数学·统计学系列

石焕南文集——受控理论与不等式研究

The Collected papers of Shi Huannan—Theory of Majorization and Inequality Research

石焕南 著

哈尔滨工业大学出版社
HARBIN INSTITUTE OF TECHNOLOGY PRESS

内 容 简 介

本书筛选石焕南教授发表的 87 篇论文，经重新编辑整理成书。主要介绍受控理论与不等式的基本内容及其新推广，重点介绍受控理论在解析不等式方面的应用，不仅包含国内外学者近年来所获得的大量研究成果，同时也包含作者近年研究的最新成果.

本书适合大学生及受控理论与不等式研究人员参考阅读.

图书在版编目（CIP）数据

石焕南文集：受控理论与不等式研究 / 石焕南著.—哈尔滨：
哈尔滨工业大学出版社，2020.9
ISBN 978-7-5603-9006 -2

Ⅰ.①石… Ⅱ.①石… Ⅲ.①不等式—研究 Ⅳ.①O178

中国版本图书馆 CIP 数据核字（2020）第 158071 号

策划编辑	刘培杰　张永芹
责任编辑	刘春雷
封面设计	孙茵艾
出版发行	哈尔滨工业大学出版社
社　　址	哈尔滨市南岗区复华四道街 10 号　　邮编 150006
传　　真	0451—86414749
网　　址	http：//hitpress.hit.edu.cn
印　　刷	哈尔滨博奇印刷有限公司
开　　本	787 mm×1 092 mm　1/16　插页 4　印张 51.25　字数 920 千字
版　　次	2020 年 9 月第 1 版　2020 年 9 月第 1 次印刷
书　　号	ISBN 978-7-5603-9006-2
定　　价	198.00 元

1973年，父母与我们六位兄弟姊妹合影

左起：我的岳父，岳母，父亲，母亲

2017年1月，我和家人在越南海滩

1969年1月，我（后排左二）在陕西省延川县居住的窑洞前留影

1976年在北京师范大学

2005年，我与王挽澜教授（左）
在广州合影

2005年，在广州全国第三届不等式学术年会上我与数学家胡克教授（右）攀谈

2008年10月，访问澳大利亚墨尔本维多利亚大学数学院时，我与
P．Cerone院长(中)和祁锋教授(左)的合影

2009年9月在内蒙古民族大学数学科学院讲学

2009年10月，我与匡继昌教授（右）和褚玉明教授（中）

在湖州师范学院合影

2009年在海宁不等式年会上作报告

2012年，我与克罗地亚Josip Pecaric 教授（中）和祁锋教授（右）
在韩国庆尚大学合影

2012年，我与美国Ravi P. Agarwal（中）和杨必成教授（右）
在韩国庆尚大学合影

2015年，北京师范大学数学学院百年院庆我与王伯英教授（左一）合影

自　　序

　　1948年12月29日，我生于广西省兴安县界首镇，老家是湖南省祁东县新塘冲.

　　为了完成国民经济第一个五年计划，中央从各地调集了大批科技人才到北京. 1953 年，父亲只身由广西调到北京有色冶金设计总院工作，两年后我们举家迁往北京. 从1956 年开始，我先后在复兴门外大街小学、青龙桥小学和北京有色冶金设计总院子弟小学(海淀区羊坊店第四小学的前身)续读完六年小学. 1962年中考时，学校期望我考上市重点中学,但我临场发挥不理想，只考上了当时区准重点的北京市第五十七中学. 不过，非常幸运的是我的初中第一位班主任萧剑桥老师就住在我家楼上，她的丈夫和我父亲在同一单位. 初一年级，我被萧老师指派为班长，这是我第一次当班长，也是最后一次当班长. 半年后因魄力不足改任学习委员，以后每年都当学习委员，一直到高中毕业. 现在想来，当时萧老师指派我为班长的决定，对我的成长影响深远，无形中约束了我，促使我不断地追求上进，完善自我. 在中小学阶段，老师委派学生做一些社会工作，对其成长的影响力不可小视. 萧老师的语文教学是出类拔萃的，她讲课绘声绘色，生动感人. 因为是邻居，我常去萧老师家串门，借阅她订的《人民文学》《儿童文学》等杂志. 萧老师常将我的作文在班上作示范，使我对作文的兴趣愈加浓厚，为我今天的创作打下了良好的基础. 我感觉在小学、初中、高中这三个初等教育阶段中，初中教育是最关键的，能不能遇到好的教师，对一个人的成长至关重要. 值得庆幸的是我遇到了一批优秀的初中老师，除萧剑桥老师之外，还有代数老师谈文荣，几何老师耿俊杰，历史老师李文琦，俄语老师周嗣宝和政治老师卓理泉. 除了语文外，我还喜欢数学，尤其钟爱几何证明题，它启蒙、训练了我的逻辑思维能力，但是我始终对三角学这门课程不感兴趣，因为厌烦要背那么多的三角公式. 我高中继续就读于北京市第五十七中学. 在我的印象中，高中只读了一学期就停课了.

　　1968年1月，北京市第五十七中学的41名同学一起来到了陕西省延川县文安驿公社下驿大队落户劳动. 当年，虽然饥饿穷困，但淳朴善良的下驿村老乡接纳了我们这群还不大懂事也不会做事的年轻娃娃，像对待自己的儿女一样关心我们, 爱护我们, 使我们能够比较顺利地度过那段艰难的日子. 由于我劳动不惜力, 肯吃苦, 1970年我出席了延安地区第一届先进知识青年代表大会. 一年后, 也就是1971年3月, 我被提拔为国家干部, 分配到延川县通讯组工作. 当时的通讯组组长曹谷溪带着我到全县各个公社去采访, 手把手地教我新闻写作和摄影. 我在谷溪手下两年半新闻报道的学习与实践, 对我今天的数学创作有着重要的影响. 二者似乎风牛马不相及, 但文理是相通的. 比如选题, 都要新, 要符合潮流; 又如语言都要简洁, 要准确. 好文章, 都要深思熟虑, 都要字斟句酌, 反复推敲. 不论文理, 要干出一番事业, 都需锲而不舍, 百折不回. 在这期间, 我有幸结识了延川众多的文人志士和有为青年, 例如本土的路遥、谷溪、北京青年孙立哲、陶海粟等. 他们强烈的事业心和不断进取的精神深深地感染了我, 影响着我后来的人生轨迹.

1973年, 全国各大学开始招收学生, 采取的是个人申请, 群众推荐, 领导审批, 文化考查及学校复审相结合的办法. 那一年我被北京师范大学数学系录取, 回到了北京, 这一步是我人生的一个重要转折点. 我一辈子都不会忘记在这关键时刻朋友对我的帮助.

从北京师范大学毕业后, 我被分配到位于北京远郊区房山县大山深处的一所煤矿中学任教, 我是全系毕业生被分得最远的一个. 1977年恢复高考, 但已毕业的学生不能报考, 我只能报考中科院数学所的研究生, 两次均"名落孙山". 失望之余, 获悉北京师范大学1978年将举办高校师资进修班. 我便白天上课, 晚上熬夜复习, 拼搏了两三个月, 终于如愿以偿, 于而立之年重返北京师范大学读书. 1980年进修班结业后, 调入北京联合大学师范学院任教. 1984年, 我又拼搏了一次, 考取了北京师范大学首届助教进修班, 进修数学研究生课程, 时年我36岁, 女儿5岁. 在助教班随机选修了王伯英教授开设的有关受控理论的硕士研究生课程"矩阵与控制不等式", 想不到此选择竟决定了我日后的科研方向, 至今我已发表了80余篇有关受控理论的论文. 三进三出北京师范大学, 累计在北京师范大学学习近七载, 从而奠定了我在高校创业的基础. 后来, 我三度申报副教授方成果, 两度冲刺教授才如愿. 在这期间我身患脑溢血, 手术开颅, 大难不死. 病休几年后, 我重登讲台, 并以更大的热情投入创作. 至今, 我在国内外所发表的160余篇论文的绝大部分, 以及两部数学专著都是大病后创作的. 2000年我晋升为教授, 2008年我晋升为三级教授. 担任院学术委员会委员、《北京联合大学学报》编委和全国不等式研究会副理事长, 所授"概率论与数理统计"课程被评为校级精品课程, 多次获学院优秀科研成果一等奖, 被评为北京联合大学2005~2007年度优秀教师. 2008年退休后, 我仍担任全国不等式研究会顾问和全国初等数学研究会常务理事, 以及美国《数学评论》的评论员, 并继续着割舍不下的不等式研究和创作. 拙著《受控理论与解析不等式》自2012年4月经哈尔滨工业大学出版社出版后, 受到国内同行的关注. 五年间, 书中所涉及的所有问题几乎都有了后续的研究成果. 我有心再版该书, 以便补充新的内容并弥补原书的疏漏, 但因退休多年, 很难获得出版资金支持, 故羞于向哈尔滨工业大学出版社启齿此事, 但意想不到的是哈尔滨工业大学出版社的刘培杰副社长竟主动建议我修订拙著, 还建议出版我的文集, 且一切费用由该出版社负担. 出版这两本书, 对于年已七旬的我来说意义重大, 它是对我一生学术研究的一个总结、一份厚礼、一份褒奖. 我不知用什么语言来表达对哈尔滨工业大学出版社, 对刘培杰数学工作室, 对刘培杰副社长本人的感激之情, 只是通过微信向刘培杰副社长表示: "太感谢了!" 他却说: "您是真正的学者, 应得的." 不久, 哈尔滨工业大学出版社便与我签订了两本书的出版合同. 目前我的新版书《Schur- 凸函数与不等式》作为国家出版基金资助项目已于2017年出版.

一个出版社有懂行的、敬业的数学编辑, 难得; 有一个懂行的、敬业的数学编辑团队, 更难得. 哈尔滨工业大学出版社的数学图书编辑工作成绩卓著, 就是因为有这样一个出色的团队 "刘培杰数学工作室". 我衷心地祝愿刘培杰数学工作室的出版事业兴旺发达.

对于热情地给予我学术帮助和鼓励的王伯英教授、胡克教授、刘绍学教授、王挽澜教授、续铁权教授、刘证教授、祁锋教授、刘国瑞教授表示衷心的谢意!

感谢祁锋教授对本书的Tex排版给予的悉心指导和帮助!

特别感谢相濡以沫四十年的妻子张景晶, 没有她一路的陪伴、呵护和付出, 我能健康地走到今天并取得一点成绩是不可想象的.

谨以此书献给逝去的四位慈祥老人: 我的父亲石承忠、母亲桂挹芬、岳父张云尧、岳母庄竹青, 愿他们在天之灵安息!

石焕南

2020 年 6 月

于北京

前　　言

1980年我到高校任教, 因教学任务繁重且无师指导, 十年间我未曾发表过一篇论文. 直到1990年秋, 我41岁时才在创刊不久的《数理统计与管理》杂志上发表了我的处女作《也谈巧算"百分比"》. 这是置于"趣味概率"栏目不到半个版面的短文, 现在看来这只不过是一道"习题解答". 但当时着实让我兴奋了一把, 并从此点燃了我的创作热情. 1991年至1992 年, 两年间我一鼓作气发表了五篇论文, 其中在核心期刊《数学通报》上发表了三篇. 正当我踌躇满志地准备要大干一场时, 1992 年年底由于突发脑溢血中断了我的创作, 差不多四年后才恢复写作. 2008 年退休后, 我的创作兴趣不减, 发表了70多篇论文并出版了两本专著.

我的论文主要包含三个方面的内容: 一是早期的初等数学研究论文, 二是结合教学的概率统计论文, 三是受控理论研究论文, 研究的主题是各类解析不等式. 本书筛选出87 篇论文, 此外还收录了我的《追念胡克教授》一文及我与胡克教授的两封通信和原文影印件, 以表达我对这位已故尊贵慈祥的数学家的怀念之情.

回顾我的科研之路, 由于没有考取研究生, 缺少导师的系统指导, 只好向书刊学习, 向同行请教, 不断地摸索、钻研, 一步步地提升自己, 向更高的台阶迈进. 这里我想向年轻的研究者及与我有类似经历的学者提几点建议:

1. 经常性地阅读文献、期刊, 要养成习惯, 要成为一种爱好. 可以说这是做好科研工作的必要条件, 是科研工作者应该具备的基本素质.

2. 在研究的起步阶段, 不要贪大求全, 而应从易到难, 别介意小问题, 可能小中见大; 别介意问题初等, 可能转向高深; 别介意教学类问题, 可能蕴含着深刻的学术问题.

3. 华罗庚先生对他的学生说过: 要做出好的文章, 关键是要有几手自己的"招路"和"拿手好戏". 因此, 一定要学点最新的东西, 至少是较新的东西.

4. 注意逐步选定方向. 只有集中目标, 深入研究, 才能有所突破, 才能上层次. 为了及时把握该方向的前沿动向, 紧跟潮流, 要搞清:

(1)该方向的发展历史及现状;

(2)该方向国内外的主要专著;

(3)该方向国内外的主要期刊(包括检索期刊);

(4)该方向国内外的主要代表人物;

(5)该方向国内外的主要组织机构.

居里夫人说过:"科学的探讨与研究,其本身就含有至美,其本身给人的愉快就是报酬."搞科研有趣,因为这是创造性的活动.只要坚持不懈,谁都会成功.

数学是千真万确的,容不得一点假,故它教人诚实、老实;数学视简洁为美,鄙视花里胡哨,故它教人朴实、踏实;学数学要心平气和,做数学要心静如水,故它教人养心向善.总之,数学真、善、美,是一门净化灵魂的科学.搞了一辈子数学,这是我的一点感受.

目　　录

第1篇　概率方法在级数求和中的应用

(石焕南. 数学通报, 1992 (3): 34-35,18)

本文的目的是通过一些例子说明如何应用概率方法求某些无穷级数的和.

例 1

$$\sum_{k=1}^{\infty}\left\{[(k-1)n+m](kn+m)\cdots[(r-k-1)n+m]\right\}^{-1}$$
$$=\{m(n+m)\cdots[(r-1)n+m]rn\}^{-1}$$

其中 m, n, r 为任意自然数.

考虑这样的随机试验: 一个口袋中装有 rn 个红球和 m 个白球, 每次从中任取一个球, 取后放回. 若取到红球, 则停止取球; 若取到白球, 则在口袋中再装入 n 个白球，然后继续按上述规则取球, 直到取得红球为止.

令 $A=\{$停止取球$\}$, $A_k=\{$取了 k 次球后停止取球$\}$, $k=1,2,3,\cdots$, 则

$$P(A_1)=\frac{rn}{rn+m}$$

$$P(A_2)=\frac{m}{rn+m}\cdot\frac{rn}{(r+1)n+m}$$

$$\vdots$$

$$P(A_{r+3})=\frac{m}{rn+m}\cdot\frac{n+m}{(r+1)n+m}\cdot\cdots\cdot$$
$$\frac{rn+m}{2rn+m}\cdot\frac{(r+1)n+m}{(2r+1)n+m}\cdot\frac{rn}{(2r+2)n+m}$$

$$\vdots$$

一般地

$$P(A_k) = \{m(n+m)\cdots[(r-1)n+m]rn\}\cdot$$
$$\{[(k-1)n+m](kn+m)\cdots[(r+k-2)n]+$$
$$m[(r+k-1)n+m]\}^{-1}$$
$$(k=1,2,3,\cdots) \tag{1}$$

由于诸A_k两两互不相容, 且$A = \bigcup\limits_{k=1}^{\infty} A_k$, 所以$P(A) = \sum\limits_{k=1}^{\infty} P(A_k)$. 另外, A 的对立事件$\bar{A} = \{$取球不止$\}$. 若令$B_k = \{$前k 次取的都是白球$\}$, 易见$\bar{A} = \bigcap\limits_{k=1}^{\infty} B_k$, 且$B_k \supset B_{k+1}, k=1,2,3,\cdots$, 由概率的连续性定理, 有

$$P(\bar{A}) = \lim_{k\to\infty} P(B_k) = \lim_{k\to\infty} \{m(m+n)\cdots[(r-1)n+m]\}\cdot$$
$$\{(kn+m)[(k+1)n+m]\cdots[(r-k-1)n+m]\}^{-1} = 0 \tag{2}$$

从而$P(A) = 1 - P(\bar{A}) = 1$, 结合(1)(2) 两式即证得例1 的结论.

当赋予m, n, r 适当的值时可得到一些常见的级数的和, 例如:

当$r = m = n = 1$时, 有

$$\frac{1}{1\cdot 2} + \frac{1}{2\cdot 3} + \frac{1}{3\cdot 4} + \cdots + \frac{1}{k\cdot(k+1)} + \cdots = 1$$

当$m = 1, n = 2, r = 1$ 时, 有

$$\frac{1}{1\cdot 3} + \frac{1}{3\cdot 5} + \frac{1}{5\cdot 7} + \cdots + \frac{1}{(2k-1)\cdot(2k+1)} + \cdots = \frac{1}{2}$$

当$m = 1, n = 1, r = 2$ 时, 有

$$\frac{1}{1\cdot 2\cdot 3} + \frac{1}{2\cdot 3\cdot 4} + \frac{1}{3\cdot 4\cdot 5} + \cdots +$$
$$\frac{1}{k\cdot(2k-1)\cdot(2k+1)} + \cdots = \frac{1}{4}$$

当$m = 1, n = 2, r = 2$ 时, 有

$$\frac{1}{1\cdot 3\cdot 5} + \frac{1}{3\cdot 5\cdot 7} + \frac{1}{5\cdot 7\cdot 9} + \cdots +$$
$$\frac{1}{(2k-1)\cdot(2k+1)\cdot(2k+3)} + \cdots = \frac{1}{12}$$

等. 仅通过建立一个简单的随机模型, 就解决了一类级数的求和问题, 概

率方法在这里所显示出来的高效率是普通的分析方法所无法比拟的.

例 2

$$\sum_{k=1}^{\infty} \frac{(k+r-1)m^{k-1}}{(k+m+r-1)!} = \frac{1}{(r+m-1)!}$$

其中 m, r 为任意自然数.

将例1 中所考虑的随机试验的条件"一个口袋中装有 rn 个红球和 m 个白球" 改为"一个口袋中装有 r 个红球和 m 个白球",并将条件"若取到白球,则在口袋中再装入一个白球" 改为"若取到白球, 则在口袋中再装入1个红球",则与例1 的证明类似,可证得例2 的结论.

特别地,当 $m = r = 1$ 时,有

$$\frac{1}{2!} + \frac{2}{3!} + \frac{3}{4!} + \cdots + \frac{k}{(k+1)!} + \cdots = 1$$

当 $m = 2, r = 1$ 时,有

$$\frac{2}{3!} + \frac{2 \cdot 2^2}{4!} + \frac{3 \cdot 2^3}{5!} + \cdots + \frac{k \cdot 2^k}{(k+2)!} + \cdots = 1$$

这些也是常见的无穷级数.

例 3

$$\sum_{k=1}^{\infty} \frac{(r-1)k+r}{k(k+1)r^k} = 1$$

其中 r 为任意自然数.

将例2 中所提及的两个条件分别改为"一个口袋中装有 $2r-1$ 个红球和1 个白球" 和"若取到白球, 则在口袋中再装入 $r-1$ 个红球和1 个白球", 则类似地可证明该例成立.

特别地, 当 $r = 1$ 时, 有

$$\frac{1}{1 \cdot 2} + \frac{1}{2 \cdot 3} + \frac{1}{3 \cdot 4} + \cdots + \frac{1}{k \cdot (k+1)} + \cdots = 1$$

当 $r = 2$ 时, 有

$$\frac{4}{1 \cdot 2 \cdot 2} + \frac{5}{2 \cdot 3 \cdot 2^2} + \frac{6}{3 \cdot 4 \cdot 2^3} + \cdots + \frac{k+3}{k \cdot (k+1) \cdot 2^k} + \cdots = 1$$

例 4

$$\sum_{k=1}^{\infty} \frac{2k-1}{(2k)!!} = 1$$

将例2 中所提及的两个条件分别改为"一个口袋中装有一个红球和一个白

球" 和"若取得白球, 则在口袋中再装入两个红球" 即可得证.

若将例4 的前一条件改为"一个口袋中装有两个红球和一个白球", 则可得到:

例 5

$$\sum_{k=1}^{\infty} \frac{k}{(2k+1)!!} = \frac{1}{2}$$

下面的例子结构稍复杂些.

例 6

$$\frac{1}{3} + \left(1 - \frac{2}{2 \cdot 3}\right) \frac{2}{3 \cdot 4} + \left(1 - \frac{2}{2 \cdot 3}\right) \left(1 - \frac{2}{3 \cdot 4}\right) \frac{2}{4 \cdot 5} + \cdots +$$

$$\left(1 - \frac{2}{2 \cdot 3}\right) \left(1 - \frac{2}{3 \cdot 4}\right) \cdot \cdots \cdot$$

$$\left(1 - \frac{2}{k(k+1)}\right) \left(1 - \frac{2}{(k+1)(k+2)}\right) + \cdots = \frac{2}{3}$$

考虑这样的随机试验有两个口袋, 其中一个口袋中装有两个红球, 另一个口袋中装有一个红球和两个白球. 有放回地从两个口袋中各取一个球, 若取到的两个球均为红球, 则停止取球, 否则在两个口袋中各加进一个白球, 然后继续按上述规则取球, 直到取到的两个球均为红球为止.

令 $A = \{停止取球\}$, $A_k = \{取了k次球后停止取球\}$, $k = 1, 2, 3, \cdots$, 则

$$P(A_1) = \frac{2}{2} \cdot \frac{1}{3} = \frac{1}{3}$$

$$P(A_2) = \left(1 - \frac{2}{2 \cdot 3}\right) \frac{2}{3 \cdot 4}$$

$$P(A_3) = \left(1 - \frac{2}{2 \cdot 3}\right) \left(1 - \frac{2}{3 \cdot 4}\right) \frac{2}{4 \cdot 5}$$

$$\vdots$$

一般地

$$P(A_k) = \left(1 - \frac{2}{2 \cdot 3}\right) \left(1 - \frac{2}{3 \cdot 4}\right) \cdots \left(1 - \frac{2}{k(k+1)}\right) \frac{2}{(k+1)(k+2)}$$

$$(k = 2, 3, 4, \cdots)$$

由于诸 A_k 两两互不相容, 且 $A = \bigcup_{k=1}^{\infty} A_k$, 所以 $P(A) = \sum_{k=1}^{\infty} P(A_k)$. 另外,

A 的对立事件为 $\bar{A} = \{$取球不止$\}$. 易见

$$P(\bar{A}) = \lim_{k \to \infty} P(B_k)$$

$$= \lim_{k \to \infty} \left(1 - \frac{2}{2 \cdot 3}\right)\left(1 - \frac{2}{3 \cdot 4}\right) \cdots \left(1 - \frac{2}{k(k+1)}\right)$$

$$= \lim_{k \to \infty} \frac{4 \cdot 1}{2 \cdot 3}\frac{5 \cdot 2}{3 \cdot 4} \cdots \frac{(k+2)(k-1)}{k(k+1)} = \lim_{k \to \infty} \frac{1}{3}\frac{k+2}{k} = \frac{1}{3}$$

从而 $P(A) = 1 - P(\bar{A}) = \frac{2}{3}$, 得证.

若将例6中的条件"有两个口袋, 其中一个口袋中装有两个红球, 另一个口袋中装有一个红球和两个白球"改为"有两个口袋, 其中一个口袋中装有一个红球和一个白球, 另一个口袋中装有三个红球和一个白球" 类似地可证得:

例 7

$$\frac{3}{3^2 - 1} + \left(1 - \frac{3}{3^2 - 1}\right)\frac{3}{4^2 - 1} +$$

$$\left(1 - \frac{3}{3^2 - 1}\right)\left(1 - \frac{3}{4^2 - 1}\right)\frac{3}{5^2 - 1} + \cdots +$$

$$\left(1 - \frac{3}{3^2 - 1}\right)\left(1 - \frac{3}{4^2 - 1}\right) \cdots \frac{3}{k^2 - 1}\frac{3}{(k+1)^2 - 1} + \cdots = \frac{3}{4}$$

若将上面提及的条件改为"有两个口袋, 其中一个口袋中装有两个红球和三个白球, 另一个口袋中装有四个红球和三个白球. "并将例6 中的条件"否则在两个口袋中各加进一个白球"改为" 否则在两个口袋中各加进两个白球", 则类似地可证得:

例 8

$$\frac{8}{5 \cdot 7} + \left(1 - \frac{8}{5 \cdot 7}\right)\frac{8}{7 \cdot 9} +$$

$$\left(1 - \frac{8}{5 \cdot 7}\right)\left(1 - \frac{8}{7 \cdot 9}\right)\frac{8}{9 \cdot 11} + \cdots +$$

$$\left(1 - \frac{8}{5 \cdot 7}\right)\left(1 - \frac{8}{7 \cdot 9}\right) \cdots \left(1 - \frac{8}{(2k+3)(2k+5)}\right) \cdot$$

$$\frac{8}{[2(k+1) + 3][2(k+1) + 5]} + \cdots = \frac{4}{7}$$

第2篇　概率方法在不等式证明中的应用

(石焕南,石敏琪. 数学通报, 1992 (7): 34-37)

设 ξ 是一个只取有限个值的离散型随机变量, 其概率分布为

$$P(\xi = a_i) = p_i > 0, i = 1, 2, \cdots, n$$

则 ξ 的方差

$$D\xi = E\xi^2 - (E\xi)^2$$

即

$$\sum_{i=1}^{n} (a_i - E\xi)^2 p_i = \sum_{i=1}^{n} a_i^2 p_i - \left(\sum_{i=1}^{n} a_i p_i \right)^2$$

由此可知

$$E\xi^2 = \sum_{i=1}^{n} a_i^2 p_i \geqslant \left(\sum_{i=1}^{n} a_i p_i \right)^2 = (E\xi)^2 \qquad (*)$$

且上式等号成立当且仅当 $a_1 = a_2 = \cdots = a_n$.

本文试用初等概率论中的这一简单结论证明一类常见的代数不等式, 以显示概率方法在解决某些数学问题中所具有的独特而简洁的功效.

例 1 (平方－算术平均值不等式)　设 $a_i \geqslant 0, i = 1, 2, \cdots, n$, 则

$$\frac{1}{n} \sum_{i=1}^{n} a_i \leqslant \left(\frac{1}{n} \sum_{i=1}^{n} a_i^2 \right)^{\frac{1}{2}}$$

且等号成立当且仅当 $a_1 = a_2 = \cdots = a_n$.

证明　设 ξ 的概率分布为

$$P(\xi = a_i) = \frac{1}{n}, i = 1, 2, \cdots, n$$

由式(∗)有

$$\frac{1}{n}\sum_{i=1}^{n}a_i^2 = E\xi^2 \geqslant (E\xi)^2 = \left(\frac{1}{n}\sum_{i=1}^{n}a_i\right)^2$$

两边开方即得所证不等式, 且等号成立当且仅当 $a_1 = a_2 = \cdots = a_n$.

例 2 (调和 − 算术平均值不等式)　设 $a_i \geqslant 0, i = 1, 2, \cdots, n$, 则

$$\frac{1}{n}\sum_{i=1}^{n}a_i \geqslant \frac{n}{\sum\limits_{i=1}^{n}\frac{1}{a_i}}$$

且等号成立当且仅当 $a_1 = a_2 = \cdots = a_n$.

证明　设 ξ 的概率分布为

$$P\left(\xi = \frac{\sum\limits_{i=1}^{n}a_i}{a_k}\right) = \frac{a_k}{\sum\limits_{i=1}^{n}a_i}, k = 1, 2, \cdots, n$$

则

$$E\xi^2 = \sum_{i=1}^{n}\left[\left(\frac{\sum\limits_{i=1}^{n}a_i}{a_k}\right)^2 \cdot \frac{a_k}{\sum\limits_{i=1}^{n}a_i}\right]$$

$$= \sum_{k=1}^{n}\frac{\sum\limits_{i=1}^{n}a_i}{a_k} = \left(\sum_{i=1}^{n}a_i\right) \cdot \left(\sum_{i=1}^{n}\frac{1}{a_i}\right)$$

而

$$E\xi = \sum_{i=1}^{n}\left[\left(\frac{\sum\limits_{i=1}^{n}a_i}{a_k}\right) \cdot \frac{a_k}{\sum\limits_{i=1}^{n}a_i}\right] = n$$

由式(∗)有

$$\left(\sum_{i=1}^{n}a_i\right)\left(\sum_{i=1}^{n}\frac{1}{a_i}\right) \geqslant n^2$$

等号成立当且仅当 $\dfrac{\sum\limits_{i=1}^{n}a_i}{a_1} = \dfrac{\sum\limits_{i=1}^{n}a_i}{a_2} = \cdots = \dfrac{\sum\limits_{i=1}^{n}a_i}{a_n}$. 即

$$\frac{1}{n}\sum_{i=1}^{n}a_i \geqslant \frac{n}{\sum\limits_{i=1}^{n}\frac{1}{a_i}}$$

且等号成立当且仅当$a_1 = a_2 = \cdots = a_n$.

例 3 (Cauchy不等式) 设$a_1, a_2, \cdots, a_n, b_1, b_2, \cdots, b_n$ 为任意实数, 则

$$\left(\sum_{i=1}^{n} a_i b_i\right)^2 \leqslant \left(\sum_{i=1}^{n} a_i^2\right) \cdot \left(\sum_{i=1}^{n} b_i^2\right)$$

且等号成立当且仅当$b_i = 0, i = 1, 2, \cdots, n$ 或存在常数l 使得$a_i = lb_i, i = 1, 2, \cdots, n$.

证明 若$b_1, b_2, \cdots, b_n$ 都是零, 则等式成立.

若$b_1, b_2, \cdots, b_n$ 不全为零, 不妨设$b_1, b_2, \cdots, b_k$ 不为零, 而$b_{k+1} = b_{k+2} = \cdots = b_n = 0$, 首先证明

$$\left(\sum_{i=1}^{k} a_i b_i\right)^2 \leqslant \left(\sum_{i=1}^{k} a_i^2\right) \cdot \left(\sum_{i=1}^{k} b_i^2\right) \tag{1}$$

即

$$\left\{\sum_{j=1}^{k} \left| a_j \cdot \frac{\left(\sum_{i=1}^{k} b_i^2\right)^{\frac{1}{2}}}{b_j} \cdot \frac{b_j^2}{\sum_{i=1}^{n} b_i^2} \right|\right\}^2 \leqslant \sum_{i=1}^{k} a_i^2$$

现ξ 的概率分布为

$$P\left(\xi = a_j \cdot \frac{\left(\sum_{i=1}^{k} b_i^2\right)^{\frac{1}{2}}}{b_j}\right) = \frac{b_j^2}{\sum_{i=1}^{n} b_i^2}, j = 1, 2, \cdots, k$$

则

$$E\xi^2 = \sum_{j=1}^{k} \left(a_j^2 \cdot \frac{\sum_{i=1}^{k} b_i^2}{b_j^2} \cdot \frac{b_j^2}{\sum_{i=1}^{n} b_i^2}\right) = \sum_{i=1}^{k} a_i^2$$

$$E\xi = \sum_{j=1}^{k} \left(a_j \cdot \frac{\left(\sum_{i=1}^{k} b_i^2\right)^{\frac{1}{2}}}{b_j} \cdot \frac{b_j^2}{\sum_{i=1}^{n} b_i^2}\right) = \sum_{i=1}^{k} a_i^2$$

由式$(*)$可知式(1)成立, 且等号成立当且仅当$\frac{a_1}{b_1} = \frac{a_2}{b_2} = \cdots = \frac{a_n}{b_n} = l$, 进而

$$\left(\sum_{i=1}^{n} a_i b_i\right)^2 = \left(\sum_{i=1}^{k} a_i b_i\right)^2$$
$$\leqslant \left(\sum_{i=1}^{k} a_i^2\right) \cdot \left(\sum_{i=1}^{k} b_i^2\right) \leqslant \left(\sum_{i=1}^{n} a_i^2\right) \cdot \left(\sum_{i=1}^{n} b_i^2\right)$$

且等号成立当且仅当$\frac{a_1}{b_1} = \frac{a_2}{b_2} = \cdots = \frac{a_k}{b_k} = l, b_{k+1} = b_{k+2} = \cdots = b_n = 0$, 即当$b_1, b_2, \ldots, b_n$ 不全为零时, 等号成立的充要条件是$a_i = l b_i, i = 1, 2, \cdots, n$.

总之, 所证不等式等号成立的充要条件是$b_i = 0, i = 1, 2, \cdots, n$ 或存在常数l 使得$a_i = l b_i, i = 1, 2, \cdots, n$.

例 4 (1984年全国高中数学联赛试题)　设$a_i > 0, i = 1, 2, \cdots, n$, 求证

$$\frac{a_1^2}{a_2} + \frac{a_2^2}{a_3} + \cdots + \frac{a_{n-1}^2}{a_n} + \frac{a_n^2}{a_1} \geqslant a_1 + a_2 + \cdots + a_n$$

证明　设ξ 的概率分布为

$$P\left(\xi = \frac{a_k}{a_{k+1}}\right) = \frac{a_{k+1}}{\sum\limits_{i=1}^{n} a_i}, k = 1, 2, \cdots, n$$

(规定$a_{n+1} = a_1$) 则

$$E\xi^2 = \frac{a_1^2}{a_2^2} \cdot \frac{a_2}{\sum\limits_{i=1}^{n} a_i} + \frac{a_2^2}{a_3^2} \cdot \frac{a_3}{\sum\limits_{i=1}^{n} a_i} +$$
$$\frac{a_{n-1}^2}{a_n^2} \cdot \frac{a_n}{\sum\limits_{i=1}^{n} a_i} + \frac{a_n^2}{a_1^2} \cdot \frac{a_1}{\sum\limits_{i=1}^{n} a_i}$$
$$= \left(\frac{a_1^2}{a_2} + \frac{a_2^2}{a_3} + \cdots + \frac{a_{n-1}^2}{a_n} + \frac{a_n^2}{a_1}\right) \cdot \frac{1}{\sum\limits_{i=1}^{n} a_i}$$

$$\frac{a_1}{a_2} \cdot \frac{a_2}{\sum\limits_{i=1}^{n} a_i} + \frac{a_2}{a_3} \cdot \frac{a_3}{\sum\limits_{i=1}^{n} a_i} + \cdots + \frac{a_{n-1}}{a_n} \cdot \frac{a_n}{\sum\limits_{i=1}^{n} a_i} + \frac{a_n}{a_1} \cdot \frac{a_1}{\sum\limits_{i=1}^{n} a_i} = 1$$

由式$(*)$即得证.

例 5 (第20届IMO试题)　设$a_1, a_2, \cdots, a_n$ 是两两不等的正整数, 求证

$$\sum_{k=1}^{n} \frac{a_k}{k^2} \geqslant \sum_{k=1}^{n} \frac{1}{k}$$

证明 要证的不等式等价于

$$\sum_{k=1}^{n}\left[\frac{\left(\frac{1}{k}\right)^2}{\left(\frac{1}{a_k}\right)^2}\cdot\frac{\frac{1}{a_k}}{\left(\sum_{i=1}^{n}\frac{1}{a_i}\right)}\right]\geqslant\frac{\sum_{k=1}^{n}\frac{1}{k}}{\sum_{k=1}^{n}\frac{1}{a_k}}\qquad(2)$$

设随机变量 ξ 的概率分布为

$$P\left(\xi=\frac{\frac{1}{k}}{\frac{1}{k}}\right)=\frac{\frac{1}{a_k}}{\sum_{i=1}^{n}\frac{1}{a_i}},k=1,2,\cdots,n$$

由式 $(*)$ 有

$$\sum_{k=1}^{n}\left[\frac{\left(\frac{1}{k}\right)^2}{\left(\frac{1}{a_k}\right)^2}\cdot\frac{\frac{1}{a_k}}{\sum_{i=1}^{n}\frac{1}{a_i}}\right]\geqslant\left[\sum_{k=1}^{n}\left(\frac{\frac{1}{k}}{\frac{1}{a_k}}\cdot\frac{\frac{1}{a_k}}{\sum_{i=1}^{n}\frac{1}{a_i}}\right)\right]^2$$

$$=\left(\frac{\sum_{k=1}^{n}\frac{1}{k}}{\sum_{k=1}^{n}\frac{1}{a_k}}\right)^2\qquad(3)$$

由于 $a_1,a_2,\cdots,a_n$ 是两两不等的正整数,易见

$$\sum_{k=1}^{n}\frac{1}{k}\geqslant\sum_{k=1}^{n}\frac{1}{a_k}$$

即

$$\frac{\sum_{k=1}^{n}\frac{1}{k}}{\sum_{k=1}^{n}\frac{1}{a_k}}\geqslant1$$

从而

$$\left(\frac{\sum_{k=1}^{n}\frac{1}{k}}{\sum_{k=1}^{n}\frac{1}{a_k}}\right)^2\geqslant\frac{\sum_{k=1}^{n}\frac{1}{k}}{\sum_{k=1}^{n}\frac{1}{a_k}}$$

结合式(3)即证得式(2).

例 6 (Schapiro不等式) 设$0 \leqslant a_i < 1, i = 1, 2, \cdots, n, \sum\limits_{i=1}^{n} a_i = a$, 则

$$\sum_{i=1}^{n} \frac{a_i}{1 - a_i} \geqslant \frac{na}{n - a}$$

且等号成立当且仅当$a_1 = a_2 = \cdots = a_n$.

证明 注意到

$$\sum_{i=1}^{n} \frac{a_i}{1 - a_i} = \sum_{i=1}^{n} \frac{1 - (1 - a_i)}{1 - a_i} = \sum_{i=1}^{n} \frac{1}{1 - a_i} - n$$

所证不等式等价于

$$\sum_{i=1}^{n} \frac{1}{1 - a_i} \geqslant n + \frac{na}{n - a} = \frac{n^2}{n - a}$$

即

$$\sum_{i=1}^{n} \frac{1}{\frac{1 - a_i}{n - a}} \geqslant n^2 \tag{4}$$

由于$0 \leqslant \frac{1 - a_i}{n - a} < 1$, 且$\sum\limits_{i=1}^{n} \frac{1 - a_i}{n - a} = 1$. 可设随机变量$\xi$ 的概率分布为

$$P\left(\xi = \frac{1}{\frac{1 - a_i}{n - a}}\right) = \frac{1 - a_i}{n - a}, i = 1, 2, \cdots, n$$

则

$$E\xi^2 = \sum_{i=1}^{n} \left[\frac{1}{\left(\frac{1 - a_i}{n - a}\right)^2} \cdot \frac{1 - a_i}{n - a} \right] = \sum_{i=1}^{n} \frac{1}{\frac{1 - a_i}{n - a}}$$

$$E\xi = \sum_{i=1}^{n} \left(\frac{1}{\frac{1 - a_i}{n - a}} \cdot \frac{1 - a_i}{n - a} \right) = n$$

由式(∗)即证得式(4), 且等号成立当且仅当$\frac{n - a}{1 - a_1} = \frac{n - a}{1 - a_2} = \cdots = \frac{n - a}{1 - a_n}$, 即$a_1 = a_2 = \cdots = a_n$.

例 7 (《数学通报》数学问题362 题) 设$a_i > 0, i = 1, 2, \cdots, n, \sum\limits_{i=1}^{n} a_i = 1$, 试证

$$\sum_{i=1}^{n} \left(a_i + \frac{1}{a_i} \right)^2 \geqslant \frac{(n^2 + 1)^2}{n}$$

证明 设随机变量 ξ 的概率分布为

$$P\left(\xi = a_i + \frac{1}{a_i}\right) = \frac{1}{n}, i = 1, 2, \cdots, n$$

则

$$E\xi^2 = \frac{1}{n}\sum_{i=1}^n \left(a_i + \frac{1}{a_i}\right)^2$$

$$E\xi = \frac{1}{n}\sum_{i=1}^n \left(a_i + \frac{1}{a_i}\right) = \frac{1}{n}\left(1 + \sum_{i=1}^n \frac{1}{a_i}\right)$$

由式 $(*)$ 有

$$\sum_{i=1}^n \left(a_i + \frac{1}{a_i}\right)^2 \geqslant \frac{1}{n}\sum_{i=1}^n \left(a_i + \frac{1}{a_i}\right)^2$$

注意到 $\sum_{i=1}^n a_i = 1$, 由例2 知 $\sum_{i=1}^n \frac{1}{a_i} \geqslant n^2$, 从而

$$\sum_{i=1}^n \left(a_i + \frac{1}{a_i}\right)^2 \geqslant \frac{1}{n}(1 + n^2)^2 = \frac{(n^2 + 1)^2}{n}$$

例 8 试证 $\sum_{k=1}^n \frac{1}{\sqrt{k}} > \sqrt{n}$, $n > 1$.

证明 设随机变量 ξ 的概率分布为

$$P\left(\xi = \frac{1}{\sqrt{k}}\right) = \frac{\sqrt{k}}{\sum\limits_{i=1}^n \sqrt{i}}, k = 1, 2, \cdots, n$$

则

$$E\xi^2 = \sum_{i=1}^n \left[\left(\frac{1}{\sqrt{k}}\right)^2 \cdot \frac{\sqrt{k}}{\sum\limits_{i=1}^n \sqrt{i}}\right] = \frac{\sum\limits_{k=1}^n \frac{1}{\sqrt{k}}}{\sum\limits_{k=1}^n \sqrt{k}}$$

$$E\xi = \sum_{i=1}^n \left[\frac{1}{\sqrt{k}} \cdot \frac{\sqrt{k}}{\sum\limits_{i=1}^n \sqrt{i}}\right] = \frac{n}{\sum\limits_{k=1}^n \sqrt{k}}$$

由式 $(*)$ 有

$$\sum_{k=1}^n \frac{1}{\sqrt{k}} \geqslant \frac{n^2}{\sum\limits_{k=1}^n \sqrt{k}}$$

故只需证

$$\frac{n^2}{\sum\limits_{k=1}^{n} \sqrt{k}} \geqslant \sqrt{k}$$

即

$$\sum_{k=1}^{n} \sqrt{k} < n\sqrt{n}$$

而此式显然成立.

　　由以上诸例可见, 根据不等式的特点构造相应的随机变量是解题的关键所在.

　　最后说明一点, 除式(∗)外, 概率论中还有其他一些有关矩的不等式, 利用它们可以在更广泛的范围内证明代数、几何、三角等其他数学领域中的不等式, 读者若有兴趣不妨一试.

第3篇　关于对称函数的一类不等式

(石焕南. 数学通报, 1996, (3): 38-40)

n个正数$x_1, x_2, \cdots, x_n$ 的初等对称函数是

$$E(\boldsymbol{x}) = E_k(x_1, \cdots, x_n) = \sum_{1 \leqslant i_1 < \cdots < i_k \leqslant n} \prod_{j=1}^{k} x_{i_j}$$

并规定$E_0(\boldsymbol{x}) = 1$, 当$k < 0$ 或$k > n$ 时$E_k(\boldsymbol{x}) = 0$. 不难验证

$$E_k(x_1, \cdots, x_n) = x_1 E_{k-1}(x_2, \cdots, x_n) + E_k(x_2, \cdots, x_n)$$
$$= x_1 x_2 E_{k-2}(x_3, \cdots, x_n) + (x_1 + x_2)E_{k-1}(x_3, \cdots, x_n) + E_k(x_3, \cdots, x_n)$$
$$\tag{1}$$

本文将证明如下两个定理

定理1 设$x_i > 0, i = 1, \cdots, n$, 且$\sum_{i=1}^{n} x_i \leqslant 1$, 则对于任何$0 \leqslant k \leqslant n$, 有

$$\frac{E_k(1 - x_1, \cdots, 1 - x_n)}{E_k(x_1, \cdots, x_n)} \geqslant (k-1)^k \tag{2}$$

定理2 设$x_i > 0, i = 1, \cdots, n$, 且$\sum_{i=1}^{n} x_i = 1$, 则对于任何$0 \leqslant k \leqslant n$, 有

$$0 \leqslant E_k(1 - x_1, \cdots, 1 - x_n) - E_k(x_1, \cdots, x_n) \leqslant \mathrm{C}_k^n \left[\left(1 - \frac{1}{n}\right)^k - \left(\frac{1}{n}\right)^k \right] \tag{3}$$

引理 设 $x_i > 0, i = 1, \cdots, n$, $\sum_{i=1}^{n} x_i \leqslant 1$, 则

$$\prod_{i=1}^{n} \frac{1 - x_i}{x_i} \geqslant (n-1)^n \tag{4}$$

证明　因 $\sum\limits_{i=1}^{n} x_i \leqslant 1$, 则

$$\prod_{i=1}^{n} \frac{1-x_i}{x_i} \geqslant \frac{1}{x_i} \sum_{j \neq i} x_j$$

由算术-几何平均值不等式, 有

$$\sum_{j \neq i} x_j \geqslant (n-1) \left(\prod_{j \neq i} x_j \right)^{\frac{1}{n-1}}$$

从而

$$\prod_{i=1}^{n} \frac{1-x_i}{x_i} \geqslant \prod_{i=1}^{n} \left(\frac{1}{x_i} \sum_{j \neq i} x_j \right) \geqslant \prod_{i=1}^{n} \left[\frac{1}{x_i}(n-1) \left(\prod_{j \neq i} x_j \right)^{\frac{1}{n-1}} \right] = (n-1)^n$$

引理放宽了Newman不等式(见[1 ,p.168] 第94题(7)) 的条件, 但结论保持不变.

定理1 的证明

注意

$$E_k(1-x_1,\cdots,1-x_n) - (k-1)^k E_k(x_1,\cdots,x_n)$$

$$= \sum_{1 \leqslant i_1 < \cdots < i_k \leqslant n} \left[\prod_{j=1}^{k}(1-x_{i_j}) - \prod_{j=1}^{k} x_{i_j} \right]$$

由引理知上述和式中的每一项均非负, 由此即得证.

注意式(4)可以写作

$$\frac{E_n(1-x_1,\cdots,1-x_n)}{E_n(E_k(x_1,\cdots,x_n))} \geqslant (n-1)^n$$

所以定理1 实际上是Newman 不等式的一种推广.

定理2 的证明

由定理1 知式(3)左边的不等式成立, 现采用逐步调整法证明式(3)右边的不等式, 记

$$f(1-x_1,\cdots,1-x_n) = E_k(1-x_1,\cdots,1-x_n)$$

若$x_1 = \cdots = x_n$, 则等式成立, 否则不妨设$x_1 < \frac{1}{n} < x_2$, 令$y_1 = \frac{1}{n}, y_2 =$

$x_1 + x_2 - \frac{1}{n}$, 由式(1)有

$$E_k(y_1, y_2, x_3, \cdots, x_n)$$
$$= y_1 y_2 E_{k-2}(x_3, \cdots, x_n) + (y_1 + y_2) E_{k-1}(x_3, \cdots, x_n) + E_k(x_3, \cdots, x_n)$$

$$E_k(x_1, x_2, x_3, \cdots, x_n)$$
$$= x_1 x_2 E_{k-2}(x_3, \cdots, x_n) + (x_1 + x_2) E_{k-1}(x_3, \cdots, x_n) + E_k(x_3, \cdots, x_n)$$

注意 $x_1 + x_2 = y_1 + y_2$, 有

$$E_k(y_1, y_2, x_3, \cdots, x_n) - E_k(x_1, x_2, x_3, \cdots, x_n) = (y_1 y_2 - x_1 x_2) E_{k-2}(x_3, \cdots, x_n)$$

同样不难验证

$$E_k(1 - y_1, 1 - y_2, 1 - x_3, \cdots, 1 - x_n) - E_k(1 - x_1, 1 - x_2, 1 - x_3, \cdots, 1 - x_n)$$
$$= (y_1 y_2 - x_1 x_2) E_{k-2}(1 - x_3, \cdots, 1 - x_n)$$

于是

$$f(y_1, y_2, x_3, \cdots, x_n) - f(x_1, x_2, x_3, \cdots, x_n)$$
$$= (y_1 y_2 - x_1 x_2)[E_{k-2}(1 - x_3, \cdots, 1 - x_n) - E_{k-2}(x_3, \cdots, x_n)]$$
$$= \left(x_2 - \frac{1}{n}\right)\left(\frac{1}{n} - x_1\right)[E_{k-2}(1 - x_3, \cdots, 1 - x_n) - E_{k-2}(x_3, \cdots, x_n)] \geqslant 0$$

最后一步用到了定理1, 这样最多经过 $n - 1$ 次调整得

$$f(x_1, x_2, \cdots, x_n) \leqslant f\left(\frac{1}{n}, x_1 + x_2 - \frac{1}{n}, x_3, \cdots, x_n\right)$$
$$\leqslant f\left(\frac{1}{n}, \frac{1}{n}, \frac{1}{n}, \cdots, \frac{1}{n}\right) = \mathrm{C}_n^k\left[\left(1 - \frac{1}{n}\right)^k - \left(\frac{1}{n}\right)^k\right]$$

由定理2 不难得到如下推论.

推论 设 $x_i > 0, i = 1, \cdots, n$, 且 $\sum_{i=1}^n x_i = s$, 则对于任何 $0 \leqslant k \leqslant n$, 有

$$0 \leqslant E_k(s - x_1, \cdots, s - x_n) - E_k(x_1, \cdots, x_n) \leqslant \mathrm{C}_n^k\left[\left(s - \frac{s}{n}\right)^k - \left(\frac{s}{n}\right)^k\right] \quad (5)$$

我们知道若 $P(x) = \prod_{i=1}^n (x - x_i)$, 则

$$P(x) = x^n - E_1 x^{n-1} + \cdots + (-1)^n E_n$$

其中 $E_k = E_k(x_1, \cdots, x_n), k = 1, \cdots, n$, 利用此性质不难验证

$$E_k(s - x_1, \cdots, s - x_n) \sum_{1 \leqslant i_1 < \cdots < i_k \leqslant n} \prod_{j=1}^{k}(s - x_{i_j})$$

$$= \sum_{j=0}^{k}(-1)^j C_{n-j}^{k-j} E_j(x_1, \cdots, x_n) s^{k-j}$$

于是式 (5) 也可写成如下形式

$$0 \leqslant \sum_{j=0}^{k}(-1)^j C_{n-j}^{k-j} E_j(x_1, \cdots, x_n) s^{k-j} - E_k(x_1, \cdots, x_n)$$

$$\leqslant C_n^k \left[\left(s - \frac{s}{n} \right)^k - \left(\frac{s}{n} \right)^k \right] \tag{6}$$

例如, 取 $s = 1, n = 4, k = 3$, 式 (6) 可化为

$$\frac{1}{2} \geqslant E_3(x_1, x_2, x_3, x_4) - E_2(x_1, x_2, x_3, x_4) \geqslant -\frac{5}{16}$$

又如取 $s = 1, n = k = 3$, 式 (6) 可化为

$$\frac{1}{2} \geqslant x_1 x_2 + x_1 x_3 + x_2 x_3 - 2 x_1 x_2 x_3 \geqslant \frac{7}{27}$$

此为第 25 届 IMO 的第 1 题, 最后对于定理 1, 笔者猜想式 (2) 可加强为

$$\frac{E_k(1 - x_1, \cdots, 1 - x_n)}{E_k(x_1, \cdots, x_n)} \geqslant (n - 1)^k$$

参考文献

[1] 匡继昌. 常用不等式 [M]. 2 版. 长沙: 湖南教育出版社, 1993.

[2] 张永红. 关于 Newman 不等式的两种推广 [J]. 湖南数学通讯, 1995(1).

第4篇　对称平均值基本定理应用数例

(石焕南, 石敏琪. 数学通报, 1996 (10): 44-45)

命题 1　设 a 和 $x_1, \cdots, x_n$ 是正数, 且 $\sum\limits_{i=1}^{n} x_i = 1, k = 1, \cdots, n$, 则

$$\prod_{1 \leqslant i_1 < \cdots < i_k \leqslant n} \left(\sum_{j=1}^{k} x_{i_j} \right) \leqslant \left(\frac{k}{n} \right)^{C_n^k} \tag{1}$$

$$\prod_{1 \leqslant i_1 < \cdots < i_k \leqslant n} \left[\sum_{j=1}^{k} (1 - x_{i_j}) \right] \leqslant \left(\frac{(n-1)k}{n} \right)^{C_n^k} \tag{2}$$

$$\sum_{1 \leqslant i_1 < \cdots < i_k \leqslant n} \left(\prod_{j=1}^{k} \frac{1 - x_{i_j}}{x_{i_j}} \right)^a \geqslant C_n^k (n-1)^{ka} \tag{3}$$

$$\sum_{1 \leqslant i_1 < \cdots < i_k \leqslant n} \prod_{j=1}^{k} \left(\frac{1}{1 - x_{i_j}} \right)^a \geqslant C_n^k \left(\frac{n}{n-1} \right)^{ka} \tag{4}$$

$$\sum_{1 \leqslant i_1 < \cdots < i_k \leqslant n} \left(\prod_{j=1}^{k} \frac{1}{x_{i_j}} \right)^a \geqslant C_n^k n^{ka} \tag{5}$$

在所有这些情形中当且仅当 $x_1 = \cdots = x_n = \frac{1}{n}$ 时等号成立.

文献[1]指出上述这组形式优美的不等式属于V. Volenec, 但没有给出证明. 文献[2]利用优超理论证明了上述不等式. 本文将利用对称平均值基本定理给出V. Volenec 不等式的一个简单证明. 为此先将文献[3]中介绍的有关定义和结论引述如下.

定理　设 $x_1, \cdots, x_n$ 是正数, $k = 1, \cdots, n$. 记

$$\prod_{n=1}^{k}(\boldsymbol{x}) = \prod_{n=1}^{k}(x_1, \ldots, x_n) = \left[\prod_{1 \leqslant i_1 < \cdots < i_k \leqslant n} \left(\frac{1}{k} \sum_{j=1}^{k} x_{i_j} \right) \right]^{\frac{1}{C_n^k}}$$

$$\sum_{n=1}^{k}(\boldsymbol{x}) = \sum_{n=1}^{k}(x_1, \cdots, x_n) = \left(\frac{1}{C_n^k} \sum_{1 \leqslant i_1 < \cdots < i_k \leqslant n} \prod_{j=1}^{k} x_{i_j} \right)^{\frac{1}{k}}$$

则:

(i)

$$G_n(\boldsymbol{x}) = \prod_{n=1}^{1}(\boldsymbol{x}) \leqslant \prod_{n=1}^{2}(\boldsymbol{x}) \leqslant \cdots \leqslant \prod_{n=1}^{n}(\boldsymbol{x}) = A_n(\boldsymbol{x})$$

(ii)

$$A_n(\boldsymbol{x}) = \sum_{n=1}^{1}(\boldsymbol{x}) \geqslant \sum_{n=1}^{2}(\boldsymbol{x}) \geqslant \cdots \geqslant \sum_{n=1}^{n}(\boldsymbol{x}) = G_n(\boldsymbol{x})$$

在所有这些情形中当且仅当 $x_1 = \cdots = x_n = \frac{1}{n}$ 时等号成立.

定理给出了算术-几何平均值不等式的两种加细.

命题1 的证明　现利用定理证明命题1 的(2), (3)两式, 其余各式可类似证得.

式(2)的证明: 由定理的情形(i)有

$$\prod_{1 \leqslant i_1 < \cdots < i_k \leqslant n} \left[\sum_{j=1}^{k}(1 - x_{i_j}) \right] = \left[k \prod_{n=1}^{k}(1 - x) \right]^{C_n^k}$$

$$\leqslant [kA_n(1 - x)]^{C_n^k} = \left[k \cdot \frac{1}{n}\sum_{i=1}^{n}(1 - x_i) \right]^{C_n^k} = \left[\frac{(n-1)k}{n} \right]^{C_n^k}$$

式(3)的证明: 由定理的情形(ii)有

$$\sum_{1 \leqslant i_1 < \cdots < i_k \leqslant n} \left(\prod_{j=1}^{k} \frac{1 - x_{i_j}}{x_{i_j}} \right)^a = C_n^k \left[\sum_{n=1}^{k} \left(\left(\frac{1-x}{x} \right)^a \right) \right]^k$$

$$\geqslant C_n^k \left[G_n \left(\left(\frac{1-x}{x} \right)^a \right) \right]^k = C_n^k \left[\left(\prod_{i=1}^{n} \frac{1-x_i}{x_i} \right)^{\frac{a}{n}} \right]^k$$

$$\geqslant C_n^k [(n-1)^a]^k = C_n^k (n-1)^{ka}$$

最后一步用到了Newman不等式[4]

$$\prod_{i=1}^{n} \frac{1 - x_i}{x_i} \geqslant (n-1)^k$$

式(3)给出了Newman不等式的一种推广.

利用定理也可以将其他一些我们所熟悉的不等式作类似的推广. 例如, 设 $x_1, \ldots, x_n$ 是正数, 则

$$\prod_{i=1}^{n}(1+x_i) \geqslant \left(1+\sqrt[n]{\prod_{i=1}^{n}x_i}\right)^n$$

这是Chrystal不等式[1], 它可推广成:

命题 2 设 a 和 $x_1, \cdots, x_n$ 是正数, $k = 1, \cdots, n$, 则

$$\sum_{1 \leqslant i_1 < \cdots < i_k \leqslant n} \prod_{j=1}^{k}(1+x_{i_j})^a \geqslant C_n^k \left(1+\sqrt[n]{\prod_{i=1}^{n}x_i}\right)^{ka} \tag{6}$$

又如, 设 $x_1, \cdots, x_n$ 是正数, 且 $\sum\limits_{i=1}^{n} x_i = 1, k = 1, \cdots, n, m \in \mathbf{N}$, 则:

(i)[5]

$$\prod_{i=1}^{n}\left(x_i^m + \frac{1}{x_i^m}\right) \geqslant \left(n^m + \frac{1}{n^m}\right)^n$$

(ii)[5]

$$\prod_{i=1}^{n}\left(1 + \frac{1}{x_i^m}\right) \geqslant (n^m + 1)^n$$

(iii)

$$\prod_{i=1}^{n}\frac{1+x_i}{1-x_i} \geqslant \left(\frac{n+1}{n-1}\right)^n$$

它们可推广成:

命题 3 设 a 和 $x_1, \cdots, x_n$ 是正数, 且 $\sum\limits_{i=1}^{n} x_i = 1, m \in \mathbf{N}, k = 1, \cdots, n$, 则

$$\sum_{1 \leqslant i_1 < \cdots < i_k \leqslant n} \prod_{j=1}^{k}\left(x_{i_j}^m + \frac{1}{x_{i_j}^m}\right)^a \geqslant C_n^k \left(n^m + \frac{1}{n^m}\right)^{ka} \tag{7}$$

$$\sum_{1 \leqslant i_1 < \cdots < i_k \leqslant n} \prod_{j=1}^{k}\left(1 + \frac{1}{x_{i_j}^m}\right)^a \geqslant C_n^k (n^m + 1)^{ka} \tag{8}$$

$$\sum_{1 \leqslant i_1 < \cdots < i_k \leqslant n} \prod_{j=1}^{k}\left(\frac{1+x_{i_j}}{1-x_{i_j}}\right)^a \geqslant C_n^k \left(\frac{n+1}{n-1}\right)^{ka} \tag{9}$$

式(7)是Mitrinovic-Djokovic不等式[1]的一种推广, 而(8), (9)两式分别是两个Klamkin不等式[4] 的推广.

参考文献

[1]　密特里诺维奇，瓦西奇. 分析不等式[M]. 赵汉宾,译. 南宁：广西人民出版社, 1986: 455, 378-379.

[2]　MARSHALL A W, OLKIN I. Inequalities: Theary of Majorization and Its Applieations[M]. New York: Academic Press, 1979, 85-91.

[3]　彭秀平. 对称平均数及其基本定理[J]. 湖南数学通讯, 1991, 9(3):39-40.

[4]　匡继昌. 常用不等式[M]. 2版. 长沙: 湖南教育出版社, 1993: 168-169.

[5]　惠州人. 也谈一组不等式的证明[J]. 数学教育研究, 1988(4)：4-7.

第5篇 代数不等式概率证法举例

(石焕南. 工科数学, 1996,12 (3): 146-149)

摘 要: 本文仅借助初等概率论中的一些简单性质, 简洁巧妙地证明一些代数不等式.

关键词: 概率; 代数不等式

用概率论的方法解决其他学科中的一些问题是一个非常重要而且十分有趣的课题. 本文将通过构造适当的概率模型, 仅借助初等概率的一些简单性质, 巧妙简洁地证明一些常见的代数不等式.

例 1 设自然数 $m_j < n_j, j = 1, 2, \cdots, k$, 令

$$A = \sum_{j=1}^{k} n_j, \ B = \sum_{j=1}^{k} m_j$$

则

$$0 < \prod_{j=1}^{k} \binom{n_j}{m_j} \leqslant \binom{A}{B} \tag{1}$$

(见[1]第 112 页第 115 题)

证明 假设盒中装有 $A = \sum\limits_{j=1}^{k} n_j$ 个球, 其中有 n_j 个球标有号码 $j, j = 1, 2, \cdots, k$. 现从中任取 $B = \sum\limits_{j=1}^{k} m_j$ 个球, 则恰好取得 m_1 个 1 号球, m_2 个 2 号球, $\cdots$, m_k 个 k 号球的概率为 $\dfrac{\prod\limits_{j=1}^{k} \binom{n_j}{m_j}}{\binom{A}{B}}$, 而任何事件的概率总是不超过 1 的, 由此即证得式 (1) 右边的不等式, 而左边的不等式显然成立.

例 2 设 $0 \leqslant a, b, c, d \leqslant 1$, 则

$$(a + b - ab)(c + d - cd) \geqslant ac + bd - abcd$$

22

证明　设事件 A, B, C, D 相互独立, 其概率依次为 a, b, c, d, 因

$$(A + B)(C + D) = AC + AD + BC + BD \supset AC + BD$$

由概率的单调性有

$$P[(A + B)(C + D)] \geqslant P(AC + BD)$$

注意到 A, B, C, D 的独立性, 有

$$P(A + B) \cdot P(C + D) \geqslant P(AC + BD) \tag{2}$$

由概率的一般加法公式, 有

$$P(A + B) = P(A) + P(B) - P(A) \cdot P(B) = a + b - ab$$

同理有

$$P(C + D) = c + d - cd, \; P(AC + BD) = ac + bd - abcd$$

将以上三式代入式(2)即得证.

例 3　设 $a_i \in \mathbf{R}_+, i = 1, 2, \ldots, 6$, 且 $a_1 + a_4 = a_2 + a_5 = a_3 + a_6 = k$, 试证

$$a_1 a_5 + a_2 a_6 + a_3 a_4 + \frac{a_1 a_2 a_3}{k} < k^2$$

(《数学通报》(1992 年8 月)问题786)

证明　设事件 A_1, A_2, A_3 相互独立, 且 $P(A_1) = \frac{a_1}{k}, P(A_2) = \frac{a_2}{k}, P(A_3) = \frac{a_3}{k}$, 则

$$P(\overline{A_1}) = 1 - \frac{a_1}{k} = \frac{a_4}{k}, \; P(\overline{A_2}) = 1 - \frac{a_2}{k} = \frac{a_5}{k}$$
$$P(\overline{A_3}) = 1 - \frac{a_3}{k} = \frac{a_6}{k}$$

由概率的一般加法公式, 有

$$P(A_1 + A_2 + A_3) = P(A_1) + P(A_2) + P(A_3) -$$
$$P(A_1)P(A_2) - P(A_2)P(A_3) - P(A_3)P(A_1) +$$
$$P(A_1)P(A_2)P(A_3)$$
$$= \frac{a_1}{k} + \frac{a_2}{k} + \frac{a_3}{k} - \frac{a_1 a_2}{k^2} - \frac{a_2 a_3}{k^2} - \frac{a_3 a_1}{k^2} + \frac{a_1 a_2 a_3}{k^3}$$

$$=\frac{a_1}{k}\left(1-\frac{a_2}{k}\right)+\frac{a_2}{k}\left(1-\frac{a_3}{k}\right)+\frac{a_3}{k}\left(1-\frac{a_1}{k}\right)+\frac{a_1a_2a_3}{k^3}$$

$$=\frac{a_1a_5}{k^2}+\frac{a_2a_6}{k^2}+\frac{a_3a_4}{k^2}+\frac{a_1a_2a_3}{k^3}$$

另外

$$P(A_1+A_2+A_3)=1-P(\overline{A}_1)P(\overline{A}_2)P(\overline{A}_3)=1-\frac{a_4a_5a_6}{k^3}<1$$

例 4　设 $a_k \geqslant 0, k=1,2,\cdots,n$, 且 $\sum_{k=1}^{n} a_k \leqslant \frac{1}{2}$, 则

$$\prod_{k=1}^{n}(1-a_k)\geqslant\frac{1}{2}$$

(波兰 6 5–6 6 中学数学竞赛题)

证明　设 $A_1, A_2, \cdots, A_n$ 相互独立, 且 $P(A_k)=a_k, k=1,2,\cdots,n$, 因

$$P\left(\sum_{k=1}^{n}A_k\right)=1-P\left(\prod_{k=1}^{n}\overline{A}_k\right)=1-\prod_{k=1}^{n}(1-a_k)$$

另外

$$P\left(\sum_{k=1}^{n}A_k\right)\leqslant\sum_{k=1}^{n}P(A_k)=\sum_{k=1}^{n}a_k$$

从而

$$1-\prod_{k=1}^{n}(1-a_k)\leqslant\sum_{k=1}^{n}a_k$$

即

$$\prod_{k=1}^{n}(1-a_k)\geqslant1-\sum_{k=1}^{n}a_k\geqslant\frac{1}{2}$$

例 5　(Weierstrass 不等式) 设 $0<a_k<1, k=1,2,\cdots,n$, 则

$$1-\sum_{k=1}^{n}a_k<\prod_{k=1}^{n}(1-a_k)<\left(1+\sum_{k=1}^{n}a_k\right)^{-1} \tag{3}$$

(见文献 [1] 第 158 页第 88 题).

证明　设 $A_1, A_2, \ldots, A_n$ 相互独立, 且 $P(A_k)=a_k, k=1,2,\ldots,n$, 由于必然事件

$$\Omega=A_1+\overline{A}_1A_2+\overline{A}_1\overline{A}_2A_3+\cdots+\overline{A}_1\cdots\overline{A}_{n-1}A_n+\overline{A}_1\overline{A}_2\cdots\overline{A}_n$$

所以

$$
\begin{aligned}
P(\Omega) = &P(A_1) + P(\overline{A}_1)P(A_2) + P(\overline{A}_1)P(\overline{A}_2)P(A_3) + \cdots + \\
&P(\overline{A}_1)\cdots P(\overline{A}_{n-1})P(A_n) + P(\overline{A}_1)P(\overline{A}_2)\cdots P(\overline{A}_n)
\end{aligned}
$$

即

$$
\begin{aligned}
1 = &a_1 + (1-a_1)a_2 + (1-a_1)(1-a_2)a_3 + \cdots + \\
&(1-a_1)\cdots(1-a_{n-1})a_n + (1-a_1)(1-a_2)\cdots(1-a_n) \\
> &\prod_{k=1}^{n}(1-a_k)(a_1 + a_2 + \cdots + a_n + 1) \\
= &\prod_{k=1}^{n}(1-a_k)\left(1 + \sum_{k=1}^{n}a_k\right)
\end{aligned}
$$

由此即证得式(3)右边不等式. 又

$$
\begin{aligned}
1 = &a_1 + (1-a_1)a_2 + (1-a_1)(1-a_2)a_3 + \cdots + \\
&(1-a_1)\cdots(1-a_{n-1})a_n + (1-a_1)(1-a_2)\cdots(1-a_n) \\
< &a_1 + a_2 + \cdots + a_n + (1-a_1)(1-a_2)\cdots(1-a_n) \\
= &\sum_{k=1}^{n}a_k + \prod_{k=1}^{n}(1-a_k)
\end{aligned}
$$

由此即证得式(3)左边的不等式.

例 6　(Schweitzer 不等式) 设 $0 < p \leqslant a_k \leqslant q, k = 1, 2, \ldots, n$, 则

$$
\left(\frac{1}{n}\sum_{k=1}^{n}a_k\right)\left(\frac{1}{n}\sum_{k=1}^{n}a_k^{-1}\right) \leqslant \frac{(p+q)^2}{4pq}
$$

(见文献[1]第173页第205题(2)).

证明　设随机变量 ξ 的分布列为

$$
P(\xi = a_k) = \frac{1}{n}, k = 1, 2, \ldots, n
$$

则

$$
E\xi = \frac{1}{n}\sum_{k=1}^{n}a_k, \quad E(\xi^{-1}) = \frac{1}{n}\sum_{k=1}^{n}a_k^{-1}
$$

令

$$\eta = (q - \xi)(p^{-1} - \xi^{-1})$$

由于 $\eta \geqslant 0$, 所以 $E\eta \geqslant 0$, 而

$$E\eta = E(q - \xi)(p^{-1} - \xi^{-1}) = E(qp^{-1} - q\xi^{-1} - \xi p^{-1} + 1)$$
$$= qp^{-1} + 1 - qE(\xi^{-1}) - p^{-1}E\xi$$

从而

$$qE(\xi^{-1}) + p^{-1}E\xi \leqslant 1 + qp^{-1}$$

又

$$qE(\xi^{-1}) + p^{-1}E\xi \geqslant 2(qp^{-1}E\xi^{-1}E\xi)^{\frac{1}{2}}$$

则

$$2(qp^{-1}E(\xi^{-1})E\xi)^{\frac{1}{2}} \leqslant 1 + qp^{-1}$$

即

$$E(\xi^{-1})E\xi \leqslant \frac{p}{4q}\left(1 + \frac{q}{p}\right)^2 = \frac{(p+q)^2}{4pq}$$

由此即得证.

例 7 令 $a_1, a_2, \ldots, a_n$ 与 $b_1, b_2, \ldots, b_n$ 为正实数, 且 $\sum\limits_{k=1}^{n} a_k = \sum\limits_{k=1}^{n} b_k$, 求证

$$\sum_{k=1}^{n} \frac{a_k^2}{a_k + b_k} \geqslant \frac{1}{2}\sum_{k=1}^{n} a_k$$

(1991 年亚太地区数学奥林匹克试题).

证明 设随机变量 ξ 的分布列为

$$P\left(\xi = \frac{a_k}{a_k + b_k}\right) = \frac{a_k + b_k}{\sum\limits_{i=1}^{n}(a_i + b_i)}, \ k = 1, 2, \ldots, n$$

则

$$E\xi^2 = \sum_{k=1}^{n} \frac{a_k^2}{(a_k + b_k)^2} \cdot \frac{a_k + b_k}{\sum\limits_{i=1}^{n}(a_i + b_i)}$$
$$= \sum_{k=1}^{n} \frac{a_k^2}{a_k + b_k} \cdot \frac{1}{\sum\limits_{i=1}^{n}(a_i + b_i)}$$

$$E\xi = \sum_{k=1}^{n} \frac{a_k}{a_k + b_k} \cdot \frac{a_k + b_k}{\sum\limits_{i=1}^{n}(a_i + b_i)} = \frac{\sum\limits_{k=1}^{n} a_k}{\sum\limits_{i=1}^{n}(a_i + b_i)} = \frac{1}{2}$$

由于

$$E\xi^2 \geqslant (E\xi)^2$$

则

$$\sum_{k=1}^{n} \frac{a_k^2}{(a_k + b_k)^2} \cdot \frac{1}{\sum\limits_{i=1}^{n}(a_i + b_i)} \geqslant \frac{1}{4}$$

即

$$\sum_{k=1}^{n} \frac{a_k^2}{(a_k + b_k)^2} \geqslant \frac{1}{4} \sum_{i=1}^{n}(a_i + b_i) = \frac{1}{2} \sum_{k=1}^{n} a_k$$

参考文献

[1] 匡继昌. 常用不等式[M]. 2 版.长沙：湖南教育出版社，1993.

第6篇　一类对称函数不等式的控制证明

(石焕南. 成都大学学报(自然科学版), 1998, 17 (4): 22-24)

摘　要:　利用控制不等式理论证明了一类对称函数不等式.

关键词:　不等式; 对称函数; 控制理论

文献[1]用数学分析的方法证明了笔者在文献[2] 末提出的猜想, 即：

定理 1　设 $x_i > 0, i = 1, \cdots, n$, 且 $\sum\limits_{i=1}^{n} x_i \leqslant 1$, 则

$$\frac{E_k(1-\boldsymbol{x})}{E_k(\boldsymbol{x})} \geqslant (n-1)^k \tag{1}$$

其中

$$E_k(\boldsymbol{x}) = E_k(x_1, \cdots, x_n) = \sum_{1 \leqslant i_1 < \ldots < i_k \leqslant n} \prod_{j=1}^{k} x_{i_j}, \quad k = 1, \cdots, n$$

并规定 $E_0(\boldsymbol{x}) = 1$, 当 $k < 0$ 或 $k > n$ 时 $E_k(\boldsymbol{x}) = 0$.

　　本文利用控制不等式理论给出式(1)的一个简洁证明,并建立与式(1)类似的另一个不等式,文中所涉及的有关控制不等式的概念及记号请参见文献[3].

　　引理　　设 $x_i \geqslant 0, i = 1, \cdots, n$, 且 $\sum\limits_{i=1}^{n} x_i = 1$, 则

$$\frac{1-\boldsymbol{x}}{n-1} = \left(\frac{1-x_1}{n-1}, \frac{1-x_2}{n-1}, \cdots, \frac{1-x_n}{n-1} \right) \prec (x_1, x_2, \cdots, x_n) = \boldsymbol{x}$$

　　证明　　不妨设 $x_1 \geqslant x_2 \geqslant \cdots \geqslant x_n$, 则 $\frac{1-x_1}{n-1} \leqslant \frac{1-x_2}{n-1} \leqslant \cdots \leqslant \frac{1-x_n}{n-1}$, 显然 $\sum\limits_{i=1}^{n} \frac{1-x_i}{n-1} = \sum\limits_{i=1}^{n} x_i = 1$, 只需证

$$\sum_{i=1}^{k} x_i \geqslant \sum_{i=1}^{k} \frac{1-x_{n-i+1}}{n-1}, k = 1, 2, \cdots, n-1$$

而此式等价于

$$(n-1)\sum_{i=1}^{k}x_i + \sum_{i=1}^{k}x_{n-i+1} \geqslant k, k=1,2,\cdots,n-1 \tag{2}$$

$$(n-k)x_1 + \sum_{i=1}^{k}x_{n-i+1} + \sum_{j=2}^{k}[(n-k)x_j + \sum_{i=1}^{k}x_i]$$

注意上式左边可写作

$$(n-k)x_1 + s\sum_{i=1}^{k}x_{n-i+1} \geqslant \sum_{i=1}^{n-k}x_i + \sum_{i=1}^{k}x_{n-i+1} = \sum_{i=1}^{n}x_i = 1$$

$$(n-k)x_j + \sum_{i=1}^{k}x_i \geqslant \sum_{i=k+1}^{n}x_i + \sum_{i=1}^{k}x_i = \sum_{i=1}^{n}x_i = 1, j=2,3,\cdots,k$$

因此式(2)成立.

定理 1 的证明　不妨设 $\sum_{i=1}^{n}x_i = 1$, 由文献[3]知 $E_k(x)$ 在 $\mathbf{R}_+^n$ 上是Schur凹函数, 结合引理有

$$\frac{E_k(1-\boldsymbol{x})}{(n-1)^k} = E_k\left(\frac{1-\boldsymbol{x}}{n-1}\right) \geqslant E_k(\boldsymbol{x})$$

即式(1)成立.

注意到当 $x_i \geqslant 0, i=1,...,n$, 且 $\sum_{i=1}^{n}x_i = 1$ 时, 有

$$\frac{1-\boldsymbol{x}}{n+1} = \left(\frac{1+x_1}{n-1}, \frac{1+x_2}{n-1}, ..., \frac{1+x_n}{n-1}\right) \prec (x_1, x_2, ..., x_n) = \boldsymbol{x}$$

类似地可证得:

定理 2　设 $x_i > 0, i=1,\cdots,n$, 且 $\sum_{i=1}^{n}x_i \leqslant 1$, 则

$$\frac{E_k(1+\boldsymbol{x})}{E_k(\boldsymbol{x})} \geqslant (n+1)^k, k=1,2,\cdots,n$$

一般地,有:

定理 3　设 $x_i > 0, i=1,\cdots,n$, 且 $\sum_{i=1}^{n}x_i = s, c \geqslant 0$, 则

$$\frac{E_k(c+\boldsymbol{x})}{E_k(\boldsymbol{x})} \geqslant \left(\frac{nc}{s}+1\right)^k, \ k=1,2,\cdots,n$$

证明 不难验证

$$\frac{c+\boldsymbol{x}}{s+nc} = \left(\frac{c+x_1}{s+nc}, \frac{c+x_2}{s+nc}, ..., \frac{c+x_n}{s+nc}\right) \prec \left(\frac{x_1}{s}, \frac{x_2}{s}, ..., \frac{x_n}{s}\right) = \frac{\boldsymbol{x}}{s}$$

事实上, 显然有 $\sum\limits_{i=1}^{n} \frac{c+x_i}{s+nc} = \sum\limits_{i=1}^{n} \frac{x_i}{s} = 1$, 又

$$\sum_{i=1}^{k} \frac{x_i}{s} \geqslant \sum_{i=1}^{k} \frac{c+x_i}{s+nc}$$

$$\Leftrightarrow (s+nc) \sum_{i=1}^{k} x_i \geqslant s \left(kc + \sum_{i=1}^{k} x_i\right)$$

$$\Leftrightarrow \sum_{i=1}^{k} x_i \geqslant \frac{sk}{n}$$

而此式成立, 假若不然, 即存在 k 使得 $\sum\limits_{i=1}^{k} x_i < \frac{sk}{n}$, 则

$$\frac{s}{n} > \frac{1}{k} \sum_{i=1}^{k} x_i \geqslant x_k \geqslant x_{k+1} \geqslant \cdots \geqslant x_n$$

从而

$$s = \sum_{i=1}^{n} x_i = \sum_{i=1}^{k} x_i + \sum_{i=k+1}^{n} x_i < \frac{sk}{n} + (n-k)\frac{s}{n} = s$$

矛盾. 利用此控制不等式及 $E_k(\boldsymbol{x})$ 的Schur凹性即可证得定理了.

参考文献

[1] 汤子赓. 一个初等对称函数不等式的加强[J].数学通报, 1997,10：44-46.

[2] 石焕南. 关于对称函数的一类不等式[J]. 数学通报, 1996,3:38-40.

[3] 王伯英. 控制不等式基础[M]. 北京:北京师范大学出版社, 1990.

[4] 赖立. 一类对称函数的Schur – 凹性[J]. 成都大学学报(自然科学版), 1997, 16(3):5-7.

第7篇　一类对称函数不等式的加细与推广

(石焕南. 数学的实践与认识, 1999, 29 (4): 81-85)

　摘　要:　利用控制不等式理论加细和推广了一类对称函数不等式, 并给出一个几何应用.

　关键词:　不等式; 对称函数; 控制理论; 单形

7.1　引言

文献[1]用数学分析的方法证明了笔者在文献[2]末提出的猜想,即:

　命题 1　设 $x_i > 0, i = 1, \cdots, n$, 且 $\sum\limits_{i=1}^{n} x_i \leqslant 1$, 则

$$\frac{E_k(1 - \boldsymbol{x})}{E_k(\boldsymbol{x})} \geqslant (n-1)^k \tag{1}$$

其中

$$E_k(\boldsymbol{x}) = E_k(x_1, \ldots, x_n) = \sum_{1 \leqslant i_1 < \cdots < i_k \leqslant n} \prod_{j=1}^{k} x_{i_j}, \quad k = 1, \cdots, n$$

并规定 $E_0(\boldsymbol{x}) = 1$, 当 $k < 0$ 或 $k > n$ 时, $E_k(\boldsymbol{x}) = 0$.

　　最近笔者在文献[3]中利用 "近几年才形成一门新兴学科的控制不等式理论"[4]给出式(1)一个简洁的证明, 并建立了与式(1)类似的另一个不等式, 即:

　命题 2　设 $x_i > 0. i = 1, \cdots, n$, 且 $\sum\limits_{i=1}^{n} x_i = s, c \geqslant 0$, 则

$$\frac{E_k(1 + \boldsymbol{x})}{E_k(\boldsymbol{x})} \geqslant \left(\frac{nc}{s} + 1\right)^k, k = 1, \cdots, n \tag{2}$$

本文将继续利用这一理论对(1),(2)两式作加细和推广, 并给出式(1)的推

广式的一个几何应用.

7.2 两个控制不等式

仿照文献[3]的引理的证明可得:

引理 1   设 $x_i \in \mathbf{R}, i = 1, \cdots, n, n \geqslant 2$, 且 $\sum_{i=1}^{n} x_i = s$, 则

$$\frac{s - \boldsymbol{x}}{n-1} = \left(\frac{s - x_1}{n-1}, \frac{s - x_2}{n-1}, ..., \frac{s - x_n}{n-1} \right) \prec (x_1, x_2, \cdots, x_n) = \boldsymbol{x} \qquad (3)$$

与文献[3]中引理比较, 这里不仅将 $\sum_{i=1}^{n} x_i = 1$ 放宽为 $\sum_{i=1}^{n} x_i = s$, 而且取消了诸 x_i 为正数的限制.

引理2   设 $x_i \in \mathbf{R}, i = 1, \cdots, n, n \geqslant 2$, 且 $\sum_{i=1}^{n} x_i = s > 0, c \geqslant s$, 则

$$\frac{c - \boldsymbol{x}}{\frac{nc}{s} - 1} = \left(\frac{c - x_1}{\frac{nc}{s} - 1}, \frac{c - x_2}{\frac{nc}{s} - 1}, \frac{c - x_n}{\frac{nc}{s} - 1} \right) \prec (x_1, x_2, \cdots, x_n) = \boldsymbol{x} \qquad (4)$$

证明   不妨设 $x_1 \geqslant x_2 \geqslant \cdots \geqslant x_n$, 则

$$\frac{c - x_1}{\frac{nc}{s} - 1} \leqslant \frac{c - x_2}{\frac{nc}{s} - 1} \leqslant \cdots \leqslant \frac{c - x_n}{\frac{nc}{s} - 1}$$

显然

$$\sum_{i=1}^{n} \frac{c - x_i}{\frac{nc}{s} - 1} = \sum_{i=1}^{n} x_i = s$$

为证式(4)只需证

$$\sum_{i=1}^{k} x_i \geqslant \sum_{i=1}^{k} \frac{c - x_i}{\frac{nc}{s} - 1}, k = 1, 2, \cdots, n - 1$$

上式等价于

$$(nc - s) \sum_{i=1}^{k} x_i + s \sum_{i=1}^{k} x_{n-i+1} \geqslant kcs \qquad (5)$$

类似于文献[3]中引理的证明可得

$$(n - 1) \sum_{i=1}^{k} x_i + \sum_{i=1}^{k} x_{n-i+1} \geqslant ks \qquad (6)$$

再有显然的不等式

$$(n-1)\sum_{i=1}^{k} x_i - \sum_{i=1}^{k} x_{n-i+1} \geqslant 0 \tag{7}$$

于是经 $(6) \times c + (7) \times (c-s)$ 即得式 (5),所以式 (4) 成立. 当 $c = s$ 时, 式 (4) 化为式 (3), 不过式 (4) 限制 $\sum_{i=1}^{n} x_i = s > 0$.

7.3　主要结果及其证明

定理 1　设 $x_i \in \mathbf{R}, i = 1, ..., n$, 且 $\sum_{i=1}^{n} x_i = s, c > s, 0 < \alpha \leqslant 1$, 则

$$\frac{E_k[(c-\boldsymbol{x})^{\alpha}]}{E_k(\boldsymbol{x}^{\alpha})} \geqslant \left(\frac{nc}{s} - 1\right)^{\alpha} \cdot \frac{E_{k-1}[(c-\boldsymbol{x})^{\alpha}]}{E_{k-1}(\boldsymbol{x}^{\alpha})}, \quad k = 1, 2, ..., n \tag{8}$$

证明　由文献[5]知 $\varphi(\boldsymbol{x}) = \dfrac{E_k(\boldsymbol{x})}{E_{k-1}(\boldsymbol{x})}$ 是 $\mathbf{R}_{++}^n = \{\boldsymbol{x} | x_1 > 0, ..., x_n > 0\}$ 上的 Schur 凹函数, 又由文献[6]的定理 1 知 $\varphi(\boldsymbol{x})$ 是增函数, 由于当 $0 < \alpha \leqslant 1$ 时, x^{α} 是凹函数, 据文献[4]的命题 6.16(b) 知 $\varphi(\boldsymbol{x}^{\alpha})$ 亦是 $\mathbf{R}_{++}^n$ 上的 Schur 凹函数, 由式 (4) 有

$$\varphi\left[\left(\frac{c-\boldsymbol{x}}{\frac{nc}{s}-1}\right)^{\alpha}\right] \geqslant \varphi(\boldsymbol{x}^{\alpha})$$

由此即可得式 (8).

定理 2　设 $x_i \in \mathbf{R}, i = 1, \cdots, n, n \geqslant 2$, 且 $\sum_{i=1}^{n} x_i = s, c > 0, 0 < \alpha \leqslant 1$, 则

$$\frac{E_k[(c+\boldsymbol{x})^{\alpha}]}{E_k(\boldsymbol{x}^{\alpha})} \geqslant \left(\frac{nc}{s} + 1\right)^{\alpha} \cdot \frac{E_{k-1}[(c-\boldsymbol{x})^{\alpha}]}{E_{k-1}(\boldsymbol{x}^{\alpha})}, \quad k = 1, 2, \cdots, n \tag{9}$$

证明　文献[3]中的定理 3 证明了控制不等式

$$\frac{c+\boldsymbol{x}}{s+nc} = \left(\frac{c+x_1}{s+nc}, \frac{c+x_2}{s+nc}, ..., \frac{c+x_n}{s+nc}\right) \prec \left(\frac{x_1}{s}, \frac{x_2}{s}, ..., \frac{x_n}{s}\right) = \frac{\boldsymbol{x}}{s}$$

由定理 1 中的 $\varphi(\boldsymbol{x}^{\alpha})$ 的 Schur 凹性, 有

$$\varphi\left[\left(\frac{c+\boldsymbol{x}}{s+nc}\right)^{\alpha}\right] \geqslant \varphi\left[\left(\frac{\boldsymbol{x}}{s}\right)^{\alpha}\right]$$

由此即得式 (9).

由定理 1 和定理 2 立得如下推论.

推论 1　设 $x_i \in \mathbf{R}, i = 1, \cdots, n, n \geqslant 2$, 且 $\sum_{i=1}^{n} x_i = s, c > s, 0 < \alpha \leqslant 1$,

则

$$\frac{E_k[(c-\boldsymbol{x})^\alpha]}{E_k(\boldsymbol{x}^\alpha)} \geqslant \left(\frac{nc}{s}-1\right)^\alpha \cdot \frac{E_{k-1}[(c-\boldsymbol{x})^\alpha]}{E_{k-1}(\boldsymbol{x}^\alpha)}$$

$$\geqslant \left(\frac{nc}{s}-1\right)^{2\alpha} \cdot \frac{E_{k-2}[(c-\boldsymbol{x})^\alpha]}{E_{k-2}(\boldsymbol{x}^\alpha)}$$

$$\geqslant \cdots \geqslant \left(\frac{nc}{s}-1\right)^{(k-1)\alpha} \cdot \frac{E_1[(c-\boldsymbol{x})^\alpha]}{E_1(\boldsymbol{x}^\alpha)}$$

$$\geqslant \left(\frac{nc}{s}-1\right)^{k\alpha} \cdot \frac{E_0[(c-\boldsymbol{x})^\alpha]}{E_0(\boldsymbol{x}^\alpha)} \tag{10}$$

取 $s=c=\alpha=1$, 则式(10)给出了式(1)的加细.

推论 2 设 $x_i \in \mathbf{R}, i=1,\cdots,n, n \geqslant 2$, 且 $\sum\limits_{i=1}^{n} x_i = s, c>0, 0<\alpha \leqslant 1$, 则

$$\frac{E_k[(c+\boldsymbol{x})^\alpha]}{E_k(\boldsymbol{x}^\alpha)} \geqslant \left(\frac{nc}{s}+1\right)^\alpha \cdot \frac{E_{k-1}[(c+\boldsymbol{x})^\alpha]}{E_{k-1}(\boldsymbol{x}^\alpha)}$$

$$\geqslant \left(\frac{nc}{s}+1\right)^{2\alpha} \cdot \frac{E_{k-2}[(c+\boldsymbol{x})^\alpha]}{E_{k-2}(\boldsymbol{x}^\alpha)}$$

$$\geqslant \cdots \geqslant \left(\frac{nc}{s}-1\right)^{(k-1)\alpha} \cdot \frac{E_1[(c+\boldsymbol{x})^\alpha]}{E_1(\boldsymbol{x}^\alpha)}$$

$$\geqslant \left(\frac{nc}{s}+1\right)^{k\alpha} \cdot \frac{E_0[(c+\boldsymbol{x})^\alpha]}{E_0(\boldsymbol{x}^\alpha)} \tag{11}$$

若取 $k=n, s=c=1$, 则由式(1)便得到Klamkin 不等式[7]

$$\prod_{i=1}^{n} \frac{1+x_i}{x_i} \geqslant (n+1)^n$$

的加细.

定理 3 设 $x_i \in \mathbf{R}, i=1,\cdots,n, n \geqslant 2$, 且 $\sum\limits_{i=1}^{n} x_i = s$, 则

$$\frac{E_2[(s-\boldsymbol{x})]}{E_2(\boldsymbol{x})} \geqslant (n-1)^2 \tag{12}$$

等式成立的充要条件是 $x_1 = x_2 = \cdots = x_n$.

证明 由文献[4]的命题6.7 知 $E_2(\boldsymbol{x})$ 在 $\mathbf{R}^n$ 上是严格Schur凹函数, 结合式(3)即得证.

7.4　一个几何应用

下面利用式(8)给出与单形内点有关的一个不等式, 采用文献[8]中的记号. 设 $A = A_1 A_2 \cdots A_{n+1}$ 是 $E_n(n \geqslant 3)$ 中的一个单形, P 是 A 内任一点, B_i 是直线 $A_i P$ 与超平面 $a_i = A_1 \cdots A_{i-1} A_{i+1} \cdots A_{n+1}$ 的交点, 易知

$$\sum_{i=1}^{n+1} \frac{PB_i}{A_i B_i} = 1, \frac{A_i P}{A_i B_i} = 1 - \frac{PB_i}{A_i B_i}, i = 1, 2, \ldots, n+1$$

取 $s = c = 1$, 由式(10)有

$$E_k \left[\left(\frac{A_1 P}{A_1 B_1} \right)^\alpha, \cdots, \left(\frac{A_{n+1} P}{A_{n+1} B_{n+1}} \right)^\alpha \right]$$

$$\geqslant n^{k\alpha} E_k \left[\left(\frac{PB_1}{A_1 B_1} \right)^\alpha, \cdots, \left(\frac{PB_{n+1}}{A_{n+1} B_{n+1}} \right)^\alpha \right] \tag{13}$$

由文献[8]的2.7和2.3, 分别有

$$E_k \left[\left(\frac{A_1 P}{A_1 B_1} \right)^\alpha, \cdots, \left(\frac{A_{n+1} P}{A_{n+1} B_{n+1}} \right)^\alpha \right] \leqslant C_{n+1}^k \left(\frac{n}{n+1} \right)^{k\alpha} \tag{14}$$

和

$$E_k \left[\left(\frac{PB_1}{A_1 B_1} \right)^\alpha, \cdots, \left(\frac{PB_{n+1}}{A_{n+1} B_{n+1}} \right)^\alpha \right] \leqslant \left(\frac{1}{n+1} \right)^{k\alpha} \tag{15}$$

注意式(13)可写作

$$E_k \left[\left(\frac{PB_1}{A_1 B_1} \right)^\alpha, \cdots, \left(\frac{PB_{n+1}}{A_{n+1} B_{n+1}} \right)^\alpha \right]$$

$$\leqslant C_{n+1}^k \left(\frac{1}{n+1} \right)^{k\alpha} \cdot \frac{E_k \left[\left(\frac{A_1 P}{A_1 B_1} \right)^\alpha, \cdots, \left(\frac{A_{n+1} P}{A_{n+1} B_{n+1}} \right)^\alpha \right]}{C_{n+1}^k \left(\frac{n}{n+1} \right)^{k\alpha}} \tag{16}$$

由式(14)知式(13)加强了式(15).

参考文献

[1]　汤子赓. 一个初等对称函数不等式的加强[J]. 数学通报, 1997,10: 44-46.

[2]　石焕南. 关于对称函数的一类不等式[J]. 数学通报, 1996, 3: 38-40.

[3]　石焕南. 一类对称函数不等式的控制证明[J]. 成都大学学报(自然科学版), 1998, 4: 22-24.

[4]　王伯英. 控制不等式基础[M]. 北京:北京师范大学出版社, 1990.

[5] MARSHALL A W, OLKIN I. Inequalities: Theory of Mjorization and Its Applications[M]. New York: Academic Press, 1979.

[6] 密特里诺维奇,瓦西奇. 分析不等式[M]. 赵汉滨,译. 南宁:广西人民出版社, 1986.

[7] 匡继昌. 常用不等式[M]. 2版. 长沙:湖南教育出版社, 1993.

[8] MITRINOVIĆ D S, PEČARIĆ J E, VOLENEC V. Recent Advances in Geometric Inequalities[M]. Dordrecht: Kluwer Academic Publishers, 1989.

第8篇 初等对称函数差的Schur凸性

(石焕南. 湖南教育学院学报, 1999, 17 (5): 135-138)

摘　要: 本文讨论了初等对称函数差$E_k(\boldsymbol{x}) - E_{k-1}(\boldsymbol{x})$ 在n 维单形 $\Omega_n = \{\boldsymbol{x} = (x_1, \cdots, x_n) \in \mathbf{R}_+^n : E_1(\boldsymbol{x}) \leqslant 1\}$ 和n 维立方体$\Omega_n' = \{\boldsymbol{x} = (x_1, \cdots, x_n) \in \mathbf{R}_+^n : 0 \leqslant x_i \leqslant 1, i = 1, \cdots, n\}$ 上的Schur 凸性.
关键词: 初等对称函数; Schur凸性; 不等式; 单形

初等对称函数

$$E_k(\boldsymbol{x}) = E_k(x_1, \ldots, x_n) = \sum_{1 \leqslant i_1 < \ldots < i_k \leqslant n} \prod_{j=1}^k x_{i_j}, \quad k = 1, \cdots, n$$

是$\mathbf{R}_+^n = \{\boldsymbol{x} = (x_1, \cdots, x_n) : x_i \geqslant 0, i = 1, \cdots, n\}$ 上的一类重要的Schur 凹函数[1], 其商函数$\frac{E_k(\boldsymbol{x})}{E_{k-1}(\boldsymbol{x})}$ 是$\mathbf{R}_{++}^n = \{\boldsymbol{x} = (x_1, \cdots, x_n) : x_i > 0, i = 1, \cdots, n\}$ 上的Schur 凹函数[2], 由文献[1] 的推论6.15 知积函数 $E_k(\boldsymbol{x}) \cdot E_{k-1}(\boldsymbol{x})$ 是$\mathbf{R}_+^n$ 上的Schur 凹函数, 此外易知和函数$E_k(\boldsymbol{x}) + E_{k-1}(\boldsymbol{x})$ 也是$\mathbf{R}_+^n$ 上的Schur 凹函数, 由此很自然地猜想差函数 $E_k(\boldsymbol{x}) - E_{k-1}(\boldsymbol{x})$ 亦是$\mathbf{R}_+^n$ 的Schur 凹函数. 但实际上此猜想不成立. 例如, 在$\mathbf{R}_+^3$ 中取$\boldsymbol{x} = (1, 1, 1), \boldsymbol{y} = (2, 0.5, 0.5)$, 则$\boldsymbol{x} \prec \boldsymbol{y}$, 而$E_3(\boldsymbol{x}) - E_2(\boldsymbol{x}) = -2 < E_3(\boldsymbol{y}) - E_2(\boldsymbol{y}) = -1.75$, 若取$\boldsymbol{x} = (5, 5, 5), \boldsymbol{y} = (10, 3, 2)$, 则$\boldsymbol{x} \prec \boldsymbol{y}$, 但$E_3(\boldsymbol{x}) - E_2(\boldsymbol{x}) = 50 > E_3(\boldsymbol{y}) - E_2(\boldsymbol{y}) = 4$. 这说明在$\mathbf{R}_+^n$ 中$E_k(\boldsymbol{x}) - E_{k-1}(\boldsymbol{x})$ 的Schur 凸性不确定. 退而求其次, 本文将在$\mathbf{R}_+^n$ 中的两个重要对称凸集上讨论$E_k(\boldsymbol{x}) - E_{k-1}(\boldsymbol{x})$ 的Schur 凸性, 一个是n 维单形$\Omega_n = \{\boldsymbol{x} = (x_1, ..., x_n) \in \mathbf{R}_+^n : E_1(\boldsymbol{x}) \leqslant 1\}$, 另一个是$n$ 维立方体 $\Omega_n' = \{\boldsymbol{x} = (x_1, ..., x_n) \in \mathbf{R}_+^n : 0 \leqslant x_i \leqslant 1, i = 1, ..., n\}$. 文中规定$E_0(\boldsymbol{x}) = 1$, 当$k > n$ 或$k < 0$ 时, $E_k(\boldsymbol{x}) = 0$.

8.1 主要结果及证明

引理 设$\boldsymbol{x} \in \mathbf{R}_+^n, k = 1, 2, \cdots, n$. 若$E_1(\boldsymbol{x}) \leqslant \frac{nk}{n-k+1}$, 则$E_{k-1}(\boldsymbol{x}) \geqslant E_k(\boldsymbol{x})$.

证明 由Maclaurin 不等式[3] 有

$$\frac{E_{k-1}(\boldsymbol{x})}{\mathrm{C}_n^{k-1}} \geqslant \left[\frac{E_k(\boldsymbol{x})}{\mathrm{C}_n^k}\right]^{\frac{k-1}{k}}$$

和

$$\frac{E_1(\boldsymbol{x})}{\mathrm{C}_n^1} \geqslant \left[\frac{E_k(\boldsymbol{x})}{\mathrm{C}_n^k}\right]^{\frac{1}{k}}$$

两式对应相乘得

$$\frac{E_{k-1}(\boldsymbol{x})E_1(\boldsymbol{x})}{\mathrm{C}_n^{k-1}\mathrm{C}_n^1} \geqslant \frac{E_k(\boldsymbol{x})}{\mathrm{C}_n^k}$$

即

$$E_{k-1}(\boldsymbol{x}) \geqslant \frac{\mathrm{C}_n^{k-1}\mathrm{C}_n^1}{\mathrm{C}_n^k E_1(\boldsymbol{x})} E_k(\boldsymbol{x})$$

注意到

$$E_1(\boldsymbol{x}) \leqslant \frac{nk}{n-k+1} \Leftrightarrow \frac{\mathrm{C}_n^{k-1}\mathrm{C}_n^1}{\mathrm{C}_n^k E_1(\boldsymbol{x})} \geqslant 1$$

即知引理成立.

定理 1 当$3 \leqslant k \leqslant n$ 时, $\varphi(\boldsymbol{x}) = E_k(\boldsymbol{x}) - E_{k-1}(\boldsymbol{x})$ 是Ω_n 上的减的Schur 凸函数, 当$k = 2$ 时, $\varphi(\boldsymbol{x})$ 是Ω_n 上的减的Schur 凹函数.

证明 利用恒等式

$$E_k(x_1,\cdots,x_n) = x_1 E_{k-1}(x_2,\cdots,x_n) + E_k(x_2,\cdots,x_n)$$

可得

$$\frac{\partial \varphi(\boldsymbol{x})}{\partial x_1} = E_{k-1}(x_2,\cdots,x_n) + E_{k-2}(x_2,\cdots,x_n) \tag{1}$$

由于当$2 \leqslant k \leqslant n$ 时

$$E_1(x_2,\cdots,x_n) \leqslant 1 + \frac{n(k-2)}{n-k+1} = \frac{(n-1)(k-1)}{(n-1)-(k-1)+1}$$

据引理知$\frac{\partial \varphi(\boldsymbol{x})}{\partial x_1} \leqslant 0$, 同理$\frac{\partial \varphi(\boldsymbol{x})}{\partial x_i} \leqslant 0, i = 1,\cdots,n$, 据文献[1]的命题4.3 (a) 知, 当$2 \leqslant k \leqslant n$ 时, $\varphi(\boldsymbol{x})$ 是Ω_n 上的减函数. 利用恒等式

$$E_k(x_1,\cdots,x_n) = x_1 x_2 E_{k-2}(x_3,\cdots,x_n) + (x_1 + x_2) E_{k-1}(x_3,\cdots,x_n) + $$
$$E_k(x_3,\cdots,x_n)$$

易得Schur 条件

$$(x_1-x_2)\left(\frac{\partial\varphi(\boldsymbol{x})}{\partial x_1}-\frac{\partial\varphi(\boldsymbol{x})}{\partial x_2}\right)=(x_1-x_2)^2[E_{k-3}(x_2,\cdots,x_n)-E_{k-2}(x_3,\cdots,x_n)]$$

$$(2)$$

当$k=2$ 时, 上式$=(x_1-x_2)^2(0-1)\leqslant 0$, 当$3\leqslant k\leqslant n$ 时, 因

$$E_1(x_3,\cdots,x_n)\leqslant 1+\frac{(k-3)(n-1)}{n-k+1}=\frac{(n-2)(k-2)}{(n-2)-(k-2)+1}$$

由引理知式(2)非负, 由此得证.

定理 2 设$n\geqslant 2$, 当$n\geqslant k\geqslant (n+3)/2$ 时, $\varphi(\boldsymbol{x})=E_k(\boldsymbol{x})-E_{k-1}(\boldsymbol{x})$ 是Ω_n' 上的减的Schur 凸函数, 当$k=2$ 时, $\varphi(\boldsymbol{x})$ 是Ω_n' 上的减的Schur 凹函数.

证明 易见当$n\geqslant k\geqslant(n+3)/2$ 时, 有

$$E_1(x_2,\cdots,x_n)\leqslant n-1\leqslant\frac{(n-1)(k-1)}{(n-1)-(k-1)+1}$$

和

$$E_1(x_3,\cdots,x_n)\leqslant n-2\leqslant\frac{(n-2)(k-2)}{(n-2)-(k-2)+1}$$

由引理知式(1)非正, 而式(2)非负, 由此得证.

推论 1 当$\lambda\geqslant 1,3\leqslant k\leqslant n$ 时, $\varphi(\boldsymbol{x},\lambda)=E_k(\boldsymbol{x})-\lambda E_{k-1}(\boldsymbol{x})$ 是Ω_n 上的减的Schur 凸函数, 当$0<\lambda\leqslant 1,k=2$ 时, $\varphi(\boldsymbol{x},\lambda)$ 是Ω_n 上的Schur 凹函数.

证明 由定理1, 当$3\leqslant k\leqslant n$ 时, $\varphi(\boldsymbol{x})=E_k(\boldsymbol{x})-E_{k-1}(\boldsymbol{x})$ 是Ω_n 上的减的Schur 凸函数, 而由文献[1]知$E_k(\boldsymbol{x})$ 是$\mathbf{R}_+^n$ 上的增的Schur 凹函数, 故$-E_k(\boldsymbol{x})$ 是$\mathbf{R}_+^n$上的减的Schur 凸函数. 从而, 当$\lambda\geqslant 1$ 时, $\varphi(\boldsymbol{x},\lambda)=E_k(\boldsymbol{x})-\lambda E_{k-1}(\boldsymbol{x})=\lambda[E_k(\boldsymbol{x})-E_{k-1}(\boldsymbol{x})]+(\lambda-1)[-E_k(\boldsymbol{x})]$ 是Ω_n 上的两个减的Schur 凸函数的非负线性组合, 因而也是$\mathbf{R}_+^n$ 上的减的Schur 凸函数. 当$k=2$ 时, 由定理1, $E_k(\boldsymbol{x})-E_{k-1}(\boldsymbol{x})$ 是Ω_n 上的Schur 凹函数, 而当$0<\lambda\leqslant 1,k=2$ 时, $\varphi(\boldsymbol{x},\lambda)=E_k(\boldsymbol{x})-\lambda E_{k-1}(\boldsymbol{x})=\lambda[E_k(\boldsymbol{x})-E_{k-1}(\boldsymbol{x})]+(1-\lambda)E_k(\boldsymbol{x})$ 是Ω_n 上的两个Schur 凹函数的非负线性组合, 因而亦是Ω_n 上的Schur 凹函数.

类似地由定理2 可得:

推论 1' 设$n\geqslant 2,\lambda\geqslant 1$, 当$n\geqslant k\geqslant(n+3)/2$ 时, $\varphi(\boldsymbol{x},\lambda)=E_k(\boldsymbol{x})-\lambda E_{k-1}(\boldsymbol{x})$ 是Ω_n' 上的减的Schur 凸函数, 当$0<\lambda\leqslant 1,k=2$ 时, $\varphi(\boldsymbol{x},\lambda)$ 是Ω_n' 上的Schur 凹函数.

推论 2 当$\lambda\geqslant 1,n\geqslant k>j\geqslant 2$ 时, $\varphi(\boldsymbol{x},\lambda)=E_k(\boldsymbol{x})-\lambda E_j(\boldsymbol{x})$ 是Ω_n

上的减的Schur 凸函数.

证明 注意

$$\varphi(\boldsymbol{x},\lambda) = [E_k(\boldsymbol{x})-\lambda E_{k-1}(\boldsymbol{x})]+[E_{k-1}(\boldsymbol{x})-E_{k-2}(\boldsymbol{x})]+\cdots+[E_{j+1}(\boldsymbol{x})-\lambda E_j(\boldsymbol{x})]$$

由推论1 即得证.

类似地由推论1′ 可得:

推论 2′ 设$n \geqslant 2, \lambda \geqslant 1$, 当$n \geqslant k > j \geqslant (n+3)/2$ 时, $\varphi(\boldsymbol{x},\lambda) = E_k(\boldsymbol{x})-\lambda E_j(\boldsymbol{x})$ 是Ω_n' 上的减的Schur 凸函数, 当$0 < \lambda \leqslant 1, k = 2$ 时, $\varphi(\boldsymbol{x},\lambda)$ 是Ω_n' 上的减的Schur 凹函数.

推论 3 设$\lambda \geqslant 1$, $\boldsymbol{A} = \{\boldsymbol{x} \in \mathbf{R}_+^n : 0 < E_1(\boldsymbol{x}) \leqslant s\}$. 当$n \geqslant k > j \geqslant 2$ 时, $\varphi(\boldsymbol{x},\lambda) = E_k(\boldsymbol{x}) - \lambda s^{k-j} E_j(\boldsymbol{x})$ 是$\boldsymbol{A}$ 上的Schur 凸函数.

证明 由推论2, $\psi(\boldsymbol{x},\lambda) = \lambda E_j(\boldsymbol{x}) - E_k(\boldsymbol{x})$ 是Ω_n 上的增的Schur 凹函数, 而$g(x) = \frac{x}{s}$ 是凹函数, 由文献[2]的命题6.16 (c) 知

$$\psi[g(\boldsymbol{x}),\lambda] = \lambda E_j\left(\frac{\boldsymbol{x}}{s}\right) - E_k\left(\frac{\boldsymbol{x}}{s}\right) = \lambda \frac{E_j(\boldsymbol{x})}{s^j} - \frac{E_k(\boldsymbol{x})}{s^k}$$

是$\boldsymbol{A}$ 上的Schur 凹函数, 从而$-s^k\psi[g(\boldsymbol{x}),\lambda] = E_k(\boldsymbol{x})-\lambda s^{k-j} E_j(\boldsymbol{x}) = \varphi(\boldsymbol{x},\lambda)$ 是$\boldsymbol{A}$ 上的Schur 凸函数.

8.2 几例应用

例 1 设$\boldsymbol{x} \in \mathbf{R}_+^n$. 若$E_1(\boldsymbol{x}) = s, 0 < \alpha \leqslant 1, 3 \leqslant k \leqslant n$ 则

$$0 \leqslant Q_k(\boldsymbol{x}) \leqslant Q_{k-1}(\boldsymbol{x}) \leqslant \cdots \leqslant Q_3(\boldsymbol{x}) \leqslant Q_2(\boldsymbol{x}) \tag{3}$$

其中

$$Q_m(\boldsymbol{x}) = s^{k-m}(n \pm 1)^{(k-m)\alpha}[E_m((s \pm \boldsymbol{x})^\alpha) - (n \pm 1)^{m\alpha}E_m(\boldsymbol{x}^\alpha)]$$

证明 由定理1, 当$3 \leqslant k \leqslant n$ 时, $\varphi(\boldsymbol{x}) = E_k(\boldsymbol{x}) - E_{k-1}(\boldsymbol{x})$ 是Ω_n 上的减的Schur 凸函数, 又当$0 < \alpha \leqslant 1$ 时, x^α 是凹函数, 据文献[1]的命题6.16 (c), $\varphi(\boldsymbol{x}^\alpha)$ 亦是Ω_n 上的Schur 凸函数, 记

$$\frac{\boldsymbol{x}}{s} = \left(\frac{x_1}{s}, \frac{x_2}{s}, \cdots, \frac{x_n}{s}\right), \quad \boldsymbol{y} = \left(\frac{1 \pm \frac{x_1}{s}}{n \pm 1}, \frac{1 \pm \frac{x_2}{s}}{n \pm 1}, \cdots, \frac{1 \pm \frac{x_n}{s}}{n \pm 1}\right)$$

则$\frac{\boldsymbol{x}}{s} \in \Omega_n$, $\boldsymbol{y} \in \Omega_n$, 且由文献[4]知$\boldsymbol{y} \prec \frac{\boldsymbol{x}}{s}$, 因而有$\varphi(\boldsymbol{y}^\alpha) \leqslant \varphi\left(\left(\frac{\boldsymbol{x}}{s}\right)^\alpha\right)$, 稍加整理即得$Q_k(\boldsymbol{x}) \leqslant Q_{k-1}(\boldsymbol{x})$. 反复应用上式即可得式(3), 式(3)左端不等式对

于$k = 1, 2$ 亦成立, 它实际上是文献[4]的定理1 和定理2 的直接结果.

例 2　沿用文献[5]中的记号. 设$A = A_1 A_2 \cdots A_{n+1}$ 是$E_n (n \geqslant 3)$ 中的一个单形, P 是A 内任一点, B_i 是直线 $A_i P$ 与超平面

$$a_i = A_1 \cdots A_{i-1} A_{i+1} \cdots A_{n+1}$$

的交点, 则对于$3 \leqslant k \leqslant n+1, 0 < \alpha \leqslant 1$, 有

$$E_k \left(\left(\frac{A_1 P}{A_1 B_1} \right)^\alpha, \cdots, \left(\frac{A_{n+1} P}{A_{n+1} B_{n+1}} \right)^\alpha \right) \geqslant C_{n+1}^k \left(\frac{1}{n+1} \right)^{k\alpha} -$$

$$\left[C_{n+1}^{k+1} \left(\frac{1}{n+1} \right)^{(k-1)\alpha} - E_{k-1} \left(\left(\frac{PB_1}{A_1 B_1} \right)^\alpha, \cdots, \left(\frac{PB_{n+1}}{A_{n+1} B_{n+1}} \right)^\alpha \right) \right] \quad (4)$$

证明　记$x_i = \frac{PB_i}{A_i B_i}, i = 1, \cdots, n+1$, 则$\sum\limits_{i=1}^{n+1} x_i = 1$, 与例1 同理, $\varphi(\boldsymbol{x}^\alpha)$ 是Ω_{n+1} 上的Schur 凸函数, 记

$$\boldsymbol{x} = (x_1, x_2, \cdots, x_{n+1}), \boldsymbol{y} = \left(\frac{1}{n+1}, \frac{1}{n+1}, \cdots, \frac{1}{n+1} \right)$$

则$\boldsymbol{x} \in \Omega_{n+1}$, $\boldsymbol{y} \in \Omega_{n+1}$, 且$\boldsymbol{y} \prec \boldsymbol{x}$, 从而有$\varphi(\boldsymbol{y}^\alpha) \leqslant \varphi(\boldsymbol{x}^\alpha)$. 由此稍加整理即得式(4), 当$k \geqslant 3$ 时, 式(4)给出文献[5]的式2.3的一个反向不等式.

例 3　在$\triangle ABC$ 中, 有

$$\frac{9}{4} \geqslant \prod \cos \frac{A}{2} + \frac{9}{4} - \frac{3\sqrt{3}}{8}$$

$$\geqslant \sum \cos \frac{A}{2} \cos \frac{B}{2} \geqslant \sin A \sin B + \left(\prod \cos \frac{A}{2} - \prod \sin A \right) \quad (5)$$

其中$\prod$ 表循环积, $\sum$ 表循环和.

证明　当$n = k = 3$ 时, 由定理2 知$\varphi(\boldsymbol{x}) = E_3(\boldsymbol{x}) - E_2(\boldsymbol{x})$ 是Ω_3' 上的Schur 凸函数, 又$\sin x$ 是$I = (0, \pi)$ 上的凹函数, 据文献[1]的命题6.16 (c) 知$\varphi(\sin \boldsymbol{x})$ 是I^3 上的Schur 凸函数, 记$\boldsymbol{x} = \left(\frac{\pi}{3}, \frac{\pi}{3}, \frac{\pi}{3} \right)$, $\boldsymbol{y} = \left(\frac{\pi-A}{2}, \frac{\pi-B}{2}, \frac{\pi-C}{2} \right)$, $\boldsymbol{z} = (A, B, C)$, 易见$\boldsymbol{x} \prec \boldsymbol{y} \prec \boldsymbol{z}$, 因而$\varphi(\sin \boldsymbol{x}) \leqslant \varphi(\sin \boldsymbol{y}) \leqslant \varphi(\sin \boldsymbol{z})$, 由此稍加整理即得式(5)中第2个和第3个不等式, 而第1 个不等式由熟知的不等式$\prod \cos \frac{A}{2} \leqslant \frac{3\sqrt{3}}{8}$ 即得证, 由已知不等式$\prod \cos \frac{A}{2} \geqslant \prod \sin A$[6] 知式(5)的第3 个不等式是已知不等式$\sum \cos \frac{A}{2} \cos \frac{B}{2} \geqslant \sum \sin A \cos B$[6] 的加强, 而式(5)第1个和第2个不等式给出已知不等式$\cos \frac{A}{2} \cos \frac{B}{2} \leqslant \frac{9}{4}$[6] 的加细.

参考文献

[1] 王伯英. 控制不等式基础[M]. 北京:北京师范大学出版社, 1990.

[2] MARSHALL A W, OLKIN I. Inequalities: Theory of Mjorization and Its Applications[M]. New York: Academic Press, 1979.

[3] HARDY G H, LITTLEWOOD J E, POLYA G. Inequalities [M]. Cambridge: Cambridge Univ. Press, 1952: 52

[4] 石焕南. 一类对称函数不等式的控制证明[J]. 工科数学, 1999, 15 (3): 140-142.

[5] MITRINOVIĆ D S, PEČARIĆ J E, VOLENEC V. Recent Advances in Geometric Inequalities[M]. Dordrecht: Kluwer Academic Publishers, 1989:473-515.

[6] 匡继昌. 常用不等式[M]. 2版. 长沙:湖南教育出版社, 1993: 259-261.

第9篇 积分不等式概率证法举例

(石焕南. 高等数学研究, 1999年专刊: 31-33)

摘　要: 本文仅借助初等概率论中的一些简单性质, 简洁巧妙地证明一些积分不等式.

关键词: 概率; 积分不等式

设 ξ 为定义在概率空间 (Ω, F, P) 上的实随机变量, 其分布函数为 $F(x)$, $g(x)$ 为一 Borel 可测函数, 则当积分 $\int_{-\infty}^{+\infty} |g(x)| \mathrm{d}F(x) < \infty$ 时, 则称 $g(\xi)$ 的均值存在, 记为 $Eg(\xi)$

$$Eg(\xi) = \int_{-\infty}^{+\infty} g(x) \mathrm{d}F(x) \tag{1}$$

若 $g(x)$ 是定义在某区间上的连续向上凹的函数, 则有

$$g(E\xi) \leqslant Eg(\xi) \tag{2}$$

若 $g(x)$ 是向上凸的函数, 则上述不等式反向[1].

式(1)是概率论中一个非常深刻的结果, 式(2)是著名的 Jensen 不等式, 利用 (1),(2) 两式及概率的一些基本性质可以简洁地证明一些常见的积分不等式, 本文略举数例.

例 1　若 $f(x)$ 在 $[0,1]$ 上连续, 且 $f(x) > 0$, 证明

$$\int_0^1 f(x)\mathrm{d}x \cdot \int_0^1 \frac{1}{f(x)}\mathrm{d}x \geqslant 1$$

(1988年北京信息学院数学竞赛试题).

证明　设 ξ 在区间 $[0,1]$ 上服从均匀分布, $\eta = f(\xi)$, $g(x) = \frac{1}{x}$ 在 $(0, +\infty)$ 上凹, 则由式(2)有

$$g(E\eta) = g[Ef(\xi)] = \frac{1}{\int_0^1 f(x)\mathrm{d}x}$$

43

$$\leqslant Eg(\eta) = E\left(\frac{1}{n}\right) = \int_0^1 \frac{1}{f(x)}\mathrm{d}x$$

例 2　设 $f(x), g(x)$ 是区间 $[a,b]$ 上的正值可积函数, 且 $\int_a^b f(x)\mathrm{d}x = 1$, 则有

$$\int_a^b f(x)\ln g(x)\mathrm{d}x \leqslant \ln\int_a^b f(x)g(x)\mathrm{d}x$$

证明　设 ξ 的密度为

$$p(x) = \begin{cases} f(x), & \text{当 } a \leqslant x \leqslant b \\ 0, & \text{其他} \end{cases}$$

$\eta = g(\xi), \varphi(x) = \ln x$ 在 $(0,+\infty)$ 上凸, 则

$$\varphi E(\eta) = \ln E[g(\xi)] = \ln\int_a^b f(x)g(x)\mathrm{d}x$$
$$\geqslant E\varphi(\eta) = E[\ln g(\xi)] = \int_a^b f(x)\ln g(x)\mathrm{d}x$$

例 3　设 f 在区间 $[0,1]$ 上连续, 且 $0 \leqslant f(x) \leqslant 1$, 则

$$\int_0^1 \frac{f(x)}{1-f(x)}\mathrm{d}x \geqslant \frac{\int_0^1 f(x)\mathrm{d}x}{1-\int_0^1 f(x)\mathrm{d}x}$$

证明　设 ξ 在 $[0,1]$ 上服从均匀分布, $\eta = f(\xi), g(x) = \frac{x}{1-x}$ 在 $(0,1)$ 上凹, 则

$$g(E\eta) = \frac{E\eta}{1-E\eta} = \frac{\int_0^1 f(x)\mathrm{d}x}{1-\int_0^1 f(x)\mathrm{d}x} \leqslant Eg(\eta) = \int_0^1 \frac{f(x)}{1-f(x)}\mathrm{d}x$$

例 4　设 f, g 是区间 $[0,1]$ 上的正值可积函数, 则

$$\int_a^b f(x)g(x)\mathrm{d}x \geqslant \frac{1}{b-a}\left(\int_a^b \sqrt{f(x)g(x)}\mathrm{d}x\right)^2$$

(见文献 [2] 第 76 题 (1))

证明　设 ξ 在 $[a,b]$ 上服从均匀分布, $\eta = f(\xi)g(\xi), \varphi(x) = \sqrt{x}$ 在 $(0,+\infty)$ 上凸, 则

$$\varphi(E\eta) = \sqrt{\frac{1}{b-a}\int_a^b f(x)g(x)\mathrm{d}x}$$

false
44

$$\geqslant E\varphi(\eta) = \frac{1}{b-a}\int_a^b \sqrt{f(x)g(x)}\mathrm{d}x$$

由此即得证.

例 5[3]　设 $f(x)$ 在 $x > 0$ 时为非负连续函数, 且 $\int_{-\infty}^{+\infty} f(x)\mathrm{d}x = 1$, 若定义 g 为

$$g(t) = \int_{-\infty}^{+\infty} f(x)\cos tx\mathrm{d}x, t\text{为实数}$$

则

$$g(2t) > 2[g(t)]^2 - 1, t \neq 0$$

证明　设 ξ 的密度函数为 $f(x)$, 注意 x^2 在 $(0, +\infty)$ 上凹, 有

$$g(2t) = \int_{-\infty}^{+\infty} f(x)(2\cos^2 tx - 1)\mathrm{d}x = 2\int_{-\infty}^{+\infty} f(x)\cos^2 tx\mathrm{d}x -$$

$$\int_{-\infty}^{+\infty} f(x)\mathrm{d}x = 2E(\cos^2 t\xi) - 1$$

$$\geqslant 2E(\cos t\xi)^2 - 1 = 2[g(t)]^2 - 1$$

例 6　若 f 在 $[x_1, x_2]$ 上可积, 且 $f > 0$, 则

$$\frac{x_2 - x_1}{\int_{x_1}^{x_2}\frac{\mathrm{d}x}{f(x)}} \leqslant \exp\left\{\frac{1}{x_2 - x_1}\int_{x_1}^{x_2}\ln f(x)\mathrm{d}x\right\}$$

$$\leqslant \frac{1}{x_2 - x_1}\int_{x_1}^{x_2} f(x)\mathrm{d}x$$

证明　设 ξ 在 $[x_1, x_2]$ 上服从均匀分布, $\eta_1 = f(\xi), \eta_2 = \frac{1}{f(\xi)}$, 因 $\varphi(x) = \ln x$ 在 $(0, +\infty)$ 上凸, 则

$$\varphi(E\eta_1) = \ln(E\eta_1) = \ln\left(\frac{1}{x_2 - x_1}\int_{x_1}^{x_2} f(x)\mathrm{d}x\right)$$

$$\geqslant E\varphi(\eta_1) = \frac{1}{x_2 - x_1}\int_{x_1}^{x_2} \ln f(x)\mathrm{d}x$$

$$\varphi(E\eta_2) = \ln(E\eta_2) = \ln\left(\frac{1}{x_2 - x_1}\int_{x_1}^{x_2} f^{-1}(x)\mathrm{d}x\right)$$

$$\geqslant E\varphi(\eta_2) = \frac{1}{x_2 - x_1}\int_{x_1}^{x_2} \ln[f(x)]^{-1}\mathrm{d}x$$

稍加整理便可由以上两式证得例6.

例 7 证明对 $a > 0$, 有

$$\frac{\sqrt{2\pi\left(1 - e^{\frac{a^2}{2}}\right)}}{2} \leqslant \int_0^a e^{-\frac{x^2}{2}}\mathrm{d}x \leqslant \frac{\sqrt{2\pi\left(1 - e^{\frac{2a^2}{\pi}}\right)}}{2}$$

证明 设二维随机变量 (ξ, η) 的联合密度函数为 $\frac{1}{2\pi}e^{-\frac{1}{2}(x^2+y^2)}$.

令

$$D = \{(x,y) : 0 \leqslant x \leqslant a, 0 \leqslant y \leqslant a\}$$

$$D_1 = \left\{(x,y) : x \geqslant 0, y \geqslant 0, x^2 + y^2 \leqslant a^2\right\}$$

$$D_2 = \left\{(x,y) : x \geqslant 0, y \geqslant 0, x^2 + y^2 \leqslant \frac{4a^2}{\pi}\right\}$$

则

$$P((\xi,\eta) \in D) = \frac{1}{2\pi}\int_0^a\int_0^a e^{-\frac{1}{2}(x^2+y^2)}\mathrm{d}x\mathrm{d}y$$

$$= \frac{1}{2\pi}\left[\int_0^a e^{-\frac{x^2}{2}}\mathrm{d}x\right]^2$$

$$P((\xi,\eta) \in D_1) = \frac{1}{2\pi}\int_0^a\int_0^a e^{-\frac{1}{2}(x^2+y^2)}\mathrm{d}x\mathrm{d}y$$

$$= \frac{1}{4}\left(1 - e^{-\frac{a^2}{2}}\right)$$

$$P((\xi,\eta) \in D_2) = \frac{1}{2\pi}\int_0^a\int_0^a e^{-\frac{1}{2}(x^2+y^2)}\mathrm{d}x\mathrm{d}y$$

$$= \frac{1}{4}\left(1 - e^{-\frac{2a^2}{\pi}}\right)$$

我们有

$$P((\xi,\eta) \in D_1) \leqslant P((\xi,\eta) \in D) \leqslant P((\xi,\eta) \in D_2).$$

第一个不等式是因 $D_1 \subset D$, 而第二个不等式是因 D 和 D_2 的面积相等且当 $(x,y) \in D - D_2$ 时, $e^{-\frac{1}{2}(x^2+y^2)} \leqslant e^{-\frac{2a^2}{\pi}}$, 当 $(x,y) \in D_2 - D$ 时, $e^{-\frac{1}{2}(x^2+y^2)} \geqslant e^{-\frac{2a^2}{\pi}}$.

参考文献

[1] 周概容. 概率论与数理统计[M]. 北京: 高等教育出版社, 1984 : 275-276.

[2] 匡继昌. 常用不等式[M]. 2版. 长沙：湖南教育出版社，1993.

[3]　MITRIONOVIĆ D S, VASIĆ P M. 分析不等式[M]. 赵汉滨,译. 南宁: 广西人民出版社, 1986: 396.

[4]　波利亚 G, 舍贵 G. 数学分析中的问题与定理[M]. 张奠宙,等,译. 上海: 上海科学技术出版社, 1981: 67.

第10篇 一类对称函数不等式的加强, 推广及应用

(石焕南. 北京联合大学学报(自然科学版), 1999, 13 (2): 51-55)

摘 要: $E_k(\boldsymbol{x})$, $E_k(1-\boldsymbol{x})$ 和 $E_k(1+\boldsymbol{x})$ 分别表示 $x_1, ..., x_n$; $1-x_1, \cdots, 1-x_n$ 和 $1 + x_1, \cdots, 1 + x_n$ 的第 k 个初等对称函数, 借助控制不等式理论加强和推广了关于对称函数的一类不等式的结论, 主要结果为:

$(1)(C_{n-1}^k)^r s^{rk\alpha} \leqslant E_k^r(s - \boldsymbol{x}^\alpha) - \lambda E_k^r(\boldsymbol{x}^\alpha) \leqslant (C_n^k)^r s^{rk\alpha}[(1 - \frac{1}{n^\alpha})^{kr} - \lambda(\frac{1}{n^\alpha})^{kr}], r \geqslant 1$

$(2)(C_{n-1}^k)^r s^{rk\alpha} \leqslant E_k^r(s - \boldsymbol{x}^\alpha) + \lambda E_k^r(\boldsymbol{x}^\alpha) \leqslant (C_n^k)^r s^{rk\alpha}[(1 - \frac{1}{n^\alpha})^{kr} + \lambda(\frac{1}{n^\alpha})^{kr}], r > 0$ 其中 $n \geqslant 3, x_i > 0, i = 1, \cdots, n, \sum\limits_{i=1}^{n} x_i = s, \alpha \geqslant 1, k = 2, \cdots, n, 0 \leqslant \lambda \leqslant \max\{1, (n-1)^{k(r-1)}\}$.

关键词: 不等式; 对称函数; 控制理论; 单形

10.1 问题的提出

笔者在文献[1]中用初等方法建立了如下不等式

$$0 \leqslant E_k(1 - \boldsymbol{x}) - E_k(\boldsymbol{x}) \leqslant C_n^k \left[\left(1 - \frac{1}{n}\right)^k - \left(\frac{1}{n}\right)^k \right] \tag{1}$$

其中 $\boldsymbol{x} \in \Omega = \{\boldsymbol{x} | x_i > 0, i = 1, \cdots, n, \sum\limits_{i=1}^{n} x_i = 1\}$, $k = 1, \cdots, n$, $E_k(\boldsymbol{x})$ 是初等对称函数, 即

$$E_k(\boldsymbol{x}) = E_k(x_1, \cdots, x_n) = \sum_{1 \leqslant i_1 < \cdots < i_k \leqslant n} \prod_{j=1}^{k} x_{i_j}$$

并在文末提出了一个猜想

$$\frac{E_k(1-\boldsymbol{x})}{E_k(\boldsymbol{x})} \geqslant (n-1)^k \tag{2}$$

其中 $\boldsymbol{x} \in \Omega_1 = \{\boldsymbol{x}|x_i > 0, i = 1, \cdots, n, \sum_{i=1}^{n} x_i \leqslant 1\}$, $k = 1, \cdots, n$.

　　本文首先用初等方法证明此猜想成立, 然后利用控制不等式理论对式(1)作进一步的加强和推广, 最后给出(1)和(2) 两式在几何上的一个应用。

10.2　猜想的证明

引理 1

$$E_j(\boldsymbol{x}) \geqslant n^{k-j} \frac{\mathrm{C}_n^j}{\mathrm{C}_n^k} E_k(\boldsymbol{x})$$

其中 $\boldsymbol{x} \in \Omega, 0 \leqslant j \leqslant k \leqslant n$.

　　证明　规定 $E_0(\boldsymbol{x}) = 1$, 由Marclaurin 不等式[2]有

$$\frac{E_{k-1}(\boldsymbol{x})}{\mathrm{C}_n^{k-1}} \geqslant \left[\frac{E_k(\boldsymbol{x})}{\mathrm{C}_n^k}\right]^{\frac{k-1}{k}}$$

$$\frac{E_1(\boldsymbol{x})}{\mathrm{C}_n^1} \geqslant \left[\frac{E_k(\boldsymbol{x})}{\mathrm{C}_n^k}\right]^{\frac{1}{k}}$$

两式对应相乘得

$$\frac{E_{k-1}(\boldsymbol{x})}{\mathrm{C}_n^{k-1}} \cdot \frac{E_1(\boldsymbol{x})}{\mathrm{C}_n^1} \geqslant \frac{E_k(\boldsymbol{x})}{\mathrm{C}_n^k}$$

连续应用上式并注意 $E_1(\boldsymbol{x}) = 1$ 即得证.

　　引理 2

$$\mathrm{C}_{n-j}^{k-j} E_j(\boldsymbol{x}) - \mathrm{C}_{n-(j+1)}^{k-(j+1)} E_{j+1}(\boldsymbol{x})$$
$$\geqslant \mathrm{C}_k^j n^{k-j} E_k(\boldsymbol{x}) - \mathrm{C}_k^{j+1} n^{k-(j+1)} E_k(\boldsymbol{x})$$

其中 $\boldsymbol{x} \in \Omega, 0 \leqslant j < j+1 \leqslant k \leqslant n$.

　　证明　由引理1有

$$E_j(\boldsymbol{x}) \geqslant \frac{n\mathrm{C}_n^j}{\mathrm{C}_n^{j+1}} E_{j+1}(\boldsymbol{x})$$

因而

$$\mathrm{C}_{n-j}^{k-j} E_j(\boldsymbol{x}) - \mathrm{C}_{n-(j+1)}^{k-(j+1)} E_{(j+1)}(\boldsymbol{x})$$

$$\geqslant \left[\mathrm{C}_{n-j}^{k-j} \frac{n\mathrm{C}_n^j}{\mathrm{C}_n^{j+1}} - \mathrm{C}_{n-(j+1)}^{k-(j+1)} \right] E_{j+1}(\boldsymbol{x})$$

注意

$$\mathrm{C}_{n-j}^{k-j} \frac{\mathrm{C}_n^j}{\mathrm{C}_n^{j+1}} - \mathrm{C}_{n-(j+1)}^{k-(j+1)} = \mathrm{C}_{n-j}^{k-j} \left[\frac{n(j+1)}{n-j} - \frac{k+j}{n-j} \right] \geqslant 0$$

再由引理1 有

$$\mathrm{C}_{n-j}^{k-j} E_j(\boldsymbol{x}) - \mathrm{C}_{n-(j+1)}^{k-(j+1)} E_{(j+1)}(\boldsymbol{x})$$
$$\geqslant \left[\mathrm{C}_{n-j}^{k-j} \frac{n\mathrm{C}_n^j}{\mathrm{C}_n^{j+1}} - \mathrm{C}_{n-(j+1)}^{k-(j+1)} \right] n^{k-(j+1)} \frac{\mathrm{C}_n^{j+1}}{\mathrm{C}_n^k} E_k(\boldsymbol{x})$$
$$= \mathrm{C}_k^j n^{k-j} E_k(\boldsymbol{x}) - \mathrm{C}_k^{j+1} n^{k-(j+1)} E_k(\boldsymbol{x})$$

猜想的证明　不妨设 $\sum\limits_{i=1}^{n} x_i = 1$, 我们知道

$$\prod_{j=1}^{k}(1-a_j) = 1 - E_1(\boldsymbol{a}) + E_2(\boldsymbol{a}) + \cdots + (-1)^k E_k(\boldsymbol{a})$$

利用此性质不难验证

$$E_k(1-\boldsymbol{x}) = \sum_{1 \leqslant i_1 < \cdots < i_k \leqslant n} \prod_{j=1}^{k}(1-x_{i_j}) = \sum_{j=0}^{k}(-1)^j \mathrm{C}_{n-j}^{k-j} E_j(\boldsymbol{x})$$

若 k 为奇数, 设 $k = 2m+1$, 由引理2 有

$$E_k(1-\boldsymbol{x})$$
$$= \sum_{j=0}^{m} \left[\mathrm{C}_{n-2j}^{k-2j} E_{2j}(\boldsymbol{x}) - \mathrm{C}_{n-(2j+1)}^{k-(2j+1)} E_{2j+1}(\boldsymbol{x}) \right]$$
$$\geqslant \sum_{j=0}^{m} \left[\mathrm{C}_k^{2j} n^{k-2j} E_k(\boldsymbol{x}) - \mathrm{C}_k^{2j+1} n^{k-(2j+1)} E_k(\boldsymbol{x}) \right]$$
$$= \left[\sum_{i=0}^{k}(-1)^i \mathrm{C}_k^i n^{k-i} \right] E_k(\boldsymbol{x})$$
$$= (n-1)^k E_k(\boldsymbol{x})$$

若 k 为偶数, 设 $k = 2m$, 则

$$E_k(1-\boldsymbol{x})$$

$$= \sum_{j=0}^{m-1} \left[C_{n-2j}^{k-2j} E_{2j}(\boldsymbol{x}) - C_{n-(2j+1)}^{k-(2j+1)} E_{2j+1}(\boldsymbol{x}) \right] +$$

$$C_{n-2m}^{k-2m} E_{2m}(\boldsymbol{x})$$

$$\geqslant \sum_{j=0}^{m-1} \left[C_k^{2j} n^{k-2j} E_k(\boldsymbol{x}) - C_k^{2j+1} n^{k-(2j+1)} E_k(\boldsymbol{x}) \right] + E_k(\boldsymbol{x})$$

$$= \left[\sum_{i=0}^{k} (-1)^i C_k^i n^{k-i} \right] E_k(\boldsymbol{x}) = (n-1)^k E_k(\boldsymbol{x})$$

10.3　式(1)的加强与推广

定理 1

$$(C_{n-1}^k)^r \leqslant E_k^r(1-\boldsymbol{x}^\alpha) - \lambda E_k^r(\boldsymbol{x}^\alpha) \leqslant (C_n^k)^r [(1-\frac{1}{n^\alpha})^{kr} - \lambda(\frac{1}{n^\alpha})^{kr}] \quad (3)$$

其中 $n \geqslant 3, x_i > 0, i = 1, \cdots, n, \sum_{i=1}^n x_i = 1, r \geqslant 1, \alpha \geqslant 1, k = 2, \cdots, n, 0 \leqslant \lambda \leqslant \max\{1, (n-1)^{k(r-1)}\}$. 当 $k = 1$ 时, 式(3) 左端为 $(C_{n-1}^r)^r - 1$.

证明　当 $k > n$ 时, 规定 $C_n^k = 0$. 考虑函数

$$\phi(\boldsymbol{x}) = E_k^r(1-\boldsymbol{x}) - \lambda E_k^r(\boldsymbol{x})$$

易知

$$E_k(\boldsymbol{x}) = x_1 E_{k-1}(x_2, \cdots, x_n) + E_k(x_2, \cdots, x_n)$$

因而

$$\frac{\partial \phi(\boldsymbol{x})}{\partial x_1} = -r[E_k^{r-1}(1-\boldsymbol{x})E_{k-1}(1-x_2, \cdots, 1-x_n) +$$

$$\lambda E_k^{r-1}(\boldsymbol{x})E_{k-1}(x_2, \cdots, x_n)] \leqslant 0$$

同理, $\frac{\partial \phi(\boldsymbol{x})}{\partial x_i} \leqslant 0, i = 1, 2, \cdots, n$. 据文献[1]的命题4.3 (a) 知 $\phi(\boldsymbol{x})$ 在 Ω 上是减函数, 利用关系式

$$E_k(\boldsymbol{x}) = x_1 x_2 E_{k-2}(x_3, \cdots, x_n) +$$

$$(x_1 + x_2) E_{k-1}(x_3, \cdots, x_n) + E_k(x_3, \cdots, x_n)$$

可得

$$\frac{\partial \phi(\boldsymbol{x})}{\partial x_1} = rE_k^{r-1}(1-\boldsymbol{x})\cdot$$
$$[(x_2-1)E_{k-2}(1-x_3,\cdots,1-x_n)-E_{k-1}(1-x_3,\cdots,1-x_n)]-$$
$$\lambda rE_k^{r-1}(\boldsymbol{x})[x_2E_{k-2}(x_3,\cdots,x_n)+E_{k-1}(x_3,\cdots,x_n)]$$

$$\frac{\partial \phi(\boldsymbol{x})}{\partial x_2} = rE_k^{r-1}(1-\boldsymbol{x})\cdot$$
$$[(x_1-1)E_{k-2}(1-x_3,\cdots,1-x_n)-E_{k-1}(1-x_3,\cdots,1-x_n)]-$$
$$\lambda rE_k^{r-1}(\boldsymbol{x})[x_1E_{k-2}(x_3,\cdots,x_n)+E_{k-1}(x_3,\cdots,x_n)]$$

从而Schur 条件是

$$(x_1-x_2)\left(\frac{\partial \phi(\boldsymbol{x})}{\partial x_1}-\frac{\partial \phi(\boldsymbol{x})}{\partial x_2}\right)=-r(x_1-x_2)^2\cdot$$
$$\left[E_k^{r-1}(1-\boldsymbol{x})E_{k-2}(1-x_3,\cdots,1-x_n)-\lambda E_k^{r-1}(\boldsymbol{x})E_{k-2}(x_3,\cdots,x_n)\right]$$

由式(2) 知上式不大于零, 据文献[1]的定理6.5 (a) 知$\phi(\boldsymbol{x})$ 是Ω 上的Schur 凹函数, 又x^α 当$\alpha \geqslant 1$ 时是凸函数, 由文献[1]的命题6.16 (d) 知

$$\psi(\boldsymbol{x}) = \phi(x_1^\alpha, x_2^\alpha, \cdots, x_n^\alpha)$$

亦是Ω 上的Schur 凹函数, 从而由

$$\left(\frac{1}{n},\frac{1}{n},\cdots,\frac{1}{n}\right) \prec (x_1,x_2,\cdots,x_n) \prec (1,0,\cdots,0)$$

有

$$\psi\left(\frac{1}{n},\frac{1}{n},\cdots,\frac{1}{n}\right) \geqslant \psi(x_1,x_2,\cdots,x_n) \geqslant \psi(1,0,\cdots,0)$$

即式(3)成立.

式(3)不仅从两方面对式(1)作了指数推广, 而且还加强了左边的不等式. 由定理1 立得:

推论

$$(C_{n-1}^k)^r s^{rk\alpha} \leqslant E_k^r(s-\boldsymbol{x}^\alpha)-\lambda E_k^r(\boldsymbol{x}^\alpha) \leqslant (C_n^k)^r s^{rk\alpha}[(1-\frac{1}{n^\alpha})^{kr}-\lambda(\frac{1}{n^\alpha})^{kr}] \quad (4)$$

其中$n \geqslant 3, x_i > 0, i = 1,\cdots,n, \sum_{i=1}^n x_i = s, r \geqslant 1, \alpha \geqslant 1, k = 2,\cdots,n, 0 \leqslant$

$\lambda \leqslant \max\{1, (n-1)^{k(r-1)}\}$. 当$k = 1$ 时, 式(4)左端为$[(\mathrm{C}_{n-1}^r)^r - 1]s^{r\alpha}$.

在式(4)中令$k = r = \lambda = 1$, 可得幂平均不等式

$$\frac{1}{n}\sum_{i=1}^{n} x_i^\alpha \geqslant \left(\frac{1}{n}\sum_{i=1}^{n} x_i\right)^\alpha, \alpha \geqslant 1$$

采用类似的证法可得:

定理2

$$(\mathrm{C}_{n-1}^k)^r s^{rk\alpha} \leqslant E_k^r(s-\boldsymbol{x}^\alpha) + \lambda E_k^r(\boldsymbol{x}^\alpha) \leqslant (\mathrm{C}_n^k)^r s^{rk\alpha}[(1-\frac{1}{n^\alpha})^{kr} + \lambda(\frac{1}{n^\alpha})^{kr}] \quad (5)$$

其中$n \geqslant 3, x_i > 0, i = 1, \cdots, n, \sum_{i=1}^{n} x_i = s, r > 0, \alpha \geqslant 1, k = 2, \cdots, n, 0 \leqslant \lambda \leqslant \max\{1, (n-1)^{k(r-1)}\}$. 当$k = 1$ 时, 式(5) 左端为$[(\mathrm{C}_{n-1}^r)^r + 1]s^{r\alpha}$. 注意不同于式(4), 这里的$r$ 可放宽为$r > 0$.

10.4　一个几何应用

本节利用(2),(3) 两式给出与单形内点有关的两个不等式, 采用文献[4]中的记号. 设$A = A_1 A_2 \cdots A_{n+1}$ 是$E_n(n \geqslant 3)$ 中的一个单形, P 是A 内任一点, B_i 是直线 $A_i P$ 与超平面$a_i = A_1 \cdots A_{i-1}A_{i+1}\cdots A_{n+1}$ 的交点, 易知

$$\sum_{i=1}^{n+1}\frac{PB_i}{A_iB_i} = 1, \frac{A_iP}{A_iB_i} = 1 - \frac{PB_i}{A_iB_i}, i = 1, 2, \cdots, n+1$$

从而当$k = 2, 3, \cdots, n$ 时, 由式(3)有

$$\mathrm{C}_n^k \leqslant E_k\left(\frac{A_1P}{A_1B_1}, \cdots, \frac{A_{n+1}P}{A_{n+1}B_{n+1}}\right) -$$
$$E_k\left(\frac{PB_1}{A_1B_1}, \cdots, \frac{PB_{n+1}}{A_{n+1}B_{n+1}}\right)$$
$$\leqslant \mathrm{C}_{n+1}^k\left[\left(1-\frac{1}{n+1}\right)^k - \left(\frac{1}{n+1}\right)^k\right]$$

即

$$\mathrm{C}_n^k + E_k\left(\frac{PB_1}{A_1B_1}, \cdots, \frac{PB_{n+1}}{A_{n+1}B_{n+1}}\right) \leqslant E_k\left(\frac{A_1P}{A_1B_1}, \cdots, \frac{A_{n+1}P}{A_{n+1}B_{n+1}}\right)$$

$$\leqslant \mathrm{C}_{n+1}^{k} \left(\frac{n}{n+1} \right)^{k} - \left[\mathrm{C}_{n+1}^{k} \left(\frac{1}{n+1} \right)^{k} - E_{k} \left(\frac{PB_1}{A_1 B_1}, \cdots, \frac{PB_{n+1}}{A_{n+1} B_{n+1}} \right) \right] \quad (6)$$

由文献[4, 479]的2.7 ($\alpha = 1$) 和[4, 476]的2.3 ($\alpha = 1$) 分别有

$$E_{k} \left(\frac{A_1 P}{A_1 B_1}, \cdots, \frac{A_{n+1} P}{A_{n+1} B_{n+1}} \right) \leqslant \mathrm{C}_{n+1}^{k} \left(\frac{n}{n+1} \right)^{k} \quad (7)$$

$$E_{k} \left(\frac{PB_1}{A_1 B_1}, \cdots, \frac{PB_{n+1}}{A_{n+1} B_{n+1}} \right) \leqslant \mathrm{C}_{n+1}^{k} \left(\frac{1}{n+1} \right)^{k} \quad (8)$$

由此可见式(6)不仅加强了式(7), 并给出了式(7)一个下界估计.

参考文献

[1] 石焕南. 关于对称函数的一类不等式[J]. 数学通报, 1996,3:38-40.

[2] 匡继昌. 常用不等式[M].2版. 长沙: 湖南教育出版社, 1993.

[3] 王伯英, 控制不等式基础[M]. 北京:北京师范大学出版社, 1990.

[4] MITRINOVIĆ D S, PEČARIĆ J E.VOLENEC V. Recent Advances in Geometric Inequalities[M]. Dordrecht: Kluwer Academic Publishers, 1989.

第11篇　整值随机变量期望的一个表示式的应用与推广

(石焕南. 辽宁师范大学学报(自然科学版), 1999, 13 (2): 102-105)

摘　要:　本文建立了整值随机变量高阶矩的一个递推公式

$$E\xi^{r+1} = (r+1)\left[\sum_{k=1}^{\infty} k^r P(\xi \geqslant k) - \sum_{k=1}^{r} a_k E\xi^k\right]$$

其中a_k 满足等式$\sum_{k=1}^{n} k^r = \sum_{k=1}^{r+1} a_k n^k$, 此公式表示式$E\xi = \sum_{k=1}^{\infty} P(\xi \geqslant k)$ 的推广, 此外给出此公式的一些应用.

关键词:　整值随机变量; 矩; 递推公式

在离散型随机变量中整值随机变量, 即只取非负整数的随机变量, 占有重要的地位, 事实上, 像二项分布、泊松分布、超几何分布、几何分布这些常见分布均是整值随机变量, 对于整值随机变量有一个很有用的期望计算公式: $E\xi = \sum_{k=1}^{\infty} P(\xi \geqslant k)$. 该公式一般教科书很少提及, 有的只是作为习题出现, 本文将给出该公式的一些应用实例, 并讨论整值随机变量一般原点矩的此类计算公式.

引理 1[1]　　若整值随机变量ξ 的期望存在, 则

$$E\xi = \sum_{k=1}^{\infty} P(\xi \geqslant k) \tag{1}$$

注　若ξ只取有限个值$0, 1, \cdots, n$, 则

$$E\xi = \sum_{k=1}^{n} P(\xi \geqslant k)$$

有时候$P(\xi \geqslant k)$ 比$P(\xi = k)$ 易于求得且形式简单, 因此用式(1)计算期

望比直接用定义 $E\xi = \sum\limits_{k=1}^{\infty} kP(\xi = k)$ 计算要来得容易些, 下面略举数例.

例 1 从 1 到 n 这 n 个自然数中有放回地随机抽取 r 个数 $(r \leqslant n)$, ξ 表示取出的最大数, 求 $E\xi$.

解 $P(\xi \geqslant k) = 1 - P(\xi < k) = 1 - \frac{(k-1)^r}{n^r}, k = 1, \cdots, n$

$$E\xi = \sum_{k=1}^{n} P(\xi \geqslant k) = \sum_{k=1}^{n}\left[1 - \frac{(k-1)^r}{n^r}\right] = n - \frac{1}{n^r}\sum_{k=1}^{n-1} k^r$$

例 2 从装有 m 个红球与 n 个白球的袋中不放回地逐个取球, 直到取得红球为止, 求所取出的球数 ξ 的期望值.

解 不难算得

$$P(\xi = k) = \frac{C_{m+n-k}^{m-1}}{C_{m+n}^{m}}, k = 1, 2, \cdots, n+1$$

由 $\sum\limits_{k=1}^{n+1} P(\xi = k) = 1$, 可得

$$\sum_{k=1}^{n+1} C_{m+n-k}^{m-1} = C_{m+n}^{m} \tag{2}$$

$$\begin{aligned}
P(\xi \geqslant k) &= \sum_{r=k}^{n+1} \frac{C_{m+n-r}^{m-1}}{C_{m+n}^{m}} \\
&= \sum_{j=1}^{(n-k+1)+1} \frac{C_{m+(n-k+1)-j}^{m-1}}{C_{m+n}^{m}} (\diamondsuit r = k+j-1) \\
&= \frac{C_{m+n-k+1}^{m}}{C_{m+n}^{m}}, \ k = 1, 2, \cdots, n+1 (由(2))
\end{aligned}$$

从而

$$\begin{aligned}
E\xi &= \sum_{k=1}^{\infty} P(\xi \geqslant k) \\
&= \sum_{k=1}^{n+1} \frac{C_{m+n-k+1}^{m}}{C_{m+n}^{m}} \\
&= \sum_{k=1}^{n+1} C_{(m+1)+n-k}^{(m+1)-1}
\end{aligned}$$

$$= \frac{C_{m+n+1}^{m+1}}{C_{m+n}^m} = \frac{m+n+1}{m+1}(由(2))$$

定理 1　设 $1 < r \leqslant n$, 考虑集合 $\{1,2,3,\cdots,n\}$ 的所有含 r 个元素的子集及每个这样子集的最小元素, 设 $F(n,r)$ 表示这样子集各自的最小元素的算术平均数, 则 $F(n,r) = \frac{n+1}{r+1}$.

证明　从 1 到 n 这 n 个自然数中不放回地随机抽取 r 个数 $(r \leqslant n)$, ξ 表示取出的最小数, 则相当于证明 $E\xi = \frac{n+1}{r+1}$. 易见

$$P(\xi = k) = \frac{C_{n-k}^{r-1}}{C_n^r},\ k = 1,2,\cdots,n-r+1$$

由 $\sum\limits_{k=1}^{n-r+1} P(\xi = k) = 1$ 可得 $\sum\limits_{k=1}^{n-r+1} C_{n-k}^{r-1} = C_n^r$. 易见

$$P(\xi \geqslant k) = \frac{C_{n-k+1}^r}{C_n^r} \tag{3}$$

于是

$$E\xi = \sum_{k=1}^{n-r+1} P(\xi \geqslant k)$$

$$= \sum_{k=1}^{n-r+1} \frac{C_{n-k+1}^r}{C_n^r}$$

$$= \sum_{k=1}^{(n+1)-(r+1)+1} \frac{C_{(n+1)-k}^{(r+1)-1}}{C_n^r}$$

$$= \frac{C_{n+1}^{r+1}}{C_n^r} = \frac{n+1}{r+1}(由(3))$$

定理 2

$$\sum_{k=1}^n k C_{m+k-1}^m = \frac{(m+1)n+1}{m+2} C_{m+n}^{m+1}$$

证明　考虑随机试验: 从 1 到 $m+n$ 这 $m+n$ 个自然数中任取 $m+1$ 个数, ξ 表示取出的最大数

$$\sum_{k=1}^n C_{m+k-1}^m = C_{m+n}^{m+1} \tag{4}$$

$$E\xi = \sum_{k=1}^n k \frac{C_{m+k-1}^m}{C_{m+n}^{m+1}} \tag{5}$$

又

$$P(\xi \geqslant k) = 1 - P(\xi < k) = 1 - \frac{C_{m+k-1}^{m+1}}{C_{m+n}^{m+1}}, k = 2, \cdots, n$$

所以

$$E\xi = \sum_{k=1}^{n} P(\xi \geqslant k) = n - \sum_{k=1}^{n} P(\xi < k) = n - \sum_{k=2}^{n} \frac{C_{m+k-1}^{m+1}}{C_{m+n}^{m+1}} \qquad (6)$$

结合(5), (6) 两式有

$$\begin{aligned}
\sum_{k=1}^{n} k C_{m+k-1}^{m} &= n C_{m+n}^{m+1} - \sum_{k=2}^{n} C_{m+k-1}^{m+1} \\
&= n C_{m+n}^{m+1} - \sum_{k=1}^{n-1} C_{m+k}^{m+1} \\
&= n C_{m+n}^{m+1} - C_{m+n}^{m+2} (由(4)) \\
&= n C_{m+n}^{m+1} - \frac{n-1}{m+2} C_{m+n}^{m+1} \\
&= \frac{(m+1)n+1}{m+2} C_{m+n}^{m+1}
\end{aligned}$$

下面我们考虑整值随机变量一般原点矩的此类计算问题, 我们知道自然数的方幂和总可表示成n 的多项式的形式, 即有:

引理 2[2]

$$\sum_{k=1}^{n} k^r = \sum_{k=1}^{r+1} a_k n^k, \ r = 1, 2, \cdots \qquad (7)$$

其中系数$a_k, k = 1, 2, \cdots, r+1$ 由下述方程组确定

$$\begin{cases}
a_1 + a_2 + \cdots + a_{r+1} = 1 \\
(r+1)a_{r+1} + ra_r + \cdots + 2a_2 = r \\
(r+1)ra_{r+1} + r(r-1)a_r + \cdots + 3 \cdot 2a_3 = r(r-1) \\
\vdots \\
(r+1)r \cdot \cdots \cdot 4 \cdot 3 a_{r+1} + r(r-1) \cdot \cdots \cdot 3 \cdot 2 a_r = r(r-1) \cdots 3 \cdot 2 \\
(r+1)! a_{r+1} = r!
\end{cases} \qquad (8)$$

定理 3 若整值随机变量的$r+1$ 阶原点矩存在, 则

$$E\xi^{r+1} = (r+1)\left(G\xi^r - \sum_{k=1}^{r} a_k E\xi^k \right) \qquad (9)$$

其中$G\xi^r = \sum\limits_{k=1}^{\infty} k^r P(\xi \geqslant k), r = 0, 1, 2, \cdots, a_k, k = 1, 2, \cdots, r+1$, 由式(9)确定.

证明　由式(8)的最后一个方程知$a_{r+1} = \frac{1}{r+1}$, 由式(7)有

$$n^{r+1} = (r+1)\left(\sum_{k=1}^{n} k^r - \sum_{k=1}^{r} a_k n^k\right)$$

从而

$$E\xi^{r+1} = \sum_{n=1}^{\infty} n^{r+1} P(\xi = n)$$

$$= (r+1)[\sum_{n=1}^{\infty}\left(\sum_{k=1}^{n} k^r\right) P(\xi = n) -$$

$$\sum_{n=1}^{\infty}\left(\sum_{k=1}^{r} a_k n^k\right) P(\xi = n)]$$

$$= (r+1)\left[\sum_{n=1}^{\infty}\sum_{n=k}^{\infty} k^r P(\xi = n) - \sum_{k=1}^{r} a_k \sum_{n=1}^{\infty} n_k P(\xi = n)\right]$$

$$= (r+1)\left[\sum_{k=1}^{\infty} k^r P(\xi \geqslant n) - \sum_{k=1}^{r} a_k E\xi^k\right]$$

例如, 由

$$\sum_{k=1}^{n} k = \frac{1}{2}n + \frac{1}{2}n^2$$

$$\sum_{k=1}^{n} k^2 = \frac{1}{6}n + \frac{1}{2}n^2 + \frac{1}{3}n^3$$

和

$$\sum_{k=1}^{n} k^3 = \frac{1}{4}n^2 + \frac{1}{2}n^3 + \frac{1}{4}n^4$$

经式(9)可分别得

$$E\xi^2 = 2G\xi - E\xi$$

$$E\xi^3 = 3G\xi^2 - \frac{1}{2}E\xi - \frac{3}{2}E\xi^2$$

和

$$E\xi^4 = 4G\xi^3 - E\xi^2 - 2E\xi^3$$

注意当$r = 0$时, 式(9)变成

$$E\xi = G\xi^0 = \sum_{k=1}^{\infty} P(N \geqslant k)$$

因此式(9)可以看作是式(1)的推广.

例 3 设ξ服从几何分布, 其分布列为

$$P(\xi = n) = pq^{n-1}, n = 1, 2, \cdots, 0 < p < 1, q = 1 - p$$

则

$$\sum_{n=1}^{\infty} pq^{n-1} = 1, E\xi = \sum_{n=1}^{\infty} npq^{n-1}$$

又由

$$P(\xi \geqslant k) = \sum_{n=k}^{\infty} pq^{n-1} = q^{k-1} \sum_{n=1}^{\infty} pq^{n-1} = q^{k-1}$$

有

$$E\xi = \sum_{k=1}^{\infty} P(\xi \geqslant k) = \sum_{k=1}^{\infty} q^{k-1} = \frac{1}{p}$$

$$G\xi = \sum_{k=1}^{\infty} kP(\xi \geqslant k) = \sum_{k=1}^{\infty} kq^{k-1}$$

$$= \frac{1}{p} \sum_{k=1}^{\infty} kpq^{k-1}$$

$$= \frac{1}{p} E\xi = \frac{1}{p^2}$$

$$E\xi^2 = 2G\xi - E\xi = \frac{2}{p^2} - \frac{1}{p}$$

$$G\xi^2 = \sum_{k=1}^{\infty} k^2 P(\xi \geqslant k) = \sum_{k=1}^{\infty} k^2 q^{k-1}$$

$$= \frac{1}{p} \sum_{k=1}^{\infty} k^2 pq^{k-1} = \frac{1}{p} E\xi^2$$

$$= \frac{1}{p} \left(\frac{2}{p^2} - \frac{1}{p} \right) = \frac{2}{p^3} - \frac{1}{p^2}$$

$$E\xi^3 = 3G\xi^2 - \frac{1}{2}E\xi - \frac{3}{2}E\xi^2$$
$$= \frac{6}{p^3} - \frac{3}{p^2} - \frac{1}{2p} - \frac{3}{p^2} + \frac{3}{2p}$$
$$= \frac{6}{p^3} - \frac{6}{p^2} + \frac{1}{p}$$

$$G\xi^3 = \sum_{k=1}^{\infty} k^3 P(\xi \geqslant k) = \sum_{k=1}^{\infty} k^3 q^{k-1}$$
$$= \frac{1}{p} \sum_{k=1}^{\infty} k^3 p q^{k-1} = \frac{1}{p} E\xi^3 = \frac{6}{p^4} - \frac{6}{p^3} + \frac{1}{p^2}$$

$$E\xi^4 = 4G\xi^3 - E\xi^2 - 2E\xi^3$$
$$= \frac{24}{p^4} - \frac{24}{p^3} + \frac{4}{p^2} - \frac{2}{p^2} + \frac{1}{p} - \frac{12}{p^3} + \frac{12}{p^2} - \frac{2}{p}$$
$$= \frac{24}{p^4} - \frac{26}{p^3} + \frac{14}{p^2} - \frac{1}{p}$$

例 4　设 $(p_1, \cdots, p_n)$ 是 $(1, 2, \cdots, n)$ 的一个随机排列, 假定每个随机排列的概率都相同, 令

$$\xi = \max(p_1, \cdots, p_r), 1 \leqslant r \leqslant n$$

求 $E\xi, E\xi^2, E\xi^3$.

解　易见

$$P(\xi = k) = \frac{C_{k-1}^{r-1}}{C_n^r}, k = r, r+1, \cdots, n$$

为了便于使用式(9), 补充定义 $P(\xi = k) = 0, k = 1, 2, \cdots, r-1$.

$$P(\xi \geqslant k) = 1 - P(\xi < k) = 1 - \frac{C_{k-1}^r}{C_n^r}$$
$$= 1 - \frac{1}{r}(k-r)\frac{C_{k-1}^{r-1}}{C_n^r} = 1 - \left(\frac{k}{r} - 1\right)P(\xi = k)$$

$$k = 1, 2, \cdots, n$$

$$E\xi = \sum_{k=1}^{n} P(\xi \geqslant k) = \sum_{k=1}^{n} \left[1 - \left(\frac{k}{r} - 1\right)P(\xi = k)\right]$$

$$= n - \frac{1}{r}\sum_{k=1}^{n} kP(\xi = k) + \sum_{k=1}^{n} P(\xi = k) = n - \frac{1}{r}E\xi + 1$$

从而

$$\left(1 + \frac{1}{r}\right)E\xi = n + 1$$

即

$$E\xi = \frac{r(n+1)}{r+1}$$

$$\begin{aligned}
G\xi &= \sum_{k=1}^{n} k^2 P(\xi \geqslant k) \\
&= \sum_{k=1}^{n} k^2 \left[1 - (\frac{k}{r} - 1)P(\xi = k)\right] \\
&= \sum_{k=1}^{n} k^2 - \frac{1}{r}E\xi^3 + E\xi^2
\end{aligned}$$

另外, 由

$$E\xi^3 = 3G\xi^2 - \frac{1}{2}E\xi - \frac{3}{2}E\xi^2$$

有

$$G\xi^2 = \frac{1}{3}E\xi^3 + \frac{1}{6}E\xi + \frac{1}{2}E\xi^2$$

因此

$$\sum_{k=1}^{n} k^2 - \frac{1}{r}E\xi^3 + E\xi^2 = \frac{1}{3}E\xi^3 + \frac{1}{6}E\xi + \frac{1}{2}E\xi^2$$

即

$$\begin{aligned}
E\xi^3 &= \frac{3r}{3+r}\left(\frac{1}{2}E\xi^2 - \frac{1}{6}E\xi + \sum_{k=1}^{n} k^2\right) \\
&= r(n+1) \cdot [3(nr + n + r)r - r(r+2) + n(1+2n)(r+1)(r+2)] \cdot \\
&\quad [2(r+1)(r+2)(r+3)]^{-1}
\end{aligned}$$

以上诸例如果从原点矩的定义去计算要烦琐些、困难些.

参考文献

[1] 朱秀娟, 洪再吉. 概率统计150 题[M].修订本. 长沙: 湖南科学技术出版社, 1987: 169-170.

[2] 余炯沛.化自然数方幂和为多项式的方法[J].数学通报, 1980, 4: 24-26, 34.

第12篇　Bonferroni不等式的推广及应用

(石焕南. 北京联合大学学报(自然科学版), 2000, 14 (2): 51-55)

摘　要:　本文利用数学归纳法推广了概率论中的Bonferroni 不等式, 并给出该不等式的一些应用.

关键词:　概率; Bonferroni 不等式; 数学归纳法

对于 n 个事件 $A_1, \cdots, A_n (n > 1)$, 有

$$\sum_{i=1}^{n} P(A_i) - \sum_{1 \leqslant i < j \leqslant n} P(A_i A_j) \leqslant P(\sum_{i=1}^{n} A_i) \leqslant \sum_{i=1}^{n} P(A_i) \tag{1}$$

这是概率论中著名的Bonferroni 不等式[1],[2]. 联系概率的一般加法公式

$$P(\sum_{i=1}^{n} A_i) = s_1 - s_2 + s_3 - \cdots + (-1)^n s_n \tag{2}$$

其中

$$
\begin{aligned}
s_m &= s_m(A_1, \cdots, A_n) \\
&= \sum_{1 \leqslant i_1 < \cdots < i_m \leqslant n} P(A_{i1} A_{i2} \cdots A_{im}), m = 1, \cdots, n
\end{aligned}
$$

(当 $m > n$ 时, 规定 $s_m = 0$).

Bonferroni不等式告诉我们:若式(2)右边只保留 s_1, 将不少于式(2)左边; 若式(2)右边保留到 s_2, 将不大于式(2)左边, 由此很自然地联想到: 若式(2)右边保留到 s_k 结果将如何? 本文将建立如下一般的结论, 并给出Bonferroni不等式的一些应用.

定理 1　设 $A_1, \cdots, A_n$ 是一随机试验中的 n 个事件, $n > 1, k \leqslant n$, 则当 k 为偶数时

$$P(\sum_{i=1}^{n} A_i) \geqslant \sum_{i=1}^{k} (-1)^{i+1} s_i \tag{3}$$

当k为奇数时, 上述不等式反向.

证明　现对n用数学归纳法. 当$n = 2$时, 命题显然成立, 假定当$n = m$时命题成立, 当$n = m+1$时, 对于$k \leqslant m+1$, 若$k = m+1$, 由式(2)知式(3)等号成立; 若$k < m+1$且为偶数, 因

$$P(\sum_{i=1}^{m+1} A_i) = P(A_{m+1}) + P(\sum_{i=1}^{m} A_i) - P(\sum_{i=1}^{m} A_i A_{m+1}).$$

由归纳假设有

$$P(\sum_{i=1}^{m} A_i) \geqslant \sum_{i=1}^{k} (-1)^{i+1} s_i(A_1, \cdots, A_m)$$

注意到$k-1$为奇数, 又有

$$P(\sum_{i=1}^{m} A_i A_{m+1}) \leqslant \sum_{i=1}^{k-1} (-1)^{i+1} s_i(A_1 A_{m+1}, \cdots, A_m A_{m+1})$$

从而

$$P(\sum_{i=1}^{m+1} A_i) \geqslant P(A_{m+1}) + \sum_{i=1}^{k} (-1)^{i+1} s_i(A_1, \cdots, A_m) -$$

$$\sum_{i=1}^{k-1} (-1)^{i+1} s_i(A_1 A_{m+1}, \cdots, A_m A_{m+1})$$

$$= P(A_{m+1}) + s_1(A_1, \cdots, A_m) + \sum_{i=2}^{k} (-1)^{i+1} s_i(A_1, \cdots, A_m) -$$

$$\sum_{i=2}^{k} (-1)^{i+1} s_{i-1}(A_1 A_{m+1}, \cdots, A_m A_{m+1})$$

$$= s_1(A_1, \cdots, A_{m+1}) +$$

$$\sum_{i=2}^{k} (-1)^{i+1} [s_i(A_1, \cdots, A_m) s_{i-1}(A_1 A_{m+1}, \cdots, A_m A_{m+1})]$$

$$= \sum_{i=1}^{k} (-1)^{i+1} s_i(A_1, \cdots, A_{m+1})$$

若$k < m+1$且为奇数, 上述不等式反向, 证毕.

显然当k取1时, 由式(3)得式(1)式右边不等式, 当k取2时, 由式(3)得式(1)左边不等式, 所以该定理推广了Bonferroni不等式.

推论 1　条件同定理1, 有

$$\sum_{i=k}^{n}(-1)^{i-k}s_i \geqslant 0 \tag{4}$$

证明　若k为偶数, 则式(4)等价于

$$\sum_{i=k}^{n}(-1)^{i+1}s_i \leqslant 0$$

注意因$k-1$为奇数, 由式(3)有

$$P(\sum_{i=1}^{n}A_i) = \sum_{i=1}^{n}(-1)^{i+1}s_i \leqslant \sum_{i=1}^{k-1}(-1)^{i+1}s_i$$

即

$$\sum_{i=1}^{n}(-1)^{i+1}s_i - \sum_{i=1}^{k-1}(-1)^{i+1}s_i = \sum_{i=k}^{n}(-1)^{i+1}s_i \leqslant 0$$

当k为奇数时类似可证.

下面利用(3), (4)两式建立下述代数不等式.

定理 2　设$0 \leqslant x_i \leqslant 1, i = 1, \cdots, n, n > 1, k \leqslant n$.

(1) 若k为偶数, 则

$$\prod_{i=1}^{n}(1-x_i) \leqslant \sum_{i=0}^{k}(-1)^i\sigma_i \tag{5}$$

若k为奇数, 上述不等式反向.

(2)

$$\sum_{i=k}^{n}(-1)^{i-k}\sigma_i \geqslant 0 \tag{16}$$

其中

$$\sigma_m = \sigma_m(x_1, \cdots, x_n) = \sum_{1 \leqslant i_1 < \cdots < i_m \leqslant n} x_{i_1} x_{i_2} \cdots x_{i_m}$$

$$m = 1, \cdots, n, \ \sigma_0 = 1$$

证明　设$A_1, \cdots, A_n$为一随机试验中的n个相互独立的事件, 且$P(A_i) =$

$x_i, i = 1, \cdots, n$, 注意到诸 A_i 的独立性, 有 $s_i = \sigma_i, i = 1, \cdots, n$, 并结合

$$P\left(\sum_{i=1}^{n} A_i\right) = 1 - P\left(\prod_{i=1}^{n} \overline{A_i}\right) = 1 - \prod_{i=1}^{n}(1 - x_i)$$

由式(3)即可证得式(5), 类似地利用式(4)可证得式(6).

推论 2 设 $x_i \geqslant 0, i = 1, \cdots, n, n > 1, k \leqslant n$. 若 k 为偶数,则

$$\prod_{i=1}^{n}(\sigma_1 - x_i) \leqslant \sum_{i=0}^{k}(-1)\sigma_1^{n-1}\sigma_i \tag{7}$$

若 k 为奇数, 则上述不等式反向.

注意 $0 \leqslant \frac{x_i}{\sigma_1} \leqslant 1$, 由定理2 即可得推论2.

当 k 取1 时, 由式(5)得Weierstrass 不等式[1]

$$\prod_{i=1}^{n}(1 - x_i) \geqslant 1 - \sum_{i=1}^{n} x_i$$

当 k 取2 时,由式(5)得

$$\sum_{i=1}^{n} x_i - \sum_{1 \leqslant i < j \leqslant n} x_i x_j \leqslant 1 - \prod_{i=1}^{n}(1 - x_i)$$

它加强了1994 年罗马尼亚数学奥林匹克竞赛题十年级第一题: 对于每个整数 $n \geqslant 2$ 和 $x_1, \cdots, x_n \in [0, 1]$, 求证

$$\sum_{i=1}^{n} x_i - \sum_{1 \leqslant i < j \leqslant n} x_i x_j \leqslant 1$$

最后, 我们利用Bonferroni 不等式证明第30 届国际数学奥林匹克竞赛的压轴题:

设 n 是正整数, 我们说集合 $\{1, 2, \cdots, 2n\}$ 的一个排列 $\{x_1, x_2, \cdots, x_{2n}\}$ 具有性质 P: 如果 $\{1, 2, \cdots, 2n-1\}$ 中至少有一个 i 使 $|x_i - x_{i+1}| = n$ 成立. 求证:对于任何 n, 具有性质 P 的排列比不具有性质 P 的排列个数多.

证明 设 $\{x_1, x_2, \cdots, x_{2n}\}$ 是 $\{1, 2, \cdots, 2n\}$ 的任一排列, 则此排列具有性质 P, 当且仅当存在 $i(1 \leqslant i \leqslant n)$ 使得 i 与 $n+i$ 相邻. 若用 A 表示此排列具有性质 P, A_i 表示在此排列中, i 与 $n+i$ 相邻, $i = 1, \cdots, n$, 则 $A = \sum_{i=1}^{n} A_i$.

设具有性质P 的排列数为M, 而总排列数为$(2n)!$, 则$P(A) = \frac{M}{(2n)!}$, 易见

$$P(A_i) = \frac{2(2n-1)!}{(2n)!}, i = 1, \cdots, n$$

$$P(A_iA_j) = \frac{2^2(2n-2)!}{(2n)!}, 1 \leqslant i < j \leqslant n$$

由Bonferroni 不等式有

$$\frac{M}{(2n)!} = P(A) = P(\sum_{i=1}^{n} A_i)$$

$$\geqslant \sum_{i=1}^{n} P(A_i) - \sum_{i \leqslant i < j \leqslant n} P(A_iA_j)$$

$$= \frac{2C_n^1(2n-1)!}{(2n)!} - \frac{2^2C_n^2(2n-2)!}{(2n)!}$$

$$= 1 - \frac{n-1}{2n-1} > \frac{1}{2}$$

即$M > \frac{(2n)!}{2}$, 这说明具有性质P 的排列个数比不具有性质P 的排列个数多.

利用Bonferroni 不等式的推广式(3) 可对此问题作更精细的估计. 例如, 若取$k = 3$, 由式(3)有

$$\frac{M}{(2n)!} = P(A) = P(\sum_{i=1}^{n} A_i)$$

$$\leqslant \sum_{i=1}^{n} P(A_i) - \sum_{1 \leqslant i < j \leqslant n} P(A_iA_j) + \sum_{1 \leqslant i < j < k \leqslant n} P(A_iA_jA_k)$$

$$= \frac{2C_n^1(2n-1)!}{(2n)!} - \frac{2^2C_n^2(2n-2)!}{(2n)!} + \frac{2^3C_n^3(2n-3)!}{(2n)!} = \frac{2}{3}$$

这说明虽然具有性质P的排列个数比不具有性质P的排列个数多,但前者不会多过后者的两倍. 若取$k = 4$, 由式(3)有

$$\frac{M}{(2n)!} = P(A) = P\left(\sum_{i=1}^{n} A_i\right)$$

$$\geqslant \frac{2C_n^1(2n-1)!}{(2n)!} - \frac{2^2C_n^2(2n-2)!}{(2n)!} +$$

$$\frac{2^3C_n^3(2n-3)!}{(2n)!} - \frac{2^4C_n^4(2n-4)!}{(2n)!}$$

$$= 1 - \frac{n-1}{2n-1} + \frac{n-2}{3(2n-1)} - \frac{(n-2)(n-3)}{6(2n-1)(2n-3)}$$

$$= \frac{15n^2 - 27n + 6}{24n^2 - 48n + 18} = \frac{5n^2 - 9n + 2}{2(4n^2 - 8n + 3)}$$

$$= \frac{5}{8} + \frac{4n - 7}{8(4n^2 - 8n + 3)} = \frac{5}{8} + \frac{4n - 7}{8(2n - 1)(2n - 3)}$$

由此可见当 $n \geqslant 2$ 时, $\frac{M}{(2n)!} > \frac{5}{8}$, 这说明具有性质 P 的排列个数的3 倍将超过不具有性质 P 的排列个数的5 倍. 显然此结论要强于原题的结论.

参考文献

[1]　匡继昌. 常用不等式[M]. 2版. 长沙: 湖南教育出版社,1993: 724.

[2]　洛哈吉 V K. 概率论及数理统计导论: 上册[M]. 高尚华, 译. 北京: 高等教育出版社,1983: 33 - 34.

[3]　黄宣国. 数学奥林匹克大集1994[M]. 上海:上海教育出版社, 1997: 732 - 733.

第13篇 一个分析不等式的推广

(石焕南. 成都大学学报(自然科学版), 2001, 20 (1): 5-7)

摘 要: 本文利用逐步调整法将一个分析不等式推广到一般初等对称函数上.

关键词: 分析不等式; 初等对称函数; 逐步调整法

记 n 个非负实数的初等对称函数为

$$E_k(x_1,\cdots,x_n) = \sum_{1 \leqslant i_1 < \cdots < i_k \leqslant n} \prod_{j=1}^{k} x_{i_j}$$

其中, $k = 1,\cdots,n$. 规定当 $k = 0$ 时, $E_k(x_1,\cdots,x_n) = 1$, 当 $k < 0$ 或 $k > n$ 时, $E_k(x_1,\cdots,x_n) = 0$. 不难验证

$$E_k(x_1,\cdots,x_n) = x_1 E_{k-1}(x_2,\cdots,x_n) + E_k(x_2,\cdots,x_n)$$

$$\begin{aligned} E_k(x_1,\cdots,x_n) =&\, x_1 x_2 E_{k-2}(x_3,\cdots,x_n)+ \\ &(x_1 + x_2)E_{k-1}(x_3,\cdots,x_n) + E_k(x_3,\cdots,x_n) \end{aligned} \qquad (1)$$

文献[1] 为了实现著名的Pedoe不等式的高维推广, 提出了一个分析不等式.

设 $0 < x_i < \frac{1}{2}, i = 1,\cdots,n$, 且 $E_1(x_1,\cdots,x_n) = 1$, 则

$$E_{n-1}\left(\frac{1}{x_1} - 2,\cdots,\frac{1}{x_n} - 2\right) \geqslant n(n-2)^{n-1} \qquad (2)$$

在文献[1]中试图用Lagrange乘法证明式(2), 但我们认为文献[1]未能证明驻点方程解的唯一性, 因此证而未明, 于是, 在文献[2]中给出另一种分析证明, 其证明富于技巧. 最近文献[3]利用逐步调整法给出式(2)一个简洁的初等证明, 并将条件放宽为 $x_i > 0, i = 1,\cdots,n, n \geqslant 2$, 且 $E_1(x_1,\cdots,x_n) = 1$. 本文采用文献[3]的方法, 将式(2)推广到一般初等对称函数上, 即建立了如下

定理.

定理 1　设 $x_i > 0, i = 1, \cdots, n, n \geqslant 2$，且 $E_1(x_1, \cdots, x_n) = 1$，则对于 $k = 1, 2, \cdots, n - 1$，有

$$E_k\left(\frac{1}{x_1} - 2, \cdots, \frac{1}{x_n} - 2\right) \geqslant C_n^2 (n - 2)^k \tag{3}$$

引理 1　设 $x_i > 0, i = 1, \cdots, n, n \geqslant 2$，且 $E_1(x_1, \cdots, x_n) \leqslant 1$，且存在 $x_i > \frac{1}{2}$，则

$$E_1\left(\frac{x_1}{1 - 2x_1}, \cdots, \frac{x_n}{1 - 2x_n}\right) < 0$$

证明　不妨设 $1 > x_1 > \frac{1}{2} > x_2 \geqslant \cdots \geqslant x_n > 0$，则

$$\frac{1}{1 - 2x_2} \geqslant \frac{1}{1 - 2x_3} \geqslant \cdots \geqslant \frac{1}{1 - 2x_n} > 0, 2x_1 - 1 > 0$$

从而

$$
\begin{aligned}
& E_k\left(\frac{x_1}{1 - 2x_1}, \cdots, \frac{x_n}{1 - 2x_n}\right) \\
=& \frac{x_1}{1 - 2x_1} + \frac{x_2}{1 - 2x_2} + \cdots + \frac{x_n}{1 - 2x_n} \\
\leqslant& \frac{x_1}{1 - 2x_1} + \frac{x_2 + \cdots + x_n}{1 - 2x_2} \\
\leqslant& \frac{x_1}{1 - 2x_1} + \frac{1 - x_1}{1 - 2x_2} \\
=& \frac{-1 + 2x_1 - 2x_1^2 + 2x_1 x_2}{(2x_1 - 1)(1 - 2x_2)} \\
\leqslant& -\frac{2x_1 - 1}{1 - 2x_2} < 0
\end{aligned}
$$

引理 2　设 $x_i > 0, i = 1, \cdots, n, n \geqslant 2$，且 $E_1(x_1, \cdots, x_n) \leqslant 1$，且存在 $x_i > \frac{1}{2}$，则对于 $k = 1, 2, \cdots, n$，有

$$E_k\left(\frac{x_1}{1 - 2x_1}, \cdots, \frac{x_n}{1 - 2x_n}\right) < 0 \tag{4}$$

证明　对 n 用数学归纳法，当 $n = 2$ 时，设 $x_1 > 0, x_2 > 0, x_1 + x_2 \leqslant 1$，不妨设 $x_1 > \frac{1}{2}$，则由引理 1 知

$$E_1\left(\frac{x_1}{1 - 2x_1}, \frac{x_2}{1 - 2x_2}\right) < 0$$

而

$$E_2\left(\frac{x_1}{1-2x_1},\frac{x_2}{1-2x_2}\right)=\frac{x_1}{1-2x_1}\cdot\frac{x_2}{1-2x_2}<0$$

故当 $n=2$ 时, 命题成立. 假定当 $n=m(m\geqslant 2)$ 时, 命题成立, 现设 $x_i>0,i=1,\cdots,m+1,E_1(x_1,\cdots,x_{m+1})\leqslant 1$, 且存在 $x_i>\frac{1}{2}$, 当 $k=1$ 时, 由引理1 知式(4)成立, 当 $2\leqslant k\leqslant m+1$ 时, 不妨设 $x_2>\frac{1}{2},x_1<\frac{1}{2}$, 由式(1)有

$$\begin{aligned}
&E_k\left(\frac{x_1}{1-2x_1},\cdots,\frac{x_{m+1}}{1-2x_{m+1}}\right)\\
&=\frac{x_1}{1-2x_1}E_{k-1}\left(\frac{x_2}{1-2x_2},\cdots,\frac{x_{m+1}}{1-2x_{m+1}}\right)+\\
&\quad E_k\left(\frac{x_2}{1-2x_2},\cdots,\frac{x_{m+1}}{1-2x_{m+1}}\right)
\end{aligned}$$

注意 $\frac{x_1}{1-2x_1}>0$, 由归纳假设以及引理1 知

$$E_{k-1}\left(\frac{x_2}{1-2x_2},\cdots,\frac{x_{m+1}}{1-2x_{m+1}}\right)<0$$

$$E_k\left(\frac{x_2}{1-2x_2},\cdots,\frac{x_{m+1}}{1-2x_{m+1}}\right)<0$$

故

$$E_k\left(\frac{x_1}{1-2x_1},\cdots,\frac{x_{m+1}}{1-2x_{m+1}}\right)<0$$

引理 3　设 $x_i>0,i=1,\cdots,n,n\geqslant 2$, 且 $E_1(x_1,\cdots,x_n)\leqslant 1$, 则对于 $k=1,2,\cdots,n-1$, 有

$$E_k\left(\frac{1}{x_1}-2,\cdots,\frac{1}{x_n}-2\right)\geqslant 0 \tag{5}$$

证明　当 $0<x_i\leqslant\frac{1}{2},i=1,\cdots,n$ 时, 上式显然成立. 否则只有一个 $x_i>\frac{1}{2}$, 其余均小于 $\frac{1}{2}$, 不妨设 $1>x_1>\frac{1}{2}\geqslant\cdots\geqslant x_n>0$, 由于

$$\begin{aligned}
&E_k\left(\frac{1}{x_1}-2,\cdots,\frac{1}{x_n}-2\right)=\left(\frac{1}{x_1}-2\right)\cdots\left(\frac{1}{x_n}-2\right)\cdot\\
&E_{n-k}\left(\frac{x_1}{1-2x_1},\cdots,\frac{x_{m+1}}{1-2x_{m+1}}\right)
\end{aligned}$$

注意 $\frac{1}{x_1-2}<0,\frac{1}{x_i-2}>0,i=2,\cdots,n$, 并结合引理2 知上式非负, 引理3 得证.

定理的证明　当 $n=2$ 时, 由引理3 知式(3)成立. 现采用逐步调整法证明 $n\geqslant 3$ 的情形, 若 $x_1=x_2=\cdots=x_n=\frac{1}{n}$, 则等式成立, 否则, 不妨

设 $x_1 < \frac{1}{n} < x_2$, 令 $y_1 = \frac{1}{n}$, $y_2 = x_1 + x_2 - \frac{1}{n}$, 则 $x_1 + x_2 = y_1 + y_2$, 由式(1)有

$$E_k(\frac{1}{x_1} - 2, \frac{1}{x_2} - 2, \cdots, \frac{1}{x_n} - 2)-$$

$$E_k(\frac{1}{y_1} - 2, \frac{1}{y_2} - 2, \cdots, \frac{1}{y_n} - 2)$$

$$=[(\frac{1}{x_1} - 2)(\frac{1}{x_2} - 2) - (\frac{1}{y_1} - 2)(\frac{1}{y_2} - 2)]\cdot$$

$$E_{k-2}(\frac{1}{x_3} - 2, \cdots, \frac{1}{x_n} - 2)+$$

$$[(\frac{1}{x_1} - 2) + (\frac{1}{x_2} - 2) - (\frac{1}{y_1} - 2) - (\frac{1}{y_2} - 2)]\cdot$$

$$E_{k-1}(\frac{1}{x_3} - 2, \cdots, \frac{1}{x_n} - 2)$$

$$=(\frac{1}{x_1 x_2} - \frac{1}{y_1 y_2})\{[1 - 2(x_1 + x_2)]\cdot$$

$$E_{k-2}(\frac{1}{x_3} - 2, \cdots, \frac{1}{x_n} - 2)+$$

$$(x_1 + x_2)E_{k-1}(\frac{1}{x_3} - 2, \cdots, \frac{1}{x_n} - 2)\}$$

$$=\frac{x_1 + x_2}{x_1 x_2 y_1 y_2}(x_2 - \frac{1}{n})(\frac{1}{n} - x_1)[(\frac{1}{x_1 + x_2} - 2)\cdot$$

$$E_{k-2}(\frac{1}{x_3} - 2, \cdots, \frac{1}{x_n} - 2) + E_{k-1}(\frac{1}{x_3} - 2, \cdots, \frac{1}{x_n} - 2)]$$

$$=\frac{x_1 + x_2}{x_1 x_2 y_1 y_2}(x_2 - \frac{1}{n})(\frac{1}{n} - x_1)\cdot$$

$$E_{k-1}(\frac{1}{x_1 + x_2} - 2, \frac{1}{x - 3} - 2, \cdots, \frac{1}{x_n} - 2) \geqslant 0$$

最后一步运用了引理3, 这样最多经 $n - 1$ 次调整有

$$E_k(\frac{1}{x_1} - 2, \frac{1}{x_2} - 2, \cdots, \frac{1}{x_n} - 2) \geqslant E_k(\frac{1}{y_1} - 2, \frac{1}{y_2} - 2, \cdots, \frac{1}{y_n} - 2)$$

$$\geqslant \cdots \geqslant E_k(n - 2, n - 2, \cdots, n - 2) = C_n^k(n - 2)^k$$

定理证毕.

值得注意的是当 $k = n$ 时, 式(3)不成立. 例如, 当 $k = n = 3$ 时, 式(3)等价于

$$(1 - 2x_1)(1 - 2x_2)(1 - 2x_3) \geqslant x_1 x_2 x_3$$

而这与熟知的不等式

$$x_1 x_2 x_3 \geqslant (x_1 + x_2 - x_3)(x_2 + x_3 - x_1)(x_3 + x_1 - x_2)$$

矛盾.

参考文献

[1]　冷岗松, 唐立华. 再论Pedoe 不等式的高维推广及应用[J]. 数学学报, 1997, 40 (1): 14 - 211.

[2]　陈计, 黄勇, 夏时洪. 关于Neuberg-Pedoe 不等式的高维推广的一个注记[J]. 四川大学学报(自然科学版), 1999, 36 (2): 197 - 2001.

[3]　樊益武. Neuberg-Pedoe 不等式的单形推广[J]. 延安大学学报(自然科学版), 1999, 18 (3): 19 - 221.

第14篇　凸数列的一个等价条件及其应用

(石焕南, 李大矛. 曲阜师范大学学报(自然科学版), 2001, 27(4)：4-6)

摘　要:　本文利用控制不等式理论建立了凸数列的一个等价条件, 并应用其推广了一类加权对称平均不等式.

关键词:　凸数列; 控制不等式; 对称平均; 权

14.1　引言

定义 1　若实数列 $\{a_i\}$ 满足条件 $a_{i-1} + a_{i+1} \geqslant 2a_i, i = 2, 3, \ldots$, 则称数列 $\{a_i\}$ 是一个凸数列.

定义 2　若实数列 $\{a_i\}$ 满足条件 $a_{i-1}a_{i+1} \geqslant a_i^2, i = 2, 3, \cdots$, 则称数列 $\{a_i\}$ 是一个对数凸数列.

对于 $\boldsymbol{x} = (x_1, \cdots, x_n) \in \mathbf{R}^n$, 把 $\boldsymbol{x}$ 的分量排成递减的次序后, 记作 $x_{[1]} \geq \cdots \geq x_{[n]}$.

定义 3[3-4]　设 $\boldsymbol{x}, \boldsymbol{y} \in \mathbf{R}^n$ 满足:

(1) $\sum\limits_{i=1}^{k} x_{[i]} \leq \sum\limits_{i=1}^{k} y_{[i]}$, $k = 1, 2, \cdots, n-1$;

(2) $\sum\limits_{i=1}^{n} x_i = \sum\limits_{i=1}^{n} y_i$.

则称 $\boldsymbol{x}$ 被 $\boldsymbol{y}$ 所控制, 记作 $\boldsymbol{x} \prec \boldsymbol{y}$.

凸数列是凸函数的离散形式, 文献[1]探讨了凸数列的一些特征性质, 给出凸数列的若干等价条件, 其中有:

定理 1　数列 $\{a_i\}$ 是凸数列的充要条件为: 对任意4 个非负整数 m, n, p, q, 当 $p < m < q, p < n < q$, 且 $m + n = p + q$ 时, 恒有

$$a_p + a_q \geqslant a_m + a_n \tag{1}$$

注　由文献[1] 的证明过程可见, 条件 $p < m < q, p < n < q$ 可放宽为 $p \leqslant m \leqslant q, p \leqslant n \leqslant q$. 从控制不等式的观点来看, 条件 $p \leqslant m \leqslant q, p \leqslant$

$n \leqslant q$, 且$m + n = p + q$ 意味着$(m, n) \prec (p, q)$, 即二维向量(m, n) 被二维向量(p, q) 控制. 很自然地想到上述结果是否可推广到n 维情形? 结论是肯定的. 本文将利用控制不等式理论建立凸数列的一个等价条件, 并应用其对肖振钢等人[1]建立的一类加权对称平均不等式作实质性的推广.

14.2　主要结果及其证明

定理 2　数列$\{a_i\}$ 是凸数列的充要条件为: 对任意的$\boldsymbol{p}, \boldsymbol{q} \in \mathbf{Z}_+^n = \{\, \boldsymbol{p} = (p_1, \cdots, p_n) : p_i$ 是非负整数, $i = 1, 2, \cdots, n, n \geqslant 2\,\}$ 时, 恒有

$$a_{q_1} + \cdots + a_{q_n} \geqslant a_{p_1} + \cdots + a_{p_n} \tag{2}$$

证明　必要性. 对n 用数学归纳法. 由定理1 知, 当$n = 2$ 时命题成立, 假定当$n = m(m \geqslant 2)$ 时命题成立.考察$n = m + 1$ 的情形. 设$\boldsymbol{p}, \boldsymbol{q} \in \mathbf{Z}_+^{m+1}$, $\boldsymbol{p} = (p_1, \cdots, p_{m+1}) \prec (q_1, \cdots, q_{m+1}) = \boldsymbol{q}$, 不妨设$p_1 \geqslant p_2 \geqslant \cdots \geqslant p_{m+1}, q_1 \geqslant q_2 \geqslant \cdots \geqslant q_{m+1}$, 它们满足$(1), (2)$, 下面分两种情形证明.

第一种情形, 存在$r, 1 \leqslant r \leqslant m + 1$, 使得$p_r = q_r$, 去掉$p_r$ 和q_r, 显然仍有

$$(p_1, \cdots, p_{r-1}, p_{r+1}, \cdots, p_{m+1}) \prec (q_1, \cdots, q_{r-1}, q_{r+1}, \cdots, p_{m+1})$$

由归纳假设有

$$a_{q_1} + \cdots + a_{q_{r-1}} + a_{q_{r+1}} + \cdots + a_{p_{m+1}}$$
$$\geqslant a_{p_1} + \cdots + a_{p_{r-1}} + a_{p_{r+1}} + \cdots + a_{p_{m+1}}$$

两边同加上a_{p_r}, 则

$$a_{q_1} + \cdots + a_{p_{m+1}} \geqslant a_{p_1} + \cdots + a_{p_{m+1}}$$

第二种情形, $p_i \neq q_i, i = 1, 2, \cdots, m + 1$, 由(1) 有$p_1 < q_1$, 由(2) 不可能对一切$i$, $p_i < q_i$, 故必存在$r, 2 \leqslant r \leqslant m + 1$, 使得$p_r > q_r$, 不妨设这个$r$ 是使$p_r > q_r$ 的最小下标, 于是$p_1 < q_1, p_2 < q_2, \cdots, p_{r-1} < q_{r-1}, p_r > q_r$, 把两个自然数$q_{r-1} - p_{r-1}, p_r - q_r$ 中较小的一个记作$h, h > 0$, 再记$q_{r-1}' = q_{r-1} - h$, $q_r' = q_r + h$, 则

$$q_{r-1}' \geqslant p_{r-1}, q_r' \leqslant p_r \tag{3}$$

并且这两个不等式中至少有一个等式成立(当$h = q_{r-1} - p_r - 1$ 时, $q_{r-1}' = q_{r-1} - h = p_{r-1}$, 当$h = p_r - q_r$ 时, $q_r' = q_r + h = p_r$). 又因$p_{r-1} \geqslant p_r$, 由(3)

有 $q_{r-1}' \geqslant q_r'$，于是 $q_{r-1} > q_{r-1}' \geqslant q_r' > q_r$ 且 $q_{r-1}' + q_r' = q_{r-1} + q_r$，由定理1 及其注有

$$a_{q_{r-1}'} + a_{q_r'} \leqslant a_{q_{r-1}} + a_{q_r} \tag{4}$$

规定当 $i \neq r-1, r$ 时，$q_i' = q_i$，在式 (4) 两边再加上 $\sum_{i \neq r-1, r} a_{q_i}$，得

$$\sum_{i=1}^{m+1} a_{q_i'} \leqslant \sum_{i=1}^{m+1} a_{q_i} \tag{5}$$

再考察 $(q_1', \cdots, q_{m+1}')$ 和 $(p_1, \cdots, p_{m+1})$，由 $q_{r-2}' = q_{r-2} \geqslant q_{r-1} > q_{r-1}' \geqslant q_r' > q_r \geqslant q_{r+1} = q_{r+1}'$ 知 $(q_1', \cdots, q_n') = (q_{[1]}', \cdots, q_{[n]}')$，容易验证 $(p_1, \cdots, p_n) \prec (q_1', \cdots, q_n')$，前面已证(3) 的两个不等式中至少有一个等式成立，即 $(p_1, \cdots, p_n)$ 与 $(q_1', \cdots, q_n')$ 至少有一对同下标的分量相等，由已证明的第一种情形，$\sum_{i=1}^{m+1} a_{q_i} \leqslant \sum_{i=1}^{m+1} a_{q_i'}$，结合式 (5) 得 $\sum_{i=1}^{m+1} a_{q_i} \leqslant \sum_{i=1}^{m+1} a_{q_i'}$。

总结以上两种情形，由归纳假设当 $n = m(m \geqslant 2)$ 时命题成立. 可以证明当 $n = m+1$ 时命题正确，从而对一切 $n \geqslant 2$ 命题正确.

充分性. 对任意正整数 i，于式 (2) 中取 $q_1 = i-1, q_2 = i+1, q_3 = \cdots = q_n = 0, p_1 = p_2 = i, p_3 = \cdots = p_n = 0$，则 $(p_1, \cdots, p_n) \prec (q_1, \cdots, q_n)$，且式 (2) 化为 $a_{i-1} + a_{i+1} \geqslant 2a_i (i = 1, 2, 3, \cdots)$，故 $\{a_n\}$ 是一凸数列，定理证毕.

推论 1 设 $f : I \subset \mathbf{R} \to \mathbf{R}$ 是增的凸函数，$\{a_n\}$ 是凸集 I 上的凸数列，$\boldsymbol{p}, \boldsymbol{q} \in \mathbf{Z}_+^n$，若 $\boldsymbol{p} \prec \boldsymbol{q}$，则

$$\sum_{i=1}^n f(a_{q_i}) \geqslant \sum_{i=1}^n f(a_{p_i}) \tag{6}$$

证明 因 f 是 I 上的增的凸函数，且 $\{a_n\}$ 凸，则

$$f(a_{i-1}) + f(a_{i+1}) \geqslant 2f\left[\frac{1}{2}(a_{i-1} + a_{i+1})\right]$$
$$\geqslant 2f\left(\frac{2a_i}{2}\right) = 2f(a)$$

这说明 $\{f(a_i)\}$ 是 $\mathbf{R}$ 上的凸数列，从而由定理2 知(6)成立.

推论 2 非负数列 $\{a_n\}$ 是对数凸数列的充要条件为：对任意 $\boldsymbol{p}, \boldsymbol{q} \in \mathbf{Z}_+^n$，若 $\boldsymbol{p} \prec \boldsymbol{q}$，则

$$\prod_{i=1}^n a_{q_i} \geqslant \prod_{i=1}^n a_{p_i} \tag{7}$$

证明 注意 $\{a_n\}$ 是对数凸数列等价于 $\{\ln a_n^{-1}\}$ 是 $\mathbf{R}$ 上的凸数列，从而

由定理2 知该推论成立.

推论 3　若$\{a_n\}$是凸数列, 则对于$0 \leqslant i \leqslant n$ 有

$$a_i \leqslant \left(1 - \frac{i}{n}\right) a_0 + \frac{i}{n} a_n \tag{8}$$

证明　注意

$$\left(\underbrace{i, \cdots, i}_{n}\right) \prec \left(\underbrace{n, \cdots, n}_{i}, \underbrace{0, \cdots, 0}_{n-i}\right)$$

由定理2 有$na_i \leqslant (n-i)a_0 + ia_n$, 即式(8)成立.

推论 4　若$\{a_n\}$是凸数列, 则有

$$\frac{1}{2^n} \sum_{i=0}^{n} a_i C_n^i \leqslant \frac{1}{n+1} \sum_{i=0}^{n} a_i \leqslant \frac{1}{2}(a_0 + a_n) \tag{9}$$

证明　注意

$$\left(\frac{2n}{n+1}, \frac{2n}{n+1}, \cdots, \frac{2n}{n+1}\right) \prec \left(C_n^0, C_n^1, \cdots, C_n^n\right)$$

和

$$(0, 0, 1, 1, \cdots, n, n) \prec \left(\underbrace{0, \cdots, 0}_{n+1}, \underbrace{n, \cdots, n}_{n+1}\right)$$

由定理2 即知式(9)中的两个不等式成立.

推论 5　(E.J.Nanson 不等式)[4] 若$\{a_n\}$是凸数列, 则

$$\frac{1}{n+1} \sum_{i=0}^{n} a_{2i+1} \leqslant \frac{1}{n} \sum_{i=1}^{n} a_{2i}, \ n \in \mathbf{N} \tag{10}$$

证明　注意

$$\left(\underbrace{2, \cdots, 2}_{n+1}, \underbrace{4, \cdots, 4}_{n+1}, \cdots, \underbrace{2n, \cdots, 2n}_{n+1}\right)$$

$$\prec \left(\underbrace{1, \cdots, 1}_{n}, \underbrace{3, \cdots, 3}_{n}, \cdots, \underbrace{2n+1, \cdots, 2n+1}_{n}\right)^{[4]}$$

由定理2 即可证得式(10).

14.3 应用

定义 4[6]   设 $\boldsymbol{a} = (a_1, a_2, \cdots, a_n)$ 是一组非负实数, $\boldsymbol{\lambda} = (\lambda_1, \lambda_2, \cdots, \lambda_n)$ 是一组正的权, $r \in \mathbf{N}$

$$E_\lambda(\boldsymbol{a}, \boldsymbol{\lambda}) = \sum_{\substack{i_1+i_2+\cdots+i_n=r}}^{i_1,i_2,\cdots,i_n \geqslant 0} \left(\sum_{k=1}^{n} (1+i_k)\lambda_k \right) a_1^{i_1} a_2^{i_2} \cdots a_n^{i_n} \tag{11}$$

和式中的 i_k 取非负整数,令

$$\Omega_r(\boldsymbol{a}, \boldsymbol{\lambda}) = \begin{cases} \left[\left(\mathrm{C}_{n+r}^r \sum_{k=1}^{n} \lambda_k \right)^{-1} E_r(\boldsymbol{a}, \boldsymbol{\lambda}) \right]^{\frac{1}{r}}, \ r \in \mathbf{N} \\ 1, \ r = 0 \end{cases}$$

则称 $\Omega_r(\boldsymbol{a}, \boldsymbol{\lambda})$ 为非负实数组 $\boldsymbol{a}$ 关于权 $\boldsymbol{\lambda}$ 的广义 r 次加权平均.

关于此类加权对称平均,文献[6]证得:

定理 3   对任意 $r \in \mathbf{N}$, 有

$$\left(\Omega_{r-1}(\boldsymbol{a}, \boldsymbol{\lambda}) \right)^{r-1} \left(\Omega_{r+1}(\boldsymbol{a}, \boldsymbol{\lambda}) \right)^{r+1} \geqslant \Omega_r(\boldsymbol{a}, \boldsymbol{\lambda})^{2r} \tag{12}$$

上式说明 $\{ \Omega_r(\boldsymbol{a}, \boldsymbol{\lambda}) \}$ 是对数凸数列, 从而由推论2, 有:

定理 4   对任意的 $\boldsymbol{p}, \boldsymbol{q} \in \mathbf{Z}_+^n$, 若 $\boldsymbol{p} \prec \boldsymbol{q}$, 则

$$\prod_{i=1}^{n} \left(\Omega_{q_i}(\boldsymbol{a}, \boldsymbol{\lambda}) \right)^{q_i} \geqslant \prod_{i=1}^{n} \left(\Omega_{p_i}(\boldsymbol{a}, \boldsymbol{\lambda}) \right)^{p_i} \tag{13}$$

取 $\boldsymbol{p} = (r, r, 0, \cdots, 0), \boldsymbol{q} = (r+1, r-1, 0, \cdots, 0)$, 则式 (14) 即化为式(13).

参考文献

[1]   朱秀娟, 洪再吉. 概率统计150 题[M].修订本. 长沙: 湖南科学技术出版社, 1987: 169-170.

[2]   MITRINOVIĆ D S, VASIC P M. 分析不等式[M]. 赵汉宾, 译. 南宁:广西人民出版社, 1986.

[3]   续铁权. 关于凸函数的一个控制不等式[J]. 数学通报, 1995(7): 42-46.

[4]   哈代 G H, 李特伍德 J E, 波利亚 G. 不等式[M]. 越民义, 译. 北京: 科学出版社, 1965: 107.

[5]   王伯英. 控制不等式基础[M]. 北京: 北京师范大学出版社, 1990.

[6]　肖振钢, 张志华, 卢小宁. 一类加权对称平均不等式[J]. 湖南教育学院学报, 1999, 17 (5)：130-134.

第15篇　Weierstrass不等式的新推广

(石焕南. 数学的实践与认识, 2002, 32 (1):132-135)

摘　要: 本文利用控制不等式理论将Weierstrass 不等式 $\prod_{i=1}^{n}(1 \pm x_i) >$ $1 \pm \sum_{i=1}^{n} x_i$ 推广到一般初等对称函数上, 并且给出一个上界估计.

关键词: Weierstrass 不等式; 初等对称函数; 控制理论

15.1　引言

设 $0 < x_i < 1, i = 1, ..., n$, 则

$$\prod_{i=1}^{n}(1 + x_i) > 1 + \sum_{i=1}^{n} x_i, \quad \prod_{i=1}^{n}(1 - x_i) > 1 - \sum_{i=1}^{n} x_i \tag{1}$$

上两式均称为Weierstrass 不等式[1], 关于Weierstrass 不等式有不少推广[2-6], 如1983 年, Pečarić J. E (克罗的亚)[3] 得到的如下结果: 设 $0 < x_i < 1, a_i \geqslant 1, i = 1, ..., n$, 则

$$\prod_{i=1}^{n}(1 + x_i)^{a_i} > 1 + \sum_{i=1}^{n} a_i x_i, \quad \prod_{i=1}^{n}(1 - x_i)^{a_i} > 1 - \sum_{i=1}^{n} a_i x_i \tag{2}$$

本文将利用控制不等式理论将上述各式推广到一般初等对称函数上, 并给出一个上界估计, 最后给出两例应用.

80

15.2　主要结果及其证明

定理 1　设 $\boldsymbol{x} \in \mathbf{R}_{++}^n = \{\boldsymbol{x} = (x_1, ..., x_n) : x_i > 0, i = 1, ..., n\}$, $n \geqslant 2, 1 < k \leqslant n, p > 0, 0 \leqslant \alpha \leqslant 1$, 则

$$\mathrm{C}_n^k \left(p + \frac{1}{n} \sum_{i=1}^n x_i \right)^{k\alpha} \geqslant E_k \left[(p + x_1)^\alpha, (p + x_2)^\alpha, ..., (p + x_n)^\alpha \right]$$

$$:= \sum_{1 \leqslant i_1 < \cdots < i_k \leqslant n} \prod_{j=1}^k (p + x_{i_j})^\alpha \geqslant \mathrm{C}_{n-1}^k p^{k\alpha} + \mathrm{C}_{n-1}^{k-1} p^{(k-1)\alpha} \left(p + \sum_{i=1}^n x_i \right)^\alpha \tag{3}$$

当 $\alpha = 1$ 时右边不等式为严格不等式.

证明　　不难证明

$$(p + \bar{x}, p + \bar{x}, ..., p + \bar{x}) \prec (p + x_1, p + x_2, ..., p + x_n) \prec\prec \left(p + \sum_{i=1}^n x_i, p, ..., p \right) \tag{4}$$

其中 $\bar{x} = \frac{1}{n} \sum_{i=1}^n x_i$. 由文[3]的命题6.7 知 $E_k(\boldsymbol{x})$ 是 $\mathbf{R}_{++}^n$ 上的增的且严格的 Schur 凹函数, 又当 $0 < \alpha \leqslant 1$ 时, x^α 是凹函数, 由文[3] 命题6.16 (b) 知 $E_k(\boldsymbol{x}^\alpha)$ 亦是 $\mathbf{R}_{++}^n$ 上的Schur 凹函数, 从而由(4) 式有

$$E_k \left[(p + \bar{x})^\alpha, (p + \bar{x}), ..., (p + \bar{x})^\alpha \right]$$
$$\geqslant E_k \left[(p + x_1)^\alpha, (p + x_2)^\alpha, ..., (p + x_n)^\alpha \right]$$
$$\geqslant E_k \left[(p + \sum_{i=1}^n x_i)^\alpha, p^\alpha, ..., p^\alpha \right]$$

当 $\alpha = 1$ 时右边不等式为严格不等式, 由此证得式(3). 当 $\alpha = p = 1, k = n$ 时, 式 (3) 右边不等式即化为式(1)中第一个不等式(注意 $\mathrm{C}_{n-1}^n = 0$).

定理 2　设 $\boldsymbol{x} \in \mathbf{R}_{++}^n$, 且 $a_i \geqslant 1, i = 1, ..., n, 1 \leqslant k \leqslant n$, 则

$$\mathrm{C}_n^k \left[\frac{1}{n} \sum_{i=1}^n (1 + x_i)^{a_i} \right]^k \geqslant E_k \left[(1 + x_1)^{a_1}, (1 + x_2)^{a_2}, ..., (1 + x_n)^{a_n} \right]$$

$$:= \sum_{1 \leqslant i_1 < \cdots < i_k \leqslant n} \prod_{j=1}^k (1 + x_{i_j})^{a_{i_j}} > \mathrm{C}_{n-1}^k + \mathrm{C}_{n-1}^{k-1} \left(1 + \sum_{i=1}^n a_i x_i \right) \tag{5}$$

证明　　记 $y = \frac{1}{n} \sum_{i=1}^n (1 + x_i)^{a_i}$, 由Bernoulli 不等式[1] $(1 + x_i)^{a_i} \geqslant 1 +$

$a_i x_i, i = 1, ..., n,$ 有

$$
\begin{aligned}
&(y, y, ..., y) \\
&\prec ((1 + x_1)^{a_1}, (1 + x_2)^{a_2}, ..., (1 + x_n)^{a_n}) \\
&\geqslant (1 + a_1 x_1, 1 + a_2 x_2, ..., 1 + a_n x_n) \\
&\prec\prec (1 + \sum_{i=1}^{n} a_i x_i, 1, ..., 1)
\end{aligned}
$$

从而由 $E_k(\boldsymbol{x})$ 的增的严格Schur 凹性, 有

$$
\begin{aligned}
&E_k(y, y, ..., y) \\
&\geqslant E_k \left[(1 + x_1)^{a_1}, (1 + x_2)^{a_2}, ..., (1 + x_n)^{a_n} \right] \\
&\geqslant E_k (1 + a_1 x_1, 1 + a_2 x_2, ..., 1 + a_n x_n) \\
&> E_k(1 + \sum_{i=1}^{n} a_i x_i, 1, ..., 1)
\end{aligned}
$$

由此证得式(5). 当 $k = n$ 时, 式(5)右边不等式化为式(2)中第一个不等式. 注意

$$
\begin{aligned}
&(p - \bar{x}, p - \bar{x}, ..., p - \bar{x}) \\
&\prec (p - x_1, p - x_2, ..., p - x_n) \\
&\prec\prec \left(p - \sum_{i=1}^{n} x_i, p, ..., p \right)
\end{aligned}
$$

其中 $\bar{x} = \frac{1}{n} \sum_{i=1}^{n} x_i$, 类似于定理1 可证得

定理3 设 $\boldsymbol{x} \in \mathbf{R}_{++}^n, n \geqslant 2, 1 < k \leqslant n, 0 \leqslant \alpha \leqslant 1, p > 0, \sum_{i=1}^{n} x_i \leqslant p,$ 则

$$
\mathrm{C}_n^k \left(p - \frac{1}{n} \sum_{i=1}^{n} x_i \right)^{k\alpha} \geqslant E_k \left[(p - x_1)^\alpha, (p - x_2)^\alpha, ..., (p - x_n)^\alpha \right]
$$

$$
:= \sum_{1 \leqslant i_1 < \cdots < i_k \leqslant n} \prod_{j=1}^{k} (p - x_{i_j})^\alpha \geqslant \mathrm{C}_{n-1}^k p^{k\alpha} + \mathrm{C}_{n-1}^{k-1} p^{(k-1)\alpha} \left(p - \sum_{i=1}^{n} x_i \right)^\alpha \quad (6)
$$

当 $\alpha = 1$ 时右边不等式为严格不等式.

对于式(6)左边不等式以及当 $k = 2$ 或 $k = n$ 时的右边不等式, 条

件$\sum\limits_{i=1}^{n} x_i \leqslant p$ 可放宽为$0 < x_i < p, i = 1, ..., n$, 当$\alpha = p = 1, k = n$ 时, 式(6)右边不等式即化为式(1)中第二个不等式. 类似于定理2, 可证得

定理 4　设$\boldsymbol{x} \in \mathbf{R}_{++}^n$, $\sum\limits_{i=1}^{n} a_i x_i \leqslant 1, a_i \geqslant 1, i = 1, ..., n, n \geqslant 2, 1 \leqslant k \leqslant n$, 则

$$
\mathrm{C}_n^k \left[\frac{1}{n} \sum_{i=1}^{n} (1 - x_i)^{a_i} \right]^k \geqslant E_k \left[(1 - x_1)^{a_1}, (1 - x_2)^{a_2}, ..., (1 - x_n)^{a_n} \right]
$$

$$
:= \sum_{1 \leqslant i_1 < \cdots < i_k \leqslant n} \prod_{j=1}^{k} (1 - x_{i_j})^{a_{i_j}} > \mathrm{C}_{n-1}^k + \mathrm{C}_{n-1}^{k-1} \left(1 - \sum_{i=1}^{n} a_i x_i \right) \tag{7}
$$

对于式(7)左边不等式以及当$k = 2$或$k = n$ 时的右边不等式, 条件$\sum_{i=1}^{n} a_i x_i \leqslant 1$ 可放宽为$0 < x_i < 1, i = 1, ..., n$. 当$k = n$ 时, 式(7)右边不等式化为式(2)中第二个不等式.

15.3　两例应用

定理 5　设$x_i \geqslant 1$, $a_i \geqslant 1$, $i = 1, \cdots, n, n \geqslant 2, 1 \leqslant k \leqslant n, A_k = \min\limits_{1 \leqslant i_1 < \cdots < i_k \leqslant n} \sum\limits_{i=j}^{k} a_{i_j}$, 则

$$
E_k \left[(1 + x_1)^{a_1}, (1 + x_2)^{a_2}, ..., (1 + x_n)^{a_n} \right]
$$

$$
\geqslant \frac{\mathrm{C}_{n-1}^{k-1} 2^{A_k}}{1 + A_k} \left[\frac{n}{k} (1 + A_k) - A_n + \sum_{i=1}^{n} a_i x_i \right] \tag{8}
$$

证明　式(8)左边

$$
= \sum_{1 \leqslant i_1 < \cdots < i_k \leqslant n} \prod_{j=1}^{k} \left(1 + x_{i_j} \right)^{a_{i_j}}
$$

$$
= \sum_{1 \leqslant i_1 < \cdots < i_k \leqslant n} 2^{\sum\limits_{j=1}^{k} a_{i_j}} \prod_{j=1}^{k} \left(1 + \frac{x_{i_j} - 1}{2} \right)^{a_{i_j}}
$$

$$
\geqslant 2^{A_k} E_k \left[\left(1 + \frac{x_1 - 1}{2} \right)^{a_1}, \left(1 + \frac{x_2 - 1}{2} \right)^{a_2}, ..., \left(1 + \frac{x_n - 1}{2} \right)^{a_n} \right]
$$

$$
\geqslant 2^{A_k} E_k \left[\mathrm{C}_{n-1}^k + \mathrm{C}_{n-1}^{k-1} \left(1 + \sum_{i=1}^{n} \frac{a_i (x_i - 1)}{2} \right) \right] \quad （由定理2）
$$

$$= 2^{A_k} C_{n-1}^k \left[\frac{k}{n} + \sum_{i=1}^n \frac{a_i(x_i - 1)}{2} \right]$$

$$\geqslant 2^{A_k} C_{n-1}^k E_k \left[\frac{k}{n} + \sum_{i=1}^n \frac{a_i(x_i - 1)}{1 + A_k} \right]$$

$$= \frac{C_{n-1}^{k-1} 2^{A_k}}{1 + A_k} \left[\frac{n}{k}(1 + A_k) - A_n + \sum_{i=1}^n a_i x_i \right]$$

当 $k = n$ 时, 式(8)化为文[2]式(7.5)

$$\prod_{i=1}^n (1 + x_i)^{a_i} \geqslant \frac{2^{A_n}}{1 + A_n} \left(1 + \sum_{i=1}^n a_i x_i \right)$$

定理 6 设 $A = A_1 A_2 \cdots A_{n+1}$ 是 $\mathbf{R}^n (n \geqslant 3)$ 中的一个单形, P 是 A 内任一点, B_i 是直线 $A_i P$ 与超平面 $a_i = A_1 \cdots A_{i-1} A_{i+1} \cdots A_{n+1}$ 的交点, 则

$$C_{n+1}^k \left(\frac{n}{n+1} \right)^{k\alpha} \geqslant \sum_{1 \leqslant i_1 < \cdots < i_k \leqslant n+1} \prod_{j=1}^k \left(\frac{A_{i_j} P}{A_{i_j} B_{i_j}} \right)^\alpha \geqslant C_n^k \ (0 < \alpha \leqslant 1) \quad (9)$$

证明 由文[4] 知

$$\sum_{i=1}^{n+1} \frac{PB_i}{A_i B_i} = 1, \frac{A_i P}{A_i B_i} = 1 - \frac{PB_i}{A_i B_i}, i = 1, 2, \cdots, n+1$$

从而由定理3 即可得证. 式(9)左边不等式即为文[4] 的式2.7, 这里给出一个下界估计.

参考文献

[1] 匡继昌. 常用不等式[M]. 2版. 长沙: 湖南教育出版社, 1993.

[2] MITRINOVIĆ D S, PEČARIĆ J E, FLNK A M. Classical and new inequalitis in analysis [M]. Dordrecht: Kluwer Academic Publishers, 1993: 69-72.

[3] PEČARIĆ J E. On Bernoulli′s inequality[M]. ANUB IH Radovi-74, Odj Prir. Nauka knj., 1983, 22: 61-65.

[4] KLAMKIN M S. NEWMAN D J. Extensions of the Weierstrass product inequalities[J]. Math. Mag, 1970, 43: 137-140.

[5] KLAMKIN M S. Extensions of the Weierstrass product inequalities II [J]. Amer. Math. Monthly, 1975, 82: 741-742.

[6]　PEČARIĆ J E. KLAMKIN M S. Extensions of the Weierstrass product inequalities III [J]. SEA Bull. Math. II, 1988(2) : 123-126.

[7]　王伯英. 控制不等式基础[M]. 北京: 北京师范大学出版社, 1990.

[8]　MITRINOVI D S, PEČARIĆ J E, VOLENEC V. Recent Advances in Geometric Inequalities[M]. Dordrecht: Kluwer Academic Publishers, 1989.

第16篇 Turner-Conway不等式的概率证明

(石焕南. 商丘师范学院学报，2002，(2)：51-52)

摘 要： 本文用概率方法给出Turner－Conway不等式的一个简洁证明.

关键词： 概率方法；独立事件；Turner－Conway不等式

著名数学家王梓坤院士在文[3]中指出："用概率的方法来证明一些关系式或解决其他数学分析中的问题, 是概率论的重要研究方向之一."

定理 若 $a_{j_k} + b_{j_k} = 1, 0 < a_{j_k} < 1, j = 1, \cdots, m, k = 1, \cdots, n, (m, n > 1)$, 则

$$\prod_{j=1}^{m}\left(1 - \prod_{k=1}^{n} a_{j_k}\right) + \prod_{j=1}^{n}\left(1 - \prod_{k=1}^{m} b_{j_k}\right) > 1 \tag{1}$$

上式为著名的Turner－Conway不等式[1]. Carlitz I. [2]曾用数学归纳法证明了式(1), 本文用概率方法给出式(1)的一个简洁证明.

证明 设事件 $A_{j_k}, j = 1, \cdots, m, k = 1, \cdots, n$ 相互独立, 且 $P(A_{j_k}) = a_{j_k}$, 则 $(P(\overline{A}_{j_k}) = b_{j_k}$. 注意

$$P\left[\sum_{j=1}^{m}\left(\prod_{k=1}^{n} A_{j_k}\right)\right] = 1 - P\left[\prod_{j=1}^{m}\left(\overline{\prod_{k=1}^{n} A_{j_k}}\right)\right]$$

$$= 1 - P\left[\prod_{j=1}^{m}\left(\sum_{k=1}^{n} \overline{A}_{j_k}\right)\right] = 1 - \prod_{j=1}^{m} P\left(\prod_{k=1}^{n} \overline{A}_{j_k}\right)$$

$$= 1 - \prod_{j=1}^{m}\left[1 - P\left(\overline{\prod_{k=1}^{n} \overline{A}_{j_k}}\right)\right] = 1 - \prod_{j=1}^{m}\left[1 - P\left(\prod_{k=1}^{n} A_{j_k}\right)\right]$$

$$= 1 - \prod_{j=1}^{m}\left[1 - \prod_{k=1}^{n} P(A_{j_k})\right] = 1 - \prod_{j=1}^{m}\left(1 - \prod_{k=1}^{n} a_{j_k}\right)$$

而

$$P\left[\prod_{k=1}^{n}\sum_{j=1}^{m}A_{j_k}\right]=\prod_{k=1}^{n}P\left(\prod_{j=1}^{m}A_{j_k}\right)$$

$$=\prod_{k=1}^{n}\left[1-P\left(\overline{\sum_{j=1}^{m}A_{j_k}}\right)\right]=\prod_{k=1}^{n}\left[1-\prod_{j=1}^{m}P\left(\overline{A}_{j_k}\right)\right]$$

$$=\prod_{k=1}^{n}\left(1-\prod_{j=1}^{m}b_{j_k}\right)$$

由式(1)等价于

$$P\left(\sum_{j=1}^{m}\prod_{k=1}^{n}A_{j_k}\right)<P\left(\prod_{k=1}^{n}\sum_{j=1}^{m}A_{j_k}\right)\tag{2}$$

不妨设$m\geqslant n$, 因$A_{j_k}\subset\prod\limits_{j=1}^{m}A_{j_k},k=1,\cdots,n$, 则$\prod\limits_{k=1}^{n}A_{i_k}\subset\prod\limits_{k=1}^{n}\sum\limits_{j=1}^{m}A_{j_k}$,

$j=1,\cdots,m$, 进而有

$$\sum_{j=1}^{m}\prod_{k=1}^{n}A_{j_k}\subset\prod_{k=1}^{n}\prod_{j=1}^{m}A_{j_k}$$

于是

$$P\left(\prod_{k=1}^{n}\sum_{j=1}^{m}A_{j_k}\right)-P\left(\sum_{j=1}^{m}\prod_{k=1}^{n}A_{j_k}\right)=P\left(\prod_{k=1}^{n}\sum_{j=1}^{m}A_{j_k}-\sum_{j=1}^{m}\prod_{k=1}^{n}A_{j_k}\right)\tag{3}$$

注意

$$\prod_{k=1}^{n}\sum_{j=1}^{m}A_{j_k}-\sum_{j=1}^{m}\prod_{k=1}^{n}A_{j_k}=\prod_{k=1}^{n}\sum_{j=1}^{m}A_{j_k}\cdot\overline{\sum_{j=1}^{m}\prod_{k=1}^{n}A_{j_k}}$$

$$=\prod_{k=1}^{n}\sum_{j=1}^{m}A_{j_k}\cdot\overline{\prod_{j=1}^{m}\prod_{k=1}^{n}A_{j_k}}=\prod_{k=1}^{n}\sum_{j=1}^{m}A_{j_k}\cdot\prod_{j=1}^{m}\sum_{k=1}^{n}\overline{A}_{j_k}\tag{4}$$

易见

$$\prod_{k=1}^{n}A_{k_k}\subset\prod_{k=1}^{n}\sum_{j=1}^{m}A_{j_k},\ \overline{A}_{1_2}\cdot\prod_{j=2}^{m}\overline{A}_{j_1}\subset\prod_{j=1}^{m}\sum_{k=1}^{n}\overline{A}_{j_k}$$

则

$$\prod_{k=1}^{n}A_{k_k}\subset\prod_{k=1}^{n}\sum_{j=1}^{m}A_{j_k},\ \overline{A}_{1_2}\cdot\prod_{j=2}^{m}\overline{A}_{j_1}\subset\prod_{k=1}^{n}\sum_{j=1}^{m}A_{j_k}\cdot\prod_{j=1}^{m}\sum_{k=1}^{n}\overline{A}_{j_k}$$

由(4)式有

$$P\left(\prod_{k=1}^{n}\sum_{j=1}^{m}A_{j_k} - \sum_{j=1}^{m}\prod_{k=1}^{n}A_{j_k}\right)$$

$$\geqslant P\left(\prod_{k=1}^{n}A_{k_k}\cdot\overline{A}_{1_2}\cdot\prod_{j=1}^{m}\overline{A}_{j_1}\right)$$

$$=\prod_{k=1}^{n}b_{k_k}\cdot(1-b_{1_2})\cdot\prod_{j=2}^{m}(1-b_{j_1}) > 0$$

从而由式(3)即知式(2)成立, 定理证毕.

若取 $a_{j_k} = x_j, b_{j_k} = y_j, j = 1, \cdots, m$, 则由定理可得

推论 1 对任意 $m, n \in \mathbf{N}$ 与满足 $x_i + y_i = 1, i = 1, \cdots, n$ 的 $x_1, \cdots, x_n$, $y_1, \cdots y_n \in (0, 1)$, 有

$$(1 - x_1 \cdots x_n)^m + (1 - y_1^m) \cdots (1 - y_n^m) > 1 \tag{5}$$

进一步再取 $x_j = x, y_j = y, j = 1, \cdots, n$, 则由推论1 可得

推论 2 对任意 $m, n \in \mathbf{N}$ 与满足 $x + y = 1$ 的 $x, y \in (0, 1)$, 有

$$(1 - x^n)^m + (1 - y^m)^n > 1 \tag{6}$$

参考文献

[1] 匡继昌. 常用不等式[M].2版. 长沙: 湖南教育出版社, 1993.

[2] CARLITZ L. SIAN Review[J]. Philadephia, 1969(11): 402-406.

[3] 王梓坤. 概率论基础及其应用[M]. 北京:科学出版社, 1979: 24.

第17篇　两个组合恒等式的概率证明

(石焕南, 范淑香. 山东师范大学学报(自然科学版), 2002, 17(1): 12-14)

摘　要:　本文给出了两个组合恒等式的概率证明,扩展了已知的结果.

关键词:　概率论; 组合恒等式; 概率方法

组合数学中的一个重要恒等式

$$\sum_{k=1}^{n}(-1)^k C_n^k (n-k)^r = \begin{cases} 0, & \text{当 } 1 \leqslant r < n \\ n!, & \text{当 } r = n \end{cases} \tag{1}$$

有着多种证法, 文[1] 、 [2]利用指母函数和积和式证明了式(1), 最近文[3]给出了一个代数证明. 本文用概率方法给出式(1)一个直观简洁的证明, 并扩展到 $r = n+1$ 的情形, 此外还建立了一个类似的恒等式.

引理 1　设随机事件 $A_1, \cdots, A_n$ 满足

$$P(A_i) = p_1 \, (i = 1, \cdots, n)$$

$$P(A_{i_1} A_{i_2}) = p_2 \, (1 \leqslant i_1 < i_2 \leqslant n)$$

$$P(A_{i_1} A_{i_2} A_{i_3}) = p_2 \, (1 \leqslant i_1 < i_2 < i_3 \leqslant n)$$

$$\cdots\cdots$$

$$P(A_1 A_2 \cdots A_n) = p_n$$

则

$$P(\bigcup_{k=1}^{n} A_k) = \sum_{k=1}^{n}(-1)^{k-1} C_n^k p_k \tag{2}$$

证明 由概率的一般加法公式

$$P(\bigcup_{i=1}^{n} A_i) = \sum_{i=1}^{n} P(A_i) - \sum_{1 \leqslant i_1 < i_2 \leqslant n} P(A_{i_1} A_{i_2}) +$$

$$\sum_{1 \leqslant i_1 < i_2 < i_3 \leqslant n} P(A_{i_1} A_{i_2} A_{i_3}) + \cdots + (-1)^{n-1} P(A_1 A_2 \cdots A_n)$$

并结合引理的条件易知式(2)成立.

定理 1 对于 $r \in \mathbf{N}$

$$\sum_{k=0}^{n} (-1)^k C_n^k (n-k)^r = \begin{cases} 0, & \text{当 } 1 \leqslant r < n \\ n!, & \text{当 } r = n \\ \frac{n(n+1)}{2} n!, & r = n+1 \end{cases} \tag{3}$$

证明 考虑随机试验:从1 到 n 这 n 个自然数中每次任取一数, 有放回地抽取 r 次, 令 $A_i = \{$取出的 r 个数均不等于 $i\}$, $i = 1, \cdots, n$, 则

$$p_k = P(A_{i_1} A_{i_2} \cdots A_{i_k}) = \left(\frac{n-k}{n}\right)^r$$

$$(1 \leqslant i_1 < i_2 < \cdots < i_k \leqslant n, k = 1, 2, \cdots, n)$$

由式(2)有

$$P(\bigcup_{k=1}^{n} A_k) = \sum_{k=1}^{n} (-1)^{k-1} C_n^k \left(\frac{n-k}{n}\right)^r \tag{4}$$

当 $1 \leqslant r < n$ 时, 必存在 i 使得取出的 r 个数均不等于 i, 因此 $\bigcup_{i=1}^{n} A_i$ 是必然事件, 于是, 由式(4)有

$$\sum_{k=1}^{n} (-1)^{k-1} C_n^k \left(\frac{n-k}{n}\right)^r = P(\bigcup_{i=1}^{n} A_i) = 1 = C_n^0$$

稍加整理, 即得

$$\sum_{k=0}^{n} (-1)^k C_n^k (n-k)^r = 0$$

当 $r = n$ 时, 注意 $\bar{A}_i = \{$取出的 n 个数中至少有一个等于 $i\}$, $i = 1, \cdots, n$,

于是, $\prod\limits_{i=1}^{n} \bar{A}_i = \{取出的n 个数均不相同\}$, 其概率为 $\frac{n!}{n^n}$, 从而

$$P(\bigcup_{i=1}^{n} A_i) = 1 - P(\prod_{i=1}^{n} \bar{A}_i) = 1 - \frac{n!}{n^n}$$

代入式(4)并稍加整理, 即得

$$\sum_{k=0}^{n}(-1)^k \mathrm{C}_n^k (n-k)^r = n!$$

当 $r = n+1$ 时, 注意 $\prod\limits_{i=1}^{n} \bar{A}_i = \{$ 取出的$n+1$ 个数恰有两个数相同$\}$, 其概率为 $\frac{n!}{n^{n+1}}\mathrm{C}_{n+1}^2$, 从而

$$P(\bigcup_{i=1}^{n} A_i) = 1 - P(\prod_{i=1}^{n} \bar{A}_i) = 1 - \frac{n!}{n^{n+1}}\mathrm{C}_{n+1}^2$$

代入式(4)并稍加整理, 即得

$$\sum_{k=0}^{n}(-1)^k \mathrm{C}_n^k (n-k)^r = n!\mathrm{C}_{n+1}^2 = \frac{n(n+1)}{2}n!$$

最后单独讨论$r = 0$ 的情形, 考虑随机试验:从大于n 的自然数中任取一数,令$A_i = \{$ 取出的数大于$i\}$, $i = 1, \cdots, n$, 则显然

$$p_k = P(A_{i_1} A_{i_2} \cdots A_{i_k}) = 1$$

$$(1 \leqslant i_1 < i_2 < \cdots < i_k \leqslant n, k = 1, 2, \cdots, n)$$

且 $\sum\limits_{i=1}^{n} P(\bigcup\limits_{i=1}^{n} A_i) = 1 = \mathrm{C}_n^0$, 代入式(2)并稍加整理, 即得 $\sum\limits_{k=0}^{n}(-1)^k \mathrm{C}_n^k = 0$. 证毕.

上述定理1 通过有放回取数这一简单概率模型证得式(3),下面的定理2 将利用一个不放回抽球的概率模型建立一个与式(3)类似的恒等式.

定理 2　对于$r \in \mathbf{N}$

$$\sum_{k=0}^{n}(-1)^k \mathrm{C}_n^k \mathrm{C}_{mn-mk}^r = \begin{cases} 0, & 当 0 < r < n \\ m^n, & 当 r = n \\ \frac{n(m-1)}{2}m^n, & r = n+1, r \geqslant 2 \end{cases} \tag{5}$$

证明 考虑随机试验:一袋中装有标号为 $1,2,\cdots,n$ 的球各 m 个, 从中不放回地任取 r 个球, 令

$$A_i = \{ \text{取出的} r \text{ 个球均不为} i \text{ 号球} \}, i = 1, \cdots, n$$

则

$$p_k = P(A_{i_1} A_{i_2} \cdots A_{i_k}) = \frac{C_{mn-mk}^r}{C_{mn}^r}$$

$$(1 \leqslant i_1 < i_2 < \cdots < i_k \leqslant n, k = 1, 2, \cdots, n)$$

由式(2)有

$$P(\bigcup_{k=1}^{n} A_k) = \sum_{k=1}^{n} (-1)^{k-1} C_n^k \frac{C_{mn-mk}^r}{C_{mn}^r} \tag{6}$$

当 $0 < r < n$ 时, 易见 $\sum\limits_{i=1}^{n} P(\bigcup\limits_{i=1}^{n} A_i) = 1 = C_n^0$, 代入式(6)稍加整理,即得

$$\sum_{k=0}^{n} (-1)^k C_n^k C_{mn-mk}^r = 0$$

当 $r = n$ 时, 注意

$$\prod_{i=1}^{n} \bar{A}_i = \{ \text{取出的} r \text{ 个数均不相同} \}$$

其概率为 $\frac{m^n}{C_{mn}^n}$, 从而

$$P(\bigcup_{i=1}^{n} A_i) = 1 - P(\prod_{i=1}^{n} \bar{A}_i) = 1 - \frac{m^n}{C_{mn}^n}$$

代入式(6)稍加整理, 即得

$$\sum_{k=0}^{n} (-1)^k C_n^k C_{mn-mk}^k = m^n$$

当 $r = n + 1$ 时, 注意

$$\prod_{i=1}^{n} \bar{A}_i = \{ \text{取出的} n + 1 \text{ 个数恰有两个数相同} \}$$

其概率为 $\dfrac{nm^{n-1}C_m^2}{C_{mn}^n}$, 从而

$$P(\bigcup_{i=1}^{n} A_i) = 1 - P(\prod_{i=1}^{n} \bar{A}_i) = 1 - \frac{nm^{n-1}C_m^2}{C_{mn}^n}$$

代入式(6)并稍加整理, 即得

$$\sum_{k=0}^{n}(-1)^k C_n^k C_{mn-mk}^{n+1} = nm^{n-1}C_m^2 = \frac{n(m-1)}{2}m^n$$

证毕.

著名数学家王梓坤院士在文[4]中指出:"用概率的方法来证明一些关系式或解放其他数学分析中的问题, 是概率论的重要研究方向之一." 用概率的方法解决一些问题, 思路别开生面, 过程常常简洁直观, 由本例可领略一斑. 最后提出一个问题:

当 $r > n + 1$ 时

$$\sum_{k=0}^{n}(-1)^k C_n^k (n-k)^r =?, \quad \sum_{k=0}^{n}(-1)^k C_n^k C_{mn-mk}^r =?$$

参考文献

[1]　BRUALDI R A. Introductory combinatorics [M]. New York : North - Holland, 1997: 1-50.

[2]　柯召, 魏万迪. 组合论: 上册[M]. 北京:科学出版社, 1981: 30-57.

[3]　王俊青. 用代数方法证明一个恒等式[J]. 山东师范大学学报(自然科学版), 2001, 16 (2) : 129-130.

[4]　王梓坤. 概率论基础及其应用[M]. 北京:科学出版社, 1979: 24.

第18篇　整幂函数不等式的控制证明

(石焕南，续铁权，顾春. 商丘师范学院学报，2003,19(2):46-48)

摘　要:　本文利用控制不等式理论简洁地证明了一些整幂函数不等式,大部分结果是一些常用不等式的推广.

关键词:　凸函数; Schur凸函数; 整幂函数; 不等式; 控制

18.1　定义与引理

定义 1[1]　设 $x, y \in \mathbf{R}^n$ 满足

$$(1) \sum_{i=1}^{k} x_{[i]} \le \sum_{i=1}^{k} y_{[i]}, k = 1, 2, ..., n-1, \quad (2) \sum_{i=1}^{n} x_i = \sum_{i=1}^{n} y_i$$

则称 x 被 y 所控制, 记作 $x \prec y$. 其中 $x_{[1]} \ge \cdots \ge x_{[n]}$ 和 $y_{[1]} \ge \cdots \ge y_{[n]}$ 分别是 x 和 y 的分量的递减重排. 又若 x 不是 y 的重排, 则称 x 被 y 严格控制, 记作 $x \prec\prec y$.

定义 2　设 $\Omega \subset \mathbf{R}^n, \phi : \Omega \to \mathbf{R}$, 若在 Ω 上 $x \prec y \Rightarrow \phi(x) \le \phi(y)$, 则称 ϕ 为 Ω 上的 Schur 凸函数; 若在 Ω 上 $x \prec\prec y \Rightarrow \phi(x) \le \phi(y)$, 则称 ϕ 为 Ω 上的严格 Schur 凸函数.

本文利用控制不等式理论证明了一类涉及整幂函数的不等式, 所获得的大部分结果是文献[2]中的不等式的推广. 我们要用到如下引理.

引理 1[2]　设 $I \subset \mathbf{R}$ 为一个区间, g 为 $I \to \mathbf{R}$ 上的凸(严格凸)函数, 则

$$\phi(x) = \sum_{i=1}^{n} g(x_i)$$

为 I^n 上的 Schur 凸(严格 Schur 凸)函数.

引理 2　设 $x = (x_1.x_2,...,x_n) \in \mathbf{R}^n$, 则

$$(\bar{x},\bar{x},\cdots,\bar{x}) \prec (x_1.x_2,...,x_n)$$

其中 $\bar{x} = \frac{1}{n}\sum_{i=1}^n x_i$.

引理 3　设 $x,y \in \mathbf{R}^n, x_1 \geqslant x_2 \geqslant \cdots \geqslant x_n, \sum_{i=1}^n x_i = \sum_{i=1}^n y_i$, 若存在 $k, 1 \leqslant k \leqslant n$, 使得 $x_i \leqslant y_i, i = 1,...,k, x_i \geqslant y_i, i = k+1,...,n$, 则 $x \prec y$. (参见文[3]第234页定理8.1.10)

引理 4　设 $g(t) = \ln \frac{x^t-1}{t}$, 有

(1) 若 $x > 1$, 则 $g(t)$ 是 $(0,+\infty)$ 上的严格凸函数;

(2) 若 $0 < x < 1$, 则 $-g(t)$ 是 $(0,+\infty)$ 上的严格凸函数.

证明　经计算

$$g''(t) = -\frac{x^t \ln^2 x}{(x^t-1)^2} + \frac{1}{t^2}$$

欲证 $g''(t) > 0$, 它等价于 $t^2 x^t (\ln x)^2 < (x^t-1)^2$, 两边开方并同除以 x^t, 亦等价于 $f(t) := x^{\frac{t}{2}} - x^{-\frac{t}{2}} - t\ln x > 0$. 当 $x > 1$ 时, $f'(t) = \frac{1}{2}\ln x(x^{\frac{t}{2}} + x^{-\frac{t}{2}} - 2) > 0$, 所以 $f(t)$ 在 $(0,+\infty)$ 上严格单调增. 从而当 $t > 0$ 时, $f(t) > f(0) = 0$, 即 $g''(t) > 0$, (1) 得证, 仿此可证得(2).

18.2　主要结论与证明

定理 1　设 $x > 0, x \neq 1, m,n,k \in \mathbf{N}$, 则

$$x^{kn-1} + x < x^{kn} + 1 \tag{1}$$

$$x^{n-(k-2)m} + x^{-(k-1)n} < x^{m-(k-2)n} + x^{-(k-1)m}(m > n, k \geqslant 2) \tag{2}$$

$$x^k(n - kx^{n-k}) < n - k(k < n) \tag{3}$$

$$\frac{x^m-1}{m} \geqslant \frac{x^n-1}{n}(m > n)^{[2]} \tag{4}$$

证明　对于 $x > 0, x \neq 1$, 由于 $g''(t) = x^t(\ln x)^2 > 0$, 所以 $g(t) = x^t$ 是 $[0,+\infty)$ 上的严格凸函数. 易见

$$(kn-1,1) \prec\prec (kn,0)$$

由引理1 即可证得式(1); 而由

$$(kn+m,(k-1)m) \prec\prec (km+n,(k-1)n)$$

有

$$x^{kn+m} + x^{(k-1)m} < x^{km+n} + x^{(k-1)n}$$

而此式与式(2)等价; 由

$$(\underbrace{k,\cdots,k}_{n},\underbrace{0,\cdots,0}_{(k-1)n}) \prec\prec (\underbrace{n,\cdots,n}_{k},\underbrace{0,\cdots,0}_{(n-1)k})$$

有

$$nx^k + (k-1)n < kx^n + (n-1)k$$

而此式与式(3)等价; 由

$$(\underbrace{n,\cdots,n}_{m},\underbrace{0,\cdots,0}_{n}) \prec (\underbrace{m,\cdots,m}_{n},\underbrace{0,\cdots,0}_{m})$$

有

$$nx^m + m > mx^n + n$$

而此式与式(4)等价.

定理 2 设 $x > 1, n \geqslant 2, n, k \in \mathbf{N}, k < n$, 则

$$k(x^n - x^{-n}) \geqslant n(x^k - x^{-k}) \tag{5}$$

证明 令 $g(t) = x^t - x^{-t}$, 对于 $x > 1$, 当 $t > 0$ 时, 由于

$$g''(t) = (\ln x)^2 (x^t - x^{-t}) = (\ln x)^2 \cdot \frac{x^{2t} - 1}{x^t} \geqslant 0$$

所以 $g(t)$ 是 $(0, +\infty)$ 上的凸函数. 不难验证

$$(\underbrace{k,\cdots,k}_{n}) \prec (\underbrace{n,\cdots,n}_{k},\underbrace{0,\cdots,0}_{n-k})$$

从而由引理1 即知式(5)成立.

注 1 取 $k = 2$, (1)和(2)两式分别化为文[2]第 117 页第 1 题 (2) 和 (3), 取 $k = 1$, (3) 和 (5) 两式分别化为文 [2] 第117页第 1 题 (4) 和 [2] 第118页第 2 题.

定理 3　设 $x > 0, n \in \mathbf{N}$, 则

$$\frac{1}{n-2m}\sum_{k=m+1}^{n-m} x^k \leqslant \frac{1}{n}\sum_{k=1}^{n} x^k \leqslant \frac{1}{2m}\left(\sum_{k=1}^{m} x^k + \sum_{k=n-m+1}^{n-m} x^k\right) \ (n > 2m)$$

(6)

证明　记

$$\boldsymbol{x} = \left(\underbrace{n-m, \cdots, n-m}_{2m}, \underbrace{m+1, \cdots, m+1}_{2m}\right)$$

$$\boldsymbol{y} = \left(\underbrace{n, \cdots, n}_{2m}, \underbrace{n-m+1, \cdots, n-m+1}_{n-m}, \underbrace{m, \cdots, m}_{n-2m}, \cdots, \underbrace{1, \cdots, 1}_{n-2m}\right)$$

注意 $\boldsymbol{y}$ 的前 $m(n-m)$ 个分量的每一个均大于 $\boldsymbol{x}$ 的任一分量, 而 $\boldsymbol{y}$ 的后 $m(n-m)$ 个分量的每一个均小于 $\boldsymbol{x}$ 的任一分量, 由引理3 可知 $\boldsymbol{x} \prec \boldsymbol{y}$, 从而有

$$2m\sum_{k=m+1}^{n-m} x^k \leqslant (n-2m)\left(\sum_{k=1}^{m} x^k + \sum_{k=n-m+1}^{n} x^k\right)$$

而不难验证式(6)的左右两个不等式均等价于上式, 因此式(6)成立.

定理 4[2]　设 $x > 0, x \neq 1, n \in \mathbf{N}$, 则

$$\sum_{k=0}^{m} x^k > \frac{n+1}{n-1}\sum_{k=1}^{n-1} x^k$$

(7)

证明　当 $x > 0, x \neq 1$ 时, 由 $g(t) = x^t$ 在 $[0, +\infty)$ 上的严格凸性, 并注意

$$\left(\underbrace{1, \cdots, 1}_{n+1}, \underbrace{2, \cdots, 2}_{n+1}, \cdots, \underbrace{n-1, \cdots, n-1}_{n+1}\right)$$

$$\prec \left(\underbrace{0, \cdots, 0}_{n-1}, \underbrace{1, \cdots, 1}_{n-1}, \cdots, \underbrace{n, \cdots, n}_{n-1}\right)$$

由引理1 有

$$\sum_{k=0}^{n}(n-1)x^k > \sum_{k=1}^{n-1}(n+1)x^k$$

而此式与式(7)等价.

定理5   若$x > 0, x \neq 1, n, k \in \mathbf{N}, n \geqslant 2, n > k$, 则

$$\left(1 - \frac{k^2}{n^2}\right)(x^n - 1)^2 < (x^{n-k} - 1)(x^{n+k} - 1) < (x^n - 1)^2 \qquad (8)$$

证明   因$(n, n) \prec\prec (n+k, n-k)$, 由$x^t$ 在$[0 + \infty)$ 上的严格凸性, 有$2x^n < x^{n+k} + x^{n-k}$, 而此式与式(8)右边不等式等价.

对于左边不等式分两种情况讨论:

(1) 当$x > 1$ 时, 由引理4 (1) 知$g(t) = \ln\frac{x^t - 1}{t}$ 是$(0 + \infty)$ 上的严格凸函数, 因$(n, n) \prec\prec (n+k, n-k)$, 由引理1 有

$$2\ln\frac{x^n - 1}{n} < \ln\frac{x^{n-k} - 1}{n-1} + \ln\frac{x^{n+k} - 1}{n+1}$$

而此式与式(8)左边不等式等价;

(2) 当$0 < x < 1$ 时, 由引理 4 (2) 同样可知式(8)左边不等式成立.

注2   取$k = 1$, 式(8)化为文[2] 第 120 页第4题(4).

定理6 [2]   设$0 \leqslant x \leqslant 1$, 则

$$(2n + 1)x^n(1 - x) \leqslant 1 - x^{2n+1} \qquad (9)$$

且当$x \neq 1$ 时, 不等式严格成立.

证明   当$x = 0$ 或1时, 式(9)显然成立, 当$0 < x < 1$ 时, 注意

$$\Big(\underbrace{n, \cdots, n}_{2n+1}\Big) \prec\prec (0, 1, \cdots, 2n)$$

由x^t 在$[0 + \infty)$ 上的严格凸性, 有

$$(2n + 1)x^n < x^{2n} + x^{2n-1} + \cdots + x + 1$$

两边同乘以$1 - x$, 并注意

$$1 - x^{2n+1} = (1 - x)(x^{2n} + x^{2n-1} + \cdots + x + 1)$$

即得证.

通过以上不等式的证明, 我们可以初步领略控制不等式一个鲜明的特点: "能把许多已有的从不同方法得来的不等式, 用一种统一的方法简便地推导出来"[1].

参考文献

[1]　王伯英. 控制不等式基础[M]. 北京: 北京师范大学出版社,1990.

[2]　匡继昌. 常用不等式[M]. 2 版. 长沙: 湖南教育出版社, 1993.

[3]　王松桂, 贾忠贞. 矩阵论中不等式[M]. 合肥: 安徽教育出版社, 1994: 312.

[4]　PEČARIĆ J E. Convex functions, partial orderings, and statistical applications[M]. New York: Academic Press, Inc., 1992.

第19篇　一类无理不等式的控制证明

(吴善和, 石焕南. 首都师范大学学报(自然科学版), 2003, 24(3):13-16)

摘　要:　本文利用控制不等式理论建立一类新的无理不等式,它们或推广或加强了已知不等式或给出已知不等式的反向估计.

关键词:　无理不等式; Schur 凸函数; 控制理论; 推广; 加强

19.1　定义与引理

关于控制不等式理论, 王伯英先生在他的专著《控制不等式基础》一书中有一段精辟的论述: "控制不等式理论作为一门新兴学科, 近几年得到迅速的发展, 它能深刻地描述许多数学量之间的内在关系, 成为推广已有不等式和发现新不等式的一种强有力手段, 具有广泛的应用价值". 本文利用控制不等式理论建立一类新的无理不等式, 它们或推广或加强了已知不等式或给出已知不等式的反向估计.

定义 1[1]　设 $x, y \in \mathbf{R}^n$, x 与 y 分量的递减重排分别记为

$$x_{[1]} \geqslant x_{[2]} \geqslant \cdots \geqslant x_{[n]}$$

和

$$y_{[1]} \geqslant y_{[2]} \geqslant \cdots \geqslant y_{[n]}$$

若满足

$$(1) \sum_{i=1}^{k} x_{[i]} \leq \sum_{i=1}^{k} y_{[i]}, k = 1, 2, ..., n-1; (2) \sum_{i=1}^{n} x_i = \sum_{i=1}^{n} y_i$$

则称 x 被 y 所控制, 记作 $x \prec y$. 又若 x 不是 y 的重排, 则称 x 被 y 严格控制, 记作 $x \prec\prec y$.

定义 2　设 $\Omega \subset \mathbf{R}^n, \phi: \Omega \to \mathbf{R}$, 若在 Ω 上 $x \prec y \Rightarrow \phi(x) \leqslant \phi(y)$, 则称 ϕ

为 Ω 上的 Schur 凸函数; 若在 Ω 上 $\boldsymbol{x} \prec\prec \boldsymbol{y} \Rightarrow \phi(\boldsymbol{x}) \leqslant \phi(\boldsymbol{y})$, 则称 ϕ 为 Ω 上的严格 Schur 凸函数.

引理 1[2]　设 $\boldsymbol{x}, \boldsymbol{y} \in \mathbf{R}^n, x_1 \geqslant x_2 \geqslant \cdots \geqslant x_n, \sum\limits_{i=1}^{n} x_i = \sum\limits_{i=1}^{n} y_i$, 若存在 $k, 1 \leqslant k \leqslant n$, 使得 $x_i \leqslant y_i, i = 1, \cdots, k, x_i \geqslant y_i, i = k+1, \cdots, n$, 则 $\boldsymbol{x} \prec \boldsymbol{y}$.

引理 2　设 $a \leqslant x_i \leqslant b, i = 1, 2, \cdots, n, n \geqslant 2, x_1 + x_2 + \cdots + x_n = s$, 则

$$\boldsymbol{x} = (x_1, x_2, \cdots, x_n) \prec \Big(\underbrace{b, \cdots, b}_{n-1-u}, c, \underbrace{a, \cdots, a}_{u} \Big) \tag{1}$$

其中 $u = \left[\frac{nb-s}{b-a} \right], c = s - b(n-1) + (b-a)u$.

证明　根据题设条件, 有 $x_1 + x_2 + \cdots + x_n = y_1 + y_2 + \cdots + y_n$, 从而欲证式 (1), 只需证明

$$x_{[1]} + x_{[2]} + \cdots + x_{[k]} \leqslant y_{[1]} + y_{[2]} + \cdots + y_{[k]} \quad (k = 1, 2, \cdots, n-1) \quad (*)$$

不难验证 $b \geqslant c \geqslant a$. 由 $a \leqslant x_i \leqslant b (i = 1, 2, \cdots, n)$ 得

$$x_{[1]} + x_{[2]} + \cdots + x_{[k]} \leqslant kb \leqslant y_{[1]} + y_{[2]} + \cdots + y_{[k]} \quad (k = 1, 2, \cdots, n-1-u)$$

$$x_{[k+1]} + x_{[k+2]} + \cdots + x_{[n]} \geqslant (n-k)a$$
$$= y_{[k+1]} + y_{[k+2]} + \cdots + y_{[n]}(k = n-u, n-u+1, \cdots, n-1)$$

由

$$x_1 + x_2 + \cdots + x_n = y_1 + y_2 + \cdots + y_n$$

知上式等价于

$$x_{[1]} + x_{[2]} + \cdots + x_{[k]} \leqslant y_{[1]} + y_{[2]} + \cdots + y_{[k]}, k = n-u, n-u+1 \cdots, n-1$$

这样就证得式 (*), 从而式 (1) 成立.

引理 3[3]　设 $x_i \in \mathbf{R}, i = 1, ..., n, n \geqslant 2$, 且 $\sum\limits_{i=1}^{n} x_i = s > 0, c \geqslant s$, 则

$$\frac{(c-\boldsymbol{x})s}{nc-s} = \Big(\frac{(c-x_1)s}{nc-s}, \frac{(c-x_2)s}{nc-s}, \cdots, \frac{(c-x_n)s}{nc-s} \Big) \prec (x_1, x_2, ..., x_n) = \boldsymbol{x} \tag{2}$$

引理 4　设 $a_1 \geqslant a_2 \geqslant \cdots \geqslant a_{2n} \geqslant 0$, 则

$$\left(\sum_{i=1}^{2n-1}(-1)^{i-1}a_i, a_2, a_4, \cdots, a_{2n-2}\right) \prec (a_1, a_3, \cdots, a_{2n-1}) \quad (n \geqslant 2) \quad (3)$$

$$\left(\sum_{i=1}^{2n}(-1)^{i-1}a_i, a_2, a_4, \cdots, a_{2n}\right) \prec (a_1, a_3, \cdots, a_{2n-1}, 0) \quad (n \geqslant 1) \quad (4)$$

证明　下面利用引理1 证式(3). 显然, 式(3)中两个向量的分量之和相等, 记 $s = \sum_{i=1}^{2n-1}(-1)^{i-1}a_i$

以下分4 种情况讨论:

（Ⅰ）若 $a_2 \geqslant a_4 \geqslant \cdots \geqslant a_{2n-2} \geqslant s$.

因为 $a_{2i} \leqslant a_{2i-1}, i=1,2,\cdots,n-1$, 而

$$s = (a_1 - a_2) + (a_3 - a_4) + \cdots + (a_{2n-3} - a_{2n-2}) + a_{2n-1} \geqslant a_{2n-1}$$

由引理1 知式(3)成立.

（Ⅱ）若 $s \geqslant a_2 \geqslant a_4 \geqslant \cdots \geqslant a_{2n-2}$, 因为

$$s = a_1 - (a_2 - a_3) - (a_4 - a_5) - \cdots - (a_{2n-2} - a_{2n-1}) \leqslant a_1$$

而 $a_{2i} \geqslant a_{2i+1}, i=1,2,\cdots,n-1$, 由引理1 知式(3)成立.

（Ⅲ）若 $a_2 \geqslant a_4 \geqslant \cdots \geqslant a_{2k} \geqslant a_{2k+1} \geqslant s \geqslant a_{2k+2} \geqslant \cdots \geqslant a_{2n-2}$.

此时 $a_{2i} \leqslant a_{2i-1}, i=1,2,\cdots,k; s \leqslant a_{2k+1}$, 而 $a_{2i} \geqslant a_{2i+1}, i=k+1,k+2,\cdots,n-1$; 由引理1 知式(3) 成立.

（Ⅳ）若 $a_2 \geqslant a_4 \geqslant \cdots \geqslant a_{2k} \geqslant s \geqslant a_{2k+1} \geqslant a_{2k+2} \geqslant \cdots \geqslant a_{2n-2}$.

此时, $a_{2i} \leqslant a_{2i-1}, i=1,2,\cdots,k; s \geqslant a_{2k+1}$, 而 $a_{2i} \geqslant a_{2i+1}, i=k+1,k+2,\cdots,n-1$; 由引理1 知式(3)成立.

根据（Ⅰ）~（Ⅳ）得, 引理4 中式(3)成立. 同理可证, 引理4 中式(4)成立.

引理 5 [1]　设 $I \subset \mathbf{R}$ 为一个区间, f 为 $I \to \mathbf{R}$ 上凹(严格凹) 函数, $x=(x_1,x_2,\cdots,x_n)$, 则 $\varphi x = \sum_{i=1}^{n} f(x_i)$ 为 I^n 上的Schur 凹(严格Schur 凹) 函数.

19.2　主要结论与证明

定理 1　设 $x_i \geqslant 0, i = 1, 2, \cdots, n, n \geqslant 2, x_1 + x_2 + \cdots + x_n = 1, k \in$ $\mathbf{N}, k > 1, m > 0, 1 - mx^{k_i} \geqslant 0$，则

$$\sqrt[k]{n^k - m} \geqslant \sum_{i=1}^{n} \sqrt[k]{1 - mx_i^k} \geqslant \sqrt[k]{1 - mc^k} + u \qquad (5)$$

其中 $c = 1 - (n-1)m^{-\frac{1}{k}} + um^{-\frac{1}{k}}, u = [n - m^{\frac{1}{k}}]$．

证明　令 $f(x) = (1 - mx^k)^{\frac{1}{k}}$，则

$$f''(x)$$
$$= -(k-1)\left[\frac{1}{k^2}(mkx^{k-1})^2(1 - mx^k)^{\frac{1}{k}-2} + mk^{k-2}(1 - mx^k)^{\frac{1}{k}}\right]$$

由 $k > 1$ 知 $f''(x) \leqslant 0$，即 $f(x)$ 是其定义域 $I = [0, m^{-\frac{1}{k}}]$ 上的凹函数，由引理 5 知 $\varphi(x_1, x_2, \cdots, x_n) = f(x_1) + f(x_2) + \cdots + f(x_n)$ 是 I^n 上的 Schur 凹函数；令 $b = m^{-\frac{1}{k}}$，则 $0 \leqslant x_i \leqslant b, i = 1, 2, \cdots, n$；运用引理 2，得

$$\left(\frac{1}{n}, \frac{1}{n}, \cdots, \frac{1}{n}\right) \prec (x_1, x_2, \cdots, x_n) \prec \left(\underbrace{b, \cdots, b}_{n-1-u}, c, \underbrace{a, \cdots, a}_{u}\right)$$

其中

$$u = \left[\frac{nb - s}{b - a}\right], c = s - b(n-1) + (b-a)u$$

所以

$$\varphi\left(\frac{1}{n}, \frac{1}{n}, \cdots, \frac{1}{n}\right) \geqslant \varphi(x_1, x_2, \cdots, x_n) \geqslant \varphi\left(\underbrace{b, \cdots, b}_{n-1-u}, c, \underbrace{a, \cdots, a}_{u}\right)$$

将上面不等式整理即得式 (5)．

定理 2　设 $x_i \geqslant 0, i = 1, 2, \cdots, n, n \geqslant 2, x_1 + x_2 + \cdots + x_n = 1, k \in$ $\mathbf{N}, k > 1, p > 0, q \geqslant 0$，则

$$n^{1-\frac{1}{k}}\sqrt[k]{np + q} \geqslant \sum_{i=1}^{n} \sqrt[k]{p + qx_i} \geqslant \sqrt[k]{p + q} + (n-1)\sqrt[k]{p} \qquad (6)$$

证明　令 $f(x) = (p + qx)^{\frac{1}{k}}$, 则

$$f''(x) = \frac{1}{k}\left(\frac{1}{k} - 1\right)q^2(p + qx)^{\frac{1}{k} - 2}$$

由 $k > 1$ 知 $f''(x) < 0$, 即 $f(x)$ 是 $\mathbf{R}_+^1$ 上的严格凹函数, 由引理5 知

$$\varphi(x_1, x_2, \cdots, x_n) = f(x_1) + f(x_2) + \cdots + f(x_n)$$

是 $\mathbf{R}_+^n$ 上的严格Schur 凹函数; 因为

$$\left(\frac{1}{n}, \frac{1}{n}, \cdots, \frac{1}{n}\right) \prec (x_1, x_2, \cdots, x_n) \prec (1, 0, \cdots 0)$$

所以

$$\varphi\left(\frac{1}{n}, \frac{1}{n}, \cdots, \frac{1}{n}\right) \geqslant \varphi(x_1, x_2, \cdots, x_n) \geqslant \varphi(1, 0, \cdots 0)$$

经整理, 上面不等式即为式(6).

在不等式(6)中, 令 $x_1 = a, x_2 = b, x_3 = c, n = 3, p = 1, q = 4$, 得

$$3^{1 - \frac{1}{k}}\sqrt[k]{7} \geqslant \sqrt[k]{4a + 1} + \sqrt[k]{4b + 1} + \sqrt[k]{4c + 1} \geqslant 2 + \sqrt[k]{5} \tag{7}$$

式(7)左边是文[4]中问题65 - 3 的一个上界估计.

定理 3　设 $m \in \mathbf{N}, m > 1$, 则

$$\frac{1}{2}\prod_{j=0}^{m-1}(n^2 + n + 2j)^{\frac{1}{m}} < \sum_{k=1}^{n}\prod_{j=0}^{m-1}(k + j)^{\frac{1}{m}} < \frac{n}{2}\prod_{j=0}^{m-1}(n + 1 + 2j)^{\frac{1}{m}} \tag{8}$$

证明　令 $f(x) = \prod_{j=0}^{m-1}(x + j)^{\frac{1}{m}}$, 则

$$f''(x) = f(x)\left[\left(\frac{1}{m}\sum_{j=0}^{m-1}\frac{1}{x + j}\right)^2 - \frac{1}{m}\sum_{j=0}^{m-1}\frac{1}{(x + j)^2}\right]$$

由平方－算术平均值不等式知 $f''(x) < 0$, 因此 $f(x)$ 是 $\mathbf{R}_{++}^1$ 上的严格凹函数, 从而 $\varphi(x_1, x_2, \cdots, x_n) = f(x_1) + f(x_2) + \cdots + f(x_n)$ 是 $\mathbf{R}_{++}^n$ 上的严格Schur凹函数; 因为

$$\left(\frac{n+1}{2}, \frac{n+1}{2}, \cdots, \frac{n+1}{2}\right) \prec\prec (1, 2, \cdots, n)$$

$$\prec\prec \left(\frac{n(n+1)}{2}, 0, \cdots 0 \right)$$

所以

$$\varphi\left(\frac{n+1}{2}, \frac{n+1}{2}, \cdots, \frac{n+1}{2} \right) > \varphi(1, 2, \cdots, n)$$

$$> \varphi\left(\frac{n(n+1)}{2}, 0, \cdots 0 \right)$$

经整理,上面不等式即为式(8).

在定理3 中,令$m = 2$, 得

$$\sqrt{\frac{n(n+1)(n^2+n+2)}{2}} < \sum_{k=1}^{n} \sqrt{k(k+1)} < \frac{n\sqrt{(n+1)(n+3)}}{2} \qquad (9)$$

不难验证

$$\frac{n(n+1)}{2} < \frac{\sqrt{n(n+1)(n^2+n+2)}}{2}$$

$$\frac{n\sqrt{(n+1)(n+3)}}{2} < \frac{n(n+2)}{2}$$

于是, 我们得到如下不等式链

$$\frac{n(n+1)}{2} < \frac{\sqrt{n(n+1)(n^2+n+2)}}{2}$$

$$< \sum_{k=1}^{n} \sqrt{k(k+1)} < \frac{n\sqrt{(n+1)(n+3)}}{2} < \frac{n(n+2)}{2} \qquad (10)$$

式(10)给出文[4]中不等式(42) 的一个加细.

定理 4　设$a_i \geqslant 0, i = 1, 2, \cdots, n, n \geqslant 2, r \geqslant 1, p = \sum_{i=1}^{n} a_i^r$, 则

$$\sum_{i=1}^{n} (p - a_i^r)^{\frac{1}{r}} \geqslant (n-1)^{\frac{1}{r}} \sum_{i=1}^{n} a_i \qquad (11)$$

证明　在式(2)中, 令$x_i = (n-1)a_i^r, i = 1, 2, \cdots, n, c = s = \sum_{i}^{n} (n-1)a_i^r$, 得

$$(p - a_1^r, p - a_2^r, \cdots, p - a_n^r)$$

$$\prec ((n-1)a_1^r, (n-1)a_2^r, \cdots, (n-1)a_n^r)$$

设 $f(x) = x^{\frac{1}{r}}$, 则 $f''(x) = \frac{1}{r}\left(\frac{1}{r} - 1\right)x^{\frac{1}{r}-2}$; 由 $r \geqslant 1$ 知 $f''(x) \leqslant 0$. 故 $f(x) = x^{\frac{1}{r}}$ 是 $[0, +\infty)$ 上的凹函数, 从而

$$\varphi(x_1, x_2, \cdots, x_n) = f(x_1) + f(x_2) + \cdots + f(x_n)$$

是 $[0, +\infty)$ 上的Schur 凹函数; 所以

$$\varphi\left(p - a_1^r, p - a_2^r, \cdots, p - a_n^r\right)$$
$$\geqslant \varphi\left((n-1)a_1^r, (n-1)a_2^r, \cdots, (n-1)a_n^r\right)$$

将上面不等式整理即得式(11).

在式(11)中,令 $r = 2, n = 3$, 便得到文[4]中的不等式(59).

定理 5 设 $a_1 > a_2 > \cdots > a_m > 0 (m \geqslant 2)$, 则

$$\sum_{i=1}^{m}(-1)^{i-1}\sqrt{a_i} < \sqrt{\sum_{i=1}^{m}(-1)^{i-1}a_i} \qquad (12)$$

证明 当 $m = 2n - 1$ 时, 根据引理4 中的式(3), 有

$$\left(\sum_{i=1}^{2n-1}(-1)^{i-1}a_i, a_2, a_4, \cdots, a_{2n-2}\right) \prec\prec (a_1, a_3, \cdots, a_{2n-1})$$

设 $f(x) = \sqrt{x}$, 则 $f''(x) = -\frac{1}{4}x^{\frac{3}{2}} < 0$, 因此 $f(x) = \sqrt{x}$ 是 $(0, +\infty)$ 上的严格凹函数, $\varphi(x_1, x_2, \cdots, x_n) = f(x_1) + f(x_2) + \cdots + f(x_n)$ 是 $[0, +\infty)$ 上的严格Schur凹函数; 所以

$$\varphi\left(\sum_{i=1}^{2n-1}(-1)^{i-1}a_i, a_2, a_4, \cdots, a_{2n-2}\right) > \varphi(a_1, a_3, \cdots, a_{2n-1})$$

从而

$$\sqrt{\sum_{i=1}^{2n-1}(-1)^{i-1}a_i} + \sum_{i=1}^{n-1}\sqrt{a_{2i}} > \sum_{i=1}^{n}\sqrt{a_{2i-1}}$$

即

$$\sum_{i=1}^{2n-1}(-1)^{i-1}\sqrt{a_i} < \sqrt{\sum_{i=1}^{2n-1}(-1)^{i-1}a_i}$$

所以式(12)成立; 同理可证, 当 $m = 2n$ 时, 式(12)成立, 定理5 得证.

参考文献

[1] 王伯英. 控制不等式基础[M]. 北京:北京师范大学出版社, 1990.

[2] 王松桂,贾忠贞. 矩阵论中不等式[M]. 合肥:安徽教育出版社, 1994.

[3] 石焕南. 一类对称函数不等式的加细与推广[J]. 数学的实践与认识,1999, 29(4):81-84.

[4] 匡继昌. 常用不等式[M]. 2版. 长沙:湖南教育出版社,1993

[5] 杜家祥. 一个不等式的推广[J]. 安徽教育学院学报,2002, 20(3) :7-8.

[6] MARSHALL A M,OLKIN I. Inequalities :Theory of Majorization and Its Application[M]. New York : Academies Press,1979.

[7] 吴善和. 两个无理不等式的推广[J]. 甘肃教育学院学报(自然科学版), 2002, 16(3): 9-11.

[8] 吴善和. 一个条件不等式与两道竞赛题的推广[J]. 铁道师范学院学报(自然科学版), 2002, 19(1) :39-42.

第20篇　凸序列不等式的控制证明

(吴善和, 石焕南. 数学的实践与认识, 2003, 33(12)：132-137)

摘　要:　本文利用控制不等式理论简洁地证明了一类凸序列不等式(包括著名的Nanson不等式的几个推广)并给出若干应用.

关键词:　凸序列; 控制不等式; Nanson 不等式; 推广

20.1　引言

定义 1　若实数列$\{a_i\}$满足条件$a_{i-1} + a_{i+1} \geqslant 2a_i, i = 2, 3, \cdots$, 则称数列$\{a_i\}$是一个凸数列.

定义 2　若实数列$\{a_i\}$满足条件$a_{i-1}a_{i+1} \geqslant a_i^2, i = 2, 3, \cdots$, 则称数列$\{a_i\}$是一个对数凸数列.

定义 3[1]　设$\boldsymbol{x}, \boldsymbol{y} \in \mathbf{R}^n$满足

$$(1) \sum_{i=1}^{k} x_{[i]} \leq \sum_{i=1}^{k} y_{[i]}, k = 1, 2, \cdots, n-1; (2) \sum_{i=1}^{n} x_i = \sum_{i=1}^{n} y_i$$

则称$\boldsymbol{x}$被$\boldsymbol{y}$所控制, 记作$\boldsymbol{x} \prec \boldsymbol{y}$. 其中$x_{[1]} \geqslant \cdots \geqslant x_{[n]}$和$y_{[1]} \geqslant \cdots \geqslant y_{[n]}$分别是$\boldsymbol{x}$和$\boldsymbol{y}$的分量的递减重排.

本文将利用控制不等式理论证明一类凸序列不等式, 其中包括著名的Nanson不等式的推广, 并给出某些应用. 为此给出如下引理.

引理 1[2]　数列$\{a_i\}$是凸数列的充要条件为: 对任意的$\boldsymbol{p}, \boldsymbol{q} \in \mathbf{Z}_+^n = \{\boldsymbol{p} = (p_1, \cdots, p_n) : p_i$ 是非负整数, $i = 1, 2, \cdots, n, n \geqslant 2 \}$, 若$\boldsymbol{p} = (p_1, \cdots, p_n) \prec (q_1, \cdots, q_n) = \boldsymbol{q}$,恒有

$$a_{q_1} + \cdots + a_{q_n} \geqslant a_{p_1} + \cdots + a_{p_n}$$

引理 2　设$\{a_n\}$是一个凸序列, $a_n \in I, n = 1, 2, \cdots$, 若$\varphi$ 是I上的增的凸函数, 则$\{\varphi(a_n)\}$也是凸序列.

108

证明　由$\{a_n\}$及φ的凸性，有

$$\varphi(a_{n-1}) - 2\varphi(a_n) + \varphi(a_{n+1})$$
$$\geqslant 2\left[\frac{1}{2}\varphi(a_{n-1}) + \frac{1}{2}\varphi(a_{n+1}) - \varphi\left(\frac{1}{2}a_{n-1} + \frac{1}{2}a_{n+1}\right)\right] \geqslant 0$$

这说明$\{\varphi(a_n)\}$也是凸序列.

引理 3[3]　若$\{a_n\}$是一个凸序列，$A_n = \frac{1}{n}\sum_{i=1}^{n} a_i$，则$\{A_n\}$也是一个凸序列.

引理 4　设$x, y \in \mathbf{R}^n, x_1 \geqslant x_2 \geqslant \cdots \geqslant x_n, \sum_{i=1}^{n} x_i = \sum_{i=1}^{n} y_i$，若存在$k, 1 \leqslant k \leqslant n$，使得$x_i \leqslant y_i, i = 1, ..., k, x_i \geqslant y_i, i = k+1, ..., n$，则$x \prec y$. (参见文[4] p. 234 定理8.1.10)

引理 5[1]　设$x, y \in \mathbf{R}^n, u, v \in \mathbf{R}^m$，且$x \prec y, u \prec v$，则$(x, u) \prec (y, v)$.

引理 6

$$\left(\underbrace{2, \cdots, 2}_{n+1}, \underbrace{4, \cdots, 4}_{n+1}, ..., \underbrace{2n, \cdots, 2n}_{n+1}\right)$$
$$\prec \left(\underbrace{1, \cdots, 1}_{n}, \underbrace{3, \cdots, 3}_{n}, ..., \underbrace{2n+1, \cdots, 2n+1}_{n}\right) \tag{1}$$

证明　用数学归纳法，当$n = 1$时式(1)化为$(2,2) \prec (1,3)$，此式显然成立. 若$n = k$时式(1)成立，即

$$x = \left(\underbrace{2, \cdots, 2}_{k+1}, \underbrace{4, \cdots, 4}_{k+1}, ..., \underbrace{2k, \cdots, 2k}_{k+1}\right)$$
$$\prec \left(\underbrace{1, \cdots, 1}_{k}, \underbrace{3, \cdots, 3}_{k}, ..., \underbrace{2k+1, \cdots, 2k+1}_{k}\right) = y$$

由引理1 易证

$$u = \left(2, 4, ..., 2k, \underbrace{2k+2, \cdots, 2k+2}_{k+2}\right)$$
$$\prec \left(1, 3, ..., 2k+1, \underbrace{2k+3, \cdots, 2k+3}_{k+1}\right) = v$$

从而由引理2 有$(\boldsymbol{x},\boldsymbol{u}) \prec (\boldsymbol{y},\boldsymbol{v})$, 即

$$\left(\underbrace{2,\cdots,2}_{k+2},\underbrace{4,\cdots,4}_{k+2},...,\underbrace{2k+2,\cdots,2k+2}_{k+2}\right)$$

$$\prec \left(\underbrace{1,\cdots,1}_{k+1},\underbrace{3,\cdots,3}_{k+1},...,\underbrace{2k+3,\cdots,2k+3}_{k+1}\right)$$

亦即当$n=k+1$ 时式(1)成立, 所以式(1)对所有$n \in \mathbf{N}$ 成立. 从而, 由引理1 即知式(1)成立.

引理 7　设$x_1 \geqslant x_2 \geqslant \cdots \geqslant x_n \geqslant 0$, 则

$$\left(\sum_{i=1}^{2n-1}(-1)^{i-1}x_i, x_2, x_4, ..., x_{2n-2}\right) \prec (x_1, x_3, ..., x_{2n-1}) \tag{2}$$

$$\left(\sum_{i=1}^{2n}(-1)^{i-1}x_i, x_2, x_4, ..., x_{2n}\right) \prec (x_1, x_3, ..., x_{2n-1}, 0) \tag{3}$$

证明　下面利用引理4 证(2).

显然, 式(2)中两个向量的分量之和相等. 记$s = \sum\limits_{i=1}^{2n-1}(-1)^{i-1}x_i$. 我们分四种情况讨论.

(I)若$x_2 \geqslant x_4 \geqslant ... \geqslant x_{2n-2} \geqslant s$, 因为$x_{2i} \leqslant x_{2i-1}, i=1,2,...,n-1$, 而

$$s = (x_1-x_2)+(x_3-x_4)+\cdots+(x_{2n-3}-x_{2n-2})+x_{2n-1} \geqslant x_{2n-1}$$

由引理4 知式(2)成立.

(II)若$s \geqslant x_2 \geqslant x_4 \geqslant ... \geqslant x_{2n-2}$, 因为

$$s = x_1-(x_2-x_3)-(x_4-x_5)-\cdots-(x_{2n-2}-x_{2n-1}) \leqslant x_1$$

而$x_{2i} \geqslant x_{2i+1}, i=1,2,...,n-1$, 由引理4 知式(2)成立.

(III)若$x_2 \geqslant x_4 \geqslant \cdots \geqslant x_{2k} \geqslant x_{2k+1} \geqslant s \geqslant x_{2k+2} \geqslant \cdots \geqslant x_{2n-2}$, 此时$x_{2i} \leqslant x_{2i-1}, i=1,2,\cdots,k, s \leqslant x_{2k+1}$, 而$x_{2i} \geqslant x_{2i+1}, i=k+1,k+2,\cdots,n-1$, 由引理4 知式(2)成立.

(IV) 若$x_2 \geqslant x_4 \geqslant \cdots \geqslant x_{2k} \geqslant s \geqslant x_{2k+1} \geqslant x_{2k+2} \geqslant \cdots \geqslant x_{2n-2}$, 此时$x_{2i} \leqslant x_{2i-1}, i=1,2,\cdots,k, s \geqslant x_{2k+1}$, 而$x_{2i} \geqslant x_{2i+1}, i=k+1,k+2,\cdots,n-1$, 由引理4 知式(2)成立.

同理可证引理7 中式(3)也成立.

引理 8　设 $\boldsymbol{y} = (y_1, y_2, ..., y_n) \in \mathbf{R}^n$, $\bar{y} = \frac{1}{n}\sum\limits_{i=1}^{n} y_i$, 则

$$(\bar{y}, \bar{y}, ..., \bar{y}) \prec (y_1, y_2, ..., y_n)$$

20.2　主要结果

定理 1　设 $\{a_n\}$ 是一个凸数列, m, k 为非负整数, 则

$$(n-2m)(a_{k+1}+a_{k+3}+\cdots+a_{k+2n+1})+(2m-n-1)(a_{k+2}+a_{k+4}+\cdots+a_{k+2n})+$$

$$2m(a_{k+1} + a_{k+2n+1}) - m(a_{k+2} + a_{k+2n}) \geqslant 0 \tag{4}$$

证明　根据定义 2 及引理 6, 有

$$\left(\underbrace{k+2,\cdots,k+2}_{n+1}, \underbrace{k+4,\cdots,k+4}_{n+1}, ..., \underbrace{k+2n,\cdots,k+2n}_{n+1}\right)$$

$$\prec \left(\underbrace{k+1,\cdots,k+1}_{n}, \underbrace{k+3,\cdots,k+3}_{n}, ..., \underbrace{k+(2n+1),\cdots,k+(2n+1)}_{n}\right)$$

运用引理 1, 得

$$(n+1)(a_{k+2} + a_{k+4} + \cdots + a_{k+2n}) \leqslant n(a_{k+1} + a_{k+3} + \cdots + a_{k+(2n+1)}) \tag{5}$$

又由引理 4 易证

$$(\underbrace{k+1,\cdots,k+1}_{2m}, \underbrace{k+2,\cdots,k+2}_{m}, \underbrace{k+3,\cdots,k+3}_{2m},$$

$$\underbrace{k+5,\cdots,k+5}_{2m}, ..., \underbrace{k+2n-1,\cdots,k+2n-1}_{2m},$$

$$\underbrace{k+2n,\cdots,k+2n}_{m}, \underbrace{k+2n+1,\cdots,k+2n+1}_{2m})$$

$$\prec (\underbrace{k+1,\cdots,k+1}_{2m}, \underbrace{k+2,\cdots,k+2}_{m},$$

$$\underbrace{k+4,\cdots,k+4}_{2m}, ..., \underbrace{k+(2n-2),\cdots,k+(2n-2)}_{2m},$$

$$\underbrace{k+2n,\cdots,k+2n}_{m}, \underbrace{k+2n+1,\cdots,k+2n+1}_{2m})$$

从而由引理1 有

$$2m(a_{k+1} + a_{k+3} + \cdots + a_{k+2n+1}) + m(a_{k+2} + a_{k+2n})$$

$$\leqslant 2m(a_{k+2} + a_{k+4} + \cdots + a_{k+2n}) + 2m(a_{k+1} + a_{k+2n+1}) \tag{6}$$

式(5)与式(6)的两边对应相加, 并稍加变形即得式(4).

当 $m = 0$ 时, 由式(4)得

推论 1　设非负数列 $\{a_n\}$ 是一个凸数列, k 为非负整数, 则

$$\frac{a_{k+1} + a_{k+3} + \cdots + a_{k+2n+1}}{n+1} \geqslant \frac{a_{k+2} + a_{k+4} + \cdots + a_{k+2n}}{n} \tag{7}$$

当 $k = 0$ 时, 式(7)为著名的Nanson不等式[3]

$$n \sum_{i=0}^{n} a_{2i+1} \geqslant (n+1) \sum_{i=1}^{n} a_{2i}$$

推论 2　设非负数列 $\{a_n\}$ 是一个凸数列, $n = 1, \cdots, k$ 为非负整数, 若 φ 是 I 上的增的凸函数, 则

$$\frac{\varphi(a_{k+1}) + \varphi(a_{k+3}) + \cdots + \varphi(a_{k+2n+1})}{n+1}$$

$$\geqslant \frac{\varphi(a_{k+2}) + \varphi(a_{k+4}) + \cdots + \varphi(a_{k+2n})}{n} \tag{8}$$

特别, 若 $\{a_n\}$ 是正的凸序列, φ 是 $(0, \infty)$ 上的凸函数, 则当 $t \geqslant 1$ 时, 有

$$\frac{a'_{k+1} + a'_{k+3} + \cdots + a'_{k+2n+1}}{n+1} \geqslant \frac{a'_{k+2} + a'_{k+4} + \cdots + a'_{k+2n}}{n} \tag{9}$$

证明　由引理2 知 $\{\varphi(a_n)\}$ 为凸序列, 从而由式(7) 即得式(8). 当 $t \geqslant 1$ 时, 是区间 $(0, +\infty)$ 上的增的凸函数, 由式(8) 即得式(9).

推论 3　设非负数列 $\{a_n\}$ 是一个凸数列, 则

$$\sum_{i=1}^{2n+1} (-1)^{i+1} a_{k+i} \geqslant \frac{1}{n+1} \sum_{i=0}^{n} a_{k+2i+1} \geqslant \frac{1}{2n+1} \sum_{i=1}^{2n+1} a_{k+i} \frac{1}{n} \sum_{i=1}^{n} a_{k+2i} \tag{10}$$

证明　不难验证式(10)中的三个不等式均等价于式(7).

定理 2　设非负数列 $\{a_n\}$ 是一个凸数列, $m \in \mathbf{N}$, 则

$$a_2 + a_4 + \cdots + a_{2m} \geqslant a_3 + a_5 + \cdots + a_{2m-1} + a_{m+1} \tag{11}$$

$$a_1 + a_3 + a_5 + \cdots + a_{2m+1} \geqslant a_2 + a_4 + \cdots + a_{2m} + a_{m+1} \qquad (12)$$

证明　令$x_i = 2m - (i-1), i = 1, 2, \cdots, m$, 则$x_1 \geqslant x_2 \geqslant \cdots \geqslant x_{2m} \geqslant 0$, 且

$$\sum_{i=1}^{2m-1} (-1)^{i-1} x_i = m + 1, \quad \sum_{i=1}^{2m} (-1)^{i-1} x_i = m$$

运用引理7, 得

$$(m+1, 2m-1, 2m-3, \cdots, 5, 3) \prec (2m, 2m-2, \cdots, 4, 2) \qquad (13)$$

$$(m, 2m-1, 2m-3, \cdots, 3, 1) \prec (2m, 2m-2, \cdots, 4, 2, 0) \qquad (14)$$

由式(14)显然有

$$(m+1, 2m, 2m-3, 2m-2, \cdots, 4, 2) \prec (2m+1, 2m-1, \cdots, 5, 3, 1) \quad (15)$$

从而由引理1 并结合式(13)和式(15)知式(11)和式(12)成立.

定理 3　设$\{a_n\}$ 是一个凸数列, $h, m, n \in \mathbf{N}$, 则

$$\frac{1}{n-2m} \sum_{k=m+1}^{n-m} a_k \leqslant \frac{1}{n} \sum_{k=1}^{n} a_k \leqslant \frac{1}{2m} \left(\sum_{k=1}^{m} a_k + \sum_{k=n-m+1}^{n} a_k \right) \quad (n \geqslant 2m) \tag{16}$$

$$\frac{n-m}{h} \sum_{k=1}^{h} a_k + \frac{h-n}{m} \sum_{k=1}^{m} a_k + \frac{m-h}{n} \sum_{k=1}^{n} a_k \geqslant 0 \quad (h < m < n) \tag{17}$$

$$\frac{n+m}{n-m} \left(\sum_{k=1}^{n} a_k - \sum_{k=1}^{m} a_k \right) \leqslant \sum_{k=1}^{m+n} a_k \quad (m \neq n) \tag{18}$$

证明　因$\{a_n\}$ 是凸数列, 由引理3 知$\{A_n\}$ 也是一个凸数列, 其中$A_n = \frac{1}{n} \sum_{i=1}^{n} a_i$, 不难验证, 对于$n > 2m$, 有

$$\left(\underbrace{n-m, \cdots, n-m}_{n-m}, \underbrace{n, \cdots, n}_{2m} \right) \prec \left(\underbrace{m, \cdots, m}_{m}, \underbrace{n, \cdots, n}_{n} \right)$$

从而由引理1 可得

$$(n-m)A_{n-m} + 2mA_n \leqslant mA_m + nA_n$$

而易见式(16)中左右两个不等式均等价于上式.

对于 $h < m < n$, 有

$$\left(\underbrace{m, \cdots, m}_{n-h}\right) \prec \left(\underbrace{n, \cdots, n}_{m-h}, \underbrace{h, \cdots, h}_{n-m}\right)$$

从而由引理1可得

$$(n-h)A_m \leqslant (m-h)A_n + (n-m)A_h$$

而式(17)与上式等价, 从而式(17)得证.

由于式(18)关于 m, n 对称, 可不妨设 $m < n$. 注意到

$$\left(\underbrace{n, \cdots, n}_{n}\right) \prec \left(\underbrace{m, \cdots, m}_{m}, \underbrace{n+m, \cdots, n+m}_{n-m}\right)$$

由引理1 可得

$$nA_n \leqslant (nm)A_{n+m} + mA_m,$$

而此式与式(18)等价, 从而式(18)得证.

定理 4　设 $\{a_n\}$ 是一个凸数列, $n \in \mathbf{N}, n > 1$, 则

$$\frac{1}{n-1}\sum_{k=1}^{n-1}a_k \leqslant \frac{1}{n+1}\sum_{k=0}^{n}a_k \leqslant \frac{1}{2}(a_0 + a_n) \tag{19}$$

证明　根据引理 4 知

$$\left(\underbrace{1, \cdots, 1}_{n+1}, \underbrace{2, \cdots, 2}_{n+1}, \cdots, \underbrace{n+1, \cdots, n+1}_{n+1}\right)$$

$$\prec \left(\underbrace{0, \cdots, 0}_{n-1}, \underbrace{1, \cdots, 1}_{n-1}, \cdots, \underbrace{n, \cdots, n}_{n-1}\right)$$

$$(0, 0, 1, 1, ..., n, n) \prec \left(\underbrace{0, \cdots, 0}_{n+1}, \underbrace{n, \cdots, n}_{n+1}\right)$$

而由引理 1 和上面两个控制关系式可分别证得式(19)左右两边不等式.

20.3　若干应用

定理 5　(Wilson不等式的加强) 设$x > 0, n \in \mathbf{N}$, 则

$$\frac{1 + x^2 + \cdots + x^{2n}}{x + x^3 + \cdots + x^{2n-1}} \geqslant \frac{n+1}{n} + \left(\sqrt{x} - \frac{1}{\sqrt{x}}\right)^2 \tag{20}$$

证明　因为式(3)右边

$$\frac{n+1}{n} + \left(\sqrt{x} - \frac{1}{\sqrt{x}}\right)^2 = \frac{nx^2 - (n-1)x + n}{nx}$$

所以

$$\begin{aligned}
\text{式(20)} &\Leftrightarrow nx(1 + x^2 + \cdots + x^{2n}) \\
&\geqslant (x + x^3 + \cdots + x^{2n-1})[nx^2 - (n-1)x + n] \\
&\Leftrightarrow n(x + x^3 + \cdots + x^{2n+1}) + 2n(x^2 + x^4 + \cdots + x^{2n}) \\
&\geqslant n(x^{2n+1} + x) + (n+1)(x^2 + x^4 + \cdots + x^{2n}) + \\
&\quad 2n(x^3 + x^5 + \cdots + x^{2n-1}) - \\
&\Leftrightarrow x[(n-1)(x + x^3 + \cdots + x^{2n-1}) \\
&\quad n(x^2 + x^4 + \cdots + x^{2n-2})] \geqslant 0
\end{aligned}$$

记$a_n = x^n, n \in \mathbf{N}$, 则

$$(n-1)(x + x^3 + \cdots + x^{2n-1}) - n(x^2 + x^4 + \cdots + x^{2n-2})$$

$$(n-1)(a_1 + a_3 + \cdots + a_{2n-1}) - n(a_2 + a_4 + \cdots + a_{2n-2})$$

不难验证$\{a_n\}$ 为凸序列, 由Nanson不等式, 知

$$(n-1)(a_1 + a_3 + \cdots + a_{2n-1}) - n(a_2 + a_4 + \cdots + a_{2n-2}) \geqslant 0$$

所以不等式(20)成立.

不等式(20)是著名的Wilson不等式

$$\frac{1 + x^2 + \cdots + x^{2n}}{x + x^3 + \cdots + x^{2n-1}} \geqslant \frac{n+1}{n}$$

的一个加强形式.

定义 4[5]   称函数 f 在 $(0, +\infty)$ 上绝对单调,如果它的各阶导数存在且

$$f^{(k)}(t) \geqslant 0, t \in (0, \infty), k = 0, 1, 2, \cdots$$

引理 8[5]   设函数 f 在 $(0, +\infty)$ 上绝对单调, 则

$$f^{(n-1)}(t) f^{(n+1)}(t) \geqslant [f^{(n)}(t)]^2, n = 1, 2, \cdots$$

上式说明 $\{f^{(k)}(t)\}$ 是对数凸数列, 从而由推论2 有

定理 6   设函数 f 在 $(0, +\infty)$ 上绝对单调, 则

$$\left(f^{(k+1)}(x) f^{(k+3)}(x) \cdots f^{(k+2n+1)}(x) \right)^{\frac{1}{n+1}}$$

$$\geqslant \left(f^{(k+2)}(x) f^{(k+4)}(x) \cdots f^{(k+2n)}(x) \right)^{\frac{1}{n}} \quad (k = 0, 1, \cdots) \tag{21}$$

证明   由引理8, $\{f^{(k)}(t)\}$ 是对数凸序列, 因而 $\{\ln(f^{(k)}(t))\}$ 是凸序列, 于是由式(7)有

$$\frac{\ln f^{(k+1)}(t) + \ln f^{(k+3)}(t) + \cdots + \ln f^{(k+2n+1)}(t)}{n+1}$$
$$\geqslant \frac{\ln f^{(k+2)}(t) + \ln f^{(k+4)}(t) + \cdots + \ln f^{(k+2n)}(t)}{n}$$

由此即得式(21).

综观全文, 运用控制不等式理论我们建立了一组关于凸序列的重要不等式, 证明方法新颖、简洁; 这些不等式若用常规方法推导, 则颇为困难. 这表明控制不等式确为发现新不等式和推广已有不等式的强有力工具, 控制不等式的理论和应用定然有美好的发展前景.

致谢: 衷心感谢鞍山科技大学的刘证教授给予本文的热情帮助.

参考文献

[1]   王伯英. 控制不等式基础[M]. 北京:北京师范大学出版社,1990.

[2]   石焕南, 李大矛. 凸数列的一个等价条件及其应用[J]. 曲阜师范大学学报(自然科学版), 2001, 27 (4):4-6.

[3]   王松桂,贾忠贞. 矩阵论中不等式[M]. 合肥:安徽教育出版社, 1994 : 312.

[4]   匡继昌. 常用不等式[M]. 2 版. 长沙:湖南教育出版社, 1993.

[5]   QI FENG, XU SENLIN. Refinements and extensions of an inequality, II [J]. J. Wath. Anal. Appl., 1997, 211: 616-620.

[6]   MITRINOVIC D S, VASIC P M. 分析不等式[M]. 赵汉宾, 译. 南宁: 广西人民出版社, 1986: 128.

[7]　张垚. 关于多胞形的一类不等式[J]. 湖南教育学院学报, 1999, 17(5): 99-105.

[8]　肖振钢. 数列与不等式的一个连接点——凸数列[J]. 湖南数学年刊, 1995,15(4) :62-69.

[9]　哈代 G H, 李特伍德 J E, 波利亚 G. 不等式[M]. 越民义,译. 北京: 科学出版社, 1965.

[10]　MITRINOVIC D S, PEČARIĆ J E, FLNK A M. Classical and New Inequalitis in Analysis[M]. Dordrecht: Kluwer Academic Publishers, 1993.

第21篇 Extensions and Refinements of Adamovic's Inequality

(SHI Huan-Nan, LI Da-mao. Chin. Quart. J. of Math, 2004,19(1): 35-40)

Abstract: In this article, by means of the theory of majorization, Adamovic's inequality is extended to the cases of the general elementary symmetric functions and its duals, and the refined and reversed forms are also given. As applications, some new inequalities for simplex are established.

Keywords: Adamovic's inequality; elementary symmetric function; majorization; simplex

21.1 Introduction

Theorem A The following inequality holds if all the factors are positive

$$\prod_{i=1}^{n} a_i \geqslant \prod_{i=1}^{n}(s - (n-1)a_i) \tag{1}$$

where $s = \sum_{i=1}^{n} a_i$. (1) is the well-known Adamovic's inequality[1].

In 1996, an extension of (1) was given by Klamkin M S as follows:

Theorem B[2] Let $a_1, \cdots, a_n$ an be positive real numbers and let $s = \sum_{i=1}^{n} a_i$. Then for every convex function f

$$\sum_{i=1}^{n} f(s - (n-1)a_i) \geqslant \sum_{i=1}^{n} f(a_i) \tag{2}$$

Taking $f(x) = -\log x$, the special case of (2) is the above inequality (1). In [3], the following (3) which is similar to (1) is given:

118

Theorem C Let $a_i \geqslant 0$, and $\sum\limits_{i=1}^{n} a_i = s, y_i = a_i + \frac{n-3}{n-1}(s - a_i), i = 1, \cdots, n$, then

$$\prod_{i=1}^{n} y_i \geqslant \prod_{i=1}^{n}(s - 2a_i) \qquad (3)$$

In this article, by means of the theory of majorization , inequalities (2) and (3) are extended to the cases involving general elementary symmetric functions and its duals , and the related refinements and reversed forms are also established. As some applications, we obtain some inequalities for a simplex. The following notation and definitions are used.

We denote the set of n-dimensional row vector on real number field by $\mathbf{R}^n$.

$$\mathbf{R}_+^n = \{a = (a_1, \cdots, a_n) \in \mathbf{R}^n : a_i \geqslant 0, i = 1, \cdots, n\}$$

$$\mathbf{R}_{++}^n = \{a = (a_1, \cdots, a_n) \in \mathbf{R}^n : a_i > 0, i = 1, \cdots, n\}$$

Let $a = (a_1, \cdots, a_n) \in \mathbf{R}^n$. Its elementary symmetric functions are

$$E_k(a) = E_k(a_1, a_2, \cdots, a_n)$$

$$= \sum_{1 \leqslant i_1 < \cdots < i_k \leqslant n} \prod_{j=1}^{k} a_{i_j}, \quad k = 1, \cdots, n$$

Let $E_0(a) = 1$. Duals of elementary symmetric functions are

$$G_k(a) = G_k(a_1, a_2, \cdots, a_n)$$

$$= \prod_{1 \leqslant i_1 < \cdots < i_k \leqslant n} \sum_{j=1}^{k} a_{i_j}, \quad k = 1, \cdots, n$$

For any $a = (a_1, \cdots, a_n) \in \mathbf{R}^n$, let $a_{[1]} \geqslant a_{[2]} \geqslant \cdots \geqslant a_{[n]}$ denote the components of a in decreasing order, and we write $a \downarrow = (a_{[1]}, a_{[2]}, \cdots, a_{[n]})$. The elementwise vector ordering $a_i \leqslant b_i, i = 1, 2, \cdots, n$, is denoted by $a \leqslant b$.

Definition 1[4] For any $a, b \in \mathbf{R}^n$, a is said to majors b (written $a \prec b$) if

(i) $\sum\limits_{i=1}^{k} a_{[i]} \leqslant \sum\limits_{i=1}^{k} b_{[i]}, \quad k = 1, 2, \cdots, n - 1$;

(ii) $\sum\limits_{i=1}^{n} a_i = \sum\limits_{i=1}^{n} b_i$.

and a is said to strictly majors b (written $a \prec\prec b$) if a is not permutation of b.

Definition 2[4] Let $\Omega \subset \mathbf{R}^n, \phi : \Omega \to \mathbf{R}$, (i) ϕ is said to be an increasing function on Ω if $\boldsymbol{a} \leqslant \boldsymbol{b}$ on $\Omega \Rightarrow \phi(\boldsymbol{a}) \leqslant \phi(\boldsymbol{b})$,(ii)$\phi$ is said to be a Schur-convex function on Ω if $\boldsymbol{a} \prec \boldsymbol{b}$ on $\Omega \Rightarrow \phi(\boldsymbol{a}) \leqslant \phi(\boldsymbol{b})$, ϕ is said to be a Schur-concave function on Ω if and only $-\phi$ is Schur- convex function, ϕ is said to be a strictly Schur-convex function on Ω if $\boldsymbol{a} \prec\prec \boldsymbol{b}$ on $\Omega \Rightarrow \phi(\boldsymbol{a}) < \phi(\boldsymbol{b})$, ϕ is said to be a strictly Schur-concave function on Ω if and only $-\phi$ is a strictly Schur-convex function on Ω.

The following results established by author of the present paper are also used in this paper.

Theorem D[5] Let $x_i \in \mathbf{R}, i = 1, 2, \cdots, n, n \geqslant 2, \sum_{i=1}^{n} x_i = s > 0, c \geqslant s$. Then

$$\frac{(c-x)s}{nc-s} = \left(\frac{(c-x_1)s}{nc-s}, \frac{(c-x_2)s}{nc-s}, \cdots, \frac{(c-x_n)s}{nc-s} \right)$$

$$\prec (x_1, x_2, \cdots, x_n) = s \tag{4}$$

Theorem E Let $\Omega = \{ \boldsymbol{x} = (x_1, x_2, \cdots, x_n) : x \in \mathbf{R}_+^n, \sum_{i=1}^{n} x_i \leqslant 1 \}$ and $\lambda \geqslant 1$. Then, for $n \geqslant k > p \geqslant 2, \phi(x) = E_k(x) - \lambda E_p(x)$ is a decreasing and Schur-convex function on Ω.(See consequence 2 of [7])

21.2　Main results

Theorem Let $a \in \mathbf{R}_+^n, \sum_{i=1}^{n} a_i = s, x_i = s - ma_i \geqslant 0, y_i = \frac{(c-x_i)(n-m)s}{nc-(n-m)s}, i = 1, 2, \cdots, n, m \leqslant n, c \geqslant (n-m)s$, and let f be concave function on $\mathbf{R}_+^1, k = 1, 2, \cdots, n$. We have

(I)

$$C_n^k \left[f\left(\frac{(n-m)s}{n} \right) \right]^k \geqslant E_k[f(y_1), f(y_2), \cdots, f(y_n)]$$

$$\geqslant E_k[f(x_1), f(x_2), \cdots, f(x_n)] \tag{5}$$

$$\left[k \cdot f\left(\frac{(n-m)s}{n} \right) \right]^{C_n^k} \geqslant G_k[f(y_1), f(y_2), \cdots, f(y_n)]$$

$$\geqslant G_k[f(x_1), f(x_2), \cdots, f(x_n)] \tag{6}$$

(II) If $s - ma_i > 0, i = 1, 2, \cdots, n, 1 \leqslant p \leqslant k \leqslant n$, then

$$\frac{E_k[f(y_1), f(y_2), \cdots, f(y_n)]}{E_k[f(x_1), f(x_2), \cdots, f(x_n)]} \geqslant \frac{E_{k-p}[f(y_1), f(y_2), \cdots, f(y_n)]}{E_{k-p}[f(x_1), f(x_2), \cdots, f(x_n)]} \quad (7)$$

(III) If $\lambda \geqslant 1, n \geqslant k > p \geqslant 2$, then

$$E_k\left[f\left(\frac{y_1}{(n-m)s}\right), f\left(\frac{y_2}{(n-m)s}\right), \cdots, f\left(\frac{y_n}{(n-m)s}\right)\right]$$

$$-E_k\left[f\left(\frac{x_1}{(n-m)s}\right), f\left(\frac{x_2}{(n-m)s}\right), \cdots, f\left(\frac{x_n}{(n-m)s}\right)\right]$$

$$\leqslant \lambda \cdot E_p\left[f\left(\frac{y_1}{(n-m)s}\right), f\left(\frac{y_2}{(n-m)s}\right), \cdots, f\left(\frac{y_n}{(n-m)s}\right)\right]$$

$$-\lambda \cdot E_p\left[f\left(\frac{x_1}{(n-m)s}\right), f\left(\frac{x_2}{(n-m)s}\right), \cdots, f\left(\frac{x_n}{(n-m)s}\right)\right] \quad (8)$$

Proof　From Theorem D we have

$$\boldsymbol{y} = (y_1, y_2, \cdots, y_n) \prec (x_1, x_2, \cdots, x_n) = \boldsymbol{x} \quad (*)$$

(I) From proposition 6.7 in monograph [4] we know that $E_k(a)$ is an increasing and Schur- concave function on $\mathbf{R}_+^n$, since f is a concave function on $\mathbf{R}_+^1$, $E_k[f(a_1), \cdots, f(a_n)]$ and it also is Schur-concave function on $\mathbf{R}_+^n$ (See [4, Proposition 6.16(b)]. Accordingly from $(*)$ and

$$(\bar{y}, \bar{y}, \cdots, \bar{y}) \prec (y_1, y_2, \cdots, y_n) \quad (\text{where } \bar{y} = \frac{1}{n}\sum_{i=1}^{n} y_i = \frac{(n-m)s}{n})$$

we have

$$E_k[f(\bar{y}), f(\bar{y}), \cdots, f(\bar{y})] \geqslant E_k[f(y_1), f(y_2), \cdots, f(y_n)]$$
$$\geqslant E_k[f(x_1), f(x_2), \cdots, f(x_n)]$$

namely (5) holds. From paper[8, p.86]or monograph [9] we know $G_k(a)$ also is an increasing and Schur-concave function on $\mathbf{R}_+^n$, analogously we can prove (6).

(II) From paper[10] we know $\frac{E_k(a)}{E_{k-p}(a)}$ is an increasing and Schur-concave function on $\mathbf{R}_{++}^n$, thus $\frac{E_k[f(a_1), f(a_2), \cdots, f(a_n)]}{E_{k-p}[f(a_1), f(a_2), \cdots, f(a_n)]}$ also is a Schur-concave function on $\mathbf{R}_{++}^n$, accordingly from $(*)$, we have

$$\frac{E_k[f(y_1), f(y_2), \cdots, f(y_n)]}{E_{k-p}[f(y_1), f(y_2), \cdots, f(y_n)]} \geqslant \frac{E_k[f(x_1), f(x_2), \cdots, f(x_n)]}{E_{k-p}[f(x_1), f(x_2), \cdots, f(x_n)]}$$

This shows that (7) holds.

(III) From theorem E we known $-\phi(x) = \lambda E_p(x) - E_k(x)$ is a increasing and Schur-concave function on $\Omega = \{ \boldsymbol{x} = (x_1, x_2, \cdots, x_n) : \boldsymbol{x} \in \mathbf{R}_+^n, \sum\limits_{i=1}^n x_i \leqslant 1 \}$ .It is clear from $(*)$ that

$$\frac{\boldsymbol{y}}{(n-m)s} = \left(\frac{y_1}{(n-m)s}, \frac{y_2}{(n-m)s}, \cdots, \frac{y_n}{(n-m)s} \right)$$
$$\prec \left(\frac{x_1}{(n-m)s}, \frac{x_2}{(n-m)s}, \cdots, \frac{x_n}{(n-m)s} \right) = \frac{\boldsymbol{x}}{(n-m)s}$$

and $\frac{\boldsymbol{y}}{(n-m)s} \in \Omega, \frac{\boldsymbol{x}}{(n-m)s} \in \Omega$, thus we have

$$- \phi \left(f \left(\frac{y_1}{(n-m)s} \right), f \left(\frac{y_2}{(n-m)s} \right), \cdots, f \left(\frac{y_n}{(n-m)s} \right) \right)$$
$$\geqslant - \phi \left(f \left(\frac{x_1}{(n-m)s} \right), f \left(\frac{x_2}{(n-m)s} \right), \cdots, f \left(\frac{x_n}{(n-m)s} \right) \right)$$

namely (8) holds. This completes the proof of theorem.

Repeatedly using (7) with $p = 1$, we obtain

Corollary Let $a \in \mathbf{R}_+^n, \sum\limits_{i=1}^n a_i = s, x_i = s - ma_i > 0, y_i = \frac{(c-x_i)(n-m)s}{nc-(n-m)s}$, $i = 1, \cdots, n, m \leqslant n, c \geqslant (n-m)s$, and let f be concave function on $\mathbf{R}_+^1, 1 \leqslant k \leqslant n$. Then

$$\frac{E_n[f(y_1), f(y_2), \cdots, f(y_n)]}{E_n[f(x_1), f(x_2), \cdots, f(x_n)]} \geqslant \frac{E_{n-1}[f(y_1), f(y_2), \cdots, f(y_n)]}{E_{n-1}[f(x_1), f(x_2), \cdots, f(x_n)]} \geqslant \cdots$$

$$\geqslant \frac{E_1[f(y_1), f(y_2), \cdots, f(y_n)]}{E_1[f(x_1), f(x_2), \cdots, f(x_n)]} \geqslant \frac{E_0[f(y_1), f(y_2), \cdots, f(y_n)]}{E_0[f(x_1), f(x_2), \cdots, f(x_n)]} = 1 \quad (9)$$

Remark 1 For any convex function $f, -f$ is concave function. taking $k = 1, m = n-1, c = (n-m)s$, from the right inequality of (5), we obtain (2).

Remark 2 Setting $f(x) = x, k = n, m = 2, c = (n-m)s$ in the right inequality of (5), we have $y_i = \frac{[(n-2)s-(s-2a_i)]}{n-1} = a_i + \frac{(n-3)(s-a_i)}{n-1}$, and obtain (3), but in (3) we need not restrict $s - ma_i \geqslant 0, i = 1, \cdots, n$. Setting $f(x) = x, k = 1, m = 2, c = (n-m)s$ in (6), we obtain (3).

Remark 3 (9) is a refinement of equivalent form of the right inequality of (5).

21.3　Applications

Applying (5), some new inequalities for simplex are established in this section.

Using notations of monograph [11] , let V be the volume and r inradius of a given n-dimensional simplex $A = A_1 A_2 \cdots A_{n+1}$ in n-dimensional Euclidean space $E^n (n \geqslant 2)$. For any $i \in \{1, 2, \cdots, n + 1\}$ let F_i be $(n - 1)$dimensional content of the $(n - 1)$dimensional simplex $a_i = A_1 \cdots A_{i-1} \cdot A_{i+1} \cdots A_{n+1}$, let $F = F_1 + \cdots + F_{n+1}$ be the 'total area' of A, let ρ_i be the radius of i-th escribedhypersphere of A and let h_i be the altitude of A from vertex A_i ,we have

$$\rho_i = \frac{nV}{F - 2F_i} \tag{10}$$

$$h_i = \frac{nV}{F_i} \tag{11}$$

$$r = \frac{nV}{F} \tag{12}$$

Replacing $n + 1$ by n, a_i by F_i, s by F in theorem respectively ,then (5) is reduces to

$$C_{n+1}^k \left[f\left(\frac{n + 1 - m}{n + 1} \right) F \right]^k \geqslant E_k(f(y_1), f(y_2), \cdots, f(y_n))$$

$$\geqslant E_k[f(F - mF_1), f(F - mF_2), \cdots, f(F - mF_n)] \tag{13}$$

where $y_i = \frac{[c - (F - mF_i)](n - m + 1)F}{(n+1)c - (n - m + 1)F}$.

In particular, setting $f(x) = x, k = n + 1, m = 2, c = (n - m + 1)F = (n - 1)F$, (13) is reduces to

$$\left(\frac{n(n - 1)F}{n + 1} \right)^{n+1} \geqslant \prod_{i=1}^{n+1} [(n - 2)(F - 2F_i) + 2(n - 1)F_i]$$

$$\geqslant n^{n+1} \prod_{i=1}^{n+1} (F - 2F_i) \tag{14}$$

Dividing(14) by $\prod\limits_{i=1}^{n+1}(F-2F_i)$,and noting

$$\frac{F_i}{F-2F_i}=\frac{\rho_i}{h_i},\ \frac{F_i}{F-2F_i}=\frac{\rho_i}{r}$$

by (10) and (11), then (14) is reduces to

$$\left(\frac{n(n-1)}{n+1}\right)^{n+1}\cdot\prod_{i=1}^{n+1}\frac{\rho_i}{r}\geqslant\prod_{i=1}^{n+1}\left[n-2+2(n-1)\frac{\rho_i}{h_i}\right]\geqslant n^{n+1}\tag{15}$$

If note that we have $\frac{\rho_i-r}{r}=\frac{2F_i}{F-2F_i}$ by (10) and (12), then (14) is also reduces to

$$\left(\frac{n(n-1)}{n+1}\right)^{n+1}\cdot\prod_{i=1}^{n+1}\frac{\rho_i}{r}$$
$$\geqslant\prod_{i=1}^{n+1}\left[n-2+(n-1)\frac{\rho_i-r}{r}\right]\geqslant n^{n+1}$$

namely

$$\prod_{i=1}^{n+1}\frac{n(n-1)}{n+1}\rho_i\geqslant\prod_{i=1}^{n+1}[(n-1)\rho_i-r]\geqslant(nr)^{n+1}\tag{16}$$

Setting $f(x)=x,k=n+1,m=1,c=(n-m+1)=nF$, then (13) is reduces to

$$\left(\frac{n^2F}{n+1}\right)^{n+1}\geqslant\prod_{i=1}^{n+1}[nF_i+(n-1)(F-F_i)]$$
$$\geqslant n^{n+1}\prod_{i=1}^{n+1}(F-F_i)\tag{17}$$

Dividing (14) by $\prod\limits_{i=1}^{n+1}(F-F_i)$, and noting $\frac{\rho_i-r}{\rho_i+r}=\frac{F_i}{F-F_i}$ by (10)and (12), and $\frac{h_i}{h_i-r}=\frac{F_i}{F-F_i}$ by (11) and (12), then (17)is reduces to

$$\left(\frac{n^2}{n+1}\right)^{n+1}\prod_{i=1}^{n+1}\frac{h_i}{h_i-r}$$
$$\geqslant\prod_{i=1}^{n+1}\left[\frac{n(\rho_i-r)}{\rho_i+r}+n-1\right]\geqslant n^{n+1}\tag{18}$$

21.4　Acknowledgements

The authors are indebted to associate professor Zhang Han-fang, professors Qi Feng and Wang Wan-lan for many helpful and valuable comments and suggestions.

参考文献

[1]　MITRINOVIC D S. Analytic Inequalities[M]. Berlin: Springer-Verlag, 1970.

[2]　KLAMKIN M S. Extensions of an inequality[J]. Publ. Elektroteh. Fak., Univ. Beogr., Ser.Mat. 1996(7): 72-73. (2000; Zbl 946:26009).

[3]　KUANG JI-CHANG. Applied Inequalities[M]. 2nd ed. Changsha: Hunan Education Press , 1993.(in Chinese)

[4]　WANG BOYING. Elements of majorization inequalities[M]. Beijing: Beijing Normal University Press, 1990. (in Chinese)

[5]　SHI HUAN-NAN. Refinement and generalization of a class of inequalities for symmetric function[J]. Chinese Math. in Practice and Theory，1999,29 (4): 81-84.

[6]　SHI HUAN-NAN. Majorized proof of a kind of inequalities for symmetric function[J]. Chinese J. Math. for Techology，1999, 15 (3): 140-142.

[7]　SHI HUAN-NAN. Schur-convexity of difference for the elementrary symmetric function [J]. J. Hunan Educational Institute，1999, 17(5): 135-138.

[8]　MARSHALL A M, OlLKIN I. Inequalities:theory of majorization and its application[M]. New York : Academies Press, 1979.

[9]　SHI HUAN-NAN. Schur - concavity of dual form for elementary symmetric function with applications[J]. J. Northeast Normal University (Natural Science), 2001, 33 Supplement: 24-27.

[10]　MA TONG-YI. Schur-concavity of quotient for the elementrary symmetric function[J]. J. LANZHOU University (Natural Science), 2001, 37(4): 19-24.

[11]　MITRINOVIC D S, PECARIC J E, VOLENEC V. Recent advances in geometric inequalities [M]. Dordrecht: Kluwer Academic Publishers, 1989: 463-473.

第22篇 极限 $\lim\limits_{n\to\infty}\left(1+\dfrac{1}{n}\right)^n$ 存在的控制证明

(石焕南. 云南师范大学学报(自然科学版), 2004, 24 (2)：13-15)

摘 要： 本文利用初等对称函数的Schur 凹性及向量的简单的控制关系, 建立了一类关于凹函数的不等式, 作为推论, 给出极限 $\lim\limits_{n\to\infty}\left(1+\frac{1}{n}\right)^n$ 存在的一种简洁的证明.

关键词： 控制; 不等式; 单调有界; 极限

证明极限 $\lim\limits_{n\to\infty}\left(1+\frac{1}{n}\right)^n$ 存在是单调有界定理的一个应用. 传统证法是利用二项式定理推出它单调且有上界. 由于此极限的重要性, 不断有人探索它的新的证法.文[1] 给出三种证法:

I.利用Bernoulli 不等式 $(1+x)^n \geqslant 1+nx(x>-1, n \in \mathbf{N})$;

II.利用关系式 $\frac{b^{n+1}-a^{n+1}}{b-a} < (n+1)b^n(0 \leqslant a < b)$;

III.利用算术-几何平均值不等式 $\frac{1}{n}\sum\limits_{i=1}^{n} x_i \geqslant \sqrt[n]{\prod\limits_{i=1}^{n} x_i}$, $x_i \geqslant 0(i = 1,\cdots, n)$.

本文利用初等对称函数的Schur 凹性及向量的简单的控制关系, 建立了一类关于凹函数的不等式, 作为推论, 给出此极限存在的一种简洁的证明.

定义 1[1] 设$\boldsymbol{x}, \boldsymbol{y} \in \mathbf{R}^n$ 满足

(1) $\sum\limits_{i=1}^{k} x_{[i]} \leqslant \sum\limits_{i=1}^{k} y_{[i]}$, $k = 1, 2, ..., n-1$;

(2) $\sum\limits_{i=1}^{n} x_i = \sum\limits_{i=1}^{n} y_i$.

则称$\boldsymbol{x}$ 被$\boldsymbol{y}$ 所控制, 记作$\boldsymbol{x} \prec \boldsymbol{y}$. 其中$x_{[1]} \geqslant \cdots \geqslant x_{[n]}$ 和$y_{[1]} \geqslant \cdots \geqslant y_{[n]}$ 分别是$\boldsymbol{x}$ 和$\boldsymbol{y}$ 的递减重排. 又若$\boldsymbol{x}$ 不是$\boldsymbol{y}$ 的重排, 则称$\boldsymbol{x}$ 被$\boldsymbol{y}$ 严格控制, 记作$\boldsymbol{x} \prec\!\prec \boldsymbol{y}$.

定义 2 [3]　设 $\Omega\subset\mathbf{R}^{n},\varphi:\Omega\to\mathbf{R}$, 若在 Ω 上, $\boldsymbol{x}\prec\boldsymbol{y}$, 恒有 $\varphi(\boldsymbol{x})\leqslant\varphi(\boldsymbol{y})$, 则称 φ 为 Ω 上的 Schur 凸函数; 若 $-\varphi$ 是 Ω 上的 Schur 凸函数, 则称 φ 为 Ω 上的 Schur 凹函数. 若在 Ω 上, $\boldsymbol{x}\prec\prec\boldsymbol{y}$, 恒有 $\varphi(\boldsymbol{x})<\varphi(\boldsymbol{y})$, 则称 φ 为 Ω 上的严格 Schur 凸函数; 若 $-\varphi$ 是 Ω 上的 Schur 凸函数, 则称 φ 为 Ω 上的严格 Schur 凹函数.

引理 [3]　设 $\boldsymbol{x}=(x_{1},...,x_{n})\in\mathbf{R}^{n},\bar{x}=\frac{1}{n}\sum\limits_{i=1}^{n}x_{i}$, 则

$$(\bar{x},\bar{x},\cdots,\bar{x})\prec(x_{1},x_{2},\cdots,x_{n})$$

定理 1　设 $\lambda\geqslant 1,t\leqslant 1,n\geqslant 2,f(x)$ 是 $\mathbf{R}_{++}=\{x:x>0\}$ 上的凹函数, 对于 $k=1,2,\cdots,n$, 有

$$\left(f\left(\lambda-\frac{t}{n}\right)\right)^{k}$$
$$\geqslant\left(\frac{k}{n}\right)f(\lambda)\left(f\left(\lambda-\frac{t}{n-1}\right)\right)^{k-1}+\left(1-\frac{k}{n}\right)\left(f\left(\lambda-\frac{t}{n-1}\right)\right)^{k}\quad(1)$$

证明　由引理知

$$\boldsymbol{u}=\left(\underbrace{\lambda-\frac{t}{n},\cdots,\lambda-\frac{t}{n}}_{n}\right)\prec\left(\lambda,\underbrace{\lambda-\frac{t}{n-1},\cdots,\lambda-\frac{t}{n-1}}_{n-1}\right)=\boldsymbol{v}\quad(2)$$

由[3]命题6.7 知初等对称函数

$$E_{k}(\boldsymbol{x})=E_{k}(x_{1},x_{2},\cdots,x_{n})=:\sum\limits_{1\leqslant i_{1}<\cdots<i_{k}\leqslant n}\prod\limits_{j=1}^{k}x_{i_{j}}\ (k=1,2,\cdots,n)$$

是 $\mathbf{R}_{++}^{n}=\{\boldsymbol{x}=(x_{1},x_{2},\cdots,x_{n}):x_{i}>0,i=1,\cdots,n\}$ 上的增的 Schur 凹函数, 由[3] 命题6.16 (b) 知 $E_{k}(f(\boldsymbol{x}))$ 也是 $\mathbf{R}_{++}^{n}$ 上的 Schur 凹函数, 从而当 $k\geqslant 1$ 时, 由式(2)有 $E_{k}(f(\boldsymbol{u}))\geqslant E_{k}(f(\boldsymbol{v}))$, 即

$$\mathrm{C}_{n}^{k}\left(f\left(\lambda-\frac{t}{n}\right)\right)^{k}$$
$$\geqslant\mathrm{C}_{n-1}^{k-1}f(\lambda)\left(f\left(\lambda-\frac{t}{n-1}\right)\right)^{k-1}+\mathrm{C}_{n-1}^{k}\left(f\left(\lambda-\frac{t}{n-1}\right)\right)^{k}$$

上式两端同除以 C_{n}^{k} 即得式(1), 类似地可证.

定理 2　设 $\lambda\geqslant 1,t\geqslant 0,n\geqslant 2,f(x)$ 是 $\mathbf{R}_{++}=\{x:x>0\}$ 上的凹函

数, 对于 $k = 1, 2, \cdots, n$, 有

$$\left(f\left(\lambda + \frac{t}{n} \right) \right)^k$$

$$\geqslant \left(\frac{k}{n} \right) f(\lambda) \left(f\left(\lambda + \frac{t}{n-1} \right) \right)^{k-1} + \left(1 - \frac{k}{n} \right) \left(f\left(\lambda + \frac{t}{n-1} \right) \right)^k \quad (3)$$

推论 1　设 $\lambda \geqslant 1, t \leqslant 1$, 则数列 $\left\{ \left(\lambda - \frac{t}{n} \right)^n \right\}$ 单调增加.

证明　取 $f(x) = x, k = n$, 由式(1)有

$$\left(\lambda - \frac{t}{n} \right)^n \geqslant \lambda \left(\lambda - \frac{t}{n-1} \right)^{n-1} \geqslant \left(\lambda - \frac{t}{n-1} \right)^{n-1} \quad (n \geqslant 2)$$

即数列 $\left\{ \left(\lambda - \frac{t}{n} \right)^n \right\}$ 单调增加, 类似地由式(3)可得

推论 2　设 $\lambda \geqslant 1, t \geqslant 0$, 则数列 $\left\{ \left(\lambda + \frac{t}{n} \right)^n \right\}$ 单调增加.

定理 3　极限 $\lim\limits_{n \to \infty} \left(1 + \frac{1}{n} \right)^n$ 存在.

证明　由推论 2, $\left\{ \left(\lambda + \frac{1}{n} \right)^n \right\}$ 单调增, 只需证 $\left\{ \left(1 + \frac{1}{n} \right)^n \right\}$ 有上界. 由推论 1, $\left\{ \left(1 - \frac{1}{n} \right)^n \right\} = \left\{ \left(\frac{n-1}{n} \right)^n \right\}$ 单调增, 则 $\left\{ \left(\frac{n}{n-1} \right)^n \right\} = \left\{ \left(1 + \frac{1}{n-1} \right)^n \right\}$ 单调减, 即 $\left\{ \left(1 + \frac{1}{n} \right)^{n+1} \right\}$ 单调减, 于是

$$\left(1 + \frac{1}{n} \right)^n < \left(1 + \frac{1}{n} \right)^n \cdot \left(1 + \frac{1}{n} \right)$$

$$= \left(1 + \frac{1}{n} \right)^{n+1} < \left(1 + \frac{1}{1} \right)^{1+1} = 4$$

即数列 $\left(1 + \frac{1}{n} \right)^n$ 有上界4, 证毕.

参考文献

[1]　刘泮振. 关于极限 $\lim\limits_{n \to \infty} \left(1 + \frac{1}{n} \right)^n$ 存在的三种证明方法[J]. 数学的实践与认识, 2001, 31 (2): 248- 250.

[2]　徐传胜. 数列 $\lim\limits_{n \to \infty} \left(1 + \frac{1}{n} \right)^{n+p}$ 的性质及应用探究[J]. 山东师范大学学报(自然科学版),2002, 17 (1) : 86-67.

[3]　王伯英. 控制不等式基础[M]. 北京: 北京师范大学出版社, 1990.

[4]　李文荣. 分析中的问题研究[M]. 北京: 中国工人出版社, 2001. 79-84.

[5]　林谦. 重要极限 $\lim\limits_{n \to \infty} \left(1 + \frac{1}{n} \right)^n = \mathrm{e}$ 的推广及应用[J]. 云南师范大学学报(自然科学版), 2000(1) : 53 - 56.

第23篇　凸数列的一个等价条件及其应用II

(石焕南. 数学杂志, 2004, 24(4):390-394)

摘　要: 本文将涉及控制向量的凸数列的一个等价条件扩展到弱控制的情形,并给出此等价条件在代数、分析、凸体几何、概率论等诸多方面的应用.

关键词: 凸数列; 对数凸数列; 控制不等式; 绝对单调函数; 多胞形; 矩

23.1　引言

定义 1　若实数列 $\{a_i\}$ 满足条件 $a_{i-1} + a_{i+1} \geqslant 2a_i, i = 2, 3, \cdots$, 则称数列 $\{a_i\}$ 是一个凸数列.

定义 2　若实数列 $\{a_i\}$ 满足条件 $a_{i-1}a_{i+1} \geqslant a_i^2, i = 2, 3, \cdots$, 则称数列 $\{a_i\}$ 是一个对数凸数列.

对于 $\boldsymbol{x} = (x_1, \cdots, x_n) \in \mathbf{R}^n$, 把 $\boldsymbol{x}$ 的分量排成递减的次序后,记作 $x_{[1]} \geqslant \cdots \geqslant x_{[n]}$.

定义 3[1]　设 $\boldsymbol{x}, \boldsymbol{y} \in \mathbf{R}^n$ 满足

$$(1) \sum_{i=1}^{k} x_{[i]} \leqslant \sum_{i=1}^{k} y_{[i]}, k = 1, 2, \cdots, n-1; (2) \sum_{i=1}^{n} x_i = \sum_{i=1}^{n} y_i$$

则称 $\boldsymbol{x}$ 被 $\boldsymbol{y}$ 所控制, 记作 $\boldsymbol{x} \prec \boldsymbol{y}$. 若只满足 $\sum_{i=1}^{k} x_{[i]} \leqslant \sum_{i=1}^{k} y_{[i]}, k = 1, 2, \cdots, n$, 则称 $\boldsymbol{x}$ 被 $\boldsymbol{y}$ 下弱控制, 记作 $\boldsymbol{x} \prec_w \boldsymbol{y}$. $\boldsymbol{x} \leqslant \boldsymbol{y}$ 表示 $x_i \leqslant y_i, i = 1, \cdots, n$.

笔者于文[2] 利用控制不等式理论建立了凸数列的一个等价条件及其若干相关结论, 主要结果是:

定理 A　数列 $\{a_i\}$ 是凸数列的充要条件为: 对任意的 $\boldsymbol{p}, \boldsymbol{q} \in \mathbf{Z}_+^n = \{\boldsymbol{p} = (p_1, \cdots, p_n) : p_i$ 是非负整数, $i = 1, 2, \cdots, n, n \geqslant 2 \}$, 若 $\boldsymbol{p} = (p_1, ..., p_n) \prec$

$(q_1, ..., q_n) = \boldsymbol{q}$, 恒有

$$a_{q_1} + \cdots + a_{q_n} \geqslant a_{p_1} + \cdots + a_{p_n}$$

推论 B   非负数列 $\{a_n\}$ 是对数凸数列的充要条件为: 对任意 $\boldsymbol{p}, \boldsymbol{q} \in \mathbf{Z}_n^+$, 若 $\boldsymbol{p} \prec \boldsymbol{q}$, 则

$$\prod_{i=1}^{n} a_{p_i} \leqslant \prod_{i=1}^{n} a_{q_i}$$

本文是文[2] 的继续, 除了将定理A 和推论B 扩展到弱控制的情形以外, 主要展示定理A 和推论B 在分析、代数、凸体几何、概率论等诸多方面的应用.

23.2   定理A的扩展

定理 1   若数列 $\{a_i\}$ 是增的凸数列, 对于任意的 $\boldsymbol{p}, \boldsymbol{q} \in \mathbf{Z}_+^n$, 若 $\boldsymbol{p} = (p_1, \cdots, p_n) \prec_w (q_1, \cdots, q_n) = \boldsymbol{q}$,   则

$$a_{p_1} + \cdots + a_{p_n} \leqslant a_{q_1} + \cdots + a_{q_n}$$

证明   因 $\boldsymbol{p} = (p_1, \cdots, p_n) \prec_w (q_1, \cdots, q_n) = \boldsymbol{q}$, 由文[1] 定理1. 20 (a) 知存在 $\boldsymbol{u} = (u_1, \cdots, u_n) \in \mathbf{R}^n$, 使得 $\boldsymbol{p} \leqslant \boldsymbol{u}, \boldsymbol{u} \prec \boldsymbol{q}$, 由该定理的证明过程可见, $\boldsymbol{u} \in \mathbf{Z}_+^n$, 从而由定理A 及 $\{a_n\}$ 是增数列的条件, 有

$$a_{p_1} + \cdots + a_{p_n} \leqslant a_{u_1} + \cdots + a_{u_n} \leqslant a_{q_1} + \cdots + a_{q_n}$$

推论 1   设非负数列 $\{a_n\}$ 是增的对数凸数列, 对于任意 $\boldsymbol{p}, \boldsymbol{q} \in \mathbf{Z}_n^+$, 若 $\boldsymbol{p} = (p_1, \cdots, p_n) \prec_w (q_1, \cdots, q_n) = \boldsymbol{q}$, 则

$$\prod_{i=1}^{n} a_{p_i} \leqslant \prod_{i=1}^{n} a_{q_i}$$

证明   若增的对数凸数列中有等于零的项, 易证此数列各项均为零, 这时推论1 显然成立, 若各项均不为零, 注意此时 $\{\ln a_n\}$ 是增的凸数列, 由定理1 即得证.

23.3　应用

引理 1　设 $x, y \in \mathbf{R}^n, x_1 \geqslant x_2 \geqslant \cdots \geqslant x_n, \sum\limits_{i=1}^{n} x_i = \sum\limits_{i=1}^{n} y_i$, 若存在 $k, 1 \leqslant k \leqslant n$, 使得 $x_i \leqslant y_i, i = 1, \cdots, k, x_i \geqslant y_i, i = k+1, \cdots, n$, 则 $x \prec y$. (参见文[3] 定理8.1.10)

引理 2[1]　设 $x, y \in \mathbf{R}^n, u, v \in \mathbf{R}^m$, 且 $x \prec y, u \prec v$, 则 $(x, u) \prec (y, v)$.

引理 3

$$\left(\underbrace{2, \cdots, 2}_{n+1}, \underbrace{4, \cdots, 4}_{n+1}, \cdots, \underbrace{2n, \cdots, 2n}_{n+1} \right)$$
$$\prec \left(\underbrace{1, \cdots, 1}_{n}, \underbrace{3, \cdots, 3}_{n}, \cdots, \underbrace{2n+1, \cdots, 2n+1}_{n} \right) \tag{1}$$

证明　用数学归纳法, 当 $n = 1$ 时式(1)化为 $(2, 2) \prec (1, 3)$, 此式显然成立. 若 $n = k$ 时式 (1) 成立, 即

$$x = \left(\underbrace{2, \cdots, 2}_{k+1}, \underbrace{4, \cdots, 4}_{k+1}, \cdots, \underbrace{2k, \cdots, 2k}_{k+1} \right)$$
$$\prec \left(\underbrace{1, \cdots, 1}_{k}, \underbrace{3, \cdots, 3}_{k}, \cdots, \underbrace{2k+1, \cdots, 2k+1}_{k} \right) = y$$

由引理1 易证

$$u = \left(2, 4, \cdots, 2k, \underbrace{2k+2, \cdots, 2k+2}_{k+2} \right)$$
$$\prec \left(1, 3, \cdots, 2k+1, \underbrace{2k+3, \cdots, 2k+3}_{k+1} \right) = v$$

从而由引理2 有 $(x, u) \prec (y, v)$, 即

$$\left(\underbrace{2, \cdots, 2}_{k+2}, \underbrace{4, \cdots, 4}_{k+2}, \cdots, \underbrace{2k+2, \cdots, 2k+2}_{k+2} \right)$$

$$\prec \left(\underbrace{1,\cdots,1}_{k+1}, \underbrace{3,\cdots,3}_{k+1}, \cdots, \underbrace{2k+3,\cdots,2k+3}_{k+1} \right)$$

亦即当 $n = k+1$ 时式(1)成立, 所以式(1)对所有 $n \in \mathbf{N}$ 成立.

定理 2 设 $a > 0$, 则

$$\sum_{k=1}^{2n+1} (-1)^{k+1} a^k \geqslant \frac{1}{n+1} \sum_{k=0}^{n} a^{2k+1}$$

$$\geqslant \frac{1}{2n+1} \sum_{k=1}^{2n+1} a^k \geqslant \frac{1}{n} \sum_{k=1}^{n} a^{2k} \tag{2}$$

$$\frac{1 + a^2 + \cdots + a^{2n}}{a + a^3 + \cdots + a^{2n-1}}$$

$$\geqslant \frac{n+1}{n} + \left(\sqrt{a} - \frac{1}{\sqrt{a}} \right)^2 \quad {}^{[4]} \tag{3}$$

证明 易见式(2)中的三个不等式均等价于

$$n(a + a^3 + \cdots + a^{2n+1}) \geqslant (n+1)(a^2 + a^4 + \cdots + a^{2n}) \tag{$*$}$$

注意 $\{a_n\}$ 是凸数列, 结合式(1), 由定理A 知式($*$)成立, 进而式(2)亦成立;

式(3)右边 $= \frac{n+1}{n} + a + \frac{1}{a} - 2 = \frac{na^2 - (n-1)a + n}{na}$, 于是

$$式(3) \Leftrightarrow na(1 + a^2 + \cdots + a^{2n})$$

$$\geqslant (a + a^3 + \cdots + a^{2n-1})[na^2 - (n-1)a + n]$$

$$\Leftrightarrow n(a + a^3 + \cdots + a^{2n+1}) + 2n(a^2 + a^4 + \cdots + a^{2n})$$

$$\geqslant n(a^{2n+1} + a) + (n+1)(a^2 + a^4 + \cdots + a^{2n}) +$$

$$2n(a^3 + a^5 + \cdots + a^{2n-1})$$

$$\Leftrightarrow (n-1)(a^2 + \cdots + a^{2n}) \geqslant n(a^3 + \cdots + a^{2n-1})$$

而上式与式($*$)等价, 由此知式(3)成立.

定理 3 设 $x \geqslant 0, m > n, f(x) = \frac{1 + x + \cdots + x^m}{1 + x + \cdots + x^n}, g(x) = f(x)x^{n-m}$, 则

$$g(x) \leqslant \frac{m+1}{n+1} \leqslant f(x) \ (1 < x < +\infty) \tag{4}$$

$$f(x) \leqslant \frac{m+1}{n+1} \leqslant g(x) \ (0 < x < 1) \tag{5}$$

证明　不难验证

$$\left(\underbrace{n,\cdots,n}_{n+1},\underbrace{n+1,\cdots,n+1}_{n+1},\cdots,\underbrace{n+m,\cdots,n+m}_{n+1}\right)$$

$$\prec_w \left(\underbrace{m,\cdots,m}_{m+1},\underbrace{m+1,\cdots,m+1}_{m+1},\cdots,\underbrace{m+n,\cdots,m+n}_{m+1}\right)$$

$$\left(\underbrace{0,\cdots,0}_{m+1},\underbrace{1,\cdots,1}_{m+1},\cdots,\underbrace{n,\cdots,n}_{m+1}\right)$$

$$\prec \left(\underbrace{0,\cdots,0}_{n+1},\underbrace{1,\cdots,1}_{n+1},\cdots,\underbrace{m,\cdots,m}_{n+1}\right)$$

当$1 < x < +\infty$ 时,$\{x^n\}$ 是增的凸数列, 根据定理1, 由上述两个弱控制不等式,可依次得

$$(n+1)(x^n + x^{n+1} + \cdots + x^{m+n}) \leqslant (m+1)(x^m + x^{m+1} + \cdots + x^{n+m})$$

$$(m+1)(1 + x + \cdots + x^n) \leqslant (n+1)(1 + x + \cdots + x^m)$$

而这两个不等式分别与式(4)中的两个不等式等价, 故式(4)得证. 注意$g(x) = f(\frac{1}{x})$, 由式(4)即可得式(5).

引理 4 [4]　设f, g 是$[a, b]$ 上的正的连续函数, 记$I_n = \int_a^b [f(x)]^n g(x)\mathrm{d}x$,则

$$I_{n-1}^2 \leqslant I_n I_{n-2} \quad (n \geqslant 2) \tag{6}$$

定理 4　在引理4 的条件下,设$m \leqslant n, n \geqslant 2$, 对于任意$p, q \in \mathbf{Z}_+^n$,若$\boldsymbol{p} \prec \boldsymbol{q}$, 有

$$\prod_{i=1}^m I_{p_i} \leqslant \prod_{i=1}^m I_{q_i} \tag{7}$$

证明　条件$I_{n-1}^2 \leqslant I_n I_{n-2}$ 说明$\{I_n\}$ 是对数凸数列, 从而由推论B 即知结论成立.

推论 1　设f, g 是$[a, b]$ 上的正的连续函数, 则对任意自然数m 都有

$$\left(\int_a^b [f(x)]^m g(x)\mathrm{d}x\right)^{m+1}$$

$$\leqslant \int_a^b g(x)\mathrm{d}x \left(\int_a^b [f(x)]^{m+1} g(x)\mathrm{d}x \right)^m \tag{8}$$

证明 取

$$\boldsymbol{p} = \Big(\underbrace{m, \cdots, m}_{m+1} \Big), \boldsymbol{q} = \Big(\underbrace{m+1, \cdots, m+1}_{m}, 0 \Big)$$

则 $\boldsymbol{p} \prec \boldsymbol{q}$, 由定理4 有 $I_m^{m+1} \leqslant I_{m+1}^m I_0$, 即式(8) 成立.

特别取 $m = 1, g(x) = 1$, 由式(8)可得熟悉的不等式

$$\left[\int_a^b f(x)\mathrm{d}x \right]^2 \leqslant \frac{1}{b-a} \left[\int_a^b (f(x))^2 \mathrm{d}x \right]$$

引理 5 [4] (指数积分不等式) 设 $E_n(z) = \int_1^\infty \frac{\mathrm{e}^{-zt}}{t^n} \mathrm{d}t, \mathrm{Re}z > 0, n = 0, 1, 2, \cdots$, 则当 $x > 0$ 时, 对于 $n = 1, 2, 3, \cdots$, 有

$$E_n^2(x) \leqslant E_{n-1}(x) E_{n+1}(x)$$

上式说明 $\{E_n(x)\}$ 是对数凸数列, 从而由推论B 有

定理 5 记号同引理5, 对于任意 $\boldsymbol{p}, \boldsymbol{q} \in \mathbf{Z}_+^n$, 若 $\boldsymbol{p} \prec \boldsymbol{q}$, 有

$$\prod_{i=1}^m E_{p_i}(x) \leqslant \prod_{i=1}^m E_{q_i}(x) \tag{9}$$

定义 4 [5] 称函数 f 在 $(0, +\infty)$ 上绝对单调, 如果它的各阶导数存在且

$$f^{(k)}(t) \geqslant 0, t \in (0, \infty), k = 0, 1, 2, \cdots \tag{10}$$

引理 6 [5] 设函数 f 在 $(0, +\infty)$ 上绝对单调, 则

$$f^{(k-1)}(t) f^{(k+1)}(t) \geqslant [f^{(k)}(t)]^2, k = 0, 1, 2, \cdots \tag{11}$$

上式说明 $\{f^{(k)}(t)\}$ 是对数凸数列, 从而由推论B 有

定理 6 设函数 f 在 $(0, +\infty)$ 上绝对单调, 对于任意 $\boldsymbol{p}, \boldsymbol{q} \in \mathbf{Z}_+^n$, 若 $\boldsymbol{p} \prec \boldsymbol{q}$, 有

$$\prod_{i=1}^m f^{(p_i)}(t) \leqslant \prod_{i=1}^m f^{(q_i)}(t) \tag{12}$$

推论 2 设 $b > a > 0, x > 0, q(x) = \frac{b^x - a^x}{x}$, 则对于任意 $\boldsymbol{p}, \boldsymbol{q} \in \mathbf{Z}_+^n$,

若 $p \prec q$, 有

$$\prod_{i=1}^{m} g^{(p_i)}(t) \leqslant \prod_{i=1}^{m} g^{(q_i)}(t) \tag{13}$$

文[5] 证明了 $g(x)$ 是 $(0, +\infty)$ 上绝对单调函数, 从而由定理6 知式(13)成立.

特别对任意非负整数 i, j, k, 取

$$\boldsymbol{p} = (p_1, p_2) = (2(i+k)+1, 2(j+k)+1)$$

$$\boldsymbol{q} = (q_1, q_2) = (2k, 2(i+j+k+1))$$

则 $\boldsymbol{p} \prec \boldsymbol{q}$.

由式(13)得文[5]中的式(28), 即

$$g^{(2(j+k)+1)}(t) g^{(2(j+k)+1)}(t)$$
$$\leqslant g^{(2k)}(t) g^{(2(i+j+k+1))}(t), t > 0$$

引理 7 [6] 若 $f(x) = \sum\limits_{i=0}^{n} c_i x^i$ 具有 n 个实根且 $c_i = \binom{n}{i} d_i$, 则

$$d_i^2 - d_{i-1} d_{i+1} \geqslant 0, i = 1, \cdots, n-1 \tag{14}$$

定理 7 设 $f(x) = \sum\limits_{i=0}^{n} c_i x^i$ 具有 n 个实根且 $c_i = \binom{n}{i} d_i$, 则对于任意 $p, q \in \mathbf{Z}_+^n$, 若 $\boldsymbol{p} \prec \boldsymbol{q}$, 有

$$\prod_{i=1}^{n} d_{q_i} \leqslant \prod_{i=1}^{n} d_{p_i} \tag{15}$$

证明 由式(14) 知 $\{d_i^{-1}\}$ 是对数凸数列, 从而由推论B 即得证.

引理 8[3] 设 X 为随机变量, 则

$$E(X^{p+q}) E(X^{p-1}) \geqslant E(X^{p-1+q}) E(X^p), p > 0, q \geqslant 0 \tag{16}$$

取 $q = 1$, 式(16)化为

$$[E(X^p)]^2 \leqslant E(X^{p+1}) E(X^{p-1})$$

上式说明 $\{E(X^k)\}$ 是对数凸数列, 从而由推论B 有

定理 8 设 X 为正随机变量, 则对于任意 $p, q \in \mathbf{Z}_+^n$, 若 $\boldsymbol{p} \prec \boldsymbol{q}$, 有

$$\prod_{i=1}^{n} E(X^{p_i}) \leqslant \prod_{i=1}^{n} E(X^{q_i}) \tag{17}$$

定义 5 [7] 设 $\mathbf{R}^n$ 中 n 维多胞形 Ω 有 $N(\geqslant n+1)$ 个 $n-1$ 维表面 $F_1, \cdots, F_N$, 用 e_i 表示 F_i 的, 我们称 $\alpha_{i_1 i_2 \cdots i_s} = \arcsin |\det\Gamma(e_{i_1}, e_{i_2}, \cdots, e_{i_s})|^{\frac{1}{2}}$ 为 Ω 的 s 个表面 $F_{i_1}, F_{i_2}, \cdots, F_{i_s}$ 形成的 s 面空间角, 其中 $\Gamma(e_{i_1}, e_{i_2}, \cdots, e_{i_s})$ 是 s 个法向量 $e_{i_1}, e_{i_2}, \cdots, e_{i_s}$ 的 Gram 矩阵, $1 \leqslant i_1 < i_2 < \cdots < i_s \leqslant N$.

引理 9 [7] 设 $\mathbf{R}^n$ 中 n 维多胞形 Ω 有 $N(\geqslant n+1)$ 个 $n-1$ 维表面 $F_1, \cdots, F_N$, 其中任意 $s(2 \leqslant s \leqslant n)$ 个表面 $F_{i_1}, F_{i_2}, \cdots, F_{i_s}$ 形成的 s 面空间角为 $\alpha_{i_1 i_2 \cdots i_s}$, 又 $x_1, \cdots, x_N$ 为 N 个实数, 令 $N_0 = 1, N_1 = \sum\limits_{i=1}^{N} x_i \neq 0$

$$N_s = \sum_{1 \leqslant i_1 < \cdots < i_s \leqslant N} x_{i_1} \cdots x_{i_s} \sin^2{}_{i_1 \cdots i_s} \quad (2 \leqslant s \leqslant n).$$

则

$$N_k^2 \geqslant \frac{k+1}{k} \cdot \frac{n+1-k}{n-k} N_{k-1} N_{k+1} \quad (1 \leqslant k \leqslant n-1) \tag{18}$$

上式等价于

$$\left(\frac{N_k}{C_n^k}\right)^2 \geqslant \left(\frac{N_{k-1}}{C_n^{k-1}}\right)\left(\frac{N_{k+1}}{C_n^{k+1}}\right)$$

这说明 $\left\{\left(\dfrac{N_k}{C_n^k}\right)^{-1}\right\}$ 是对数凸数列, 从而由推论 B 有

定理 9 条件同引理 9, 则对任意 $p, q \in \mathbf{Z}_+^n$, 若 $\boldsymbol{p} \prec \boldsymbol{q}$, 有

$$\prod_{i=1}^{n} \frac{N_{q_i}}{C_n^{q_i}} \leqslant \prod_{i=1}^{n} \frac{N_{p_i}}{C_n^{p_i}} \tag{19}$$

推论 3 条件同定理 9, 对于 $k \leqslant n$ 有

$$N_k N_{n-k} \geqslant (C_n^k)^2 N_n \tag{20}$$

$$(N_k)^n \geqslant (C_n^k)^n (N_n)^k \tag{21}$$

证明 注意

$$\left(k, n-k, \underbrace{0, \cdots, 0}_{n-2}\right) \prec \left(n, \underbrace{0, \cdots, 0}_{n-1}\right)$$

和

$$\left(\underbrace{k, \cdots, k}_{n}\right) \prec \left(\underbrace{n, \cdots, n}_{k}, \underbrace{0, \cdots, 0}_{n-k}\right)$$

由定理 9 即可得证.

致谢: 衷心感谢山东青岛教育学院续铁权教授及审稿人给予本文的宝贵的修改意见.

参考文献

[1]　王伯英. 控制不等式基础[M]. 北京:北京师范大学出版社,1990.

[2]　石焕南,李大矛. 凸数列的一个等价条件及其应用[J]. 曲阜师范大学学报(自然科学版), 2001, 27 (4):4-6.

[3]　王松桂,贾忠贞. 矩阵论中不等式[M]. 合肥:安徽教育出版社, 1994 :312

[4]　匡继昌. 常用不等式[M]. 2 版. 长沙:湖南教育出版社, 1993.

[5]　QI FENG, XU SENLIN. Refinements and extensions of an inequality, II [J]. J. Wath. Anal. Appl., 1997, 211: 616-620.

[6]　MITRINOVIĆ D S, VASIC P M. 分析不等式[M]. 赵汉宾,译.南宁: 广西人民出版社, 1986: 128.

[7]　张垚. 关于多胞形的一类不等式[J]. 湖南教育学院学报,1999, 17(5): 99-105.

[8]　肖振钢. 数列与不等式的一个连接点——凸数列[J]. 湖南数学年刊,1995, 15(4): 62-69.

[9]　哈代 G H, 李特伍德 J E, 波利亚 G. 不等式[M]. 越民义, 译. 北京:科学出版社, 1965.

第24篇　一类积分不等式的控制证明

(石焕南, 张鉴, 徐坚. 首都师范大学学报(自然科学版), 2004,25(4):11-13)

摘　要:　本文利用控制不等式理论建立了一类积分不等式.

关键词:　对数凸序列; 积分不等式; 控制

记号 $R[a,b]$ 表示有界闭区间 $[a,b]$ 上所有黎曼可积函数所组成的集合, 黎曼积分 $\int_a^b g(x)f^n(x)\mathrm{d}x$ 简记作 $\int_a^b gf^n\mathrm{d}x$. 文[1] 利用二重积分证明了如下结论:

定理 A　对于 $f,g \in R[a,b], m,n > 0$, 且 $f,g \geqslant 0, m \leqslant n, \int_a^b gf^n\mathrm{d}x \neq 0, \int_a^b gf^m\mathrm{d}x \neq 0$, 则

$$\frac{\int_a^b gf^{m+1}\mathrm{d}x}{\int_a^b gf^m\mathrm{d}x} \leqslant \frac{\int_a^b gf^{n+1}\mathrm{d}x}{\int_a^b gf^n\mathrm{d}x} \tag{1}$$

本文将利用控制不等式理论证明并推广(1), 进而建立一系列此类积分不等式, 以显示控制不等式理论 "成批生产不等式" 的鲜明特征. 为此, 引入如下定义和引理.

定义 1[2]　设 $\boldsymbol{x},\boldsymbol{y} \in \mathbf{R}^n$, $\boldsymbol{x}$ 和 $\boldsymbol{y}$ 的分量的递减重排分别记为 $x_{[1]} \geqslant \cdots \geqslant x_{[n]}$ 和 $y_{[1]} \geqslant \cdots \geqslant y_{[n]}$, 若满足

(1) $\sum\limits_{i=1}^k x_{[i]} \leqslant \sum\limits_{i=1}^k y_{[i]}, k = 1, 2, ..., n-1$;

(2) $\sum\limits_{i=1}^n x_i = \sum\limits_{i=1}^n y_i$. 则称 $\boldsymbol{x}$ 被 $\boldsymbol{y}$ 所控制, 记作 $\boldsymbol{x} \prec \boldsymbol{y}$.

定义 2[2]　设 $\Omega \subset \mathbf{R}^n, \phi: \Omega \to \mathbf{R}$, 若在 Ω 上 $\boldsymbol{x} \prec \boldsymbol{y} \Rightarrow \phi(\boldsymbol{x}) \leqslant \phi(\boldsymbol{y})$, 则称 ϕ 为 Ω 上的Schur凸函数.

引理 1[2]　$I \subset \mathbf{R}$ 为一个区间, g 为 $I \to \mathbf{R}$ 上的凸函数, 则

$$\phi(\boldsymbol{x}) = \sum_{i=1}^n g(x_i)$$

为 I^n 上的Schur凸函数.

138

引理 2[2] 设 $\boldsymbol{x}=(x_1.x_2,\cdots,x_n)\in\mathbf{R}^n$, $\bar{x}=\frac{1}{n}\sum\limits_{i=1}^{n}x_i$, 则

$$(\bar{x},\bar{x},\cdots,\bar{x})\prec(x_1,x_2,\cdots,x_n)$$

引理 3 设 $\boldsymbol{x},\boldsymbol{y}\in\mathbf{R}^n,x_1\geqslant x_2\geqslant\cdots\geqslant x_n,\sum\limits_{i=1}^{n}x_i=\sum\limits_{i=1}^{n}y_i$, 若存在 $k,1\leqslant k\leqslant n$, 使得 $x_i\leqslant y_i,i=1,\cdots,k,x_i\geqslant y_i,i=k+1,\cdots,n$, 则 $\boldsymbol{x}\prec\boldsymbol{y}$. (参见文[3]定理8.1.10)

引理 4 设 $f,g\in R[a,b]$ 且 $f,g\geqslant0$, 记 $I(r)=\int_a^b gf^r\mathrm{d}x,r\geqslant0$, 则 $\ln I(r)$ 是 $[0+\infty)$ 上的凸函数.

证明 设 $\alpha,\beta\geqslant0,0<t<1$, 利用积分的Hölder 不等式, 有

$$I(t\alpha+(1-t)\beta)=\int_a^b gf^{t\alpha+(1-t)\beta}\mathrm{d}x=\int_a^b [gf^\alpha]^t[gf^\beta]^{1-t}\mathrm{d}x$$
$$\leqslant\left[\int_a^b gf^\alpha\mathrm{d}x\right]^t\left[\int_a^b gf^\beta\mathrm{d}x\right]^{1-t}=I^t(\alpha)I^{1-t}(\beta)$$

$$\ln I(t\alpha+(1-t)\beta)\leqslant t\ln I(\alpha)+(1-t)\ln I(\beta)$$

这说明 $\ln I(r)$ 是 $[0+\infty)$ 上的凸函数.

定理 1 设 $f,g\in R[a,b]$ 且 $f,g\geqslant0,n\geqslant2$. 记 $I(r)=\int_a^b gf^r\mathrm{d}x,r\geqslant0$, 对于任意 $\boldsymbol{p},\boldsymbol{q}\in\mathbf{R}_+^n=\{\boldsymbol{x}=(x_1,\cdots,x_n)\in\mathbf{R}^n:x_i\geqslant0,i=1,\cdots,n\}$, 若 $\boldsymbol{p}\prec\boldsymbol{q}$, 有

$$\prod_{i=1}^{n}I(p_i)\leqslant\prod_{i=1}^{n}I(q_i) \tag{2}$$

证明 由引理4 知 $\ln I(r)$ 是 $[0+\infty)$ 上的凸函数, 从而由引理1 有

$$\sum_{i=1}^{n}\ln I(p_i)\leqslant\sum_{i=1}^{n}\ln I(q_i)$$

即(2)成立.

定理 A 的别证 取 $\boldsymbol{p}=(m+1,n),\boldsymbol{q}=(m,n+1)$, 易见 $\boldsymbol{p}\prec\boldsymbol{q}$, 从而由(2)有 $I(n)I(m+1)\leqslant I(m)I(n+1)$, 即

$$\int_a^b gf^n\mathrm{d}x\int_a^b gf^{m+1}\mathrm{d}x\leqslant\int_a^b gf^m\mathrm{d}x\int_a^b gf^{n+1}\mathrm{d}x$$

而此式与式(1)等价.

下面我们通过构造各种控制关系, 建立一系列此类不等式.

定理 2 设 $f,g \in R[a,b]$ 且 $f \geqslant 0, g > 0, m, n \in \mathbf{N}, m \leqslant n$, 则

$$\left(\frac{\int_a^b gf^m \mathrm{d}x}{\int_a^b g\mathrm{d}x}\right)^n \leqslant \left(\frac{\int_a^b gf^n \mathrm{d}x}{\int_a^b g\mathrm{d}x}\right)^m \tag{3}$$

证明 取 $\boldsymbol{p} = (\underbrace{m,\cdots,m}_{n}), \boldsymbol{q} = (\underbrace{n,\cdots,n}_{m},\underbrace{0,\cdots,0}_{n-m})$ 易见 $\boldsymbol{p} \prec \boldsymbol{q}$, 从而由定理1 有 $[I(m)]^n \leqslant [I(n)]^m[I(0)]^{n-m}$, 即

$$\left(\int_a^b gf^m \mathrm{d}x\right)^n \leqslant \left(\int_a^b gf^n \mathrm{d}x\right)^m \left(\int_a^b g\mathrm{d}x\right)^{n-m}$$

此式等价于式(3).

特别取 $n = m+1$, 则式(3)化为

$$\left(\int_a^b gf^m \mathrm{d}x\right)^{m+1} \leqslant \int_a^b g\mathrm{d}x \left(\int_a^b g^{m+1}\mathrm{d}x\right)^m \tag{4}$$

此为文[6]的定理4.

取 $g(x) = 1$, 则式(3)化为

$$\left(\frac{1}{b-a}\int_a^b f^m \mathrm{d}x\right)^n \leqslant \left(\frac{1}{b-a}\int_a^b f^n \mathrm{d}x\right)^m \quad (m \leqslant n) \tag{5}$$

进一步取 $n=2, m=1$, 则式(5)化为熟悉的不等式

$$\left(\int_a^b f\mathrm{d}x\right)^2 \leqslant (b-a)\int_a^b f^2\mathrm{d}x \tag{6}$$

而取 $n=3, m=2$, 则式(5)化为

$$\left(\frac{1}{b-a}\int_a^b f^2\mathrm{d}x\right)^3 \leqslant \left(\frac{1}{b-a}\int_a^b f^3\mathrm{d}x\right)^2 \tag{7}$$

类似地, 利用 $(m,n-m) \prec (n,0)$, 可得

定理 3 设 $f,g \in R[a,b]$ 且 $f,g \geqslant 0, m,n \in \mathbf{N}, m \leqslant n$, 则

$$\int_a^b gf^m \mathrm{d}x \int_a^b gf^{n-m}\mathrm{d}x \leqslant \int_a^b g\mathrm{d}x \int_a^b gf^n \mathrm{d}x \tag{8}$$

若 $f > 0$, 取 $n=2, m=1, g=\frac{1}{f}$, 则由式(8)可得

推论 1　设 $f, g \in R[a, b]$, 且 $f > 0$, 则

$$\left(\int_a^b f \mathrm{d}x\right)\left(\int_a^b \frac{1}{f} \mathrm{d}x\right) \geqslant (b - a)^2 \tag{9}$$

这也是熟知的结果. 利用

$$(\underbrace{n - m, \cdots, n - m}_{n-m}, \underbrace{n, \cdots, n}_{2m}) \prec (\underbrace{m, \cdots, m}_{m}, \underbrace{n, \cdots, n}_{n})$$

可得

定理 4　设 $f, g \in R[a, b]$ 且 $f, g \geqslant 0, m, n \in \mathbf{N}, 2m < n$, 则

$$\left(\int_a^b g f^{n-m} \mathrm{d}x\right)^{n-m}\left(\int_a^b g f^n \mathrm{d}x\right)^{2m} \leqslant \left(\int_a^b g f^m \mathrm{d}x\right)^m\left(\int_a^b g f^n \mathrm{d}x\right)^n \tag{10}$$

利用

$$(\underbrace{m, \cdots, m}_{n-h}) \prec (\underbrace{n, \cdots, n}_{m-h}, \underbrace{h, \cdots, h}_{n-m}) \quad (h < m < n)$$

可得

定理 5　设 $f, g \in R[a, b]$ 且 $f, g \geqslant 0, h, m, n \in \mathbf{N}, h < m < n$, 则

$$\left(\int_a^b g f^m \mathrm{d}x\right)^{n-h} \leqslant \left(\int_a^b g f^n \mathrm{d}x\right)^{m-h}\left(\int_a^b g f^h \mathrm{d}x\right)^{n-m} \tag{11}$$

由引理 3 容易验证

$$(0, 0, 1, 1, \cdots, n, n) \prec (\underbrace{0, \cdots, 0}_{n+1}, \underbrace{n, \cdots, n}_{n+1})$$

利用上式可得

定理 6　设 $f, g \in R[a, b]$ 且 $f, g \geqslant 0, m, n \in \mathbf{N}$, 则

$$\prod_{i=0}^n \left(\int_a^b g f^i \mathrm{d}x\right)^2 \leqslant \left(\int_a^b g \mathrm{d}x\right)^{n+1}\left(\int_a^b g f^n \mathrm{d}x\right)^{n+1} \tag{12}$$

取 $g(x) = 1$, 由式 (12) 可得

$$\prod_{i=0}^n \left(\frac{1}{b-a}\int_a^b f^i \mathrm{d}x\right)^2 \leqslant \left(\frac{1}{b-a}\int_a^b f^n \mathrm{d}x\right)^{\frac{n+1}{2}} \tag{13}$$

特别取$n = 3$, 由式(13)可得

$$\left(\frac{1}{b-a}\int_a^b f\mathrm{d}x\right)\left(\frac{1}{b-a}\int_a^b f^2\mathrm{d}x\right) \leqslant \frac{1}{b-a}\int_a^b f^3\mathrm{d}x \qquad (14)$$

由引理3 容易验证

$$(\underbrace{1,\cdots,1}_{n+1},\underbrace{2,\cdots,2}_{n+1},\underbrace{n-1,\cdots,n-1}_{n+1}) \prec (\underbrace{0,\cdots,0}_{n-1},\underbrace{1,\cdots,1}_{n-1},\underbrace{n,\cdots,n}_{n-1})$$

利用上式可得

定理 7 设$f, g \in R[a,b]$ 且$f, g \geqslant 0, n \in \mathbf{N}$, 则

$$\prod_{i=0}^{n-1}\left(\int_a^b gf^i\mathrm{d}x\right)^{n+1} \leqslant \prod_{i=0}^{n}\left(\int_a^b gf^i\mathrm{d}x\right)^{n-1}$$

即

$$\prod_{i=0}^{n-1}\left(\int_a^b gf^i\mathrm{d}x\right)^2 \leqslant \left(\int_a^b g\mathrm{d}x\right)^{n-1}\left(\int_a^b gf^n\mathrm{d}x\right)^{n-1} \qquad (15)$$

利用

$$(\underbrace{2,\cdots,2}_{n+1},\underbrace{4,\cdots,4}_{n+1},\underbrace{2n,\cdots,2n}_{n+1}) \prec (\underbrace{1,\cdots,1}_{n},\underbrace{3,\cdots,3}_{n},\underbrace{2n+1,\cdots,2n+1}_{n})^{[7]}$$

可得

定理 8 设$f, g \in R[a,b]$ 且$f, g \geqslant 0, n \in \mathbf{N}$, 则

$$\prod_{i=1}^{n}\left(\int_a^b gf^{2i}\mathrm{d}x\right)^{n+1} \leqslant \prod_{i=0}^{n}\left(\int_a^b gf^{2i+1}\mathrm{d}x\right)^{n}$$

由引理2 有

$$(\underbrace{n,\cdots,n}_{2n+1}) \prec (0,1,\cdots,2n)$$

利用上式可得

定理 9 设$f, g \in R[a,b]$ 且$f, g \geqslant 0, n \in \mathbf{N}$, 则

$$\left(\int_a^b gf^n\mathrm{d}x\right)^{2n+1} \leqslant \prod_{i=0}^{2n}\left(\int_a^b gf^i\mathrm{d}x\right)$$

需要指出的是, 由于对于可测集上的Lebesgue积分, Hölder不等式亦成

立, 进而引理4 和定理1 亦成立. 因此上述结论不局限于有限区间上的黎曼可积函数. 例如, 下面是有关无穷广义积分的一个例子.

对于 $g(t) = \exp\left(-\frac{t^2}{2}\right)$, $f(t) = t$, 注意 $(1,1) \prec (0,2)$, 由定理1 可得

定理 10[5]　对于 $x > 0$, 有

$$\left[\int_x^{+\infty} t\exp\left(-\frac{t^2}{2}\right)\mathrm{d}t\right]^2 \leqslant \int_x^{+\infty} \exp\left(-\frac{t^2}{2}\right)\mathrm{d}t \times \int_x^{+\infty} t^2\exp\left(-\frac{t^2}{2}\right)\mathrm{d}t \tag{16}$$

致谢　作者衷心感谢青岛职业技术学院续铁权教授给予本文的宝贵的修改意见.

参考文献

[1]　陈欢. 定积分的一个不等式及其应用[J]. 福州大学学报(自然科学版), 2003, 31(6): 649–651.

[2]　王伯英. 控制不等式基础[M]. 北京: 北京师范大学出版社, 1990.

[3]　王松桂, 贾忠贞. 矩阵论中不等式[M]. 合肥:安徽教育出版社, 1994.

[4]　石焕南, 李大矛. 凸数列的一个等价条件及其应用[J]. 曲阜师范大学学报(自然科学版), 2001, 27(4): 4–6.

[5]　匡继昌. 常用不等式[M]. 3 版. 济南: 山东科技出版社, 2003.

[6]　梁新健, 刘碧秋, 林菊芳. 关于一些积分不等式的注记[J]. 宁夏大学学报(自然科学版), 1998, 19(2):113–116.

[7]　吴善和, 石焕南. 凸序列不等式的控制证明[J]. 数学的实践与认识, 2003, 33(12): 132–137.

[8]　吴善和, 石焕南. 一类无理不等式的控制证明[J]. 首都师范大学学报(自然科学版), 2003, 24(3): 13–16.

[9]　石焕南, 续铁权, 顾春. 整幂函数不等式的控制证明[J]. 商丘师范学院学报, 2003, 19(2): 46–48.

第25篇 Exponential Generalization of Newman's Inequality and Klamkin's Inequality

(SHI HUAN-NAN. Northeast Math. J., 2005, 21(4): 431-438)

Abstract: Exponential generalizations of Newman's inequality and Klamkin's inequality are established by the Wang Wan-lan's inequality, and they are extended to the cases involving general elementary symmetric functions. As an application, some new inequalities for a simplex are established. In the final, an open problem is posed.

Keywords: Newman's inequality; Klamkin's inequality; Wan wan-lan's inequality; elementary symmetric function; exponent; simplex.

25.1 Introduction

Let $x_1, x_2, \cdots, x_n \in \mathbf{R}^+, \sum\limits_{i=1}^{n} x_i = 1, n \in \mathbf{N}$. Then

$$\prod_{i=1}^{n}\left(\frac{1}{x_i} - 1\right) \geqslant (n-1)^n \quad \text{(Newman)} \tag{1}$$

$$\prod_{i=1}^{n}\left(\frac{1}{x_i} + 1\right) \geqslant (n+1)^n \quad \text{(Klamkin)} \tag{2}$$

$$\prod_{i=1}^{n}\frac{1+x_i}{1-x_i} \geqslant \left(\frac{n+1}{n-1}\right)^n \quad \text{(Klamkin)} \tag{3}$$

The forms of the above inequalities are beautiful and succinct which have evoked the interest of a number of scholars. Various extensions of these inequalities have been published to this day. For example, the following extensions of(1) were recorded in [1]:

Theorem A Let $s_n = \sum_{i=1}^{n} x_i, x_i > 0, i = 1, 2, \cdots, n$. Then

$$\prod_{i=1}^{n} \left(\frac{1}{x_i} - t \right) \geq \left(\frac{n}{s_n} - t \right)^n, \quad \forall\, t \leq \frac{1}{s_n} \tag{4}$$

$$\prod_{i=1}^{n} \left(\frac{s_n}{x_i} - \frac{1}{m} \right) \geq \left(n - \frac{1}{m} \right)^n, \quad m \in \mathbf{N} \tag{5}$$

In [2], (1)-(3) have been extended to the following (6)-(8), respectively.

Theorem B Let $x_i > 0, \alpha_i > 0, i = 1, 2, \cdots, n, \sum_{i=1}^{n} \alpha_i = 1,$ and let $x_i + x_j \leq 1, i, j = 1, 2, \cdots, n, i \neq j$. Then

$$\prod_{i=1}^{n} \left(\frac{1}{x_i} - 1 \right)^{\alpha_i} \geq \frac{1}{\sum_{i=1}^{n} \alpha_i x_i} - 1 \tag{6}$$

Theorem C Let $x_i > 0, \ \alpha_i > 0, \ i = 1, 2, \cdots, n, \ \sum_{i=1}^{n} \alpha_i = 1, \ m > 0.$ Then

$$\prod_{i=1}^{n} \left(\frac{1}{x_i^m} + 1 \right)^{\alpha_i} \geq \frac{1}{\left(\sum_{i=1}^{n} \alpha_i x_i \right)^m} + 1 \tag{7}$$

Theorem D Let $0 < x_i < a, \ \alpha_i > 0, \ i = 1, 2, \cdots, n, \ \sum_{i=1}^{n} \alpha_i = 1.$ Then

$$\prod_{j=1}^{n} \left(\frac{a + x_i}{a - x_i} \right)^{\alpha_i} \geq \frac{a + \sum_{i=1}^{n} \alpha_i x_i}{a - \sum_{i=1}^{n} \alpha_i x_i} \tag{8}$$

Remark From the proof process in [2], it is easy to see that in (6)-(8) equalities holding if and only if $x_1 = x_2 = \cdots = x_n$.

Volenec V extended (1) to the cases involving general elementary symmetric functions(see [3]):

Theorem E Let $x_i > 0, \ i = 1, 2, \cdots, n, \ \beta > 0,$ and $\sum_{i=1}^{n} x_i = 1$. Then

$$\sum_{1 \leq i_1 < \cdots < i_k \leq n} \prod_{j=1}^{k} \left(\frac{1}{x_{i_j}} - 1 \right)^{\beta} \geq C_n^k (n-1)^{k\beta} \quad (k = 1, \cdots, n) \tag{9}$$

In 1999, by using the theory of majorization , the author[4] extended

(1) and (2) to the following (10) and (11) respectively:

Theorem F Let $x_i > 0, i = 1, 2, \cdots, n, n \geqslant 2$, and $\sum\limits_{i=1}^{n} x_i = s, c \geqslant s, 0 < \beta \leqslant 1$. Then

$$\frac{\sum\limits_{1\leqslant i_1 < \cdots < i_k \leqslant n} \prod\limits_{j=1}^{k} \left(c - x_{i_j} \right)^{\beta}}{\sum\limits_{1\leqslant i_1 < \cdots < i_k \leqslant n} \prod\limits_{j=1}^{k} x_{i_j}^{\beta}} \geqslant \left(\frac{nc}{s} - 1 \right)^{k\beta} \quad (k = 1, \cdots, n) \qquad (10)$$

Theorem G Let $x_i > 0, i = 1, 2, \cdots, n, n \geqslant 2$, and $\sum\limits_{i=1}^{n} x_i = s, c \geqslant 0, 0 < \beta \leqslant 1$. Then

$$\frac{\sum\limits_{1\leqslant i_1 < \cdots < i_k \leqslant n} \prod\limits_{j=1}^{k} \left(c + x_{i_j} \right)^{\beta}}{\sum\limits_{1\leqslant i_1 < \cdots < i_k \leqslant n} \prod\limits_{j=1}^{k} x_{i_j}^{\beta}} \geqslant \left(\frac{nc}{s} + 1 \right)^{k\beta} \quad (k = 1, \cdots, n) \qquad (11)$$

In 2000, by using the theory of majorization, Xu Tie-quan[5] extended (3) to the following (12):

Theorem H Let $0 \leqslant x_i < 1$, $i = 1, 2, \cdots, n$, $n \geqslant 2$, and let $\sum\limits_{i=1}^{n} x_i = s < n$. Then

$$\frac{\sum\limits_{1\leqslant i_1 < \cdots < i_k \leqslant n} \prod\limits_{j=1}^{k} \left(1 + x_{i_j} \right)}{\sum\limits_{1\leqslant i_1 < \cdots < i_k \leqslant n} \prod\limits_{j=1}^{k} \left(1 - x_{i_j} \right)} \geqslant \left(\frac{n+s}{n-s} \right)^{k} \quad (k = 2, \cdots, n) \qquad (12)$$

The aim of this paper is to establish the exponential and weighted generalizations of (1)-(3), and they are extended to the cases involving general elementary symmetric functions. As an application, we obtain some new inequalities for a simplex.

25.2　Main results and Proof

Lemma 1 (Jensen's inequality)[1] If f is a strictly convex function on an interval $I \subset \mathbf{R}$, $\boldsymbol{x} = (x_1, \cdots x_n) \in I^n$ $(n \geqslant 2)$, $\boldsymbol{p}$ is a positive n- tuple,

$P_n = \sum_{i=1}^{n} p_i$, then

$$f\left(\frac{1}{P_n}\sum_{i=1}^{n} p_i x_i\right) \leqslant \frac{1}{P_n}\sum_{i=1}^{n} p_i f\left(x_i\right) \tag{13}$$

with equality holding if and only if $x_1 = x_2 = \cdots = x_n$.

Using lemma 1, we establish the following weighted form of Wang Wan-lan's inequality see [1]).

Lemma 2 Let $x_i > 0, p_i > 0, i = 1, 2, \cdots, n, P_n = \sum_{i=1}^{n} p_i$. Then

$$\prod_{i=1}^{n}\left[(x_i + 1)^t - 1\right]^{\frac{p_i}{P_n}} \geqslant \left[\left(\prod_{i=1}^{n} x_i^{\frac{p_i}{P_n}}\right) + 1\right]^t - 1 \quad (t \geqslant 1) \tag{14}$$

$$\prod_{i=1}^{n}\left[(x_i + 1)^t + 1\right]^{\frac{p_i}{P_n}} \geqslant \left[\left(\prod_{i=1}^{n} x_i^{\frac{p_i}{P_n}}\right) + 1\right]^t + 1 \quad (t > 0) \tag{15}$$

with both equalities holding if and only if $x_1 = x_2 = \cdots = x_n$.

Proof Let $f(y) = \ln\left[(e^y + 1)^t - 1\right]$. Then

$$f'(y) = \frac{te^y (e^y + 1)^{t-1}}{(e^y + 1)^t - 1}, \quad f''(y) = \frac{te^y (e^y + 1)^{t-2}\left[(e^y + 1)^t - te^y - 1\right]}{\left[(e^y + 1)^t - 1\right]^2}$$

When $t \geqslant 1$, by the Bernoulli's inequality(see[1]), we have

$$(e^y + 1)^t > 1 + te^y$$

So $f''(y) > 0$, i.e. $f(y)$ is a strictly convex function. Then from (13) we have

$$\frac{1}{P_n}\sum_{i=1}^{n} p_i f\left(\ln x_i\right) \geqslant f\left(\frac{1}{P_n}\sum_{i=1}^{n} p_i \ln x_i\right)$$

i.e.

$$\frac{1}{P_n}\sum_{i=1}^{n} p_i \ln\left[\left(e^{\ln x_i} + 1\right)^t - 1\right] \geqslant \ln\left[\left(e^{\frac{1}{P_n}\sum_{i=1}^{n} p_i \ln x_i} + 1\right)^t - 1\right]$$

which is equivalent to (14).

Similarly, let $g(y) = \ln\left[(e^y + 1)^t + 1\right]$. Then

$$g''(y) = \frac{te^y(e^y+1)^{t-2}\left[(e^y+1)^t + te^y + 1\right]}{\left[(e^y+1)^t + 1\right]^2}$$

When $t > 0, g''(y) > 0$, i.e., $g(y)$ is a strictly convex function. Then from (13) we obtain (15). In (14) and (15) both equalities holding if and only if $x_1 = x_2 = \cdots = x_n$.

Theorem 1 Let $x_i > 0, p_i > 0, i = 1, 2, \cdots, n, n \geqslant 2, P_n = \sum\limits_{i=1}^{n} p_i$, and let $x_i + x_j \leqslant 1, i \neq j, i, j = 1, 2, \cdots, n$. Then for $t \geqslant 1$, we have

$$\prod_{i=1}^{n}\left(\frac{1}{x_i^t} - 1\right)^{\frac{p_i}{P_n}} \geqslant \left(\frac{P_n}{\sum\limits_{i=1}^{n} p_i x_i}\right)^t - 1 \tag{16}$$

with equality holding if and only if $x_1 = x_2 = \cdots = x_n$.

Proof Let $y_i = \frac{1}{x_i} - 1$, $i = 1, 2, \cdots, n$. From Lemma 2 we have

$$\prod_{i=1}^{n}\left[(y_i+1)^t - 1\right]^{\frac{p_i}{P_n}} \geqslant \left[\left(\prod_{i=1}^{n} y_i^{\frac{p_i}{P_n}}\right) + 1\right]^t - 1$$

i.e.

$$\prod_{i=1}^{n}\left(\frac{1}{x_i^t} - 1\right)^{\frac{p_i}{P_n}} \geqslant \left[\prod_{i=1}^{n}\left(\frac{1}{x_i} - 1\right)^{\frac{p_i}{P_n}} + 1\right]^t - 1$$

Moreover, from Theorem B, we have

$$\prod_{i=1}^{n}\left(\frac{1}{x_i} - 1\right)^{\frac{p_i}{P_n}} \geqslant \frac{1}{\sum\limits_{i=1}^{n} \frac{p_i}{P_n} x_i} - 1 = \frac{P_n}{\sum\limits_{i=1}^{n} p_i x_i} - 1$$

Hence

$$\prod_{i=1}^{n}\left(\frac{1}{x_i^t} - 1\right)^{\frac{p_i}{P_n}} \geqslant \left[\left(\frac{P_n}{\sum\limits_{i=1}^{n} p_i x_i} - 1\right) + 1\right]^t - 1 = \left(\frac{P_n}{\sum\limits_{i=1}^{n} p_i x_i}\right)^t - 1$$

with equality holding if and only if $x_1 = x_2 = \cdots = x_n$.

Similarly, using Theorem C, we can obtain

Theorem 2 Let $x_i > 0$, $p_i > 0$, $i = 1, 2, \cdots, n$, $P_n = \sum\limits_{i=1}^{n} p_i$. Then

$$\prod_{i=1}^{n} \left(\frac{1}{x_i^t} + 1 \right)^{\frac{p_i}{P_n}} \geqslant \left(\frac{P_n}{\sum\limits_{i=1}^{n} p_i x_i} \right)^t + 1 \quad (t > 0) \tag{17}$$

with equality holding if and only if $x_1 = x_2 = \cdots = x_n$.

Corollary 1 Let $x_i > 0, i = 1, 2, \cdots, n, n \geqslant 2, A_n = \frac{1}{n} \sum\limits_{i=1}^{n} x_i$, and let $x_i + x_j \leqslant 1, i \neq j, i, j = 1, 2, \cdots, n$. Then for $t \geqslant 1, \beta > 0$, we have

$$\sum_{1 \leqslant i_1 < \cdots < i_k \leqslant n} \prod_{j=1}^{k} \left(\frac{1}{x_{i_j}^t} - 1 \right)^{\beta} \geqslant C_n^k \left(\frac{1}{A_n^t} - 1 \right)^{k\beta} \quad (k = 1, \cdots, n) \tag{18}$$

with equality holding if and only if $x_1 = x_2 = \cdots = x_n$.

Proof Taking $p_i = 1, i = 1, 2, \cdots, n$, from Theorem 1 we have

$$\prod_{i=1}^{n} \left(\frac{1}{x_i^t} - 1 \right) \geqslant \left(\frac{1}{A_n^t} - 1 \right)^n$$

Then by the Maclaurin's inequality [1], we have

$$\sum_{1 \leqslant i_1 < \cdots < i_k \leqslant n} \prod_{j=1}^{k} \left(\frac{1}{x_{i_j}^t} - 1 \right)^{\beta} \geqslant C_n^k \left[\prod_{i=1}^{n} \left(\frac{1}{x_i^t} - 1 \right)^{\frac{\beta}{n}} \right]^k \geqslant C_n^k \left(\frac{1}{A_n^t} - 1 \right)^{k\beta}$$

with equality holding if and only if $x_1 = x_2 = \cdots = x_n$. This completes the proof of Corollary 1.

Similarly, from Theorem 2 and the Maclaurin's inequality, we have

Corollary 2 Let $x_i > 0, i = 1, 2, \cdots, n, A_n = \frac{1}{n} \sum\limits_{i=1}^{n} x_i$. Then for $t > 0, \beta > 0$, we have

$$\sum_{1 \leqslant i_1 < \cdots < i_k \leqslant n} \prod_{j=1}^{k} \left(\frac{1}{x_{i_j}^t} + 1 \right)^{\beta} \geqslant C_n^k \left(\frac{1}{A_n^t} + 1 \right)^{k\beta} \quad (k = 1, \cdots, n) \tag{19}$$

with equality holding if and only if $x_1 = x_2 = \cdots = x_n$.

Corollary 3 Let $0 < x_i < 1, i = 1, 2, \cdots, n, A_n = \frac{1}{n} \sum\limits_{i=1}^{n} x_i$. Then for

$\beta > 0$, we have

$$\sum_{1 \leqslant i_1 < \cdots < i_k \leqslant n} \prod_{j=1}^{k} \left(\frac{1 + x_{i_j}}{1 - x_{i_j}}\right)^{\beta} \geqslant C_n^k \left(\frac{1 + A_n}{1 - A_n}\right)^{k\beta} \tag{20}$$

with equality holding if and only if $x_1 = x_2 = \cdots = x_n$.

Proof   Notice that

$$\frac{1 + x}{1 - x} = \frac{1}{(1 - x)/2} - 1, \quad \frac{1}{n}\sum_{i=1}^{n} \frac{1}{2}(1 - x_i) = \frac{1}{2}(1 - A_n)$$

since

$$\frac{1}{2}(1 - x_i) > 0, \quad \frac{1}{2}(1 - x_j) > 0$$

and

$$x_i + x_j \geqslant 0 \Leftrightarrow \frac{1}{2}(1 - x_i) + \frac{1}{2}(1 - x_j) \leqslant 1, \ i, j = 1, 2, \cdots, n, \quad i \neq j$$

From Corollary 1, we get that

$$\sum_{1 \leqslant i_1 < \cdots < i_k \leqslant n} \prod_{j=1}^{k} \left(\frac{1 + x_{i_j}}{1 - x_{i_j}}\right)^{\beta} = \sum_{1 \leqslant i_1 < \cdots < i_k \leqslant n} \prod_{j=1}^{k} \left(\frac{1}{(1 - x_{i_j})/2} - 1\right)^{\beta}$$

$$\geqslant C_n^k \left(\frac{2}{(1 - A_n)} - 1\right)^{k\beta} = C_n^k \left(\frac{1 + A_n}{1 - A_n}\right)^{k\beta}$$

with equality holding if and only if $\frac{1}{2}(1 - x_1) = \frac{1}{2}(1 - x_2) = \cdots = \frac{1}{2}(1 - x_n)$, i.e. $x_1 = x_2 = \cdots = x_n$. This completes the proof of Corollary 3.

25.3   Applications

Theorem 3 In isosceles tetrahedron $ABCD$, if γ_1, γ_2, γ_3 are dihedral angles between $\triangle ABC$, $\triangle ABD$, $\triangle ACD$ and the base $\triangle BCD$ respectively, $\beta > 0$. Then

$$\tan^{2\beta} \gamma_1 + \tan^{2\beta} \gamma_2 + \tan^{2\beta} \gamma_3 \geqslant 3 \times 8^{\beta} \tag{21}$$

$$\tan^{2\beta} \gamma_1 \tan^{2\beta} \gamma_2 + \tan^{2\beta} \gamma_2 \tan^{2\beta} \gamma_3 + \tan^{2\beta} \gamma_3 \tan^{2\beta} \gamma_1 \geqslant 3 \times 8^{2\beta} \tag{22}$$

$$\tan \gamma_1 \tan \gamma_2 \tan \gamma_3 \geqslant 16\sqrt{2} \tag{23}$$

with equalities holding in (21)-(23) if and only if $\gamma_1 = \gamma_2 = \gamma_3$.

Proof Theorem 4 in [6] showed that

$$\cos \gamma_1 + \cos \gamma_2 + \cos \gamma_3 = 1$$

Let

$$x_1 = \cos \gamma_1, x_2 = \cos \gamma_2, x_3 = \cos \gamma_3$$

It is clear that $\gamma_1, \gamma_2, \gamma_3$ are all acute angles. So that x_1, x_2, x_3 are all greater than zero. Taking $t = 2$, and let $k = 1, 2, 3$ respectively, from Corollary 1, we obtain (21)-(23).

In general, we have

Theorem 4 Let $0 < \gamma_i < \frac{\pi}{2}, i = 1, 2, \cdots, n$, and let $\sum\limits_{i=1}^{n} \cos \gamma_i = 1$, $\beta > 0$. Then

$$\sum_{1 \leqslant i_1 < \cdots < i_k \leqslant n} \prod_{j=1}^{k} \tan^{2\beta} \gamma_{i_j} \geqslant C_n^k \left(n^2 - 1\right)^{k\beta} \tag{24}$$

with equality holding if and only if $\gamma_1 = \gamma_2 = \cdots = \gamma_n$.

Theorem 5 Let A be an n-dimensional simplex in n-dimensional Euclidean space $E^n (n \geqslant 2)$ and $\{A_1, A_2, \cdots, A_{n+1}\}$ is the set of vertices. Let P be an arbitrary point in the interior of A. If B_i is the intersection point of the extension line of $A_i P$ and the n-1-dimensional hyperplane opposite to the point A_i and write

$$|A_i P| = R_i, |P B_i| = r_i \ (1 \leqslant i \leqslant n + 1)$$

then for $\beta > 0$ and $m \in \mathbf{N}$, we have

$$\sum_{1 \leqslant i_1 < \cdots < i_k \leqslant n+1} \prod_{j=1}^{k} \left(\sum_{l=1}^{m} C_m^l \left(\frac{R_{i_j}}{r_{i_j}} \right)^l \right)^{\beta} \geqslant C_{n+1}^k \left((n+1)^m - 1 \right)^{k\beta} \tag{25}$$

with equality holding if and only if when P is centroid of the simplex A.

$$\sum_{1 \leqslant i_1 < \cdots < i_k \leqslant n+1} \prod_{j=1}^{k} \left[1 + 2 \left(\frac{r_{i_j}}{R_{i_j}} \right) \right]^{\beta} \geqslant C_{n+1}^k \left(1 + \frac{2}{n} \right)^{k\beta} \tag{26}$$

Proof Suppose that the barycentric coordinates of the point P are

$P(\lambda_1, \lambda_2, \cdots, \lambda_{n+1})$. In [7] one showed that $\lambda_i = \frac{PB_i}{A_i B_i} = \frac{r_i}{R_i + r_i}$, and then

$$\frac{1}{\lambda_i} - 1 = \frac{R_i}{r_i}, \quad \frac{1 + \lambda_i}{1 - \lambda_i} = 1 + \frac{2r_i}{R_i}$$

Since

$$\frac{1}{\lambda_i^m} = \left(\left(\frac{1}{\lambda_i} - 1\right) + 1\right)^m = \sum_{i=0}^{m} C_m^i \left(\frac{1}{\lambda_i} - 1\right)^i = 1 + \sum_{i=1}^{m} C_m^i \left(\frac{1}{\lambda_i} - 1\right)^i$$

therefore

$$\frac{1}{\lambda_i^m} - 1 = \sum_{i=1}^{m} C_m^i \left(\frac{1}{\lambda_i} - 1\right)^i = \sum_{i=1}^{m} C_m^i \left(\frac{R_i}{r_i}\right)^i$$

From the above equality and Corollary 1, and notice that $\sum_{i=1}^{n} \lambda_i = 1$, we have

$$\sum_{1 \leqslant i_1 < \cdots < i_k \leqslant n+1} \prod_{j=1}^{k} \left(\sum_{l=1}^{m} C_m^l \left(\frac{R_{i_j}}{r_{i_j}}\right)^l\right)^\beta = \sum_{1 \leqslant i_1 < \cdots < i_k \leqslant n+1} \prod_{j=1}^{k} \left(\frac{1}{\lambda_{i_j}^m} - 1\right)^\beta$$
$$\geqslant C_{n+1}^k \left((n+1)^m - 1\right)^{k\beta}$$

i.e., (25) holds. It is clear that equality holds if and only if when P is centroid of the simplex A.

Moreover, from Corollary 3, we can obtain

$$\sum_{1 \leqslant i_1 < \cdots < i_k \leqslant n+1} \prod_{j=1}^{k} \left(1 + \frac{2r_{i_j}}{R_{i_j}}\right)^\beta = \sum_{1 \leqslant i_1 < \cdots < i_k \leqslant n} \prod_{j=1}^{k} \left(\frac{1 + \lambda_{i_j}}{1 - \lambda_{i_j}}\right)^\beta$$
$$\geqslant C_{n+1}^k \left(\frac{n+2}{n}\right)^{k\beta}$$

i.e. (26) holds. This completes the proof of Theorem 3.

In the final, the author supposes that(1) also has another kind of exponential generalization as follows:

Let $x_i > 0, i = 1, \cdots, n$, $n \geqslant 2$, and $\sum_{i=1}^{n} x_i = 1$. Then for $k = 1, 2, \cdots, n$ and $\beta \geqslant 1$, we have

$$\frac{\displaystyle\sum_{1 \leqslant i_1 < \cdots < i_k \leqslant n} \prod_{j=1}^{k} \left(1 - x_{i_j}^\beta\right)}{\displaystyle\sum_{1 \leqslant i_1 < \cdots < i_k \leqslant n} \prod_{j=1}^{k} x_{i_j}^\beta} \geqslant \left(n^\beta - 1\right)^k \tag{27}$$

Acknowledgements The author is indebted to professor Xu Tie-quan,
Liu zheng, Zhang Han-fang professor and referee for their many helpful and
valuable comments and suggestions.

References

[1]　KUANG JI-CHANG. Applied Inequalities (Changyong budengshi)[M].
3rd ed. Jinan: Shandong Press of science and technology, 2002.
(Chinese).

[2]　HUANG ZHI-XING. A new function inequality [J]. J. Fuzhou Teachers
College, 2000,19 (2): 20-24.

[3]　MINTRINVIC D S . Analytic Inequalities[M]. Berlin: Springer-Verlag,
1970.

[4]　HUAN-NAN SHI. Refinement and Generalization of a Class of Inequalities for symmetric function[J]. Chinese Math. in Practice and Theory,
1999, 29(4): 81-84.

[5]　TIE-QUAN XU. Several inequalities for the symmetric function [J]. J.
Chengdu University(Natural Science), 2000, 19 (2): 25-18.

[6]　HUA-MING SU. Some properties of the isosceles tetrahedron and the
right angle tetrahedron [J]. Hunan Annals of Mathematics, 1995, 15
(4): 20-28.

[7]　ZHANG HAN-FANG. Geometry inequalities guide [M]. Beijing: China
science and Culture Press, 2003.

第26篇　An Alternative Note on Schur-Convexity of Extended Mean Values

(HUAN-NAN SHI, SHAN-HE WU, FENG QI. Mathematical Inequal-

ities and Applications, 2006, 9 (2): 219-224)

Abstract:　The Schur-convex and Schur-concave properties with (x, y) in $(0, +\infty) \times (0, +\infty)$ for fixed (r, s) of the extended mean values $E(r, s; x, y)$ are researched again, some errors in (F. Qi, J. Sándor, S. S. Dragomir, and A. Sofo, *Notes on the Schur-convexity of the extended mean values*, Taiwanese J. Math. **9** (2005), no. 3, in press. RGMIA Res. Rep. Coll. **5** (2002), no. 1, Art. 3, 19–27. Available online at `http://rgmia.vu.edu.au/v5n1.html`.) are corrected.

Keywords:　Schur-convexity; extended mean values

26.1　Introduction

Let $\boldsymbol{x} = (x_0, \cdots, x_n)$ and $\boldsymbol{y} = (y_0, \cdots, y_n)$ denote two real $(n + 1)$-tuples. $\boldsymbol{x}$ is said to be not greater than $\boldsymbol{y}$ (in symbols, $\boldsymbol{x} \leqslant \boldsymbol{y}$) if $x_i \leqslant y_i$ for $0 \leqslant i \leqslant n + 1$. $\boldsymbol{x}$ is said to majorize $\boldsymbol{y}$ (in symbols, $\boldsymbol{x} \succ \boldsymbol{y}$) if

$$\sum_{i=0}^{k} x_{[i]} \geqslant \sum_{i=0}^{k} y_{[i]} \tag{1}$$

for $k = 0, 1, \cdots, n - 1$ and

$$\sum_{i=0}^{n} x_i = \sum_{i=0}^{n} y_i \tag{2}$$

154

where

$$x_{[0]} \geqslant x_{[1]} \geqslant \cdots \geqslant x_{[n]} \tag{3}$$

and

$$y_{[0]} \geqslant y_{[1]} \geqslant \cdots \geqslant y_{[n]} \tag{4}$$

are the decreasingly ordered components of $\boldsymbol{x}$ and $\boldsymbol{y}$. See [10].

A function $\psi : \mathbf{R}^{n+1} \to \mathbf{R}$ is said to be increasing if $\boldsymbol{x} \leqslant \boldsymbol{y}$ implies $\psi(\boldsymbol{x}) \leqslant \psi(\boldsymbol{y})$. It has been proved in [21, Proposition 4.3] that the function $\psi(\boldsymbol{x})$ is increasing if and only if $\nabla\psi(\boldsymbol{x}) \geqslant 0$ for $\boldsymbol{x} \in A$, where $A \subset \mathbf{R}^{n+1}$ is an open set, $\psi : A \to \mathbf{R}$ is differentiable, and

$$\nabla\psi(\boldsymbol{x}) = \left(\frac{\partial\psi(\boldsymbol{x})}{\partial x_0}, \ldots, \frac{\partial\psi(\boldsymbol{x})}{\partial x_n} \right) \in \mathbf{R}^{n+1} \tag{5}$$

A function $g : \mathbf{R}^{n+1} \to \mathbf{R}$ is said to be Schur-convex if $\boldsymbol{x} \succ \boldsymbol{y}$ implies $g(\boldsymbol{x}) \geqslant g(\boldsymbol{y})$. A function f is Schur-concave if and only if $-f$ is Schur-convex. See [10].

The extended mean values $E(r, s; x, y)$ were defined in [20] by

$$E(r, s; x, y) = \left[\frac{r}{s} \cdot \frac{y^s - x^s}{y^r - x^r} \right]^{1/(s-r)}, rs(r-s)(x-y) \neq 0 \tag{6}$$

$$E(r, 0; x, y) = \left[\frac{1}{r} \cdot \frac{y^r - x^r}{\ln y - \ln x} \right]^{1/r}, r(x-y) \neq 0 \tag{7}$$

$$E(r, r; x, y) = \frac{1}{e^{1/r}} \left[\frac{x^{x^r}}{y^{y^r}} \right]^{1/(x^r - y^r)}, r(x-y) \neq 0 \tag{8}$$

$$E(0, 0; x, y) = \sqrt{xy}, x \neq y \tag{9}$$

$$E(r, s; x, x) = x, x = y$$

where $x, y > 0$ and $(r, s) \in \mathbf{R}^2$.

Leach and Sholander[6] showed that $E(r, s; x, y)$ are increasing with both r and s, or with both x and y. Later, the monotonicity of the extended mean values E was also researched in [2,3,5,16,18,19] by using different ideas and simpler approaches.

Leach and Sholander[7] and Páles [9] respectively solved the problem of comparison of E. They found necessary and sufficient conditions for the parameters r, s and u, v in order that $E(r, s; x, y) \leqslant E(u, v; x, y)$ be satisfied for all positive x and y.

The logarithmic convexity of the extended mean values $E(r, s; x, y)$ with two parameters r and s was obtained in [12,13].

The Schur-convexities of the extended mean values $E(r, s; x, y)$ with (r, s) and (x, y) were presented in [11,14,17] as follows.

Theorem A[11] For fixed (x, y) with $x > 0$, $y > 0$ and $x \neq y$, the extended mean values $E(r, s; x, y)$ are Schur-concave on $[0, +\infty) \times [0, +\infty)$ and Schur-convex on $(-\infty, 0] \times (-\infty, 0]$ with (r, s).

Theorem B[17] For given (r, s) with $r, s \notin (0, \frac{3}{2})$ (or $r, s \in (0, 1]$, resp.), the extended mean values $E(r, s; x, y)$ are Schur-concave (or Schur-convex, resp.) with (x, y) on the domain $(0, +\infty) \times (0, +\infty)$.

For more information on the extended mean values E, please refer to [1,15] and the references therein.

The following two counterexamples of Theorem B tell us that there must exist some errors about the proof of the above Theorem B.

Example 1 Let $(r, s) = (4, 2)$. It is clear that $(4, 2) \notin \left(0, \frac{3}{2}\right) \times \left(0, \frac{3}{2}\right)$. For $(2, 2) \succ (1, 3)$, directly calculating yields

$$E(4, 2; 1, 3) = \left(\frac{4}{2} \cdot \frac{3^2 - 1^2}{3^4 - 1^4}\right)^{1/(2-4)} = \sqrt{5} > E(4, 2; 2, 2) = 2$$

This leads to a contradiction with Theorem B.

Example 2 Take $(r, s) = (1, 1)$. It is clear that $(r, s) \in (0, 1] \times (0, 1]$. For $(2, 2) \succ (1, 3)$, straightforward computation gives

$$E(1, 1; 1, 3) = e^{-1/1} \left(\frac{1^{1^1}}{3^{3^1}}\right)^{1/(1^1 - 3^1)} = \frac{3\sqrt{3}}{e} < E(1, 1; 2, 2) = 2$$

This also leads to a contradiction with Theorem B.

These two contradictions motivate us to reconsider to find a new approach to prove the Schur-convexity of the extended mean values $E(r, s; x, y)$ and obtain the following

Theorem 1 For fixed $(r, s) \in \mathbf{R}^2$.

1. If $2 < 2r < s$ or $2 \leqslant 2s \leqslant r$, then the extended mean values $E(r, s; x, y)$ is Schur-convex with $(x, y) \in (0, +\infty) \times (0, +\infty)$,

2. If $(r, s) \in \{r < s \leqslant 2r, 0 < r \leqslant 1\} \cup \{s < r \leqslant 2s, 0 < s \leqslant 1\} \cup \{0 < s < r \leqslant 1\} \cup \{0 < r < s \leqslant 1\} \cup \{s \leqslant 2r < 0\} \cup \{r \leqslant 2s < 0\}$, then the extended mean values $E(r, s; x, y)$ is Schur-concave with $(x, y) \in$

$(0, +\infty) \times (0, +\infty)$.

26.2　Lemmas

In order to verify Theorem 1, the following lemmas are necessary.

Lemma 1 [21] Let $g : I \to \mathbf{R}$, $\phi : \mathbf{R}^n \to \mathbf{R}$ and $\psi(x) = \phi(g(x_1), \ldots, g(x_n))$.

1. If g is convex (concave) and ϕ is increasing and Schur-convex (Schur-concave), then ψ is Schur-convex (Schur-concave);

2. If g is concave (convex) and ϕ is decreasing and Schur-convex (Schur-concave), then ψ is Schur-convex (Schur-concave).

Lemma 2 [21] Let $A \subset \mathbf{R}^n$, $\phi_i : A \to \mathbf{R}$ for $1 \leqslant i \leqslant k$, $h : \mathbf{R}^k \to \mathbf{R}$ and $\psi(x) = h(\phi_1(x), \ldots, \phi_k(x))$.

1. If each of ϕ_i for $1 \leqslant i \leqslant k$ is Schur-convex and h is increasing (decreasing), then ψ is Schur-convex (Schur-concave);

2. If each of ϕ_i for $1 \leqslant i \leqslant k$ is Schur-concave and h is increasing (decreasing), then ψ is Schur-concave (Schur-convex).

Remark 1　These two lemmas can also be found in [8] and [10].

Lemma 3 [17] Let f be a continuous function and p a positive continuous weight on I. Then the weighted arithmetic mean of function indexweighted arithmetic mean of function f with weight p defined by

$$F(x, y) = \begin{cases} \dfrac{\int_x^y p(t) f(t) \, \mathrm{d}t}{\int_x^y p(t) \, \mathrm{d}t}, & x \neq y \\ f(x), & x = y \end{cases} \tag{10}$$

is Schur-convex (Schur-concave) on I^2 if and only if inequality

$$\frac{\int_x^y p(t) f(t) \, \mathrm{d}t}{\int_x^y p(t) \, \mathrm{d}t} \leqslant \frac{p(x) f(x) + p(y) f(y)}{p(x) + p(y)} \tag{11}$$

holds (reverses) for $(x, y) \in I^2$.

Remark 2　A corresponding result for the special case $p \equiv 1$ of Lemma 3 has been showed in [1, Theorem 21, p. 384] and [4].

Lemma 4 Let f be a continuous function and p a positive continuous weight on I. Then the function $F(x, y)$ defined by (10) is increasing (decreasing) on I^2 if f is an increasing (decreasing) function on I.

Proof Direct calculation yields

$$\frac{\partial F(x, y)}{\partial y} = -\frac{p(y)}{\left[\int_x^y p(t)\, \mathrm{d}t\right]^2} \int_x^y p(t) f(t)\, \mathrm{d}t + \frac{p(y) f(y)}{\int_x^y p(t)\, \mathrm{d}t}$$

$$= \frac{p(y)}{\left[\int_x^y p(t)\, \mathrm{d}t\right]^2} \int_x^y p(t)\left[f(y) - f(t)\right] \mathrm{d}t \tag{12}$$

Hence, if f is increasing (decreasing), then $\frac{\partial F(x,y)}{\partial y}$ is positive (negative), this means that the function $F(x, y)$ is increasing (decreasing) with $y \in I$.

Since $F(x, y) = F(y, x)$ is symmetric, if f is increasing (decreasing), then $F(x, y)$ is also increasing (decreasing) with $x \in I$. The proof is complete.

Remark 3 A special case of Lemma 4 for $p \equiv 1$ has been proved in [1, Theorem 5] and [19].

26.3　Proof of Theorem 1

The extended mean values $E(r, s; x, y)$ can be expressed for $r(s - r) \neq 0$ as

$$E(r, s; x, y) = \left(\frac{1}{y^r - x^r} \int_{x^r}^{y^r} t^{s/r - 1}\, \mathrm{d}t\right)^{1/(s-r)} \triangleq \left[\phi(x^r, y^r)\right]^{1/(s-r)} \tag{13}$$

In $(0, \infty)$, the function $f(t) = t^{s/r - 1}$ is increasing and convex if $\frac{s}{r} - 1 \geqslant 1$, decreasing and convex if $\frac{s}{r} - 1 \leqslant 0$, and increasing and concave if $0 < \frac{s}{r} - 1 \leqslant 1$. Utilizing Lemma 3 for a special case $p \equiv 1$ and Lemma 4, it follows that the function $\phi(x, y)$ defined in (13) is increasing and Schur-convex in $(0, \infty) \times (0, \infty)$ if $0 < 2r \leqslant s$ or $s \leqslant 2r < 0$, decreasing and Schur-convex in $(0, \infty) \times (0, \infty)$ if $0 < s < r$ or both $r \leqslant s$ and $r < 0$, and increasing and Schur-concave in $(0, \infty) \times (0, \infty)$ if $r < s \leqslant 2r$ or $0 > r > s \geqslant 2r$.

In $(0, \infty)$, the function $g(t) = t^r$ is increasing and convex if $r \geqslant 1$, decreasing and convex if $r < 0$, and increasing and concave if $r \in (0, 1]$. From Lemma 1, it is deduced that the function

$$\psi(x, y) = \phi(g(x), g(y)) = \phi(x^r, y^r) \tag{14}$$

is increasing and Schur-convex in $(0, +\infty) \times (0, +\infty)$ if $1 \leqslant r \leqslant \frac{s}{2}$, decreasing and Schur-convex in $(0, +\infty) \times (0, +\infty)$ if $0 < s < r \leqslant 1$ or $s \leqslant 2r < 0$, and increasing and Schur-concave in $(0, +\infty) \times (0, +\infty)$ if $r < s \leqslant 2r$ and $0 < r \leqslant 1$.

Further, since the function $h(t) = t^{1/(s-r)}$ is increasing when $s > r$ and decreasing when $s < r$ on $(0, +\infty)$, then from Lemma 2 and formula (13), it is deduced that the function $h\big(\psi(x, y)\big) = E(r, s; x, y)$ is increasing and Schur-convex with $(x, y) \in (0, +\infty) \times (0, +\infty)$ if $1 \leqslant r \leqslant \frac{s}{2}$ and increasing and Schur-concave with $(x, y) \in (0, +\infty) \times (0, +\infty)$ if $\{r < s \leqslant 2r, 0 < r \leqslant 1\} \cup \{0 < s < r \leqslant 1\} \cup \{s \leqslant 2r < 0\}$.

Since the extended mean values $E(r, s; x, y) = E(s, r; x, y)$, then $E(r, s; x, y)$ is also increasing and Schur-convex with $(x, y) \in (0, +\infty) \times (0, +\infty)$ if $1 \leqslant s \leqslant \frac{r}{2}$ and increasing and Schur-concave with $(x, y) \in (0, +\infty) \times (0, +\infty)$ if $\{s < r \leqslant 2s, 0 < s \leqslant 1\} \cup \{0 < r < s \leqslant 1\} \cup \{r \leqslant 2s < 0\}$.

If $r = 0$ and $s \neq 0$, the extended mean values $E(r, s; x, y)$ can be rewritten as

$$E(0, s; x, y) = \left[\frac{1}{\ln y - \ln x} \int_{\ln x}^{\ln y} t^{st} \, \mathrm{d}\, t \right]^{1/s} \triangleq \big[\theta(\ln x, \ln y) \big]^{1/s}$$

Since the function $f(t) = \mathrm{e}^{st}$ is decreasing and convex in $(0, +\infty)$ for $s < 0$, from Lemma 3 and Lemma 4, it follows that the function $\theta(x, y)$ is decreasing and Schur-convex in $(0, +\infty) \times (0, +\infty)$ for $s < 0$. Furthermore, since $g(t) = \ln t$ is increasing and convex in $(0, +\infty)$, by Lemma 1, it is found that $\theta(\ln x, \ln y)$ for $s < 0$ is increasing and Schur-convex in $(0, +\infty)$. Since the function $h(t) = \frac{1}{t}$ is decreasing for $t > 0$, by Lemma 2, it is deduced that $h[\psi(x, y)] = E(0, s; x, y)$ is increasing and Schur-concave in $(x, y) \in (0, +\infty) \times (0, +\infty)$. The proof is complete.

References

[1]　P S BULLEN. Handbook of Means and Their Inequalities[M]//Mathematics and its Applications 560. Dordrecht: Kluwer Academic Publishers, 2003.

[2]　CH-P CHEN, F QI. An alternative proof of monotonicity for the extended mean values[J/OL]. Austral. J. Math. Anal. Appl., 2004, 1 (2): Art. 11. http://ajmaa.org/cgi-bin/paper.pl?string=v1n2/V1I2P11.tex.

[3]　SU-LING ZHANG, CH-P CHEN, F QI. Another proof of monotonicity

for the extended mean values[J]. Tamkang J. Math., 2006, 37 (3): 207-209.

[4] N ELEZOVIĆ, J PEČARIĆ. A note on Schur-convex functions[J]. Rocky Mountain J. Math., 2000, 30 (3): 853–856.

[5] B-N GUO, SH-Q ZHANG, F QI. Elementary proofs of monotonicity for extended mean values of some functions with two parameters[J]. Shùxué de Shíjiàn yǔ Rènshi (Math. Practice Theory), 1999, 29 (2): 169–174. (Chinese)

[6] E B LEACH, M C SHOLANDER. Extended mean values[J]. Amer. Math. Monthly, 1978, 85 : 84–90.

[7] E LEACH, M SHOLANDER. Extended mean values II[J]. J. Math. Anal. Appl., 1983, 92: 207–223.

[8] A W MAESHALL, I OLKIN. Inequalities: Theory of Majorization and its Applications[M]. New York: Academic Press, 1979.

[9] Z PÁLES. Inequalities for differences of powers[J]. J. Math. Anal. Appl., 1988, 131: 271–281.

[10] J PEČARIĆ, F PROSCHAN, Y L TONG. Convex Functions, Partial Orderings, and Statistical Applications[M]. Mathematics in Science and Engineering 187, Academic Press, 1992.

[11] F QI. A note on Schur-convexity of extended mean values[J]. Rocky Mountain J. Math., 2005 35(5): 1787 – 1793.

[12] F QI. Logarithmic convexity of extended mean values[J]. Proc. Amer. Math. Soc., 2002, 130(6) : 1787 – 1796.

[13] F QI. Logarithmic convexities of the extended mean values[J]. RGMIA Res. Rep. Coll., 1999, 2(5): 643 – 652.

[14] F QI. Schur-convexity of the extended mean values[J]. RGMIA Res. Rep. Coll., 2001, 4 (4): 529 – 533.

[15] F QI. The extended mean values: definition, properties, monotonicities, comparison, convexities, generalizations, and applications[J]. Cubo Mat. Educ., 2003, 5 (3): 63 – 90. RGMIA Res. Rep. Coll., 2002, 5 (1): 57 – 80.

[16] F QI, Q-M LUO. A simple proof of monotonicity for extended mean values[J]. J. Math. Anal. Appl., 1998, 224 (2): 356–359.

[17] F QI, J SÁNDOR, S S DRAGOMIR, A SOFO. Notes on the Schur-convexity of the extended mean values[J]. Taiwanese J. Math., 2005,9 (3):411-420.

[18] F QI S-L XU. The function $(b^x - a^x)/x$: Inequalities and properties[J]. Proc. Amer. Math. Soc., 1998, 126 (11): 3355–3359.

[19] F QI, S-L XU, L DEBNATH. A new proof of monotonicity for extended mean values[J]. Internat. J. Math. Math. Sci., 1999, 22 (2):415–420.

[20] K B STOLARSKY. Generalizations of the logarithmic mean[J]. Mag. Math., 1975, 48: 87–92.

[21] B-Y WANG. Kongzhi Budengshi Jichu[M]. Beijing: Beijing Normal University Press, 1990.

第27篇　Refinements of an Inequality for the Rational Fraction

(HUAN-NAN SHI. Pure and Applied Mathematics, 2006, 22(2): 256-262)

Abstract: By using methods on the theory of majorization , two refinements of an inequality for the rational fraction are established. Comparison between these two refinements is considered.

Keywords: Rational Fraction; inequality; majorization.

27.1　Introduction

Let $x \geqslant 0, m > n$ and $P_n(x) = \sum\limits_{k=0}^{n} x^k$ and let $f(x) = \frac{P_m(x)}{P_n(x)}, g(x) = f(x)x^{n-m}$. Then

$$1 < g(x) < \frac{m+1}{n+1} < f(x) \quad (1 < x < \infty) \tag{1}$$

$$\max\left\{1, x^{m-n}\frac{m+1}{n+1}\right\} < f(x) < \frac{m+1}{n+1} < g(x) \quad (0 < x < 1) \tag{2}$$

The aim of this paper is to establish two refinements of (1) and (2),i.e. the following Theorem 1 and Theorem 2, by using methods on the theory of majorization. Also, comparison between these two refinements is given.

Theorem 1 Let $x \geqslant 0, m > n$ and $P_n(x) = \sum\limits_{k=0}^{n} x^k$ and let

$$f(x) = \frac{P_m(x)}{P_n(x)}, g(x) = f(x)x^{n-m}$$

Then

$$1 < g(x) < \frac{m+1}{n+1} - \frac{x^{n-m}\left(1 - x^{-\frac{1}{2}(m+1)(n+1)(m-n)}\right)}{(n+1)P_n(x)} < \frac{m+1}{n+1}$$

$$< \frac{m+1}{n+1} + \frac{1 - x^{-\frac{1}{2}(m+1)(n+1)(m-n)}}{(n+1)P_n(x)} < f(x) \quad (1 < x < \infty) \qquad (3)$$

$$\max\left\{1, x^{m-n}\frac{m+1}{n+1}\right\} < f(x) < \frac{m+1}{n+1} - \frac{x^m\left(1 - x^{-\frac{1}{2}(m+1)(n+1)(m-n)}\right)}{(n+1)P_n(x)}$$

$$< \frac{m+1}{n+1} < \frac{m+1}{n+1} + \frac{x^n\left(1 - x^{-\frac{1}{2}(m+1)(n+1)(m-n)}\right)}{(n+1)P_n(x)} < g(x) \quad (0 < x < 1) \quad (4)$$

Theorem 2 Let $x \geqslant 0, m > n, P_n(x) = \sum\limits_{k=0}^{n} x^k$ and let

$$f(x) = \frac{P_m(x)}{P_n(x)}, g(x) = f(x)x^{n-m}$$

Then

$$1 < g(x) < \frac{m+1}{n+1} - \frac{(m+1)\left(x^{\frac{m-n}{2}} - 1\right)}{x^m P_n(x)} < \frac{m+1}{n+1}$$

$$< \frac{m+1}{n+1} + \frac{(m+1)\left(x^{\frac{m-n}{2}} - 1\right)}{P_n(x)} < f(x) \quad (1 < x < \infty) \qquad (5)$$

$$\max\left\{1, x^{m-n}\frac{m+1}{n+1}\right\} < f(x) < \frac{m+1}{n+1} - \frac{(m+1)x^{\frac{m+3n}{2}}\left(1 - x^{\frac{m-n}{2}}\right)}{P_n(x)} < \frac{m+1}{n+1}$$

$$< \frac{m+1}{n+1} + \frac{(m+1)x^{\frac{3n-m}{2}}\left(1 - x^{\frac{m-n}{2}}\right)}{P_n(x)} < g(x) \quad (0 < x < 1) \qquad (6)$$

27.2　Definitions and Lemmas

Definition[2]　 Let $\boldsymbol{x} = (x_1, \ldots, x_n)$ and $\boldsymbol{y} = (y_1, \ldots, y_n)$ be two real n-tuples. Then $\boldsymbol{x}$ is said to be majorized by $\boldsymbol{y}$ (in symbols $\boldsymbol{x} \prec \boldsymbol{y}$) if

$$(\text{i}) \sum_{i=1}^{k} x_{[i]} \leqslant \sum_{i=1}^{k} y_{[i]} \ for \ k = 1, 2, \ldots, n-1; \ (\text{ii}) \sum_{i=1}^{n} x_i = \sum_{i=1}^{n} y_i.$$

where $x_{[1]} \geqslant x_{[2]} \cdots \geqslant x_{[n]}$ and $y_{[1]} \geqslant y_{[2]} \cdots \geqslant y_{[n]}$ are components of $\boldsymbol{x}$ and $\boldsymbol{y}$ rearranged in descending order, $\boldsymbol{x}$ is said to be strictly majorized by $\boldsymbol{y}$

(written $\boldsymbol{x} \prec\prec \boldsymbol{y}$) if $\boldsymbol{x}$ is not a permutation of $\boldsymbol{y}$. And $\boldsymbol{x}$ is said to be weakly majorized by $\boldsymbol{y}$(written $\boldsymbol{x} \prec_w \boldsymbol{y}$) if

$$\sum_{i=1}^{k} x_{[i]} \leqslant \sum_{i=1}^{k} y_{[i]} \; for \; k = 1, 2, \ldots, n$$

Lemma 1 [2] For $\boldsymbol{x}, \boldsymbol{y} \in \mathbf{R}^n$, let $x_{n+1} = \min\{x_1, \cdots, x_n\}, y_{n+1} = \sum_{i=1}^{k} x_i - \sum_{i=1}^{n} y_i$. If $\boldsymbol{x} \prec_w \boldsymbol{y}$, then

$$(\boldsymbol{x}, x_{n+1}) \prec (\boldsymbol{y}, y_{n+1})$$

Lemma 2 [2] For $\boldsymbol{x} \in \mathbf{R}_+^n, \boldsymbol{y} \in \mathbf{R}^n$, let $W = \sum_{i=1}^{n} (y_i - x_i)$. If $\boldsymbol{x} \prec_w \boldsymbol{y}$, then

$$\left(\boldsymbol{x}, \underbrace{\frac{W}{n}, \cdots, \frac{W}{n}}_{n} \right) \prec \left(\boldsymbol{y}, \underbrace{0, \cdots, 0}_{n} \right)$$

Lemma 3 [2] Let interval $I \subset \mathbf{R}, \boldsymbol{x}, \boldsymbol{y} \in I^n \subset \mathbf{R}^n$. Then

$$\boldsymbol{x} \prec \boldsymbol{y} \Leftrightarrow \sum_{i=1}^{n} g(x_i) < \sum_{i=1}^{n} g(y_i)$$

for all strictly convex function $g : I \to \mathbf{R}$.

Lemma 4 [3] Let $\boldsymbol{x}, \boldsymbol{y} \in \mathbf{R}^n, x_1 \geqslant x_2 \geqslant \cdots \geqslant x_n$ and $\sum_{i=1}^{n} x_i = \sum_{i=1}^{n} y_i$. If for some $k, 1 \leqslant k \leqslant n, x_i \leqslant y_i, i = 1, \cdots, k, x_i \geqslant y_i, i = k+1, \cdots, n$, then $\boldsymbol{x} \prec \boldsymbol{y}$.

27.3 Proofs of Theorem

Proof of Theorem 1

We first prove (3). It is clear that we only need to prove the second and fifth inequalities in (3). It is not difficult to verify that

$$\boldsymbol{x} = \left(\underbrace{n, \cdots, n}_{n+1}, \underbrace{n+1, \cdots, n+1}_{n+1}, \ldots, \underbrace{n+m, \cdots, n+m}_{n+1} \right)$$

$$\prec_w \left(\underbrace{m, \cdots, m}_{m+1}, \underbrace{m+1, \cdots, m+1}_{m+1}, \cdots, \underbrace{m+n, \cdots, m+n}_{m+1} \right) = \boldsymbol{y} \quad (7)$$

In fact, since from $m > n$ we have $x_{[i]} \leqslant y_{[i]}, i = 1, \ldots, (m+1)(n+1)$, and so (7) holds.

For (7), let

$$y_{(m+1)(n+1)+1} = \sum_{i=1}^{(m+1)(n+1)+1} x_i - \sum_{i=1}^{(m+1)(n+1)} y_i = n - \frac{1}{2}(m+1)(n+1)(m-n)$$

and

$$x_{(m+1)(n+1)+1} = \min\{x_1, \cdots, x_{(m+1)(n+1)}\} = n$$

Then from Lemma 1 we have

$$\left(n, \underbrace{n, \cdots, n}_{n+1}, \underbrace{n+1, \cdots, n+1}_{n+1}, \ldots, \underbrace{n+m, \cdots, n+m}_{n+1} \right) \prec\prec$$

$$\left(\underbrace{m, \cdots, m}_{m+1}, \underbrace{m+1, \cdots, m+1}_{m+1}, \ldots, \underbrace{m+n, \cdots, m+n}_{m+1}, n - \frac{1}{2}(m+1)(n+1)(m-n) \right) \tag{8}$$

since the left side of (8) is not a permutation of the right side.

Let $h(t) = x^t$. When $x \neq 1, x > 0$, $\frac{d^2 h(t)}{dt^2} = x^t (\log x)^2 > 0$. So that $h(t)$ is a strictly convex function on $\mathbf{R}$. Hence by Lemma 3 and from (8) we have

$$x^n + (n+1)P_m(x)x^n < (m+1)x^m P_n(x) + x^{n-\frac{1}{2}(m+1)(n+1)(m-n)}$$

Dividing both sides in the above inequality by $(n+1)x^m P_n(x)$, we get

$$\frac{x^n}{(n+1)x^m P_n(x)} + g(x) < \frac{m+1}{n+1} + \frac{x^{n-\frac{1}{2}(m+1)(n+1)(m-n)}}{(n+1)x^m P_n(x)}$$

i. e.

$$g(x) < \frac{m+1}{n+1} - \frac{x^{n-m}\left(1 - x^{-\frac{1}{2}(m+1)(n+1)(m-n)}\right)}{(n+1)P_n(x)}$$

Thus the second inequality in (3) holds.

Similarly, it is not difficult to verify that

$$\left(\underbrace{0, \cdots, 0}_{m+1}, \underbrace{1, \cdots, 1}_{m+1}, \ldots, \underbrace{n, \cdots, n}_{m+1} \right) \prec_w \left(\underbrace{0, \cdots, 0}_{n+1}, \underbrace{1, \cdots, 1}_{n+1}, \ldots, \underbrace{m, \cdots, m}_{n+1} \right) \tag{9}$$

For (9), let

$$y_{(m+1)(n+1)+1} = \sum_{i=1}^{(m+1)(n+1)+1} x_i - \sum_{i=1}^{(m+1)(n+1)} y_i = n - \frac{1}{2}(m+1)(n+1)(m-n)$$

and

$$x_{(m+1)(n+1)+1} = \min\{x_1, \cdots, x_{(m+1)(n+1)}\} = 0$$

Then from Lemma 1 we have

$$\left(0,0,\cdots,0,\underbrace{1,\cdots,1}_{m+1},\cdots,\underbrace{n,\cdots,n}_{m+1} \right)$$
$$\underbrace{}_{m+1}$$
$$\prec\prec \left(\underbrace{0,\cdots,0}_{n+1},\underbrace{1,\cdots,1}_{n+1},\cdots,\underbrace{m,\cdots,m}_{n+1},-\frac{1}{2}(m+1)(n+1)(m-n) \right) \quad (10)$$

So from Lemma 3 we have

$$1 + (m+1)P_n(x) < (n+1)P_m(x) + x^{-\frac{1}{2}(m+1)(n+1)(m-n)}$$

Dividing both sides in the above inequality by $(n+1)P_n(x)$, we get

$$\frac{1}{(n+1)P_n(x)} + \frac{m+1}{n+1} < f(x) + \frac{x^{-\frac{1}{2}(m+1)(n+1)(m-n)}}{(n+1)P_n(x)}$$

i. e.

$$f(x) > \frac{m+1}{n+1} + \frac{1 - x^{-\frac{1}{2}(m+1)(n+1)(m-n)}}{(n+1)P_n(x)}$$

Thus the fifth inequality in (3) holds.

Now we prove(4). It is sufficient to prove the second and fifth inequalities in (4) hold. Notice that when $0 < x < 1$, we have $1 < \frac{1}{x} < \infty$, and $g(x) = f(\frac{1}{x})$. Replacing x by $\frac{1}{x}$ in the second and fifth inequalities of (3), we get the second and fifth inequalities of (4) correspondingly. This completes the proof of Theorem 1.

Proof of Theorem 2 We first prove (5). It is clear that we only need to prove the second and fifth inequalities in (5). For (7), let

$$W = \sum_{i=1}^{(m+1)(n+1)} (y_i - x_i) = \frac{1}{2}(m+1)(n+1)(m-n)$$

Then from Lemma 2 we have

$$
\left(\underbrace{n,n,\cdots,n,}_{n+1}\underbrace{n+1,\cdots,n+1,}_{n+1}\ldots,\underbrace{n+m,\cdots,n+m,}_{n+1}\underbrace{\frac{m-n}{2},\cdots,\frac{m-n}{2}}_{(m+1)(n+1)}\right)
$$

$$
\prec\prec\left(\underbrace{m,\cdots,m,}_{m+1}\underbrace{m+1,\cdots,m+1,}_{m+1}\cdots,\underbrace{m+n,\cdots,m+n,}_{m+1}\underbrace{0,\cdots,0}_{(m+1)(n+1)}\right)
$$

$$(11)$$

Since when $x \neq 1, x > 0, h(t) = x^t$ is a strictly convex function on $\mathbf{R}$, from Lemma 3 we have

$$(n+1)x^n P_m(x) + (n+1)(m+1)x^{\frac{m-n}{2}} < (m+1)x^m P_n(x) + (n+1)(m+1)$$

Dividing both sides in the above inequality by $(n+1)x^n P_m(x)$, we get

$$g(x) + \frac{(m+1)x^{\frac{m-n}{2}}}{x^m P_n(x)} < \frac{m+1}{n+1} + \frac{n+1}{x^m P_n(x)}$$

i. e.

$$g(x) < \frac{m+1}{n+1} - \frac{(m+1)\left(x^{\frac{m-n}{2}} - 1\right)}{x^m P_n(x)}$$

Thus the second inequality in (5) holds.

Similarly, for (7), let

$$W = \sum_{i=1}^{(m+1)(n+1)} (y_i - x_i) = \frac{1}{2}(m+1)(n+1)(m-n)$$

Then from Lemma 2 we have

$$
\left(\underbrace{0,\cdots,0,}_{m+1}\underbrace{1,\cdots,1,}_{m+1}\cdots,\underbrace{n,\cdots,n,}_{m+1}\underbrace{\frac{m-n}{2},\cdots,\frac{m-n}{2}}_{(m+1)(n+1)}\right)
$$

$$
\prec\prec\left(\underbrace{0,\cdots,0,}_{n+1}\underbrace{1,\cdots,1,}_{n+1}\cdots,\underbrace{m,\cdots,m,}_{n+1}\underbrace{0,\cdots,0}_{(m+1)(n+1)}\right)
$$

Hence from Lemma 3 we have

$$(m+1)P_n(x) + (n+1)(m+1)x^{\frac{m-n}{2}} < (n+1)P_m(x) + (m+1)(n+1)$$

Dividing both sides in the above inequality by $(n+1)P_n(x)$, we get

$$\frac{m+1}{n+1} + \frac{(m+1)x^{\frac{m-n}{2}}}{P_n(x)} < f(x) + \frac{m+1}{P_n(x)}$$

i. e.

$$f(x) > \frac{m+1}{n+1} + \frac{(m+1)\left(x^{\frac{m-n}{2}} - 1\right)}{P_n(x)}$$

Thus the fifth inequality in (5) holds.

Now we secondly prove (6). It is sufficient to prove the second and fifth inequalities in (6) holds. Replacing x by $\frac{1}{x}$ in the second and fifth inequalities of (5), we get the second and fifth inequalities of (6) correspondingly. This completes the proof of Theorem 2.

27.4　Comparison between two refinements

Remark 1 When $m \geqslant 3n$, i. e. $n \leqslant \frac{m-n}{2}$, notice that

$$n - \frac{1}{2}(m+1)(n+1)(m-n) \leqslant 0$$

from Lemma 4 we have

$$\left(n, \underbrace{0, \cdots, 0}_{(m+1)(n+1)} \right)$$

$$\prec \left(\underbrace{\frac{m-n}{2}, \cdots, \frac{m-n}{2}}_{(m+1)(n+1)}, n - \frac{1}{2}(m+1)(n+1)(m-n) \right)$$

and then

$$(m+1)(n+1)x^{\frac{m-n}{2}} + x^{n-\frac{1}{2}(m+1)(n+1)(m-n)} \geqslant x^n + (m+1)(n+1) \qquad (*)$$

$$\Leftrightarrow (m+1)(n+1)\left(x^{\frac{m-n}{2}} - 1\right) \geqslant x^n \left(1 - x^{-\frac{1}{2}(m+1)(n+1)(m-n)}\right)$$

$$\Leftrightarrow \frac{m+1}{n+1} - \frac{(m+1)\left(x^{\frac{m-n}{2}} - 1\right)}{x^m P_n(x)} \leqslant \frac{m+1}{n+1} - \frac{x^{n-m}\left(1 - x^{-\frac{1}{2}(m+1)(n+1)(m-n)}\right)}{(n+1)P_n(x)}$$

This shows that when $m \geqslant 3n$, then second inequality in (5) is stronger than the second inequality in (3). However, when $m < 3n$, they are not comparable. For example, taking $m = 4, n = 2$, we have

$$(*) \Leftrightarrow 15x + x^{-13} \geqslant x^2 + 15$$
$$\Leftrightarrow h(x) =: 15x^{14} - x^{15} - 15x^{13} + 1 \geqslant 0$$

It follows
$$h'(x) = -15x^{12}(x - 13)(x - 1)$$

When $1 \leqslant x \leqslant 13, h'(x) \geqslant 0$, and then $h(x) \geqslant h(1) = 0$, i. e. $(*)$ holds. When $x > 15, h'(x) < 0$, and then $h(x) < h(15) = 1 - 15^{14} < 0$, i. e. the reverse of $(*)$ holds.

Remark 2 From Lemma 4 we have

$$\left(\underbrace{0, \cdots, 0}_{(m+1)(n+1)+1}\right) \prec \left(\underbrace{\frac{m-n}{2}, \cdots, \frac{m-n}{2}}_{(m+1)(n+1)}, -\frac{1}{2}(m+1)(n+1)(m-n)\right)$$

and then

$$(m+1)(n+1)x^{\frac{m-n}{2}} + x^{-\frac{1}{2}(m+1)(n+1)(m-n)} \geqslant (m+1)(n+1) + 1$$
$$\Leftrightarrow (m+1)(n+1)\left(x^{\frac{m-n}{2}} - 1\right) \geqslant 1 - x^{-\frac{1}{2}(m+1)(n+1)(m-n)}$$

$$\Leftrightarrow \frac{m+1}{n+1} + \frac{1 - \frac{1}{2}(m+1)(n+1)(m-n)}{(n+1)P_n(x)} \leqslant \frac{m+1}{n+1} - \frac{(n+1)\left(x^{\frac{m-n}{2}} - 1\right)}{P_n(x)}$$

This shows that the fifth inequality in (5) is always stronger than the fifth inequality in (3).

27.5　Acknowledgements

The author is indebted to professor XU tie-quan and the referee for his many helpful and valuable comment s and suggestions.

References

[1]　KUANG JICHANG. Applied Inequalities(Chang yong Budengshi)[M].

3rd ed. Jinan: Shandong Press of science and technology, 2002.

[2]   WANG BOYING. Elements of Majorization Inequalities [M]. Beijing: Beijing Normal University Press, 1990.

[3]   WANG SONGGUI, JIA ZONGZHEN. Inequalities in Theory of Matrix [M]. Anhui: Anhui Education Press, 1994.

第28篇　Heron平均幂型推广的Schur凸性

(李大矛, 顾春, 石焕南. 数学的实践与认识, 2006, 36(9):387-390)

摘　要:　本文讨论了两个正数a, b的Heron平均幂型推广在$\mathbf{R}_+^2$上的单调性和Schur凸性, 并得到了两个新的不等式.

关键词:　Heron平均; 单调性; Schur凸性; 不等式

28.1　引言

设$a, b > 0$, 令$A = A(a,b) = \frac{a+b}{2}, G = G(a,b) = \sqrt{ab}, H = H(a,b) = \frac{a+\sqrt{ab}+b}{3}$, 则

$$L = L(a,b) = \begin{cases} \dfrac{a-b}{\ln a - \ln b}, & a \neq b \\ a, & a = b \end{cases}$$

$$M_q = M_q(a,b) = \begin{cases} \left(\dfrac{b^q + a^q}{2} \right)^{\frac{1}{q}}, & q \neq 0 \\ \sqrt{ab}, & q = 0 \end{cases}$$

它们分别称为算术平均, 几何平均, Heron平均[1], 对数平均和幂平均.

1986年, 匡继昌建立了如下不等式链[1]

$$G \leqslant L \leqslant M_{\frac{1}{3}} \leqslant M_{\frac{1}{2}} \leqslant H \leqslant M_{\frac{2}{3}} \leqslant A$$

文[2] 研究了Heron平均的幂型推广

$$H_p = H_p(a,b) = \begin{cases} \left[\dfrac{a^p + (ab)^{\frac{p}{2}} + b^p}{3} \right]^{\frac{1}{p}}, & p \neq 0 \\ \sqrt{ab}, & p = 0 \end{cases}$$

并得到不等式

$$L \leqslant H_p \leqslant M_q$$

其中$p \geqslant \frac{1}{2}, q \geqslant \frac{2}{3}p$, 并且$p = \frac{1}{2}, q = \frac{1}{3}$ 是最佳常数.

本文将从一个新的视角研究Heron平均, 讨论$H_p(a,b)$ 关于(a,b) 在$\mathbf{R}_+^2 = \{(a,b)|a \geqslant 0, b \geqslant 0\}$ 上的单调性及Schur凸性, 并进而建立相应的不等式.为此我们需要如下的定义和引理.

对于$\boldsymbol{x} = (x_1, ..., x_n) \in \mathbf{R}^n$, 将$\boldsymbol{x}$ 的分量排成递减的次序后, 记作$x_{[1]} \geqslant x_2 \geqslant \cdots \geqslant x_{[n]}$. $\boldsymbol{x} \leqslant \boldsymbol{y}$ 表示$x_i \leqslant y_i, i = 1, \ldots, n$.

定义 1[3]　　设$\boldsymbol{x}, \boldsymbol{y} \in \mathbf{R}^n$ 满足

(1) $\sum\limits_{i=1}^{k} x_{[i]} \leq \sum\limits_{i=1}^{k} y_{[i]}$, $k = 1, 2, ..., n-1$; (2) $\sum\limits_{i=1}^{n} x_i = \sum\limits_{i=1}^{n} y_i$, 则称$\boldsymbol{x}$ 被$\boldsymbol{y}$ 所控制, 记作$\boldsymbol{x} \prec \boldsymbol{y}$.

定义 2[3]　　设$\Omega \subset \mathbf{R}^n, \varphi : \Omega \to \mathbf{R}$.

(1) 若在Ω 上$\boldsymbol{x} \leqslant \boldsymbol{y} \Rightarrow \varphi(\boldsymbol{x}) \leqslant \varphi(\boldsymbol{y})$, 则称$\varphi$ 为Ω 上的增函数; 若$-\varphi$ 是Ω 上的增函数, 则称φ 为Ω 上的减函数;

(2) 若在Ω 上$\boldsymbol{x} \prec \boldsymbol{y} \Rightarrow \varphi(\boldsymbol{x}) \leqslant \varphi(\boldsymbol{y})$, 则称$\varphi$ 为Ω 上的Schur凸函数; 若$-\varphi$ 是Ω 上的Schur 凸函数, 则称φ 为Ω 上的Schur 凹函数.

引理 1[3]　　设φ 在开凸集$\Omega \subset \mathbf{R}^n$ 上可微, 则φ 是Ω 上的增函数的充要条件是$\nabla\varphi(\boldsymbol{x}) \geqslant 0, \forall \boldsymbol{x} \in \Omega$, 其中

$$\nabla\varphi(\boldsymbol{x}) = \left(\frac{\partial\varphi(\boldsymbol{x})}{\partial x_1}, \ldots, \frac{\partial\varphi(\boldsymbol{x})}{\partial x_n} \right) \in \mathbf{R}^n$$

引理 2 [3]　　设$\Omega \subset \mathbf{R}^n$ 是有内点的对称凸集, $\varphi : \Omega \to \mathbf{R}$ 在Ω 上连续, 在Ω 的内部Ω° 可微, 则在Ω 上Schur-凸(Schur-凹)的充要条件是φ 在Ω 上对称且$\forall \boldsymbol{x} \in \Omega$, 有

$$(x_1 - x_2) \left(\frac{\partial\varphi}{\partial x_1} - \frac{\partial\varphi}{\partial x_2} \right) \geq 0 (\leqslant 0)$$

上式称为Schur条件.

引理 3 [1]　　广义对数平均(Stolarsky平均)

$$S_p(a,b) = \begin{cases} \left(\dfrac{b^p - a^p}{p(b-a)} \right)^{\frac{1}{p-1}}, & a \neq b, p \neq 0, 1 \\ b, & a = b \end{cases}$$

当$a \neq b$ 时是p 的严格递增函数.

引理 4 [4]　设$a, b > 0$且$a \neq b$, 若$x > 0, y \leqslant 0, x + y \geqslant 0$, 则

$$\frac{b^{x+y} - a^{x+y}}{b^x - a^x} \leqslant \frac{x+y}{x}(ab)^{\frac{y}{2}}$$

引理 5　设$a \leqslant b, u(t) = tb + (1-t)a, v(t) = ta + (1-t)b$. 若$\frac{1}{2} \leqslant t_2 \leqslant t_l \leqslant 1$, 则

$$(u(t_2), v(t_2)) \prec (u(t_1), v(t_1)) \prec (a, b)$$

证明　由$a \leqslant b$和$\frac{1}{2} \leqslant t_2 \leqslant t_l \leqslant 1$, 易见$u(t_1) \geqslant v(t_1), u(t_2) \geqslant v(t_2)$, 且$b \geqslant u(t_1) \geqslant u(t_2)$, 又$u(t_2) + v(t_2) = u(t_1) + v(t_1) = a + b$, 故引理5 成立.

引理 6　设$0 \leqslant a \leqslant b, c \geqslant 0$, 则

$$\left(\frac{a+c}{a+b+2c}, \frac{b+c}{a+b+2c} \right) \prec \left(\frac{a}{a+b}, \frac{b}{a+b} \right)$$

证明　因$0 \leqslant a \leqslant b, c \geqslant 0$, 有$\frac{a+c}{a+b+2c} \leqslant \frac{b+c}{a+b+2c}$, $\frac{a}{a+b} \leqslant \frac{b}{a+b}$且$\frac{b+c}{a+b+2c} \leqslant \frac{b}{a+b}$, 又$\frac{a+c}{a+b+2c} + \frac{b+c}{a+b+2c} = \frac{a}{a+b} + \frac{b}{a+b} = 1$, 故结论成立.

28.2　主要结果及其证明

定理 1　$H_p(a, b)$关于(a, b)在$\mathbf{R}_+^2$上单调增. 当$p \leqslant \frac{3}{2}$时, $H_p(a, b)$关于(a, b)在$\mathbf{R}_+^2$上Schur 凹; 当$p \geqslant 2$时, $H_p(a, b)$关于(a, b)在$\mathbf{R}_+^2$上Schur 凸, 而当$\frac{3}{2} < p < 2$时, $H_p(a, b)$关于(a, b)在$\mathbf{R}_+^2$上的Schur凸性不确定.

证明　记$\varphi(a, b) = \frac{a^p + (ab)^{\frac{p}{2}} + b^p}{3}$, 则当$p \neq 0$时, $H_p(a, b) = \varphi^{\frac{1}{p}}(a, b)$. $H_p(a, b)$显然是$\mathbf{R}_+^2$上的对称函数

$$\frac{\partial H_p(a, b)}{\partial a} = \frac{1}{3} \left[a^{p-1} + \frac{b}{2}(ab)^{\frac{b}{2}-1} \right] \varphi^{\frac{1}{p}-1}(a, b) \geqslant 0$$

$$\frac{\partial H_p(a, b)}{\partial b} = \frac{1}{3} \left[b^{p-1} + \frac{b}{2}(ab)^{\frac{a}{2}-1} \right] \varphi^{\frac{1}{p}-1}(a, b) \geqslant 0$$

由引理1 知$H_p(a, b)$在$\mathbf{R}_+^2$上单调增. 记

$$\Lambda := (b - a) \left(\frac{\partial H_p(a, b)}{\partial b} - \frac{\partial H_p(a, b)}{\partial a} \right)$$

则当$a = b$时, $\Lambda = 0$, 而当$a \neq b$时

$$\Lambda = \frac{(b-a)^2}{3} \varphi^{\frac{1}{p}-1}(a, b)Q$$

其中

$$Q = \frac{b^{p-1} - a^{p-1}}{b-a} - \frac{1}{2}(ab)^{\frac{p}{2}-1}$$

当 $p = 2$ 时, $Q = \frac{1}{2} > 0 \Rightarrow \Lambda \geqslant 0$, 由引理2 知 $H_p(a,b)$ Schur-凸.

当 $p \leqslant 1$ 时, x^{p-1} 在 $(0, +\infty)$ 上递减

$$\frac{b^{p-1} - a^{p-1}}{b-a} \leqslant 0 \Rightarrow Q \leqslant 0 \Rightarrow \Lambda \leqslant 0$$

即 $H_p(a,b)$ Schur-凹.

当 $p > 2$ 时, 注意

$$Q = (p-1)[S_{p-1}(a,b)]^{p-2} - \frac{1}{2}[S_{-1}(a,b)]^{p-2}$$

由引理3, $S_p(a,b)$ 关于 p 递增, 故 $S_{p-1}(a,b) \geqslant S_{-1}(a,b) = G(a,b)$, 进而因 $p - 2 > 0$, 有 $[S_{p-1}(a,b)]^{p-2} \geqslant [S_{-1}(a,b)]^{p-2}$, 又 $p - 1 > \frac{1}{2}$, 所以 $Q \geqslant 0 \Rightarrow \Lambda \geqslant 0$, 即 $H_p(a,b)$ Schur-凸.

当 $1 < p < \frac{3}{2}$ 时, 令 $x = 1, y = p - 2$, 则 $x > 0, y \leqslant 0, x + y \geqslant 0$ 由引理4 并注意 $p - 1 \leqslant \frac{1}{2}$ 有

$$\frac{b^{p-1} - a^{p-1}}{b-a} \leqslant (p-1)(ab)^{\frac{p-2}{2}} \leqslant \frac{1}{2}(ab)^{\frac{p}{2}-1} \Rightarrow Q \leqslant 0 \Rightarrow \Lambda \leqslant 0$$

即 $H_p(a,b)$ Schur-凹

当 $\frac{3}{2} < p < 2$ 时, 取 $a = 4, b = 2, p = 1.6$, 则 $Q = 0,0609630938 > 0 \Rightarrow \Lambda > 0$, 而取 $a = 21, b = 12221, p = 1.6$, 则 $Q = -0.01869701065 < 0$, 这说明当 $\frac{3}{2} < p < 2$ 时, $H_p(a,b)$ 的Schur-凸性不确定. (这里的两个实例属于重庆市长寿区第四中学的江永明老师)

28.3 应用

结合引理5 和定理1 即可得

定理 2 设 $a \leqslant b, u(t) = tb + (1-t)a, v(t) = ta + (1-t)b, \frac{1}{2} \leqslant t_2 \leqslant t_l \leqslant 1$, 则当 $p \leqslant \frac{3}{2}$ 时, 有

$$A(a,b) \geqslant H_p(u(t_2), v(t_2)) \geqslant H_p(u(t_1), v(t_1)) \geqslant H_p(a,b) \qquad (1)$$

而当 $p \geqslant 2$ 时, 上述不等式反向.

推论 设 $a \leqslant b, u(t) = tb + (1-t)a, v(t) = ta + (1-t)b, k \in \mathbf{N}$,

$\frac{1}{2} \leqslant t_k \leqslant t_{k-l} \leqslant \cdots \leqslant t_2 \leqslant t_1 \leqslant 1$, 则当 $p \leqslant \frac{3}{2}$ 时, 有

$$A(a,b) \geqslant H_p(u(t_k),v(t_k)) \geqslant H_p(u(t_{k-1}),v(t_{k-1})) \geqslant \cdots$$

$$\geqslant H_p(u(t_2),v(t_2)) \geqslant H_p(u(t_1),v(t_1)) \geqslant H_p(a,b) \tag{2}$$

选取特殊的 r,t_i, 由(1)和(2)两式可导出许多有趣的不等式. 例如, 取 $p = \frac{1}{2}, t_1 = \frac{3}{4}, t_2 = \frac{1}{2}$, 由式(1)可得

$$\frac{a+b}{2} \geqslant \frac{1}{36}[\sqrt{a+3b}+\sqrt[4]{(a+3b)(3a+b)}+\sqrt{3a+b}]^2 \geqslant \frac{1}{9}\left(\sqrt{a}+\sqrt[4]{ab}+\sqrt{b}\right)^2$$

取 $p = 2, t_i = \frac{3}{4}, t_2 = \frac{1}{2}$, 由式(1)可得

$$\frac{a+b}{2} \leqslant \sqrt{\frac{(a+3b)^2+(a+3b)(3a+b)+(3a+b)^2}{3}} \leqslant \sqrt{\frac{a^2+ab+b^2}{3}}$$

取 $p = \frac{1}{2}, t_7 = \frac{9}{16}, t_6 = \frac{10}{16}, \ldots, t_2 = \frac{14}{16}, t_1 = \frac{15}{16}$, 由式(2)可得

$$\frac{a+b}{2} \geqslant \frac{1}{144}\left[\sqrt{9a+7b}+\sqrt[4]{(9a+7b)(9b+7a)}+\sqrt{9b+7a}\right]^2$$
$$\geqslant \frac{1}{144}\left[\sqrt{10a+6b}+\sqrt[4]{(10a+6b)(10b+6a)}+\sqrt{10b+6a}\right]^2$$
$$\geqslant \cdots$$
$$\geqslant \frac{1}{144}\left[\sqrt{14a+2b}+\sqrt[4]{(14a+2b)(14b+2a)}+\sqrt{14b+2a}\right]^2$$
$$\geqslant \frac{1}{144}\left[\sqrt{15a+b}+\sqrt[4]{(15a+b)(15b+a)}+\sqrt{15b+a}\right]^2$$
$$\geqslant \frac{1}{9}(\sqrt{a}+\sqrt{ab}+\sqrt{b})^2$$

结合引理6 和定理1 还可得

定理 3　设 $0 \leqslant a \leqslant b, c \geqslant 0$ 则当 $p \leqslant \frac{3}{2}$ 时, 有

$$\frac{H_p(a+c,b+c)}{a+b+2c} \geqslant \frac{H_p(a,b)}{a+b} \tag{3}$$

而当 $p \geqslant 2$ 时, 上述不等式反向.

参考文献

[1]　匡继昌. 常用不等式[M].3版. 济南: 山东科技出版社, 2003.

[2]　JIA G, CAO J-D. A new upper bound of the logarithmic mean[J]. JIPAM. J. Inequal. Pure Appl. Math., 2003, 4 (4), Article 80.

[3]　王伯英. 控制不等式基础[M]. 北京: 北京师范大学出版社, 1990.

[4]   LIU ZHENG. A note on an inequality [J], 纯粹数学与应用数学, 2001, 17 (4): 349-351.

第29篇 Majorized Proof and Improvement of the Discrete Steffensen's Inequality

(SHI HUAN-NAN. Taiwanese Journal of Mathematics, 2007, 11(4): 1203-1208)

Abstract: We enlarge two weak majorization relations of the vectors to strong majorization relations of the vectors. An improvement of the discrete Steffensen's inequalities is established by the related propositions in the theory of majorization.

Keywords: Steffensen's inequality; majorization; weak majorization; refinement

29.1   Introduction

Let $\{x_i\}_{i=1}^n$ be a nonincreasing finite sequence of nonnegative real numbers, and let $\{y_i\}_{i=1}^n$ be a finite sequence of real numbers such that for every i, $0 \leqslant y_i \leqslant 1$. Let $k_1, k_2 \in \{1, 2, \cdots, n\}$ be such that $k_2 \leqslant \sum_{i=1}^n y_i \leqslant k_1$. Then

$$\sum_{i=n-k_2+1}^n x_i \leqslant \sum_{i=1}^n x_i y_i \leqslant \sum_{i=1}^{k_1} x_i \qquad (1)$$

The inequality (1) is called the discrete Steffensen's inequality. It was first given in [1] and then cited repeatedly in [2-6]. Recently, a new proof which is very simple and clear is given in [7]. The purpose of this note is to establish an improved Steffensen's inequality by means of the theory of majorization. The following definitions and lemmas will be used:

Definition[8] Let $\boldsymbol{x}= (x_1,\ldots,x_n)$ and $\boldsymbol{y} = (y_1,\ldots,y_n)$ be two real n-tuples. Then $\boldsymbol{x}$ is said to be majorized by $\boldsymbol{y}$ (in symbols $\boldsymbol{x} \prec \boldsymbol{y}$) if

(i) $\sum\limits_{i=1}^{k} x_{[i]} \leqslant \sum\limits_{i=1}^{k} y_{[i]}$ for $k = 1, 2, ..., n-1$; (ii) $\sum\limits_{i=1}^{n} x_i = \sum\limits_{i=1}^{n} y_i$.

where $x_{[1]} \geqslant x_{[2]} \geqslant \cdots \geqslant x_{[n]}$ and $y_{[1]} \geqslant y_{[2]} \geqslant \cdots \geqslant y_{[n]}$ are components of x and y rearranged in descending order. And x is said to be weakly submajorized by y (written $x \prec_w y$) if

$$\sum_{i=1}^{k} x_{[i]} \leqslant \sum_{i=1}^{k} y_{[i]}, k = 1, 2, ..., n$$

And x is said to be weakly supermajorized by y (written $x \prec^w y$) if

$$\sum_{i=1}^{k} x_{(i)} \geqslant \sum_{i=1}^{k} y_{(i)}, k = 1, 2, ..., n$$

where $x_{(1)} \leqslant x_{(2)} \leqslant \cdots \leqslant x_{(n)}$ and $y_{(1)} \leqslant y_{(2)} \leqslant \cdots \leqslant y_{(n)}$ are components of x and y rearranged in increasing order.

Relatively, the majorization is also said to be the strong majorization .

Lemma 1 [9] Let $x, y \in \mathbf{R}^n$.

(a) if $x \prec_w y$, then

$$(\boldsymbol{x}, x_{n+1}) \prec (\boldsymbol{y}, y_{n+1})$$

where $x_{n+1} = \min\{x_1, x_2, \ldots, x_n, y_1, y_2, \ldots, y_n\}, y_{n+1} = \sum\limits_{i=1}^{n+1} x_i - \sum\limits_{i=1}^{n} y_i$.

(b) if $x \prec^w y$, then

$$(x_0, \boldsymbol{x}) \prec (y_0, \boldsymbol{y})$$

where $x_0 = \max\{x_1, x_2, \ldots, x_n, y_1, y_2, \ldots, y_n\}, y_0 = \sum\limits_{i=0}^{n} x_i - \sum\limits_{i=1}^{n} y_i$.

Lemma 2 For $x, y \in \mathbf{R}^n$, we have

$$\sum_{i=1}^{n} x_{[i]} y_{(i)} \leqslant \sum_{i=1}^{n} x_i y_i \leqslant \sum_{i=1}^{n} x_{[i]} y_{[i]}.$$

Lemma 3 [8] For $x, y \in \mathbf{R}^n$, we have

(a) $\boldsymbol{x} \prec \boldsymbol{y} \Leftrightarrow \sum\limits_{i=1}^{n} x_{[i]} u_{[i]} \leqslant \sum\limits_{i=1}^{n} y_{[i]} u_{[i]}, \forall \boldsymbol{u} \in \mathbf{R}^n$;

(b) $\boldsymbol{x} \prec \boldsymbol{y} \Leftrightarrow \sum\limits_{i=1}^{n} x_{(i)} u_{[i]} \leqslant \sum\limits_{i=1}^{n} y_{(i)} u_{[i]}, \forall \boldsymbol{u} \in \mathbf{R}^n$;

(c) $\boldsymbol{x} \prec \boldsymbol{y} \Leftrightarrow \sum\limits_{i=1}^{n} x_{[i]} u_{(i)} \leqslant \sum\limits_{i=1}^{n} y_{[i]} u_{(i)}, \forall \boldsymbol{u} \in \mathbf{R}^n$.

29.2　Main results and proofs

Theorem 1 Let $\{y_i\}_{i=1}^n$ be a finite sequence of real numbers such that $0 \leqslant y_i \leqslant 1$ for $i = 1, 2, \cdots, n$, and let $k_1, k_2 \in \{1, 2, \cdots, n\}$ be such that $k_2 \leqslant \sum_{i=1}^n y_i \leqslant k_1$. Then

$$y = (y_1, y_2, \cdots, y_n) \prec^w \Big(\underbrace{0, \cdots, 0}_{n-k_2}, \underbrace{1, \cdots, 1}_{k_2} \Big) = z \tag{2}$$

$$y = (y_1, y_2, \cdots, y_n) \prec_w \Big(\underbrace{1, \cdots, 1}_{k_1}, \underbrace{0, \cdots, 0}_{n-k_1} \Big) = v \tag{3}$$

Proof we prove that $y \prec^w z$ by the definition of the weak supermajorization . When $1 \leqslant k \leqslant n - k_2$, clearly

$$\sum_{i=1}^k y_{(i)} \geqslant \sum_{i=1}^k z_{(i)} = 0$$

When $n \geqslant k > n - k_2$, using reduction to absurdity, we prove that $\sum_{i=1}^k y_{(i)} \geqslant \sum_{i=1}^k z_{(i)}$, namely, if there exist $k(n \geqslant k > n - k_2)$ such that $\sum_{i=1}^k y_{(i)} < \sum_{i=1}^k z_{(i)}$, then by $0 \leqslant y_i \leqslant 1, i = 1, \cdots, n$, we have

$$\sum_{i=1}^n y_{(i)} = \sum_{i=1}^k y_{(i)} + \sum_{i=k+1}^n y_{(i)} < k - (n - k_2) + (n - k) = k_2$$

It contradicts with $k_2 \leqslant \sum_{i=1}^n y_i$.

Secondly, we prove that $y \prec_w v$ by the definition of the weak submajorizzation . Note that $0 \leqslant y_i \leqslant 1, i = 1, 2, \cdots, n$, when $1 \leqslant k \leqslant k_1$, we have

$$\sum_{i=1}^k y_{[i]} \leqslant k = \sum_{i=1}^k v_{[i]}$$

When $k_1 + 1 \leqslant k \leqslant n$, we have

$$\sum_{i=1}^k y_{[i]} \leqslant \sum_{i=1}^n y_{[i]} \leqslant k_1 = \sum_{i=1}^k v_{[i]}$$

This completes the proof of Theorem 1.

Theorem 2 Let $\{x_i\}_{i=1}^n$ be a nonincreasing finite sequence of real numbers and let $\{y_i\}_{i=1}^n$ be a finite sequence of real numbers such that for every $i, 0 \leqslant y_i \leqslant 1$. Let $k_1, k_2 \in \{1, 2, \ldots, n\}$ be such that $k_2 \leqslant \sum\limits_{i=1}^n y_i \leqslant k_1$. Then

$$\sum_{i=n-k_2+1}^n x_i + \left(\sum_{i=1}^n y_i - k_2\right) x_n \leqslant \sum_{i=1}^n x_i y_i \leqslant \sum_{i=1}^{k_1} x_i - \left(k_1 - \sum_{i=1}^n y_i\right) x_n \quad (4)$$

Proof By Theorem 1, we have

$$\boldsymbol{y} = (y_1, y_2, \cdots, y_n) \prec_w \left(\underbrace{1, \cdots, 1}_{k_1}, \underbrace{0, \cdots, 0}_{n-k_1}\right) = \boldsymbol{v}$$

using Lemma 1 (a), we obtain

$$\boldsymbol{y} = (y_1, y_2, \cdots, y_n, y_{n+1}) \prec \left(\underbrace{1, \cdots, 1}_{k_1}, \underbrace{0, \cdots, 0}_{n-k_1}, v_{n+1}\right)$$

where $y_{n+1} = \min\{y_1, y_2, \cdots, y_n, v_1, v_2, \cdots, v_n\}, v_{n+1} = \sum\limits_{i=1}^{n+1} y_i - \sum\limits_{i=1}^n v_i = \sum\limits_{i=1}^n y_i + y_{n+1} - k_1$. It is clear that $y_{n+1} = 0$, and then $v_{n+1} \leqslant 0$.

Choosing $\boldsymbol{u} = (x_1, x_2, \cdots, x_n, x_n)$, from Lemma 2 and Lemma 3 (a) we have

$$\sum_{i=1}^n x_i y_i + x_n y_{n+1} \leqslant \sum_{i=1}^n x_{[i]} y_{[i]} + x_n y_{n+1} \leqslant \sum_{i=1}^{k_1} x_i + \left(\sum_{i=1}^n y_i + y_{n+1} - k_1\right) x_n$$

hence

$$\sum_{i=1}^n x_i y_i \leqslant \sum_{i=1}^{k_1} x_i - \left(k_1 - \sum_{i=1}^n y_i\right) x_n$$

Also, from Theorem 1 and Lemma 1 (b) we have

$$\boldsymbol{y} = (y_1, y_2, \cdots, y_n) \prec^w \left(\underbrace{0, \cdots, 0}_{n-k_2}, \underbrace{1, \cdots, 1}_{k_2}\right) = \boldsymbol{z}$$

and

$$y = (y_1, y_2, \cdots, y_n, y_0) \prec \left(\underbrace{0, \cdots, 0}_{n-k_2}, \underbrace{1, \cdots, 1}_{k_2}, z_0 \right)$$

where $y_0 = \max\{y_1, y_2, \cdots, y_n, z_1, \cdots, z_n\}$, $z_0 = \sum_{i=0}^{n} y_i - \sum_{i=1}^{n} v_i = \sum_{i=1}^{n} y_i + y_0 - k_2$. It is clear that $y_0 \geqslant 1$, and then $z_0 \geqslant 1$.

Choosing $\boldsymbol{u} = (x_1, x_2, \cdots, x_n, x_n)$, from Lemma 2 and Lemma 3 (b) we have

$$\sum_{i=1}^{n} y_i x_i + y_0 x_n \geqslant \sum_{i=1}^{n} y_{(i)} x_{[i]} + y_0 x_n \geqslant \sum_{i=n-k_2+1}^{n} x_i + \left(\sum_{i=1}^{n} y_i + y_0 - k_2 \right) x_n$$

thus

$$\sum_{i=1}^{n} x_i y_i \geqslant \sum_{i=n-k_2+1}^{n} x_i + \left(\sum_{i=1}^{n} y_i - k_2 \right) x_n$$

This completes the proof of Theorem 2.

As consequence, a refinement of the discrete Steffensen's inequality follows from Theorem 2 directly:

Corollary Let $\{x_i\}_{i=1}^{n}$ be a nonincreasing finite sequence of real numbers and let $\{y_i\}_{i=1}^{n}$ be a finite sequence of real numbers such that for every $i, 0 \leqslant y_i \leqslant 1$. Let $k_1, k_2 \in \{1, 2, \cdots, n\}$ be such that $k_2 \leqslant \sum_{i=1}^{n} y_i \leqslant k_1$. Then

$$\sum_{i=n-k_2+1}^{n} x_i \leqslant \sum_{i=n-k_2+1}^{n} x_i + \left(\sum_{i=1}^{n} y_i - k_2 \right) x_n$$

$$\leqslant \sum_{i=1}^{n} x_i y_i \leqslant \sum_{i=1}^{k_1} x_i - \left(k_1 - \sum_{i=1}^{n} y_i \right) x_n \leqslant \sum_{i=1}^{k_1} x_i \tag{5}$$

29.3　Acknowledgment

The authors are indebted to professors Zheng Liu, Tie-quan Xu and Wan-lan Wang for their many helpful and valuable comments and suggestions.

References

[1]　J C EVARD, H GAUCHMAN. Steffensen type inequalities over general measure spaces[J]. Analysis, 1997, 17 : 301-322.

[2] H GAUCHMAN. On a further generalization of Steffensen's inequality[J]. J. Inequal. Appl., 2000, 5 : 505-513.

[3] H GAUCHMAN. A Steffensen type inequality[J]. J. Inequal. Pure Appl. Math., 2000, 1: Art. 3.

[4] J E PECARI' C. On the Bellman generalization of steffensen's inequality, II[J]. J. Math. Anal. Appl., 1984, 104 : 432-434.

[5] F QI, J X CHENG, G WANG. New Steffensen pairs[J]. Inequality Theory and Applications, 2002, 1 : 273-279.

[6] F QI, B N GUO. On Steffensen pairs[J]. J. Math. Anal.Appl., 2002, 271 : 534-541.

[7] Z LIU. Simple proof of the discrete steffensen's inequality[J]. Tamkang J. Math., 2004, 35(4) : 281-282.

[8] B Y WANG. Elements of Majorization Inequalities[M]. Beijing: Beijing Normal University Press, 1990.

[9] A W MARSHALL, I OLKIN. Inequalities: Theory of majorization and its application[M]. New York: Academies Press, 1979.

第30篇　Schur-Convex Functions Relate to Hadamard-Type Inequalities

(SHI HUAN-NAN. Journal Mathematical Inequalities, Volum 1, Number 1, (2007): 127-136)

Abstract:　The Schur-convexity on the upper and the lower limit of the integral for a mean of the convex function is researched. As applications, a generalized logarithmic mean with a parameter is obtained and a relevant double inequality that is a extension of the known inequality is established.

Keywords:　Schur-convex function; inequality; Hadamard's inequality; generalized logarithmic mean

30.1　Introduction

Let f be a convex function defined on the interval $I \subseteq \mathbf{R} \to \mathbf{R}$ of real numbers and $a, b \in I$ with $a < b$. Then

$$f\left(\frac{a+b}{2}\right) \leqslant \frac{1}{b-a} \int_a^b f(x)\,\mathrm{d}x \leqslant \frac{f(a)+f(b)}{2} \tag{1}$$

is known as the Hadamard's inequality for convex function.

In [1], S. S. Dragomir established the following two theorems which are refinements of the first inequality of (1).

Theorem A[1] If $f : [a, b] \to \mathbf{R}$ is a convex function, and H is defined on $[0, 1]$ by

$$H(t) = \frac{1}{b-a} \int_a^b f\left(tx + (1-t)\frac{a+b}{2}\right)\,\mathrm{d}x$$

then H is convex, increasing on $[0, 1]$, and for all $t \in [0, 1]$, we have

$$f\left(\frac{a+b}{2}\right) = H(0) \leqslant H(t) \leqslant H(1) = \frac{1}{b-a} \int_a^b f(x)\,\mathrm{d}x \tag{2}$$

Theorem B[1] If $f : [a, b] \to \mathbf{R}$ is a convex function, and F is defined on $[0, 1]$ by

$$F(t) = \frac{1}{(b-a)^2} \int_a^b \int_a^b f\left(tx + (1-t)y\right) \mathrm{d}\,x\,\mathrm{d}\,y$$

then

(i) F is convex on $[0, 1]$, symmetric about $\frac{1}{2}$, (i.e.$F(t) = F(1 - t)$ for all $t \in [0, 1]$), F is increasing on $[0, \frac{1}{2}]$ and increasing on $[\frac{1}{2}, 1]$, and for all $t \in [0, 1]$, we have

$$F(t) \leqslant F(1) = \frac{1}{b-a} \int_a^b f(x) \,\mathrm{d}\,x \tag{3}$$

and

$$F(t) \geqslant F\left(\frac{1}{2}\right) = \frac{1}{(b-a)^2} \int_a^b \int_a^b f\left(\frac{x+y}{2}\right) \mathrm{d}\,x\,\mathrm{d}\,y \geqslant f\left(\frac{a+b}{2}\right) \tag{4}$$

(ii) for all $t \in [0, 1]$, we have

$$F(t) \geqslant \max\{H(t), H(1 - t)\} \tag{5}$$

where $H(t)$ is defined in Theorem A.

In [2], S. S. Dragomir established the following theorem which is a extension of the relevant conclusion in [3].

Theorem C[2] If $f : [a, b] \to \mathbf{R}$ is a convex function, and G is defined on $[0, 1]$ by

$$G(t) = \frac{1}{2(b-a)} \int_a^b \left[f\left(ta + (1-t)x\right) + f\left(tb + (1-t)x\right)\right] \mathrm{d}\,x$$

then G is convex on $[0, 1]$, and for all $t \in [0, 1]$, we have

$$\frac{1}{b-a} \int_a^b f(x) \,\mathrm{d}\,x = G(0) \leqslant G(t) \leqslant G(1) = \frac{f(a) + f(b)}{2} \tag{6}$$

Remark 1 If f is concave, then (6) is reversed. (notice that $-f$ is convex)

In [4], N. Elezovic and J. Pecaric researched the Schur-convexity on the upper and the lower limit of the integral for the mean of the convex function and established the following important result by using the Hadamard's inequality.

Theorem D[4] Let I be an interval with nonempty interior on $\mathbf{R}$ and f be a continuous function on I. Then

$$\Phi(a,b) = \begin{cases} \frac{1}{b-a}\int_a^b f(t)\mathrm{d}t, & a,b \in I,\ a \neq b \\ f(a), & a = b \end{cases}$$

is Schur-convex (Schur-concave) on I^2 if and only if f is convex(concave) on I.

The aim of this paper is to establish the following results which are similar to Theorem D. As applications, a generalized logarithmic mean with a parameter is obtained and a relevant double inequality which is a extension of the known inequality is established.

Theorem 1 Let I be an interval with nonempty interior on $\mathbf{R}$ and define a function of two variables as follows

$$P(a,b) = \begin{cases} G(t), & a,b \in I,\ a \neq b \\ f(a), & a = b \end{cases}$$

(i) For $\frac{1}{2} \leqslant t \leqslant 1$, if f is convex on I, then $P(a,b)$ is Schur-convex on I^2.

(ii) For $0 \leqslant t \leqslant \frac{1}{2}$, if f is concave on I, then $P(a,b)$ is Schur-concave on I^2.

Theorem 2 Let I be an interval with nonempty interior on $\mathbf{R}$ and f be a continuous function on I. For any $t \in [0,1]$, we define a function of two variables as follows

$$Q(a,b) = \begin{cases} F(t), & a,b \in I,\ a \neq b \\ f(a), & a = b \end{cases}$$

if f is convex (concave) on I, then $P(a,b)$ is Schur-convex (Schur-concave) on I^2.

30.2　Definitions and Lemmas

We need the following definitions and lemmas.

Definition 1[5,7] Let $\Omega \subseteq \mathbf{R}^n, \boldsymbol{x} = (x_1,\cdots,x_n)$ and $\boldsymbol{y} = (y_1,\cdots,y_n) \in \Omega$, and let $\varphi : \Omega \to \mathbf{R}$.

1. $\boldsymbol{x}$ is said to be majorized by $\boldsymbol{y}$ (in symbols $\boldsymbol{x} \prec \boldsymbol{y}$) if $\sum\limits_{i=1}^{k} x_{[i]} \leqslant \sum\limits_{i=1}^{k} y_{[i]}$ for $k = 1, 2, \cdots, n-1$ and $\sum\limits_{i=1}^{n} x_i = \sum\limits_{i=1}^{n} y_i$, where $x_{[1]} \geqslant \cdots \geqslant x_{[n]}$ and $y_{[1]} \geqslant \cdots \geqslant y_{[n]}$ are rearrangements of $\boldsymbol{x}$ and $\boldsymbol{y}$ in a descending order.

2. $\boldsymbol{x} \geqslant \boldsymbol{y}$ means $x_i \geqslant y_i$ for all $i = 1, 2, \cdots, n$. φ is said to be increasing if $\boldsymbol{x} \geqslant \boldsymbol{y}$ implies $\varphi(\boldsymbol{x}) \geqslant \varphi(\boldsymbol{y})$. φ is said to be decreasing if and only if $-\varphi$ is increasing.

3. φ is said to be a Schur-convex function on Ω if $\boldsymbol{x} \prec \boldsymbol{y}$ on Ω implies $\varphi(\boldsymbol{x}) \leqslant \varphi(\boldsymbol{y})$, φ is said to be a Schur-concave function on Ω if and only if $-\varphi$ is Schur- convex function.

Lemma 1[5] Let $\boldsymbol{x} \in \mathbf{R}^n$ and $\bar{x} = \frac{1}{n} \sum\limits_{i=1}^{n} x_i$. Then

$$(\bar{x}, \ldots, \bar{x}) \prec \boldsymbol{x}$$

Lemma 2 Let $a \leqslant b, u(t) = tb + (1-t)a, v(t) = ta + (1-t)b$.
(i) If $\frac{1}{2} \leqslant t_2 \leqslant t_1 \leqslant 1$, then

$$(u(t_2), v(t_2)) \prec (u(t_1), v(t_1)) \tag{7}$$

(ii) If $0 \leqslant t_2 \leqslant t_1 \leqslant \frac{1}{2}$, then

$$(u(t_1), v(t_1)) \prec (u(t_2), v(t_2)) \tag{8}$$

Proof From $a < b, \frac{1}{2} \leqslant t_2 \leqslant t_1 \leqslant 1$, it is easy to see that $u(t_1) \geqslant v(t_1), u(t_2) \geqslant v(t_2), u(t_1) \geqslant u(t_2)$ and $u(t_2) + v(t_2) = u(t_1) + v(t_1) = a + b$. By Definition 1, it follows that (7) holds. It is similarly to prove that (8) holds.

Lemma 3[5] Let $\Omega \subseteq \mathbf{R}^n$ is symmetric and has a nonempty interior set. Ω^0 is the interior of Ω. $\varphi : \Omega \to \mathbf{R}$ is continuous on Ω and differentiable in Ω^0. Then φ is the Schur-convex(Schur-concave)function, if and only if φ is symmetric on Ω and

$$(x_1 - x_2) \left(\frac{\partial \varphi}{\partial x_1} - \frac{\partial \varphi}{\partial x_2} \right) \geqslant 0 (\leqslant 0)$$

holds for any $\boldsymbol{x} = (x_1, x_2, \cdots, x_n) \in \Omega^0$.

Lemma 4[4] Let $\Omega \subseteq \mathbf{R}^n$, $\varphi : \Omega \to \mathbf{R}, h : \mathbf{R} \to \mathbf{R}, \psi(x) = h(\varphi(x))$.

1. If φ is Schur-convex and h is increasing, then ψ is also Schur-convex.

2. If φ is Schur-convex and h is decreasing, then ψ is also Schur-concave.

3. If φ is Schur-concave and h is increasing, then ψ is also Schur-concave.

Lemma 5 Let $F(\alpha, \beta) = \int_\alpha^\beta \int_\alpha^\beta f(x, y) \, \mathrm{d}x \, \mathrm{d}y$, where $f(x, y)$ is continuous on the rectangle $[a, p; a, q]$, $\alpha = \alpha(b)$ and $\beta = \beta(b)$ are differentiable with b, $a \leqslant \alpha(b) \leqslant p$ and $a \leqslant \beta(b) \leqslant q$. Then

$$\frac{\partial F}{\partial b} = \left(\int_\alpha^\beta f(\alpha, y) \, \mathrm{d}y \right) \alpha'(b) + \left(\int_\alpha^\beta f(x, \beta) \, \mathrm{d}x \right) \beta'(b) \qquad (9)$$

Proof Since $F(\alpha, \beta) = \int_\alpha^\beta \int_\alpha^\beta f(x, y) \, \mathrm{d}x \, \mathrm{d}y$, by the derivation rule for the composite functions, we have

$$\frac{\partial F}{\partial b} = \frac{\partial F}{\partial \alpha} \frac{\mathrm{d}\alpha}{\mathrm{d}b} + \frac{\partial F}{\partial \beta} \frac{\mathrm{d}\beta}{\mathrm{d}b}$$

which is the equation[9].

30.3 Proofs of main results

Proof of Theorem 1

It is sufficient prove that (i), the proof of (ii) is similar with (i). We need only consider the case of $\frac{1}{2} \leqslant t < 1$. It is clear that $P(a, b)$ is symmetric. When $a \neq b$, let

$$P_1(a, b) = \int_a^b f(ta + (1 - t)x) \, \mathrm{d}x$$

and

$$P_2(a, b) = \int_a^b f(tb + (1 - t)x) \, \mathrm{d}x$$

Then

$$P(a, b) = \frac{1}{2(b - a)} [P_1(a, b) + P_2(a, b)] = G(t), a \neq b$$

By the transformation $s = ta + (1 - t)x$, we get

$$P_1(a, b) = \frac{1}{1 - t} \int_a^{ta + (1 - t)b} f(s) \, \mathrm{d}s$$

$$= \frac{1}{1 - t} \left[\int_0^{ta + (1 - t)b} f(s) \, \mathrm{d}s - \int_0^a f(s) \, \mathrm{d}s \right]$$

$$\frac{\partial P_1(a,b)}{\partial a} = \frac{1}{1-t}\left[f\left(ta+(1-t)b\right)t - f(a)\right]$$

$$= \frac{t}{1-t}f\left(ta+(1-t)b\right) - \frac{f(a)}{1-t} \tag{10}$$

$$\frac{\partial P_1(a,b)}{\partial b} = \frac{1}{1-t}\left[(1-t)f\left(ta+(1-t)b\right)\right] = f\left(ta+(1-t)b\right) \tag{11}$$

Notice that $P_2(a,b) = -P_1(b,a)$, from (11), we get

$$\frac{\partial P_2(a,b)}{\partial a} = -\frac{\partial P_1(b,a)}{\partial a} = -f\left(tb+(1-t)a\right) \tag{12}$$

and from (10), we get

$$\frac{\partial P}{\partial b} = -\frac{\partial P_1(b,a)}{\partial b} = \frac{f(b)}{1-t} + \frac{t}{1-t}f\left(tb+(1-t)a\right) \tag{13}$$

And then

$$\frac{\partial P_1(a,b)}{\partial b} - \frac{\partial P_1(a,b)}{\partial a} = f\left(ta+(1-t)b\right) - \frac{t}{1-t}f\left(ta+(1-t)b\right) - \frac{f(a)}{1-t}$$

$$= \frac{1-2t}{1-t}f\left(ta+(1-t)b\right) + \frac{f(a)}{1-t}$$

$$\frac{\partial P_2(a,b)}{\partial b} - \frac{\partial P_2(a,b)}{\partial a} = \frac{f(b)}{1-t} - \frac{t}{1-t}f\left(tb+(1-t)a\right) + f\left(tb+(1-t)a\right)$$

$$= \frac{f(b)}{1-t} + \frac{1-2t}{1-t}f\left(tb+(1-t)a\right)$$

Since

$$\frac{\partial P(a,b)}{\partial b} = \left\{-\frac{1}{2(b-a)^2}[P_1(a,b)+P_2(a,b)] + \frac{1}{2(b-a)}\left[\frac{\partial P_1(a,b)}{\partial b} + \frac{\partial P_2(a,b)}{\partial b}\right]\right\}$$

and

$$\frac{\partial P(a,b)}{\partial a} = \left\{\frac{1}{2(b-a)^2}[P_1(a,b)+P_2(a,b)] + \frac{1}{2(b-a)}\left[\frac{\partial P_1(a,b)}{\partial a} + \frac{\partial P_2(a,b)}{\partial a}\right]\right\}$$

then

$$(b-a)\left(\frac{\partial P(a,b)}{\partial b} - \frac{\partial P(a,b)}{\partial a}\right)$$

$$= \frac{1}{2}\left[\left(\frac{\partial P_1(a,b)}{\partial b} - \frac{\partial P_1(a,b)}{\partial a}\right) + \left(\frac{\partial P_2(a,b)}{\partial b} - \frac{\partial P_2(a,b)}{\partial a}\right)\right] - G(t)$$

$$= \frac{1}{2(1-t)} \left[f(a) + f(b) + (1-2t) \left(f(ta + (1-t)b) + f(tb + (1-t)a) \right) \right] - G(t)$$

$$\geqslant \frac{1}{2(1-t)} \left[f(a) + f(b) + (1-2t)(tf(a) + (1-t)f(b) + tf(b) + (1-t)f(a)) \right] - G(t)$$

(notice that f is convex and from $\frac{1}{2} \leqslant t < 1$, we have $1 - 2t \leqslant 0$)

$$= f(a) + f(b) - G(t) \geqslant 0 \quad \text{(by the right inequality in (6))}$$

According to Lemma 3, it follows that $P(a, b)$ is Schur-convex on I^2. The proof of Theorem 1 is completed.

Proof of Theorem 2

Taking $\alpha = \beta = b$ in Lemma 5, when $a \neq b$, we have

$$\frac{\partial Q(a, b)}{\partial b} = \frac{-2}{(b-a)^3} \int_a^b \int_a^b f(tx + (1-t)y) \, dx \, dy$$

$$+ \frac{1}{(b-a)^2} \left[\int_a^b f(tb + (1-t)y) \, dy + \int_a^b f(tx + (1-t)b) \, dx \right]$$

$$\frac{\partial Q(a, b)}{\partial a} = \frac{2}{(b-a)^3} \int_a^b \int_a^b f(tx + (1-t)y) \, dx \, dy$$

$$+ \frac{1}{(b-a)^2} \left[\int_a^b f(ta + (1-t)y) \, dy + \int_a^b f(tx + (1-t)a) \, dx \right]$$

Now we only consider the case of convexity, the case of concavity is similar.

$$(b-a) \left(\frac{\partial Q(a, b)}{\partial b} - \frac{\partial Q(a, b)}{\partial a} \right)$$

$$= \frac{1}{b-a} \left[\int_a^b (f(tb + (1-t)y) + f(ta + (1-t)y)) \, dy \right.$$

$$+ \int_a^b (f(tx + (1-t)b) + f(tx + (1-t)a)) \, dx \right]$$

$$- \frac{4}{(b-a)^2} \int_a^b \int_a^b f(tx + (1-t)y) \, dx \, dy$$

$$\geqslant \frac{4}{b-a} \int_a^b f(x) \, dx - \frac{4}{(b-a)^2} \int_a^b \int_a^b f(tx + (1-t)y) \, dx \, dy$$

(by the left inequality in (6))

$$\geqslant 0 \quad \text{(by (3))}$$

According to Lemma 3, it follows that $Q(a, b)$ is Schur-convex on I^2. The proof of Theorem 2 is completed.

30.4 Applications

Theorem 3 Let $t \in [0, 1)$, $a, b \in \mathbf{R}_+ = [0, +\infty)$, and let

$$L_r(a, b; t) = \left[\frac{(b^r - a^r) - (u^r - v^r)}{r(1 - t)(b - a)} \right]^{\frac{1}{r-1}} \quad (a \neq b)$$

$$L_r(a, a; t) = a$$

where $u = tb + (1 - t)a$, $v = ta + (1 - t)b$.

(i) If $r > 2$ and $\frac{1}{2} \leqslant t \leqslant 1$, then $L_r(a, b; t)$ is Schur-convex on $\mathbf{R}_+^2$,

(ii) If $1 < r < 2$ and $0 \leqslant t \leqslant \frac{1}{2}$, then $L_r(a, b; t)$ is Schur-concave on $\mathbf{R}_+^2$,

(iii) If $r \leqslant 1, r \neq 0$ and $\frac{1}{2} \leqslant t \leqslant 1$, then $L_r(a, b; t)$ is Schur-concave on $\mathbf{R}_+^2$.

Proof Taking $f(x) = x^{r-1}, r \neq 0$, then for $a \neq b$, from Theorem 1, we have

$$G(t) = \frac{1}{2(b - a)} \int_a^b \left[(ta + (1 - t)x)^{r-1} + (tb + (1 - t)x)^{r-1} \right] \mathrm{d}\, x$$

$$= \frac{1}{2r(1 - t)(b - a)} \left[(ta + (1 - t)x)^r \Big|_a^b + (tb + (1 - t)x)^r \Big|_a^b \right]$$

$$= \frac{(b^r - a^r) + (ta + (1 - t)b)^r - (tb + (1 - t)a)^r}{2r(1 - t)(b - a)}$$

$$= \frac{(b^r - a^r) + (u^r - v^r)}{2r(1 - t)(b - a)}$$

(i) If $r > 2$ and $\frac{1}{2} \leqslant t \leqslant 1$, since $f(x) = x^{r-1}$ is convex on $\mathbf{R}_+$, from Theorem 1 we obtain that $P(a, b)$ is Schur-convex on $\mathbf{R}_+^2$. Furthermore, since $h : t \to t^{\frac{1}{r-1}}$ is increasing on $\mathbf{R}_+$, then from (1) in Lemma 4, $L_r(a, b; t) = [P(a, b)]^{\frac{1}{r-1}}$ is Schur-convex on $\mathbf{R}_+^2$,

(ii) If $1 < r < 2$ and $0 \leqslant t \leqslant \frac{1}{2}$, since $f(x) = x^{r-1}$ is concave on $\mathbf{R}_+$, from Theorem 1 we obtain that $P(a, b)$ is Schur-concave on $\mathbf{R}_+^2$. Furthermore, since $h : t \to t^{\frac{1}{r-1}}$ is increasing on $\mathbf{R}_+$, then from (3) in Lemma 4, $L_r(a, b; t) = [P(a, b)]^{\frac{1}{r-1}}$ is Schur-concave on $\mathbf{R}_+^2$,

(iii) If $r \leqslant 1, r \neq 0$ and $\frac{1}{2} \leqslant t \leqslant 1$, since $f(x) = x^{r-1}$ is convex on $\mathbf{R}_+$, from Theorem 1 we obtain that $P(a, b)$ is Schur-convex on $\mathbf{R}_+^2$. Furthermore, since $h : t \rightarrow t^{\frac{1}{r-1}}$ is decreasing on $\mathbf{R}_+$, then from (2) in Lemma 4, $L_r(a, b; t)$ is Schur-concave on $\mathbf{R}_+^2$. Setting $r \rightarrow 1$, it is deduced that $L_r(a, b; t)$ is still Schur-concave on $\mathbf{R}_+^2$ for $r = 1$. The proof of Theorem 3 is completed.

Corllary 1 If $(r, t) \in \{r > 2, \frac{1}{2} \leqslant t \leqslant 1\}$, then

$$\frac{a+b}{2} \leqslant L_r(a, b; t) \leqslant (a+b) \left(\frac{(t^r - 1)^r + (1 - t)^r}{2r(t-1)} \right)^{\frac{1}{r-1}} \tag{14}$$

if $(r, t) \in \{1 < r < 2, 0 \leqslant t \leqslant 1/2\} \cup \{r \leqslant 1, r \neq 0, 1/2 \leqslant t \leqslant 1\}$, then the two inequalities in (14) are all reversed.

Proof Since

$$\left(\frac{a+b}{2}, \frac{a+b}{2} \right) \prec (a, b) \prec (a+b, 0)$$

then from Theorem 3, when $r > 2$ and $\frac{1}{2} \leqslant t \leqslant 1$, we have

$$L_r \left(\frac{a+b}{2}, \frac{a+b}{2} \right) \leqslant L_r(a, b) \leqslant L_r(a+b, 0)$$

i.e. (14) is holds. When $(r, t) \in \{1 < r < 2, 0 \leqslant t \leqslant 1/2\} \cup \{r \leqslant 1, r \neq 0, 1/2 \leqslant t \leqslant 1\}$, the two inequalities in (14) are all reversed.

Remark 2 $L_r(a, b; 0)$ is the generalized logarithmic mean (or Stolarsky's mean)

$$S_r(a, b) = \left(\frac{b^r - a^r}{r(b-a)} \right)^{\frac{1}{r-1}}$$

Taking $t = 0$, from (14), we can obtain known inequality[5]

$$\frac{a+b}{2} \leqslant S_r(a, b) \leqslant \frac{a+b}{r^{\frac{1}{r-1}}}, \quad r > 2 \tag{15}$$

Theorem 4 Let $\frac{1}{2} \leqslant t < 1, a, b \in \mathbf{R}_+$, and let

$$L(a, b; t) = \frac{(\ln b - \ln a) - (\ln u - \ln v)}{2(1-t)(b-a)}, a \neq b$$

$$L(a, a; t) = a^{-1}$$

where $u = tb + (1 - t)a, v = ta + (1 - t)b$. Then $L(a, b; t)$ is Schur-convex on $\mathbf{R}_+^2$.

Proof Taking $f(x) = x^{-1}$, then for $a \neq b$, from Theorem 1, we have

$$G(t) = \frac{1}{2(b-a)} \int_a^b \left[(ta + (1-t)x)^{-1} + (tb + (1-t)x)^{-1} \right] dx$$

$$= \frac{1}{2(1-t)(b-a)} \left[\ln(ta + (1-t)x) \Big|_a^b + \ln(tb + (1-t)x) \Big|_a^b \right]$$

$$= \frac{(\ln b - \ln a) - [\ln(tb + (1-t)a) - \ln(ta + (1-t)b)]}{2(1-t)(b-a)}$$

$$= \frac{(\ln b - \ln a) - (\ln u - \ln v)}{2(1-t)(b-a)}$$

Since $f(x) = x^{-1}$ is convex on $\mathbf{R}_+$, when $\frac{1}{2} \leqslant t < 1$, by Theorem 1, it follows that $L(a, b; t)$ is Schur-convex on $\mathbf{R}_+^2$.

From Theorem 4 and $\left(\frac{a+b}{2}, \frac{a+b}{2} \right) \prec (a, b)$, we get

Corllary 2 Let $\frac{1}{2} \leqslant t < 1, a, b \in \mathbf{R}_+$. Then

$$L(a, b; t) \leqslant \frac{2}{a+b} \tag{16}$$

Remark 3 $L(a, b; 0)$ is the logarithmic mean $L(a, b) = \frac{\ln b - \ln a}{b-a}$. Taking $t = 0$, from (16), we can obtain the Ostle-Terwilliger inequality [7]

$$\frac{\ln b - \ln a}{b - a} \leqslant \frac{2}{a+b} \tag{17}$$

Theorem 5 Let $t \in (0, 1)$, $a, b \in \mathbf{R}_+ = [0, +\infty)$, and let

$$Q_r(a, b; t) = \left[\frac{a^{r+1} + b^{r+1} - (u^{r+1} - v^{r+1})}{r(r+1)t(1-t)(b-a)^2} \right]^{\frac{1}{r-1}}, a \neq b$$

$$Q_r(a, a; t) = a$$

where $u = tb + (1-t)a$, $v = ta + (1-t)b$, if $r \geqslant 2$, then $Q_r(a, b; t)$ is Schur-convex on $\mathbf{R}_+^2$, if $r \in \{1 \leqslant r < 2\} \cup \{r < 1, r \neq 0, -1\}$, then $Q_r(a, b; t)$ is Schur-concave on $\mathbf{R}_+^2$.

Proof Taking $f(x) = x^{r-1}, r \neq 0$, then for $a \neq b$

$$Q(a, b) = F(t) = \frac{1}{(b-a)^2} \int_a^b \int_a^b f(tx + (1-t)y) \, dx \, dy$$

$$= \frac{1}{rt(b-a)^2} \int_a^b (tx + (1-t)y)^r \Big|_a^b \, dy$$

$$= \frac{1}{rt(b-a)^2} \int_a^b \left[(tb + (1-t)y)^r - (ta + (1-t)y)^r \right] \mathrm{d}\,y$$

$$= \frac{1}{r(r+1)t(1-t)(b-a)^2} \left[(tb+(1-t)y)^{r+1} - (ta+(1-t)y)^{r+1} \right]_a^b$$

$$= \frac{b^{r+1} - (ta+(1-t)y)^{r+1} - (tb+(1-t)y)^{r+1} + a^{r+1}}{\cdot\, r(r+1)t(1-t)(b-a)^2}$$

$$= \frac{a^{r+1} + b^{r+1} - (u^{r+1} - v^{r+1})}{r(r+1)t(1-t)(b-a)^2}$$

The following discussions are similar with Theorem 3, hence it is omitted. Theorem 5 is completed.

Corllary 3 When $r \geqslant 2$, we have

$$\frac{a+b}{2} \leqslant Q_r(a,b;t) \leqslant (a+b) \left(\frac{1 - (1-t)^{r+1} - r^{r+1}}{r(r+1)t(1-t)} \right)^{\frac{1}{r-1}} \tag{18}$$

when $r \in \{1 \leqslant r < 2\} \cup \{r < 1, r \neq 0, -1\}$, the two inequalities in (17) are all reversed.

Proof Since

$$\left(\frac{a+b}{2}, \frac{a+b}{2} \right) \prec (a,b) \prec (a+b, 0)$$

then from Theorem 5, when $r \geqslant 2$ we have

$$Q_r \left(\frac{a+b}{2}, \frac{a+b}{2} \right) \leqslant Q_r(a,b) \leqslant Q_r(a+b, 0)$$

i.e. (18) is holds. when $r \in \{1 \leqslant r < 2\} \cup \{r < 1, r \neq 0, -1\}$, the two inequalities in (18) are all reversed.

Acknowledgements

The author is indebted to associate professor Xiao-Ming Zhang for his valuable comments and suggestions.

References

[1]　S S DRAGOMIR. Two mappings in connection to Hadamard's inequalities[J]. J. Math. Anal. Appl. 1992,167(1): 49-56.

[2]　G S YANG, M C HHONG. A note on Hadamard's inequality[J]. Tamkang.J. Math. 1997, 28 (1): 33-37.

[3] S S DRAGOMIR. Further properties of some mappings associated with Hermite- Hadamard's inequalities[J]. Tamkang. J. Math., 2003, 34(1): 45-57.

[4] N ELEZOVIC, J PECARIC. A note on Schur-convex fuctions[J]. Rocky Mountain J. Math. 2000, 30 (3): 853-856.

[5] B Y WANG. Foundations of Majorization Inequalities[M]. Beijing: Beijing Normal Univ. Press, 1990. (Chinese)

[6] J C KUANG. Applied Inequalities (Chang yong bu deng shi)[M]. 3rd ed. Jinan: Shandong Press of science and technology, 2002. (Chinese).

[7] A W MARSHALL, I OLKIN. Inequalities: Theory of majorization and its application[M]. New York: Academies Press, 1979.

[8] B OSTLE, H L TERWILLIGER. A companion of two means[J]. Proc. Montana Acad.Sci. 1957, 17(1): 69-70.

第31篇 两个凸函数单调平均不等式的改进

(续铁权，石焕南. 数学的实践与认识, 2007, 37(19)：150-154)

摘 要： 本文改进了有关凹函数和凸函数的算术平均值单调性的某些已知结论. 作为应用, 加强了Minc-Sathre 不等式和Alzer 不等式.

关键词： 不等式; 凸函数; 凹函数; 单调性; Minc-Sathre 不等式; Alzer 不等式

31.1 引言

定理 A[1] 设 $f(t)$ 是 $[0,1]$ 上增加的凸(或凹) 函数, $\{a_n\}$ 是递增的正数列, 使得 $\left\{i\left(\frac{a_i}{a_{i+1}}-1\right)\right\}$ 递减(或 $\left\{i\left(\frac{a_{i+1}}{a_i}-1\right)\right\}$ 递增), 则

$$\frac{1}{n}\sum_{i=1}^{n}f\left(\frac{a_i}{a_n}\right) \geqslant \frac{1}{n+1}\sum_{i=1}^{n+1}f\left(\frac{a_i}{a_{n+1}}\right) \geqslant \int_0^1 f(x)\mathrm{d}x \tag{1}$$

定理 B[2] 设 $f(t)$ 是 $[0,1]$ 上增加的凸(或凹) 函数, $\{a_n\}$ 是递增的正数列, 使得 $\left\{i\left(\frac{a_{i+1}}{a_i}-1\right)\right\}$ 递减(或 $\left\{i\left(\frac{a_i}{a_{i+1}}-1\right)\right\}$ 递增), 规定 $a_0 = 0$, 则

$$\int_0^1 f(x)\mathrm{d}x \geqslant \frac{1}{n+1}\sum_{i=0}^{n}f\left(\frac{a_i}{a_{n+1}}\right) \geqslant \frac{1}{n}\sum_{i=0}^{n-1}f\left(\frac{a_i}{a_n}\right) \tag{2}$$

本文将这两个定理改进为下面一组定理.

定理 1 设 $f(t)$ 是 $[0,1]$ 上增加的凹函数, $\{a_n\}$ 是递增的正数列, 使得 $\left\{i\left(\frac{a_{i+1}}{a_i}-1\right)\right\}$ 递增, 则

$$\frac{1}{n}\sum_{i=1}^{n}f\left(\frac{a_i}{a_{n+1}}\right) \geqslant \frac{1}{n+1}\sum_{i=1}^{n+1}f\left(\frac{a_i}{a_{n+2}}\right) \geqslant f(0) \tag{3}$$

若$f(t)$在$t = 0$右连续, 则下界$f(0)$是最好的. 即当$f(t)$给定之后, 对于一切满足定理1要求的数列$\{a_n\}$和自然数n, $f(0)$是数集$\left\{\frac{1}{n}\sum_{i=1}^{n}f\left(\frac{a_i}{a_{n+1}}\right)\right\}$的最大下界.

定理 2 设$f(t)$是$[0,1]$上递增的凸函数, $\{a_n\}$是递增的正数列, 使得$\left\{i\left(\frac{a_i}{a_{i+1}} - 1\right)\right\}$递减, 则

$$\frac{1}{n}\sum_{i=1}^{n}f\left(\frac{a_i}{a_n}\right) \geqslant \frac{1}{n+1}\sum_{i=1}^{n+1}f\left(\frac{a_i}{a_{n+1}}\right) \geqslant f(0) \tag{4}$$

下界$f(0)$是最好的.

定理 3 设$f(t)$是$[0,1]$上递增的凹函数, $\{a_n\}$是递增的正数列, 使得$\left\{i\left(\frac{a_{i-1}}{a_i} - 1\right)\right\}$递增, 规定$a_0 = 0$, 则

$$\frac{1}{n}\sum_{i=0}^{n-1}f\left(\frac{a_i}{a_{n-1}}\right) \leqslant \frac{1}{n+1}\sum_{i=0}^{n}f\left(\frac{a_i}{a_n}\right) \leqslant f(1) \tag{5}$$

上界$f(1)$是最好的.

定理 4 设$f(t)$是$[0,1]$上增加的凸函数, $\{a_n\}$是递增的正数列, 使得$\left\{i\left(\frac{a_{i+1}}{a_i} - 1\right)\right\}$递减, 规定$a_0 = 0$, 则

$$\frac{1}{n}\sum_{i=0}^{n-1}f\left(\frac{a_i}{a_n}\right) \leqslant \frac{1}{n+1}\sum_{i=0}^{n}f\left(\frac{a_i}{a_{n+1}}\right) \leqslant f(1) \tag{6}$$

若$f(t)$在$t = 1$左连续, 则上界$f(1)$是最好的.

应用上述定理, 本文将加强Minc-Sathre不等式和Alzer不等式. 本文还将说明式(1)的第二个不等式和式(2)的第一个不等式不成立.

31.2 定理的证明

定理 1 的证明 先证明第一个不等式. 因为$\{a_n\}$是递增的正数列, 使得$\left\{i\left(\frac{a_{i+1}}{a_i} - 1\right)\right\}$递增, 所以当$1 \leqslant i \leqslant n$时, 有

$$(n+1)\left(\frac{a_{n+2}}{a_{n+1}} - 1\right) \geqslant i\left(\frac{a_{i+1}}{a_i} - 1\right)$$

即

$$\frac{(n+1)a_{n+2}}{a_{n+1}} \geqslant \frac{i(a_{i+1} - a_i)}{a_i} + (n+1)$$

即

$$\frac{ia_{i+1} + (n-i+1)a_i}{(n+1)a_{n+2}} \leqslant \frac{a_i}{a_{n+1}}$$

因为 $f(t)$ 是 $[0,1]$ 上增加的凹函数, 所以

$$\frac{i}{n+1}f\left(\frac{a_{i+1}}{a_{n+2}}\right) + \left(1 - \frac{i}{n+1}\right)f\left(\frac{a_i}{a_{n+2}}\right)$$

$$\leqslant f\left(\frac{ia_{i+1} + (n-i+1)a_i}{(n+1)a_{n+2}}\right) \leqslant f\left(\frac{a_i}{a_{n+1}}\right)$$

$$\sum_{i=1}^{n}\left[\frac{i}{n+1}f\left(\frac{a_{i+1}}{a_{n+2}}\right) + \left(\frac{n}{n+1} - \frac{i-1}{n+1}\right)f\left(\frac{a_i}{a_{n+2}}\right)\right] \leqslant \sum_{i=1}^{n}f\left(\frac{a_i}{a_{n+1}}\right)$$

注意

$$\sum_{i=1}^{n}\left[\frac{i}{n+1}f\left(\frac{a_{i+1}}{a_{n+2}}\right) - \frac{i-1}{n+1}f\left(\frac{a_i}{a_{n+2}}\right)\right] = \frac{n}{n+1}f\left(\frac{a_{n+1}}{a_{n+2}}\right)$$

有

$$\frac{n}{n+1}\sum_{i=1}^{n+1}f\left(\frac{a_i}{a_{n+2}}\right) \leqslant \sum_{i=1}^{n}f\left(\frac{a_i}{a_{n+1}}\right)$$

由此即得式(3)的第一个不等式. 第二个不等式是显然的. 现在我们证明: 若 $f(t)$ 在 $t = 0$ 处右连续, 则下界 $f(0)$ 是最好的.

因为 $f(t)$ 在 $t = 0$ 右连续, 任取 $\varepsilon > 0$, 存在 $m > 1$, 使得当 $0 \leqslant t \leqslant \frac{1}{m}$, 有 $f(0) \leqslant f(t) < f(0) + \varepsilon$. 令 $a_n = m^n$, $\{a_n\}$ 是递增的正数列, $i\left(\frac{a_{i+1}}{a_i} - 1\right) = (m-1)i$ 递增, 这时

$$\frac{1}{n}\sum_{i=1}^{n}f\left(\frac{a_i}{a_{n+1}}\right) = \frac{1}{n}\left[f\left(\frac{1}{m}\right) + \left(\frac{1}{m^2}\right) + \cdots + \left(\frac{1}{m^n}\right)\right] < f(0) + \varepsilon$$

这就证明了 $f(0)$ 是数集 $\left\{\frac{1}{n}\sum\limits_{i=1}^{n}f\left(\frac{a_i}{a_{n+1}}\right)\right\}$ 的最大下界.

注 1　由式(3) 第一个不等式可推出式(1) 的第一个不等式. 这是因为, 由式(3) 得

$$(n+1)\sum_{i=1}^{n}f\left(\frac{a_i}{a_{n+1}}\right) \geqslant n\sum_{i=1}^{n+1}f\left(\frac{a_i}{a_{n+2}}\right) \tag{7}$$

由 $\{a_n\}$ 递增和 $f(t)$ 在 $[0,1]$ 上增加可得 $f\left(\frac{a_i}{a_{n+1}}\right) \geqslant f\left(\frac{a_i}{a_{n+2}}\right), i = 1, 2, \ldots, n+$

1, 所以

$$\sum_{i=1}^{n+1} f\left(\frac{a_i}{a_{n+1}}\right) \geqslant \sum_{i=1}^{n+1} f\left(\frac{a_i}{a_{n+2}}\right) \tag{8}$$

最后, 显然有

$$(n+1)f\left(\frac{a_{n+1}}{a_{n+1}}\right) = (n+1)f\left(\frac{a_{n+2}}{a_{n+2}}\right) \tag{9}$$

将(7), (8), (9) 三式相加, 并把 n 换成 $n-1$, 就得到

$$(n+1)\sum_{i=1}^{n} f\left(\frac{a_i}{a_n}\right) \geqslant n\sum_{i=1}^{n+1} f\left(\frac{a_i}{a_{n+1}}\right)$$

它等价于式(1) 的第一个不等式.

定理 2 的证明　文[1]的定理1 已证明式(4)的第一个不等式, 式(4)的第二个不等式是显然的. 本文的第四部分将证明: 若 $f(t)$ 是 $[0,1]$ 上增加的凸函数, 则 $f(t)$ 在 $t=0$ 处右连续. 于是仿定理1 可证下界 $f(0)$ 是最好的.

定理 3 的证明　先证明第一个不等式. 因为 $\{a_n\}$ 是递增的正数列, 使得 $\left\{i\left(\frac{a_{i-1}}{a_i}-1\right)\right\}$ 递增, 所以当 $1 \leqslant i \leqslant n$ 时, 有

$$n\left(\frac{a_{n-1}}{a_n}-1\right) \geqslant i\left(\frac{a_{i-1}}{a_i}-1\right)$$

即

$$\frac{na_{n-1}}{a_n} \geqslant \frac{i(a_{i-1}-a_i)}{a_i} + n$$

即

$$\frac{ia_{i-1}+(n-i)a_i}{na_{n-1}} \leqslant \frac{a_i}{a_n}$$

因为 $f(t)$ 是 $[0,1]$ 上增加的凹函数, 所以

$$\frac{i}{n}f\left(\frac{a_{i-1}}{a_{n-1}}\right) + \left(1-\frac{i}{n}\right)f\left(\frac{a_i}{a_{n-1}}\right) \leqslant f\left(\frac{ia_{i-1}+(n-i)a_i}{na_{n-1}}\right) \leqslant f\left(\frac{a_i}{a_n}\right)$$

即

$$\frac{i}{n}f\left(\frac{a_{i-1}}{a_{n-1}}\right) + \left(\frac{n+1}{n}-\frac{i+1}{n}\right)f\left(\frac{a_i}{a_{n-1}}\right) \leqslant f\left(\frac{a_i}{a_n}\right)$$

于是

$$\sum_{i=1}^{n-1}\left[\frac{i}{n}f\left(\frac{a_{i-1}}{a_{n-1}}\right) + \left(\frac{n+1}{n}-\frac{i+1}{n}\right)f\left(\frac{a_i}{a_{n-1}}\right)\right] \leqslant \sum_{i=1}^{n-1} f\left(\frac{a_i}{a_{n-1}}\right)$$

注意

$$\sum_{i=1}^{n-1}\left[\frac{i}{n}f\left(\frac{a_{i-1}}{a_{n-1}}\right)-\frac{i+1}{n}f\left(\frac{a_i}{a_{n-1}}\right)\right]=\frac{1}{n}f(0)-f(1)$$

有

$$\frac{n+1}{n}\sum_{i=0}^{n-1}f\left(\frac{a_i}{a_{n-1}}\right)\leqslant\sum_{i=0}^{n-1}f\left(\frac{a_i}{a_n}\right)$$

由此即得式(5)的第一个不等式. 第二个不等式是显然的. 本文的第四部分将证明: 若$f(t)$ 是$[0,1]$ 上增加的凹函数, 则$f(t)$ 在$t=1$ 处左连续. 现在我们证明: 上界$f(1)$ 是最好的.

因为$f(t)$ 在$t=1$ 左连续. 任取$\varepsilon>0$, 存在$m>1$, 使得当$1-\frac{1}{m}\leqslant t\leqslant1$, 有$f(1)-\frac{\varepsilon}{2}\leqslant f(t)\leqslant f(1)$. 不妨假定$m>\mathrm{e}$. 令$a_n=1-\frac{1}{m^n}$, $\{a_n\}$ 是递增的正数列

$$i\left(\frac{a_{i-1}}{a_i}-1\right)=i\cdot\frac{\frac{1}{m_i}-\frac{1}{m^{i-1}}}{1-\frac{1}{m^i}}=i\cdot\frac{1-m}{m^i-1}$$

为了证明它是递增的, 注意$1-m<0$, 只须证明函数$g(t)=\frac{t}{m^t-1}$ 在$[1,+\infty)$ 上减少. 微分可得$(m^t-1)^2g'(t)=m^t-1-tm^t\ln m$, 因$m>\mathrm{e},\ln m>1$, 所以当$t\geqslant1$ 时, $m^t-1-tm^t\ln m=m^t(1-t\ln m)-1<0$, 函数$g(t)=\frac{t}{m^t-1}$ 在$[1,+\infty)$ 上减少. 这时

$$\begin{aligned}\frac{1}{n}\sum_{i=0}^{n-1}f\left(\frac{a_i}{a_{n-1}}\right)&=\frac{1}{n}\sum_{i=0}^{n-1}f\left(\frac{1-\frac{1}{m^i}}{1-\frac{1}{m^{n-1}}}\right)\geqslant\frac{1}{n}\sum_{i=0}^{n-1}f\left(1-\frac{1}{m^i}\right)\\&\geqslant\frac{1}{n}\left[f(0)+(n-1)\left(f(1)-\frac{\varepsilon}{2}\right)\right]\\&\geqslant f(1)-\frac{1}{n}(f(1)-f(0))-\frac{\varepsilon}{2}\end{aligned}$$

取n 充分大, 使得$\frac{f(1)-f(0)}{n}<\frac{\varepsilon}{2}$, 于是$\frac{1}{n}\sum_{i=0}^{n-1}f\left(\frac{a_i}{a_{n-1}}\right)>f(1)-\varepsilon$, 这就证明了$f(1)$ 是数集$\left\{\frac{1}{n}\sum_{i=0}^{n-1}f\left(\frac{a_i}{a_{n-1}}\right)\right\}$的最小上界.

注 2　从$\left\{i\left(\frac{a_i}{a_{i+1}}-1\right)\right\}$ 递增可以推出当$i\geqslant2$时, $\left\{i\left(\frac{a_{i-1}}{a_i}-1\right)\right\}$ 递增. 因为$\left\{i\left(\frac{a_i}{a_{i+1}}-1\right)\right\}$ 递增$\Rightarrow i\left(\frac{a_{i-1}}{a_i}-1\right)\leqslant(i+1)\left(\frac{a_{i+1}}{a_{i+2}}-1\right)\leqslant i\left(\frac{a_{i+1}}{a_{i+2}}-1\right)\Rightarrow\frac{a_i}{a_{i+1}}\leqslant\frac{a_{i+1}}{a_{i+2}}$, 所以当$\left\{i\left(\frac{a_i}{a_{i+1}}-1\right)\right\}$ 递增时, $\left\{\frac{a_i}{a_{i+1}}\right\}$ 递增. 于是有$\frac{a_i}{a_{i+1}}-1\leqslant\frac{a_{i+1}}{a_{i+2}}-1$, 两式相加, 得$(i+1)\left(\frac{a_i}{a_{i+1}}-1\right)\leqslant(i+2)\left(\frac{a_{i+1}}{a_{i+2}}-1\right)$, 这就说明当$i\geqslant2$时, $\left\{i\left(\frac{a_{i-1}}{a_i}-1\right)\right\}$ 递增.

注 3　由式(5)第一个不等式可推出式(2)的第一个不等式. 这是因为,

由式(5)得

$$(n+1)\sum_{i=0}^{n-1} f\left(\frac{a_i}{a_{n-1}}\right) \leqslant n\sum_{i=0}^{n} f\left(\frac{a_i}{a_n}\right) \qquad (10)$$

由$\{a_n\}$递增和$f(t)$在$[0,1]$上增加可得$f\left(\frac{a_i}{a_{n-1}}\right) \geqslant f\left(\frac{a_i}{a_n}\right), i = 0,1,...,n-1,$ 所以

$$-\sum_{i=0}^{n-1} f\left(\frac{a_i}{a_{n-1}}\right) \leqslant -\sum_{i=0}^{n-1} f\left(\frac{a_i}{a_n}\right) \qquad (11)$$

最后, 显然有

$$-nf\left(\frac{a_{n-1}}{a_{n-1}}\right) = -nf\left(\frac{a_n}{a_n}\right) \qquad (12)$$

将(10), (11), (12) 三式相加, 并把n换成$n+1$, 就得到

$$(n+1)\sum_{i=0}^{n-1} f\left(\frac{a_i}{a_n}\right) \leqslant n\sum_{i=0}^{n} f\left(\frac{a_i}{a_{n+1}}\right)$$

它等价于式(2)的第一个不等式.

定理 4 的证明 文[2]的定理2 已证明式(6)的第一个不等式, 式(6)的第二个不等式是显然的. 仿定理3 可证: 若$f(t)$在$t = 1$处左连续, 则上界$f(1)$是最好的.

注 4 定理4 的条件"$\{a_n\}$是递增的正数列, 使得$\left\{i\left(\frac{a_{i+1}}{a_i}-1\right)\right\}$递减, 规定$a_0 = 0$"可以修改为"$\{a_n\}$是递增的正数列, 使得$\left\{i\left(\frac{a_i}{a_{i-1}}-1\right)\right\}$递减, 且$a_0 \geqslant 0$", 但当$a_0 > 0$时, 应当在$i = 1,2,...$时$\left\{i\left(\frac{a_i}{a_{i-1}}-1\right)\right\}$递减, 修改的定理可以仿照定理3 证明之, 此处从略.

31.3 应用

下面应用这些定理证明一些不等式. 对于自然数n, 有Minc-Sathre 不等式[3]

$$\frac{n}{n+1} < \frac{\sqrt[n]{n!}}{\sqrt[n+1]{(n+1)!}} < 1 \qquad (13)$$

令$f(t) = \ln t, t \in (0,1]$, $f(t)$是$(0,1]$上增加的凹函数. 取$a_i = i$, 使用定理A 的(1), 得

$$\frac{1}{n}\sum_{i=1}^{n}(\ln i - \ln n) \geqslant \frac{1}{n+1}\sum_{i=1}^{n+1}[\ln i - \ln(n+1)]$$

由此得到式(13)的第一个不等式. 如果使用定理1 的式(3), 得

$$\frac{1}{n}\sum_{i=1}^{n}[\ln i-\ln(n+1)]\geqslant\frac{1}{n+1}\sum_{i=1}^{n+1}[\ln i-\ln(n+2)]$$

由此得到

推论 1　对于自然数n, 有

$$\frac{n+1}{n+2}\leqslant\frac{\sqrt[n]{n!}}{\sqrt[n+1]{(n+1)!}}\tag{14}$$

显然式(14) 强于式(13) 的第一个不等式

设$r>0$, 有Alzer 不等式[1-3]

$$\frac{n}{n+1}\leqslant\left[\frac{\dfrac{1}{n}\sum\limits_{i=1}^{n}i^r}{\dfrac{1}{n+1}\sum\limits_{i=1}^{n+1}i^r}\right]^{\frac{1}{r}}\tag{15}$$

文[1], [2] 指出用定理A 的式(1) 可证明式(15).

令$f(t)=t^r, t\in[0,1]$, 当$0<r\leqslant 1, f(t)$ 是$[0,1]$ 上增加的凹函数. 取$a_i=i$, 由定理1 的式(3) 知

$$\frac{1}{n(n+1)^r}\sum_{i=1}^{n}i^r\geqslant\frac{1}{(n+1)(n+2)}\sum_{i=1}^{n+1}i^r$$

由此得到

推论 2　当$0<r\leqslant 1$, 有

$$\frac{n+1}{n+2}\leqslant\left[\frac{\dfrac{1}{n}\sum\limits_{i=1}^{n}i^r}{\dfrac{1}{n+1}\sum\limits_{i=1}^{n+1}i^r}\right]^{\frac{1}{r}}\tag{16}$$

显然当$0<r\leqslant 1$ 时, 式(16)强于式(15).

文[2]用定理B 的式(2)证明了当$r>0$ 有

$$\left[\frac{\dfrac{1}{n}\sum\limits_{i=1}^{n}i^r}{\dfrac{1}{n+1}\sum\limits_{i=1}^{n+1}i^r}\right]^{\frac{1}{r}}\leqslant\frac{n}{n+1}$$

令 $f(t) = t^r, t \in [0,1]$, 当 $0 < r \leqslant 1$, $f(t)$ 是 $[0,1]$ 上增加的凹函数. 取 $a_i = i$, 由定理3 的式(5)知

$$\frac{1}{n(n-1)^r} \sum_{i=1}^{n-1} i^r \geqslant \frac{1}{(n+1)n^r} \sum_{i=1}^{n} i^r$$

由此得到

推论 3 当 $0 < r \leqslant 1$, 有

$$\left[\frac{\frac{1}{n} \sum_{i=1}^{n-1} i^r}{\frac{1}{n+1} \sum_{i=1}^{n} i^r} \right]^{\frac{1}{r}} \leqslant \frac{n-1}{n}$$

显然当 $0 < r \leqslant 1$, 式(18) 强于式(17).

31.4 几点说明

(I) **命题**   若 $f(t)$ 是 $[0,1]$ 上增加的凸函数, 则 $f(t)$ 在 $t = 0$ 右连续; 若 $f(t)$ 是 $[0,1]$ 上增加的凹函数, 则 $f(t)$ 在 $t = 1$ 处左连续.

证明   只证明第一个命题. 因为 $f(t)$ 增加, 所以 $f(0+0)$ 存在, 记做 A.

若 $f(0) < A$, 取 $\varepsilon = A - f(0) > 0$, 存在 $\delta > 0$, 当 $0 < t < \delta, A \leqslant f(t) < A + \varepsilon$ 时, 令 $t_1 = \frac{\delta}{2}, t_2 = \frac{\delta}{4}$, 则 $f(t_1) < A + \varepsilon$, 由凸函数定义及 $f(0) = A - \varepsilon$, 得到

$$f(t_2) = f\left(\frac{t_1 + 0}{2} \right) \leqslant \frac{1}{2}[f(t_1) + f(0)] < \frac{1}{2}[(A+\varepsilon) + (A-\varepsilon)] = A$$

这与 $f(t_2) \geqslant A$ 矛盾. 若 $f(0) > A$, 取 $\varepsilon = f(0) - A > 0$, 存在 $\delta > 0$, 当 $0 < t < \delta, A \leqslant f(t) < A + \varepsilon = f(0)$, 这与 $f(t)$ 增加矛盾. 所以 $f(0) = A$, 即 $f(t)$ 在 $t = 0$ 右连续.

(II) 现在说明式(1)的第二个不等式和式(2) 的第一个不等式不成立. 我们只说明凸函数的情形. 任取在 $[0,1]$ 上严格增加的连续的凸函数 $f(t)$, 由Hadamard 不等式得

$$f(0) < f\left(\frac{0+1}{2} \right) \leqslant \int_0^1 f(t)\,\mathrm{d}t \leqslant (f(0) + f(1)) < f(1)$$

根据定理2知, 对于一切满足定理2 要求的数列 $\{a_n\}$ 和自然数 n, 数集 $\left\{ \frac{1}{n} f\left(\frac{a_i}{a_n} \right) \right\}$ 的最大下界是 $f(0)$, 因为 $f(0) < \int_0^1 f(t)\mathrm{d}t$, 所以 $\int_0^1 f(t)\mathrm{d}t$ 不是数集 $\left\{ \frac{1}{n} f\left(\frac{a_i}{a_n} \right) \right\}$

的下界, 即存在满足定理2 要求的数列$\{a_n\}$ 和自然数n, 使$\frac{1}{n}\sum_{i=1}^{n} f\left(\frac{a_i}{a_n}\right) < \int_0^1 f(t)\mathrm{d}t$, 从而式(1) 的第二个不等式不成立. 同样, 对于一切满足定理4 要求的数列$\{a_n\}$ 和自然数n, $\int_0^1 f(t)\,\mathrm{d}t$不是数集$\left\{\frac{1}{n}\sum_{i=0}^{n-1} f\left(\frac{a_i}{a_n}\right)\right\}$ 的上界, 式(2)的第一个不等式不成立.

(III) 前面的定理$1-4$ 中的$f(t)$ 均是$[0,1]$ 上增加的凸函数(凹函数), 如果把$f(t)$ 改为$[0,1]$ 上减少的凹函数(凸函数) , 可以得到一组对偶的定理. 我们只写出定理1 的对偶定理, 其证明与定理1 相同, 从略.

定理 5　设$f(t)$ 是$[0,1]$ 上减少的凸函数, $\{a_n\}$ 是递增的正数列, 使得$\left\{i\left(\frac{a_{i+1}}{a_i}-1\right)\right\}$ 递增, 则式(3)反向成立, 即

$$\frac{1}{n}\sum_{i=1}^{n} f\left(\frac{a_i}{a_{n+1}}\right) \leqslant \frac{1}{n+1}\sum_{i=1}^{n+1} f\left(\frac{a_i}{a_{n+2}}\right) \leqslant f(0) \tag{3'}$$

若$f(t)$ 在$t=0$ 右连续, 则上界$f(0)$ 是最好的.

利用这一定理, 可以证明当$r<0$, 不等式(16)仍成立.

参考文献

[1]　QI F, GUO B N. Monotonicity of sequences involving conex function and sequence [J]. RGMIA Res Rep Coll, 2000, 3 (2): 321-329.

[2]　CHEN C P, QI F, Cerone P, D ragomir S S. Monotonicity of sequences involving conex and concave functions[J]. Math Inequal App l, 2003, 6 (2): 229-239. RGMIA Res. Rep. Co ll, 2002, 5 (1): 3-13.

[3]　KUANG J C. Some extensions and refinements of Minc-Sathre inequality [J]. Math Gaz, 1999, 83 (17) : 123-127.

第32篇　Refinement of an Inequality for the Generalized Logarithmic Mean

(SHI HUAN-NAN, WU SHANHE. Chin. Quart. J of Math., 2008,

23(4): 594-599)

Abstract:　In this article, we show that the generalized logarithmic mean is strictly Schur-convex function for $p > 2$ and strictly Schur-concave function for $p < 2$ on $\mathbf{R}_+^2$. And then we give a refinement of an inequality for the generalized logarithmic mean inequality using a simple majoricotion relation of the vector.

Keywords:　inequality; refinement; generalized logarithmic mean; strictly Schur-convex function; strictly Schur-concave function

32.1　Introduction

The generalized logarithmic mean (or Stolarsky's mean) [1] of two positive numbers a and b is defined as follows

$$
S_p(a,b) = \begin{cases}
\left(\frac{b^p - a^p}{p(b-a)}\right)^{\frac{1}{p-1}}, & p \neq 0, 1, \ a \neq b \\
\mathrm{e}^{-1}\left(a^a/b^b\right)^{1/(a-b)}, & p = 1, \ a \neq b \\
\frac{b-a}{\ln b - \ln a}, & p = 0, \ a \neq b \\
b, & a = b
\end{cases}
$$

In particular, $S_1(a,b)$ is also called the exponent mean of a and b, in symbols $E(a,b)$. $S_0(a,b)$ is also called the logarithmic mean of a and b, in symbols $L(a,b)$.

By using different methods respectively, Stolarsky K. B. [2] and Liu Zheng [3] proved that, when $a \neq b$, $S_p(a,b)$ is a strictly increasing function for p.

In 1984, Yang Zhenhang [4] obtained an upper bound of $S_p(a,b)$

$$S_p(a,b) < M_{p-1}(a,b) = \left[\frac{1}{2} \left(a^{p-1} + b^{p-1} \right) \right]^{1/(p-1)}, \quad p > 2 \qquad (1)$$

In 1997, Shi Minqi and Shi Huanan [5] obtained other upper bound of $S_p(a,b)$

$$S_p(a,b) < \frac{a+b}{p^{\frac{1}{p-1}}}, \quad p > 2 \qquad (2)$$

and showed that two upper bounds in (1) and (2) can not compare.

In 1999, Li Kanghai[6] further proved that, if $p > 2$ then

$$\frac{a+b}{2} < S_p(a,b) < \frac{a+b}{p^{\frac{1}{p-1}}}, \quad a \neq b \qquad (3)$$

if $0 < p < 2$ and $p \neq 1$ then then two inequalities in (3) are reversed.

In 2000, by using Hermite-Hadamard inequality [1], i.e.

$$\frac{1}{b-a} \int_a^b f(t)\,\mathrm{d}t \leqslant \frac{f(a)+f(b)}{2}$$

holds for all $a, b \in I$ if and if f is convex function.

Elezovic N and Pecaric J [7-8] proved the following theorem:

Theorem A　Let I be an interval with nonempty interior on R and f be a continuous function on I. Then

$$F(a,b) = \begin{cases} \frac{1}{b-a} \int_a^b f(t)\,\mathrm{d}t, & a, b \in I, \ a \neq b \\ f(a), & a = b \end{cases} \qquad (4)$$

is Schur-convex (Schur-concave) on I^2 if and only if f is convex (concave) on I.

Remark 1　The equality in Hermite-Hadamard's inequality holds[1] only if f is a linear function, therefore if f is a nonlinear convex (concave) function on I, then $F(a,b)$ is strictly Schur-convex (Schur-concave) function on I^2.

In 2000, Qi Feng [9] established the following weighted form of inequality(4).

Theorem B　Let f be a continuous function and p a positive continuous weight on I. Then the weighted arithmetic mean of function f with weight p defined by

$$F(a,b) = \begin{cases} \dfrac{\int_a^b p(t)f(t)\,dt}{\int_a^b p(t)\,dt}, & a \neq b \\ f(a), & a = b \end{cases}$$

is Schur-convex(Schur-concave) on I^2 if and only if the inequality

$$\frac{\int_a^b p(t)f(t)\,dt}{\int_a^b p(t)\,dt} \leqslant \frac{p(a)f(a)+p(b)f(b)}{p(a)+p(b)}$$

holds (reverses) for $(a,b) \in I^2$.

As a corollary of Theorem A, the following result is given in [7]:

Theorem C $S_p(a,b)$ is Schur- convex for $p > 2$ or Schur-concave for $p < 2$ on $\mathbf{R}_+^2$.

In this article, we show that Schur-concavity (convexity) of $S_p(a,b)$ on $\mathbf{R}_+^2$ is strictly and give a refinement of (3). Our main results are as follows.

Theorem 1 $S_p(a,b)$ is an increasing and strictly Schur-convex function for $p > 2$, or an increasing and strictly Schur-concave function for $p < 2$ on $\mathbf{R}_+^2$.

Theorem 2 Let $0 < c < a < b$, $u(t) = tb + (1-t)\frac{a+b}{2}$, $v(t) = ta + (1-t)\frac{a+b}{2}$ and $0 < t_2 < t_1 < 1$.

(I) If $p > 2$, then

$$\frac{a+b}{2} < \left(\frac{u^p(t_2)-v^p(t_2)}{pt_2(b-a)}\right)^{\frac{1}{p-1}} < \left(\frac{u^p(t_1)-v^p(t_1)}{pt_1(b-a)}\right)^{\frac{1}{p-1}}$$

$$< \left(\frac{b^p-a^p}{p(b-a)}\right)^{\frac{1}{p-1}} < \left(\frac{(a+b-c)^p-c^p}{p(a+b-2c)}\right)^{\frac{1}{p-1}} < \frac{a+b}{p^{\frac{1}{p-1}}} \tag{5}$$

If $0 < p < 2$ and $p \neq 1$, then five inequalities in (5) are all reversed. If $p \leqslant 0$, then the very right inequality in (5) be taken out and the other inequalities in (5) are reversed.

(II) When $p = 1$, we have

$$\frac{a+b}{2} > E[u(t_2), v(t_2)] > E[u(t_1), v(t_1)]$$

$$> E(a,b) > E(a+b-c,c) > \frac{a+b}{e} \tag{6}$$

(III) When $p = 0$, we have

$$\frac{a+b}{2} > L[u(t_2), v(t_2)] > L[u(t_1), v(t_1)]$$

$$> L(a,b) > L(a+b-c,c) > 0 \qquad (7)$$

Theorem 3　Let $0 < a < b$, $u(t) = tb + (1-t)\frac{a+b}{2}$ and $v(t) = ta + (1-t)\frac{a+b}{2}$. If $f(x)$ is apositive convex function on $[0, a+b]$ and $0 < t_2 < t_1 < 1$, then for $p > 2$, we have

$$f\left(\frac{a+b}{2}\right) \leqslant \left(\frac{f^p(u(t_2)) - f^p(v(t_2))}{p(f(u(t_2)) - f(v(t_2)))}\right)^{\frac{1}{p-1}} \leqslant \left(\frac{f^p(u(t_1)) - f^p(v(t_1))}{p(f(u(t_1)) - f(v(t_1)))}\right)^{\frac{1}{p-1}}$$

$$\leqslant \left(\frac{f^p(b) - f^p(a)}{p(f(b) - f(a))}\right)^{\frac{1}{p-1}} \leqslant \left(\frac{f^p(a+b) - f^p(0)}{p(f(a+b) - f(0))}\right)^{\frac{1}{p-1}} \qquad (8)$$

If $f(x)$ is a positive concave function on $[0, a+b]$, then for $p < 2$ and $p \neq 1$, four inequalities in (8) are all reversed.

32.2　Definitions and Lemmas

Definition 1[10] Let $x = (x_1, x_2, \ldots, x_n)$ and $y = (y_1, y_2, \ldots, y_n)$ be two real n-tuples. Then x is said to be majorized by y (in symbols $x \prec y$) if

(i) $\sum_{i=1}^{k} x_{[i]} \leqslant \sum_{i=1}^{k} y_{[i]}$, for $k = 1, 2 \cdots, n-1$; (ii) $\sum_{i=1}^{n} x_i = \sum_{i=1}^{n} y_i$.

where $x_{[1]} \geqslant x_{[2]} \geqslant \cdots \geqslant x_{[n]}$ and $y_{[1]} \geqslant y_{[2]} \geqslant \cdots \geqslant y_{[n]}$ are components of x and y rearranged in descending order, and x is said to strictly majors by y (written $x \prec\prec y$) if x is not permutation of y.

Definition 2[10]　Let $\Omega \subset \mathbf{R}^n$, $\varphi: \Omega \to \mathbf{R}$.

(i) φ is said to be an increasing function on Ω if $x \leqslant y$ on $\Omega \Rightarrow \varphi(x) \leqslant \varphi(y)$.

(ii) φ is said to be a Schur-convex function on Ω if $x \prec y$ on $\Omega \Rightarrow \varphi(x) \leqslant \varphi(y)$, φ is said to be a Schur-concave function on Ω if and only if $-\varphi$ is Schur-convex, φ is said to be a strictly Schur-convex function on Ω if $x \prec\prec y$ on $\Omega \Rightarrow \phi(x) < \phi(y)$, φ is said to be a strictly Schur-concave function on Ω if and only if $-\varphi$ is strictly Schur-convex on Ω.

Lemma 1[10] Let $g: I \to \mathbf{R}$, $\varphi: \mathbf{R}^n \to \mathbf{R}$, $\psi(x) = \varphi(g(x_1), g(x_2), \cdots, g(x_n))$.

(i) If g is a convex function and φ is an increasing function and Schur-convex function, then ψ is Schur-convex.

(ii) If g is a concave function and φ is an increasing function and Schur-concave function, then ψ is Schur-concave.

Lemma 2[8,11]　Let interval $I \subset \mathbf{R}$ and f is the continuous function on I. If f is a increasing (decreasing) function on I, then $F(a,b)$ is the increasing (decreasing) function on I^2.

32.3   Proof of Theorems

Proof of Theorem 1   Let $Q(a,b) = \frac{1}{b-a}\int_a^b t^{p-1}\,\mathrm{d}t$, it is easy to see that, when $p \neq 1$, for $a,b \in \mathbf{R}_+$, $a \neq b$, we have

$$S_p(a,b) = [Q(a,b)]^{\frac{1}{p-1}}$$

Now we consider four cases:

Case 1 When $p > 2$, since $f : t \to t^{p-1}$ is the nonlinear convex function on $\mathbf{R}_+$, then from Remark 1 we obtain that $Q(a,b)$ is strictly Schur-convex on $\mathbf{R}_+^2$. Furthermore, since $g : t \to t^{\frac{1}{p-1}}$ is the strictly increasing function on $\mathbf{R}_+$, then $S_p(a,b)$ is also strictly Schur-convex on $\mathbf{R}_+^2$.

Case 2 When $1 < p < 2$, since $f : t \to t^{p-1}$ is the nonlinear concave function on $\mathbf{R}_+$, from Remark 1 we know that $Q(a,b)$ is strictly Schur-concave on $\mathbf{R}_+^2$. Furthermore, since $g : t \to t^{\frac{1}{p-1}}$ is strictly increasing function on $\mathbf{R}_+$, then $S_p(a,b)$ is also strictly Schur-concave on $\mathbf{R}_+^2$.

Case 3 When $p < 1$, since $f : t \to t^{p-1}$ is the nonlinear convex function on $\mathbf{R}_+$, from Remark 1 we know that $Q(a,b)$is strictly Schur-convex on $\mathbf{R}_+^2$. Furthermore, since $g : t \to t^{\frac{1}{p-1}}$ is strictly descending function on $\mathbf{R}_+$, then $S_p(a,b)$ is strictly Schur-concave on $\mathbf{R}_+^2$.

Case 4 When $p = 1$, since $E(a,b) = \exp\left\{\frac{1}{b-a}\int_a^b \ln x\,\mathrm{d}x\right\}$, by similar argument to discussing with Case 1-3, we conclude that $E(a,b)$ is strictly Schur-concave on $\mathbf{R}_+^2$.

Combining Lemma 2, the proof of Theorem 1 is complete.

Proof of Theorem 2   First we show that for $0 \leqslant t_2 < t_1 \leqslant 1$,

$$(u(t_2), v(t_2)) \prec\prec (u(t_1), v(t_1)) \tag{9}$$

In fact, since $a \leqslant b$, $u(t_2) \geqslant v(t_2)$, $u(t_1) \geqslant v(t_1)$, then

$$u(t_1) - u(t_2) = (t_1 - t_2)\left(b - \frac{a+b}{2}\right) \geqslant 0$$

Furthermore, it is clear that $a + b = u(t_2) + v(t_2) = u(t_1) + v(t_1)$, and so (9) holds.

In particular, since

$$(u\,(0)\,,\ v\,(0)) = \left(\frac{a+b}{2},\ \frac{a+b}{2}\right),\quad (u\,(1)\,,\ v\,(1)) = (a,\ b)$$

and

$$(a,\ b) \prec\prec (a+b-c,\ c)$$

then for $0 < t_2 < t_1 < 1$, we have

$$\left(\frac{a+b}{2},\ \frac{a+b}{2}\right) \prec\prec (u\,(t_2),\ v\,(t_2)) \prec\prec (u\,(t_1),\ v\,(t_1))$$

$$\prec\prec (a,\ b) \prec\prec (a+b-c,\ c) \qquad (10)$$

and for $p > 2$, from Theorem 1, we have

$$S_p\left(\frac{a+b}{2},\ \frac{a+b}{2}\right) < S_p\,(u\,(t_2),\ v\,(t_2)) < S_p\,(u\,(t_1),\ v\,(t_1))$$

$$< S_p\,(a,\ b) < S_p\,(a+b-c,\ c)$$

Since

$$S_p\,(a+b-c,\ c) = \left(\frac{(a+b-c)^p - c^p}{p(a+b-2c)}\right)^{\frac{1}{p-1}} \to \frac{a+b}{p^{\frac{1}{p-1}}}\ (c \to 0)$$

then the fifth inequality in (5) also holds. If $0 < p < 2$ and $p \neq 1$, then the five inequalities in (5) are all reversed. If $p \leqslant 0$, then the very right inequality in (5)can be taken out and other qualities in (5) are reversed.

When $p = 1$, using L'Hospital's law, we can prove that

$$E(a+b-c, c) \to \frac{a+b}{e}, c \to 0$$

similarly, we can prove that (6) and (7) holds.

Proof of Theorem 3　Since $f(x)$ is the positive convex function on $[0,\ a+b]$, from (a) in Lemma 1 we obtain that $S_p\,(f\,(a),\,f\,(b))$ is Schur-convex function. And combining (10), we get (8). If $f(x)$ is the positive concave function on $[0,\ a+b]$, from (b) in Lemma 1 we obtain that four inequalities in (8) are all reversed.

Acknowledgements The author is indebted to professor XU Tie-quan and the referee for their many helpful and valuable comments and suggestions.

References

[1] KUANG JI-CHANG. Applied Inequalities(Chang yong bu deng shi) [M].3rd ed.Jinan: Shandong Press of science and technology, 2004. (Chinese)

[2] STOLARSKY K B. Generalizations of the Logarithmic mean [J]. Math, Mag, 1875, 48:87–92.

[3] LIU ZHENG. On generalized Logarithmic Mean [J]. Journal of Anshan Institute of I. & S. Technology, 1998, 21 (5):1–4.

[4] YANG ZHEN-HANG. Other property of the convex function [J]. Chinese Mathematics Bulletin, 1984(2): 31–32.

[5] SHI MIN-QI, SHI HUAN-NAN. An inequality for the generalized Logarithmic Mean [J]. Chinese Mathematics Bulletin, 1997(5): 37–38.

[6] LI KANG-HAI. Two inequality for the generalized Logarithmic Mean [M]//Xue-zhi Yang. Inequalities. People's Press of Tibet ,2000: 117-118.

[7] ELEZOVIC N, PECARIC J. A note on Schur-convex functions [J]. Rocky Mountain J. Math. 2000, 30 (3):853–856.

[8] BULLEN P S. Handbook of Means and Their inequalities Mathematics and its Applications[M]// Dordrecht: Kluwer Academic Publishers, 2003.

[9] QI F, SANDOR J, DRAGOMIR S S, SOFO A. Notes on the Schur-convexity of the extended mean values[J]. Taiwansese J. Math. 2005,9 (3): 411-420. RGMIA Res. Rep. Coll., 2002, 5 (1): 19-27.

[10] WANG BO-YING. Elements of Majorization Inequalities [M]. Beijing: Beijing Normal University Press, 1990 (Chinese).

[11] QI F, XU SEN-LIN, DEBNATH L. A new poof of monotonicity for extended mean values[J]. Internat. J. Math. Math. Sci. 1999, 22(2):415-420.

第33篇　Bernoulli不等式的控制证明及推广

(石焕南. 北京联合大学学报(自然科学版) 2008, 22 (2)：58-61)

摘　要： 本文利用控制不等式理论并结合分析的方法, 给出了经典Bernoulli不等式的3 种已知的推广的新的证明. 此外, 利用初等对称函数的Schur凹性和简单的控制关系建立了该不等式的一种新椎广.

关键词： Bernoulli不等式; 凸函数; Schur凹性; 初等对称函数; 控制理论

33.1　引言

设$x > -1$, n 是正整数, 则

$$(1+x)^n \geqslant 1 + nx \tag{1}$$

这是著名的Bernoulli不等式. 该不等式在数学分析中占有非常重要的地位. 因此, 不断有人探索它的变形、推广、证明和应用. 例如, 文献[1]记录了如下推广和变形:

定理 A　对于$x > -1$, 若$\alpha > 1$ 或$\alpha < 0$, 则

$$(1+x)^\alpha \geqslant 1 + \alpha x \tag{2}$$

若$0 < \alpha < 1$, 则

$$(1+x)^\alpha \leqslant 1 + \alpha x \tag{3}$$

式(2)和式(3)中的等式成立当且仅当$x = 0$.

定理 B 设 $a_i \geqslant 0, x_i > -1, i = 1, \ldots, n$, 且 $\sum\limits_{i=1}^{n} a_i \leqslant 1$, 则

$$\prod_{i=1}^{n}(1+x_i)^{a_i} \leqslant 1 + \sum_{i=1}^{n} a_i x_i \qquad (4)$$

若 $a_i \geqslant 1, x_i > 0$ 或 $a_i \leqslant 0, x_i < 0, i = 1, \ldots, n$, 则

$$\prod_{i=1}^{n}(1+x_i)^{a_i} \geqslant 1 + \sum_{i=1}^{n} a_i x_i \qquad (5)$$

定理 C 设 $x > 0, x \neq 1$, 则当 $0 < \alpha < 1$ 时

$$\alpha x^{\alpha-1}(x-1) < x^{\alpha} - 1 < \alpha(x-1) \qquad (6)$$

当 $\alpha > 1$ 或 $\alpha < 0$时, 两个不等式均反向.

文献[1-3]给出了定理A 的7 种证法, 包括数学归纳法、幂级数方法、积分方法以及利用算术-几何平均值不等式、利用幂平均不等式、利用函数的单调性、利用中值定理等方法. 本文另辟新径, 利用控制不等式理论证明定理A 和定理C 并推广Bernoulli不等式(1)和定理B.

33.2 定义和引理

在本文中, $\mathbf{R}^n$ 表示实数域上n 维行向量的集合, 并记

$$\mathbf{R}_+^n = \{\boldsymbol{x} = (x_1, \ldots, x_n) \in \mathbf{R}^n | x_i \geqslant 0, i = 1, \ldots, n\}$$

$$\mathbf{R}_{++}^n = \{\boldsymbol{x} = (x_1, \ldots, x_n) \in \mathbf{R}^n | x_i > 0, i = 1, \ldots, n\}$$

$\boldsymbol{x}$ 的第k 个初等对称函数为

$$E_k(\boldsymbol{x}) = E_k(x_1, \ldots, x_n) = \sum_{1 \leqslant i_1 < \ldots < i_k \leqslant n} \prod_{j=1}^{k} x_{i_j}, \ k = 1, \ldots, n$$

特别, $E_n(\boldsymbol{x}) = \prod\limits_{i=1}^{n} x_i, E_1(\boldsymbol{x}) = \sum\limits_{i=1}^{n} x_i$. 当 $k > n$ 时, 规定$E_k(\boldsymbol{x}) = 0$, 并规定$E_0(\boldsymbol{x}) = 1$.

对于$\boldsymbol{x} = (x_1, \ldots, x_n) \in \mathbf{R}^n$, 将$\boldsymbol{x}$ 的分量排成递减的次序后, 记作$x_{[1]} \geqslant x_2 \geqslant \ldots \geqslant x_{[n]}$. $\boldsymbol{x} \leqslant \boldsymbol{y}$ 表示$x_i \leqslant y_i, i = l, \ldots, n$.

定义 1[4] 设$\boldsymbol{x}, \boldsymbol{y} \in \mathbf{R}^n$ 满足

(1) $\sum\limits_{i=1}^{k} x_{[i]} \leqslant \sum\limits_{i=1}^{k} y_{[i]}$, $k = 1, 2, ..., n-1$;

(2) $\sum\limits_{i=1}^{n} x_i = \sum\limits_{i=1}^{n} y_i$.

则称$\boldsymbol{x}$ 被$\boldsymbol{y}$ 所控制, 记作$\boldsymbol{x} \prec \boldsymbol{y}$. 又若$\boldsymbol{x}$ 不是$\boldsymbol{y}$ 的重排, 则称$\boldsymbol{x}$ 被$\boldsymbol{y}$ 严格控制, 记作$\boldsymbol{x} \prec\prec \boldsymbol{y}$.

定义 2[4]　设$\Omega \subset \mathbf{R}^n$, $\varphi : \Omega \to \mathbf{R}$

(1) 若在Ω 上$\boldsymbol{x} \leqslant \boldsymbol{y} \Rightarrow \varphi(\boldsymbol{x}) \leqslant \varphi(\boldsymbol{y})$, 则称$\varphi$ 为Ω 上的增函数; 若$-\varphi$ 是Ω 上的增函数, 则称φ 为Ω 上的减函数;

(2) 若在Ω 上$\boldsymbol{x} \prec \boldsymbol{y} \Rightarrow \varphi(\boldsymbol{x}) \leqslant \varphi(\boldsymbol{y})$, 则称$\varphi$ 为Ω 上的Schur凸函数; 若$-\varphi$ 是Ω 上的Schur 凸函数, 则称φ 为Ω 上的Schur 凹函数.

(3) 若在Ω 上$\boldsymbol{x} \prec\prec \boldsymbol{y} \Rightarrow \varphi(\boldsymbol{x}) < \varphi(\boldsymbol{y})$, 则称$\varphi$ 为Ω 上的严格Schur凸函数; 若$-\varphi$ 是Ω 上的严格Schur 凸函数, 则称φ 为dΩ 上的严格Schur 凹函数.

引理 1[4]　设$\boldsymbol{x} = (x_1, ..., x_n) \in \mathbf{R}^n$, $\bar{x} = \frac{1}{n} \sum\limits_{i=1}^{n} x_i$, 则

$$(\bar{x}, \bar{x}, \ldots, \bar{x}) \prec (x_1, x_2, \ldots, x_n)$$

引理 2[5]　设$\boldsymbol{x}, \boldsymbol{y} \in \mathbf{R}^n$, $x_1 \geqslant x_2 \geqslant ... \geqslant x_n$, $\sum\limits_{i=1}^{n} x_i = \sum\limits_{i=1}^{n} y_i$, 若存在$k(1 \leqslant k \leqslant n)$, 使得$x_i \leqslant y_i(i = 1, ..., k)$, $x_i \geqslant y_i(i = k+1, ..., n)$, 则$\boldsymbol{x} \prec \boldsymbol{y}$.

引理 3[4]　设$I \subset \mathbf{R}$ 为一个区间, $\boldsymbol{x}, \boldsymbol{y} \in I^n \subset \mathbf{R}^n$, 则

(1)$\boldsymbol{x} \prec \boldsymbol{y} \Leftrightarrow \sum\limits_{i=1}^{n} g(x_i) \leqslant \sum\limits_{i=1}^{n} g(y_i)$, 对于任意凸函数$g : I \to \mathbf{R}$;

(2)$\boldsymbol{x} \prec \boldsymbol{y} \Leftrightarrow \sum\limits_{i=1}^{n} g(x_i) \geqslant \sum\limits_{i=1}^{n} g(y_i)$, 对于任意凹函数$g : I \to \mathbf{R}$;

(3)$\boldsymbol{x} \prec\prec \boldsymbol{y} \Leftrightarrow \sum\limits_{i=1}^{n} g(x_i) < \sum\limits_{i=1}^{n} g(y_i)$, 对于任意严格凸函数$g : I \to \mathbf{R}$;

(4)$\boldsymbol{x} \prec\prec \boldsymbol{y} \Leftrightarrow \sum\limits_{i=1}^{n} g(x_i) > \sum\limits_{i=1}^{n} g(y_i)$, 对于任意严格凹函数$g : I \to \mathbf{R}$.

33.3　主要结果及其证明

定理 1　设m, n 是正整数, 若$m \geqslant n$, 对于$x > -1$ 有

$$\mathrm{C}_m^k \left(1 + \frac{n}{m}x\right)^k \geqslant \sum_{i=0}^{k} \mathrm{C}_n^i \mathrm{C}_{m-n}^{k-i} (1+x)^i, \quad k = 1, \ldots, m \tag{7}$$

若$m < n$ 对于$x > -\frac{m}{n}$ 有

$$\mathrm{C}_n^k (1+x)^k \geqslant \sum_{i=0}^{k} \mathrm{C}_m^i \mathrm{C}_{m-n}^{k-i} (1 + \frac{n}{m}x)^i, \quad k = 1, \ldots, n \tag{8}$$

且等式(7)和(8)成立当且仅当$x = 0$, 这里$C_n^k = \frac{n!}{k!(n-k)!}$ 是组合数, 规定$C_n^0 = 1$, 当$k > n$ 时, $C_n^k = 0$.

证明 由引理1 知

$$\boldsymbol{p} := \left(\underbrace{1 + \frac{n}{m}x, \cdots, 1 + \frac{n}{m}x}_{m}\right) \prec \left(\underbrace{1 + x, \cdots, 1 + x}_{n}, \underbrace{1, \cdots, 1}_{m-n}\right) =: \boldsymbol{q}$$

当$x \neq 0$ 时, 此控制是严格的. 若$m \geqslant n$, 对于$x > -1$ 有$1 + x > 0$, $1 + \frac{n}{m}x > 1 - \frac{n}{m} \geqslant 0$, 即$\boldsymbol{p}, \boldsymbol{q} \in \mathbf{R}_{++}^n$. 据文献[4]命题6.7, $E_k(\boldsymbol{x})$ 是$\mathbf{R}_+^n$ 上的增的Schur凹函数, 且当$k > 1$ 时, $E_k(\boldsymbol{x})$ 是$\mathbf{R}_+^n$ 上的增的严格Schur凹函数, 因此

$$E_k\left(\underbrace{1 + \frac{n}{m}x, \cdots, 1 + \frac{n}{m}x}_{m}\right) \geqslant E_k\left(\underbrace{1 + x, \cdots, 1 + x}_{n}, \underbrace{1, \cdots, 1}_{m-n}\right)$$

即式(7)成立, 且等式成立当且仅当$x = 0$.

若$m < n$, 由引理1 知

$$\left(\underbrace{1 + x, \cdots, 1 + x}_{n}\right) \prec \left(\underbrace{1 + \frac{n}{m}x, \cdots, 1 + \frac{n}{m}x}_{m}, \underbrace{1, \ldots, 1}_{n-m}\right)$$

当$x \neq 0$ 时, 此控制是严格的. 因此

$$E_k\left(\underbrace{1 + x, \cdots, 1 + x}_{n}\right) \geqslant E_k\left(\underbrace{1 + \frac{n}{m}x, \cdots, 1 + \frac{n}{m}x}_{m}, \underbrace{1, \ldots, 1}_{n-m}\right)$$

即式(8)成立且等式成立当且仅当$x = 0$, 定理1 证毕.

定理A 的控制证明 分三种情形讨论.

(1) $0 < \alpha < 1$ 的情形. 当$k = m$ 时, 注意对于$k > n$, $C_n^k = 0$, 由式(7)有

$$\left(1 + \frac{n}{m}x\right)^m \geqslant (1 + x)^n$$

即

$$(1 + x)^{\frac{n}{m}} \leqslant 1 + \frac{n}{m}x \tag{9}$$

且等式成立当且仅当$x = 0$. 式(9)表明对于满足$0 < \alpha < 1$ 的有理数α, 式(3)

成立. 若α 是无理数，存在有理数列 $\{r_k\}$ 满足$0 < r_k < 1, k = 1, 2, \ldots,$ 且$r_k \to \alpha, (k \to \infty)$. 对于有理数$r_k$, 由式(9) 有$(1+x)^{r_k} \leqslant 1+r_k x$, 令$k \to \infty$, 即得式(3).

(2) $\alpha > 1$ 的情形. 若$\alpha x \leqslant -1$, 即$1 + \alpha x \leqslant 0$, 式(2) 显然成立; 若$\alpha x > -1$, 因$0 < \frac{1}{\alpha} < 1$, 由式(3) 有$(1 + \alpha x)^{\frac{1}{\alpha}} \leqslant 1 + \frac{1}{\alpha}(\alpha x)$, 即式(2)成立.

(3) $\alpha < 0$ 的情形. 此时$1 - \alpha > 1, -\frac{x}{1+x} > -1$, 由式(2)有

$$\left(\frac{1}{1+x}\right)^{1-\alpha} = \left(1 - \frac{x}{1+x}\right)^{1-\alpha} \geqslant 1 - \frac{(1-\alpha)x}{1+x}$$

即$(1+x)^\alpha \geqslant 1 + \alpha x$, 这样我们利用定理1 证得定理A, 证毕.

定理 2　设$a_i \geqslant 1, x_i > 0$, 或$a_i \leqslant 0, -1 < x_i \leqslant 0$, 则对于$k = 1, \cdots, n$, 有

$$C_n^k \left[\frac{1}{n}\sum_{i=1}^n (1+x_i)^{a_i}\right]^k \geqslant \sum_{1 \leqslant i_1 < \cdots < i_k \leqslant n} \prod_{j=1}^k (1+x_{i_j})^{a_{i_j}}$$

$$\geqslant \sum_{1 \leqslant i_1 < \cdots < i_k \leqslant n} \prod_{j=1}^k (1+a_{i_j}x_{i_j}) > C_{n-1}^k + C_{n-1}^{k-1}\left(1 + \sum_{i=1}^n a_i x_i\right) \tag{10}$$

证明　记$y = \frac{1}{n}\sum_{i=1}^n (1+x_i)^{a_i}$, 由式(2)有$(1+x_i)^{a_i} \geqslant 1+a_i x_i, i = 1, \cdots, n$, 于是由引理1 有

$$(y, y, \cdots, y) \prec ((1+x_1)^{a_1}, (1+x_2)^{a_2}, \cdots, (1+x_n)^{a_n})$$
$$\geqslant (1+a_1 x_1, 1+a_2 x_2, \cdots, 1+a_n x_n)$$
$$\prec\prec (1 + \sum_{i=1}^n a_i x_i, 1, \cdots, 1)$$

从而由$E_k(\boldsymbol{x})$ 的增的严格Schur凹性, 有

$$E_k(y, y, \cdots, y) \geqslant E_k\left[(1+x_1)^{a_1}, (1+x_2)^{a_2}, \cdots, (1+x_n)^{a_n}\right]$$
$$\geqslant E_k(1+a_1 x_1, 1+a_2 x_2, \cdots, 1+a_n x_n)$$
$$> E_k(1 + \sum_{i=1}^n a_i x_i, 1, \cdots, 1)$$

由此证得式(10). 证毕.

注　当$k = n$ 时，由式(10)即可得式(5).

定理C 的控制证明　由于式(6)右边的不等式等价于式(3)，故只需证式(6)左边的不等式，分三种情形讨论.

(1) $\alpha > 1$ 的情形. 设 $m < n$, 欲证

$$\frac{n}{m}x^{\frac{n}{m}-1}(x-1) > x^{\frac{n}{m}} - 1 \tag{11}$$

对于 $x > 1$, 令

$$\boldsymbol{u} := \left(\underbrace{\frac{n}{m}\ln x, \cdots, \frac{n}{m}\ln x}_{m}, \underbrace{(\frac{n}{m}-1)\ln x, \cdots, (\frac{n}{m}-1)\ln x}_{n}\right)$$

$$\boldsymbol{v} := \left(\underbrace{\frac{n}{m}\ln x, \cdots, \frac{n}{m}\ln x}_{n}, \underbrace{(0,\cdots,(0}_{m}\right)$$

因 $m < n$, 有 $0 < (\frac{n}{m}-1)\ln x < \frac{n}{m}\ln x$, 从而 $u_i \leqslant v_i, i = 1,\cdots,n$, 而 $u_i \geqslant v_i, i = n+1,\cdots,m+n$, 又 $\sum_{i=1}^{m+n} u_i = \frac{n^2}{m}\ln x = \sum_{i=1}^{m+n} v_i$, 由引理2 知 $\boldsymbol{u} \prec \boldsymbol{v}$, 且因 $x \neq 1$, 此控制是严格的. 从而因 e^x 是 $\mathbf{R}$ 上的严格凸函数, 由引理3 有 $\sum_{i=1}^{m+n} e^{u_i} < \sum_{i=1}^{m+n} e^{v_i}$, 即

$$mx^{\frac{n}{m}} + nx^{\frac{n}{m}-1} < nx^{\frac{n}{m}} + m$$

而此式等价于式(11).

对于 $0 < \alpha < 1$, 令

$$\boldsymbol{u}' := \left(\underbrace{\left(\frac{n}{m}-1\right)\ln x, \cdots, \left(\frac{n}{m}-1\right)\ln x}_{n}, \underbrace{\frac{n}{m}\ln x, \cdots, \frac{n}{m}\ln x}_{m}\right)$$

$$\boldsymbol{v}' := \left(\underbrace{0, \cdots, 0}_{m}, \underbrace{\frac{n}{m}\ln x, \cdots, \frac{n}{m}\ln x}_{n}\right)$$

因 $m < n$, 有 $0 > (\frac{n}{m}-1)\ln x > \frac{n}{m}\ln x$, 从而 $u_i' \leqslant v_i', i = 1,\cdots,m$, 而 $u_i' \geqslant v_i', i = m+1,\cdots,m+n$, 又 $\sum_{i=1}^{m+n} u_i' = \frac{n^2}{m}\ln x = \sum_{i=1}^{m+n} v_i'$, 由引理2 知 $\boldsymbol{u}' \prec \boldsymbol{v}'$, 又因 $x \neq 1$, 此控制是严格的. 从而因 e^x 是 $\mathbf{R}$ 上的严格凸函数, 有 $\sum_{i=1}^{m+n} e^{u_i'} <$

$\sum\limits_{i=1}^{m+n} e^{v_i'}$, 即

$$mx^{\frac{n}{m}} + nx^{\frac{n}{m}-1} < nx^{\frac{n}{m}} + m \Leftrightarrow \frac{n}{m}x^{\frac{n}{m}-1}(x-1) > x^{\frac{n}{m}} - 1$$

总之, 对于 $x > 0, x \neq 1, m < n$, 式(11)成立. 这说明对于满足 $x > 1$ 的有理数有 $\alpha x^{\alpha-1}(x-1) > x^{\alpha} - 1$, 通过有理逼近知, 对于满足 $\alpha > l$ 的任意实数均有 $\alpha x^{\alpha-1}(x-1) > x^{\alpha} - 1$.

(2) $\alpha < 0$ 的情形. 此时 $1-\alpha > 1$, 对于 $x > 0, x \neq 1$, 有 $x^{-1} > 0, x^{-1} \neq 1$, 由(1) 有

$$(1-\alpha)(x^{-1})^{(1-\alpha)-1}(x^{-1} - 1) > (x^{-1})^{1-\alpha} - 1$$

即 $\alpha x^{\alpha-1}(x-1) > x^{\alpha} - 1$.

(3) $0 < \alpha < 1$的情形. 此时 $\frac{1}{\alpha} > 1$, 对于 $x > 0, x \neq 1$, 由(1)有

$$\frac{1}{\alpha}(x^{\alpha})^{\frac{1}{\alpha}-1}((x^{\alpha})^{-1} - 1) > (x^{\alpha})^{\frac{1}{\alpha}} - 1$$

即 $\alpha x^{\alpha-1}(x-1) < x^{\alpha} - 1$, 证毕.

致谢 作者衷心感谢匡继昌教授给予本文的宝贵的修改意见.

参考文献

[1] 匡继昌. 常用不等式[M]. 3 版. 济南: 山东科技出版社, 2003.

[2] 王向东, 苏化明, 王方汉. 不等式·理论·方法[M]. 郑州: 河南教育出版社, 1994.

[3] ALFRED W. A new of the monotonicity of power menas[J]. J Ineq Pure. Appl. Math., 2004, 5(1): Article 6.

[4] 王伯英. 控制不等式基础[M]. 北京: 北京师范大学出版社, 1990.

[5] 王松桂, 吴密霞, 贾忠贞. 矩阵论中不等式[M]. 2版. 北京: 科学出版社, 2006.

[6] MITRINOVIĆ D S, PEČARIĆ J E, FINK A M. Classical and new inequalities in analysis[M]. London: Khwer Academic Publishers, 1993.

[7] 文家金, 罗钊. Bernoulli不等式的优化推广及其应用[J]. 成都大学学报(自然科学版), 2001, 20(4): 1-8.

[8] 席华昌. Bernoulli不等式及其应用[J]. 高等数学研究, 2005, 4(8): 42-44.

[9] 苏灿荣, 禹春福. 也谈Bernoulli不等式的证明与应用[J]. 高等数学研究, 2006, 9(6)：39-41.

[10] 崔丽鸿, 吴守玲, 杨士俊. Bernoulli型不等式[J]. 杭州师范学院学报, 1998, 6(6)：33-35.

[11] 葛健芽. Bernoulli不等式的几个注记[J]. 浙江师范大学学报(自然科学版), 2003, 26(2)：119-122

[12] 薛昌兴. 关于Bernoulli不等式的隔离和推广[J]. 甘肃教育学院学报(自然科学版), 1999, 13(3)：5-7.

[13] 刘广文. 伯努利不等式的推广[J]. 聊城师院学报(自然科学版), 1997, 10(3)：23-24.

第34篇 反向Chrystal不等式

(顾春, 石焕南. 数学的实践与认识, 2008, 38 (13):163-167)

摘 要: 本文利用控制不等式理论建立了三个反向Chrystal不等式. 并给出若干应用.

关键词: 反向Chrystal不等式; 控制理论; 初等对称函数; 算术平均; 几何平均

34.1 引言

在本文中, $\mathbf{R}^n$ 表示实数域上 n 维行向量的集合，并记

$$\mathbf{R}^n_+ = \{\boldsymbol{x} = (x_1, \cdots, x_n) \in \mathbf{R}^n | x_i \geqslant 0, i = 1, \cdots, n\}$$

$\boldsymbol{x}$ 的第 k 个初等对称函数为

$$E_k(\boldsymbol{x}) = E_k(x_1, \cdots, x_n) = \sum_{1 \leqslant i_1 < \cdots < i_k \leqslant n} \prod_{j=1}^{k} x_{i_j}, \ k = 1, \cdots, n$$

特别, $E_n(\boldsymbol{x}) = \prod_{i=1}^{n} x_i, E_1(\boldsymbol{x}) = \sum_{i=1}^{n} x_i$. 当 $k > n$ 时, 规定 $E_k(\boldsymbol{x}) = 0$, 并规定 $E_0(\boldsymbol{x}) = 1$.

对于 $\boldsymbol{x} = (x_1, \cdots, x_n) \in \mathbf{R}^n, A_n(\boldsymbol{x}) = \frac{1}{n} \sum_{i=1}^{n} x_i = \frac{1}{n} E_1(\boldsymbol{x})$ 表示 $x_1, \cdots, x_n$ 的算术平均. 对于 $\boldsymbol{x} = (x_1, \cdots, x_n) \in \mathbf{R}^n_+, G_n(\boldsymbol{x}) = (\prod_{i=1}^{n} x_i)^{\frac{1}{n}} = (E_n(\boldsymbol{x}))^{\frac{1}{n}}$ 表示 $x_1, \cdots, x_n$ 的几何平均.

著名的Chrystal不等式[1] 为

$$\prod_{i=1}^{n}(1 + x_i) \geqslant (1 + G_n(\boldsymbol{x}))^n \tag{1}$$

219

仅当 $x_1 = \cdots = x_n = G_n(\boldsymbol{x})$ 时等号成立.

1996年, 文[2]将式(1)推广到初等对称函数的情形

$$E_k((1+\boldsymbol{x})^a) \geqslant \mathrm{C}_n^k (1 + G_n(\boldsymbol{x}))^{ka} \tag{2}$$

其中 $\boldsymbol{x} = (x_1, \cdots, x_n) \in \mathbf{R}_{++}^n = \{\boldsymbol{x} = (x_1, \cdots, x_n) \in \mathbf{R}^n | x_i > 0, i = 1, \cdots, n\}$, $a > 0, k = 1, \cdots, n$.

$$E_k((1+\boldsymbol{x})^a) = E_k((1+x_1)^a, \cdots, (1+x_n)^a) := \sum_{1 \leqslant i_1 < \cdots < i_k \leqslant n} \prod_{j=1}^{k} (1+x_{i_j})^a$$

当 $a = 1, k = n$ 时, 式(2)化为式(1).

最近文[3]将式(1)推广到 m 组 n 维向量的情形

$$\prod_{i=1}^{n} \sum_{j=1}^{m} x_{i_j} \geqslant \left(\sum_{j=1}^{m} \left(\prod_{i=1}^{n} x_{i_j} \right)^{\frac{1}{n}} \right)^n \tag{3}$$

其中 $x_{i_j} \geqslant 0, i = 1, \cdots, n, j = 1, \cdots, m$. 当 $m = 2, a_{i_1} = 1, a_{i_2} > 0, (i = 1, \cdots, n)$ 时, 式(3)化为式(1).

本文将利用控制不等式理论建立如下三个反向Chrystal不等式.

定理 1 设 $\boldsymbol{x} = (x_1, \cdots, x_n) \in \mathbf{R}_+^n$, 则

$$(1 + G_n(\boldsymbol{x}))^{n+1} \geqslant (1 + G_n(\boldsymbol{x}) - n(A_n(\boldsymbol{x}) - G_n(\boldsymbol{x}))) \prod_{i=1}^{n} (1+x_i) \tag{4}$$

定理 2 设 $\boldsymbol{x} = (x_1, \cdots, x_n) \in \mathbf{R}_+^n, n \in \mathbf{N}, x_1 \leqslant \cdots \leqslant x_n$, 则

$$(1 + G_n(\boldsymbol{x}))^n \geqslant (1 + x_n - n(A_n(\boldsymbol{x}) - G_n(\boldsymbol{x}))) \prod_{i=1}^{n-1} (1+x_i) \tag{5}$$

定理 3 设 $\boldsymbol{x} = (x_1, \cdots, x_n) \in \mathbf{R}_+^n, n \in \mathbf{N}, k = 1, \cdots, n, 0 \leqslant \alpha \leqslant 1$, 则

$$\sum_{k=0}^{n} (\mathrm{C}_n^k)^2 (1 + G_n(\boldsymbol{x}))^{\alpha k} (A_n(\boldsymbol{x}) - G_n(\boldsymbol{x}))^{\alpha(n-k)} \geqslant \prod_{i=1}^{n} (1+x_i)^\alpha \tag{6}$$

对于 $x_1 = 1, x_2 = 2, \cdots, x_n = n$ 和 $x_1 = 1, x_2 = 4, \cdots, x_n = n^2$ 由(4), (5), (6)三式不难得

推论 1 设 $n \in \mathbf{N}, k = 1, \cdots, n, 0 \leqslant \alpha \leqslant 1$, 则

$$\left(1 + \sqrt[n]{n!}\right)^{n+1} \geqslant \left(1 + \sqrt[n]{n!} - n\left(\frac{n+1}{2} - \sqrt[n]{n!}\right)\right) \prod_{i=1}^{n}(1+i) \tag{7}$$

$$\left(1 + \sqrt[n]{n!}\right)^{n} \geqslant \left(1 + n - n\left(\frac{n+1}{2} - \sqrt[n]{n!}\right)\right) \prod_{i=1}^{n-1} \tag{8}$$

$$\sum_{k=0}^{n}(\mathrm{C}_n^k)^2 \left(1 + \sqrt[n]{n!}\right)^{\alpha k} \left(\frac{n+1}{2} - \sqrt[n]{n!}\right)^{\alpha(n-k)} \geqslant \prod_{i=1}^{n}(1+i)^{\alpha} \tag{9}$$

$$\left(1 + \left(\sqrt[n]{n!}\right)^2\right)^{n+1} \geqslant \left(1 + \left(\sqrt[n]{n!}\right)^2 - mn\right) \prod_{i=1}^{n}(1+i^2) \tag{10}$$

$$\left(1 + \left(\sqrt[n]{n!}\right)^2\right)^{n} \geqslant \left(1 + n^2 - mn\right) \prod_{i=1}^{n-1}(1+i^2) \tag{11}$$

$$\sum_{k=0}^{n}(\mathrm{C}_n^k)^2 \left(1 + \left(\sqrt[n]{n!}\right)^2\right)^{\alpha k} m^{\alpha(n-k)} \geqslant \prod_{i=1}^{n}(1+i^2)^{\alpha} \tag{12}$$

其中

$$m = \frac{(n+1)(2n+1)}{6} - \left(\sqrt[n]{n!}\right)^2$$

对于 $x_1 = 1, x_2 = x, \cdots, x_n = x^n$, 由(4), (5), (6)三式不难得

推论 2 设 $n \in \mathbf{N}, k = 1, \cdots, n, 0 \leqslant \alpha \leqslant 1, x \geqslant 0$, 则

$$\left(1 + x^{\frac{n}{2}}\right)^{n+2} \geqslant \left(1 + x^{\frac{n}{2}} - (n+1)s\right) \prod_{i=0}^{n}(1+x^i) \tag{13}$$

$$\left(1 + x^{\frac{n}{2}}\right)^{n+1} \geqslant \left(1 + x^n - (n+1)s\right) \prod_{i=0}^{n}(1+x^i), \ x \geqslant 1 \tag{14}$$

$$\left(1 + x^{\frac{n}{2}}\right)^{n+1} \geqslant \left(2 - (n+1)s\right) \prod_{i=0}^{n}(1+x^i), \ 0 \leqslant x < 1 \tag{15}$$

$$\sum_{k=0}^{n}(\mathrm{C}_n^k)^2 \left(1 + x^{\frac{n}{2}}\right)^{\alpha k} s^{\alpha(n-k)} \geqslant \prod_{i=0}^{n}(1+x^i)^{\alpha} \tag{16}$$

其中

$$s = \frac{x^{n+1} - 1}{(n+1)(x-1)} - x^{\frac{n}{2}}$$

注 在式(14)和式(15) 的证明中需要注意, 当 $x \geqslant 1$ 时,有 $1 \leqslant x \leqslant x^2 \leqslant \cdots \leqslant x^n$, 而当 $0 \leqslant x < 1$ 时, 有 $x^n \leqslant x^{n-1} \leqslant \cdots \leqslant x \leqslant 1$.

对于 $x = \frac{1}{1\cdot 2}, x_2 = \frac{1}{2\cdot 3}, \cdots, \frac{1}{n\cdot(n+1)}$, 由(4), (5), (6) 三式不难得

推论 3 设 $n \in \mathbf{N}, k = 1, \cdots, n, 0 \leqslant \alpha \leqslant 1$, 则

$$\left(1 + \frac{1}{\sqrt[n]{(n+1)(n!)^2}}\right)^{n+1} \geqslant \left(1 + \frac{1}{\sqrt[n]{(n+1)(n!)^2}} - nt\right)\prod_{i=1}^{n}\left(1 + \frac{1}{i(i+1)}\right) \tag{17}$$

$$\left(1 + \frac{1}{\sqrt[n]{(n+1)(n!)^2}}\right)^{n} \geqslant \left(\frac{3}{2} - nt\right)\prod_{i=1}^{n}\left(1 + \frac{1}{i(i+1)}\right) \tag{18}$$

$$\sum_{k=0}^{n}(C_n^k)^2\left(1 + \frac{1}{\sqrt[n]{(n+1)(n!)^2}}\right)^{\alpha k} t^{\alpha(n-k)} \geqslant \prod_{i=1}^{n}(1 + \frac{1}{i(i+1)})^{\alpha} \tag{19}$$

其中

$$t = \frac{1}{n+1} - \frac{1}{\sqrt[n]{(n+1)(n!)^2}}$$

对于 $x_1 = C_n^0, x_2 = C_n^1, \cdots, x_{n+1} = C_n^n$, 由(4), (5), (6) 三式不难得

推论 4 设 $n \in \mathbf{N}, k = 1, \cdots, n, 0 \leqslant \alpha \leqslant 1$, 则

$$\left(1 + \sqrt[n+1]{\prod_{k=0}^{n}C_n^k}\right)^{n+2} \geqslant \left(1 + \sqrt[n+1]{\prod_{k=0}^{n}C_n^k} - (n+1)p\right)\prod_{i=1}^{n}(1 + C_n^i) \tag{20}$$

$$\sum_{k=0}^{n}(C_n^k)^2\left(1 + \sqrt[n+1]{\prod_{k=0}^{n}C_n^k}\right)^{\alpha k} p^{\alpha(n-k)} \geqslant \prod_{i=1}^{n}(1 + C_n^i)^{\alpha} \tag{21}$$

其中

$$p = \frac{2^n}{n+1} - \sqrt[n+1]{\prod_{k=0}^{n}C_n^k}$$

34.2 定义和引理

定义 1[4] 设 $\boldsymbol{x}, \boldsymbol{y} \in \mathbf{R}^n$ 满足

$$(1)\sum_{i=1}^{k}x_{[i]} \leqslant \sum_{i=1}^{k}y_{[i]}, \quad k = 1, 2, \cdots, n-1; (2)\sum_{i=1}^{n}x_i = \sum_{i=1}^{n}y_i.$$

称$\boldsymbol{x}$ 被$\boldsymbol{y}$ 所控制, 记作$\boldsymbol{x} \prec \boldsymbol{y}$. 其中$x_{[1]} \geqslant \cdots \geqslant x_{[n]}$ 是$\boldsymbol{x}$ 的分量的递减重排. 若满足

$$\sum_{i=1}^{k} x_{[i]} \leqslant \sum_{i=1}^{k} y_{[i]}, \quad k = 1, 2, \cdots, n$$

称$\boldsymbol{x}$ 被$\boldsymbol{y}$ 下(弱)控制, 记作$\boldsymbol{x} \prec_w \boldsymbol{y}$.

定义 2[4]　设$\Omega \subset \mathbf{R}^n, \varphi : \Omega \to \mathbf{R}$.

(1) 若在Ω 上$\boldsymbol{x} \leqslant \boldsymbol{y} \Rightarrow \varphi(\boldsymbol{x}) \leqslant \varphi(\boldsymbol{y})$, 则称$\varphi$ 为Ω 上的增函数; 若$-\varphi$ 是Ω 上的增函数, 则称φ 为Ω 上的减函数;

(2) 若在Ω 上$\boldsymbol{x} \prec \boldsymbol{y} \Rightarrow \varphi(\boldsymbol{x}) \leqslant \varphi(\boldsymbol{y})$, 则称$\varphi$ 为Ω 上的Schur凸函数; 若$-\varphi$ 是Ω 上的Schur-凸函数, 则称φ 为Ω 上的Schur-凹函数.

引理 1[4]　设$\boldsymbol{x} = (x_1, \cdots, x_n) \in \mathbf{R}^n, \bar{x} = \frac{1}{n} \sum_{i=1}^{n} x_i$, 则

$$(\bar{x}, \bar{x}, \cdots, \bar{x}) \prec (x_1, x_2, \cdots, x_n) \tag{22}$$

引理 2[4]　设$\boldsymbol{x}, \boldsymbol{y} \in \mathbf{R}^n, 令x_{n+1} = \min\{x_1, x_2, \cdots, x_n\}, y_{n+1} = \sum_{i=1}^{n+1} x_i - \sum_{i=1}^{n} y_i$. 若$\boldsymbol{x} \prec_w \boldsymbol{y}$, 则

$$(\boldsymbol{x}, x_{n+1}) \prec (\boldsymbol{y}, y_{n+1})$$

引理 3[4]　设$\boldsymbol{x}, \boldsymbol{y} \in \mathbf{R}^n, y_1 \leqslant y_2 \leqslant \cdots \leqslant y_n, 令\tilde{y} = y_n - (\sum_{i=1}^{n} y_i - \sum_{i=1}^{n} x_i)$. 若$\boldsymbol{x} \prec_w \boldsymbol{y}$, 则

$$(x_1, x_2, \cdots, x_n) \prec (y_1, \cdots, y_{n-1}, \tilde{y})$$

引理 4[4]　设$\boldsymbol{x} \in \mathbf{R}_+^n, \boldsymbol{y} \in \mathbf{R}^n, 令\delta = \sum_{i=1}^{n} (y_i - x_i)$. 若$\boldsymbol{x} \prec_w \boldsymbol{y}$, 则

$$\left(\boldsymbol{x}, \underbrace{\frac{\delta}{n}, \ldots, \frac{\delta}{n}}_{n} \right) \prec (\boldsymbol{y}, \underbrace{0, \cdots, 0}_{n})$$

引理 5[4]　初等对称函数$E_k(\boldsymbol{x})$ 在$\mathbf{R}_+^n$ 上是递增且Schur 凹函数.

引理 6[2]　设$g : I \to \mathbf{R}, \varphi : \mathbf{R}^n \to \mathbf{R}, \psi(\boldsymbol{x}) = \varphi(g(x_1), \cdots, g(x_n))$. 若$g$ 是凹函数, φ 是递增且Schur凹函数, 则ψ 是Schur凹函数.

34.3　定理的证明

定理 1 的证明　由引理1 有

$$\left(\underbrace{1+G(\boldsymbol{x}),\cdots,1+G(\boldsymbol{x})}_{n}\right) \leqslant \left(\underbrace{1+A(\boldsymbol{x}),\cdots,1+A(\boldsymbol{x})}_{n}\right) \prec (1+x_1,\ldots,1+x_n)$$

故

$$\left(\underbrace{1+G(\boldsymbol{x}),\cdots,1+G(\boldsymbol{x})}_{n}\right) \prec_w (1+x_1,\cdots,1+x_n) \qquad (*)$$

由引理2 有

$$\left(\underbrace{1+G(\boldsymbol{x}),\cdots,1+G(\boldsymbol{x})}_{n+1}\right) \prec (1+x_1,\cdots,1+x_n,1+G(\boldsymbol{x})-n(A(\boldsymbol{x})-G(\boldsymbol{x})))$$

进而由引理5, 有

$$E_{n+1}\left(\underbrace{1+G(\boldsymbol{x}),\cdots,1+G(\boldsymbol{x})}_{n+1}\right)$$
$$\geqslant E_{n+1}(1+x_1,\cdots,1+x_n,1+G(\boldsymbol{x})-n(A(\boldsymbol{x})-G(\boldsymbol{x})))$$

即式(4)成立.

定理 2 的证明　据引理3, 由式(∗)有

$$\left(\underbrace{1+G(\boldsymbol{x}),\cdots,1+G(\boldsymbol{x})}_{n}\right) \prec (1+x_1,\cdots,1+x_{n-1},1+x_n-n(A(\boldsymbol{x})-G(\boldsymbol{x})))$$

进而由引理5, 有

$$E_{n+1}\left(\underbrace{1+G(\boldsymbol{x}),\cdots,1+G(\boldsymbol{x})}_{n}\right)$$
$$\geqslant E_{n+1}(1+x_1,\cdots,1+x_{n-1},1+x_n-n(A(\boldsymbol{x})-G(\mathbf{x})))$$

即式(5)成立.

定理 3 的证明　据引理4, 由$(*)$有

$$\left(\underbrace{1+G(\boldsymbol{x}),\cdots,1+G(\boldsymbol{x})}_{n},\underbrace{A(\boldsymbol{x})-G(\boldsymbol{x}),\cdots,A(\boldsymbol{x})-G(\boldsymbol{x})}_{n}\right)$$

$$\prec\left(1+x_1,\cdots,1+x_n,\underbrace{0,\cdots,0}_{n}\right)$$

进而由引理5 和引理6, 有

$$E_n\left(\underbrace{(1+G(\boldsymbol{x}))^{\alpha},\cdots,(1+G(\boldsymbol{x}))^{\alpha}}_{n},\underbrace{(A(\boldsymbol{x})-G(\boldsymbol{x}))^{\alpha},\cdots,(A(\boldsymbol{x})-G(\boldsymbol{x}))^{\alpha}}_{n}\right)$$

$$\geqslant E_n\left((1+x_1)^{\alpha},\cdots,(1+x_n)^{\alpha},\underbrace{0,\cdots,0}_{n}\right)$$

即式(6)成立.

参考文献

[1]　匡继昌. 常用不等式[M]. 2 版. 长沙: 湖南教育出版社, 1993.

[2]　石焕南, 石敏琪. 对称平均值基本定理应用数例[J]. 数学通报, 1996 (10): 14-45.

[3]　郭要红. Chrystal不等式的一个推广及其应用[J]. 大学数学, 2005, 21 (5): 81-83.

[4]　王伯英. 控制不等式基础[M]. 北京: 北京师范大学出版社, 1990.

第35篇　Generalizations of Bernoulli's Inequality with Applications

(SHI HUAN-NAN. Journal Mathematical Inequalities, Volum 2, Number 1, (2008): 101-107)

Abstract: By using methods on the theory of majorization, some new generalizations of Bernoulli's inequality are established and some applications of the generalizations are given.

Keywords: Bernoulli's inequality; Schur-concavity; elementary symmetric function; majorization

35.1　Introduction

Let $x > -1$ and n is a positive integer. Then

$$(1+x)^n \geqslant 1 + nx \tag{1}$$

(1) is known as the Bernoulli's inequality which play an important role in analysis and its applications. So, during the past few years, many researchers obtained various generalizations, extensions of inequality (1). For example, the following generalizations and variants of (1) were recorded in [1]:

Theorem A Let $x > -1$. If $\alpha > 1$ or $\alpha < 0$, then

$$(1+x)^\alpha \geqslant 1 + \alpha x \tag{2}$$

if $0 < \alpha < 1$, then

$$(1+x)^\alpha \leqslant 1 + \alpha x \tag{3}$$

In (2) and (3), equalities holding if and only if $x = 0$.

Theorem B Let $a_i \geqslant 0$, $x_i > -1, i = 1, \cdots, n$, and $\sum\limits_{i=1}^{n} a_i \leqslant 1$. Then

$$\prod_{i=1}^{n}(1+x_i)^{a_i} \leqslant 1 + \sum_{i=1}^{n} a_i x_i \qquad (4)$$

if $a_i \geqslant 1$ or $a_i \leqslant 0$, and if $x_i > 0$ or $-1 < x_i < 0, i = 1, \cdots, n$, then

$$\prod_{i=1}^{n}(1+x_i)^{a_i} \geqslant 1 + \sum_{i=1}^{n} a_i x_i \qquad (5)$$

For more information on the Bernoulli's inequality, please refer to [4, 6-10] and the references therein.

In this paper, some new generalizations of Bernoulli's inequality are established by the Schur-concatity of the elementary symmetric functions and the dual form of the elementary symmetric functions, and some applications of the generalizations are given. We obtain the following results.

Theorem 1 Let m, n is a positive integer, $k = 1, \cdots, n$.

(i)If $m \geqslant n$ and $x > -1$, then

$$C_m^k \left(1 + \frac{n}{m}x\right)^k \geqslant \sum_{i=0}^{k} C_n^i C_{m-n}^{k-i}(1+x)^i \qquad (6)$$

and

$$k^{C_m^k} \left(1 + \frac{n}{m}x\right)^{C_m^k} \geqslant \prod_{i=0}^{k}(ix+k)^{C_n^i C_{m-n}^{k-i}} \qquad (7)$$

(ii)If $m < n$ and $x > -\frac{m}{n}$, then

$$C_m^k (1+x)^k \geqslant \sum_{i=0}^{k} C_m^i C_{n-m}^{k-i} \left(1 + \frac{m}{n}x\right)^i \qquad (8)$$

and

$$k^{C_m^k} (1+x)^{C_m^k} \geqslant \prod_{i=0}^{k} \left(\frac{m}{n}ix + k\right)^{C_m^i C_{n-m}^{k-i}} \qquad (9)$$

where $C_n^k = \frac{n!}{k!(n-k)!}$ is the number of combinations of n elements taken k at a time, defined $C_n^0 = 1$ and $C_n^k = 0$ for $k > n$. In (6), (7), (8) and (9), equalities holding if and only if $x = 0$.

Remark 1 When $x = 0$, (6), (7), (8) and (9)are deduce to Vandermonde identity

$$C_m^k = \sum_{i=0}^{k} C_n^i C_{m-n}^{k-i} \tag{10}$$

Theorem 2 If $a_i \geqslant 1$ or $a_i \leqslant 0$, and if $x_i > 0$ or $0 \geqslant x_i \geqslant -1$. Then

$$\frac{k}{n}\left(\frac{1}{n}\sum_{i=1}^{n}(1+x_i)^{a_i}\right)^k \geqslant \sum_{1\leqslant i_1<\cdots<i_k\leqslant n}\prod_{j=1}^{k}(1+x_{i_j})^{a_{i_j}}$$

$$\geqslant \sum_{1\leqslant i_1<\cdots<i_k\leqslant n}\prod_{j=1}^{k}(1+a_{i_j}x_{i_j}) \geqslant C_{n-1}^k + C_{n-1}^{k-1}\left(1+\sum_{i=1}^{n}a_i x_i\right) \tag{11}$$

$$\left(\frac{k}{n}\sum_{i=1}^{n}(1+x_i)^{a_i}\right)^{C_n^k} \geqslant \prod_{1\leqslant i_1<\cdots<i_k\leqslant n}\sum_{j=1}^{k}(1+x_{i_j})^{a_{i_j}}$$

$$\geqslant \prod_{1\leqslant i_1<\cdots<i_k\leqslant n}\sum_{j=1}^{k}(1+a_{i_j}x_{i_j}) \geqslant k^{C_{n-1}^k}\left(k+\sum_{i=1}^{n}a_i x_i\right)^{C_{n-1}^{k-1}} \tag{12}$$

Remark 2 When $k = n$, (11) is deduce to (5), and when $k = 1$, (12) is deduce to (5) too.

35.2 Proof of Theorem

For our own convenience, we introduce the following notations. we assume that the set of n-dimensional row vector on real number field by $\mathbf{R}^n$.

$$\mathbf{R}_+^n = \{x = (x_1,\cdots,x_n) \in \mathbf{R}^n : x_i \geqslant 0, i = 1,\cdots,n\}$$

$$\mathbf{R}_{++}^n = \{x = (x_1,\cdots,x_n) \in \mathbf{R}^n : x_i > 0, i = 1,\cdots,n\}$$

Let $x = (x_1,\cdots,a_n) \in \mathbf{R}^n$. Its elementary symmetric functions are

$$E_k(x) = E_k(x_1,\ldots,x_n) = \sum_{1\leqslant i_1<\cdots<i_k\leqslant n}\prod_{j=1}^{k}x_{i_j}, \quad k = 1,\cdots,n$$

In particular, $E_n(x) = \prod_{i=1}^{n}x_i$, $E_1(x) = \sum_{i=1}^{n}x_i$, and defined $E_0(x) = 1$ and $E_k(x) = 0$ for $k < 0$ or $k > n$.

The dual form of the elementary symmetric functions are

$$E_k^*(\boldsymbol{x}) = E_k^*(x_1, \ldots, x_n) = \prod_{1 \leqslant i_1 < \ldots < i_k \leqslant n} \sum_{j=1}^{k} x_{i_j}, \quad k = 1, \cdots, n$$

and defined $E_0^*(\boldsymbol{x}) = 1$, and $E_k^*(\boldsymbol{x}) = 0$ for $k < 0$ or $k > n$.

We need the following definitions and lemmas.

Definition 1[2-3] Let $\boldsymbol{x} = (x_1, \ldots, x_n)$ and $\boldsymbol{y} = (y_1, \ldots, y_n) \in \mathbf{R}^n$.

(i)$\boldsymbol{x}$ is said to be majorized by $\boldsymbol{y}$ (in symbols $\boldsymbol{x} \prec \boldsymbol{y}$) if $\sum_{i=1}^{k} x_{[i]} \leqslant \sum_{i=1}^{k} y_{[i]}$

for $k = 1, 2, \ldots, n-1$ and $\sum_{i=1}^{n} x_i = \sum_{i=1}^{n} y_i$, where $x_{[1]} \geqslant \cdots \geqslant x_{[n]}$ and $y_{[1]} \geqslant \cdots \geqslant y_{[n]}$ are rearrangements of $\boldsymbol{x}$ and $\boldsymbol{y}$ in a descending order, and $\boldsymbol{x}$ is said to strictly majorized by $\boldsymbol{y}$(in symbols $\boldsymbol{x} \prec\prec \boldsymbol{y}$) if $\boldsymbol{x}$ is not permutation of $\boldsymbol{y}$.

(ii)$\boldsymbol{x} \geq \boldsymbol{y}$ means $x_i \geqslant y_i$ for all $i = 1, 2, \ldots, n$. Let $\Omega \subset \mathbf{R}^n$, $\varphi \colon \Omega \to \mathbf{R}$ is said to be increasing if $\boldsymbol{x} \geq \boldsymbol{y}$ implies $\varphi(\boldsymbol{x}) \geqslant \varphi(\boldsymbol{y})$. φ is said to be decreasing if and only if $-\varphi$ is increasing.

(iii)$\Omega \subset \mathbf{R}^n$ is called a convex set if $(\alpha x_1 + \beta y_1, \ldots, \alpha x_n + \beta y_n) \in \Omega$ for any $\boldsymbol{x}$ and $\boldsymbol{y} \in \Omega$, where α and $\beta \in [0, 1]$ with $\alpha + \beta = 1$.

(iv)Let $\Omega \subset \mathbf{R}^n$, $\varphi \colon \Omega \to \mathbf{R}$ is said to be a Schur-convex function on Ω if $\boldsymbol{x} \prec \boldsymbol{y}$ on Ω implies $\varphi(\boldsymbol{x}) \leq \varphi(\boldsymbol{y})$. φ is said to be a Schur-concave function on Ω if and only if $-\varphi$ is Schur-convex function. φ is said to be a strictly Schur-convex function on Ω if $\boldsymbol{x} \prec\prec \boldsymbol{y}$ on Ω implies $\phi(\boldsymbol{x}) < \phi(\boldsymbol{y})$, φ is said to be a strictly Schur-concave function on Ω if and if only $-\varphi$ is strictly Schur-convex on Ω.

Lemma 1[2] Let $\boldsymbol{x} \in \mathbf{R}^n$ and $\bar{x} = \frac{1}{n} \sum_{i=1}^{n} x_i$. Then $(\bar{x}, \ldots, \bar{x}) \prec \boldsymbol{x}$.

Proof of Theorem 1

From Lemma 1, we have

$$\boldsymbol{p} := \left(\underbrace{1 + \frac{n}{m}x, \ldots, 1 + \frac{n}{m}x}_{m} \right) \prec \left(\underbrace{1 + x, \ldots, 1 + x}_{n}, \underbrace{1, \ldots, 1}_{m-n} \right) := \boldsymbol{q}$$

and $\boldsymbol{p} \prec\prec \boldsymbol{q}$ for $x \neq 0$. If $m \geqslant n$, from $x > -1$, we have $x + 1 > 0$ and $1 + \frac{n}{m}x > 1 - \frac{n}{m} > 0$, i.e. $\boldsymbol{p}, \boldsymbol{q} \in \mathbf{R}_{++}^n$. Since $E_k(\boldsymbol{x})$ be increasing and Schur-concave on $\mathbf{R}_+^n$ and it be increasing and strictly Schur-concave on $\mathbf{R}_{++}^n$ for $k > 1$ (see Proposition 6.7 in [2]), we have $E_k(\boldsymbol{p}) \geqslant E_k(\boldsymbol{q})$, i.e. (6) holds, and equality holding if and only if $x = 0$.

Since $E_k^*(\boldsymbol{x})$ be increasing and Schur-concave on $\mathbf{R}_+^n$ and it be increasing and strictly Schur-concave on $\mathbf{R}_{++}^n$ for $k > 1$ (see [3, 5]), we have $E_k^*(\boldsymbol{p}) \geqslant E_k^*(\boldsymbol{q})$, i.e. (7) holds, and equality holding if and only if $x = 0$.

If $m < n$, from Lemma 1, we have

$$\boldsymbol{p}' := \left(\underbrace{1+x,\ldots,1+x}_{n}\right) \prec \left(\underbrace{1+\frac{n}{m}x,\ldots,1+\frac{n}{m}x}_{m},\underbrace{1,\ldots,1}_{n-m}\right) := \boldsymbol{q}'$$

and $\boldsymbol{p}' \prec\prec \boldsymbol{q}'$ for $x \neq 0$. From $x > -\frac{m}{n}$, we have $1 + \frac{n}{m}x > 1 - \frac{n}{m}\cdot\frac{n}{m} > 0$, i.e. $\boldsymbol{p}', \boldsymbol{q}' \in \mathbf{R}_{++}^n$. Thus we have $E_k(\boldsymbol{p}') \geqslant E_k(\boldsymbol{q}')$ and $E_k^*(\boldsymbol{p}') \geqslant E_k^*(\boldsymbol{q}')$, i.e. (8)and (9) hold, and equality holding if and only if $x = 0$.

The proof of Theorem 1 is completed.

Proof of Theorem 2 Set $y = \frac{1}{n}\sum_{i=1}^{n}(1+x_i)^{a_i}$, from (2), we have $(1+x_i)^{a_i} \geqslant 1 + a_i x_i$, $i = 1,\ldots,n$, and by lemma 1, it follows that

$$(\underbrace{y,\ldots,y}_{n}) \prec ((1+x_1)^{a_1},\ldots,(1+x_n)^{a_n})$$

$$\geqslant (1+a_1x_1,\ldots,1+a_nx_n) \prec \left(1+\sum_{i=1}^{n}a_ix_i,\underbrace{1,\ldots,1}_{n-1}\right)$$

And then, since $E_k(\boldsymbol{x})$ and $E_k^*(\boldsymbol{x})$ are increasing and Schur-concave on $\mathbf{R}_+^n$ and are increasing and strictly Schur-concave on $\mathbf{R}_{++}^n$ for $k > 1$, we have

$$E_k(\underbrace{y,\ldots,y}_{n}) \geqslant E_k((1+x_1)^{a_1},\ldots,(1+x_n)^{a_n})$$

$$\geqslant E_k(1+a_1x_1,\ldots,1+a_nx_n) \geqslant E_k\left(1+\sum_{i=1}^{n}a_ix_i,\underbrace{1,\ldots,1}_{n-1}\right)$$

i.e. (11) is holds.

And

$$E_k^*(\underbrace{y,\ldots,y}_{n}) \geqslant E_k((1+x_1)^{a_1},\ldots,(1+x_n)^{a_n})$$

$$\geqslant E_k^*(1+a_1x_1,\ldots,1+a_nx_n) \geqslant E_k^*\left(1+\sum_{i=1}^{n}a_ix_i,\underbrace{1,\ldots,1}_{n-1}\right)$$

i.e. (12) is holds.

The proof of Theorem 2 is completed.

35.3　Applications

Theorem 3 Let $a_i \geqslant 1$ and $x_i \geqslant 1$, $i = 1, \cdots, n, n \in \mathbf{N}, n \geqslant 2$. Then for $k = 1, \cdots, n$, we have

$$\sum_{1 \leqslant i_1 < \ldots < i_k \leqslant n} \prod_{j=1}^{k} \left(1 + x_{i_j}\right)^{a_{i_j}} \geqslant \frac{2^{A_k} \mathrm{C}_{n-1}^{k-1}}{1 + A_k} \left(\frac{n}{k}(1 + A_k) - A_n + \sum_{i=1}^{n} a_i x_i\right) \tag{13}$$

where $A_k = \min\limits_{1 \leqslant i_1 < \ldots < i_k \leqslant n} \sum\limits_{i=j}^{k} a_{i_j}$.

Proof Firstly, since $x_i \geqslant 1$ implies $\frac{1+x}{2} \geqslant 0$, from Theorem 2 we have

$$\sum_{1 \leqslant i_1 < \ldots < i_k \leqslant n} \prod_{j=1}^{k} \left(1 + \frac{x_{i_j} - 1}{2}\right)^{a_{i_j}} \geqslant \mathrm{C}_{n-1}^{k} + \mathrm{C}_{n-1}^{k-1}\left(1 + \sum_{i=1}^{n} \frac{a_i(x_i - 1)}{2}\right)$$

$$= \mathrm{C}_{n-1}^{k-1}\left(\frac{n}{k} + \sum_{i=1}^{n} \frac{a_i(x_i - 1)}{2}\right) \geqslant \mathrm{C}_{n-1}^{k-1}\left(\frac{n}{k} + \sum_{i=1}^{n} \frac{a_i(x_i - 1)}{1 + A_k}\right)$$

$$= \frac{\mathrm{C}_{n-1}^{k-1}}{1 + A_k}\left(\frac{n}{k}(1 + A_k) - A_n + \sum_{i=1}^{n} a_i x_i\right) \tag{14}$$

$$\sum_{1 \leqslant i_1 < \ldots < i_k \leqslant n} \prod_{j=1}^{k} \left(1 + x_{i_j}\right)^{a_{i_j}} = \sum_{1 \leqslant i_1 < \ldots < i_k \leqslant n} 2^{\sum_{j=1}^{k} a_{i_j}} \prod_{j=1}^{k}\left(1 + \frac{x_{i_j} - 1}{2}\right)^{a_{i_j}}$$

$$\geqslant 2^{A_k} \sum_{1 \leqslant i_1 < \ldots < i_k \leqslant n} \prod_{j=1}^{k}\left(1 + \frac{x_{i_j} - 1}{2}\right)^{a_{i_j}} \tag{15}$$

Combining (14) with (15), we get (13). The proof of Theorem 3 is completed.

Remark 3 When $k = n$, (13) is deduce to (7.5) in[4]

$$\prod_{i=1}^{n} (1 + x_i)^{a_i} \geqslant \frac{2^{A_n}}{1 + A_n}\left(1 + \sum_{i=1}^{n} a_i x_i\right) \tag{16}$$

Theorem 4 Let $a_i \geqslant 1$ or $a_i \leqslant 0$ and $0 > x_i > -1$ or $x_i > 0$,

$i = 1, \cdots, n, n \in \mathbf{N}, n \geqslant 2$. Then for $k = 1, \cdots, n$, we have

$$\sum_{1 \leqslant i_1 < \ldots < i_k \leqslant n} \prod_{j=1}^{k} \left(1 + x_{i_j}\right)^{-a_{i_j}} \geqslant \sum_{1 \leqslant i_1 < \ldots < i_k \leqslant n} \prod_{j=1}^{k} \left(1 - a_{i_j} x_{i_j} (1 + x_{i_j})^{-1}\right)$$

$$\geqslant C_{n-1}^{k} + C_{n-1}^{k-1} \left(1 - a_{i_j} x_{i_j} (1 + x_{i_j})^{-1}\right) \tag{17}$$

Proof Since $0 > x_i > -1$ or $x_i > 0$ implies $-x_i (1 + x_i)^{-1}$ or $0 > -x_i (1 + x_i)^{-1} > -1$, from Theorem 2 we have

$$\sum_{1 \leqslant i_1 < \ldots < i_k \leqslant n} \prod_{j=1}^{k} \left(1 + x_{i_j}\right)^{-a_{i_j}} = \sum_{1 \leqslant i_1 < \ldots < i_k \leqslant n} \prod_{j=1}^{k} \left(1 - x_{i_j} \left(1 + x_{i_j}\right)^{-1}\right)^{a_{i_j}}$$

$$\geqslant \sum_{1 \leqslant i_1 < \ldots < i_k \leqslant n} \prod_{j=1}^{k} \left(1 + a_{i_j} x_{i_j} (1 + x_{i_j})^{-1}\right)$$

$$\geqslant C_{n-1}^{k} + C_{n-1}^{k-1} \left(1 + a_{i_j} x_{i_j} (1 + x_{i_j})^{-1}\right)$$

The proof of Theorem 4 is completed.

Remark 4 When $k = n$, from (17) we have

$$\prod_{i=1}^{n} (1 + x_i)^{-a_i} \geqslant 1 - \sum_{i=1}^{n} a_i x_i \left(1 + x_{i_j}\right)^{-1}$$

i.e.

$$\prod_{i=1}^{n} (1 + x_i)^{a_i} \leqslant \left(1 - \sum_{i=1}^{n} a_i x_i \left(1 + x_{i_j}\right)^{-1}\right)^{-1} \tag{18}$$

(18) is (7.3) in [4].

References

[1] JI-CHANG KUANG. Applied Inequalities (Chang yong bu deng shi) [M]. 3rd ed. Jinan: Shandong Press of science and technology, 2004.

[2] BO-YING WANG. Foundations of Majorization Inequalities [M]. Beijing: Beijing Normal Univ. Press, 1990. (Chinese)

[3] A M MARSHALL, I OLKIN. Inequalities:theory of majorization and its application[M]. New York: Academies Press, 1979.

[4] MITRINOVIC D S, PECARIC J E, FLNK A M. Classical and New Inequalities in Analysis[M]. Dordrecht: Kluwer Academic Publishers, 1993: 65-82.

[5] HUAN-NAN SHI. Schur-Concavity of Dual Form for Elementary symmetric function with applications[J]. Journal of Northeast Normal University (Natural Sciences Edition), 2001: 33.

[6] XIANG-DONG WANG, HUA-MING SU, FANG-UAN WANG. Theory and method of inequalities[M]. Zheng Zhou: Henan education Press, 1994.

[7] ALFRED WITKOWSKI. A new of the monotonicity of power menas[J]. J. Ineq. Pure. Appl. Math.,2004, 5 (1), Article 6.

[8] JIA-JIN WEN, ZHAO LUO. Optimal Generalization of Bernoulli's Inequality and Its Applications[J]. Journal of Chengdu University (Natural science), 2001, 20 (4):1-8.

[9] JIAN-YA GE. Some extensions on Bernoulli's inequality[J]. Journal of Zhejiang University (Natural science), 2003, 26 (2): 119-122.

[10] CHANG-XING XUE. Separations and extensions on Bernoulli's inequality[J]. Journal of Gansu Education College(Natural Scienc Edition), 1999, 13 (3): 5-7.

第36篇 涉及Schwarz 积分不等式的 Schur-凸函数

(石焕南. 湖南理工学院学报(自然科学版), 2008, 21 (4): 1-3)

摘　要: 本文讨论了由Schwarz 积分不等式生成的函数在$\mathbf{R} \times \mathbf{R}$上的Schur-凸性和在$(0, +\infty) \times (0, +\infty)$上的Schur- 几何凸性, 进而得到Schwarz 积分不等式的两个加强.

关键词: Schwarz 积分不等式; Schur-凸性; Schur-几何凸性; 不等式

设f 和g 是区间$[a,b](a < b)$ 上的可积函数, 则

$$\left(\int_a^b f(x)g(x)\mathrm{d}x \right)^2 \leqslant \int_a^b f^2(x)\mathrm{d}x \int_a^b g^2(x)\mathrm{d}x \tag{1}$$

这是著名的Schwarz 积分不等式.

文[1]通过引入一个正定二次型, 得到了Schwarz 积分不等式的如下改进:

定理 A 设$f, g \in L_2(a,b)$, 则

$$\left(\int_a^b f(x)g(x)\mathrm{d}x \right)^2 \leqslant \int_a^b f^2(x)\mathrm{d}x \int_a^b g^2(x)\mathrm{d}x - \lambda \tag{2}$$

其中

$$\lambda = \int_a^b f^2(x)\mathrm{d}x \left(\int_a^b g(x)h(x)\mathrm{d}x \right)^2 - 2 \int_a^b f(x)g(x)\mathrm{d}x \int_a^b f(x)h(x)\mathrm{d}x +$$

$$\int_a^b g^2(x)\mathrm{d}x \left(\int_a^b f(x)h(x)\mathrm{d}x \right)^2 \geqslant 0$$

$$\int_a^b h^2(x)\mathrm{d}x = 1$$

当且仅当$f = kg(k \neq 0)$ 或者存在不全为零的数k_1, k_2, 使得$h = k_1 f + k_2 g$ 时等式成立.

文[2]给出Schwarz 积分不等式的如下加强:

定理 B　设f, g 均在$[a, b]$ 上可积, 则

$$\left(\int_a^b f(x)g(x)\mathrm{d}x \right)^2$$

$$\leqslant \int_a^b f^2(x)\mathrm{d}x \int_a^b g^2(x)\mathrm{d}x + \frac{2}{b-a} \int_a^b f(x)g(x)\mathrm{d}x \int_a^b f(x)\mathrm{d}x \int_a^b g(x)\mathrm{d}x -$$

$$\frac{1}{b-a} \int_a^b f^2(x)\mathrm{d}x \left(\int_a^b g(x)\mathrm{d}x \right)^2 - \frac{1}{b-a} \int_a^b g^2(x)\mathrm{d}x \left(\int_a^b f(x)\mathrm{d}x \right)^2 \quad (3)$$

并且当存在常数k_1, k_2, 使得$f = k_1 g + k_2$ 或$g = k_1 f + k_2$ 时等式成立. 若f, g 在$[a, b]$ 上连续时, 反之亦然.

文[3]利用式(1)定义了一个二元函数$W : [a, b] \times [a, b] \to \mathbf{R}$

$$W(x, y) = W(x, y; f, g) = \int_x^y f^2(t)\mathrm{d}t \int_x^y g^2(t)\mathrm{d}t - \left(\int_x^y f(t)g(t)\mathrm{d}t \right)^2 \quad (4)$$

并研究了W 的一些性质, 得到如下结论:

定理 C　设f, g 在$[a, b]$ 上可积, 则

(1) $W(x, b)$ 关于x 在$[a, b]$ 上非负单调递减; $W(a, y)$ 关于y 在$[a, b]$ 上非负单调递增.

(2) 对于一切$x, y \in [a, b]$, 有

$$\left(\int_a^b f(x)g(x)\mathrm{d}x \right)^2 \leqslant W(x, y) + \left(\int_x^y f(x)g(x)\mathrm{d}x \right)^2 \leqslant \int_x^y f^2(x)\mathrm{d}x \int_x^y g^2(x)\mathrm{d}x$$
$$(5)$$

(3) 对于一切$x, y, z \in [a, b], x < y < z$, 有

$$W(x, y) + W(y, z) \leqslant W(x, z) \quad (6)$$

文[3]是利用函数单调性的定义证得(1), 本文将用分析的方法给出定理C (1) 的一个简洁的证明, 并考察$W(x, y)$ 的Schur-凸性和Schur-几何凸性, 进而得到Schwarz积分不等式的两个加强. 我们的主要结论是

定理 1　设f, g 在$[a, b]$ 上可积, 则$W(x, y)$ 在$[a, b] \times [a, b] \subset \mathbf{R} \times \mathbf{R}$ 上Schur-凸, 在$[a, b] \times [a, b] \subset \mathbf{R}_{++}^2 = (0, +\infty) \times (0, +\infty)$ 上Schur-几何凸.

定理 2　设$a < b$, 若$\frac{1}{2} \leqslant t_2 \leqslant t_1 \leqslant 1$ 或$0 \leqslant t_1 \leqslant t_2 \leqslant \frac{1}{2}$, 则

$$0 \leqslant W(t_2 b + (1-t_2)a, t_2 a + (1-t_2)b) \leqslant W(t_1 b + (1-t_1)a, t_1 a + (1-t_1)b) \leqslant W(a, b)$$
$$(7)$$

定理 3 设 $0 < a \leqslant b$, 若 $\frac{1}{2} \leqslant t_2 \leqslant t_1 \leqslant 1$ 或 $0 \leqslant t_1 \leqslant t_2 \leqslant \frac{1}{2}$, 则

$$0 \leqslant W(b^{t_2}a^{1-t_2}, a^{t_2}b^{1-t_2}) \leqslant W(b^{t_1}a^{1-t_1}, a^{t_1}b^{1-t_1}) \leqslant W(a,b) \qquad (8)$$

由定理2 和定理3 可分别直接得到下述推论1 和推论2.

推论 1 设 f, g 是区间 $[a,b](a < b)$ 上的可积函数, $0 \leqslant t \leqslant 1$, 则

$$\left(\int_a^b f(x)g(x)\mathrm{d}x \right)^2 \leqslant \int_a^b f^2(x)\mathrm{d}x \int_a^b g^2(x)\mathrm{d}x - W(tb+(1-t)a, ta+(1-t)b) \qquad (9)$$

推论 2 设 f, g 是区间 $[a,b](0 < a < b)$ 上的可积函数, $0 \leqslant t \leqslant 1$, 则

$$\left(\int_a^b f(x)g(x)\mathrm{d}x \right)^2 \leqslant \int_a^b f^2(x)\mathrm{d}x \int_a^b g^2(x)\mathrm{d}x - W(b^t a^{1-t}, a^t b^{1-t}) \qquad (10)$$

注意 $W(tb + (1-t)a, ta + (1-t)b) \geqslant 0$ 和 $W(b^t a^{1-t}, a^t b^{1-t}) \geqslant 0$, 式(9) 和式(10) 均给出了Schwarz积分不等式的加强.

为证明我们的主要结论, 需要用到几个概念和引理.

对于 $\boldsymbol{x} = (x_1, ..., x_n)$, 将 $\boldsymbol{x}$ 的分量排成递减的次序后, 记作 $x_{[1]} \geqslant ... \geqslant x_{[n]}$.

定义 1[4] 设 $\boldsymbol{x}, \boldsymbol{y} \in \mathbf{R}^n$, 满足

(1) 若 $\sum_{i=1}^k x_{[i]} \leqslant \sum_{i=1}^k y_{[i]}$, $k = 1, 2, ..., n-1$;

(2) $\sum_{i=1}^n x_i = \sum_{i=1}^n y_i$, 则称 $\boldsymbol{x}$ 被 $\boldsymbol{y}$ 所控制, 记作 $\boldsymbol{x} \prec \boldsymbol{y}$.

定义 2[4] 设 $\Omega \subset \mathbf{R}^n, \varphi : \Omega \to \mathbf{R}$, 若 $\forall \boldsymbol{x}, \boldsymbol{y} \in \Omega$, $\boldsymbol{x} \prec \boldsymbol{y}$, 恒有 $\varphi(\boldsymbol{x}) \leqslant \varphi(\boldsymbol{y})$, 则称 φ 为 Ω 上的Schur凸函数; 若 $-\varphi$ 是 Ω 上的Schur凸函数, 则称 φ 为 Ω 上的Schur凹函数.

定义 3[5] 设 $\Omega \subset \mathbf{R}_{++}^n, f : \Omega \to \mathbf{R}_+$, 对于任意 $\boldsymbol{x}, \boldsymbol{y} \in \Omega$, 若 $\ln \boldsymbol{x} \prec \ln \boldsymbol{y}$, 恒有 $\varphi(\boldsymbol{x}) \leqslant \varphi(\boldsymbol{y})$, 则称 φ 为 Ω 上的Schur 几何凸函数; 若 $\ln \boldsymbol{x} \prec \ln \boldsymbol{y}$, 恒有 $\varphi(\boldsymbol{x}) \geqslant \varphi(\boldsymbol{y})$, 则称 φ 为 Ω 上的Schur几何凹函数.

引理 1 [4] 设 $\Omega \subset \mathbf{R}^n$ 是有内点的对称凸集, $\varphi : \Omega \to \mathbf{R}$ 在 Ω 上连续, 在 Ω 的内部 Ω^o 可微, 则在 Ω 上Schur-凸(Schur-凹)的充要条件是 φ 在 Ω 上对称且 $\forall \boldsymbol{x} \in \Omega$, 有

$$(x_1 - x_2) \left(\frac{\partial \varphi}{\partial x_1} - \frac{\partial \varphi}{\partial x_2} \right) \geqslant 0 (\leqslant 0)$$

引理 2 [5] 设 $\Omega \subset \mathbf{R}_{++}^n$ 是有内点的对称凸集, $\varphi : \Omega \to \mathbf{R}$ 在 Ω 上连续, 在 Ω 的内部 Ω^o 可微, 则在 Ω 上Schur-几何凸(Schur-凹)的充要条件是 φ 在 Ω

上对称且$\forall \boldsymbol{x} \in \Omega$, 有

$$(\ln x_1 - \ln x_2)\left(x_1 \frac{\partial \varphi}{\partial x_1} - x_2 \frac{\partial \varphi}{\partial x_2}\right) \geqslant 0 (\leqslant 0)$$

引理 3[6]　设$a \leqslant b$, $u(t) = tb + (1-t)a, v(t) = ta + (1-t)b$. 若$\frac{1}{2} \leqslant t_2 \leqslant t_l \leqslant 1$, 或$0 \leqslant t_1 \leqslant t_2 \leqslant \frac{1}{2}$, 则

$$\left(\frac{a+b}{2}, \frac{a+b}{2}\right) \prec (u(t_2), v(t_2)) \prec (u(t_1), v(t_1)) \prec (a, b) \qquad (11)$$

定理 C 的别证　不妨设$y \geqslant x$, 由

$$\frac{\partial W}{\partial x} = -\left(f^2(x)\int_x^y g^2(t)\mathrm{d}t + g^2(x)\int_x^y f^2(t)\mathrm{d}t - 2f(x)g(x)\int_x^y f(t)g(t)\mathrm{d}t\right)$$
$$= -\int_x^y (f(x)g(t) - g(x)f(t))^2\mathrm{d}t \leqslant 0$$

$$\frac{\partial W}{\partial y} = f^2(y)\int_x^y g^2(t)\mathrm{d}t + g^2(y)\int_x^y f^2(t)\mathrm{d}t - 2f(y)g(y)\int_x^y f(t)g(t)\mathrm{d}t$$
$$= \int_x^y (f(y)g(t) - g(y)f(t))^2\mathrm{d}t \geqslant 0$$

可知定理C (1) 成立.

定理 1 的证明　易见$W(x,y)$ 关于x, y 对称. 不妨设$y \geqslant x$, 因

$$\frac{\partial W}{\partial x} - \frac{\partial W}{\partial y} = 2(f(x)g(x) + f(y)g(y))\int_x^y f(t)g(t)\mathrm{d}t -$$
$$(f^2(x) + f^2(y))\int_x^y g^2(t)\mathrm{d}t - (g^2(x) + g^2(y))\int_x^y f^2(t)\mathrm{d}t$$
$$= -\int_x^y (f(x)g(t) - g(x)f(t))^2\mathrm{d}t -$$
$$\int_x^y (f(y)g(t) - g(y)f(t))^2\mathrm{d}t \leqslant 0$$

所以

$$(x - y)\left(\frac{\partial W}{\partial x} - \frac{\partial W}{\partial y}\right) \geqslant 0$$

由引理1 知$W(x,y)$ 在$[a,b] \times [a,b] \subset \mathbf{R}^2$ 上Schur-凸.

经计算

$$x\frac{\partial W}{\partial x} - y\frac{\partial W}{\partial y} = 2(xf(x)g(x) + yf(y)g(y))\int_x^y f(t)g(t)\mathrm{d}t -$$

$$(xf^2(x) + yg^2(y))\int_x^y g^2(t)\mathrm{d}t -$$

$$(xg^2(x) + yg^2(y))\int_x^y f^2(t)\mathrm{d}t$$

$$= -x\int_x^y (f(x)g(t) - g(x)f(t))^2\mathrm{d}t -$$

$$y\int_x^y (f(y)g(t) - g(y)f(t))^2\mathrm{d}t \leqslant 0$$

所以

$$(\ln x - \ln y)\left(x\frac{\partial W}{\partial x} - y\frac{\partial W}{\partial y}\right) \geqslant 0$$

由引理2 知 $W(x,y)$ 在 $[a,b] \times [a,b] \subset \mathbf{R}_{++}^2$ 上Schur-几何凸.

定理 2 的证明　结合引理3 和定理1 即得证.

定理 3 的证明　由引理3 有

$$(\ln\sqrt{ab}, \ln\sqrt{ab}) \prec (\ln b^{t_2}a^{1-t_2}, \ln a^{t_2}b^{1-t_2}) \prec (\ln b^{t_1}a^{1-t_1}, \ln a^{t_1}b^{1-t_1}) \prec (\ln a, \ln b)$$

于是结合 $W(x,y)$ 在 $[a,b] \times [a,b]$ 上Schur-几何凸性即得证.

致谢　感谢网友鱼儿对本文主要结论的证明给予的帮助.

参考文献

[1]　谭立, 龚焰, 王文杰. 关于Cauchy-Buniakowski-Schwarz不等式的改进及其应用[J]. 常德师范学院学报(自然科学版), 2001, 13 (1): 3-5, 12.

[2]　时统业, 周本虎. Cauchy－Schwarz不等式的一个加强[J]. 高等数学研究, 2006, 9 (6): 28–30.

[3]　张敏, 杨灵, 李志伟. 由Schwarz积分不等式生成的函数[J]. 重庆工学院学报(自然科学版), 2008, 22 (5): 160–162, 167.

[4]　王伯英. 控制不等式基础[M]. 北京: 北京师范大学出版社, 1990.

[5]　张小明. 几何凸函数[M]. 合肥: 安徽大学出版社, 2004.

[6]　顾春, 石焕南. Lehme 平均的Schur 凸性和Schur 几何凸性[J]. 不等式研究通讯, 2008, 15 (1): 21–25.

第37篇　Schur Convexity of Generalized Heronian Means Involving Two Parameters

(SHI HUAN-NAN , MIHALY BENCZE, WU SHAN-HE, LI DA-MAO.

Journal of Inequalities and Applications, Volume 2008, Article ID 879273, 9 pages doi:10.1155/2008/)

Abstract: In this paper, the Schur-convexity and Schur-geometric convexity of generalized Heronian means involving two parameters are studied, the main result is then used to obtain several interesting and significantly inequalities for generalized Heronian means.

Keywords: Heronian means; generalized Heronian means; majorization; Schur-convexity; Schur-geometric convexity; inequality

37.1　Introduction

Throughout the paper, $\mathbf{R}$ denotes the set of real numbers, $\boldsymbol{x} = (x_1, x_2, \cdots, x_n)$ denotes n-tuple (n-dimensional real vector), the set of vectors can be written as

$$
\begin{cases}
\mathbf{R}^n = \{\boldsymbol{x} = (x_1, \cdots, x_n) : x_i \in \mathbf{R}, i = 1, \ldots, n\} \\
\mathbf{R}_+^n = \{\boldsymbol{x} = (x_1, \ldots, x_n) : x_i \geqslant 0, i = 1, \ldots, n\} \\
\mathbf{R}_{++}^n = \{\boldsymbol{x} = (x_1, \ldots, x_n) : x_i > 0, i = 1, \ldots, n\}
\end{cases}
\tag{1}
$$

In particular, the notations $\mathbf{R}$, $\mathbf{R}_+$ and $\mathbf{R}_{++}$ denote $\mathbf{R}^1$, $\mathbf{R}_+^1$ and $\mathbf{R}_{++}^1$, respectively.

In what follows, we assume that $(a, b) \in \mathbf{R}_+^2$.

The classical Heronian means of a and b is defined as ([1], see also [2])

$$H_e(a,b) = \frac{a + \sqrt{ab} + b}{3} \tag{2}$$

In [3], an analogue of Heronian means is defined by

$$\tilde{H}(a,b) = \frac{a + 4\sqrt{ab} + b}{6} \tag{3}$$

Janous [4] presented a weighted generalization of the above Heronian-type means, as follows

$$H_w(a,b) = \begin{cases} \dfrac{a + w\sqrt{ab} + b}{w + 2}, & 0 \leqslant w < +\infty \\ \sqrt{ab}, & w = +\infty \end{cases} \tag{4}$$

Recently, the following exponential generalization of Heronian means was considered by Jia and Cao in [5]

$$H_p = H_p(a,b) = \begin{cases} \left[\dfrac{a^p + (ab)^{p/2} + b^p}{3} \right]^{\frac{1}{p}}, & p \neq 0 \\ \sqrt{ab}, & p = 0 \end{cases} \tag{5}$$

Several variants as well as interesting applications of Heronian means can be found in the recent papers [6-11].

The weighted and exponential generalizations of Heronian means motivate us to consider a unified generalization of Heronian means (4) and (5), as follows:

$$H_{p,w}(a,b) = \begin{cases} \left[\dfrac{a^p + w(ab)^{p/2} + b^p}{w + 2} \right]^{\frac{1}{p}}, & p \neq 0 \\ \sqrt{ab}, & p = 0 \end{cases} \tag{6}$$

where $w \geqslant 0$.

In this paper, the Schur-convexity, Schur-geometric convexity and monotonicity of the generalized Heronian means $H_{p,w}(a,b)$ are discussed. As consequences, some interesting inequalities for generalized Heronian means are obtained.

37.2　Definitions and Lemmas

We begin by introducing the following definitions and lemmas.

Definition 1[12,13]　Let $\boldsymbol{x} = (x_1,\ldots,x_n)$ and $\boldsymbol{y} = (y_1,\ldots,y_n) \in \mathbf{R}^n$.

1. $\boldsymbol{x}$ is said to be majorized by $\boldsymbol{y}$ (in symbols $\boldsymbol{x} \prec \boldsymbol{y}$) if $\sum\limits_{i=1}^{k} x_{[i]} \leqslant \sum\limits_{i=1}^{k} y_{[i]}$
   for $k = 1,2,\ldots,n-1$ and $\sum\limits_{i=1}^{n} x_i = \sum\limits_{i=1}^{n} y_i$, where $x_{[1]} \geqslant \cdots \geqslant x_{[n]}$ and
   $y_{[1]} \geqslant \cdots \geqslant y_{[n]}$ are rearrangements of $\boldsymbol{x}$ and $\boldsymbol{y}$ in a descending order.

2. $\boldsymbol{x} \geqslant \boldsymbol{y}$ means that $x_i \geqslant y_i$ for all $i = 1,2,\ldots,n$. Let $\Omega \subset \mathbf{R}^n$, φ:
   $\Omega \to \mathbf{R}$ is said to be increasing if $\boldsymbol{x} \geqslant \boldsymbol{y}$ implies $\varphi(\boldsymbol{x}) \geqslant \varphi(\boldsymbol{y})$. φ is
   said to be decreasing if and only if $-\varphi$ is increasing.

3. Let $\Omega \subset \mathbf{R}^n$, $\varphi\colon \Omega \to \mathbf{R}$ is said to be a Schur-convex function on Ω if
   $\boldsymbol{x} \prec \boldsymbol{y}$ on Ω implies $\varphi(\boldsymbol{x}) \leq \varphi(\boldsymbol{y})$. φ is said to be a Schur-concave
   function on Ω if and only if $-\varphi$ is Schur-convex function.

Definition 2[14,15]　Let $\boldsymbol{x} = (x_1,\ldots,x_n)$ and $\boldsymbol{y} = (y_1,\ldots,y_n) \in \mathbf{R}^n_{++}$.

1. Ω is called a geometrically convex set if $(x_1^\alpha y_1^\beta,\ldots,x_n^\alpha y_n^\beta) \in \Omega$ for any
   $\boldsymbol{x}$ and $\boldsymbol{y} \in \Omega$. Where α and $\beta \in [0,1]$ with $\alpha + \beta = 1$.

2. Let $\Omega \subset \mathbf{R}^n_{++}$, $\varphi\colon \Omega \to \mathbf{R}_+$ is said to be a Schur-geometrically convex
   functionon Ω if $(\ln x_1,\ldots,\ln x_n) \prec (\ln y_1,\ldots,\ln y_n)$ on Ω implies
   $\varphi(\boldsymbol{x}) \leqslant \varphi(\boldsymbol{y})$. φ is said to be a Schur-geometrically concave function
   on Ω if and only if $-\varphi$ is Schur-geometrically convex function.

Lemma 1[12]　A function $\varphi(\boldsymbol{x})$ is increasing if and only if $\nabla\varphi(\boldsymbol{x}) \geq 0$
for $\boldsymbol{x} \in \Omega$, where $\Omega \subset \mathbf{R}^n$ is an open set, $\varphi : \Omega \to \mathbf{R}$ is differentiable, and

$$\nabla\varphi(\boldsymbol{x}) = \left(\frac{\partial\varphi(\boldsymbol{x})}{\partial x_1},\ldots,\frac{\partial\varphi(\boldsymbol{x})}{\partial x_n} \right) \in \mathbf{R}^n \tag{7}$$

Lemma 2[12]　Let $\Omega \subset \mathbf{R}^n$ is symmetric and has a nonempty interior
set. Ω^0 is the interior of Ω. $\varphi : \Omega \to \mathbf{R}$ is continuous on Ω and differentiable
in Ω^0. Then φ is the $Schur - convex(Schur - concave)function$, if and
only if φ is symmetric on Ω and

$$(x_1 - x_2)\left(\frac{\partial\varphi}{\partial x_1} - \frac{\partial\varphi}{\partial x_2} \right) \geqslant 0 (\leqslant 0) \tag{8}$$

holds for any $x = (x_1, x_2, \cdots, x_n) \in \Omega^0$.

Lemma 3[14] Let $\Omega \subset \mathbf{R}^n_{++}$ is a symmetric and has a nonempty interior geometrically convex set. Ω^0 is the interior of Ω. $\varphi : \Omega \to \mathbf{R}_+$ is continuous on Ω and differentiable in Ω^0. If φ is symmetric on Ω and

$$(\ln x_1 - \ln x_2) \left(x_1 \frac{\partial \varphi}{\partial x_1} - x_2 \frac{\partial \varphi}{\partial x_2} \right) \geqslant 0 (\leqslant 0) \tag{9}$$

holds for any $x = (x_1, x_2, \cdots, x_n) \in \Omega^0$, then φ is the Schur-geometrically convex (Schur-geometrically concave) function.

Lemma 4[12] Let $x \in \mathbf{R}^n$ and $\bar{x} = \frac{1}{n} \sum\limits_{i=1}^{n} x_i$. Then

$$(\bar{x}, \ldots, \bar{x}) \prec x \tag{10}$$

Lemma 5[16] The generalized logarithmic means (Stolarsky's means) of two positive numbers a and b is defined as follows

$$S_p(a, b) = \begin{cases} \left(\frac{b^p - a^p}{p(b-a)} \right)^{\frac{1}{p-1}}, & p \neq 0, 1, \ a \neq b \\ \mathrm{e}^{-1} \left(a^a / b^b \right)^{1/(a-b)}, & p = 1, \ a \neq b \\ \frac{b-a}{\ln b - \ln a}, & p = 0 \ a \neq b \\ b, & a = b \end{cases} \tag{11}$$

When $a \neq b$, $S_p(a, b)$ is a strictly increasing function for $p \in \mathbf{R}$.

Lemma 6[17] Let $a, b > 0$ and $a \neq b$. If $x > 0, y \leqslant 0$ and $x + y \geqslant 0$, then

$$\frac{b^{x+y} - a^{x+y}}{b^x - a^x} \leqslant \frac{x+y}{x} (ab)^{\frac{y}{2}} \tag{12}$$

37.3 Main results and their proofs

Our main results are stated in the Theorem 1 and Theorem 2 below.

Theorem 1 For fixed $(p, w) \in \mathbf{R}^2$.

1. $H_{p,w}(a, b)$ is increasing for $(a, b) \in \mathbf{R}^2_+$;

2. If $(p, w) \in \{p \leqslant 1, w \geqslant 0\} \cup \{1 < p \leqslant 3/2, w \geqslant 1\} \cup \{3/2 < p \leqslant 2, w \geqslant 2\}$, then $H_{p,w}(a, b)$ is Schur-concave for $(a, b) \in \mathbf{R}^2_+$;

3. If $p \geqslant 2, 0 \leqslant w \leqslant 2$, then $H_{p,w}(a, b)$ is Schur-convex for $(a, b) \in \mathbf{R}^2_+$.

Proof Let

$$\varphi(a,b) = \frac{a^p + w(ab)^{p/2} + b^p}{w + 2} \tag{13}$$

When $p \neq 0$ and $w \geqslant 0$, we have $H_{p,w}(a,b) = \varphi^{\frac{1}{p}}(a,b)$. It is clear that $H_{p,w}(a,b)$ is symmetric with $(a,b) \in \mathbf{R}_+^2$.

Since

$$\frac{\partial H_{p,w}(a,b)}{\partial a} = \frac{1}{w+2}\left[a^{p-1} + \frac{wb}{2}(ab)^{\frac{p}{2}-1}\right]\varphi^{\frac{1}{p}-1}(a,b) \geqslant 0$$

and

$$\frac{\partial H_{p,w}(a,b)}{\partial b} = \frac{1}{w+2}\left[b^{p-1} + \frac{wa}{2}(ab)^{\frac{p}{2}-1}\right]\varphi^{\frac{1}{p}-1}(a,b) \geqslant 0 \tag{14}$$

we deduce from Lemma 2 that $H_{p,w}(a,b)$ is increasing for $(a,b) \in \mathbf{R}_+^2$.

Let

$$\Lambda := (b-a)\left(\frac{\partial H_{p,w}(a,b)}{\partial b} - \frac{\partial H_{p,w}(a,b)}{\partial a}\right) \tag{15}$$

When $a = b$, then $\Lambda = 0$. We assume $a \neq b$ below.

Let $\Lambda = \frac{(b-a)^2}{w+2}\varphi^{\frac{1}{p}-1}(a,b)Q$, where

$$Q = \frac{b^{p-1} - a^{p-1}}{b-a} - \frac{w}{2}(ab)^{\frac{p}{2}-1} \tag{16}$$

We consider the following four cases.

Case I: If $p \leqslant 1, w \geqslant 0$, then $\frac{b^{p-1}-a^{p-1}}{b-a} \leqslant 0$, which implies that $\Lambda \leqslant 0$. It follows from Lemma 2 that $H_{p,w}(a,b)$ is Schur-concave.

Case II: If $1 < p \leqslant \frac{3}{2}, w \geqslant 1$, then $p - 1 \leqslant \frac{1}{2} \leqslant \frac{w}{2}$.

In Lemma 6, letting $x = 1, y = p-2$, which implies $x > 0, y < 0, x+y > 0$. By Lemma 6 we have

$$\frac{b^{p-1} - a^{p-1}}{b-a} \leqslant (p-1)(ab)^{\frac{p-2}{2}} \leqslant \frac{w}{2}(ab)^{\frac{p}{2}-1} \tag{17}$$

We conclude that $\Lambda \leqslant 0$. Therefore, $H_{p,w}(a,b)$ is Schur-concave.

Case III: If $\frac{3}{2} < p \leqslant 2, w \geqslant 2$, then $p - 1 \leqslant 1 \leqslant \frac{w}{2}$.

In Lemma 6, letting $x = 1, y = p-2$, which implies $x > 0, y \leqslant 0, x+y > 0$. By Lemma 6 we have

$$\frac{b^{p-1} - a^{p-1}}{b-a} \leqslant (p-1)(ab)^{\frac{p-2}{2}} \leqslant \frac{w}{2}(ab)^{\frac{p}{2}-1} \tag{18}$$

It follows that $\Lambda \leqslant 0$. Therefore, $H_{p,w}(a,b)$ is Schur-concave.

Case IV: If $p \geqslant 2, 0 \leqslant w \leqslant 2$. Note that

$$Q = (p-1) \left[S_{p-1}(a,b) \right]^{p-2} - \frac{w}{2} \left[S_{-1}(a,b) \right]^{p-2} \tag{19}$$

By Lemma 5, we obtain that $S_p(a,b)$ is increasing for $p \in \mathbf{R}$. Thus, we conclude that $\left[S_{p-1}(a,b) \right]^{p-2} \geqslant \left[S_{-1}(a,b) \right]^{p-2}$. Then, using $p - 1 \geqslant 1 \geqslant \frac{w}{2}$, we have $\Lambda \geqslant 0$. Therefore, $H_{p,w}(a,b)$ is Schur-convex.

This completes the proof of Theorem 1.

Theorem 2 For fixed $(p, w) \in \mathbf{R}^2$.

1. If $p < 0, w \geqslant 0$, then $H_{p,w}(a,b)$ is Schur-geometrically concave for $(a,b) \in \mathbf{R}_{++}^2$;

2. If $p > 0, w \geqslant 0$, then $H_{p,w}(a,b)$ is Schur-geometrically convex for $(a,b) \in \mathbf{R}_{++}^2$.

Proof Since

$$a \frac{\partial H_{p,w}(a,b)}{\partial a} = \frac{1}{w+2} \left[a^p + \frac{wb}{2}(ab)^{\frac{p}{2}} \right] \varphi^{\frac{1}{p}-1}(a,b)$$

and

$$b \frac{\partial H_{p,w}(a,b)}{\partial b} = \frac{1}{w+2} \left[b^p + \frac{wa}{2}(ab)^{\frac{p}{2}} \right] \varphi^{\frac{1}{p}-1}(a,b) \tag{20}$$

we have

$$\Delta := (\ln b - \ln a) \left(a \frac{\partial H_{p,w}(a,b)}{\partial b} - b \frac{\partial H_{p,w}(a,b)}{\partial a} \right)$$

$$= \frac{(\ln b - \ln a)(b^p - a^p)}{w+2} \varphi^{\frac{1}{p}-1}(a,b). \tag{21}$$

When $p < 0, w \geqslant 0$, then $(\ln b - \ln a)(b^p - a^p) \leqslant 0$, which implies that $\Delta \leqslant 0$. Therefore, $H_{p,w}(a,b)$ is Schur-geometrically concave.

When $p > 0, w \geqslant 0$, then $(\ln b - \ln a)(b^p - a^p) \geqslant 0$, which implies that $\Delta \geqslant 0$. Therefore, $H_{p,w}(a,b)$ is Schur-geometrically convex.

The proof of Theorem 2 is complete.

37.4   Some applications

In this section, we provide several interesting applications of Theorem 1 and 2.

Theorem 3 Let $0 < a \leqslant b$, $A(a,b) = \frac{a+b}{2}$, $u(t) = tb + (1-t)a$, $v(t) = ta + (1-t)b$, and let $\frac{1}{2} \leqslant t_2 \leqslant t_1 \leqslant 1$ or $0 \leqslant t_1 \leqslant t_2 \leqslant 1/2$. If $(p, w) \in \{p \leqslant$

$1, w \geqslant 0\} \cup \{1 < p \leqslant 3/2, w \geqslant 1\} \cup \{3/2 < p \leqslant 2, w \geqslant 2\}$, then

$$A(a,b) \geqslant H_{p,w}(u(t_2), v(t_2)) \geqslant H_{p,w}(u(t_1), v(t_1)) \geqslant H_{p,w}(a,b) \qquad (22)$$

If $p \geqslant 2, 0 \leqslant w \leqslant 2$, then each of the inequalities in (22) is reversed.

Proof When $1/2 \leqslant t_2 \leqslant t_1 \leqslant 1$. From $0 < a \leqslant b$, it is easy to see that $u(t_1) \geqslant v(t_1), u(t_2) \geqslant v(t_2), b \geqslant u(t_1) \geqslant u(t_2)$ and $u(t_2) + v(t_2) = u(t_1) + v(t_1) = a + b$.

We thus conclude that

$$(u(t_2), v(t_2)) \prec (u(t_1), v(t_1)) \prec (a, b) \qquad (23)$$

When $0 \leqslant t_1 \leqslant t_2 \leqslant 1/2$, then $1/2 \leqslant 1 - t_2 \leqslant 1 - t_1 \leqslant 1$, it follows that

$$(u(1 - t_2), v(1 - t_2)) \prec (u(1 - t_1), v(1 - t_1)) \prec (a, b)$$

Since $u(1 - t_2) = v(t_2), v(1 - t_2) = u(t_2), u(1 - t_1) = v(t_1), v(1 - t_1) = u(t_1)$, we also have

$$(u(t_2), v(t_2)) \prec (u(t_1), v(t_1)) \prec (a, b) \qquad (24)$$

On the other hand, it follows from Lemma 4 that $\left(\frac{a+b}{2}, \frac{a+b}{2}\right) \prec (u(t_2), v(t_2))$.

Applying Theorem 1 gives the inequalities asserted by Theorem 3.

Theorem 3 enable us to obtain a large number of refined inequalities by assigning appropriate values to the parameters p, w, t_1 and t_2. For example, putting $p = \frac{1}{2}, w = 1, t_1 = \frac{3}{4}, t_2 = \frac{1}{2}$ in (22), we obtain

$$\frac{a+b}{2} \geqslant \left(\frac{\sqrt{a+3b} + \sqrt[4]{(a+3b)(3a+b)} + \sqrt{3a+b}}{6}\right)^2 \geqslant \left(\frac{\sqrt{a} + \sqrt[4]{ab} + \sqrt{b}}{3}\right)^2 \qquad (25)$$

Putting $p = 2, w = 1, t_1 = \frac{3}{4}, t_2 = \frac{1}{2}$ in (11), we get

$$\frac{a+b}{2} \leqslant \sqrt{\frac{(a+3b)^2 + (a+3b)(3a+b) + (3a+b)^2}{48}} \leqslant \sqrt{\frac{a^2 + ab + b^2}{3}} \qquad (26)$$

Theorem 4 Let $0 < a \leqslant b, c \geqslant 0$. If $(p, w) \in \{p \leqslant 1, w \geqslant 0\} \cup \{1 < p \leqslant 3/2, w \geqslant 1\} \cup \{3/2 < p \leqslant 2, w \geqslant 2\}$, then

$$\frac{H_{p,w}(a+c, b+c)}{a+b+2c} \geqslant \frac{H_{p,w}(a,b)}{a+b} \qquad (27)$$

If $p \geqslant 2, 0 \leqslant w \leqslant 2$, then each of the inequalities in (27) is reversed.

Proof From the hypotheses $0 \leqslant a \leqslant b, c \geqslant 0$, we deduce that

$$\frac{a+c}{a+b+2c} \leqslant \frac{b+c}{a+b+2c}, \frac{a}{a+b} \leqslant \frac{b}{a+b}, \frac{b+c}{a+b+2c} \leqslant \frac{b}{a+b}$$

$$\frac{a+c}{a+b+2c} + \frac{b+c}{a+b+2c} = \frac{a}{a+b} + \frac{b}{a+b} = 1 \qquad (28)$$

We hence have

$$\left(\frac{a+c}{a+b+2c}, \frac{b+c}{a+b+2c} \right) \prec \left(\frac{a}{a+b}, \frac{b}{a+b} \right) \qquad (29)$$

Using Theorem 1 yields the inequalities asserted by Theorem 4.

Theorem 5 Let $0 < a \leqslant b, G(a,b) = \sqrt{ab}, \widetilde{u}(t) = b^t a^{1-t}, \widetilde{v}(t) = a^t b^{1-t}$, and let $\frac{1}{2} \leqslant t_2 \leqslant t_1 \leqslant 1$ or $0 \leqslant t_1 \leqslant t_2 \leqslant 1/2$. If $p > 0, w \geqslant 0$, then

$$G(a,b) \leqslant H_{p,w}(\widetilde{u}(t_2), \widetilde{v}(t_2)) \leqslant H_{p,w}(\widetilde{u}(t_1), \widetilde{v}(t_1)) \leqslant H_{p,w}(a,b) \qquad (30)$$

If $p < 0, w \geqslant 0$, then each of the inequalities in (30) is reversed.

Proof From the hypotheses $0 < a \leqslant b, \frac{1}{2} \leqslant t_2 \leqslant t_1 \leqslant 1$ (or $0 \leqslant t_1 \leqslant t_2 \leqslant 1/2$), it is easy to verify that

$$(\ln \widetilde{u}(t_2), \ln \widetilde{v}(t_2)) \prec (\ln \widetilde{u}(t_1), \ln \widetilde{v}(t_1)) \prec (\ln a, \ln b) \qquad (31)$$

In addition, from Lemma 4 we have $\left(\ln \sqrt{ab}, \ln \sqrt{ab} \right) \prec (\ln \widetilde{u}(t_2), \ln \widetilde{v}(t_2))$.

By applying Theorem 2, we obtain the desired inequalities in Theorem 5.

Combining the inequalities (22) and (30), we obtain the following refinement of arithmetic-geometric means inequality.

Theorem 6 Let $0 < a \leqslant b, u(t) = tb+(1-t)a, v(t) = ta+(1-t)b, \widetilde{u}(t) = b^t a^{1-t}, \widetilde{v}(t) = a^t b^{1-t}$, and let $\frac{1}{2} \leqslant t_2 \leqslant t_1 \leqslant 1$ or $0 \leqslant t_1 \leqslant t_2 \leqslant 1/2$. If $(p,w) \in \{0 < p \leqslant 1, w \geqslant 0\} \cup \{1 < p \leqslant 3/2, w \geqslant 1\} \cup \{3/2 < p \leqslant 2, w \geqslant 2\}$, then

$$G(a,b) \leqslant H_{p,w}(\widetilde{u}(t_2), \widetilde{v}(t_2)) \leqslant H_{p,w}(\widetilde{u}(t_1), \widetilde{v}(t_1)) \leqslant H_{p,w}(a,b)$$

$$\leqslant H_{p,w}(u(t_1), v(t_1)) \leqslant H_{p,w}(u(t_2), v(t_2)) \leqslant A(a,b) \qquad (32)$$

Acknowledgements

The present investigation was supported, in part, by the Scientific Research Common Program of Beijing Municipal Commission of Education under Grant KM200611417009, in part, by the Natural Science Foundation of Fujian province of China under Grant S0850023, and, in part, by the Science Foundation of Project of Fujian Province Education Department of China under Grant JA08231.

The authors would like to express heartily thanks to professor Kai-Zhong Guan for his useful suggestions.

References

[1] H ALZER, W JANOUS. Solution of problem 8*[J]. Crux Mathematicorum, 1987, 13: 173 – 178.

[2] P S BULLEN, D S MITRINVI′ C, P M VASI′ C. Means and Their Inequalities[M]. Dordrecht: Kluwer Academic Publishers, 1988.

[3] Q-J MAO. Dual means, logarithmic and Heronian dual means of two positive numbers[J]. Journal of Suzhou College of Education, 1999, 16:82 – 85.

[4] W JANOUS. A note on generalized Heronian means[J]. Mathematical Inequalities & Applications, 2001, 4 (3):369 – 375.

[5] G JIA, J CAO. A new upper bound of the logarithmic mean[J]. Journal of Inequalities in Pure and Applied Mathematics, 2003, 4 (4): article 80, 4 pages.

[6] K GUAN, H ZHU. The generalized Heronian mean and its inequalities[J]. Univerzitet u Beogradu. Publikacije Elektrotehničkog Fakulteta. Serija Matematika, 2006, 17:60 – 75.

[7] Z ZHANG, Y WU. The generalized Heron mean and its dual form[J]. Applied Mathematics E-Notes,2005(5):16 – 23.

[8] Z ZHANG, Y WU, A ZHAO. The properties of the generalized Heron means and its dual form[J]. RGMIA Research Report Collection, 2004, 7(2) article 1.

[9] Z LIU. Comparison of some means[J]. Journal of Mathematical Research and Exposition, 2002, 22(4): 583 – 588 .

[10] N-G ZHENG, Z-H ZHANG, X-M ZHANG. Schur-convexity of two types of one-parameter mean values in n variables[J]. Journal of Inequalities and Applications, vol. 2007, Article ID 78175, 10 pages, 2007.

[11] H-N SHI, S-H WU, F QI. An alternative note on the Schur-convexity of the extended mean values[J]. Mathematical Inequalities & Applications, 2006, 9(2):219‑224.

[12] B-Y WANG. Foundations of Majorization Inequalities[M]. Beijing: Beijing Normal University Press, 1990.

[13] A W MARSHALL, I OLKIN. Inequalities: Theory of Majorization and Its Applications vol. 143 of Mathematics in Science and Engineering[M]. New York: Academic Press, 1979.

[14] X-M ZHANG. Geometrically Convex Functions[M]. Hefei: Anhui University Press, 2004.

[15] C P NICULESCU. Convexity according to the geometricmean[M]. Mathematical Inequalities & Applications,2000, 3(2):155‑167.

[16] J-C KUANG. Applied Inequalities[M].3rd ed. Ji'nan: Shandong Science and Technology Press, 2004.

[17] Z LIU. A note on an inequality[J]. Pure and Applied Mathematics, 2001, 17(4): 349‑351.

第38篇 Schur-Convexity of a Mean of Convex Function

(SHI HUAN-NAN, LI DA-MAO, GU CHUN. Applied Mathematics Letters, 22 (2009): 932-937)

Abstract: The Schur-convexity on the upper and the lower limit of the integral for a mean of the convex function is researched. As applications, a generalized logarithmic mean with a parameter is obtained and a relevant double inequality that is a extension of the known inequality is established.

Keywords: Schur-convex function; inequality; Hadamard's inequality; generalized logarithmic mean

38.1 Introduction

Let f be a convex function defined on the interval $I \subseteq \mathbf{R} \to \mathbf{R}$ of real numbers and $a, b \in I$ with $a < b$. Then

$$f\left(\frac{a+b}{2}\right) \leqslant \frac{1}{b-a} \int_a^b f(x)\,\mathrm{d}x \leqslant \frac{f(a)+f(b)}{2} \tag{1}$$

is known as the Hadamard's inequality for convex function.

In [1], S. S. Dragomir established the following two theorems which are refinements of the first inequality of (1).

Theorem A[1] If $f : [a, b] \to \mathbf{R}$ is a convex function, and H is defined on $[0, 1]$ by

$$H(t) = \frac{1}{b-a} \int_a^b f\left(tx + (1-t)\frac{a+b}{2}\right)\mathrm{d}x$$

then H is convex, increasing on $[0, 1]$, and for all $t \in [0, 1]$, we have

$$f\left(\frac{a+b}{2}\right) = H(0) \leqslant H(t) \leqslant H(1) = \frac{1}{b-a} \int_a^b f(x)\,\mathrm{d}x \tag{2}$$

In [2], Yang and Hong established the following theorem which is a refinement of the second inequality of (1).

Theorem B[2] If $f : [a,b] \to \mathbf{R}$ is a convex function, and G is defined on $[0,1]$ by

$$G(t) = \frac{1}{2(b-a)} \int_a^b \left[f\left(\frac{1+t}{2}a + \frac{1-t}{2}x\right) + f\left(\frac{1+t}{2}b + \frac{1-t}{2}x\right) \right] \mathrm{d}\,x$$

then G is convex, increasing on $[0,1]$, and for all $t \in [0,1]$, we have

$$\frac{1}{b-a} \int_a^b f(x)\,\mathrm{d}\,x = G(0) \leqslant G(t) \leqslant G(1) = \frac{f(b) + f(a)}{2} \tag{3}$$

Remark 1 If f is concave, then (3) is reversed (notice that $-f$ is convex).

In [3], Elezovic and Pecaric researched the Schur-convexity on the upper and lower limits of the integral for the mean of a convex function and established the following important result by using Hadamard's inequality.

Theorem C[3] Let I be an interval with nonempty interior on $\mathbf{R}$ and f be a continuous function on I. Then

$$\Phi(a,b) = \begin{cases} \frac{1}{b-a}\int_a^b f(t)\mathrm{d}t, & a,b \in I, \ a \neq b \\ f(a), & a = b \end{cases}$$

is Schur-convex (Schur-concave) on I^2 if and only if f is convex(concave)on I.

In this work, the first aim is to prove once again the inequality in (2) using Theorem C. The second aim is to establish the following result which is similar to Theorem C. As applications, a form with a parameter of Stolarsky's mean is obtained and a relevant double inequality that is an extension of a known inequality is established.

Theorem 1 Let I be an interval with nonempty interior on $\mathbf{R}$ and define a function of two variables as follows

$$P(a,b) = \begin{cases} G(t), & a,b \in I, \ a \neq b \\ f(a), & a = b \end{cases}$$

(i) For $\frac{1}{2} \leqslant t \leqslant 1$, if f is convex on I, then $P(a,b)$ is Schur-convex on I^2.

(ii) For $0 \leqslant t \leqslant \frac{1}{2}$, if f is concave on I, then $P(a,b)$ is Schur-concave on I^2.

38.2　Definitions and Lemmas

We need the following definitions and lemmas.

Definition 1[5,7] Let $\Omega \subseteq \mathbf{R}^n, \boldsymbol{x} = (x_1, \ldots, x_n)$ and $\boldsymbol{y} = (y_1, \ldots, y_n) \in \Omega$, and let $\varphi : \Omega \to \mathbf{R}$.

1. $\boldsymbol{x}$ is said to be majorized by $\boldsymbol{y}$ (in symbols $\boldsymbol{x} \prec \boldsymbol{y}$) if $\sum\limits_{i=1}^{k} x_{[i]} \leqslant \sum\limits_{i=1}^{k} y_{[i]}$ for $k = 1, 2, \ldots, n-1$ and $\sum\limits_{i=1}^{n} x_i = \sum\limits_{i=1}^{n} y_i$, where $x_{[1]} \geqslant \cdots \geqslant x_{[n]}$ and $y_{[1]} \geqslant \cdots \geqslant y_{[n]}$ are rearrangements of $\boldsymbol{x}$ and $\boldsymbol{y}$ in a descending order.

2. $\boldsymbol{x} \geq \boldsymbol{y}$ means $x_i \geqslant y_i$ for all $i = 1, 2, \ldots, n$. φ is said to be increasing if $\boldsymbol{x} \geqslant \boldsymbol{y}$ implies $\varphi(\boldsymbol{x}) \geqslant \varphi(\boldsymbol{y})$. φ is said to be decreasing if and only if $-\varphi$ is increasing.

3. φ is said to be a Schur-convex function on Ω if $\boldsymbol{x} \prec \boldsymbol{y}$ on Ω implies $\varphi(\boldsymbol{x}) \leqslant \varphi(\boldsymbol{y})$, φ is said to be a Schur-concave function on Ω if and only if $-\varphi$ is Schur-convex function.

Lemma 1[24] Let $\boldsymbol{x} \in \mathbf{R}^n$ and $\bar{x} = \frac{1}{n} \sum\limits_{i=1}^{n} x_i$. Then

$$(\bar{x}, \ldots, \bar{x}) \prec \boldsymbol{x}$$

Lemma 2 Let $a \leqslant b, u = \frac{1+t}{2}b + \frac{1-t}{2}a, v = \frac{1+t}{2}a + \frac{1-t}{2}b$. Then

$$(u, v) \prec (a, b) \tag{4}$$

Proof From $a < b, t \in [0; 1]$, it is easy to see that $u \geqslant v, u \leqslant b$ and $u + v = a + b$. By Definition 1, it follows that (4) holds.

Lemma 3[4,7] Let us have that $\Omega \subseteq \mathbf{R}^n$, is symmetric and has a nonempty interior set. Ω^0 is the interior of Ω. $\varphi : \Omega \to \mathbf{R}$ is continuous on Ω and differentiable in Ω^0. Then φ is the *Schur − convex(Schur − concave)function*, if and only if φ is symmetric on Ω and

$$(x_1 - x_2)\left(\frac{\partial \varphi}{\partial x_1} - \frac{\partial \varphi}{\partial x_2}\right) \geqslant 0(\leqslant 0)$$

holds for any $\boldsymbol{x} = (x_1, x_2, \cdots, x_n) \in \Omega^0$.

Lemma 4[4] Let us have $\Omega \subseteq \mathbf{R}^n$, $\varphi : \Omega \to \mathbf{R}, h : \mathbf{R} \to \mathbf{R}, \psi(\boldsymbol{x}) = h\left(\varphi(\boldsymbol{x})\right)$.

1. If φ is Schur-convex and h is increasing, then ψ is also Schur-convex.

2. If φ is Schur-convex and h is decreasing, then ψ is also Schur-concave.

3. If φ is Schur-concave and h is increasing, then ψ is also Schur-concave.

38.3　Proofs of main results

Another proof of inequalities in (2) By the transformation $s = tx + (1-t)\frac{a+b}{2}$, we get

$$H(t) = \frac{1}{u-v} \int_v^u f(s)\,\mathrm{d}s$$

where $u = \frac{1+t}{2}b + \frac{1-t}{2}a$ and $v = \frac{1+t}{2}a + \frac{1-t}{2}b$. From Lemma 2, And by Lemma 1, it follows that

$$\Phi\left(\frac{a+b}{2}, \frac{a+b}{2}\right) \leqslant \Phi(u,v) \leqslant \Phi(a,b)$$

i.e. (2)holds.

Proof of Theorem 1

It is clear that $P(a,b)$ is symmetric. When $a \neq b$, let

$$P_1(a,b) = \int_a^b f\left(\frac{1+t}{2}a + \frac{1-t}{2}x\right)\mathrm{d}x$$

and

$$P_2(a,b) = \int_a^b f\left(\frac{1+t}{2}b + \frac{1-t}{2}x\right)\mathrm{d}x$$

then

$$P(a,b) = \frac{1}{2(b-a)}\left[P_1(a,b) + P_2(a,b)\right]$$

By the transformation $s = \frac{1+t}{2}a + \frac{1-t}{2}x$, we get

$$P_1(a,b) = \frac{2}{1-t}\int_a^{\frac{1+t}{2}a+\frac{1-t}{2}b} f(s)\,\mathrm{d}s$$

$$= \frac{2}{1-t}\left[\int_0^{\frac{1+t}{2}a+\frac{1-t}{2}b} f(s)\,\mathrm{d}s - \int_0^a f(s)\,\mathrm{d}s\right]$$

$$\frac{\partial P_1(a,b)}{\partial a} = \frac{2}{1-t}\left[f\left(\frac{1+t}{2}a+\frac{1-t}{2}b\right)\frac{1+t}{2}-f(a)\right]$$

$$= \frac{1+t}{1-t}f\left(\frac{1+t}{2}a+\frac{1-t}{2}b\right)-\frac{2f(a)}{1-t} \qquad (5)$$

$$\frac{\partial P_1(a,b)}{\partial b} = \frac{2}{1-t}\left[f\left(\frac{1+t}{2}a+\frac{1-t}{2}b\right)\frac{1-t}{2}\right] = f\left(\frac{1+t}{2}a+\frac{1-t}{2}b\right) \qquad (6)$$

Since

$$P_2(a,b) = \int_a^b f\left(\frac{1+t}{2}b+\frac{1-t}{2}x\right)\mathrm{d}x$$

$$= -\int_b^a f\left(\frac{1+t}{2}b+\frac{1-t}{2}x\right)\mathrm{d}x = -P_1(b,a)$$

then from (6), we get

$$\frac{\partial P_2(a,b)}{\partial a} = -\frac{\partial P_1(b,a)}{\partial a} = -f\left(\frac{1+t}{2}b+\frac{1-t}{2}a\right) \qquad (7)$$

and from (5), we get

$$\frac{\partial P}{\partial b} = -\frac{\partial P_1(b,a)}{\partial b} = \frac{2f(b)}{1-t}-\frac{1+t}{1-t}f\left(\frac{1+t}{2}b+\frac{1-t}{2}a\right) \qquad (8)$$

$$\frac{\partial P(a,b)}{\partial b} = \left\{-\frac{1}{2(b-a)^2}[P_1(a,b)+P_2(a,b)]+\frac{1}{2(b-a)}\left[\frac{\partial P_1(a,b)}{\partial b}+\frac{\partial P_2(a,b)}{\partial b}\right]\right\}$$

$$\frac{\partial P(a,b)}{\partial a} = \left\{\frac{1}{2(b-a)^2}[P_1(a,b)+P_2(a,b)]+\frac{1}{2(b-a)}\left[\frac{\partial P_1(a,b)}{\partial a}+\frac{\partial P_2(a,b)}{\partial a}\right]\right\}$$

If f is convex, then

$$(b-a)\left(\frac{\partial P(a,b)}{\partial b}-\frac{\partial P(a,b)}{\partial a}\right)$$

$$=\frac{1}{2}\left[\left(\frac{\partial P_1(a,b)}{\partial b}-\frac{\partial P_1(a,b)}{\partial a}\right)+\left(\frac{\partial P_2(a,b)}{\partial b}-\frac{\partial P_2(a,b)}{\partial a}\right)\right]-$$

$$\frac{1}{b-a}[P_1(a,b)-P_2(a,b)]$$

$$\geqslant\frac{1}{1-t}\left\{f(a)+f(b)-t\left[\frac{1+t}{2}f(b)+\frac{1-t}{2}f(a)+\frac{1+t}{2}f(a)+\frac{1-t}{2}f(b)\right]\right\}-$$

$$\frac{1}{b-a}[P_1(a,b) - P_2(a,b)]$$

$$=\frac{1}{1-t}\{f(a) + f(b) - t[f(a) + f(b)]\} - \frac{1}{b-a}[P_1(a,b) - P_2(a,b)]$$

$$=f(a) + f(b) - t[f(a) + f(b)] - \frac{1}{b-a}[P_1(a,b) - P_2(a,b)] \geqslant 0$$

(by the right inequality in (3))

According to Lemma 3, it follows that $P(a,b)$ is Schur-convex on I^2.

If f is concave, then

$$(b-a)\left(\frac{\partial P(a,b)}{\partial b} - \frac{\partial P(a,b)}{\partial a}\right) = f(a) + f(b) - t[f(a) + f(b)]$$

$$- \frac{1}{b-a}[P_1(a,b) - P_2(a,b)] \leqslant 0 \text{(by Remark 1)}$$

According to Lemma 3, it follows that $P(a,b)$ is Schur-concave on I^2. The proof of Theorem 1 is completed.

38.4   Applications

Theorem 2 Let $t \in [0,1), a, b \in \mathbf{R}_+ = [0,+\infty)$, and let

$$L_r(a,b;t) = \left[\frac{(b^r - a^r) - (u^r - v^r)}{r(1-t)(b-a)}\right]^{\frac{1}{r-1}}, a \neq b$$

$$L_r(a,a;t) = a$$

where $u = \frac{1+t}{2}b + \frac{1-t}{2}a, v = \frac{1+t}{2}a + \frac{1-t}{2}b$. Then $L_r(a,b;t)$ is Schur-convex, for $r > 2$ and $L_r(a,b;t)$ is Schur-concave for $r < 2$ and $r \neq 0$ on $\mathbf{R}_+^2$,

Proof Taking $f(x) = x^{r-1}, r \neq 0$, then for $a \neq b$, from Theorem 1, we have

$$G(t) = \frac{1}{2(b-a)}\int_a^b\left[\left(\frac{1+t}{2}a + \frac{1-t}{2}x\right)^{r-1} + \left(\frac{1+t}{2}b + \frac{1-t}{2}x\right)^{r-1}\right]\mathrm{d}x$$

$$= \frac{1}{r(1-t)(b-a)}\left[\left(\frac{1+t}{2}a + \frac{1-t}{2}x\right)^r\Big|_a^b + \left(\frac{1+t}{2}b + \frac{1-t}{2}x\right)^r\Big|_a^b\right]$$

$$= \frac{(b^r - a^r) + \left(\frac{1+t}{2}a + \frac{1-t}{2}b\right)^r - \left(\frac{1+t}{2}b + \frac{1-t}{2}a\right)^r}{r(1-t)(b-a)}$$

$$= \frac{(b^r - a^r) + (u^r - v^r)}{r(1-t)(b-a)}$$

When $r < 1$ or $r > 2$, $f(x) = x^{r-1}$ is convex on $\mathbf{R}_+$, and when $1 < r < 2$, $f(x) = x^{r-1}$ is concave on $\mathbf{R}_+$. From Theorem 1 we obtain that $P(a, b)$ is Schur-convex on $\mathbf{R}_+^2$ for $0 \neq r < 1$ or $r > 2$, and $P(a, b)$ is Schur-concave on $\mathbf{R}_+^2$ for $1 < r < 2$. Furthermore, since when $r > 1$, $h : t \to t^{\frac{1}{r-1}}$ is increasing on $\mathbf{R}_+$, then from (1) in Lemma 4, $L_r(a, b; t) = [P(a, b)]^{\frac{1}{r-1}}$ is Schur-convex on $\mathbf{R}_+^2$ for $r > 2$, and from (3) in Lemma 4, $L_r(a, b; t)$ is Schur-concave on $\mathbf{R}_+^2$ for $1 < r < 2$. Since when $0 \neq r < 1$, $h : t \to t^{\frac{1}{r-1}}$ is decreasing on $\mathbf{R}_+$, then from (2) in Lemma 4, $L_r(a, b; t)$ is Schur-concave on $\mathbf{R}_+^2$ for $0 \neq r < 1$. Letting $r \to 1$, we obtain that $L_r(a, b; t)$ is still Schur-concave on $\mathbf{R}_+^2$ for $r = 1$. The proof of Theorem 2 is complete.

Corollary 1 For $r > 2$, we have

$$\frac{a+b}{2} \leqslant L_r(a, b; t) \leqslant (a+b) \left(\frac{2^r + (1-t)^r - (1+t)^r}{r 2^r (1-t)} \right)^{\frac{1}{r-1}} \tag{9}$$

For $r < 2$ and $r \neq 0$, the two inequalities in (9) are reversed.

Proof Since

$$\left(\frac{a+b}{2}, \frac{a+b}{2} \right) \prec (a, b) \prec (a+b, 0)$$

then from Corollary 1, we have

$$L_r \left(\frac{a+b}{2}, \frac{a+b}{2} \right) \leqslant L_r(a, b) \leqslant L_r(a+b, 0)$$

for $r > 2$, i.e. (9) holds. For $r < 2$ and $r \neq 0$, the two inequalities in (9) are reversed.

Remark 2 $L_r(a, b; 0)$ is the generalized logarithmic mean (or Stolarsky's mean)

$$S_r(a, b) = \left(\frac{b^r - a^r}{r(b-a)} \right)^{\frac{1}{r-1}}.$$

Taking $t = 0$, from (14), from (9), we can obtain a known inequality[5]

$$\frac{a+b}{2} \leqslant S_r(a, b) \leqslant \frac{a+b}{r^{\frac{1}{r-1}}}, \quad r > 2 \tag{10}$$

Theorem 3 Let $t \in [0, 1), a, b \in \mathbf{R}_+$, and let

$$L(a, b; t) = \frac{\ln b[(1+t)a + (1-t)b] - \ln a[(1+t)b + (1-t)a]}{(1-t)(b-a)}, a \neq b$$

$$L(a, a; t) = a^{-1}$$

Then $L(a, b; t)$ is Schur-convex on $\mathbf{R}_+^2$.

Proof Taking $f(x) = x^{-1}$, then for $a \neq b$, from Theorem 1, we have

$$G(t) = \frac{1}{2(b-a)} \int_a^b \left[\left(\frac{1+t}{2}a + \frac{1-t}{2}x \right)^{-1} + \left(\frac{1+t}{2}b + \frac{1-t}{2}x \right)^{-1} \right] dx$$

$$= \frac{1}{(1-t)(b-a)} \left[\ln\left(\frac{1+t}{2}a + \frac{1-t}{2}x \right)\Big|_a^b + \ln\left(\frac{1+t}{2}b + \frac{1-t}{2}x \right)\Big|_a^b \right]$$

$$= \frac{\ln b[(1+t)a + (1-t)b] - \ln a[(1+t)b + (1-t)a]}{(1-t)(b-a)}$$

Since $f(x) = x^{-1}$ is convex on $\mathbf{R}_+$, when by Theorem 1, it follows that $L(a, b; t)$ is Schur-convex on $\mathbf{R}_+^2$.

From Theorem 3 and $\left(\frac{a+b}{2}, \frac{a+b}{2} \right) \prec (a, b)$, we get

Corollary 2 Let $\frac{1}{2} \leqslant t < 1, a, b \in \mathbf{R}_+$. Then

$$L(a, b; t) \geqslant \frac{2}{a + b} \tag{11}$$

Remark 3 Taking $t = 0$, from (11), we can obtain the Ostle-Terwilliger inequality[6]

$$\frac{\ln b - \ln a}{b - a} \leqslant \frac{2}{a + b} \tag{12}$$

Acknowledgements

The authors are indebted to the referees for their helpful suggestions. Huan-Nan Shi was supported in part by the Scientific Research Common Program of Beijing Municipal Commission of Education (KM200611417009).

References

[1] S S DRAGOMIR. Two mappings in connection to Hadamard's inequalities[J]. J. Math. Anal. Appl., 1992,167(1): 49-56.

[2] G S YANG, M C HONG. A note on Hadamard's inequality[J]. Tamkang. J. Math., 1997, 28 (1): 33-37.

[3] N ELEZOVIC, J PECARIC. A note on Schur-convex fuctions[J]. Rocky Mountain J. Math., 2000, 30 (3): 853-856.

[4]　B Y WANG. Foundations of Majorization Inequalities[M].Beijing: Bei-
　　　jing Normal Univ. Press, 1990.

[5]　J C KUANG. Applied Inequalities[M]. 3rd ed. Jinan: Shandong Press
　　　of science and technology, 2002.

[6]　B OSTLE, H L TERWILLIGER. A companion of two means[J]. Proc.
　　　Montana Acad.Sci., 1957, 17(1): 69-70.

[7]　A W MARSHALL, I OLKIN. Inequalities: Theory of majorization and
　　　its application[M]. New York: Academies Press, 1979.

第39篇　Schur-Convexity and Schur-Geometrically Concavity of Gini Means

(SHI HUAN-NAN, JIANG YONG-MING, JIANG WEI-DONG. Computers and Mathematics with Applications, 57 (2009): 266-274.)

Abstract: The Schur-convexity and the Schur-geometrically convexity with variables $(x,y) \in \mathbf{R}^2_{++}$ for fixed (s,t) of Gini means $G(r,s;x,y)$ are discussed. Some new inequalities are obtained.

Keywords: Gini means; Stolarsky means; Schur-convexity; Schur-geometric concavity; inequalities

39.1　Introduction

Throughout the paper we assume that the set of n-dimensional row vector on the real number field by $\mathbf{R}^n$.

$$\mathbf{R}^n_+ = \{\boldsymbol{x} = (x_1, \ldots, x_n) \in \mathbf{R}^n : x_i > 0, i = 1, \ldots, n\}$$

$$\mathbf{R}^n_- = \{\boldsymbol{x} = (x_1, \ldots, x_n) \in \mathbf{R}^n : x_i < 0, i = 1, \ldots, n\}$$

In particular, $\mathbf{R}^1$, $\mathbf{R}^1_+$ and $\mathbf{R}^1_-$ denoted by $\mathbf{R}$, $\mathbf{R}_+$ and $\mathbf{R}_-$ respectively.

Let $(s,t) \in \mathbf{R}^2, (x,y) \in \mathbf{R}^2_+$. The Gini means of (x,y) is defined in [1] and [2] as

$$G(r,s;x,y) = \begin{cases} \left(\dfrac{x^s + y^s}{x^r + y^r} \right)^{1/(s-r)}, & r \neq s \\ \exp\left(\dfrac{x^s \ln x + y^s \ln y}{x^r + y^r} \right), & r = s \end{cases}$$

The Gini means are also called the "sum means". Clearly, $G(0,-1;x,y)$

is the harmonic mean, $G(0,0;x,y)$ is the geometric mean, $G(1,0;x,y)$ is the arithmetic mean.

Some properties of Gini means are given in next theorem.

Theorem A[3]

(i)

$$\lim_{s \to r} G(r,s;x,y) = G(r,r;x,y)$$

$$\lim_{s \to \infty} G(r,s;x,y) = \max\{x,y\}; \quad \lim_{s \to -\infty} G(r,s;x,y) = \min\{x,y\}$$

(ii)If $s_1 \leqslant s_2, r_1 \leqslant r_2$ then

$$G(r_1,s_1;x,y) \leqslant G(r_2,s_2;x,y) \tag{1}$$

Further if $s_1 \neq s_2$ or $r_1 \leqslant r_2$, then inequality (1) is strict unless $x = y$.

(iii)If $s \geqslant 1 \geqslant r \geqslant 0$ then

$$G(r,s;x_1+x_2,y_1+y_2) \leqslant G(r,s;x_1,y_1) + G(r,s;x_2,y_2) \tag{2}$$

The Stolarsky means of (x,y) is defined in [4-5] as

$$E(r,s;x,y) = \begin{cases} \left(\dfrac{r}{s} \cdot \dfrac{y^s - x^s}{y^r - x^r}\right)^{1/(s-r)}, & rs(r-s)(x-y) \neq 0 \\[4mm] \left(\dfrac{1}{r} \cdot \dfrac{y^r - x^r}{\ln y - \ln x}\right)^{1/r}, & r(x-y) \neq 0 \\[4mm] \dfrac{1}{e^{1/r}}\left(\dfrac{x^{x^r}}{y^{y^r}}\right)^{1/(x^r - y^r)}, & r(x-y) \neq 0 \\[4mm] \sqrt{xy}, & x \neq y \\[2mm] x, & x = y \end{cases}$$

The Stolarsky means are sometimes called the "difference means", or the "extended means".

The Schur-convexities of the Stolarsky means $E(r,s;x,y)$ with (r,s) and (x,y) were presented in [6-7] as follows.

Theorem B[6] For fixed $(x,y) \in \mathbf{R}_+^2$ with $x \neq y$, $E(r,s;x,y)$ is Schur-concave on $\mathbf{R}_+^2$ and Schur-convex on $\mathbf{R}_-^2$ with (r,s).

Theorem C[7] For fixed $(r,s) \in \mathbf{R}^2$.

(i) If $2 < 2r < s$ or $2 \leq 2s \leqslant r$, then $E(r,s;x,y)$ is Schur-convex on $\mathbf{R}_+^2$ with (x,y);

(ii) If $(r, s) \in \{r < s \le 2r, 0 < r \le 1\} \cup \{s < r \le 2s, 0 < s \le 1\} \cup \{0 < s < r \le 1\} \cup \{0 < r < s \le 1\} \cup \{s \le 2r < 0\} \cup \{r \le 2s < 0\}$, then $E(r, s; x, y)$ is Schur-concave on $\mathbf{R}_+^2$ with (x, y).

In a recent paper, József Sándor[8] has proved the following result:

Theorem D For fixed $(x, y) \in \mathbf{R}_+^2$ with $x \ne y$, $G(r, s; x, y)$ is Schur-concave on $\mathbf{R}_+^2$ and Schur-convex on $\mathbf{R}_-^2$ with (r, s).

And József Sándor point out that the Schur-convexity problem of $G(r, s; x, y)$ for fixed (s, t) with respect to $(x, y) \in \mathbf{R}_+^2$ are still open.

In this paper, the Schur-convexity and the Schur-geometrically convexity with variables $(x, y) \in \mathbf{R}_+^2$ for fixed (s, t) of Gini means $G(r, s; x, y)$ are discussed, moreover, some new inequalities are obtained. We obtain the following results.

Theorem 1 For fixed $(r, s) \in \mathbf{R}^2$.

(i) If $(r, s) \in \{r \ge 0, s \ge 0, r + s \ge 1\}$, then $G(r, s; x, y)$ is the Schur-convex with $(x, y) \in \mathbf{R}_+^2$;

(ii) If $(r, s) \in \{r \le 0, r + s \le 1\} \cup \{s \le 0, r + s \le 1\}$, then $G(r, s; x, y)$ is the Schur-concave with $(x, y) \in \mathbf{R}_+^2$.

Theorem 2 If $(r, s) \in \mathbf{R}_+^2$, then $G(r, s; x, y)$ is the Schur-geometrically convex with $(x, y) \in \mathbf{R}_+^2$.

For more information on the Stolarsky means and the Gini means, please refer to [9-15] and the references therein.

39.2 Definitions and Lemmas

We need the following definitions and lemmas.

Definition 1[16-17] Let $\boldsymbol{x} = (x_1, \ldots, x_n)$ and $\boldsymbol{y} = (y_1, \ldots, y_n) \in \mathbf{R}^n$.

(i) $\boldsymbol{x}$ is said to be majorized by $\boldsymbol{y}$ (in symbols $\boldsymbol{x} \prec \boldsymbol{y}$) if $\sum_{i=1}^{k} x_{[i]} \le \sum_{i=1}^{k} y_{[i]}$ for $k = 1, 2, \ldots, n-1$ and $\sum_{i=1}^{n} x_i = \sum_{i=1}^{n} y_i$, where $x_{[1]} \ge \cdots \ge x_{[n]}$ and $y_{[1]} \ge \cdots \ge y_{[n]}$ are rearrangements of $\boldsymbol{x}$ and $\boldsymbol{y}$ in a descending order.

(ii) $\Omega \subset \mathbf{R}^n$ is called a convex set if $(\alpha x_1 + \beta y_1, \ldots, \alpha x_n + \beta y_n) \in \Omega$ for any $\boldsymbol{x}$ and $\boldsymbol{y} \in \Omega$, where α and $\beta \in [0, 1]$ with $\alpha + \beta = 1$.

(iii) Let $\Omega \subset \mathbf{R}^n$, $\varphi: \Omega \to \mathbf{R}$ is said to be a Schur-convex function on Ω if $\boldsymbol{x} \prec \boldsymbol{y}$ on Ω implies $\varphi(\boldsymbol{x}) \leq \varphi(\boldsymbol{y})$. φ is said to be a Schur-concave function on Ω if and only if $-\varphi$ is Schur-convex function.

Definition 2[18-19] Let $\boldsymbol{x} = (x_1, \ldots, x_n)$ and $\boldsymbol{y} = (y_1, \ldots, y_n) \in \mathbf{R}_+^n$.

(i) $\Omega \subset \mathbf{R}_+^n$ is called a geometrically convex set if $(x_1^\alpha y_1^\beta, \ldots, x_n^\alpha y_n^\beta) \in \Omega$ for any $\boldsymbol{x}$ and $\boldsymbol{y} \in \Omega$, where α and $\beta \in [0, 1]$ with $\alpha + \beta = 1$.

(ii) let $\Omega \subset \mathbf{R}_+^n$, $\varphi: \Omega \to \mathbf{R}_+$ is said to be a Schur-geometrically convex function on Ω if $(\ln x_1, \ldots, \ln x_n) \prec (\ln y_1, \ldots, \ln y_n)$ on Ω implies $\varphi(\boldsymbol{x}) \leqslant \varphi(\boldsymbol{y})$. φ is said to be a Schur-geometrically concave function on Ω if and only if $-\varphi$ is Schur-geometrically convex function.

Lemma 1[16] Let $\Omega \subset \mathbf{R}^n$ is symmetric with respect to permutations and convex set, and has a nonempty interior set Ω^0. Let $\varphi: \Omega \to \mathbf{R}$ is continuous on Ω and differentiable in Ω^0. Then φ is the $Schur-convex(Schur-concave)function$, if and only if it is symmetric on Ω and if

$$(x_1 - x_2) \left(\frac{\partial \varphi}{\partial x_1} - \frac{\partial \varphi}{\partial x_2} \right) \geqslant 0 (\leqslant 0)$$

holds for any $\boldsymbol{x} = (x_1, x_2, \cdots, x_n) \in \Omega^0$.

Lemma 2[18] Let $\Omega \subset \mathbf{R}_+^n$ is a symmetric with respect to permutations and geometrically convex set, and has a nonempty interior set Ω^0, Let $\varphi: \Omega \to \mathbf{R}_+$ is continuous on Ω and differentiable in Ω^0. Then φ is the Schur-geometrically convex (Schur-geometrically concave) function if φ is symmetric on Ω and

$$(\ln x_1 - \ln x_2) \left(x_1 \frac{\partial \varphi}{\partial x_1} - x_2 \frac{\partial \varphi}{\partial x_2} \right) \geqslant 0 (\leqslant 0)$$

holds for any $\boldsymbol{x} = (x_1, x_2, \cdots, x_n) \in \Omega^0$.

Lemma 3 Let $a \leqslant b, u(t) = tb + (1 - t)a, v(t) = ta + (1 - t)b$. If $1/2 \leqslant t_2 \leqslant t_1 \leqslant 1$ or $0 \leqslant t_1 \leqslant t_2 \leqslant 1$, then

$$(u(t_2), v(t_2)) \prec (u(t_1), v(t_1)) \prec (a, b) \tag{3}$$

Proof Case 1. When $1/2 \leqslant t_2 \leqslant t_1 \leqslant 1$, it is easy to see that $u(t_1) \geqslant v(t_1), u(t_2) \geqslant v(t_2), u(t_1) \geqslant u(t_2)$ and $u(t_2) + v(t_2) = u(t_1) + v(t_1) = a + b$, this is (3) holds.

Case 2. When $0 \leqslant t_1 \leqslant t_2 \leqslant 1$, then $1/2 \leqslant 1 - t_2 \leqslant 1 - t_1 \leqslant 1$, by the Case 1; it follows

$$(u(1 - t_2), v(1 - t_2)) \prec (u(1 - t_1), v(1 - t_1))$$

i.e. $(u(t_2), v(t_2)) \prec (u(t_1), v(t_1))$.

Lemma 4[20] Let $l, t, p, q \in \mathbf{R}_+, p > q$ and $p + q \leqslant 3(l + t)$. Assume also that $1/3 \leqslant l/t \leqslant 3$ or $q \leqslant l + t$. Then

$$G(l, t; x, y) \leqslant (p/q)^{1/(p-q)} E(p, q; x, y)$$

Lemma 5 Let

$$g(t, z) = \frac{z^t + 1}{t(z^{t-1} - 1)}$$

Then for fixed $z > 1$

(i) $g(t, z)$ is increasing on $(-\infty, 0)$ with t;

(ii) $g(t, z)$ is increasing on $(0, \xi_z)$ with t;

(iii) $g(t, z)$ is decreasing on $(\xi_z, 1)$ or $(1, +\infty)$ with t. Where ξ_z is an zero of the function

$$g_1(t, z) = t(z^t + z^{t-1}) \ln z + (z^t + 1)(z^{t-1} - 1)$$

with $0 < \xi_z < 1/2$.

Proof Differentiate $g(t, z)$ with respect to t to obtain

$$\frac{\partial g(t, z)}{\partial t} = \frac{tz^t(z^{t-1} - 1) \ln z - (z^t + 1)(z^{t-1} - 1) - tz^{t-1}(z^t + 1) \ln z}{t^2(z^{t-1} - 1)^2} = -\frac{g_1(t, z)}{t^2(z^{t-1} - 1)}$$

For fixed $z > 1$, $g_1(t, z) < 0$ and $\frac{\partial g(t,z)}{\partial t} > 0$ on $(-\infty, 0)$, then $g(t, z)$ increases on $(-\infty, 0)$ with t, and $g_1(t, z) > 0$ and $\frac{\partial g(t,z)}{\partial t} < 0$ on $(1, +\infty)$, then $g(t, z)$ decreases on $(1, +\infty)$ with t.

Differentiate $g_1(t, z)$ with respect to t to obtain

$$\frac{\partial g_1(t, z)}{\partial t} = [2z^{2t-1} + 2z^{t-1} + t(z^t + z^{t-1}) \ln z] \ln z$$

Since $\frac{\partial g_1(t,z)}{\partial t} > 0$ on $(0, 1)$, $g_1(t, z)$ increases on $(0, 1)$, it following that $g_1(0, z) \leqslant g_1(t, z) \leqslant g_1(1, z)$. Furthermore, $g_1(0, z) = 2(z^{-1} - 1) < 0$ and $g_1(1, z) = (z + 1) \ln z > 0$, hence there exist $\xi_z \in (0, 1)$ such that $g_1(\xi_z, z) = 0$, and $g_1(t, z) \leqslant 0$ and $\frac{\partial g(t,z)}{\partial t} \geqslant 0$ for $0 < t \leqslant \xi_z$, and $g_1(t, z) > 0$ and

$\frac{\partial g(t,z)}{\partial t} < 0$ for $\xi_z < t < 1$. This is, $g(t,z)$ increases on $(0, \xi_z)$ and decreases on $(\xi_z, 1)$.

Differentiate $g_1(t,z)$ with respect to z to obtain

$$\frac{\partial g_1(t,z)}{\partial z} = tz^{t-1}(z^{t-1} - 1) + (t-1)z^{t-2}(z^t + 1) +$$
$$t(z^{t-1} + z^{t-2}) + t[tz^{t-1} + (t-1)z^{t-2}]\ln z$$
$$= (2t-1)z^{2t-2} + t^2 z^{t-1}\ln z + (2t-1)z^{t-2} + (t^2 - t)z^{t-2}\ln z$$

For $1 > t \geqslant 1/2$, we have

$$\frac{\partial g_1(t,z)}{\partial z} \geqslant t^2 z^{t-1}\ln z + (2t-1)z^{t-2} + (t^2 - t)z^{t-2}\ln z$$
$$= (t^2 z + t^2 - t)z^{t-2}\ln z$$
$$> (2t^2 - t)z^{t-2}\ln z = t(2t-1)z^{t-2}\ln z \geqslant 0$$

Hence, for $1 > t \geqslant 1/2$, $g_1(t,z)$ increases on $(1, +\infty)$ with z, and then

$$g_1(t,z) > \lim_{z \to 1^+} g_1(t,z) = g_1(t,1) = 0$$

Thus we conclude that $0 < \xi_z < 1/2$.

Lemma 6 For fixed (x,y) with $x > y > 0$, If $(r,s) \in \{r > 1, s < 0, r+s \leqslant 1\} \cup \{1 < r \leqslant s\} \cup \{0 < r \leqslant 1 - r \leqslant s < 1\} \cup \{1/2 \leqslant r \leqslant s < 1\}$, then

$$s(x^r + y^r)(x^{s-1} - y^{s-1}) \geqslant r(x^s + y^s)(x^{r-1} - y^{r-1}) \tag{4}$$

if $(r,s) \in \{s > 1, r < 0, r+s \leqslant 1\} \cup \{r \leqslant s < 0\}$, then (4) is reversed.

Proof Let $g(t) = \frac{z^t + 1}{t(z^{t-1} - 1)}$ with $z = x/y > 1$. Notice that $y > 0$, it is easy to see that (4) equivalent to $g(r) \geqslant g(s)$. For $r > 1$, we first prove that $g(r) \geqslant g(1-r)$, i.e.

$$\frac{y(z^r + 1)}{r(z^{r-1} - 1)} \geqslant \frac{y(z^{1-r} + 1)}{(1-r)(z^{-r} - 1)} = \frac{y(z^r + z)}{(r-1)(z^r - 1)}$$

It is sufficient prove that

$$h(z) := (r-1)(z^r - 1)(z^r + 1) - r(z^{r-1} - 1)(z^r + z) \geqslant 0$$

Directly calculating yields

$$h(z) = (r-1)z^{2r} - rx^{2r-1} + rx - r + 1$$

$$h'(z) = 2r(r-1)z^{2r-1} - r(2r-1)z^{2r-2} + r$$

$$h''(z) = 2r(r-1)(2r-1)z^{2r-3}(z-1)$$

By $r > 1$, and $z > 1$, it follows $h''(z) > 0$. Therefore, $h'(z) > h'(1) = 0$, moreover, $h(z) > h(1) = 0$, i.e. $g(r) \geqslant g(1-r)$.

If $r > 1, s < 0, r + s \leqslant 1$, then $s \leqslant 1 - r < 0$, from (i) of Lemma 5, we have $g(r) \geqslant g(s)$, i.e. (4) holds.

If $s > 1, r < 0, r + s \leqslant 1$, replacing r by s and replacing s by r in the above case, it follows that $g(r) \leqslant g(s)$, i.e. (4) is reversed.

If $0 < r \leqslant 1/2 \leqslant 1 - r \leqslant s < 1$, then $h''(z) > 0$, it follows $h'(z) > h'(1) = 0$, moreover, $h(z) > h(1) = 0$, i.e. $g(r) \geqslant g(1-r)$, from (iii) of Lemma 5, we have $g(r) \geqslant g(1-r) \geqslant g(s)$, i.e. (4) holds.

If $1/2 \leqslant r \leqslant s < 1$ or $1 < r \leqslant s$, from (iii) of Lemma 5, we have $g(r) \geqslant g(s)$ i.e. (4) holds.

If $r \leqslant s < 0$, from (i) of Lemma 5, we have $g(r) \leqslant g(s)$ i.e. (4) is reversed.

39.3   Proofs of Main results

Proof of Theorem 1
Let $\varphi(x,y) = \dfrac{x^s + y^s}{x^r + y^r}$. When $r \neq s$, for fixed $(x,y) \in \mathbf{R}^2$, we have

$$\frac{\partial \varphi}{\partial x} = \frac{sx^{s-1}(x^r + y^r) - rx^{r-1}(x^s + y^s)}{(x^r + y^r)^2}$$

$$\frac{\partial \varphi}{\partial y} = \frac{sy^{s-1}(x^r + y^r) - ry^{r-1}(x^s + y^s)}{(x^r + y^r)^2}$$

$$\begin{aligned}
\frac{\partial \varphi}{\partial x} - \frac{\partial \varphi}{\partial y} &= \frac{s(x^r + y^r)(x^{s-1} - y^{s-1}) - r(x^s + y^s)(x^{r-1} - y^{r-1})}{(x^r + y^r)^2} \\
&= \frac{s(x^{r-1} - y^{r-1})}{(x^r + y^r)}\left[\frac{s-1}{r-1} \cdot \frac{(r-1)(x^{s-1} - y^{s-1})}{(s-1)(x^{r-1} - y^{r-1})} - \frac{r}{s} \cdot \frac{x^s + y^s}{x^r + y^r}\right] \\
&= \frac{s(x^{r-1} - y^{r-1})}{(x^r + y^r)}\left[\frac{s-1}{r-1} \cdot E^{s-r}(r-1, s-1; x, y) - \frac{r}{s} \cdot G^{s-r}(r, s; x, y)\right]
\end{aligned}$$

and then

$$\Delta := (x-y)\left(\frac{\partial G}{\partial x} - \frac{\partial G}{\partial y}\right) = \frac{x-y}{s-r}\left(\frac{\partial \varphi}{\partial x} - \frac{\partial \varphi}{\partial y}\right)\varphi^{\frac{1}{s-r}-1}(x,y)$$

$$= \frac{s(x-y)(x^{r-1}-y^{r-1})}{(s-r)(x^r+y^r)}.$$

$$\left[\frac{s-1}{r-1} \cdot E^{s-r}(r-1,s-1;x,y) - \frac{r}{s} \cdot G^{s-r}(r,s;x,y)\right]\varphi^{\frac{1}{s-r}-1}(x,y)$$

In Lemma 4, taking $l=r, t=s, p=r-1, q=s-1$, we have

$$\begin{cases} l>0, t>0, p>0, q>0 \\ p>q \\ p+q \leqslant 3(l+t) \\ 1/3 \leqslant l/t \leqslant 3 \end{cases} \Leftrightarrow \begin{cases} r>1, s>1 \\ r>s \\ r+s \geqslant -1 \\ s/3 \leqslant r \leqslant 3s \end{cases} \Leftrightarrow 3s \geqslant r>s>1$$

and

$$\begin{cases} l>0, t>0, p>0, q>0 \\ p>q \\ p+q \leqslant 3(l+t) \\ q \leqslant l+t \end{cases} \Leftrightarrow \begin{cases} r>1, s>1 \\ r>s \\ r+s \geqslant -1 \\ r \geqslant -1 \end{cases} \Leftrightarrow r>s>1$$

Hence, when $r>s>1$, we have

$$G(r,s;x,y) \leqslant \left(\frac{r-1}{s-1}\right)^{\frac{1}{r-s}} E(r-1,s-1;x,y)$$

i.e.

$$G^{s-r}(r,s;x,y) \geqslant \frac{s-1}{r-1} \cdot E^{s-r}(r-1,s-1;x,y) \tag{5}$$

When $r>s>1$, we have $s-r<0$ and $(x-y)(x^{r-1}-y^{r-1}) \geqslant 0$. Combining with (3), it follows that $\Delta \geqslant 0$. By Lemma 1, $G(r,s;x,y)$ is the Schur-convex with $(x,y) \in \mathbf{R}^2_{++}$.

Now we consider other cases. Notice that

$$(x-y)\left(\frac{\partial \varphi}{\partial x} - \frac{\partial \varphi}{\partial y}\right)$$

$$= \frac{s(x^r+y^r)(x-y)(x^{s-1}-y^{s-1}) - r(x^s+y^s)(x-y)(x^{r-1}-y^{r-1})}{(x^r+y^r)^2}$$

when $r \geqslant 1, 0 \leqslant s \leqslant 1$, since t^{r-1} and t^{s-1} is increasing and decreasing in $\mathbf{R}_+$ respectively, it follows that $(x-y)(x^{s-1}-y^{s-1}) \geqslant 0$ and $(x-y)(x^{r-1}-y^{r-1}) \leqslant 0$, moreover, $(x-y)\left(\frac{\partial\varphi}{\partial x}-\frac{\partial\varphi}{\partial y}\right) \leqslant 0$ and

$$\Delta = \frac{x-y}{s-r}\left(\frac{\partial\varphi}{\partial x}-\frac{\partial\varphi}{\partial y}\right)\varphi^{\frac{1}{s-r}-1}(x,y) \geqslant 0$$

That is, when $r \geqslant 1, 0 \leqslant s \leqslant 1$, $G(r,s;x,y)$ is the Schur-convex with $(x,y) \in \mathbf{R}_+^2$.

When $r < 0, 0 < s \leqslant 1$, since t^{r-1} and t^{s-1} are decreasing in $\mathbf{R}_{++}$, it follows that $(x-y)(x^{s-1}-y^{s-1}) \leqslant 0$ and $(x-y)(x^{r-1}-y^{r-1}) \leqslant 0$, moreover, $(x-y)\left(\frac{\partial\varphi}{\partial x}-\frac{\partial\varphi}{\partial y}\right) \leqslant 0$ and $\Delta \leqslant 0$, that is, when $r < 0, 0 < s \leqslant 1$, $G(r,s;x,y)$ is the Schur-concave with $(x,y) \in \mathbf{R}_+^2$.

Without loss of generality, we may assume $x > y > 0$. Notice that

$$\Delta = \frac{x-y}{s-r} \cdot \frac{s(x^r+y^r)(x^{s-1}-y^{s-1}) - r(x^s+y^s)(x^{r-1}-y^{r-1})}{(x^r+y^r)^2}\varphi^{\frac{1}{s-r}-1}(x,y)$$

When $r > 1, s < 0, r+s \leqslant 1$, from Lemma 6, it following that $\Delta \leqslant 0$, i.e. $G(r,s;x,y)$ is the Schur-concave with $(x,y) \in \mathbf{R}_+^2$.

Similarly, we can prove that when $r \leqslant s < 0$, $G(r,s;x,y)$ is the Schur-concave with $(x,y) \in \mathbf{R}_+^2$, and when $0 < r \leqslant 1-r \leqslant s$ or $1/2 \leqslant r \leqslant s < 1$, $G(r,s;x,y)$ is the Schur-convex with $(x,y) \in \mathbf{R}_+^2$.

When $r = s \geqslant 1$, let

$$\psi(x,y) = \frac{x^s \ln x + y^s \ln y}{x^r + y^r} = \frac{x^s \ln x + y^s \ln y}{x^s + y^s}$$

Then

$$\frac{\partial\psi}{\partial x} = \frac{x^{s-1}h(x,y)}{(x^s+y^s)^2}, \quad \frac{\partial\psi}{\partial y} = \frac{y^{s-1}k(x,y)}{(x^s+y^s)^2}$$

where

$$h(x,y) = (s\ln x + 1)(x^s+y^s) - s(x^s\ln x + y^s\ln y)$$

$$k(x,y) = (s\ln y + 1)(x^s+y^s) - s(x^s\ln x + y^s\ln y)$$

By computing

$$x^{s-1}h(x,y) - y^{s-1}k(x,y)$$
$$= (x^s+y^s)\left[x^{s-1}(s\ln x + 1) - y^{s-1}(s\ln y + 1)\right]$$
$$- s(x^s\ln x + y^s\ln y)(x^{s-1} - y^{s-1})$$

$$= s^{s-1}y^{s-1}(x+y)(\ln x - \ln y) + (x^{s-1} - y^{s-1})(x^s + y^s)$$

and then

$$(x-y)\left(\frac{\partial G}{\partial x} - \frac{\partial G}{\partial y}\right) = (x-y)\left(\frac{\partial \psi}{\partial x} - \frac{\partial \psi}{\partial y}\right)e^{\psi(x,y)}$$

$$= \frac{sx^{s-1}y^{s-1}(x+y)(x-y)(\ln x - \ln y) + (x-y)(x^{s-1} - y^{s-1})(x^s + y^s)}{(x^s + y^s)^2}e^{\psi(x,y)}$$

Since $\ln t$ and t^{s-1} are increasing in $\mathbf{R}_+$ with t for $s \geqslant 1$, therefore, $(x-y)(\ln x - \ln y) \geqslant 0$ and $(x-y)(x^{s-1} - y^{s-1}) \geqslant 0$, moreover, $(x-y)\left(\frac{\partial G}{\partial x} - \frac{\partial G}{\partial y}\right) \geqslant 0$. That is, when $r = s \geqslant 1$, $G(r,s;x,y)$ is the Schur-convex with $(x,y) \in \mathbf{R}_+^2$.

In conclusion, if $(r,s) \in \{r > s > 1\} \cup \{r = s \geqslant 1\} \cup \{r \geqslant 1, 0 \leqslant s \leqslant 1\} \cup \{0 < r \leqslant 1 - r \leqslant s\} \cup \{1/2 \leqslant r \leqslant s < 1\}$, then $G(r,s;x,y)$ is the Schur-convex with $(x,y) \in \mathbf{R}_+^2$, and if $(r,s) \in \{r < 0, 0 < s \leqslant 1\} \cup \{r > 1, s < 0, r + s \leqslant 1\} \cup \{r \leqslant s < 0\}$, then $G(r,s;x,y)$ is the Schur-concave with $(x,y) \in \mathbf{R}_+^2$.

Since $G(r,s;x,y)$ is symmetric with (r,s), if $(r,s) \in \{s > r > 1\} \cup \{s \geqslant 1, 0 \leqslant r \leqslant 1\} \cup \{0 < s \leqslant 1 - s \leqslant r\} \cup \{1/2 \leqslant s \leqslant r < 1\}$, then $G(r,s;x,y)$ is also the Schur-convex with $(x,y) \in \mathbf{R}_+^2$, and if $(r,s) \in \{s < 0, 0 < r \leqslant 1\} \cup \{s > 1, r < 0, r + s \leqslant 1\} \cup \{s \leqslant r < 0\}$, then $G(r,s;x,y)$ is also the Schur-concave with $(x,y) \in \mathbf{R}_+^2$.

The proof is complete.

Remark 1 The Schur-convexity of the function $G(r,s;x,y)$ on the set $\{s < 0, r + s > 1\}$ or $\{r < 0, r + s > 1\}$ or $\{r > 0, s > 0, r + s < 1\}$ with (x,y) is uncertainty.

Example 1 Let $(r,s) = (2.5, -1.2)$. It is clear that $(2.5, -1.2) \in \{s < 0, r + s > 1\}$. For $(3,3) \prec (5,1)$, directly calculating yields

$$G(2.5, -1.2; 3, 3) = 3.000000000 > G(2.5, -1.2; 5, 1) = 2.873884533$$

But, for $(1.25, 1.25) \prec (1.5, 1)$, directly calculating yields

$$G(2.5, -1.2; 1.25, 1.25) = 1.250000000 < G(2.5, -1.2; 1.5, 1) = 1.256253447$$

Example 2 Let $(r,s) = (-0.2, 1.5)$. It is clear that $(-0.2, 1.5) \in \{r <$

$0, r + s > 1\}$. For $(8,8) \prec (15,1)$, directly calculating yields

$$G(-0.2, 1.5; 8, 8) = 8.000000000 < G(-0.2, 1.5; 15, 1) = 8.412747770$$

But, for$(25.5, 25.5) \prec (50, 1)$, directly calculating yields

$$G(-0.2, 1.5; 25.5, 25.5) = 25.50000000 > G(-0.2, 1.5; 50, 1) = 25.32833093$$

Example 3 Let $(r, s) = (0.6, 0.2)$. It is clear that $(0.6, 0.2) \in \{r > 0, s > 0, r + s < 1\}$. For $(10.5, 10.5) \prec (20.9, 0.1)$, directly calculating yields

$$G(0.6, 0.2; 10.5, 10.5) = 10.50000000 < G(0.6, 0.2; 20.9, 0.1) = 11.03249418$$

But, for$(10.5, 10.5) \prec (18, 3)$, directly calculating yields

$$G(0.6, 0.2; 10.5, 10.5) = 10.50000000 > G(0.6, 0.2; 18, 3) = 9.970045812.$$

Proof of Theorem 2 Let

$$\varphi(x, y) = \frac{x^s + y^s}{x^r + y^r}$$

When $r \neq s$, for fixed $(x, y) \in \mathbf{R}^2$, we have

$$x\frac{\partial \varphi}{\partial x} = \frac{sx^s(x^r + y^r) - rx^r(x^s + y^s)}{(x^r + y^r)^2}$$

$$y\frac{\partial \varphi}{\partial y} = \frac{sy^s(x^r + y^r) - ry^r(x^s + y^s)}{(x^r + y^r)^2}$$

$$x\frac{\partial \varphi}{\partial x} - y\frac{\partial \varphi}{\partial y} = \frac{s(x^r + y^r)(x^s - y^s) - r(x^s + y^s)(x^r - y^r)}{(x^r + y^r)^2}$$

$$= \frac{s(x^r - y^r)}{x^r + y^r}\left[\frac{s}{r} \cdot \frac{r(x^s - y^s)}{s(x^r - y^r)} - \frac{r}{s} \cdot \frac{x^s + y^s}{x^r + y^r}\right]$$

$$= \frac{s(x^r - y^r)}{x^r + y^r}\left[\frac{s}{r} \cdot E^{s-r}(r, s; x, y) - \frac{r}{s} \cdot G^{s-r}(r, s; x, y)\right]$$

and then

$$(\ln x - \ln y)\left(x\frac{\partial G}{\partial x} - y\frac{\partial G}{\partial y}\right) = \frac{\ln x - \ln y}{s - r}\left(x\frac{\partial \varphi}{\partial x} - y\frac{\partial \varphi}{\partial y}\right)\varphi^{\frac{1}{s-r}-1}(x, y)$$

$$= \frac{s(\ln x - \ln y)(x^r - y^r)}{(s - r)(x^r + y^r)}\left[\frac{s}{r} \cdot E^{s-r}(r, s; x, y) - \frac{r}{s} \cdot G^{s-r}(r, s; x, y)\right]\varphi^{\frac{1}{s-r}-1}(x, y)$$

In Lemma 4, taking $l = p = r, t = q = s$, we have

$$\begin{cases} l > 0, t > 0, p > 0, q > 0 \\ p > q \\ p + q \leqslant 3(l+t) \\ 1/3 \leqslant l/t \leqslant 3 \end{cases} \Leftrightarrow \begin{cases} r > 0, s > 0 \\ r > s \\ r + s \geqslant -1 \\ s/3 \leqslant r \leqslant 3s \end{cases} \Leftrightarrow 3s \geqslant r > s > 0$$

and

$$\begin{cases} l > 0, t > 0, p > 0, q > 0 \\ p > q \\ p + q \leqslant 3(l+t) \\ q \leqslant l+t \end{cases} \Leftrightarrow \begin{cases} r > 0, s > 0 \\ r > s \\ r + s \geqslant -1 \\ r \geqslant 0 \end{cases} \Leftrightarrow r > s > 0$$

Hence, when $r > s > 0$, we have

$$G(r, s; x, y) \leqslant \left(\frac{r}{s}\right)^{\frac{1}{r-s}} E(r, s; x, y)$$

i.e.

$$G^{s-r}(r, s; x, y) \geqslant \frac{s}{r} \cdot E^{s-r}(r, s; x, y) \tag{6}$$

When $r > s > 0$, we have $s - r < 0$, and since $\ln t$ and t^r are increasing in $\mathbf{R}_+$ with t, therefore $(\ln x - \ln y)(x^r - y^r) \geqslant 0$. Combining with (6), it follows that $(\ln x - \ln y)\left(x\frac{\partial G}{\partial x} - y\frac{\partial G}{\partial y}\right) \geqslant 0$. By Lemma 2, $G(r, s; x, y)$ is the Schur-geometrically convex with (x, y) in $\mathbf{R}_+^2$. Since $G(r, s; x, y)$ is symmetric with (r, s), when $s > r > 0$, $G(r, s; x, y)$ is also the Schur-geometrically convex with $(x, y) \in \mathbf{R}_+^2$.

Now we consider other cases.

Without loss of generality, we may assume $x > y > 0$. Notice that

$$\Lambda = \frac{\ln x - \ln y}{s - r} \cdot \frac{s(x^r + y^r)(x^s - y^s) - r(x^s + y^s)(x^r - y^r)}{(x^r + y^r)^2} \varphi^{\frac{1}{s-r}-1}(x, y)$$

when $r = s > 0$, we have

$$\frac{\partial \psi}{\partial x} = \frac{x^{s-1}h(x, y)}{(x^s + y^s)^2}, \quad \frac{\partial \psi}{\partial y} = \frac{x^{s-1}k(x, y)}{(x^s + y^s)^2}$$

where $h(x, y), k(x, y)$ and $\psi(x, y)$ are same as in Theorem 2.

By computing

$$x^s h(x,y) - y^s k(x,y) = s^s y^s (x+y)(\ln x - \ln y) + (x^s - y^s)(x^s + y^s)$$

and then

$$(\ln x - \ln y)\left(x\frac{\partial G}{\partial x} - y\frac{\partial G}{\partial y}\right) = (\ln x - \ln y)\left(x\frac{\partial \psi}{\partial x} - y\frac{\partial \psi}{\partial y}\right) e^{\psi(x,y)}$$

$$= \frac{sx^s y^s (x+y)(\ln x - \ln y)^2 + (\ln x - \ln y)(x^s - y^s)(x^s + y^s)}{(x^s + y^s)^2} e^{\psi(x,y)}$$

Since when $s > 0$, $\ln t$ and t^s are increasing in $\mathbf{R}_+$, $(\ln x - \ln y)(x^s - y^s) \geqslant 0$, moreover, $(\ln x - \ln y)\left(x\frac{\partial G}{\partial x} - y\frac{\partial G}{\partial y}\right) \geqslant 0$. That is, when $r = s > 0$, $G(r,s;x,y)$ is the Schur-geometrically convex with $(x,y) \in \mathbf{R}_+^2$.

In conclusion, if $(r,s) \in \{r > s > 0\} \cup \{s > r > 0\} \cup \{r = s > 0\} = \mathbf{R}_+^2$, $G(r,s;x,y)$ is the Schur-geometrically convex with $(x,y) \in \mathbf{R}_+^2$.

The proof is complete.

39.4 Applications

Theorem 3 Let $(x,y) \in \mathbf{R}_{++}^2$, $u(t) = ty + (1-t)x$, $v(t) = tx + (1-t)y$. Assume also that $\frac{1}{2} \leqslant t_2 \leqslant t_1 \leqslant 1$ or $0 \leqslant t_1 \leqslant t_2 \leqslant 1$. If $(r,s) \in \{r \geqslant 0, s \geqslant 0, r+s \geqslant 1\} \subseteq \mathbf{R}^2$, then for fixed $(r,s) \in \mathbf{R}^2$, we have

$$G\left(r,s;\frac{x+y}{2},\frac{x+y}{2}\right) \leqslant G(r,s;u(t_2),v(t_2))$$

$$\leqslant G(r,s;u(t_1),v(t_1)) \leqslant G(r,s;x,y) \leqslant G(r,s;x+y,0) \tag{7}$$

if $(r,s) \in \{r \leqslant 0, r+s \leqslant 1\} \cup \{s \leqslant 0, r+s \leqslant 1\} \subseteq \mathbf{R}^2$, then inequalities in (7) are all reversed.

Proof From lemma 3, we have

$$\left(\frac{x+y}{2},\frac{x+y}{2}\right) \prec (u(t_2),v(t_2)) \prec (u(t_1),v(t_1)) \prec (r,s)$$

and it is clear that $(x,y) \prec (x+y-\varepsilon,\varepsilon)$, where ε is enough small positive number.

If $(r,s) \in \{r \geqslant 1, s > 0\} \cup \{0 < r < 1, s \geqslant 1\}$, by Theorem 1, and let $\varepsilon \to 0$, it follows that (7) are holds. If $(r,s) \in \{r < 0, 0 < s < 1\} \cup \{0 < r < 1, s < 0\}$, then inequalities in (7) are all reversed.

The proof is complete.

Theorem 4 Let $(x, y) \in \mathbf{R}_{++}^2$. For fixed $(r, s) \in \mathbf{R}_+^2$, we have

$$G\left(r, s; \sqrt{xy}, \sqrt{xy}\right) \leqslant G(r, s; x, y) \tag{8}$$

Proof Since $(\ln \sqrt{xy}, \ln \sqrt{xy}) \prec (\ln x, \ln y)$, by Theorem 2, it follows that (8) is holds.

The proof is complete.

References

[1]　C GINI. Di una formula compresiva delle medie[J]. Metron, 1938, 13: 3-22.

[2]　P S BULLEN, D S MITRINOVIĆ, P M VASIĆ. Means and Their Inequalities[M]. Dordecht: Reidel, 1988.

[3]　P S BULLEN. Handbook of Means and their Inequalities[M].Dordrecht: Kluwer Academic Publishers, 2003.

[4]　K B STOLARSKY. Generalizations of the logarithmic mean[J]. Math. Mag., 1975, 48 (2):87-92.

[5]　K B STOLARSKY. The power and generalized logarithmic means[J]. Amer. Math. Monthly, 1980 (87):545-548.

[6]　FENG QI, JÓZSEF SÁNDOR, SEVER S DRAGOMIR, ANTHO-NY SOFO. Schur-convexity of the extended mean values[J]. Taiwanese Journal of Mathematics, 2005, 9(3):411-420.

[7]　HUAN-NAN SHI, SHAN-HE WU, FENG QI. An alternative note on the Schur-convexity of the extended mean values[J]. Mathematical Inequalities and Applications, 2006, 9 (2): 219-224.

[8]　JÓZSEF SÁNDOR. The Schur-convexity of Stolarsky and Gini Means [J]. Banach J. Math. Anal., 2007, 1 (2): 212-215.

[9]　EDWARD NEUMAN, ZSOLT PÁLES. On comparison of Stdlarsky and Gini means[J]. J. Math. Anal. Appl.,2003, 278:274-284.

[10]　ZSOLT PÁLES. Inequalities for differences of powers[J]. J. Math. Anal. Appl., 1988, 131:265-270.

[11]　ZSOLT PÁLES. Inequalities for sums of powers[J]. J. Math. Anal. Appl.,1988, 131: 271-281.

[12]　JÓZSEF SÁNDOR. A note on the Gini means[J]. Genneral Mathematics, 2004, 12 (4): 17-21.

[13] C E M PEARCE, J PEČARIĆ, J SÁNDOR. Ageneralzation of Pólya's inequality to Stolarsky and Gini means[J]. Mathematical Inequalities & Applications, 1998, 1 (2) :211-222.

[14] PETER CZINDER, ZSOLT PÁLES. An extension of the Gini and Stdlarsky means[J]. J.Inequl. Pure and Appl. Math., 2004, 5(2), Art. 42.

[15] CZINDEER, PÉTER, PÁLES, ZSOLT. An extension of the Hermite-Hadamard inequality and an application for Gini and Stolarsky means [J]. J. Inequal. Pure Appl. Math., 2004, 5(2): 8.

[16] BO-YING WANG. Foundations of Majorization Inequalities[M]. Beijing: Beijing Normal Univ. Press, 1990.

[17] A M MARSHALL, I OLKIN. Inequalities:theory of majorization and its application[M]. New York : Academies Press, 1979.

[18] XIAO-MING ZHANG.Geometrically Convex Functions[M]. Hefei: An hui University Press, 2004.

[19] CONSTANTIN P NICULESCU. Convexity According to the Geometric Mean[J]. Mathematical Inequalities & Applications, 2000, 3(2):155-167.

[20] PETER A HASTO. Monotonicity property of ratios of symmetric homogeneous means[J]. J. Ineq. Pure Appl. Math., 2002, 3(5):1-23.

第40篇　Schur Convexity of Generalized Exponent Mean

(LI DA-MAO, SHI HUAN-NAN. Journal Mathematical Inequalities,

Volume 3, Number 2 (2009): 217 - 225)

Abstract: The monotonicity, the Schur-convexity and the Schur-geometrically convexity with variables (x, y) in $\mathbf{R}_{++}^2$ for fixed a of the generalized exponent mean $I_a(x, y)$ is proved. Besides, the monotonicity with parameters a in $\mathbf{R}$ for fixed (x, y) of $I_a(x, y)$ is discussed by using the hyperbolic composite function. Furthermore, some new inequalities are obtained.

Keywords: generalized exponent mean; monotonicity; Schur-convexity; Schur-geometrically concavity; inequality; hyperbolic function

40.1　Introduction

Throughout the paper we denote the set of the real numbers, the non-negative real numbers and the positive real numbers by $\mathbf{R}, \mathbf{R}_+$ and $\mathbf{R}_{++}$ respectively.

Let $(a, b) \in \mathbf{R}^2, (x, y) \in \mathbf{R}_{++}^2$. The extended mean (or Stolarsky mean) of (x, y) is defined in [1] as

$$
E(a, b; x, y) = \begin{cases}
\left(\dfrac{b}{a} \cdot \dfrac{y^a - x^a}{y^b - x^b} \right)^{1/(a-b)}, & ab(a-b)(x-y) \neq 0 \\[3ex]
\left(\dfrac{1}{a} \cdot \dfrac{y^a - x^a}{\ln y - \ln x} \right)^{1/a}, & a(x-y) \neq 0, b = 0 \\[3ex]
\dfrac{1}{e^{1/a}} \left(\dfrac{x^{x^a}}{y^{y^a}} \right)^{1/(x^a - y^a)}, & a(x-y) \neq 0, a = b \\[2ex]
\sqrt{xy}, & a = b = 0, x \neq y \\[1ex]
x, & x = y
\end{cases}
$$

In particular, for $a \neq 0$

$$E(a, a; x, y) = \begin{cases} \dfrac{1}{e^{1/a}} \left(\dfrac{x^{x^a}}{y^{y^a}} \right)^{1/(x^a - y^a)}, & x \neq y \\ x, & x = y \end{cases}$$

is called the generalized exponent or identric mean, in symbols $I_a(x, y)$.

The Schur-convexity of the extended mean $E(r, s; x, y)$ with (x, y) was discussed in [2] and the following conclusion is obtained:

Theorem A For fixed $(a, b) \in \mathbf{R}^2$.

(i) If $2 < 2a < b$ or $2 \leq 2b \leqslant a$, then $E(a, b; x, y)$ is Schur-convex with (x, y) on $\mathbf{R}^2_{++}$;

(ii) If $(a, b) \in \{a < b \leq 2a, 0 < a \leq 1\} \cup \{b < a \leq 2b, 0 < b \leq 1\} \cup \{0 < b < a \leq 1\} \cup \{0 < a < b \leq 1\} \cup \{b \leq 2a < 0\} \cup \{a \leq 2b < 0\}$, then $E(a, b; x, y)$ is Schur-concave with (x, y) on $\mathbf{R}^2_{++}$.

But this conclusion is not related to the case $a = b$. In other words, the Schur-convexity of the generalized exponent mean $I_a(x, y)$ with (x, y) is not discussed in [2].

In this paper, the monotonicity, the Schur-convexity and the Schur-geometrically convexity with variables (x, y) in $\mathbf{R}^2_{++}$ for fixed a of the generalized exponent mean $I_a(x, y)$ is proved. Besides, the monotonicity with parameters a in $\mathbf{R}$ for fixed (x, y) of $I_a(x, y)$ is discussed by using the hyperbolic composite function. Furthermore, some new inequalities are obtained.

40.2 Definitions and Lemmas

We need the following definitions and lemmas.

Definition 1[3-4] Let $x = (x_1, \ldots, x_n)$ and $y = (y_1, \ldots, y_n) \in \mathbf{R}^n$.

(i) x is said to be majorized by y (in symbols $x \prec y$) if $\sum\limits_{i=1}^{k} x_{[i]} \leqslant \sum\limits_{i=1}^{k} y_{[i]}$ for $k = 1, 2, \ldots, n-1$ and $\sum\limits_{i=1}^{n} x_i = \sum\limits_{i=1}^{n} y_i$, where $x_{[1]} \geqslant \cdots \geqslant x_{[n]}$ and $y_{[1]} \geqslant \cdots \geqslant y_{[n]}$ are rearrangements of x and y in a descending order.

(ii) $x \geqslant y$ means $x_i \geqslant y_i$ for all $i = 1, 2, \ldots, n$. Let $\Omega \subset \mathbf{R}^n$. The function $\varphi \colon \Omega \to \mathbf{R}$ is said to be increasing if $x \geqslant y$ implies $\varphi(x) \geqslant \varphi(y)$. φ is said to be decreasing if and only if $-\varphi$ is increasing.

(iii) $\Omega \subset \mathbf{R}^n$ is called a convex set if $(\alpha x_1 + \beta y_1, \dots, \alpha x_n + \beta y_n) \in \Omega$ for every x and $y \in \Omega$, where α and $\beta \in [0,1]$ with $\alpha + \beta = 1$.

(iv) let $\Omega \subset \mathbf{R}^n$. The function $\varphi \colon \Omega \to \mathbf{R}$ be said to be a Schur-convex function on Ω if $x \prec y$ on Ω implies $\varphi(x) \le \varphi(y)$. φ is said to be a Schur-concave function on Ω if and only if $-\varphi$ is Schur-convex.

Definition 2[5-6] Let $x = (x_1, \dots, x_n)$ and $y = (y_1, \dots, y_n) \in \mathbf{R}^n_{++}$.

(i) $\Omega \subset \mathbf{R}^n_{++}$ is called a geometrically convex set if $(x_1^\alpha y_1^\beta, \dots, x_n^\alpha y_n^\beta) \in \Omega$ for all x and $y \in \Omega$, where α and $\beta \in [0,1]$ with $\alpha + \beta = 1$.

(ii) Let $\Omega \subset \mathbf{R}^n_{++}$. The function $\varphi \colon \Omega \to \mathbf{R}_+$ is said to be Schur-geometrically convex function on Ω if $(\ln x_1, \dots, \ln x_n) \prec (\ln y_1, \dots, \ln y_n)$ on Ω implies $\varphi(x) \le \varphi(y)$. The function φ is said to be a Schur-geometrically concave on Ω if and only if $-\varphi$ is Schur-geometrically convex.

Definition 3[4]

(i) $\Omega \subset \mathbf{R}^n$ is called symmetric set, if $x \in \Omega$ implies $Px \in \Omega$ for every $n \times n$ permutation matrix P.

(ii) The function $\varphi \colon \Omega \to \mathbf{R}$ is called symmetric if for every permutation matrix P, $\varphi(Px) = \varphi(x)$ for all $x \in \Omega$.

Lemma 1[3-4] A function $\varphi(x)$ is increasing if and only if $\nabla \varphi(x) \ge 0$ for $x \in \Omega$, where $\Omega \subset \mathbf{R}^n$ is an open set, $\varphi \colon \Omega \to \mathbf{R}$ is differentiable, and

$$\nabla \varphi(x) = \left(\frac{\partial \varphi(x)}{\partial x_1}, \dots, \frac{\partial \varphi(x)}{\partial x_n} \right) \in \mathbf{R}^n$$

Lemma 2[3-4] Let $\Omega \subset \mathbf{R}^n$ be a symmetric set and with a nonempty interior Ω^0, $\varphi \colon \Omega \to \mathbf{R}$ be a continuous on Ω and differentiable in Ω^0. Then φ is the $Schur - convex(Schur - concave) function$, if and only if φ is symmetric on Ω and

$$(x_1 - x_2)\left(\frac{\partial \varphi}{\partial x_1} - \frac{\partial \varphi}{\partial x_2} \right) \ge 0 (\le 0)$$

holds for any $x = (x_1, x_2, \cdots, x_n) \in \Omega^0$.

Lemma 3[5] Let $\Omega \subset \mathbf{R}^n_{++}$ be symmetric with a nonempty interior geometrically convex set. Let $\varphi \colon \Omega \to \mathbf{R}_+$ be continuous on Ω and differentiable in Ω^0. If φ is symmetric on Ω and

$$(\ln x_1 - \ln x_2)\left(x_1\frac{\partial\varphi}{\partial x_1} - x_2\frac{\partial\varphi}{\partial x_2}\right) \geqslant 0(\leqslant 0)$$

holds for any $\boldsymbol{x} = (x_1, x_2, \cdots, x_n) \in \Omega^0$, then φ is a Schur-geometrically convex (*Schur $-$ geometrically concave*) function.

Lemma 4 Let $x \leqslant y, u(t) = tx + (1-t)y, v(t) = ty + (1-t)x$. If $1/2 \leqslant t_2 \leqslant t_1 \leqslant 1$ or $0 \leqslant t_1 \leqslant t_2 \leqslant 1/2$, then

$$(u(t_2), v(t_2)) \prec (u(t_1), v(t_1)) \prec (x, y) \tag{1}$$

Proof Case 1. When $1/2 \leqslant t_2 \leqslant t_1 \leqslant 1$, it is easy to see that $u(t_1) \geqslant v(t_1), u(t_2) \geqslant v(t_2), u(t_1) \geqslant u(t_2)$ and $u(t_2) + v(t_2) = u(t_1) + v(t_1) = x + y$, that is (1) holds.

Case 2. When $0 \leqslant t_1 \leqslant t_2 \leqslant 1$, then $1/2 \leqslant 1 - t_2 \leqslant 1 - t_1 \leqslant 1$, by the Case 1, it follows

$$(u(1 - t_2), v(1 - t_2)) \prec (u(1 - t_1), v(1 - t_1))$$

i.e. $(u(t_2), v(t_2)) \prec (u(t_1), v(t_1))$.

Lemma 5[4,7] Let $0 \leqslant x \leqslant y, c \geqslant 0$. Then

$$\left(\frac{x+c}{x+y+2c}, \frac{y+c}{x+y+2c}\right) \prec \left(\frac{x}{x+y}, \frac{y}{x+y}\right) \tag{2}$$

Lemma 6 For x in $\mathbf{R}$ with $x \neq 0$, we have

$$\sinh^2 x > x^2 \tag{3}$$

Proof Let $f(x) = \sinh^2 x - x^2$. Then $f'(x) = \sinh 2x - 2x$. Since $f''(x) = 2(\cosh 2x - 1) > 0$ for $x \in \mathbf{R}$ with $x \neq 0$, $f'(x)$ is strictly increasing. It follows that $f'(x) > f'(0) = 0$, so $f(x) > f(0) = 0$ for $x > 0$. As $f(-x) = f(x)$, (3) holds for any $x \in \mathbf{R}$ with $x \neq 0$.

Lemma 7 Let (x, y) and $(a, b) \in \mathbf{R}_{++}^2$ with $x < y, a < b, a + b = 1$. Then

$$ax + by > \frac{x+y}{2} \tag{4}$$

$$bx + ay < \frac{x+y}{2} \tag{5}$$

Proof As

$$ax + by - \frac{x+y}{2} = \left(a - \frac{1}{2}\right)x + \left(b - \frac{1}{2}\right)y$$

$$= \left(1 - b - \frac{1}{2}\right)x + \left(b - \frac{1}{2}\right)y = -\left(b - \frac{1}{2}\right)x + \left(b - \frac{1}{2}\right)y$$

$$= \left(b - \frac{1}{2}\right)(y - x) > 0$$

(4) holds. (5) can be proved similarly.

Lemma 8 Let $(x, y) \in \mathbf{R}_{++}^2$ and $(a, b) \in \mathbf{R}^2$ with $ab(a-b)(x-y) \neq 0$. Then

$$E(a, b; x, y) = \sqrt{xy}\left(\frac{b\sinh(a\ln\sqrt{u})}{a\sinh(b\ln\sqrt{u})}\right)^{\frac{1}{a-b}} \tag{6}$$

where $u = y/x$.

Proof Without loss of generality, we may assume $0 < x < y$. Then

$$E(a, b; x, y) = \left(\frac{b}{a}\cdot\frac{y^a - x^a}{y^b - x^b}\right)^{\frac{1}{a-b}} = \left(\frac{b}{a}\cdot\frac{u^a - 1}{u^b - 1}x^{a-b}\right)^{1/(a-b)}$$

$$= x\left(\frac{b}{a}\cdot\frac{e^{2a\ln\sqrt{u}} - 1}{e^{2b\ln\sqrt{u}} - 1}\right)^{\frac{1}{a-b}} = x\left(\frac{b}{a}\cdot\frac{\frac{e^{2a\ln\sqrt{u}} - 1}{2e^{a\ln\sqrt{u}}}}{\frac{e^{2b\ln\sqrt{u}} - 1}{2e^{b\ln\sqrt{u}}}}e^{(a-b)\ln\sqrt{u}}\right)^{\frac{1}{a-b}}$$

$$= x\sqrt{u}\left(\frac{b\sinh(a\ln\sqrt{u})}{a\sinh(b\ln\sqrt{u})}\right)^{\frac{1}{a-b}} = \sqrt{xy}\left(\frac{b\sinh(a\ln\sqrt{u})}{a\sinh(b\ln\sqrt{u})}\right)^{\frac{1}{a-b}}$$

Lemma 9 Let $(x, y) \in \mathbf{R}_{++}^2$ with $x \neq y$, and let $a \in \mathbf{R}$ with $a \neq 0$. Then

$$I_a(x, y) = \sqrt{xy}\exp\left\{\frac{t}{\tanh(at)} - \frac{1}{a}\right\} \tag{7}$$

where $t = \ln\sqrt{u}, u = y/x$.

Proof For $b \in \mathbf{R}$ with $b \neq a$, let

$$v = \frac{b\sinh(a\ln\sqrt{u}) - a\sinh(b\ln\sqrt{u})}{a\sinh(b\ln\sqrt{u})}$$

Then from Lemma 8 we have

$$I_a(x, y) = \lim_{b \to a} E(a, b; x, y) = \lim_{b \to a}\sqrt{xy}\left(\frac{b\sinh(a\ln\sqrt{u})}{a\sinh(b\ln\sqrt{u})}\right)^{\frac{1}{a-b}}$$

$$= \sqrt{xy}\lim_{b \to a}(1 + v)^{\frac{1}{a-b}}$$

$$= \sqrt{xy} \lim_{b \to a} \left[(1+v)^{\frac{1}{v}} \right]^{\frac{b\sinh(a\ln\sqrt{u}) - a\sinh(b\ln\sqrt{u})}{a-b} \cdot \frac{1}{a\sinh(b\ln\sqrt{u})}}$$

$$= \sqrt{xy} \exp\left\{ \lim_{b\to a} \frac{b\sinh(a\ln\sqrt{u}) - a\sinh(b\ln\sqrt{u})}{a-b} \cdot \frac{1}{a\sinh(b\ln\sqrt{u})} \right\}$$

$$= \sqrt{xy} \exp\left\{ \frac{1}{a\sinh(a\ln\sqrt{u})} \lim_{b\to a} \frac{b\sinh(a\ln\sqrt{u}) - a\sinh(b\ln\sqrt{u})}{a-b} \right\}$$

$$= \sqrt{xy} \exp\left\{ \frac{1}{a\sinh(a\ln\sqrt{u})} \lim_{b\to a} \frac{\sinh(a\ln\sqrt{u}) - a(\ln\sqrt{u})\cosh(b\ln\sqrt{u})}{-1} \right\}$$

$$= \sqrt{xy} \exp\left\{ \frac{a(\ln\sqrt{u})\cosh(a\ln\sqrt{u}) - \sinh(a\ln\sqrt{u})}{a\sinh(a\ln\sqrt{u})} \right\}$$

$$= \sqrt{xy} \exp\left\{ \frac{(at)\cosh(at) - \sinh(at)}{a\sinh(at)} \right\}$$

$$= \sqrt{xy} \exp\left\{ \frac{t}{\tanh(at)} - \frac{1}{a} \right\}$$

40.3   Main results and their proofs

Theorem 1 For fixed $(x,y) \in \mathbf{R}^2_{++}$, $I_a(x,y)$ is increasing with a on $\mathbf{R}$.

Proof For $a \neq 0$, set $f(a) = \frac{t}{\tanh(at)} - \frac{1}{a}$, where $t = \ln\sqrt{u}, u = y/x$. Then

$$f'(a) = \frac{-t^2}{\tanh^2(at)\cosh^2(at)} + \frac{1}{a^2} = \frac{-t^2}{\sinh^2(at)} + \frac{1}{a^2} = \frac{\sinh^2(at) - (at)^2}{a^2\sinh^2(at)}$$

Thus from Lemma 6 it follows that $f'(a) > 0$, that is $f(a)$ is increasing on $\mathbf{R}$ with a and

$$I_a(x,y) = \sqrt{xy} \exp\left\{ \frac{t}{\tanh(at)} - \frac{1}{a} \right\} = \sqrt{xy} e^{f(a)}$$

is increasing on $\mathbf{R}$ with a. The proof of Theorem 1 is completed.

Theorem 2 For fixed $a \in \mathbf{R}$, $I_a(x,y)$ is increasing with (x,y) on $\mathbf{R}^2_{++}$.

Proof Let $A = x^a, B = y^a$. Then

$$\ln I_a(x,y) = \frac{x^a\ln x - y^a\ln y}{x^a - y^a} - \frac{1}{a} = \frac{1}{a}\left(\frac{A\ln A - B\ln B}{A - B} - 1 \right)$$

$$\frac{\partial \ln I_a}{\partial x} = \frac{\partial \ln I_a}{\partial A}\frac{\mathrm{d}A}{\mathrm{d}x} = \frac{1}{a}\frac{\partial}{\partial a}\left(\frac{A\ln A - B\ln B}{A - B} - 1 \right)ax^{a-1}$$

$$= \frac{A}{x}\left[\frac{(A-B)-B(\ln A - \ln B)}{(A-B)^2}\right]$$

$$= \frac{A}{x(A-B)}\left(1 - \frac{\ln A - \ln B}{A-B}\cdot B\right)$$

$$= \frac{A}{x(A-B)}\left(1 - \frac{B}{\xi}\right) \quad \text{(where } \xi \text{ lies between } A \text{ and } B\text{)}$$

$$= \frac{A}{x(A-B)}\frac{\xi - B}{\xi} = \frac{A}{x\xi}\cdot\frac{\xi-B}{A-B} \geqslant 0$$

Similarly can be proved that $\frac{\partial \ln I_a}{\partial y} \geqslant 0$.

By Lemma 1, it follows that $\ln I_a(x,y)$ is increasing with (x,y) on $\mathbf{R}^2_{++}$, and then $I_a(x,y)$ is increasing with (x,y) on $\mathbf{R}^2_{++}$ too.

The proof of Theorem 2 is completed.

Theorem 3 If $0 < a \leqslant 1$, then $I_a(x,y)$ is Schur-concave with (x,y) on $\mathbf{R}^2_{++}$.

Proof For $(x,y) \in \mathbf{R}^2_{++}, 0 < a \leqslant 1$, let $A = x^a, B = y^a$. When $x \neq y$, we have

$$\frac{\partial \ln I_a}{\partial x} = \frac{A}{x}\cdot\frac{(A-B)-B(\ln A - \ln B)}{(A-B)^2}$$

$$\frac{\partial \ln I_a}{\partial y} = \frac{B}{y}\cdot\frac{A(\ln A - \ln B)-(A-B)}{(A-B)^2}$$

and then

$$\Delta := (x-y)\left(\frac{\partial \ln I_a}{\partial x} - \frac{\partial \ln I_a}{\partial y}\right)$$

$$= (x-y)\left[\frac{A}{x}\cdot\frac{(A-B)-B(\ln A - \ln B)}{(A-B)^2} - \frac{B}{y}\cdot\frac{A(\ln A - \ln B)-(A-B)}{(A-B)^2}\right]$$

$$= \frac{x-y}{(A-B)^2}\left[\frac{A}{x}(A-B) - \frac{AB}{x}(\ln A - \ln B) - \frac{AB}{y}(\ln A - \ln B) + \frac{B}{y}(A-B)\right]$$

$$= \frac{x-y}{(A-B)^2}\left[\left(\frac{A}{x}+\frac{B}{y}\right)(A-B) - AB\left(\frac{1}{x}+\frac{1}{y}\right)(\ln A - \ln B)\right]$$

$$= \frac{x-y}{A-B}\cdot\frac{\ln A - \ln B}{A-B}\left[\left(\frac{A}{x}+\frac{B}{y}\right)\frac{A-B}{\ln A - \ln B} - AB\left(\frac{1}{x}+\frac{1}{y}\right)\right]$$

$$= \frac{x-y}{A-B}\cdot\frac{\ln A - \ln B}{A-B}\left(\frac{A}{x}+\frac{B}{y}\right)\left[\frac{A-B}{\ln A - \ln B} - \frac{\left(\frac{1}{x}+\frac{1}{y}\right)AB}{\frac{A}{x}+\frac{B}{y}}\right]$$

$$= \frac{x-y}{A-B}\cdot\frac{\ln A - \ln B}{A-B}\left(\frac{A}{x}+\frac{B}{y}\right)\left(\frac{x^a-y^a}{\ln x^a - \ln y^a} - \frac{y^{a-1}x^a + x^{a-1}y^a}{x^{a-1}+y^{a-1}}\right)$$

$$= \frac{x-y}{A-B}\cdot\frac{\ln A - \ln B}{A-B}\left(\frac{A}{x}+\frac{B}{y}\right).$$

$$\left[L\left(x^a, y^a\right) - \left(\frac{y^{a-1}}{x^{a-1} + y^{a-1}} x^a + \frac{x^{a-1}}{x^{a-1} + y^{a-1}} y^a \right) \right]$$

where L denotes the logarithm mean.

Without loss of generality, we may assume $0 < x < y$. When $0 < a < 1$ we have

$$\frac{y^{a-1}}{x^{a-1} + y^{a-1}} < \frac{x^{a-1}}{x^{a-1} + y^{a-1}}$$

and

$$\frac{y^{a-1}}{x^{a-1} + y^{a-1}} + \frac{x^{a-1}}{x^{a-1} + y^{a-1}} = 1$$

and then by Lemma 7, it follows that

$$\frac{y^{a-1}}{x^{a-1} + y^{a-1}} x^a + \frac{x^{a-1}}{x^{a-1} + y^{a-1}} y^a > \frac{x^a + y^a}{2} = A\left(x^a, y^a\right)$$

Furthermore notice that $L\left(x^a, y^a\right) < A\left(x^a, y^a\right)$, we have

$$L\left(x^a, y^a\right) - \left(\frac{y^{a-1}}{x^{a-1} + y^{a-1}} x^a + \frac{x^{a-1}}{x^{a-1} + y^{a-1}} y^a \right) < L\left(x^a, y^a\right) - A\left(x^a, y^a\right) < 0$$

Hence $\Delta < 0$ for $0 < a < 1$. It is easy to see that $\Delta < 0$ for $a = 1$. By Lemma 2, it follows that for $0 < a \leqslant 1$, $\ln I_a(x, y)$ is Schur-concave on $\mathbf{R}^2_{++}$ with (x, y), and then $I_a(x, y)$ is Schur-concave on $\mathbf{R}^2_{++}$ with (x, y) too.

The proof of Theorem 3 is completed.

Theorem 4 If $a > 0$, then $I_a(x, y)$ is Schur-geometrically convex with (x, y) on $\mathbf{R}^2_{++}$; If $a < 0$, then $I_a(x, y)$ is Schur-geometrically concave with (x, y) on $\mathbf{R}^2_{++}$.

Proof For $(x, y) \in \mathbf{R}^2_{++}, a \in \mathbf{R}$, let $A = x^a, B = y^a$. When $x \neq y$, we have

$$\frac{\partial \ln I_a}{\partial x} = \frac{A}{x} \cdot \frac{(A - B) - B(\ln A - \ln B)}{(A - B)^2}$$

$$\frac{\partial \ln I_a}{\partial y} = \frac{B}{y} \cdot \frac{A(\ln A - \ln B) - (A - B)}{(A - B)^2}$$

and then

$$\Lambda := (x - y)\left(x \frac{\partial \ln I_a}{\partial x} - y \frac{\partial \ln I_a}{\partial y} \right)$$

$$= \frac{x - y}{(A - B)^2}\left[A(A - B) - AB(\ln A - \ln B) - AB(\ln A - \ln B) + B(A - B) \right]$$

$$= \frac{x - y}{(A - B)^2}\left[(A + B)(A - B) - 2AB(\ln A - \ln B) \right]$$

$$= \frac{(x-y)(A+B)(\ln A - \ln B)}{(A-B)^2} \left(\frac{A-B}{\ln A - \ln B} - \frac{2AB}{A+B} \right)$$

$$= \frac{(x-y)(A+B)(\ln A - \ln B)}{(A-B)^2} (L(A,B) - H(A,B))$$

where H denote the harmonic mean.

For $(x,y) \in \mathbf{R}_{++}^2$ with $x \neq y$ and $a \in \mathbf{R}$, we have $L(A,B) > H(A,B)$. If $a > 0 (< 0)$, then $(x-y)(\ln A - \ln B) = a(x-y)(\ln x - \ln y) > 0 (< 0)$, and then $\Lambda > 0 (< 0)$. By Lemma 3, it follows that $\ln I_a(x,y)$ is Schur-geometrically convex (concave) on $\mathbf{R}_{++}^2$ with (x,y), and then $I_a(x,y)$ is Schur-geometrically convex (concave) on $\mathbf{R}_{++}^2$ with (x,y) too.

The proof of Theorem 4 is completed.

40.4　Applications

Theorem 5 Let $0 < a \leqslant 1$, and let $x \leqslant y, u(t) = tx + (1-t)y, v(t) = ty + (1-t)x$. If $1/2 \leqslant t_2 \leqslant t_1 \leqslant 1$ or $0 \leqslant t_1 \leqslant t_2 \leqslant 1/2$, then we have

$$G(x,y) \leqslant I_a \left(x^{u(t_1)} y^{v(t_1)}, x^{v(t_1)} y^{u(t_1)} \right) \leqslant I_a \left(x^{u(t_2)} y^{v(t_2)}, x^{v(t_2)} y^{u(t_2)} \right)$$

$$\leqslant I_a(x,y) \leqslant I_a \left(u(t_2), v(t_2) \right) \leqslant I_a \left(u(t_1), v(t_1) \right) \leqslant A(x,y) \qquad (9)$$

Proof Combining Lemma 4 with Theorem 3, we have

$$I_a(x,y) \leqslant I_a \left(u(t_2), v(t_2) \right) \leqslant I_a \left(u(t_1), v(t_1) \right)$$
$$\leqslant I_a \left((x+y)/2, (x+y)/2 \right) = A(x,y)$$

On the other hand, since

$$(\ln \sqrt{xy}, \ln \sqrt{xy}) \prec \left(\ln x^{u(t_1)} y^{v(t_1)}, \ln x^{v(t_1)} y^{u(t_1)} \right)$$
$$\prec \left(\ln x^{u(t_2)} y^{v(t_2)}, \ln x^{v(t_2)} y^{u(t_2)} \right) \prec (\ln x, \ln y)$$

from Theorem 4, it follows that

$$G(x,y) = I_a \left(\sqrt{xy}, \sqrt{xy} \right) \leqslant I_a \left(x^{u(t_1)} y^{v(t_1)}, x^{v(t_1)} y^{u(t_1)} \right)$$
$$\leqslant I_a \left(x^{u(t_2)} y^{v(t_2)}, x^{v(t_2)} y^{u(t_2)} \right) \leqslant I_a(x,y)$$

The proof is complete.

Theorem 6 Let $0 \leqslant x \leqslant y, c \geqslant 0, 0 < a \leqslant 1$. Then

$$I_a \left(\frac{x+c}{x+y+2c}, \frac{y+c}{x+y+2c} \right) \geqslant I_a \left(\frac{x}{x+y}, \frac{y}{x+y} \right) \tag{9}$$

Proof By Lemma 5 and Theorem 3, it follows that (9) holds. The proof is complete.

Acknowledgements. The authors are indebted to the referees for their helpful suggestions.

References

[1] K B STOLARSKY. Generalizations of the logarithmic mean[J]. Math. Mag., 1975, 48(2):87-92.

[2] HUAN-NAN SHI, SHAN-HE WU, FENG QI. An alternative note on the Schur-convexity of the extended mean values[J]. Mathematical Inequalities and Applications, 2006, 9(2): 219-224.

[3] BO-YING WANG. Foundations of Majorization Inequalities[M]. Beijing: Beijing Normal Univ. Press, 1990.

[4] A M MARSHALL, I OLKIN. Inequalities:theory of majorization and its application[M]. New York : Academies Press, 1979.

[5] XIAO-MING ZHANG. Geometrically Convex Functions[M]. Hefei: An hui University Press, 2004.

[6] CONSTANTIN P NICULESCU. Convexity According to the Geometric Mean[J]. Mathematical Inequalities & Applications, 2000, 3(2):155-167.

[7] DA-MAO LI, CHUN GU, HUAN-NAN SHI. Schur Convexity of the Power-Type Generalization of Heronian Mean[J]. Mathematics in practice and theory(Chinese), 2006, 36 (9): 387-390.

第41篇 Lehme平均的Schur凸性和Schur几何凸性

(顾春, 石焕南. 数学的实践与认识, 2009, 39(12): 183-188)

摘　要: 本文讨论了二元Lehme平均$L_p(a,b)$关于变量(a,b) 在$\mathbf{R}^2_{++}$ 上的Schur凸性和Schur几何凸性, 并建立了相应的不等式.

关键词: Lehme平均; Schur凸性; Schur几何凸性; 不等式

41.1　引言

在本文中, $\mathbf{R}^n$ 和$\mathbf{R}^n_{++}$ 分别表示n 维实数集和n 维正实数集.

设$(a,b) \in \mathbf{R}^n_{++}$, (a,b) 的Lehme平均[1] 定义为

$$L_p(a,b) = \frac{a^p + b^p}{a^{p-1} + b^{p-1}}, -\infty < p < +\infty$$

$L_p(a,b)$ 是一类重要的二元平均, 它包含了下述常见二元平均

$$L_0(a,b) = \frac{2}{a^{-1} + b^{-1}} = H(a,b) : 调和平均$$

$$L_{\frac{1}{2}}(a,b) = \sqrt{ab} = G(a,b) : 几何平均$$

$$L_1(a,b) = \frac{a+b}{2} = A(a,b) : 算术平均$$

$$L_2(a,b) = \frac{a^2 + b^2}{a + b} : 反调和平均$$

关于$L_p(a,b)$, 文[2]给出如下结论:

(1)对于固定的a, b, $L_p(a,b)$ 是p 的严格递增函数, 且

$$L_{+\infty}(a,b) = \lim_{p \to +\infty} L_p(a,b) = \max(a,b)$$

$$L_{-\infty}(a,b) = \lim_{p \to -\infty} L_p(a,b) = \min(a,b)$$

(2)对于$a \neq b$, 当$p < (>) - \frac{1}{2}$ 时, $L_p(a,b)$ 是p 的对数凸(凹)函数. 若$p > (<) - \frac{1}{2}$,则对于任意实数t, 有

$$L_{p_0-t}(a,b)L_{p_0+t}(a,b) \leqslant (\geqslant)L_{p_0}^2(a,b)$$

本文从另一个视角考察$L_p(a,b)$, 完整解决了二元Lehme平均$L_p(a,b)$ 关于变量(a,b) 在$\mathbf{R}_{++}^n$ 上的Schur凸性和Schur几何凸性, 并对n 元Lehme平均在$\mathbf{R}_{++}^n$ 上的Schur凸性作了初步的探讨. 我们的主要结论是

定理 1 当$p \geqslant 1$ 时, $L_p(a,b)$关于(a,b) 在$\mathbf{R}_{++}^n$ 上Schur-凸; 而当$p \leqslant 1$ 时, $L_p(a,b)$ 关于(a,b) 在$\mathbf{R}_{++}^n$ 上Schur- 凹.

定理 2 当$p \geqslant \frac{1}{2}$ 时, $L_p(a,b)$关于(a,b) 在$\mathbf{R}_{++}^n$ 上Schur-几何凸; 而当$p \leqslant \frac{1}{2}$ 时, $L_p(a,b)$ 关于(a,b) 在$\mathbf{R}_{++}^n$ 上Schur-几何凹.

定理 3 设$\boldsymbol{a} \in \mathbf{R}_{++}^n = \{\boldsymbol{a} = (a_1,\ldots,a_n) : a_i > 0, i = 1,\ldots,n\}$, 则$\boldsymbol{a}$ 的Lehmer平均

$$L_p(a_1,\ldots,a_n) = \frac{\sum_{i=1}^n a_i^p}{\sum_{i=1}^n a_i^{p-1}}$$

当$1 \leqslant p \leqslant 2$ 时, 在$\mathbf{R}_{++}^n$ 上Schur-凸; 而当$0 \leqslant p \leqslant 1$ 时, 在$\mathbf{R}_{++}^n$ 上Schur-凹.

推论 1 设$0 \leqslant a \leqslant b$, $u(t) = tb + (1-t)a, v(t) = ta + (1-t)b$. 若$\frac{1}{2} \leqslant t_2 \leqslant t_1 \leqslant 1$ 或$0 \leqslant t_1 \leqslant t_2 \leqslant \frac{1}{2}$, 则当$p \geqslant 1$ 时

$$A(a,b) \leqslant L_p(u(t_2),v(t_2)) \leqslant L_p(u(t_1),v(t_1)) \leqslant L_p(a,b) \tag{1}$$

而当$p < 1$ 时, 不等式(1)反向.

推论 2 设$(a,b) \in \mathbf{R}_{++}^n$, 则当$p \geqslant \frac{1}{2}$ 时, 有

$$L_p\left(\sqrt{ab},\sqrt{ab}\right) \leqslant L_p(a,b) \tag{2}$$

而$p \leqslant \frac{1}{2}$ 时, 不等式(2)反向.

41.2　定义和引理

对于$\boldsymbol{x} = (x_1,\ldots,x_n) \in \mathbf{R}^n$, 将$\boldsymbol{x}$ 的分量排成递减的次序后, 记作$x_{[1]} \geqslant x_2 \geqslant \cdots \geqslant x_{[n]}$. $\boldsymbol{x} \leqslant \boldsymbol{y}$ 表示$x_i \leqslant y_i, i = 1,\ldots,n$.

定义 1[3] 设$\boldsymbol{x},\boldsymbol{y} \in \mathbf{R}^n$ 满足

(1) $\sum\limits_{i=1}^{k} x_{[i]} \leqslant \sum\limits_{i=1}^{k} y_{[i]}$, $k = 1, 2, ..., n-1$;

(2) $\sum\limits_{i=1}^{n} x_i = \sum\limits_{i=1}^{n} y_i$, 则称$\boldsymbol{x}$ 被$\boldsymbol{y}$ 所控制, 记作$\boldsymbol{x} \prec \boldsymbol{y}$.

定义 2[3]　设$\Omega \subset \mathbf{R}^n, \varphi : \Omega \to \mathbf{R}$,

(1) 若在Ω 上$\boldsymbol{x} \leqslant \boldsymbol{y} \Rightarrow \varphi(\boldsymbol{x}) \leqslant \varphi(\boldsymbol{y})$, 则称$\varphi$ 为Ω 上的增函数; 若$-\varphi$ 是Ω 上的增函数, 则称φ 为Ω 上的减函数;

(2) 若在Ω 上$\boldsymbol{x} \prec \boldsymbol{y} \Rightarrow \varphi(\boldsymbol{x}) \leqslant \varphi(\boldsymbol{y})$, 则称$\varphi$ 为Ω 上的Schur凸函数; 若$-\varphi$ 是Ω 上的Schur 凸函数, 则称φ 为Ω 上的Schur 凹函数.

(3) Ω 称为凸集, 如果$\forall \boldsymbol{x}, \boldsymbol{y} \in \Omega$ 均有$(\alpha x_1 + \beta y_1, \ldots, \alpha x_n + \beta y_n) \in \Omega$, 其中$\alpha, \beta \in [0,1]$, 且$\alpha + \beta = 1$.

定义 3[4]　设$\boldsymbol{x} = (x_1, x_2, \ldots, x_n) \in \mathbf{R}^n$ 和$\boldsymbol{y} = (y_1, y_2, \ldots, y_n) \in \mathbf{R}^n$.

(1) $\Omega \subset \mathbf{R}_{++}^n$ 称为几何凸集, 如果$\forall \boldsymbol{x}, \boldsymbol{y} \in \Omega$ 有$(x_1^\alpha y_1^\beta, \ldots, x_n^\alpha y_n^\beta) \in \Omega$, 其中$\alpha, \beta \in [0,1]$, 且$\alpha + \beta = 1$.

(2) $\Omega \subset \mathbf{R}_{++}^n, f : \Omega \to \mathbf{R}_+$, 对于任意$\boldsymbol{x}, \boldsymbol{y} \in \Omega$, 若$\ln \boldsymbol{x} \prec \ln \boldsymbol{y}$ 有$f(\boldsymbol{x}) \leqslant f(\boldsymbol{y})$, 则称为$f$ 为Ω 上的Schur几何凸函数. 若$\ln \boldsymbol{x} \prec \ln \boldsymbol{y}$ 有$f(\boldsymbol{x}) \geqslant f(\boldsymbol{y})$, 则称为$f$ 为Ω 上的Schur几何凹函数.

定义 4 [3]　(1) $\Omega \subset \mathbf{R}^n$ 称为对称集, 若$\boldsymbol{x} \in \Omega$, 对于任意$n \times n$ 置换矩阵$\boldsymbol{P}$ 均有$\boldsymbol{P}\boldsymbol{x} \in \Omega$. (2) $\Omega \subset \mathbf{R}^n, f : \Omega \to \mathbf{R}_+$ 称为对称函数, 若$\boldsymbol{x} \in \Omega$, 对予任意$n \times n$ 置换矩阵$\boldsymbol{P}$ 均有$f(\boldsymbol{P}\boldsymbol{x}) = f(\boldsymbol{x})$.

引理 1 [3]　设φ 在开凸集$\Omega \subset \mathbf{R}^n$ 上可微, 则φ 是Ω 上的增函数的充要条件是$\nabla\varphi(\boldsymbol{x}) \geqslant 0, \forall \boldsymbol{x} \in \Omega$, 其中

$$\nabla\varphi(\boldsymbol{x}) = \left(\frac{\partial\varphi(\boldsymbol{x})}{\partial x_1}, \ldots, \frac{\partial\varphi(\boldsymbol{x})}{\partial x_n} \right) \in \mathbf{R}^n$$

引理 2 [3]　设$\Omega \subset \mathbf{R}^n$ 是有内点的对称凸集, $\varphi : \Omega \to \mathbf{R}$ 在Ω 上连续, 在Ω 的内部Ω^0 可微, 则在Ω 上Schur-凸(Schur-凹)的充要条件是φ 在Ω 上对称且$\forall \boldsymbol{x} \in \Omega$, 有

$$(x_1 - x_2) \left(\frac{\partial\varphi}{\partial x_1} - \frac{\partial\varphi}{\partial x_2} \right) \geqslant 0 (\leqslant 0)$$

引理 3[3]　设$\Omega \subset \mathbf{R}_{++}^n$ 是有内点的对称凸集, $\varphi : \Omega \to \mathbf{R}$ 在Ω 上连续, 在Ω 的内部Ω^0 可微, 则在Ω 上Schur-几何凸(Schur-凹)的充要条件是φ 在Ω 上对称且$\forall \boldsymbol{x} \in \Omega$, 有

$$(\ln x_1 - \ln x_2) \left(x_1 \frac{\partial\varphi}{\partial x_1} - x_2 \frac{\partial\varphi}{\partial x_2} \right) \geqslant 0 (\leqslant 0)$$

引理 4　设$f(t) = 2pt^p - 2pt^{p-1} + t^{2p-1} + t^{p-1} - t^p - 1, t \geqslant 1$, 则当$p \geqslant \frac{1}{2}$

时, $f(t) \geqslant 0$; 当 $0 \leqslant p \leqslant \frac{1}{2}$ 时, $f(x) \leqslant 0$.

证明

$$
\begin{aligned}
f^{'}(t) &= 2p^2 t^{p-1} - 2p(p-1)t^{p-2} + (2p-1)t^{2p-2} + (p-1)t^{p-2} - pt^{p-1} \\
&= pt^{p-1}(2p-1) + (p-1)(1-2p)t^{p-2} + (2p-1)t^{2p-2} \\
&= (2p-1)g(t)t^{p-2}
\end{aligned}
$$

其中

$$
g(t) = pt + 1 - p + t^p
$$

对于 $t \geqslant 1, p \geqslant 0$, 有

$$
g^{'}(t) = p(1 + t^{p-1}) \geqslant 0
$$

因此, $g(t) \geqslant g(1) = 2 > 0$, 从而当 $p \geqslant \frac{1}{2}$ 时, $f(t) = (2p-1)g(t)t^{p-2} \geqslant 0$, 进而对于 $t \geqslant 1$, 有 $f(t) \geqslant f(1) = 0$; 当 $0 \leqslant p \leqslant \frac{1}{2}$ 时, $f(t) = (2p-1)g(t)t^{p-2} \leqslant 0$, 进而对于 $t \geqslant 1$, 有 $f(t) \leqslant f(1) = 0$. 引理得证.

引理 5 设 $0 \leqslant a \leqslant b$, $u(t) = tb + (1-t)a, v(t) = ta + (1-t)b$. 若 $\frac{1}{2} \leqslant t_2 \leqslant t_1 \leqslant 1$ 或 $0 \leqslant t_1 \leqslant t_2 \leqslant \frac{1}{2}$, 则

$$
\left(\frac{a+b}{2}, \frac{a+b}{2} \right) \prec (u(t_2), v(t_2)) \prec (u(t_1), v(t_1)) \prec (a, b) \tag{3}
$$

证明 只需证 $(u(t_2), v(t_2)) \prec (u(t_1), v(t_1))$. 首先 $u(t_2) + v(t_2) = u(t_1) + v(t_1) = a + b$

若 $\frac{1}{2} \leqslant t_2 \leqslant t_1 \leqslant 1$, 因 $0 \leqslant a \leqslant b$, 则 $u(t_i) - v(t_i) = (a-b)(1-2t_i) \geqslant 0$, 即 $u(t_i) \geqslant v(t_i), i = 1, 2$. 又 $u(t_1) - u(t_2) = (t_1 - t_2)(b-a) \geqslant 0$, 即 $u(t_1) \geqslant u(t_2)$, 故 $(u(t_2), v(t_2)) \prec (u(t_1), v(t_1))$ 成立.

若 $0 \leqslant t_1 \leqslant t_2 \leqslant \frac{1}{2}, u(t_i) - v(t_i) = (a-b)(1-2t_i) \leqslant 0$, 即 $u(t_i) \leqslant v(t_i), i = 1, 2$. 又 $v(t_1) - v(t_2) = (t_1 - t_2)(a-b) \geqslant 0$, 即 $v(t_1) \geqslant v(t_2)$. 故 $(u(t_2), v(t_2)) \prec (u(t_1), v(t_1))$ 成立.

41.3 定理的证明

定理 1 的证明 显然 $L_p(a, b)$ 在 $\mathbf{R}_{++}^2$ 上对称. $\forall (a, b) \in \mathbf{R}_{++}^2$, 经计算

$$
\frac{\partial L_p}{\partial a} = \frac{pa^{p-1}(a^{p-1} + b^{p-1}) - (p-1)a^{p-2}(a^p + b^p)}{(a^{p-1} + b^{p-1})^2}
$$

$$
\frac{\partial L_p}{\partial b} = \frac{pb^{p-1}(a^{p-1} + b^{p-1}) - (p-1)b^{p-2}(a^p + b^p)}{(a^{p-1} + b^{p-1})^2}
$$

于是

$$\Delta := (a - b)\left(\frac{\partial L_p}{\partial a} - \frac{\partial L_p}{\partial b}\right)$$

$$(a - b)\frac{p(a^{p-1} - b^{p-1})(a^{p-1} + b^{p-1}) - (p-1)(a^{p-2} - b^{p-2})(a^p + b^p)}{(a^{p-1} + b^{p-1})^2}$$

$$(a - b)\frac{p(a^{2p-2} - b^{2p-2}) - (p-1)(a^{2p-2} + a^{p-2}b^p - a^p b^{p-2} - b^{2p-2})}{(a^{p-1} + b^{p-1})^2}$$

$$(a - b)\frac{-pa^{p-2}b^p + pa^p b^{p-2} + a^{2p-2} + a^{p-2}b^p - a^p b^{p-2} - b^{2p-2}}{(a^{p-1} + b^{p-1})^2}$$

$$(a - b)\frac{(p-1)a^{p-2}b^{p-2}(a^2 - b^2)(a - b) + (a - b)(a^{2p-2} - b^{2p-2})}{(a^{p-1} + b^{p-1})^2}$$

注意$(a-b)(a^2-b^2) \geqslant 0$. 当$p \geqslant 1$, 即$p-1 \geqslant 0$时, 亦有$(a-b)(a^{2p-2}-b^{2p-2}) \geqslant 0$, 从而$\Delta \geqslant 0$, 即$L_p(a,b)$ 关于(a,b) 在$\mathbf{R}_{++}^2$ 上Schur-凸; 而当$p \leqslant 1$ 时, $p-1 \leqslant 0$, $(a-b)(a^{2p-2}-b^{2p-2}) \leqslant 0$, 从而$\Delta \leqslant 0$, 即$L_p(a,b)$ 关于(a,b) 在$\mathbf{R}_{++}^2$ 上Schur-凹. 定理1 得证.

定理 2 的证明　$\forall (a,b) \in \mathbf{R}_{++}^2$.

$$\Lambda : (\ln a - \ln b)\left(a\frac{\partial L_p}{\partial a} - b\frac{\partial L_p}{\partial b}\right)$$

$$(\ln a - \ln b)\frac{p(a^p - b^p)(a^{p-1} + b^{p-1}) - (p-1)(a^{p-1} - b^{p-1})(a^p + b^p)}{(a^{p-1} + b^{p-1})^2}$$

$$\frac{2pa^{p-1}b^{p-1}(\ln a - \ln b)(a - b) + (\ln a - \ln b)(a^{p-1} - b^{p-1})(a^p + b^p)}{(a^{p-1} + b^{p-1})^2}$$

注意$(\ln a - \ln b)(a - b) \geqslant 0$, 当$p \geqslant 1$ 时, 亦有$(\ln a - \ln b)(a^{p-1} - b^{p-1}) \geqslant 0$ 从而$\Lambda \geqslant 0$, 而当$p \leqslant 0$ 时, $(\ln a - \ln b)(a^{p-1} - b^{p-1}) \leqslant 0$, 从而$\Lambda \leqslant 0$. 下面考虑$0 < p < 1$ 的情形. 由对称性, 不妨设$a \geqslant b$, 令$t = \frac{a}{b}$, 不难验证

$$p(a^p - b^p)(a^{p-1} + b^{p-1}) - (p-1)(a^{p-1} - b^{p-1})(a^p + b^p) = b^{2p-1}f(t), t \geqslant 1$$

其中

$$f(t) = 2pt^p - 2pt^{p-1} + t^{2p-1} + t^{p-1} - t^p - 1, \ t \geqslant 1$$

于是

$$\Lambda = (\ln a - \ln b)\frac{b^{2p-1}f(t)}{(a^{p-1} + b^{p-1})^2}$$

因$a \geqslant b$, 即$t \geqslant 1$, 有$\ln a - \ln b \geqslant 0$, 结合引理4 知, 当$1 > p \geqslant \frac{1}{2}$ 时, $\Lambda \geqslant 0$,

当 $0 < p \leqslant \frac{1}{2}$ 时, $\Lambda \leqslant 0$. 总之, 当 $p \geqslant \frac{1}{2}$ 时, $\Lambda \geqslant 0$, 即 $L_p(a,b)$ 关于 (a,b) 在 $\mathbf{R}_{++}^2$ 上Schur-几何凸, 当 $p \leqslant \frac{1}{2}$ 时, $\Lambda \leqslant 0$, 即 $L_p(a,b)$ 关于 (a,b) 在 $\mathbf{R}_{++}^2$ 上Schur-几何凹, 定理2 得证.

定理 3 的证明 显然 $L_p(a_1,\ldots,a_n)$ 在 $\mathbf{R}_{++}^2$ 上对称. $\forall (a_1,\ldots,a_n) \in \mathbf{R}_{++}^2$, 经计算

$$\frac{\partial L_p}{\partial a_1} = \frac{pa_1^{p-1}\sum\limits_{i=1}^{n}a_i^{p-1} - (p-1)a_1^{p-2}\sum\limits_{i=1}^{n}a_i^p}{(\sum\limits_{i=1}^{n}a_i^{p-1})^2}$$

$$\frac{\partial L_p}{\partial a_2} = \frac{pa_2^{p-1}\sum\limits_{i=1}^{n}a_i^{p-1} - (p-1)a_2^{p-2}\sum\limits_{i=1}^{n}a_i^p}{(\sum\limits_{i=1}^{n}a_i^{p-1})^2}$$

于是

$$\Gamma := (a_1 - a_2)\left(\frac{\partial L_p}{\partial a_1} - \frac{\partial L_p}{\partial a_2}\right)$$
$$= (a_1 - a_2)\frac{p(a_1^{p-1}-a_2^{p-1})\sum\limits_{i=1}^{n}a_i^{p-1} - (p-1)(a_1^{p-2}-a_2^{p-2})\sum\limits_{i=1}^{n}a_i^p}{(\sum\limits_{i=1}^{n}a_i^{p-1})^2}$$

当 $1 \leqslant p \leqslant 2$ 时, $\Gamma \geqslant 0$, 而当 $0 \leqslant p \leqslant 1$ 时, $\Gamma \leqslant 0$, 定理3 得证.

推论 1 的证明 结合引理5 和定理1 并注意 $L_p\left(\frac{a+b}{2},\frac{a+b}{2}\right) = A(a,b)$ 即得证.

推论 2 的证明 注意 $\left(\ln\sqrt{xy},\ln\sqrt{xy}\right) \prec (\ln x,\ln y)$, 由定理2 即得证.

猜想 n 元Lehmer平均

$$L_p(a_1,\ldots,a_n) = \frac{\sum\limits_{i=1}^{n}a_i^p}{\sum\limits_{i=1}^{n}a_i^{p-1}}$$

当 $p \geqslant 2$ 时, 在 $\mathbf{R}_{++}^2$ 上Schur-凸; 而当 $p \leqslant 0$ 时, 在 $\mathbf{R}_{++}^2$ 上Schur-凹.

参考文献

[1] RAJENDRA BHATIA, HIDEKI KOSAKI. Mean matrices and infinite divisibility[J]. Linear Algebra and Its Applicationst, 2007, 424(1): 36-54.

[2] ALFRED WITKOWSKI. Covexity of weighted Stolarsky means[J].

J Inequal Pure Appl Math, 2006, 7 (2), Article 73.

[3]　王伯英. 控制不等式基础[M]. 北京:北京师范大学出版社,1990.

[4]　张小明. 几何凸函数[M]. 合肥：安徽大学出版社, 2004.

第42篇 一对互补对称函数的Schur凸性

(石焕南,张小明. 湖南理工学院学报(自然科学版), 2009, 22 (4)：1-5)

摘 要: 本文讨论了一对互补对称函数及它们的差在$\mathbf{R}_{++}^n$上的Schur凹凸性, 并据此建立相关的不等式.

关键词: 对称函数; 互补函数; Schur 凹凸性; 不等式

在本文中, $\mathbf{R}^n$ 和$\mathbf{R}_{++}^n$ 分别表示n 维实数集和n 维正实数集, 并记$\mathbf{R}^1 = \mathbf{R}, \mathbf{R}_{++}^1 = \mathbf{R}_{++}$. 对于$\boldsymbol{a} = (a_1, \ldots, a_n) \in \mathbf{R}_{++}^n$, 记

$$\varphi_k(\boldsymbol{a}) = \sum_{1 \leqslant i_1 < \ldots < i_k \leqslant n} \frac{\sum\limits_{j=1}^{k} a_{i_j}}{S - \sum\limits_{j=1}^{k} a_{i_j}}, \quad \overline{\varphi}_k(\boldsymbol{a}) = \sum_{1 \leqslant i_1 < \ldots < i_k \leqslant n} \frac{S - \sum\limits_{j=1}^{k} a_{i_j}}{\sum\limits_{j=1}^{k} a_{i_j}}$$

其中$S = \sum\limits_{i=1}^{n} a_i$. 若$k = 0$, 规定$\varphi_k(\boldsymbol{a}) = \overline{\varphi}_k(\boldsymbol{a}) = 1$; 若$k > n$, 规定$\varphi_k(\boldsymbol{a}) = \overline{\varphi}_k(\boldsymbol{a}) = 0$.

对于上述两个互补的对称函数 $\varphi_k(\boldsymbol{a})$与$\overline{\varphi}_k(\boldsymbol{a})$, 最近文[1] 获得了如下三个结论:

定理 A[1] 设n 和k 是自然数, 满足$n \geqslant 2$ 和$k \leqslant \left[\frac{n}{2}\right]$, 则对于任意$a \in \mathbf{R}_{++}^n$, 有

$$\varphi_k(\boldsymbol{a}) \leqslant \frac{k^2}{(n-k)^2} \overline{\varphi}_k(\boldsymbol{a}) \tag{1}$$

定理 B [1] 设n 和k 是自然数, 满足$n \geqslant 2$ 和$1 \leqslant k \leqslant n-1$, 则对于任意$a \in \mathbf{R}_{++}^n$, 有

$$\varphi_k(\boldsymbol{a}) \leqslant \frac{k}{n-k} \mathrm{C}_n^k \tag{2}$$

定理 C[1] 设n 和k 是自然数, 满足$n \geqslant 2$ 和$k \leqslant \left[\frac{n}{2}\right]$, 则对于任意$a \in \mathbf{R}_{++}^n$, 有

$$\overline{\varphi}_k(\boldsymbol{a}) - \varphi_k(\boldsymbol{a}) \geqslant \frac{(n-2k)n}{(n-k)k} \mathrm{C}_n^k \tag{3}$$

本文将研究$\varphi_k(\boldsymbol{a})$, $\overline{\varphi}_k(\boldsymbol{a})$ 及二者差的Schur 凹凸性, 并据此给出(2), (3)两式的控制证明, 并建立与(2)互补的不等式. 我们的主要结论是:

定理 1　设n 和k 是自然数, 满足$n \geqslant 2$ 和$1 \leqslant k \leqslant n-1$, 则$\varphi_k(\boldsymbol{a})$ 和$\overline{\varphi}_k(\boldsymbol{a})$ 均是$\mathbf{R}_{++}^n$ 上的Schur凸函数.

注 1　因$\left(\frac{S}{n}, \ldots, \frac{S}{n}\right) \prec (a_1, \ldots, a_n)$, 据定理1 有

$$\overline{\varphi}_k(a_1, \ldots, a_n) \geqslant \overline{\varphi}_k\left(\frac{S}{n}, \ldots, \frac{S}{n}\right) = \frac{k}{n-k} \mathrm{C}_n^k$$

由此给出(2) 的控制证明. 同理可得下述与(2)互补的不等式.

推论 1　设n 和k 是自然数, 满足$n \geqslant 2$ 和$1 \leqslant k \leqslant n-1$, 则

$$\overline{\varphi}_k(\boldsymbol{a}) \geqslant \frac{n-k}{k} \mathrm{C}_n^k \tag{4}$$

定理 2　设n 和k 是自然数, $n \geqslant 2$. 若$k \leqslant \frac{n}{2}$, 则$\overline{\varphi}_k(\boldsymbol{a}) - \varphi_k(\boldsymbol{a})$ 在$\mathbf{R}_{++}^n$ 上为Schur 凸函数; 若$k \geqslant \frac{n}{2}$, 则$\overline{\varphi}_k(\boldsymbol{a}) - \varphi_k(\boldsymbol{a})$ 在$\mathbf{R}_{++}^n$ 上为Schur 凹函数.

注 2　因$\left(\frac{S}{n}, \ldots, \frac{S}{n}\right) \prec (a_1, \ldots, a_n)$, 据定理2 可推广定理C 得到

推论 2　设n 和k 是自然数, 满足$n \geqslant 2$ 和$k \leqslant \frac{n}{2}$, 则对于任意$a \in \mathbf{R}_{++}^n$, 有

$$\overline{\varphi}_k(\boldsymbol{a}) - \varphi_k(\boldsymbol{a}) \leqslant \frac{(n-2k)n}{(n-k)k} \mathrm{C}_n^k \tag{5}$$

对于$\boldsymbol{a} = (a_1, ..., a_n) \in \mathbf{R}^n$, 将$\boldsymbol{a}$ 的分量排成递减的次序后, 记作$a_{[1]} \geqslant a_2 \geqslant ... \geqslant a_{[n]}$. $\boldsymbol{a} \leqslant \boldsymbol{b}$ 表示$a_i \leqslant b_i, i = 1, \ldots, n$.

定义 1 [3-4]　设$\boldsymbol{a}, \boldsymbol{b} \in \mathbf{R}^n$ 满足

(1) $\sum\limits_{i=1}^{k} a_{[i]} \leqslant \sum\limits_{i=1}^{k} b_{[i]}$, $k = 1, 2, ..., n-1$; (2) $\sum\limits_{i=1}^{n} a_i = \sum\limits_{i=1}^{n} b_i$, 则称$\boldsymbol{a}$ 被$\boldsymbol{b}$ 所控制, 记作$\boldsymbol{a} \prec \boldsymbol{b}$.

定义 2 [3-4]　设$\Omega \subset \mathbf{R}^n, \varphi : \Omega \to \mathbf{R}$.

若在Ω 上$\boldsymbol{a} \prec \boldsymbol{b} \Rightarrow \varphi(\boldsymbol{a}) \leqslant \varphi(\boldsymbol{b})$, 则称$\varphi$ 为Ω 上的Schur凸函数; 若$-\varphi$ 是Ω 上的Schur 凸函数, 则称φ 为Ω 上的Schur 凹函数.

引理 1 [3-4]　设$\Omega \subset \mathbf{R}^n$ 是有内点的对称凸集, $\varphi : \Omega \to \mathbf{R}$ 在Ω 上连续, 在Ω 的内部Ω^0 可微, 则在Ω 上Schur-凸(Schur-凹)的充要条件是φ 在Ω 上对称且$\forall \boldsymbol{a} \in \Omega$, 有

$$(a_1 - a_2)\left(\frac{\partial \varphi}{\partial a_1} - \frac{\partial \varphi}{\partial a_2}\right) \geqslant 0 (\leqslant 0)$$

定理 1 的证明　易见$\varphi_k(\boldsymbol{a}) = g_k(\boldsymbol{a}) - \mathrm{C}_n^k$, 其中

$$g_k(\boldsymbol{a}) = g_k(a_1, \ldots, a_n) = S \sum_{1 \leqslant i_1 < \cdots < i_k \leqslant n} \left(S - \sum_{j=1}^{k} a_{ij}\right)^{-1}$$

为证 $\varphi_k(\boldsymbol{a})$ 是 $\mathbf{R}_{++}^n$ 上的 Schur 凸函数, 只需证 $g_k(\boldsymbol{a})$ 是 $\mathbf{R}_{++}^n$ 上的 Schur 凸函数. 注意

$$
\sum_{1 \leqslant i_1 < \cdots < i_k \leqslant n} \left(S - \sum_{j=1}^{k} a_{i_j}\right)^{-1}
$$

$$
= \sum_{3 \leqslant i_1 < \cdots < i_k \leqslant n} \left[S - \left(a_1 + \sum_{j=1}^{k-1} a_{i_j}\right)\right]^{-1} + \sum_{3 \leqslant i_1 < \cdots < i_k \leqslant n} \left[S - \left(a_2 + \sum_{j=1}^{k-1} a_{i_j}\right)\right]^{-1} +
$$

$$
\sum_{3 \leqslant i_1 < \cdots < i_k \leqslant n} \left[S - \left(a_1 + a_2 \sum_{j=2}^{k-2} a_{i_j}\right)\right]^{-1} + \sum_{3 \leqslant i_1 < \cdots < i_k \leqslant n} \left(S - \sum_{j=1}^{k} a_{i_j}\right)^{-1}
$$

可得

$$
\frac{\partial g_k(\boldsymbol{a})}{\partial a_1} = \sum_{1 \leqslant i_1 < \cdots < i_k \leqslant n} \left(S - \sum_{j=1}^{k} a_{i_j}\right)^{-1} -
$$

$$
S \left\{ \sum_{3 \leqslant i_1 < \cdots < i_k \leqslant n} \left[S - \left(a_1 + \sum_{j=1}^{k-1} a_{i_j}\right)\right]^{-2} + \sum_{3 \leqslant i_1 < \cdots < i_k \leqslant n} \left(S - \sum_{j=1}^{k} a_{i_j}\right)^{-2} \right\}
$$

$$
\frac{\partial g_k(\boldsymbol{a})}{\partial a_2} = \sum_{1 \leqslant i_1 < \cdots < i_k \leqslant n} \left(S - \sum_{j=1}^{k} a_{i_j}\right)^{-1} -
$$

$$
S \left\{ \sum_{3 \leqslant i_1 < \cdots < i_k \leqslant n} \left[S - \left(a_2 + \sum_{j=1}^{k-1} a_{i_j}\right)\right]^{-2} + \sum_{3 \leqslant i_1 < \cdots < i_k \leqslant n} \left(S - \sum_{j=1}^{k} a_{i_j}\right)^{-2} \right\}
$$

$$
\Delta := (a_1 - a_2) \left(\frac{\partial g_k(\boldsymbol{a})}{\partial a_1} - \frac{\partial g_k(\boldsymbol{a})}{\partial a_2} \right)
$$

$$
= S(a_1 - a_2) \sum_{3 \leqslant i_1 < \cdots < i_k \leqslant n} \left\{ \left[S - \left(a_1 + \sum_{j=1}^{k-1} a_{i_j}\right)\right]^{-2} - \left[S - \left(a_2 + \sum_{j=1}^{k-1} a_{i_j}\right)\right]^{-2} \right\}
$$

$$
= S(a_1 - a_2) \sum_{3 \leqslant i_1 < \cdots < i_k \leqslant n} \frac{\left[S - \left(a_1 + \sum\limits_{j=1}^{k-1} a_{i_j}\right)\right]^2 - \left[S - \left(a_2 + \sum\limits_{j=1}^{k-1} a_{i_j}\right)\right]^2}{\left[S - \left(a_1 + \sum_{j=1}^{k-1} a_{i_j}\right)\right]^2 \left[S - \left(a_2 + \sum\limits_{j=1}^{k-1} a_{i_j}\right)\right]^2}
$$

$$
= \sum_{3 \leqslant i_1 < \cdots < i_k \leqslant n} \frac{S(a_1 - a_2)^2 \left[2S - \left(a_1 + a_2 + 2\sum\limits_{j=1}^{k-1} a_{i_j}\right)\right]}{\left[S - \left(a_1 + \sum\limits_{j=1}^{k-1} a_{i_j}\right)\right]^2 \left[S - \left(a_2 + \sum\limits_{j=1}^{k-1} a_{i_j}\right)\right]^2}
$$

再注意$3 \leqslant i_1 < \cdots < i_k \leqslant n$, 有$2S - (a_1 + a_2 + 2\sum\limits_{j=1}^{k-1} a_{i_j}) \geqslant 0$, 故$\Delta \geqslant 0$. 由引理1 知$g_k(\boldsymbol{a})$ Schur 凸, 进而$\varphi_k(\boldsymbol{a}) = g_k(\boldsymbol{a}) - C_n^k$ 亦Schur 凸.

易见$\overline{\varphi}_k(\boldsymbol{a}) = h_k(\boldsymbol{a}) - C_n^k$, 其中

$$h_k(\boldsymbol{a}) = h_k(a_1, \ldots, a_n) = S \sum_{1 \leqslant i_1 < \cdots < i_k \leqslant n} \left(\sum_{j=1}^{k} a_{i_j}\right)^{-1}$$

为证$\overline{\varphi}_k(\boldsymbol{a})$ 是$\mathbf{R}_{++}^n$ 上的Schur凹函数, 只需证$h_k(\boldsymbol{a})$ 是$\mathbf{R}_{++}^n$ 上的Schur凹函数. 注意

$$
\begin{aligned}
&\sum_{1 \leqslant i_1 < \cdots < i_k \leqslant n} \left(S - \sum_{j=1}^{k} a_{i_j}\right)^{-1} \\
= &\sum_{3 \leqslant i_1 < \cdots < i_k \leqslant n} \left(a_1 + \sum_{j=1}^{k-1} a_{i_j}\right)^{-1} + \sum_{3 \leqslant i_1 < \cdots < i_k \leqslant n} \left(a_2 + \sum_{j=1}^{k-1} a_{i_j}\right)^{-1} + \\
&\sum_{3 \leqslant i_1 < \cdots < i_k \leqslant n} \left(a_1 + a_2 \sum_{j=2}^{k-2} a_{i_j}\right)^{-1} + \sum_{3 \leqslant i_1 < \cdots < i_k \leqslant n} \left(\sum_{j=1}^{k} a_{i_j}\right)^{-1}
\end{aligned}
$$

可得

$$
\begin{aligned}
\frac{\partial h_k(\boldsymbol{a})}{\partial a_1} = &\sum_{1 \leqslant i_1 < \cdots < i_k \leqslant n} \left(\sum_{j=1}^{k} a_{i_j}\right)^{-1} - \\
&S\left[\sum_{3 \leqslant i_1 < \cdots < i_k \leqslant n} \left(a_1 + \sum_{j=1}^{k-1} a_{i_j}\right)^{-2} + \sum_{3 \leqslant i_1 < \cdots < i_k \leqslant n} \left(a_1 + a_2 + \sum_{j=1}^{k-2} a_{i_j}\right)^{-2}\right]
\end{aligned}
$$

$$
\begin{aligned}
\frac{\partial g_k(\boldsymbol{a})}{\partial a_2} = &\sum_{1 \leqslant i_1 < \cdots < i_k \leqslant n} \left(S - \sum_{j=1}^{k} a_{i_j}\right)^{-1} - \\
&S\left[\sum_{3 \leqslant i_1 < \cdots < i_k \leqslant n} \left(a_2 + \sum_{j=1}^{k-1} a_{i_j}\right)^{-2} + \sum_{3 \leqslant i_1 < \cdots < i_k \leqslant n} \left(a_1 + a_2 + \sum_{j=1}^{k} a_{i_j}\right)^{-2}\right]
\end{aligned}
$$

$$
\begin{aligned}
\Delta := &(a_1 - a_2)\left(\frac{\partial h_k(\boldsymbol{a})}{\partial a_1} - \frac{\partial h_k(\boldsymbol{a})}{\partial a_2}\right) \\
= &-S(a_1 - a_2) \sum_{3 \leqslant i_1 < \cdots < i_k \leqslant n} \left[\left(a_1 + \sum_{j=1}^{k-1} a_{i_j}\right)^{-2} - \left(a_2 + \sum_{j=1}^{k-1} a_{i_j}\right)^{-2}\right]
\end{aligned}
$$

$$= -S(a_1 - a_2) \frac{\left(a_2 + \sum\limits_{j=1}^{k-1} a_{i_j}\right)^2 - \left(a_1 + \sum\limits_{j=1}^{k-1} a_{i_j}\right)^2}{\left(a_1 + \sum\limits_{j=1}^{k-1} a_{i_j}\right)^2 \left(a_2 + \sum\limits_{j=1}^{k-1} a_{i_j}\right)^2}$$

$$= \frac{S(a_1 - a_2)^2 \left(a_1 + a_2 + 2\sum\limits_{j=1}^{k-1} a_{i_j}\right)}{\left(a_1 + \sum\limits_{j=1}^{k-1} a_{i_j}\right)^2 \left(a_2 + \sum\limits_{j=1}^{k-1} a_{i_j}\right)^2} \geqslant 0$$

由引理1 知$h_k(\boldsymbol{a})$ Schur凸, 进而$\overline{\varphi}_k(\boldsymbol{a}) = h_k(\boldsymbol{a}) - \mathrm{C}_n^k$ 亦Schur 凸.

定理2 的证明 因$\overline{\varphi}_k(\boldsymbol{a}) - \varphi_k(\boldsymbol{a}) = h_k(\boldsymbol{a}) - g_k(\boldsymbol{a})$, 所以

$$\Lambda := (a_1 - a_2) \left[\frac{\partial \overline{\varphi}_k(\boldsymbol{a}) - \varphi_k(\boldsymbol{a})}{\partial a_1} - \frac{\overline{\varphi}_k(\boldsymbol{a}) - \varphi_k(\boldsymbol{a})}{\partial a_2} \right]$$

$$= (a_1 - a_2) \left(\frac{\partial h_k(\boldsymbol{a})}{\partial a_1} - \frac{\partial h_k(\boldsymbol{a})}{\partial a_2} \right) - (a_1 - a_2) \left(\frac{\partial g_k(\boldsymbol{a})}{\partial a_1} - \frac{\partial g_k(\boldsymbol{a})}{\partial a_2} \right)$$

$$= \sum_{3 \leqslant i_1 < \cdots < i_k \leqslant n} \frac{S(a_1 - a_2)^2 (a_1 + a_2 + 2\sum\limits_{j=1}^{k-1} a_{i_j})}{(a_1 + \sum\limits_{j=1}^{k-1} a_{i_j})^2 (a_2 + \sum\limits_{j=1}^{k-1} a_{i_j})^2} -$$

$$\sum_{3 \leqslant i_1 < \cdots < i_k \leqslant n} \frac{S(a_1 - a_2)^2 [2S - (a_1 + a_2 + 2\sum\limits_{j=1}^{k-1} a_{i_j})]}{[S - (a_1 + \sum\limits_{j=1}^{k-1} a_{i_j})]^2 [S - (a_2 + \sum\limits_{j=1}^{k-1} a_{i_j})]^2}$$

$$= S(a_1 - a_2)^2 \sum_{3 \leqslant i_1 < \cdots < i_k \leqslant n} \left\{ \frac{1}{(a_1 + \sum\limits_{j=1}^{k-1} a_{i_j})^2 (a_2 + \sum\limits_{j=1}^{k-1} a_{i_j})} + \right.$$

$$\frac{1}{(a_1 + \sum\limits_{j=1}^{k-1} a_{i_j})(a_2 + \sum\limits_{j=1}^{k-1} a_{i_j})^2} -$$

$$\frac{1}{[S - (a_1 + \sum\limits_{j=1}^{k-1} a_{i_j})]^2 [S - (a_2 + \sum\limits_{j=1}^{k-1} a_{i_j})]} -$$

$$\left. \frac{1}{[S - (a_1 + \sum\limits_{j=1}^{k-1} a_{i_j})][S - (a_2 + \sum\limits_{j=1}^{k-1} a_{i_j})]^2} \right\} \tag{6}$$

(1) 当$k = \frac{n}{2}$ 时, 显然有$\overline{\varphi}_k(\boldsymbol{a}) = \varphi_k(\boldsymbol{a})$, 此时定理2 是平凡的.

(2) 当$k > \frac{n}{2}$时, 有$k > l$ 和$k - l > n - k - 1$. 对于(6)式中的数

组 $\{i_j \mid 3 \leqslant i_1 < i_2 < i_{k-1} \leqslant n\}$，设 I_{n-k-1} 是其中的任一含有 $n-k-1$ 个元素的子集. 因

$$\left(a_1 + \sum_{j=1}^{k-1} a_{i_j}\right)^2 > \left(a_1 + \sum_{t \in I_{n-k-1}} a_t\right)^2, \quad a_2 + \sum_{j=1}^{k-1} a_{i_j} > a_2 + \sum_{t \in I_{n-k-1}} a_t$$

所以

$$\frac{1}{\left(a_1 + \sum\limits_{j=1}^{k-1} a_{i_j}\right)^2 \left(a_2 + \sum\limits_{j=1}^{k-1} a_{i_j}\right)} < \frac{1}{\left(a_1 + \sum\limits_{t \in I_{n-k-1}} a_t\right)^2 \left(a_2 + \sum\limits_{t \in I_{n-k-1}} a_t\right)}$$

满足上式的子集 I_{n-k-1} 共有 C_{k-1}^{n-k-1} 个, 故有

$$\mathrm{C}_{k-1}^{n-k-1} \frac{1}{\left(a_1 + \sum\limits_{j=1}^{k-1} a_{i_j}\right)^2 \left(a_2 + \sum\limits_{j=1}^{k-1} a_{i_j}\right)}$$

$$< \sum_{I_{n-k-1} \subseteq \{i_1, i_2, \ldots, i_{k-1}\}} \frac{1}{\left(a_1 + \sum\limits_{t \in I_{n-k-1}} a_t\right)^2 \left(a_2 + \sum\limits_{t \in I_{n-k-1}} a_t\right)}$$

$$\mathrm{C}_{k-1}^{n-k-1} \sum_{3 \leqslant i_1 < \cdots < i_k \leqslant n} \frac{1}{\left(a_1 + \sum\limits_{j=1}^{k-1} a_{i_j}\right)^2 \left(a_2 + \sum\limits_{j=1}^{k-1} a_{i_j}\right)}$$

$$< \sum_{3 \leqslant i_1 < \cdots < i_k \leqslant n} \sum_{I_{n-k-1} \subseteq \{i_1, i_2, \ldots, i_{k-1}\}} \frac{1}{\left(a_1 + \sum\limits_{t \in I_{n-k-1}} a_t\right)^2 \left(a_2 + \sum\limits_{t \in I_{n-k-1}} a_t\right)}$$

注意 $S = a_1 + a_2 + \sum_{t=3}^n a_t$, 上式可化为

$$\mathrm{C}_{k-1}^{n-k-1} \sum_{3 \leqslant i_1 < \cdots < i_k \leqslant n} \frac{1}{\left(a_1 + \sum\limits_{j=1}^{k-1} a_{i_j}\right)^2 \left(a_2 + \sum\limits_{j=1}^{k-1} a_{i_j}\right)}$$

$$< \sum_{3 \leqslant i_1 < \cdots < i_k \leqslant n} \sum_{I_{n-k-1} \subseteq \{i_1, i_2, \ldots, i_{k-1}\}}$$

$$\frac{1}{\left[S - \left(a_2 + \sum\limits_{t \geqslant 3, t \notin I_{n-k-1}} a_t\right)\right]^2 \left[S - \left(a_1 + \sum\limits_{t \geqslant 3, t \notin I_{n-k-1}} a_t\right)\right]} \tag{7}$$

此时, 上式右边和中的各项就是和式

$$\sum_{3 \leqslant i_1 < \cdots < i_k \leqslant n} \frac{1}{\left[S - (a_2 + \sum\limits_{j=1}^{k-1} a_{i_j})\right]^2 \left[S - (a_1 + \sum\limits_{j=1}^{k-1} a_{i_j})\right]}$$

中的各项, 且二式关于诸 a_i, 显然具有对称性. 再计算各单项的数目知

$$\frac{1}{\mathrm{C}_{k-1}^{n-k-1}} \frac{1}{\mathrm{C}_{n-2}^{k-1}} \sum_{3 \leqslant i_1 < \cdots < i_k \leqslant n} \sum_{I_{n-k-1} \subseteq \{i_1, i_2, \ldots, i_{k-1}\}}$$

$$\frac{1}{\left[S - (a_2 + \sum\limits_{t \notin I_{n-k-1}} a_t)\right]^2 \left[S - (a_1 + \sum\limits_{t \notin I_{n-k-1}} a_t)\right]}$$

$$= \frac{1}{\mathrm{C}_{n-2}^{k-1}} \sum_{3 \leqslant i_1 < \cdots < i_k \leqslant n} \frac{1}{\left[S - a_1 + \sum\limits_{j=1}^{k-1} a_{i_j}\right] \left[S - (a_2 + \sum\limits_{j=1}^{k-1} a_{i_j})\right]^2} \tag{8}$$

联立(7),(8)两式, 得

$$\sum_{3 \leqslant i_1 < \cdots < i_k \leqslant n} \frac{1}{\left(a_1 + \sum\limits_{j=1}^{k-1} a_{i_j}\right)^2 \left(a_2 + \sum\limits_{j=1}^{k-1} a_{i_j}\right)}$$

$$< \sum_{3 \leqslant i_1 < \cdots < i_k \leqslant n} \frac{1}{\left[S - a_1 + \sum\limits_{j=1}^{k-1} a_{i_j}\right] \left[S - (a_2 + \sum\limits_{j=1}^{k-1} a_{i_j})\right]^2} \tag{9}$$

同理可证

$$\sum_{3 \leqslant i_1 < \cdots < i_k \leqslant n} \frac{1}{\left(a_1 + \sum\limits_{j=1}^{k-1} a_{i_j}\right) \left(a_2 + \sum\limits_{j=1}^{k-1} a_{i_j}\right)^2}$$

$$< \sum_{3 \leqslant i_1 < \cdots < i_k \leqslant n} \frac{1}{\left[S - (a_1 + \sum\limits_{j=1}^{k-1} a_{i_j})\right]^2 \left[S - (a_2 + \sum\limits_{j=1}^{k-1} a_{i_j})\right]} \tag{10}$$

联立(6), (9), (10)三式即知 $\Lambda < 0$, 故此时 $\overline{\varphi}_k(\boldsymbol{a}) - \varphi_k(\boldsymbol{a})$ 为Schur 凹函数.

(3) 当 $k < \frac{n}{2}$ 时, 对于式(1)中的数组 $\{i_j \mid 3 \leqslant i_1 < i_2 < i_{k-1} \leqslant n\}$, 设

$$L_{n-k-1} = \{3, 4, \ldots, n\} - \{i_1, i_2, \ldots, i_{k-1}\}$$

则 L_{n-k-1} 中的元素个数为 $n - k - 1$, 且大于 $k - 1$. 对于的任一具有 $k - 1$ 元

素的子集J_{k-1}, 易见

$$\left(a_2 + \sum_{t\in I_{n-k-1}} a_t\right)^2 > \left(a_2 + \sum_{t\in J_{k-1}} a_t\right)^2$$

$$a_1 + \sum_{t\in I_{n-k-1}} a_t > a_1 + \sum_{t\in J_{k-1}} a_t$$

因而

$$\cfrac{1}{\left(a_2 + \sum\limits_{t\in I_{n-k-1}} a_t\right)^2 (a_1 + \sum\limits_{t\in I_{n-k-1}} a_t)}$$
$$< \cfrac{1}{\left(a_2 + \sum\limits_{t\in J_{k-1}} a_t\right)^2 (a_1 + \sum\limits_{t\in J_{k-1}} a_t)}$$

$$\cfrac{1}{\left[S-(a_1 + \sum\limits_{j=1}^{k-1} a_{i_j})\right]^2 \left[S-(a_2 + \sum\limits_{j=1}^{k-1} a_{i_j})\right]}$$
$$< \cfrac{1}{\left(a_2 + \sum\limits_{t\in J_{k-1}} a_t\right)^2 (a_1 + \sum\limits_{t\in J_{k-1}} a_t)}$$

由于这样的子集J_{k-1} 有C_{n-k-1}^{k-1}个, 故

$$\mathrm{C}_{n-k-1}^{k-1} \cfrac{1}{\left[S-(a_1 + \sum\limits_{j=1}^{k-1} a_{i_j})\right]^2 \left[S-(a_2 + \sum\limits_{j=1}^{k-1} a_{i_j})\right]}$$
$$< \sum_{J_{k-1}\subset I_{n-k-1}} \cfrac{1}{\left(a_2 + \sum\limits_{t\in J_{k-1}} a_t\right)^2 \left(a_1 + \sum\limits_{t\in J_{k-1}} a_t\right)}$$

$$\mathrm{C}_{n-k-1}^{k-1} \sum_{3\leqslant i_1<\cdots<i_k\leqslant n} \cfrac{1}{\left[S-(a_1 + \sum\limits_{j=1}^{k-1} a_{i_j})\right]^2 \left[S-(a_2 + \sum\limits_{j=1}^{k-1} a_{i_j})\right]}$$

$$< \sum_{3 \leqslant i_1 < \cdots < i_k \leqslant n} \sum_{J_{k-1} \subset I_{n-k-1}} \frac{1}{\left(a_2 + \sum\limits_{t \in J_{k-1}} a_t\right)^2 \left(a_1 + \sum\limits_{t \in J_{k-1}} a_t\right)} \tag{11}$$

式(11)右边和式中各项就是和式

$$\sum_{3 \leqslant i_1 < \cdots < i_k \leqslant n} \frac{1}{\left(a_1 + \sum\limits_{j=1}^{k-1} a_{i_j}\right)\left(a_2 + \sum\limits_{j=1}^{k-1} a_{i_j}\right)^2}$$

中的各项, 且二式关于诸a_i 显然具有对称性, 再计算各单项的数目, 得

$$\frac{1}{C_{n-2}^{k-1}} \sum_{3 \leqslant i_1 < \cdots < i_k \leqslant n} \frac{1}{\left(a_1 + \sum\limits_{j=1}^{k-1} a_{i_j}\right)\left(a_2 + \sum\limits_{j=1}^{k-1} a_{i_j}\right)^2}$$

$$= \frac{1}{C_{n-k-1}^{k-1}} \frac{1}{C_{n-2}^{k-1}} \sum_{3 \leqslant i_1 < \cdots < i_k \leqslant n} \sum_{J_{k-1} \subset I_{n-k-1}} \frac{1}{\left(a_2 + \sum\limits_{t \in J_{k-1}} a_t\right)^2 \left(a_1 + \sum\limits_{t \in J_{k-1}} a_t\right)} \tag{12}$$

联立(11), (12)两式即知

$$\sum_{3 \leqslant i_1 < \cdots < i_k \leqslant n} \frac{1}{\left(a_1 + \sum\limits_{j=1}^{k-1} a_{i_j}\right)\left(a_2 + \sum\limits_{j=1}^{k-1} a_{i_j}\right)^2}$$

$$> \sum_{3 \leqslant i_1 < \cdots < i_k \leqslant n} \frac{1}{\left[S - \left(a_1 + \sum\limits_{j=1}^{k-1} a_{i_j}\right)\right]^2 \left[S - \left(a_2 + \sum\limits_{j=1}^{k-1} a_{i_j}\right)\right]} \tag{13}$$

同理可证

$$\sum_{3 \leqslant i_1 < \cdots < i_k \leqslant n} \frac{1}{\left(a_2 + \sum_{j=1}^{k-1} a_{i_j}\right)\left(a_1 + \sum_{j=1}^{k-1} a_{i_j}\right)^2}$$

$$> \sum_{3 \leqslant i_1 < \cdots < i_k \leqslant n} \frac{1}{\left[S - \left(a_2 + \sum_{j=1}^{k-1} a_{i_j}\right)\right]^2 \left[S - \left(a_1 + \sum_{j=1}^{k-1} a_{i_j}\right)\right]} \tag{14}$$

至此联立(6), (13), (14)三式即知$\Lambda > 0$, 故此时$\overline{\varphi}_k(\boldsymbol{a}) = \varphi_k(\boldsymbol{a})$也为Schur凸函数.

参考文献

[1]　O BAGDASAR. The Extension of A Cyclic Inequalities to the Sym-

metric Form[E].　Volume 9(2008), Issue 1, Article 10, 10 pp.

[2]　O BAGDASAR. Inequalities for Chains of Normalized Symmetric Sums[E]. J. Inequal. Pure. Appl. Math, Volume 9(2008)，Issue 1, Article 24, 7 pp.　]

[3]　王伯英. 控制不等式基础[M]. 北京:北京师范大学出版社,1990.

[4]　MARSHALL A M, OLKIN I. Inequalities: Theory of Majorization and Its Application[M]. New York：Academies Press，1979.

第43篇　A Generalization of Qi's Inequality for Sums

(SHI HUAN-NAN. Kragujevac J. Math. 33 (2010) 101-106.)

Abstract: By a majorization method, a pair of inequalities for sums of nonnegative sequences are established, and so an open problem posed by F. Qi is resolved.

Keywords: inequalities for sums; Schur-convexity; majorization

43.1　Introduction

In [4], the following inequality between the sum of squares and the exponential of sum of a nonnegative sequence was obtained: For $(x_1, x_2, \ldots, x_n) \in \mathbf{R}_+^n$ and $n \geqslant 2$, the inequality

$$\frac{e^2}{4} \sum_{i=1}^{n} x_i^2 \leqslant \exp\left(\sum_{i=1}^{n} x_i\right) \tag{1}$$

is valid, where $\mathbf{R}_+^n = \{(x_1, \ldots, x_n) \in \mathbf{R}^n : x_i \geqslant 0, i = 1, \ldots, n\}$. The equality in (1) holds if $x_i = 2$ and $x_j = 0$ for some given $1 \leqslant i \leqslant n$ and all $1 \leqslant j \leqslant n$ with $j \neq i$. The constant $\frac{e^2}{4}$ in the inequality(1) is the best possible.

The first open problem in [4] may be quoted as follows: For $(x_1, x_2, \ldots, x_n) \in \mathbf{R}_+^n$ and $n \geqslant 2$, determine the best possible constants $\alpha_n, \lambda_n \in \mathbf{R}$ and $0 < \beta_n, \mu_n < \infty$ such that

$$\beta_n \sum_{i=1}^{n} x_i^{\alpha_n} \leqslant \exp\left(\sum_{i=1}^{n} x_i\right) \leqslant \mu_n \sum_{i=1}^{n} x_i^{\lambda_n} \tag{2}$$

First of all, we claim that the right-hand side inequality in (2) is generally untenable. In fact, when $n = 2$, the right-hand side inequality in (2) becomes

$$e^{x_1+x_2} \leqslant \mu_2 \left(x_1^{\lambda_2} + x_2^{\lambda_2}\right) \tag{3}$$

Further taking $x_2 = 0$ in the inequality (3) reduces

$$\frac{e^{x_1}}{x_1^\lambda} \leqslant \mu \tag{4}$$

For any given $\lambda > 0$, the function $\frac{e^{x_1}}{x_1^\lambda}$ tends to ∞ as $x_1 \to \infty$. Hence, the inequality (4) does not hold if x_1 is large enough.

In this short note, by using a method in the theory of majorization, we give an affirmative solution to the left-hand side inequality in (2), which is also a generalization of the inequality (1), as follows.

Theorem 1 Let $(x_1, x_2, \ldots, x_n) \in \mathbf{R}_+^n$ and $n \geqslant 2$. If $\alpha \geqslant 1$, then the inequality

$$\frac{e^\alpha}{\alpha^\alpha} \left(\sum_{i=1}^n x_i^\alpha\right) \leqslant \exp\left(\sum_{i=1}^n x_i\right) \tag{5}$$

is valid. The equality in (5) holds if and only if $x_i = \alpha$ and $x_j = 0$ for some given $1 \leqslant i \leqslant n$ and all $1 \leqslant j \leqslant n$ with $j \neq i$.

Theorem 2 Let $\{x_i\}_{i=1}^{+\infty}$ be a nonnegative sequence such that $\sum_{i=1}^{+\infty} x_i < +\infty$. For $\alpha > 1$, the inequality

$$\frac{e^\alpha}{\alpha^\alpha} \left(\sum_{i=1}^{+\infty} x_i^\alpha\right) \leqslant \exp\left(\sum_{i=1}^{+\infty} x_i\right) \tag{6}$$

is valid.

43.2　Definitions and Lemmas

In order to prove our theorems, the following definitions and lemmas are needed.

Definition 1[1,6]

1. The sequence $\boldsymbol{x}$ is said to be majorized by $\boldsymbol{y}$ (in symbols $\boldsymbol{x} \preceq \boldsymbol{y}$) if $\sum_{i=1}^k x_{[i]} \leqslant \sum_{i=1}^k y_{[i]}$, for $k = 1, 2, \ldots, n-1$ and $\sum_{i=1}^n x_i = \sum_{i=1}^n y_i$, where $x_{[1]}, \ldots, x_{[n]}$ and $y_{[1]}, \ldots, y_{[[n]}$ are rearrangements of $\boldsymbol{x}$ and $\boldsymbol{y}$ in a descending order, and $\boldsymbol{x}$ is said to strictly majorized by $\boldsymbol{y}$ (in symbols $\boldsymbol{x} \prec \boldsymbol{y}$) if $\boldsymbol{x}$ is not a permutation of $\boldsymbol{y}$.

2. A function $f : \Omega \to \mathbf{R}$ is said to be a strictly Schur-convex on $\Omega \subset$

R if the relation $x \prec y$ on Ω implies $f(x) < f(y)$. A function f is said to be strictly Schur-concave on Ω if and only if $-f$ is strictly Schur-convex on Ω.

Definition 2[6] Let set $\Omega \subset \mathbf{R}^n$. Ω is said to be a convex set if $x, y \in \Omega, 0 \leqslant \alpha \leqslant 1$ implies $\alpha x + (1-\alpha)y = (\alpha x_1 + (1-\alpha)y_1, \ldots, \alpha x_n + (1-\alpha)y_n) \in \Omega$

Lemma 1[6] Let $\Omega \subset \mathbf{R}^n$ be symmetric and have a nonempty interior convex set Ω^0, and let $f : \Omega \to \mathbf{R}$ be continuous on Ω and differentiable on Ω^0. Then the function f is strictly Schur-convex (or Schur-concave respectively) on Ω if and only if f is symmetric on Ω and satisfies

$$(x_1 - x_2) \left(\frac{\partial f}{\partial x_1} - \frac{\partial f}{\partial x_2} \right) > 0 (\text{or} < 0; \text{respectively}) \tag{7}$$

for $x = (x_1, \ldots x_n) \in \Omega^0$ with $x_1 \neq x_2$.

Lemma 2 For any given positive real number s and α, we have

$$\frac{e^\alpha}{\alpha^\alpha} \leqslant \frac{e^s}{s^\alpha} \tag{8}$$

The equality in (8) holds if and only if $s = \alpha$.

proof Let $\varphi(s) = \alpha \ln s - s$. Then $\varphi'(s) = \frac{\alpha}{s} - 1 < 0$ for $s \geqslant \alpha > 0$, which means that $\varphi(s)$ is increasing, and $\varphi'(s) \geqslant 0$ for $0 < s \leqslant \alpha$ which means that $\varphi(s)$ is decreasing. Hence, for any $s > 0$, we have

$$\varphi(s) = \alpha \ln s - s \leqslant \varphi(\alpha) = \alpha \ln \alpha - \alpha$$

i.e., the inequality (8) is valid and the equality in (8) holds if and only if $s = \alpha$.

43.3　Proofs of Theorems

Proofs of Theorem 1　Let

$$f(x) = f(x_1, \ldots, x_n) = \ln \left(\sum_{i=1}^n x_i^\alpha \right) - s \tag{9}$$

where $s = \sum_{i=1}^n x_i$. Simple calculation gives

$$(x_1 - x_2) \left(\frac{\partial f}{\partial x_1} - \frac{\partial f}{\partial x_2} \right) = \frac{\alpha(x_1 - x_2)(x_1^{\alpha-1} - x_2^{\alpha-1})}{\sum\limits_{i=1}^n x_i^\alpha}$$

When $\alpha > 1$, since $x^{\alpha-1}$ is strictly increasing on $(0,+\infty)$, it follows easily that $(x_1 - x_2)(x_1^{\alpha-1} - x_2^{\alpha-1}) > 0$ for $x_1 \neq x_2$, and then $\Delta > 0$, so, by Lemma 1, $f(\boldsymbol{x})$ is strictly Schur-convex on $\mathbf{R}_+^n$. It is easy to see that

$$\boldsymbol{x} = (x_1,\ldots,x_n) \preceq \left(s, \underbrace{0,\ldots,0}_{n-1}\right) = \boldsymbol{y} \tag{10}$$

and $\boldsymbol{x} \prec \boldsymbol{y}$ unless $x_i = s$ and $x_j = 0$ for some given $1 \leqslant i \leqslant n$ and all $1 \leqslant j \leqslant n$ with $j \neq i$. Hence

$$f(x_1,\ldots,x_n) = \ln\left(\sum_{i=1}^n x_i^\alpha\right) - s \leqslant f\left(s, \underbrace{0,\ldots,0}_{n-1}\right) = \alpha \ln s - s \tag{11}$$

that is

$$\frac{e^s}{\alpha^s}\left(\sum_{i=1}^n x_i^\alpha\right) \leq \exp\left(\sum_{i=1}^n x_i\right) \tag{12}$$

is valid. The equality in (12) holds if and only if $x_i = s$ and $x_j = 0$ for some given $1 \leqslant i \leqslant n$ and all $1 \leqslant j \leqslant n$ with $j \neq i$. Combining the inequality (12) with the inequality (8) yields that the inequality (5) is valid and the equality in (5) holds if and only if $x_i = \alpha$ and $x_j = 0$ for some given $1 \leqslant i \leqslant n$ and all $1 \leqslant j \leqslant n$ with $j \neq i$. The proof of Theorem 1 is complete.

Proofs of Theorem 2　Letting $n \to +\infty$ in Theorem 1 yields Theorem 2 readily.

Remark After the preprint [5] of this paper was announced, there have been several papers such as [2-3] dedicated to discuss the open problems posed by F. Qi in [4].

References

[1]　A M MARSHALL, I OLKIN. Inequalities: Theory of Majorization and its Appli- cation[M]. New York: Academic Press, 1979.

[2]　Y MIAO, L-M LIU, F QI. Refinements of inequalities between the sum of squares and the exponential of sum of a nonnegative sequence[J]. J. Inequal. Pure Appl. Math. 2008, 9 (2), Art. 53.

[3]　T F MÓRI. On an inequality of Feng Qi[J]. J. Inequal. Pure Appl. Math. 2008, 9 (3), Art. 87.

[4]　F QI. Inequalities between the sum of squares and the exponential of sum of a nonnegative sequence[J]. J. Inequal. Pure Appl. Math., 2007, 8 (3), Art. 78.

[5] H-N SHI. Solution of an open problem proposed by Feng Qi[J]. RGMIA Res. Rep. Coll., 2007, 10 (4), Art. 9.

[6] B-Y WANG. Foundations of Majorization Inequalities[M]. Beijing: Beijing Normal Univ. Press, 1990.

第44篇 一类控制不等式及其应用

(石焕南. 北京联合大学学报(自然科学版), 2010, 24 (1)：60-64)

摘　要： 本文对于满足 $x_1 \geqslant \cdots \geqslant x_n \geqslant 0$ 的实数 $x_1, \cdots, x_n$, 建立了几个与循环和有关的控制关系, 并结合初等对称函数的Schur凹性给出若干不等式链.

关键词： 控制；不等式；循环和；循环积；初等对称函数

44.1　引言

在控制不等式理论的研究中, 发现和建立向量间的新的控制关系是一项重要的内容. 因为控制关系深刻地描述了向量间的内在联系, 一个新的控制关系与适当的Schur凹函数或Schur凸函数的结合, 常常能繁衍出许多形形色色的有趣的不等式. 本文将建立几个与循环和有关的控制关系, 并给出若干应用. 为此先给出如下定义和引理.

在本文中, $\mathbf{R}^n$ 表示实数域上 n 维行向量的集合, 并记

$$\mathbf{R}_+^n = \{\boldsymbol{x} = (x_1, \ldots, x_n) \in \mathbf{R}^n | x_i \geqslant 0, i = 1, \ldots, n\}$$

$\boldsymbol{x}$ 的第 k 个初等对称函数为

$$E_k(\boldsymbol{x}) = E_k(x_1, \ldots, x_n) = \sum_{1 \leqslant i_1 < \ldots < i_k \leqslant n} \prod_{j=1}^{k} x_{i_j}$$

特别 $E_n(\boldsymbol{x}) = \prod_{i=1}^{n} x_i$, $E_1(\boldsymbol{x}) = \sum_{i=1}^{n} x_i$. 当 $k > n$ 时, 规定 $E_n(\boldsymbol{x}) = 0$, 并规定 $E_0(\boldsymbol{x}) = 1$.

$G(\boldsymbol{x}) = \sqrt[n]{\prod_{i=1}^{n} x_i}$ 和 $A(\boldsymbol{x}) = \frac{1}{n} \sum_{i=1}^{n} x_i$ 分别为 $\boldsymbol{x}$ 的算术平均和几何平均.

对于 $\boldsymbol{x} = (x_1, ..., x_n) \in \mathbf{R}^n$, 将 $\boldsymbol{x}$ 的分量排成递减的次序后, 记作 $x_{[1]} \geqslant \ldots \geqslant x_{[n]}$.

定义 1[1]　设 x, y 满足

(1) $\sum_{i=1}^{k} x_{[i]} \leqslant \sum_{i=1}^{k} y_{[i]}, k = 1, 2, ..., n-1$; (2) $\sum_{i=1}^{n} x_i = \sum_{i=1}^{n} y_i$, 则称 x 被 y 所控制, 记作 $x \prec y$.

定义 2[1]　设 $\Omega \subset \mathbf{R}^n, \phi : \Omega \to \mathbf{R}$, 若在 Ω 上 $x \prec y \Rightarrow \phi(x) \leqslant \phi(y)$, 则称 ϕ 为 Ω 上的 Schur 凸函数.

引理 1[1]　设 $x = (x_1, ..., x_n) \in \mathbf{R}^n$, 则

$$(A(x), A(x), \ldots, A(x)) \prec (x_1, x_2, \ldots, x_n)$$

引理 2[1]　设 $\Omega \subset \mathbf{R}$ 为一个区间, $g : I \to \mathbf{R}$ 为 Ω 上的凹(凸)函数, 则

$$\varphi(x) = \sum_{i=1}^{n} g(x_i)$$

为 Ω^n 上的 Schur 凹(Schur 凸)函数.

引理 3[2]　对于 $1 \leqslant k \leqslant n$, $E_k(x)^{1/k}$ 是 $\mathbf{R}_+^n$ 上的 Schur 凹函数.

引理 4[1]　设 $x, y \in \mathbf{R}^n$, 则存在双随机矩阵 $P = (p_{ij})$ (即 $p_{ij} \geqslant 0$ 且 $\sum_j p_{ij} = 1, \forall j, \sum_j p_{ij} = 1, \forall i$) 使得 $x = yP$.

44.2　主要结果及其证明

定理 1　设 $x_1 \geqslant x_2 \geqslant x_3 \geqslant x_4 \geqslant 0$, 则

$$
\begin{aligned}
u &:= \left(\frac{x_1 + x_2 + x_3}{3}, \frac{x_2 + x_3 + x_4}{3}, \frac{x_3 + x_4 + x_1}{3}, \frac{x_4 + x_1 + x_2}{3} \right) \\
&\prec v := \left(\frac{x_1 + x_2}{2}, \frac{x_2 + x_3}{2}, \frac{x_3 + x_4}{2}, \frac{x_4 + x_1}{2} \right)
\end{aligned}
\tag{1}
$$

证明　我们根据控制的定义证明. 因 $x_1 \geqslant x_2 \geqslant x_3 \geqslant x_4 \geqslant 0$, 将 u 的分量按从大到小的顺序排列有

$$\frac{x_1 + x_2 + x_3}{3} \geqslant \frac{x_4 + x_1 + x_2}{3} \geqslant \frac{x_3 + x_4 + x_1}{3} \geqslant \frac{x_2 + x_3 + x_4}{3}$$

而 v 的分量的排列需分两种情况讨论: 由于

情况 1. 若 $x_4 + x_1 \leqslant x_2 + x_3$, 则有

$$\frac{x_1 + x_2}{2} \geqslant \frac{x_2 + x_3}{2} \geqslant \frac{x_4 + x_1}{2} \geqslant \frac{x_3 + x_4}{2}$$

由于

$$\frac{x_1 + x_2 + x_3}{3} \leqslant \frac{x_1 + x_2}{2} \Leftrightarrow 2x_3 \leqslant x_1 + x_2 \tag{2}$$

$$\frac{x_1 + x_2 + x_3}{3} + \frac{x_4 + x_1 + x_2}{3} \lessgtr \frac{x_1 + x_2}{2} + \frac{x_2 + x_3}{2} \Leftrightarrow x_1 + 2x_4 \leqslant 2x_2 + x_3 \tag{3}$$

$$\frac{x_1 + x_2 + x_3}{3} + \frac{x_4 + x_1 + x_2}{3} + \frac{x_3 + x_4 + x_1}{3}$$

$$\leqslant \frac{x_1 + x_2}{2} + \frac{x_2 + x_3}{2} + \frac{x_4 + x_1}{2} \Leftrightarrow x_3 + x_4 \leqslant 2x_2 \tag{4}$$

以及

$$\frac{x_1 + x_2 + x_3}{3} + \frac{x_4 + x_1 + x_2}{3} + \frac{x_3 + x_4 + x_1}{3} + \frac{x_2 + x_3 + x_4}{3}$$

$$= \frac{x_1 + x_2}{2} + \frac{x_2 + x_3}{2} + \frac{x_4 + x_1}{2} + \frac{x_3 + x_4}{2} = x_1 + x_2 + x_3 + x_4 \tag{5}$$

故(1)成立.

情况 2. 若 $x_2 + x_3 \leqslant x_4 + x_1$, 则有

$$\frac{x_1 + x_2}{2} \geqslant \frac{x_4 + x_1}{2} \geqslant \frac{x_2 + x_3}{2} \geqslant \frac{x_3 + x_4}{2}$$

此时, (2)、(4)和(5)三式依然成立. 又

$$\frac{x_1 + x_2 + x_3}{3} + \frac{x_4 + x_1 + x_2}{3} \lessgtr \frac{x_1 + x_2}{2} + \frac{x_4 + x_1}{2} \Leftrightarrow x_2 + 2x_3 \leqslant 2x_1 + x_4$$

故(1)成立. 定理1 证毕.

定理 2 设 $x_1 \geqslant x_2 \geqslant x_3 \geqslant x_4 \geqslant x_5 \geqslant 0$, 则

$$\boldsymbol{s} := \left(\frac{x_1 + x_2 + x_3}{3}, \frac{x_2 + x_3 + x_4}{3}, \frac{x_3 + x_4 + x_5}{3}, \frac{x_4 + x_5 + x_1}{3}, \frac{x_5 + x_1 + x_2}{3} \right)$$

$$\prec \boldsymbol{t} := \left(\frac{x_1 + x_2}{2}, \frac{x_2 + x_3}{2}, \frac{x_3 + x_4}{2}, \frac{x_4 + x_5}{2}, \frac{x_5 + x_1}{2} \right) \tag{6}$$

证明 因 $x_1 \geqslant x_2 \geqslant x_3 \geqslant x_4 \geqslant 0$, 有

$$x_1 + x_2 \geqslant x_2 + x_3 \geqslant x_3 + x_4 \geqslant x_4 + x_5$$

将 $\boldsymbol{s}, \boldsymbol{t}$ 的分量按从大到小的顺序排列, 需分3 种情况讨论:

情况 1. 若 $x_5 + x_1 \geqslant x_2 + x_3$, 则有

$$\frac{x_1 + x_2}{2} \geqslant \frac{x_5 + x_1}{2} \geqslant \frac{x_2 + x_3}{2} \geqslant \frac{x_3 + x_4}{2} \geqslant \frac{x_4 + x_5}{2}$$

和

$$\frac{x_1 + x_2 + x_3}{3} \geqslant \frac{x_5 + x_1 + x_2}{3} \geqslant \frac{x_4 + x_5 + x_1}{3} \geqslant \frac{x_2 + x_3 + x_4}{3} \geqslant \frac{x_3 + x_4 + x_5}{3}$$

此时(2)成立. 又

$$\frac{x_1+x_2+x_3}{3}+\frac{x_5+x_1+x_2}{3}\leqslant\frac{x_1+x_2}{2}+\frac{x_5+x_1}{2}\Leftrightarrow x_2+2x_3\leqslant 2x_1+x_5$$

$$\frac{x_1+x_2+x_3}{3}+\frac{x_5+x_1+x_2}{3}+\frac{x_4+x_5+x_1}{3}$$
$$\leqslant\frac{x_1+x_2}{2}+\frac{x_5+x_1}{2}+\frac{x_2+x_3}{2}\Leftrightarrow x_5+2x_4\leqslant 2x_2+x_3$$

以及

$$\frac{x_1+x_2+x_3}{3}+\frac{x_5+x_1+x_2}{3}+\frac{x_4+x_5+x_1}{3}+\frac{x_2+x_3+x_4}{3}+\frac{x_3+x_4+x_5}{3}$$
$$=\frac{x_1+x_2}{2}+\frac{x_5+x_1}{2}+\frac{x_2+x_3}{2}+\frac{x_3+x_4}{2}+\frac{x_4+x_5}{2}=x_1+x_2+x_3+x_4+x_5 \tag{7}$$

故(6)成立.

情况 2. 若$x_2+x_3\leqslant x_5+x_1\leqslant x_3+x_4$, 则有

$$\frac{x_1+x_2}{2}\geqslant\frac{x_2+x_3}{2}\geqslant\frac{x_5+x_1}{2}\geqslant\frac{x_3+x_4}{2}\geqslant\frac{x_4+x_5}{2}$$

和

$$\frac{x_1+x_2+x_3}{3}\geqslant\frac{x_5+x_1+x_2}{3}\geqslant\frac{x_2+x_3+x_4}{3}\geqslant\frac{x_4+x_5+x_1}{3}\geqslant\frac{x_3+x_4+x_5}{3}$$

此时, (2)和(7)依然成立. 又

$$\frac{x_1+x_2+x_3}{3}+\frac{x_5+x_1+x_2}{3}\leqslant\frac{x_1+x_2}{2}+\frac{x_2+x_3}{2}\Leftrightarrow x_1+2x_5\leqslant 2x_2+x_3$$

$$\frac{x_1+x_2+x_3}{3}+\frac{x_5+x_1+x_2}{3}+\frac{x_2+x_3+x_4}{3}$$
$$\leqslant\frac{x_1+x_2}{2}+\frac{x_2+x_3}{2}+\frac{x_5+x_1}{2}\Leftrightarrow x_3+2x_4\leqslant 2x_1+x_5$$

以及

$$\frac{x_1+x_2+x_3}{3}+\frac{x_5+x_1+x_2}{3}+\frac{x_2+x_3+x_4}{3}+\frac{x_4+x_5+x_1}{3}$$
$$\leqslant\frac{x_1+x_2}{2}+\frac{x_2+x_3}{2}+\frac{x_5+x_1}{2}+\frac{x_3+x_4}{2}\Leftrightarrow x_3+x_5\leqslant 2x_3$$

情况 3. 若 $x_5 + x_1 \leqslant x_3 + x_4$, 则有

$$\frac{x_1 + x_2}{2} \geqslant \frac{x_2 + x_3}{2} \geqslant \frac{x_3 + x_4}{2} \geqslant \frac{x_5 + x_1}{2} \geqslant \frac{x_4 + x_5}{2}$$

和

$$\frac{x_1 + x_2 + x_3}{3} \geqslant \frac{x_2 + x_3 + x_4}{3} \geqslant \frac{x_5 + x_1 + x_2}{3} \geqslant \frac{x_4 + x_5 + x_1}{3} \geqslant \frac{x_3 + x_4 + x_5}{3}$$

此时, (2)和(7)两式依然成立. 又

$$\frac{x_1 + x_2 + x_3}{3} + \frac{x_2 + x_3 + x_4}{3} \leqslant \frac{x_1 + x_2}{2} + \frac{x_2 + x_3}{2} \Leftrightarrow x_3 + 2x_4 \leqslant x_1 + 2x_2$$

$$\frac{x_1 + x_2 + x_3}{3} + \frac{x_2 + x_3 + x_4}{3} + \frac{x_5 + x_1 + x_2}{3}$$
$$\leqslant \frac{x_1 + x_2}{2} + \frac{x_2 + x_3}{2} + \frac{x_3 + x_4}{2} \Leftrightarrow x_1 + 2x_5 \leqslant 2x_3 + x_4$$

以及

$$\frac{x_1 + x_2 + x_3}{3} + \frac{x_2 + x_3 + x_4}{3} + \frac{x_5 + x_1 + x_2}{3} + \frac{x_4 + x_5 + x_1}{3}$$
$$\leqslant \frac{x_1 + x_2}{2} + \frac{x_2 + x_3}{2} + \frac{x_3 + x_4}{2} + \frac{x_5 + x_1}{2} \Leftrightarrow x_4 + x_5 \leqslant 2x_3$$

故式(6)成立. 定理2 证毕.

定理 3　设 $x_1 \geqslant x_2 \geqslant x_3 \geqslant x_4 \geqslant x_5 \geqslant 0$, 则

$$\boldsymbol{p} := \left(\frac{x_1 + x_2 + x_3 + x_4}{4}, \frac{x_2 + x_3 + x_4 + x_5}{4}, \frac{x_3 + x_4 + x_5 + x_1}{4}, \right.$$
$$\left. \frac{x_4 + x_5 + x_1 + x_2}{4}, \frac{x_5 + x_1 + x_2 + x_3}{4} \right)$$
$$\prec \boldsymbol{q} := \left(\frac{x_1 + x_2 + x_3}{3}, \frac{x_2 + x_3 + x_4}{3}, \frac{x_3 + x_4 + x_5}{3}, \frac{x_4 + x_5 + x_1}{3}, \frac{x_5 + x_1 + x_2}{3} \right)$$

$$(8)$$

证明　因 $x_1 \geqslant x_2 \geqslant x_3 \geqslant x_4 \geqslant x_5 \geqslant 0$, 将 $\boldsymbol{p}$ 的分量按从大到小的顺序排列有

$$\frac{x_1 + x_2 + x_3 + x_4}{4} \geqslant \frac{x_5 + x_1 + x_2 + x_3}{4} \geqslant \frac{x_4 + x_5 + x_1 + x_2}{4}$$
$$\geqslant \frac{x_3 + x_4 + x_5 + x_1}{4} \geqslant \frac{x_2 + x_3 + x_4 + x_5}{4}$$

将 $\boldsymbol{q}$ 的分量按从大到小的顺序排列, 需分3 种情况讨论:

情况 1. 若 $x_5 + x_1 \geqslant x_2 + x_3$, 则有

$$\frac{x_1 + x_2 + x_3}{3} \geqslant \frac{x_5 + x_1 + x_2}{3} \geqslant \frac{x_4 + x_5 + x_1}{3} \geqslant \frac{x_2 + x_3 + x_4}{3} \geqslant \frac{x_3 + x_4 + x_5}{3}$$

由于

$$\frac{x_1 + x_2 + x_3 + x_4}{4} \leqslant \frac{x_1 + x_2 + x_3}{3} \Leftrightarrow 3x_4 \leqslant x_1 + x_2 + x_3 \tag{9}$$

又

$$\frac{x_1 + x_2 + x_3 + x_4}{4} + \frac{x_5 + x_1 + x_2 + x_3}{4}$$

$$\leqslant \frac{x_1 + x_2 + x_3}{3} + \frac{x_5 + x_1 + x_2}{3} \Leftrightarrow 2x_3 + 3x_4 \leqslant 2x_1 + 2x_2 + x_5 \tag{10}$$

$$\frac{x_1 + x_2 + x_3 + x_4}{4} + \frac{x_5 + x_1 + x_2 + x_3}{4} + \frac{x_4 + x_5 + x_1 + x_2}{4}$$

$$\leqslant \frac{x_1 + x_2 + x_3}{3} + \frac{x_5 + x_1 + x_2}{3} + \frac{x_4 + x_5 + x_1}{3} \Leftrightarrow x_2 + 3x_3 + 2x_4 \leqslant 3x_1 + 2x_5 \tag{11}$$

$$\frac{x_1 + x_2 + x_3 + x_4}{4} + \frac{x_5 + x_1 + x_2 + x_3}{4} + \frac{x_4 + x_5 + x_1 + x_2}{4} + \frac{x_3 + x_4 + x_5 + x_1}{4}$$

$$\leqslant \frac{x_1 + x_2 + x_3}{3} + \frac{x_5 + x_1 + x_2}{3} + \frac{x_4 + x_5 + x_1}{3} + \frac{x_2 + x_3 + x_4}{3} \Leftrightarrow x_3 + x_4 + x_5 \leqslant 3x_2 \tag{12}$$

以及

$$\frac{x_1 + x_2 + x_3 + x_4}{4} + \frac{x_5 + x_1 + x_2 + x_3}{4} + \frac{x_4 + x_5 + x_1 + x_2}{4}$$

$$+ \frac{x_3 + x_4 + x_5 + x_1}{4} + \frac{x_2 + x_3 + x_4 + x_5}{4}$$

$$= \frac{x_1 + x_2 + x_3}{3} + \frac{x_5 + x_1 + x_2}{3} + \frac{x_4 + x_5 + x_1}{3} + \frac{x_2 + x_3 + x_4}{3} + \frac{x_3 + x_4 + x_5}{3}$$

$$= x_1 + x_2 + x_3 + x_4 + x_5 \tag{13}$$

故式(8)成立.

情况 2. 若 $x_2 + x_3 \leqslant x_5 + x_1 \leqslant x_3 + x_4$, 则有

$$\frac{x_1 + x_2 + x_3}{3} \geqslant \frac{x_5 + x_1 + x_2}{3} \geqslant \frac{x_2 + x_3 + x_4}{3} \geqslant \frac{x_4 + x_5 + x_1}{3} \geqslant \frac{x_3 + x_4 + x_5}{3}$$

此时(9),(13)两式依然成立

$$\frac{x_1 + x_2 + x_3 + x_4}{4} + \frac{x_5 + x_1 + x_2 + x_3}{4}$$

$$\leqslant \frac{x_1 + x_2 + x_3}{3} + \frac{x_5 + x_1 + x_2}{3} \Leftrightarrow 2x_3 + 3x_4 \leqslant 2x_1 + 2x_2 + x_5$$

$$\frac{x_1 + x_2 + x_3 + x_4}{4} + \frac{x_5 + x_1 + x_2 + x_3}{4} + \frac{x_4 + x_5 + x_1 + x_2}{4}$$
$$\leqslant \frac{x_1 + x_2 + x_3}{3} + \frac{x_5 + x_1 + x_2}{3} + \frac{x_2 + x_3 + x_4}{3} \Leftrightarrow x_1 + 2x_4 + 2x_5 \leqslant 3x_2 + 2x_3$$

$$\frac{x_1 + x_2 + x_3 + x_4}{4} + \frac{x_5 + x_1 + x_2 + x_3}{4} + \frac{x_4 + x_5 + x_1 + x_2}{4} + \frac{x_3 + x_4 + x_5 + x_1}{4}$$
$$\leqslant \frac{x_1 + x_2 + x_3}{3} + \frac{x_5 + x_1 + x_2}{3} + \frac{x_2 + x_3 + x_4}{3} + \frac{x_4 + x_5 + x_1}{3}$$
$$\Leftrightarrow x_3 + x_4 + x_5 \leqslant 3x_2$$

故式(8)成立.

情况 3. 若 $x_5 + x_1 \leqslant x_3 + x_4$, 则有

$$\frac{x_1 + x_2 + x_3}{3} \geqslant \frac{x_2 + x_3 + x_4}{3} \geqslant \frac{x_5 + x_1 + x_2}{3} \geqslant \frac{x_4 + x_5 + x_1}{3} \geqslant \frac{x_3 + x_4 + x_5}{3}$$

此时, (9)和(13)两式依然成立.

$$\frac{x_1 + x_2 + x_3 + x_4}{4} + \frac{x_5 + x_1 + x_2 + x_3}{4}$$
$$\leqslant \frac{x_1 + x_2 + x_3}{3} + \frac{x_2 + x_3 + x_4}{3} \Leftrightarrow 2x_1 + 3x_5 \leqslant 2x_1 + 2x_3 + x_4$$

$$\frac{x_1 + x_2 + x_3 + x_4}{4} + \frac{x_5 + x_1 + x_2 + x_3}{4} + \frac{x_4 + x_5 + x_1 + x_2}{4}$$
$$\leqslant \frac{x_1 + x_2 + x_3}{3} + \frac{x_2 + x_3 + x_4}{3} + \frac{x_5 + x_1 + x_2}{3}$$
$$\Leftrightarrow x_1 + 2x_4 + 2x_5 \leqslant 3x_2 + 2x_3$$

$$\frac{x_1 + x_2 + x_3 + x_4}{4} + \frac{x_5 + x_1 + x_2 + x_3}{4} +$$
$$\frac{x_4 + x_5 + x_1 + x_2}{4} + \frac{x_3 + x_4 + x_5 + x_1}{4}$$
$$\leqslant \frac{x_1 + x_2 + x_3}{3} + \frac{x_2 + x_3 + x_4}{3} + \frac{x_5 + x_1 + x_2}{3} + \frac{x_4 + x_5 + x_1}{3}$$
$$\Leftrightarrow x_3 + x_4 + x_5 \leqslant 3x_2$$

故(8)成立. 定理3 证毕.

定理 4 设 $\boldsymbol{x} = (x_1, \cdots, x_n) \in \mathbf{R}^n$, 则

$$\left(\frac{x_1 + x_2}{2}, \frac{x_2 + x_3}{2}, \ldots, \frac{x_{n-1} + x_n}{2}, \frac{x_n + x_1}{2} \right) \prec (x_1, x_2, \cdots, x_n) \qquad (14)$$

证明

$$\left(\frac{x_1 + x_2}{2}, \frac{x_2 + x_3}{2}, \cdots, \frac{x_{n-1} + x_n}{2}, \frac{x_n + x_1}{2}\right) = (x_1, x_2, \cdots, x_n)\, \boldsymbol{P}$$

其中

$$\boldsymbol{P} = \begin{pmatrix} 1/2 & 0 & \cdots & 0 & 1/2 \\ 1/2 & 1/2 & 0 & \cdots & 0 \\ 0 & 1/2 & 1/2 & \ddots & \vdots \\ \vdots & \ddots & \ddots & \ddots & 0 \\ 0 & \cdots & 0 & 1/2 & 1/2 \end{pmatrix}$$

是双随机矩阵, 由引理5 知式(14)成立.

44.3 应用

定理 5　设$x_1 \geqslant x_2 \geqslant x_3 \geqslant x_4 \geqslant 0$, 对于定理1 的$\boldsymbol{u}, \boldsymbol{v}$ 有

$$(E_k(\boldsymbol{x}))^{1/k} \leqslant (E_k(\boldsymbol{v}))^{1/k} \leqslant (E_k(\boldsymbol{u}))^{1/k} \leqslant (\mathrm{C}_n^k)^{1/k} A(\boldsymbol{x}), \ 1 \leqslant k \leqslant 4 \quad (15)$$

特别当$k = 4$ 时, 有

$$G(\boldsymbol{x}) \leqslant \frac{\sqrt[4]{\prod(x_1 + x_2)}}{2} \leqslant \frac{\sqrt[4]{\prod(x_1 + x_2 + x_3)}}{3} \leqslant A(\boldsymbol{x}) \quad (16)$$

对于$0 \leqslant \beta \leqslant 1$, 有

$$\sum x_1^\beta \leqslant \sum \left(\frac{x_1 + x_2}{2}\right)^\beta \leqslant \sum \left(\frac{x_1 + x_2 + x_3}{3}\right)^\beta \leqslant 4 A^\beta(\boldsymbol{x}) \quad (17)$$

若$\beta > 1$, 式(17)中不等式均反向. 其中$\prod, \sum$ 分别表示循环积与循环和.

　　证明　由定理4、定理1 和引理1 并结合引理3 即可证得式(15). 当$0 \leqslant \beta \leqslant 1(\beta > 1)$ 时, t^β 是$\mathbf{R}_+$ 上的凹(凸)函数, 由定理4、定理1 和引理1 并结合引理2 即可证得式(17).

　　定理 6　设$x_1 \geqslant x_2 \geqslant x_3 \geqslant x_4 \geqslant x_5 \geqslant 0$, 对于定理2 和定理3 的$\boldsymbol{s}, \boldsymbol{t}, \boldsymbol{p}$, 有

$$(E_k(\boldsymbol{x}))^{1/k} \leqslant (E_k(\boldsymbol{t}))^{1/k} \leqslant (E_k(\boldsymbol{s}))^{1/k} \leqslant (E_k(\boldsymbol{p}))^{1/k} \leqslant (\mathrm{C}_n^k)^{1/k} A(\boldsymbol{x}), \ 1 \leqslant k \leqslant 5 \quad (18)$$

特别当 $k = 5$ 时, 有

$$G(\boldsymbol{x}) \leqslant \frac{\sqrt[5]{\prod(x_1 + x_2)}}{2} \leqslant \frac{\sqrt[5]{\prod(x_1 + x_2 + x_3)}}{3}$$

$$\leqslant \frac{\sqrt[5]{\prod(x_1 + x_2 + x_3 + x_4)}}{4} \leqslant A(\boldsymbol{x}) \tag{19}$$

对于 $0 \leqslant \beta \leqslant 1$, 有

$$\sum x_1^\beta \leqslant \sum \left(\frac{x_1 + x_2}{2}\right)^\beta \leqslant \sum \left(\frac{x_1 + x_2 + x_3}{3}\right)^\beta$$

$$\leqslant \sum \left(\frac{x_1 + x_2 + x_3 + x_4}{4}\right)^\beta \leqslant 5 A^\beta(\boldsymbol{x}) \tag{20}$$

若 $\beta > 1$, 式(20)中不等式均反向. 其中 $\prod, \sum$ 分别表示循环积与循环和.

　　证明　由定理4、定理2、定理3 和引理1 并结合引理3 即可证得式(18). 当 $0 \leqslant \beta \leqslant 1 (\beta > 1)$ 时, t^β 是 $\mathbf{R}_+$ 上的凹(凸)函数, 由定理4、定理2、定理3 和引理1 并结合引理2 即可证得式(20).

　　最后提出一个猜想.

　　猜想　设 $x_1 \geqslant x_2 \geqslant x_3 \geqslant x_4 \geqslant x_5 \geqslant x_6 \geqslant 0$, 则

$$\left(\frac{x_1 + x_2 + x_3}{3}, \frac{x_2 + x_3 + x_4}{3}, \frac{x_3 + x_4 + x_5}{3}, \right.$$

$$\left. \frac{x_4 + x_5 + x_6}{3}, \frac{x_5 + x_6 + x_1}{3}, \frac{x_6 + x_1 + x_2}{3}\right)$$

$$\prec \left(\frac{x_1 + x_2}{2}, \frac{x_2 + x_3}{2}, \frac{x_3 + x_4}{2}, \frac{x_4 + x_5}{2}, \frac{x_5 + x_6}{2}, \frac{x_6 + x_1}{2}\right) \tag{21}$$

参考文献

[1]　王伯英. 控制不等式基础[M]. 北京:北京师范大学出版社, 1990.

[2]　MARSHALL A W, OLKIN I. Inequalities: theory of majorization and its application [M]. New York：Academies Press,1979.

第45篇　Schur-Geometric Convexity for Difference of Some Means

(SHI HUAN-NAN, ZHANG JIAN, LI DAMAO. Applied Math.

E-Notes, 10(2010), 275-284)

Abstract: The Schur-geometric convexity in $(0, \infty) \times (0, \infty)$ for the difference of some famous means such as arithmetic mean, geometric mean, harmonic mean, root-square mean, etc. is discussed. And some inequalities related to the difference of means are obtained.

Keywords: arithmetic mean; geometric mean, harmonic mean; root-square mean; square-root mean; Heron mean; difference; Schur-geometric convexity;inequality

45.1　Introduction

Recently, the following chain of inequalities for the binary means is given in [1]

$$H(a, b) \leqslant G(a, b) \leqslant N_1(a, b) \leqslant N_3(a, b) \leqslant N_2(a, b) \leqslant A(a, b) \leqslant S(a, b) \quad (1)$$

where

$$A(a, b) = \frac{a + b}{2}$$

$$G(a, b) = \sqrt{ab}$$

$$H(a, b) = \frac{2ab}{a + b}$$

$$N_1(a, b) = \left(\frac{\sqrt{a} + \sqrt{b}}{2} \right)^2 = \frac{A(a, b) + G(a, b)}{2}$$

$$N_3(a, b) = \frac{a + \sqrt{ab} + b}{3} = \frac{2A(a, b) + G(a, b)}{3}$$

314

$$N_2(a,b) = \left(\frac{\sqrt{a} + \sqrt{b}}{2}\right)\left(\sqrt{\frac{a+b}{2}}\right)$$

$$S(a,b) = \sqrt{\frac{a^2 + b^2}{2}}$$

The means, $A(a,b), G(a,b), H(a,b), S(a,b), N_1(a,b)$ and $N_3(a,b)$ are arithmetic, geometric, harmonic, root-square, square-root and Heron's means respectively. The $N_2(a,b)$ can be seen in Taneja [2-3].

Furthermore the following difference of means are considered in [1]

$$M_{SA}(a,b) = S(a,b) - A(a,b) \tag{2}$$

$$M_{SN_2}(a,b) = S(a,b) - N_2(a,b) \tag{3}$$

$$M_{SN_3}(a,b) = S(a,b) - N_3(a,b) \tag{4}$$

$$M_{SN_1}(a,b) = S(a,b) - N_1(a,b) \tag{5}$$

$$M_{SG}(a,b) = S(a,b) - G(a,b) \tag{6}$$

$$M_{SH}(a,b) = S(a,b) - H(a,b) \tag{7}$$

$$M_{AN_2}(a,b) = A(a,b) - N_2(a,b) \tag{8}$$

$$M_{AG}(a,b) = A(a,b) - G(a,b) \tag{9}$$

$$M_{AH}(a,b) = A(a,b) - H(a,b) \tag{10}$$

$$M_{N_2 N_1}(a,b) = N_2(a,b) - N_1(a,b) \tag{11}$$

$$M_{N_2 G}(a,b) = N_2(a,b) - G(a,b) \tag{12}$$

and the following Theorem is established:

Theorem 1 The difference of means given by (2)−(12) are nonnegative and convex in $\mathbf{R}_+^2 = (0, +\infty) \times (0, +\infty)$.

In this paper, the following Theorem is proved, and by this Theorem, some inequalities in (1) are strengthened.

Theorem 2 The difference of means given by (2) − (12) are Schur-geometrically convex in $\mathbf{R}_+^2 = (0, +\infty) \times (0, +\infty)$.

45.2 Definitions and lemmas

The Schur-convex function was introduced by I. Schur in 1923, and it has many important applications in analytic inequalities, linear regression, graphs and matrices, combinatorial optimization, information-theoretic topics, Gamma functions, stochastic orderings, reliability, and other related fields[4] and [18-32].

In 2003, Xiao-ming Zhang first propose concepts of "Schur-geometrically convex function" which is extension of "Schur-convex function" and establish corresponding decision theorem [6]. Since then, Schur-geometric convexity has evoked the interest of of many researchers and numerous applications and extensions have appeared in the literature, see [7-17]

In order to verify our Theorems, the following definitions and lemmas are necessary.

Definition 1[4−5] Let $\boldsymbol{x} = (x_1, \ldots, x_n)$ and $\boldsymbol{y} = (y_1, \ldots, y_n) \in \mathbf{R}^n$.

(i) $\boldsymbol{x}$ is said to be majorized by $\boldsymbol{y}$ (in symbols $\boldsymbol{x} \prec \boldsymbol{y}$) if $\sum\limits_{i=1}^{k} x_{[i]} \leqslant \sum\limits_{i=1}^{k} y_{[i]}$ for $k = 1, 2, \ldots, n-1$ and $\sum\limits_{i=1}^{n} x_i = \sum\limits_{i=1}^{n} y_i$, where $x_{[1]} \geqslant \cdots \geqslant x_{[n]}$ and $y_{[1]} \geqslant \cdots \geqslant y_{[n]}$ are rearrangements of $\boldsymbol{x}$ and $\boldsymbol{y}$ in a descending order.

(ii) $\Omega \subseteq \mathbf{R}^n$ is called a convex set if $(\alpha x_1 + \beta y_1, \ldots, \alpha x_n + \beta y_n) \in \Omega$ for every $\boldsymbol{x}$ and $\boldsymbol{y} \in \Omega$, where α and $\beta \in [0, 1]$ with $\alpha + \beta = 1$.

(iii) Let $\Omega \subseteq \mathbf{R}^n$. The function φ: $\Omega \to \mathbf{R}$ be said to be a Schur-convex function on Ω if $\boldsymbol{x} \prec \boldsymbol{y}$ on Ω implies $\varphi(\boldsymbol{x}) \leq \varphi(\boldsymbol{y})$. φ is said to be a Schur-concave function on Ω if and only if $-\varphi$ is Schur-convex.

Definition 2[6] Let $\boldsymbol{x} = (x_1, \ldots, x_n)$ and $\boldsymbol{y} = (y_1, \ldots, y_n) \in \mathbf{R}_+^n$.

(i) $\Omega \subseteq \mathbf{R}_+^n$ is called a geometrically convex set if $(x_1^\alpha y_1^\beta, \ldots, x_n^\alpha y_n^\beta) \in \Omega$ for all $\boldsymbol{x}$ and $\boldsymbol{y} \in \Omega$, where α and $\beta \in [0, 1]$ with $\alpha + \beta = 1$.

(ii) Let $\Omega \subseteq \mathbf{R}_+^n$. The function φ: $\Omega \to \mathbf{R}_+$ is said to be Schur-geometrically convex function on Ω if $(\ln x_1, \ldots, \ln x_n) \prec (\ln y_1, \ldots, \ln y_n)$ on Ω implies $\varphi(\boldsymbol{x}) \leqslant \varphi(\boldsymbol{y})$. The function φ is said to be a Schur-geometrically concave on Ω if and only if $-\varphi$ is Schur-geometrically convex.

Definition 3[4−5]

(i) $\Omega \subseteq \mathbf{R}^n$ is called symmetric set, if $\boldsymbol{x} \in \Omega$ implies $\boldsymbol{P}\boldsymbol{x} \in \Omega$ for every $n \times n$ permutation matrix $\boldsymbol{P}$.

(ii) The function $\varphi : \Omega \to \mathbf{R}$ is called symmetric if for every permutation matrix $\mathbf{P}$, $\varphi(\mathbf{P}\boldsymbol{x}) = \varphi(\boldsymbol{x})$ for all $\boldsymbol{x} \in \Omega$.

Lemma 1[4-5] Let $\Omega \subset \mathbf{R}^n, \varphi : \Omega \to \mathbf{R}$ is a symmetric and convex function. Then φ is Schur convex on Ω.

Remark 1 It is obvious that the difference of means given by (2) – (12) are symmetric, so by Theorem A and Lemma 1, it follows that those differences are all Schur-convex in $\mathbf{R}_+^2 = (0, +\infty) \times (0, +\infty)$.

Lemma 2[6] Let $\Omega \subseteq \mathbf{R}_+^n$ be symmetric with a nonempty interior geometrically convex set. Let $\varphi : \Omega \to \mathbf{R}_+$ be continuous on Ω and differentiable in Ω^0. If φ is symmetric on Ω and

$$(\ln x_1 - \ln x_2) \left(x_1 \frac{\partial \varphi}{\partial x_1} - x_2 \frac{\partial \varphi}{\partial x_2} \right) \geqslant 0 (\leqslant 0) \tag{13}$$

holds for any $\boldsymbol{x} = (x_1, \cdots, x_n) \in \Omega^0$, then φ is a Schur-geometrically convex (Schur-geometrically concave) function.

Lemma 3[7] Let $a \leqslant b, u(t) = ta + (1-t)b, v(t) = tb + (1-t)a$. If $1/2 \leqslant t_2 \leqslant t_1 \leqslant 1$ or $0 \leqslant t_1 \leqslant t_2 \leqslant 1/2$. Then

$$\left(\frac{a+b}{2}, \frac{a+b}{2} \right) \prec (u(t_2), v(t_2)) \prec (u(t_1), v(t_1)) \prec (a, b) \tag{14}$$

45.3   Proofs of main results

Proof of Theorem 2:

(1) For

$$M_{SA}(a, b) = S(a, b) - A(a, b) = \sqrt{\frac{a^2 + b^2}{2}} - \frac{a+b}{2}$$

we have

$$\frac{\partial M_{SA}}{\partial a} = \frac{a}{2} \left(\frac{a^2 + b^2}{2} \right)^{-1/2} - \frac{1}{2}$$

$$\frac{\partial M_{SA}}{\partial b} = \frac{b}{2} \left(\frac{a^2 + b^2}{2} \right)^{-1/2} - \frac{1}{2}$$

and then

$$\Lambda := (\ln a - \ln b) \left(a \frac{\partial M_{SA}}{\partial a} - b \frac{\partial M_{SA}}{\partial b} \right)$$

$$= (\ln a - \ln b) \left[\left(\frac{a^2 + b^2}{2} \right)^{-1/2} \frac{a^2 - b^2}{2} - \frac{a - b}{2} \right]$$

$$= \frac{(\ln a - \ln b)(a - b)}{2} \left[(a + b) \left(\frac{a^2 + b^2}{2} \right)^{-1/2} - 1 \right]$$

Since $\ln x$ is increasing, we have $(\ln a - \ln b)(a - b) \geqslant 0$, and $(a + b) \left(\frac{a^2 + b^2}{2} \right)^{-1/2} - 1 \geqslant 0$ is equivalent to $a^2 + b^2 \leqslant 2a^2 + 2b^2 + 4ab$, which is hold obviously, so $\Lambda \geqslant 0$. By the Lemma 2, it follows that $M_{SA}(a, b)$ is Schur-geometrically convex in $\mathbf{R}_+^2 = (0, +\infty) \times (0, +\infty)$.

(2) For

$$M_{AN_2}(a, b) = A(a, b) - N_2(a, b) = \frac{a + b}{2} - \left(\frac{\sqrt{a} + \sqrt{b}}{2} \right) \left(\sqrt{\frac{a + b}{2}} \right)$$

we have

$$\frac{\partial M_{AN_2}}{\partial a} = \frac{1}{2} - \frac{1}{4\sqrt{a}} \sqrt{\frac{a + b}{2}} - \frac{1}{4} \left(\frac{\sqrt{a} + \sqrt{b}}{2} \right) \left(\frac{a + b}{2} \right)^{-1/2}$$

$$\frac{\partial M_{AN_2}}{\partial b} = \frac{1}{2} - \frac{1}{4\sqrt{b}} \sqrt{\frac{a + b}{2}} - \frac{1}{4} \left(\frac{\sqrt{a} + \sqrt{b}}{2} \right) \left(\frac{a + b}{2} \right)^{-1/2}$$

and then

$$\Lambda = (\ln a - \ln b) \left(a \frac{\partial M_{AN_2}}{\partial a} - b \frac{\partial M_{AN_2}}{\partial b} \right)$$

$$= (\ln a - \ln b) \left[\frac{a - b}{2} - \frac{1}{4} \sqrt{\frac{a + b}{2}} (\sqrt{a} - \sqrt{b}) - \frac{1}{4} \left(\frac{\sqrt{a} + \sqrt{b}}{2} \right) \left(\frac{a + b}{2} \right)^{-1/2} (a - b) \right]$$

$$= \frac{(\ln a - \ln b)(a - b)}{2} \left[1 - \frac{1}{2} \sqrt{\frac{a + b}{2}} (\sqrt{a} + \sqrt{b})^{-1} - \frac{1}{2} \left(\frac{\sqrt{a} + \sqrt{b}}{2} \right) \left(\frac{a + b}{2} \right)^{-1/2} \right]$$

It is easy to check that

$$1 - \frac{1}{2} \sqrt{\frac{a + b}{2}} (\sqrt{a} + \sqrt{b})^{-1} - \frac{1}{2} \left(\frac{\sqrt{a} + \sqrt{b}}{2} \right) \left(\frac{a + b}{2} \right)^{-1/2} \geqslant 0$$

is equivalent to

$$(a + b)^2 + 2(a + b)\sqrt{ab} \geqslant ab$$

which is clearly hold, so $\Lambda \geqslant 0$. By the Lemma 2, it follows that $M_{AN_2}(a, b)$ is Schur-geometrically convex in $\mathbf{R}_+^2$.

(3) For

$$M_{SN_2}(a,b) = S(a,b) - N_2(a,b) = \sqrt{\frac{a^2+b^2}{2}} - \left(\frac{\sqrt{a}+\sqrt{b}}{2}\right)\left(\sqrt{\frac{a+b}{2}}\right)$$

notice that

$$M_{SN_2}(a,b) = M_{SA}(a,b) + M_{AN_2}(a,b)$$

by the definition of the Schur-geometrically convex function, it follows that the sum of two Schur-geometrically convex function is also the Schur-geometrically convex, so $M_{SN_2}(a,b)$ is Schur-geometrically convex in $\mathbf{R}_+^2$.

(4) For

$$M_{SN_3}(a,b) = S(a,b) - N_3(a,b) = \sqrt{\frac{a^2+b^2}{2}} - \frac{a+\sqrt{ab}+b}{3}$$

we have

$$\frac{\partial M_{SN_3}}{\partial a} = \frac{a}{2}\left(\frac{a^2+b^2}{2}\right)^{-1/2} - \frac{1}{3}\left(1+\frac{b}{2\sqrt{ab}}\right)$$

$$\frac{\partial M_{SN_3}}{\partial b} = \frac{b}{2}\left(\frac{a^2+b^2}{2}\right)^{-1/2} - \frac{1}{3}\left(1+\frac{a}{2\sqrt{ab}}\right)$$

and then

$$\Lambda = (\ln a - \ln b)\left(a\frac{\partial M_{SN_3}}{\partial a} - b\frac{\partial M_{SN_3}}{\partial b}\right)$$

$$= (\ln a - \ln b)(a-b)\left[\left(\frac{a^2+b^2}{2}\right)^{-1/2}\left(\frac{a+b}{2}\right) - \frac{1}{3}\right]$$

notice that

$$\left(\frac{a^2+b^2}{2}\right)^{-1/2}\left(\frac{a+b}{2}\right) - \frac{1}{3} \geqslant 0 \Leftrightarrow 9(a+b)^2 \geqslant 2\left(a^2+b^2\right)$$

we have $\Lambda \geqslant 0$, so $M_{SN_3}(a,b)$ is Schur-geometrically convex in $\mathbf{R}_+^2$.

(5) For

$$M_{N_2 N_1}(a,b) = N_2(a,b) - N_1(a,b) = \left(\frac{\sqrt{a}+\sqrt{b}}{2}\right)\left(\sqrt{\frac{a+b}{2}}\right) - \frac{a+b}{4} - \frac{\sqrt{ab}}{2}$$

we have

$$\frac{\partial M_{N_2 N_1}}{\partial a} = \frac{1}{4\sqrt{a}}\sqrt{\frac{a+b}{2}} + \frac{1}{4}\left(\frac{\sqrt{a}+\sqrt{b}}{2}\right)\left(\frac{a+b}{2}\right)^{-1/2} - \frac{1}{4} - \frac{b}{4\sqrt{ab}}$$

$$\frac{\partial M_{N_2 N_1}}{\partial b} = \frac{1}{4\sqrt{b}}\sqrt{\frac{a+b}{2}} + \frac{1}{4}\left(\frac{\sqrt{a}+\sqrt{b}}{2}\right)\left(\frac{a+b}{2}\right)^{-1/2} - \frac{1}{4} - \frac{a}{4\sqrt{ab}}$$

and then

$$\Lambda = (\ln a - \ln b)\left(a\frac{\partial M_{N_2 N_1}}{\partial a} - b\frac{\partial M_{N_2 N_1}}{\partial b}\right)$$

$$= (\ln a - \ln b)[\frac{1}{4}\sqrt{\frac{a+b}{2}}(\sqrt{a} - \sqrt{b}) +$$

$$\frac{1}{4}(\frac{\sqrt{a}+\sqrt{b}}{2})(\frac{a+b}{2})^{-1/2}(a-b) - \frac{1}{4}(a-b)]$$

$$= \frac{1}{4}(\ln a - \ln b)(a-b)[\sqrt{\frac{a+b}{2}}(\sqrt{a}+\sqrt{b})^{-1} +$$

$$(\frac{\sqrt{a}+\sqrt{b}}{2})(\frac{a+b}{2})^{-1/2} - 1]$$

By the AM-GM inequality, we have

$$\sqrt{\frac{a+b}{2}}\left(\sqrt{a}+\sqrt{b}\right)^{-1} + \left(\frac{\sqrt{a}+\sqrt{b}}{2}\right)\left(\frac{a+b}{2}\right)^{-1/2} - 1$$

$$\geqslant 2\left[\sqrt{\frac{a+b}{2}}\left(\sqrt{a}+\sqrt{b}\right)^{-1} \cdot \left(\frac{\sqrt{a}+\sqrt{b}}{2}\right)\left(\frac{a+b}{2}\right)^{-1/2}\right]^{1/2} - 1$$

$$= \sqrt{2} - 1 \geqslant 0$$

so $M_{N_2 N_1}(a,b)$ is Schur-geometrically convex in $\mathbf{R}_+^2$.

(6) For

$$M_{SN_1}(a,b) = S(a,b) - N_1(a,b) = \sqrt{\frac{a^2+b^2}{2}} - \left(\frac{\sqrt{a}+\sqrt{b}}{2}\right)^2$$

notice that

$$M_{SN_1}(a,b) = M_{SN_2}(a,b) + M_{N_2 N_1}(a,b)$$

i.e. $M_{SN_1}(a,b)$ is the sum of two Schur-geometrically convex function, so $M_{SN_2}(a,b)$ is Schur-geometrically convex in $\mathbf{R}_+^2$.

(7) For

$$M_{AG}(a, b) = A(a, b) - G(a, b) = \frac{a+b}{2} - \sqrt{ab}$$

we have

$$\frac{\partial M_{AG}}{\partial a} = \frac{1}{2} - \frac{b}{2\sqrt{ab}}, \quad \frac{\partial M_{AG}}{\partial b} = \frac{1}{2} - \frac{a}{2\sqrt{ab}}$$

and then

$$\Lambda = (\ln a - \ln b)\left(a\frac{\partial M_{AG}}{\partial a} - b\frac{\partial M_{AG}}{\partial b}\right) = \frac{1}{2}(\ln a - \ln b)(a - b) \geqslant 0$$

so $M_{AG}(a, b)$ is Schur-geometrically convex in $\mathbf{R}_+^2$.

(8) For

$$M_{SG}(a, b) = S(a, b) - G(a, b) = \sqrt{\frac{a^2 + b^2}{2}} - \sqrt{ab}$$

notice that

$$M_{SG}(a, b) = M_{SA}(a, b) + M_{AG}(a, b)$$

i.e. $M_{SG}(a, b)$ is the sum of two Schur-geometric convex function, so $M_{SG}(a, b)$ is Schur-geometrically convex in $\mathbf{R}_+^2$.

(9) For

$$M_{AH}(a, b) = A(a, b) - H(a, b) = \frac{a+b}{2} - \frac{2ab}{a+b}$$

we have

$$\frac{\partial M_{AH}}{\partial a} = \frac{1}{2} - \frac{2b^2}{(a+b)^2}, \quad \frac{\partial M_{AH}}{\partial b} = \frac{1}{2} - \frac{2a^2}{(a+b)^2}$$

and then

$$\Lambda = (\ln a - \ln b)\left(a\frac{\partial M_{AH}}{\partial a} - b\frac{\partial M_{AH}}{\partial b}\right)$$

$$= (\ln a - \ln b)(a - b)\left[\frac{1}{2} + \frac{2ab}{(a+b)^2}\right] \geqslant 0$$

so $M_{AH}(a, b)$ is Schur-geometrically convex in $\mathbf{R}_+^2$.

(10) For

$$M_{SH}(a, b) = S(a, b) - H(a, b) = \sqrt{\frac{a^2 + b^2}{2}} - \frac{2ab}{a+b}$$

notice that

$$M_{SH}(a,b) = M_{SA}(a,b) + M_{AH}(a,b)$$

i.e. $M_{SH}(a,b)$ is the sum of two Schur-geometrically convex function, so $M_{SH}(a,b)$ is Schur-geometrically convex in $\mathbf{R}_+^2$.

(11) For

$$M_{N_2G}(a,b) = N_2(a,b) - G(a,b) = \left(\frac{\sqrt{a}+\sqrt{b}}{2}\right)\left(\sqrt{\frac{a+b}{2}}\right) - \sqrt{ab}$$

we have

$$\frac{\partial M_{N_2G}}{\partial a} = \frac{1}{4\sqrt{a}}\left(\sqrt{\frac{a+b}{2}}\right) + \frac{1}{4}\left(\frac{\sqrt{a}+\sqrt{b}}{2}\right)\left(\frac{a+b}{2}\right)^{-1/2} - \frac{b}{2\sqrt{ab}}$$

$$\frac{\partial M_{N_2G}}{\partial b} = \frac{1}{4\sqrt{b}}\left(\sqrt{\frac{a+b}{2}}\right) + \frac{1}{4}\left(\frac{\sqrt{a}+\sqrt{b}}{2}\right)\left(\frac{a+b}{2}\right)^{-1/2} - \frac{a}{2\sqrt{ab}}$$

and then

$$\Lambda = (\ln a - \ln b)\left(a\frac{\partial M_{N_2G}}{\partial a} - b\frac{\partial M_{N_2G}}{\partial b}\right)$$

$$= (\ln a - \ln b)\left[\frac{1}{4}\left(\sqrt{\frac{a+b}{2}}\right)(\sqrt{a}-\sqrt{b}) + \frac{1}{4}\left(\frac{\sqrt{a}+\sqrt{b}}{2}\right)\left(\frac{a+b}{2}\right)^{-1/2}(a-b)\right]$$

$$= \frac{1}{4}(\ln a - \ln b)(a-b)\left[\left(\sqrt{\frac{a+b}{2}}t\right)(\sqrt{a}+\sqrt{b})^{-1} + \frac{1}{4}\left(\frac{\sqrt{a}+\sqrt{b}}{2}\right)\left(\frac{a+b}{2}\right)^{-1/2}\right] \geqslant 0$$

so $M_{N_2G}(a,b)$ is Schur-geometrically convex in $\mathbf{R}_+^2$.

Thus the proof of Theorem 2 is complete.

45.4 Applications

Theorem 3 Let $0 < a \leqslant b$. If $1/2 \leqslant t \leqslant 1$ or $0 \leqslant t \leqslant 1/2$. Then

$$0 \leqslant \sqrt{\frac{a^{t^2}b^{(1-t)^2} + a^{(1-t)^2}b^{t^2}}{2}} - \frac{a^tb^{1-t} + a^{1-t}b^t}{2} \leqslant \sqrt{\frac{a^2+b^2}{2}} - \frac{a+b}{2} \quad (15)$$

$$0 \leqslant \sqrt{\frac{a^{t^2}b^{(1-t)^2} + a^{(1-t)^2}b^{t^2}}{2}} - \left(\frac{\sqrt{a^tb^{1-t}} + \sqrt{a^{1-t}b^t}}{2}\right)\left(\sqrt{\frac{a^tb^{1-t} + a^{1-t}b^t}{2}}\right)$$

$$\leqslant \sqrt{\frac{a^2+b^2}{2}} - \left(\frac{\sqrt{a}+\sqrt{b}}{2}\right)\left(\sqrt{\frac{a+b}{2}}\right) \tag{16}$$

$$0 \leqslant \sqrt{\frac{a^{t^2}b^{(1-t)^2}+a^{(1-t)^2}b^{t^2}}{2}} - \frac{a^t b^{1-t}+\sqrt{ab}+a^{1-t}b^t}{3} \leqslant \sqrt{\frac{a^2+b^2}{2}} - \frac{a+\sqrt{ab}+b}{3} \tag{17}$$

$$0 \leqslant \frac{a^{t^2}b^{(1-t)^2}+a^{(1-t)^2}b^{t^2}}{2} - \left(\frac{\sqrt{a^t b^{1-t}}+\sqrt{a^{1-t}b^t}}{2}\right)\left(\sqrt{\frac{a^t b^{1-t}+a^{1-t}b^t}{2}}\right)$$

$$\leqslant \frac{a+b}{2} - \left(\frac{\sqrt{a}+\sqrt{b}}{2}\right)\left(\sqrt{\frac{a+b}{2}}\right) \tag{18}$$

$$0 \leqslant \left(\frac{\sqrt{a^t b^{1-t}}+\sqrt{a^{1-t}b^t}}{2}\right)\left(\sqrt{\frac{a^t b^{1-t}+a^{1-t}b^t}{2}}\right) - \left(\frac{\sqrt{a^t b^{1-t}}+\sqrt{a^{1-t}b^t}}{2}\right)^2$$

$$\leqslant \left(\frac{\sqrt{a}+\sqrt{b}}{2}\right)\left(\sqrt{\frac{a+b}{2}}\right) - \left(\frac{\sqrt{a}+\sqrt{b}}{2}\right)^2 \tag{19}$$

Proof From the Lemma 3, we have

$$\left(\ln\sqrt{ab}, \ln\sqrt{ab}\right) \prec \left(\ln(b^t a^{1-t}), \ln(a^t b^{1-t})\right) \prec (\ln a, \ln b),$$

and by Theorem 2, the difference of two means in (2)

$$M_{SA}(a,b) = S(a,b) - A(a,b) = \sqrt{\frac{a^2+b^2}{2}} - \frac{a+b}{2}$$

is Schur-geometrically convex in $\mathbf{R}_+^2$, so we have

$$M_{SA}(\sqrt{ab}, \sqrt{ab}) \leqslant M_{SA}(a^t b^{1-t}, a^{1-t}b^t) \leqslant M_{SA}(a,b)$$

i.e. (15) is hold.

Similarly, by Schur-geometric convexity of the difference of two means in (3),(4),(8) and (11), from (20) it follows that (16),(17),(18) and (19) is hold respectively.

The proof of Theorem 3 is complete.

Remark 2 (15) and is the sharpening of the inequality $A(a,b) \leqslant S(a,b)$ in (1), and (16) is the sharpening of the inequality $N_2(a,b) \leqslant A(a,b)$ in (1).

References

[1] I J TANEJA. Refinement of inequalities among means[J]. Journal of Combinatorics, Information & System Sciences,2006, Volume 31, IS-SUE 1-4, 343-364, arXiv:math/0505192v2 [math.GM] 12 Jul 2005.

[2] I J TANEJA. On a Difference of Jensen Inequality and its Applications to Mean Divergence Measures[J]. RGMIA Research Report Collection, http://rgmia.vu.edu.au, 7(4)(2004), Art. 16.

[3] I J TANEJA. On Symmetric and Non-Symmetric Divergence Measures and Their Generalizations[J]. Advances in Imaging and Electron Physics, 2005, 138: 177-250.

[4] A M MARSHALL, I OLKIN. Inequalities: Theory of Majorization and Its Application[M]. New York : Academies Press, 1979.

[5] BO-YING WANG. Foundations of Majorization Inequalities[M]. Beijing: Beijing Normal Univ. Press, 1990.

[6] XIAO-MING ZHANG. Geometrically Convex Functions[M]. Hefei: An hui University Press, 2004.

[7] HUAN-NAN SHI, YONG-MING JIANG, WEI-DONG JIANG. Schur-Convexity and Schur-Geometrically Concavity of Gini Mean[J]. Computers and Mathematics with Applications, 2009, 57: 266-274.

[8] YU-MING CHU, XIAO-MING ZHANG. The Schur geometrical convexity of the extended mean values[J]. Journal of Convex Analysis, 2008, 15(4): 869-890.

[9] CHUN GU, HUAN-NAN SHI. Schur-Convexity and Schur-Geometric Concavity of Lehme Means[J]. Mathematics in Practice and Theory, 2009, 39 (12): 183-188.

[10] KAI-ZHONG GUAN. A class of symmetric functions for multiplicatively convex function[J]. Math. Inequal. Appl., 2007, 10 (4): 745-753.

[11] HUAN-NAN SHI, MIHALY BENCZE, SHAN-HE WU, DA-MAO LI. Schur Convexity of generalized Heronian Means involving two parameters[J]. J. Inequal. Appl., Volume 2008, Article ID 879273, 9 pages doi:10.1155/2008/879273

[12] XIAO-MING ZHANG, SHI-JIE LI. Two S-Geometrically Convex Functions Involving the Elementary Symmetric Function[J]. Journal of Sichuan Normal University (Natural Science), 2007, 30 (2): 188-190.

[13] XIAO-MING ZHANG. S-geometric convexity of a function involving Maclaurin's elementary symmetric mean[J]. J. Inequal. Pure and Appl. Math., 8(2), Art. 51, 6 pp. 2007.

[14] XIAO-MING ZHANG, TIE-QUAN XU. The Definition of Generalized S-Geometrically Convex Function and One of Its Applications[J]. Journal of Qindao Vocational and Technical College, 2005, 18 (4): 60-62.

[15] YIN-HAI ZHOU, XIAO-MING ZHANG. Geometrically Convexity of Stolarsky's Mean of n Variables[J]. Journal of Beijing Union University(Natural Science), 2006, 20(2): 73-79.

[16] XIAO-MING ZHANG. The Schur geometrical convexity of integral arithmetic mean[J]. International Journal of Pure and Applied Mathematics, 2007, 41(7): 919-925.

[17] CONSTANTIN P NICULESCU. Convexity according to the geometric mean[J]. Math. Inequal. Appl. 2000, 3(2):155-167.

[18] WEI-FENG XIA, YU-MING CHU. Schur-convexity for a class of symmetric functions and its applications[J]. Journal of Inequalities and Applications Volume 2009, Article ID 493759, 15 pages doi:10.1155/2009/493759

[19] KAI-ZHONG GUAN. The Hamy symmetric function and its generalization[J]. Math. Inequal. Appl., 2006, 9(4): 797-805.

[20] KAI-ZHONG GUAN. Schur-convexity of the complete symmetric function[J]. Math. Inequal. Appl., 2006, 9(4): 567-576.

[21] KAI-ZHONG GUAN. Some properties of a class of symmetric functions[J]. J. Math. Anal. Appl., 2007, 336(1): 70-80.

[22] C STEPNIAK. An effective characterization of Schur-convex functions with applications[J]. J. Convex Anal., 2007, 14(1): 103-108.

[23] HUAN-NAN SHI. Schur-Convex Functions relate to Hadamard-type inequalities[J]. J. Math. Inequal., 2007, 1(1): 127-136.

[24] HUAN-NAN SHI, DA-MAO LI, CHUN GU. Schur-Convexity of a Mean of Convex Function[J]. Applied Mathematics Letters, 2009, 22 : 932-937.

[25] YU-MING CHU XIAO-MING ZHANG. Necessary and sufficient conditions such that extended mean values are Schur-convex or Schur-concave[J]. Journal of Mathematics of Kyoto University, 2008, 48 (1): 229–238.

[26] ELEZOVIC N, PECARIC J. Note on Schur-convex functions[J]. Rocky Mountain J. Math., 1998, 29: 853–856.

[27] JÓZSEF SÁNDOR. The Schur-convexity of Stolarsky and Gini means [J]. Banach J. Math. Anal., 2007, 1(2): 212-215.

[28] QI FENG, JÓZSEF SÁNDOR, SEVER S DRAGOMIR, ANTHONY SOFO. Notes on the Schur- convexity of the Extended Mean Values[J]. Taiwanese Journal of Mathematics, 2005, 9(3): 411–420.

[29] HUAN-NAN ShI, SHAN-HE WU, FENG QI. An alternative note on the Schur-convexity of the extended mean values[J]. Math. Inequal. Appl., 2006, 9(2): 219–224.

[30] NING-GUO ZHENG, ZHI-HUA ZHANG, XIAO-MING ZHANG. Schur-convexity of two types of one-parameter mean values in n variables[J]. J. Inequal. Appl., Volume 2007 (2007), Article ID 078175, 10 pagesdoi:10.1155/2007/78175.

[31] XIN-MIN ZHANG. Opimization of Schur-convex functions[J]. Math. Inequal. Appl., 1998,1(3): 319-330.

[32] XIN-MIN ZHANG. Schur-convex functions and isoperimetric inequalities[J]. Proceedings of the american mathematical society, 1998, 126(2): 461-470.

第46篇　Two Schur-Convex Functions Related to Hadamard-Type Integral Inequalities

(SHI HUAN-NAN. Publicationes Mathematicae Debrecen, 2011,78(2):

393-403)

Abstract: The Schur-convexity, the Schur-geometric convexity and the Schur-harmonic convexity of two mappings which related to Hadamard-type integral inequalities are researched. And three refinements of Hadamard-type integral inequality are obtained, as applications, some inequalities related to the arithmetic mean, the logarithmic mean and the power mean are established. obtained.

Keywords: Schur-convex function; Schur-geometrical convex function and the Schur-harmonic convex function; inequality; convex function Hadamard's inequality; logarithmic mean; power mean

46.1　Introduction

Throughout the paper we assume that the set of n-dimensional row vector on real number field by $\mathbf{R}^n$, and $\mathbf{R}_+^n = \boldsymbol{x} = (x_1, \ldots, x_n) \in \mathbf{R}^n : x_i > 0, i = 1, \ldots, n\}$. In particular, $\mathbf{R}^1$ and $\mathbf{R}_+^1$ denoted by $\mathbf{R}$ and $\mathbf{R}_+$ respectively.

Let f be a convex function defined on the interval $I \subseteq \mathbf{R} \to \mathbf{R}$ of real numbers and $a, b \in I$ with $a < b$. Then

$$f\left(\frac{a+b}{2}\right) \leqslant \frac{1}{b-a}\int_a^b f(x)\,\mathrm{d}x \leqslant \frac{f(a)+f(b)}{2} \tag{1}$$

is known as the Hadamard's inequality for convex function[1]. For some recent results which generalize, improve, and extend this classical inequality, see [2-8]and [15-17].

When $f, -g$ both are convex functions satisfying $\int_a^b g(x)\,\mathrm{d}x > 0$ and $f\left(\frac{a+b}{2}\right) \geqslant 0$, S.-J. Yang in [5] generalized (1) as

$$\frac{f\left(\frac{a+b}{2}\right)}{g\left(\frac{a+b}{2}\right)} \leqslant \frac{\frac{1}{b-a}\int_a^b f(x)\,\mathrm{d}x}{\frac{1}{b-a}\int_a^b g(x)\,\mathrm{d}x} \tag{2}$$

To go further in exploring (2), Lan He in [8] define two mappings L and F by $L : [a, b] \times [a, b] \to \mathbf{R}$

$$L(x, y; f, g) = \left[\int_x^y f(t)\,\mathrm{d}t - (y-x)f\left(\frac{x+y}{2}\right)\right]\left[(y-x)g\left(\frac{x+y}{2}\right) - \int_x^y g(t)\,\mathrm{d}t\right]$$

and $F : [a, b] \times [a, b] \to \mathbf{R}$,

$$F(x, y; f, g) = g\left(\frac{x+y}{2}\right)\int_x^y f(t)\,\mathrm{d}t - f\left(\frac{x+y}{2}\right)\int_x^y g(t)\,\mathrm{d}t$$

and established the following two theorems which are refinements of the inequality of (2).

Theorem A[8] Let $f, -g$ both are convex functions on $[a, b]$. Then we have

(i) $L(a, y; f, g)$ is nonnegative increasing with y on $[a, b]$, $L(x, b; f, g)$ is nonnegative decreasing with x on $[a, b]$.

(ii) When $\int_b^a g(x)\,\mathrm{d}x > 0$ and $f\left(\frac{a+b}{2}\right) \geqslant 0$, for any $x, y \in (a, b)$ and $\alpha \geqslant 0$ and $\beta \geqslant 0$ such that $\alpha + \beta = 1$, we have the following refinement of (2)

$$\frac{f\left(\frac{a+b}{2}\right)}{g\left(\frac{a+b}{2}\right)} \leqslant \frac{(b-a)f\left(\frac{a+b}{2}\right)}{2\int_a^b g(t)\,\mathrm{d}t} + \frac{\int_a^b f(t)\,\mathrm{d}t}{2(b-a)g\left(\frac{a+b}{2}\right)}$$

$$\leqslant \frac{(b-a)f\left(\frac{a+b}{2}\right)}{2\int_a^b g(t)\,\mathrm{d}t} + \frac{\int_a^b f(t)\,\mathrm{d}t}{2(b-a)g\left(\frac{a+b}{2}\right)} + \frac{\alpha L(a, y; f, g) + \beta L(x, b; f, g)}{2(b-a)g\left(\frac{a+b}{2}\right)\int_a^b g(t)\,\mathrm{d}t}$$

$$\leqslant \frac{\int_a^b f(t)\,\mathrm{d}t}{2\int_a^b g(t)\,\mathrm{d}t} + \frac{2f\left(\frac{a+b}{2}\right)}{2g\left(\frac{a+b}{2}\right)} \leqslant \frac{\int_a^b f(t)\,\mathrm{d}t}{\int_a^b g(t)\,\mathrm{d}t} \tag{3}$$

Theorem B[8] Let $f, -g$ both are nonnegative convex functions on $[a, b]$ satisfying $\int_a^b g(x)\,\mathrm{d}x > 0$. Then we have the following two results:

(i) If f and $-g$ both are increasing, then $F(a, y; f, g)$ is nonnegative increasing with y on $[a, b]$, and we have the following refinement of (2)

$$\frac{f\left(\frac{a+b}{2}\right)}{g\left(\frac{a+b}{2}\right)} \leqslant \frac{f\left(\frac{a+b}{2}\right)}{g\left(\frac{a+b}{2}\right)} + \frac{F(a, y; f, g)}{g\left(\frac{a+b}{2}\right) \int_a^b g(t)\,\mathrm{d}t} \leqslant \frac{\int_a^b f(t)\,\mathrm{d}t}{\int_a^b g(t)\,\mathrm{d}t} \qquad (4)$$

where $y \in (a, b)$.

(ii) If f and $-g$ both are decreasing, then $F(a, y; f, g)$ is nonnegative decreasing with y on $[a, b]$, and we have the following refinement of (2)

$$\frac{f\left(\frac{a+b}{2}\right)}{g\left(\frac{a+b}{2}\right)} \leqslant \frac{f\left(\frac{a+b}{2}\right)}{g\left(\frac{a+b}{2}\right)} + \frac{F(x, b; f, g)}{g\left(\frac{a+b}{2}\right) \int_a^b g(t)\,\mathrm{d}t} \leqslant \frac{\int_a^b f(t)\,\mathrm{d}t}{\int_a^b g(t)\,\mathrm{d}t} \qquad (5)$$

where $x \in (a, b)$.

The aim of this paper is to study the Schur-convexity of $L(x, y; f, g)$ and $F(x, y; f, g)$ with variables (x, y) in $[a, b] \times [a, b] \subseteq \mathbf{R}^2$, and study the Schur-geometric convexity and the Schur-harmonic convexity of $L(x, y; f, g)$ with variables (x, y) in $[a, b] \times [a, b] \subseteq \mathbf{R}_+^2$. We obtain the following results.

Theorem 1 Let f and $-g$ both be convex function on $[a, b]$. Then

(i) $L(x, y; f, g)$ is Schur-convex on $[a, b] \times [a, b] \subseteq \mathbf{R}^2$, and $L(x, y; f, g)$ is Schur-geometrically convex and Schur-harmonic convex in $[a, b] \times [a, b] \subseteq \mathbf{R}_+^2$.

(ii) If $\frac{1}{2} \leqslant t_2 \leqslant t_1 \leqslant 1$ or $0 \leqslant t_2 \leqslant t_1 \leqslant \frac{1}{2}$, then for $a < b$, we have

$$0 \leqslant L(t_1 a + (1 - t_1)b, t_1 b + (1 - t_1)a; f, g)$$

$$\leqslant L(t_2 a + (1 - t_2)b, t_2 b + (1 - t_2)a; f, g) \leqslant L(a, b; f, g) \qquad (6)$$

and for $0 < a < b$, we have

$$0 \leqslant L\left(b^{t_2} a^{1-t_2}, a^{t_2} b^{1-t_2}; f, g\right) \leqslant L\left(b^{t_1} a^{1-t_1}, a^{t_1} b^{1-t_1}; f, g\right) \leqslant L(a, b; f, g) \qquad (7)$$

and

$$0 \leqslant L\left(1/(t_2 b + (1 - t_2)a), 1/(t_2 a + (1 - t_2)b); f, g\right)$$

$$\leqslant L\left(1/(t_1 b + (1 - t_1)a), 1/(t_1 a + (1 - t_1)b); f, g\right) \leqslant L(1/a, 1/b; f, g) \qquad (8)$$

Theorem 2 Let f and $-g$ both be nonnegative convex function on $[a, b]$. Then

(i) $F(x, y; f, g)$ is Schur-convex on $[a, b] \times [a, b] \subseteq \mathbf{R}^2$;

(ii) If $\frac{1}{2} \leqslant t_2 \leqslant t_1 \leqslant 1$ or $0 \leqslant t_2 \leqslant t_1 \leqslant \frac{1}{2}$, then for $a < b$, we have

$$0 \leqslant F(t_1 a + (1-t_1)b, t_1 b + (1-t_1)a; f, g)$$

$$\leqslant F(t_2 a + (1-t_2)b, t_2 b + (1-t_2)a; f, g) \leqslant F(a, b; f, g) \qquad (9)$$

Theorem 3 Let f and $-g$ both be convex function on $[a, b] \subseteq \mathbf{R}$. If $\int_b^a g(x)\,\mathrm{d}x > 0$ and $f\left(\frac{a+b}{2}\right) \geqslant 0$, then

$$\frac{f\left(\frac{a+b}{2}\right)}{g\left(\frac{a+b}{2}\right)} \leqslant \frac{\int_a^b f(t)\,\mathrm{d}t - \int_{ta+(1-t)b}^{tb+(1-t)a} f(t)\,\mathrm{d}t}{\int_a^b g(t)\,\mathrm{d}t - \int_{ta+(1-t)b}^{tb+(1-t)a} g(t)\,\mathrm{d}t} \leqslant \frac{\int_a^b f(t)\,\mathrm{d}t}{\int_a^b g(t)\,\mathrm{d}t} \qquad (10)$$

where $\frac{1}{2} \leqslant t < 1$ or $0 \leqslant t \leqslant \frac{1}{2}$.

Theorem 4 Let $f, -g$ both are nonnegative convex functions on $[a, b]$ satisfying $\int_a^b g(x)\,\mathrm{d}x > 0$, then for $a < b$, we have

$$\frac{f\left(\frac{a+b}{2}\right)}{g\left(\frac{a+b}{2}\right)} \leqslant \frac{\int_a^b f(t)\,\mathrm{d}t}{\int_a^b g(t)\,\mathrm{d}t} - \frac{L(ta + (1-t)b, tb + (1-t)a; f, g)}{2(b-a)g\left(\frac{a+b}{2}\right)\int_a^b g(t)\,\mathrm{d}t} \leqslant \frac{\int_a^b f(t)\,\mathrm{d}t}{\int_a^b g(t)\,\mathrm{d}t} \qquad (11)$$

and for $0 < a < b$, we have

$$\frac{f\left(\frac{a+b}{2}\right)}{g\left(\frac{a+b}{2}\right)} \leqslant \frac{\int_a^b f(t)\,\mathrm{d}t}{\int_a^b g(t)\,\mathrm{d}t} - \frac{L\left(b^t a^{1-t}, a^t b^{1-t}; f, g\right)}{2(b-a)g\left(\frac{a+b}{2}\right)\int_a^b g(t)\,\mathrm{d}t} \leqslant \frac{\int_a^b f(t)\,\mathrm{d}t}{\int_a^b g(t)\,\mathrm{d}t} \qquad (12)$$

where $\frac{1}{2} \leqslant t \leqslant 1$ or $0 \leqslant t \leqslant \frac{1}{2}$.

46.2 Definitions and Lemmas

We need the following definitions and lemmas.

Definition 1[9−10] Let $\boldsymbol{x} = (x_1, \ldots, x_n)$ and $\boldsymbol{y} = (y_1, \ldots, y_n) \in \mathbf{R}^n$.

(i) $\boldsymbol{x}$ is said to be majorized by $\boldsymbol{y}$ (in symbols $\boldsymbol{x} \prec \boldsymbol{y}$) if $\sum_{i=1}^{k} x_{[i]} \leqslant \sum_{i=1}^{k} y_{[i]}$ for $k = 1, 2, \ldots, n-1$ and $\sum_{i=1}^{n} x_i = \sum_{i=1}^{n} y_i$, where $x_{[1]} \geqslant \cdots \geqslant x_{[n]}$ and $y_{[1]} \geqslant \cdots \geqslant y_{[n]}$ are of $\boldsymbol{x}$ and $\boldsymbol{y}$ in a descending order.

(ii) Let $\Omega \subseteq \mathbf{R}^n$. The function $\varphi \colon \Omega \to \mathbf{R}$ be said to be a Schur-convex function on Ω if $\boldsymbol{x} \prec \boldsymbol{y}$ on Ω implies $\varphi(\boldsymbol{x}) \leqslant \varphi(\boldsymbol{y})$. φ is said to be a Schur-concave function on Ω if and only if $-\varphi$ is Schur-convex.

Definition 1[11−12] Let $\boldsymbol{x} = (x_1, \ldots, x_n)$ and $\boldsymbol{y} = (y_1, \ldots, y_n) \in \mathbf{R}_+^n$.

(i) $\Omega \subseteq \mathbf{R}_+^n$ is called a geometrical convex set if $(x_1^\alpha y_1^\beta, \ldots, x_n^\alpha y_n^\beta) \in \Omega$ for all x and $y \in \Omega$, where α and $\beta \in [0,1]$ with $\alpha + \beta = 1$.

(ii) Let $\Omega \subseteq \mathbf{R}_+^n$. The function $\varphi: \Omega \to \mathbf{R}_+$ is said to be Schur-geometrical convex function on Ω if $(\ln x_1, \ldots, \ln x_n) \prec (\ln y_1, \ldots, \ln y_n)$ on Ω implies $\varphi(x) \leqslant \varphi(y)$. The function φ is said to be a Schur-geometrical concave on Ω if and only if $-\varphi$ is Schur-geometrical convex.

Definition 3[13] Let $x = (x_1, \ldots, x_n)$ and $y = (y_1, \ldots, y_n) \in \mathbf{R}_+^n$.

(i) $\Omega \subseteq \mathbf{R}_+^n$ is called a harmonic convex set if $(x_1 y_1/(\alpha x_1 + \beta y_1), \ldots, x_n y_n/(\alpha x_n + \beta y_n)) \in \Omega$ for all x and $y \in \Omega$, where α and $\beta \in [0,1]$ with $\alpha + \beta = 1$.

(ii) Let $\Omega \subseteq \mathbf{R}_+^n$. The function $\varphi: \Omega \to \mathbf{R}_+$ is said to be Schur-harmonic convex function on Ω if $(1/x_1, \ldots, 1/x_n) \prec (1/y_1, \ldots, 1/y_n)$ on Ω implies $\varphi(x) \leqslant (\geqslant) \varphi(y)$. The function φ is said to be a Schur-harmonic concave on Ω if and only if $-\varphi$ is Schur-harmonic convex.

Lemma 1[9-10] Let $\Omega \subseteq \mathbf{R}^n$ be a symmetric set and with a nonempty interior Ω^0, $\varphi : \Omega \to \mathbf{R}$ be a continuous on Ω and differentiable in Ω^0. Then φ is the $Schur-convex(Schur-concave)function$, if and only if φ is symmetric on Ω and

$$(x_1 - x_2)\left(\frac{\partial \varphi}{\partial x_1} - \frac{\partial \varphi}{\partial x_2}\right) \geqslant 0 (\leqslant 0) \tag{13}$$

holds for any $x = (x_1, \cdots, x_n) \in \Omega^0$.

Lemma 2[11] Let $\Omega \subseteq \mathbf{R}_+^n$ be symmetric with a nonempty interior geometrically convex set. Let $\varphi : \Omega \to \mathbf{R}_+$ be continuous on Ω and differentiable in Ω^0. If φ is symmetric on Ω and

$$(\ln x_1 - \ln x_2)\left(x_1 \frac{\partial \varphi}{\partial x_1} - x_2 \frac{\partial \varphi}{\partial x_2}\right) \geqslant 0 (\leqslant 0) \tag{14}$$

holds for any $x = (x_1, \cdots, x_n) \in \Omega^0$, then φ is a Schur-geometrical convex (Schur-geometrical concave) function.

Lemma 3[13] Let $\Omega \subseteq \mathbf{R}_+^n$ be symmetric with a nonempty interior harmonic convex set. Let $\varphi : \Omega \to \mathbf{R}_+$ be continuous on Ω and differentiable in Ω^0. If φ is symmetric on Ω and

$$(x_1 - x_2)\left(x_1^2 \frac{\partial \varphi}{\partial x_1} - x_2^2 \frac{\partial \varphi}{\partial x_2}\right) \geqslant 0 (\leqslant 0) \tag{15}$$

holds for any $x = (x_1, \cdots, x_n) \in \Omega^0$, then φ is a Schur-harmonic convex (Schur-harmonic concave) function.

Lemma 4[14] Let $a \leqslant b, u(t) = ta + (1-t)b, v(t) = tb + (1-t)a$. If $1/2 \leqslant t_2 \leqslant t_1 \leqslant 1$ or $0 \leqslant t_1 \leqslant t_2 \leqslant 1/2$, then

$$\left(\frac{a+b}{2}, \frac{a+b}{2}\right) \prec (u(t_2), v(t_2)) \prec (u(t_1), v(t_1)) \prec (a, b) \qquad (16)$$

Lemma 5[18] Let I be an interval with nonempty interior on $\mathbf{R}$ and f be a continuous function on I. Then

$$\Phi(a, b) = \begin{cases} \frac{1}{b-a} \int_a^b f(t) dt, & a, b \in I, \ a \neq b \\ f(a), & a = b \end{cases}$$

is $Schur-convex(Schur-concave)$ on I^2 if and only if f is $convex(concave)$on I.

Lemma 6 Let f and $-g$ both be convex function on $[a, b] \subseteq \mathbf{R}$. If $\int_b^a g(x) \, dx \geqslant 0$ and $f\left(\frac{a+b}{2}\right) \geqslant 0$, then

$$L(a, b; f, g) \leqslant 2(b-a)\left[g\left(\frac{a+b}{2}\right)\int_a^b f(t)\,dt - f\left(\frac{a+b}{2}\right)\int_a^b g(t)\,dt\right] \qquad (17)$$

Proof

$L(a, b; f, g)$

$$= \left[\int_a^b f(t)\,dt - (b-a)f\left(\frac{a+b}{2}\right)\right]\left[(b-a)g\left(\frac{a+b}{2}\right) - \int_a^b g(t)\,dt\right]$$

$$= (b-a)g\left(\frac{a+b}{2}\right)\int_a^b f(t)\,dt - \int_a^b f(t)\,dt\int_a^b g(t)\,dt$$

$$- (b-a)^2 f\left(\frac{a+b}{2}\right) g\left(\frac{a+b}{2}\right) + (b-a)f\left(\frac{a+b}{2}\right)\int_a^b g(t)\,dt \qquad (18)$$

Combining (18) with (3.6) and (3.7) in [10], it following that (17) is hold.

46.3　Proofs of main results

Proof of Theorem 1:

(i) It is clear that $L(x, y; f, g)$ is symmetric with x, y. Without loss of generality, we may assume $y \geqslant x$. Directly calculating yields

$$\frac{\partial L}{\partial y} = \left[f(y) - f\left(\frac{x+y}{2}\right) - \frac{y-x}{2} f'\left(\frac{x|!+y}{2}\right) \right] \left[(y-x)g\left(\frac{x+y}{2}\right) - \int_x^y g(t)\, dt \right] +$$
$$\left[\int_x^y f(t)\, dt - (y-x)f\left(\frac{x+y}{2}\right) \right] \left[g\left(\frac{x+y}{2}\right) + \frac{y-x}{2} g'\left(\frac{x+y}{2}\right) - g(y) \right]$$

$$\frac{\partial L}{\partial x} = \left[-f(x) + f\left(\frac{x+y}{2}\right) - \frac{y-x}{2} f'\left(\frac{x+y}{2}\right) \right] \left[(y-x)g\left(\frac{x+y}{2}\right) - \int_x^y g(t)\, dt \right] +$$
$$\left[\int_x^y f(t)\, dt - (y-x)f\left(\frac{x+y}{2}\right) \right] \left[-g\left(\frac{x+y}{2}\right) + \frac{y-x}{2} g'\left(\frac{x+y}{2}\right) + g(x) \right]$$

By Lagrange mean value theorem, there is $\xi \in ((x+y)/2, y)$ such that

$$f(y) - f\left(\frac{x+y}{2}\right) = \left(y - \frac{x+y}{2}\right) f'(\xi) = \frac{y-x}{2} f'(\xi)$$

Since f is convex, f' is increasing, we have $f'(\xi) \geqslant f'\left(\frac{x+y}{2}\right)$, so

$$f(y) - f\left(\frac{x+y}{2}\right) - \frac{y-x}{2} f'\left(\frac{x+y}{2}\right) \geqslant 0$$

By the same arguments, we have

$$-f(x) + f\left(\frac{x+y}{2}\right) - \frac{y-x}{2} f'\left(\frac{x+y}{2}\right) \leqslant 0$$

Similarly, since $-g$ is convex, we have

$$g\left(\frac{x+y}{2}\right) + \frac{y-x}{2} g'\left(\frac{x+y}{2}\right) - g(y) \geqslant 0$$

and

$$-g\left(\frac{x+y}{2}\right) + \frac{y-x}{2} g'\left(\frac{x+y}{2}\right) + g(x) \leqslant 0$$

And by Hadamard's inequality (1), it follows that $(y-x)g\left(\frac{x+y}{2}\right) - \int_x^y g(t)\, dt \geqslant 0$ and $\int_x^y f(t)\, dt - (y-x)f\left(\frac{x+y}{2}\right) \geqslant 0$. So $\frac{\partial L}{\partial y} \geqslant 0$ and $\frac{\partial L}{\partial x} \leqslant 0$, further $(y-x)\left(\frac{\partial L}{\partial y} - \frac{\partial L}{\partial x}\right) \geqslant 0$ and $(x-y)\left(x^2 \frac{\partial L}{\partial x} - y^2 \frac{\partial L}{\partial y}\right) \geqslant 0$. Notice that from $y \geqslant x$, we have $\ln x - \ln y \leqslant 0$, and then $(\ln x - \ln y)\left(x\frac{\partial L}{\partial x} - y\frac{\partial L}{\partial y}\right) \geqslant 0$. According to Lemma 1, Lemma 2 and Lemma 3, it follows that $L(x, y; f, g)$ is Schur-convex in $[a, b] \times [a, b] \subseteq \mathbf{R}^2$, and $L(x, y; f, g)$ is Schur-geometrical convex and Schur-harmonic convex in $[a, b] \times [a, b] \subseteq \mathbf{R}_+^2$.

(ii) From Lemma 4, we have

$$\left(\ln\sqrt{ab}, \ln\sqrt{ab}\right) \prec \left(\ln(b^{t_2}a^{1-t_2}), \ln(a^{t_2}b^{1-t_2})\right)$$

$$\prec \left(\ln(b^{t_1}a^{1-t_1}), \ln(a^{t_1}b^{1-t_1})\right) \prec (\ln a, \ln b) \tag{19}$$

By (i) in Theorem 1, from (16) and (19) it follows that (6), (8) and (7) are hold.

The proof of Theorem 1 is completed.

Proof of Theorem 2:

(i) It is clear that $F(x, y; f, g)$ is symmetric. Without loss of generality, we may assume $y \geqslant x$. Directly calculating yields

$$\frac{\partial F}{\partial y} = \frac{1}{2}g'\left(\frac{x+y}{2}\right)\int_x^y f(t)\,\mathrm{d}t + g\left(\frac{x+y}{2}\right)f(y)$$
$$- \frac{1}{2}f'\left(\frac{x+y}{2}\right)\int_x^y g(t)\,\mathrm{d}t - f\left(\frac{x+y}{2}\right)g(y)$$

$$\frac{\partial F}{\partial x} = \frac{1}{2}g'\left(\frac{x+y}{2}\right)\int_x^y f(t)\,\mathrm{d}t - g\left(\frac{x+y}{2}\right)f(x)$$
$$- \frac{1}{2}f'\left(\frac{x+y}{2}\right)\int_x^y g(t)\,\mathrm{d}t + f\left(\frac{x+y}{2}\right)g(x)$$

and then

$$(y-x)\left(\frac{\partial F}{\partial y} - \frac{\partial F}{\partial x}\right)$$
$$= (y-x)\left[g\left(\frac{x+y}{2}\right)(f(x)+f(y)) - f\left(\frac{x+y}{2}\right)(g(x)+g(y))\right]$$

Since f and $-g$ both be convex function on $[a,b]$, $f(x)+f(y) \geqslant 2f\left(\frac{x+y}{2}\right)$ and $g\left(\frac{x+y}{2}\right) \geqslant \frac{g(x)+g(y)}{2}$, and then $g\left(\frac{x+y}{2}\right)(f(x)+f(y)) - f\left(\frac{x+y}{2}\right)(g(x)+g(y)) \geqslant 0$, so $(y-x)\left(\frac{\partial F}{\partial y} - \frac{\partial F}{\partial x}\right) \geqslant 0$. From Lemma 1, it follows that $F(x, y; f, g)$ is Schur-convex on $[a,b] \times [a,b]$.

(ii) By (i) in Theorem 2, from (16) it follows that the (9) is hold.

The proof of Theorem 2 is completed.

Proof of Theorem 3:

By the Theorem 2, for $\frac{1}{2} \leqslant t < 1$ or $0 \leqslant t \leqslant \frac{1}{2}$, we have

$$F(ta + (1-t)b, tb + (1-t)a; f, g)$$

$$
= g\left(\frac{a+b}{2}\right) \int_{ta+(1-t)b}^{tb+(1-t)a} f(t)\,\mathrm{d}t - f\left(\frac{a+b}{2}\right) \int_{ta+(1-t)b}^{tb+(1-t)a} g(t)\,\mathrm{d}t
$$

$$
\leqslant g\left(\frac{a+b}{2}\right) \int_{a}^{b} f(t)\,\mathrm{d}t - f\left(\frac{a+b}{2}\right) \int_{a}^{b} g(t)\,\mathrm{d}t = F(a,b;f,g)
$$

i.e.

$$
f\left(\frac{a+b}{2}\right)\left[\int_{a}^{b} g(t)\,\mathrm{d}t - \int_{ta+(1-t)b}^{tb+(1-t)a} g(t)\,\mathrm{d}t\right]
$$

$$
\leqslant g\left(\frac{a+b}{2}\right)\left[\int_{a}^{b} f(t)\,\mathrm{d}t - \int_{ta+(1-t)b}^{tb+(1-t)a} f(t)\,\mathrm{d}t\right]
$$

which is equivalent to left inequality in (10).

Since f is convex on $[a,b]$, by Lemma 5, it follows that $\frac{1}{y-x}\int_{x}^{y} f(t)\,\mathrm{d}t$ is Schur convex on $[a,b]\times[a,b]$, and since $-g$ is convex on $[a,b]$, i.e. g is concave on $[a,b]$, by Lemma 5, it follows that $\frac{1}{y-x}\int_{x}^{y} g(t)\,\mathrm{d}t$ is Schur concave on $[a,b]\times[a,b]$, and then

$$
\frac{\frac{1}{y-x}\int_{x}^{y} f(t)\,\mathrm{d}t}{\frac{1}{y-x}\int_{x}^{y} g(t)\,\mathrm{d}t} = \frac{\int_{x}^{y} f(t)\,\mathrm{d}t}{\int_{x}^{y} g(t)\,\mathrm{d}t}
$$

is Schur convex on $[a,b]\times[a,b]$. Therefore, from (16) we have

$$
\frac{\int_{ta+(1-t)b}^{tb+(1-t)a} f(t)\,\mathrm{d}t}{\int_{ta+(1-t)b}^{tb+(1-t)a} g(t)\,\mathrm{d}t} \leqslant \frac{\int_{a}^{b} f(t)\,\mathrm{d}t}{\int_{a}^{b} g(t)\,\mathrm{d}t}
$$

The above inequality equivalent to the right inequality in (10).

The proof of Theorem 3 is completed.

Proof of Theorem 4:

By the Theorem 1, for $a < b$, we have

$$
L(ta + (1-t)b, tb + (1-t)a; f,g) \leqslant L(a,b;f,g) \tag{20}
$$

and for $0 < a < b$, we have

$$
L\left(b^{t}a^{1-t}, a^{t}b^{1-t}; f,g\right) \leqslant L(a,b;f,g) \tag{21}
$$

Combining (17) with (20) and (21) respectively, it is deduced that (11) and (12) are hold.

The proof of Theorem 4 is completed.

46.4   Applications

Corollary 1 Let $a, b \in \mathbf{R}_+$ with $a < b$, and let $u = tb + (1-t)a, v = ta + (1-t)b, \frac{1}{2} \leqslant t < 1$ or $0 \leqslant t \leqslant \frac{1}{2}$. Then for $1 \leqslant r \leqslant 2$, we have

$$\left(\frac{2}{a+b}\right)^r \leqslant \frac{r[(\ln b - \ln a) - (\ln u - \ln v)]}{2(b-a)(1-t)} \leqslant \frac{r(\ln b - \ln a)}{b-a} \tag{22}$$

Proof For $1 \leqslant r \leqslant 2$, taking $f(x) = x^{-1}$ and $g(x) = x^{r-1}$, then f and $-g$ both be convex function on $[a, b]$. From Theorem 3, it follows that (22) is hold.

The proof of Corollary 1 is completed.

Remark 1 Taking $r = 1$, from (22), we have

$$\frac{2}{a+b} \leqslant \frac{(\ln b - \ln a) - (\ln u - \ln v)}{2(b-a)(1-t)} \leqslant \frac{\ln b - \ln a}{b-a} \tag{23}$$

(23) is a refinement of the following Ostle-Terwilliger inequality

$$\frac{\ln b - \ln a}{b-a} \geqslant \frac{2}{a+b} \tag{24}$$

Corollary 2 Let $a, b \in \mathbf{R}_+$ with $a < b$, and let $u = tb + (1-t)a, v = ta + (1-t)b, \frac{1}{2} \leqslant t < 1$ or $0 \leqslant t \leqslant \frac{1}{2}$. Then for $1 \leqslant r \leqslant 2$, we have

$$\frac{a+b}{2} \leqslant \left[\frac{(b^{2r} - a^{2r}) - (u^{2r} - v^{2r})}{2(b^r - a^r) - 2(u^r - v^r)}\right]^{\frac{1}{r}} \leqslant \left(\frac{a^r + b^r}{2}\right)^{\frac{1}{r}} \tag{25}$$

Proof For $1 \leqslant r \leqslant 2$, taking $f(x) = x^{2r-1}$ and $g(x) = x^{r-1}$, then f and $-g$ both be convex function on $[a, b]$, from Theorem 3, it is easy to prove that (25) is hold.

The proof of Corollary 2 is completed.

Acknowledgements. The author is indebted to the referees for their helpful suggestions.

References

[1]   J HADAMARD. Étude sur les propriétés des fonctions entières et en particulier d'une fonction considérée par Riemann[J]. J. Math. Pures Appl., 1893, 58 : 171-215.

[2]   S S DRAGOMIR, Y J CHO, S S KIM. Inequalities of Hadamard'
s type for Lipschitzian mappings and their applications[J]. J. Math.
Anal. Appl., 2000, 245 : 489-501.

[3]   G-S YANG, K-L TSENG. Inequalities of Hadamard's type for Lipschitzian mappings[J]. J. Math. Anal. Appl., 2001, 260 : 230-238.

[4]   M MATIĆ, J PEČARIĆ. Note on inequalities of Hadamard's type for
Lipschitzian mappings[J]. Tamkang J. Math.,2001, 32(2) : 127-130.

[5]   S-J YANG. A direct proof and extensions of an inequality[J]. J. Math.
Res. Exposit., 2004, 24(4): 649-652.

[6]   S S DRAGOMIR, R P AGARWAL. Two new mappings associated with
Hadamard's inequalities for convex functions[J]. Appl. Math. Lett.,
1998, 11(3): 33-38.

[7]   L-C WANG. Some refinements of Hermite-Hadamard inequalities for
convex functions[J]. Univ. Beograd. Publ. Elektrotehn. Fak. Ser.
Mat., 2004, 15 : 40-45.

[8]   L HE. Two new mappings associated with inequalities of Hadamard-
type for convex functions[J]. J. Inequal. Pure and Appl. Math., 2009,
10(3), Art. 81, 5 pp.

[9]   A M MARSHALL, I OLKIN. Inequalities:theory of majorization and
its application[M]. New York :Academies Press, 1979.

[10]   B-Y WANG. Foundations of Majorization Inequalities[M]. Beijing:
Beijing Normal Univ. Press, 1990.

[11]   X-M ZANG. Geometrically Convex Functions[M]. Hefei: An hui University Press, 2004.

[12]   CONSTANTIN P NICULESCU. Convexity according to the geometric
mean[J]. Math. Inequal. Appl., 2000, 3(2): 155-167.

[13]   Y-M CHU, Y-P LV. The Schur harmonic convexity of the Hamy symmetric function and its applications[J]. J.Inequal. Appl., Volume 2009,
Article ID 838529, 10 pages doi:10.1155/2009/838529.

[14]   H-N SHI, Y-M JIANG, W-D JIANG. Schur-Convexity and Schur-
Geometrically Concavity of Gini Mean[J]. Comput. Math. Appl.,2009,
57 :266-274.

[15]   H-N SHI. Schur-Convex Functions relate to Hadamard-type inequalities[J]. J. Math. Inequal., 2007, 1 (1): 127-136.

[16] H-N SHI, D-M LI, CH GU. Schur-Convexity of a Mean of Convex Function[J]. Appl. Math. Lett., 2009, 22 : 932-937.

[17] X-M ZANG. Geometric convexity of integral of a geometrically concave function[J]. Int. J. Math. Inequal. Appl.,2007, 1(1): 121-130.

[18] ELEZOVIĆ N, J E PEČARIĆ. Note on Schur-convex functions[J]. Rocky Mountain J. Math., 1998, 29: 853-856.

[19] B OSTLE, H L TERWILLIGER. A companion of two means[J]. Proc. Montana Acad. Sci. 1957, 17 (1): 69-70.

第47篇　加权算术-几何平均值不等式的控制证明

(张鉴, 石焕南. 北京联合大学学报(自然科学版), 2011, 25 (4):46-47)

摘　要:　众所周知，算术－几何平均值不等式是最基本、最重要的不等式，寻求它的不同证法，一直是人们研究的热点，至今已有上百种不同的证明方法. 本文利用控制不等式的方法，并结合分析技巧给出加权算术——几何平均值不等式的一个新的证明.

关键词:　加权算术; 几何平均值不等式; 控制; Schur-凹; 初等对称函数

本文中, $\mathbf{R}^n$ 表示实数域上 n 维行向量的集合, 记

$$\mathbf{R}_+^n = \{\boldsymbol{x} = (x_1, \ldots, x_n) \in \mathbf{R}^n | x_i \geqslant 0, i = 1, \ldots, n\}$$

$$\mathbf{R}_{++}^n = \{\boldsymbol{x} = (x_1, \ldots, x_n) \in \mathbf{R}^n | x_i > 0, i = 1, \ldots, n\}$$

并记 $\mathbf{R}^1 = \mathbf{R}, \mathbf{R}_+^1 = \mathbf{R}_+$ 和 $\mathbf{R}_{++}^1 = \mathbf{R}_{++}$.

n 维向量 $\boldsymbol{x}$ 的第 k 个初等对称函数为

$$E_k(\boldsymbol{x}) = E_k(x_1, \ldots, x_n) = \sum_{1 \leqslant i_1 < \ldots < i_k \leqslant n} \prod_{j=1}^{k} x_{i_j}$$

特别 $E_n(\boldsymbol{x}) = \prod_{i=1}^{n} x_i$, $E_1(\boldsymbol{x}) = \sum_{i=1}^{n} x_i$. 当 $k > n$ 时, 规定 $E_n(\boldsymbol{x}) = 0$, 并规定 $E_0(\boldsymbol{x}) = 1$.

文献[1] 利用凸函数的性质及 Jensen 不等式

$$-\ln\left(\frac{a_j}{\sum_{i=1}^{n} a_i} \cdot \frac{x_j}{a_j}\right) \leqslant \sum_{j=1}^{n} \frac{a_j}{\sum_{i=1}^{n} a_i} \cdot \ln \frac{x_j}{a_j}$$

得到以下结论:

339

定理 1 设 $\boldsymbol{x}, \boldsymbol{a} \in \mathbf{R}_{++}^n$, 则

$$\prod_{i=1}^n x_i^{a_i} \leqslant \prod_{i=1}^n a_i^{a_i} \cdot \left(\frac{\sum\limits_{i=1}^n x_i}{\sum\limits_{i=1}^n a_i} \right)^{\sum\limits_{i=1}^n a_i} \tag{1}$$

实际上, 式(1) 是加权算术-几何平均值不等式的变形, 若以 $a_i x_i$ 置换 $a_i, i = 1, \ldots, n$, 则式(1) 就化为

$$\left(\prod_{i=1}^n x_i^{a_i} \right)^{\frac{1}{\sum_{i=1}^n a_i}} \leqslant \frac{1}{\sum_{i=1}^n a_i} \sum_{i=1}^n a_i x_i \tag{2}$$

文献[2]利用Lagrange数乘法给出式(1) 的另一证明. 本文结合初等对称函数的Schur 凹性与一个简单的受控关系给出式(1), 即加权算术-几何平均值不等式(2) 的控制(majorization) 证明. 等权算术-几何平均值不等式两种不同的控制证明请参见文献[3]和文献[4].

对于 $\boldsymbol{x} = (x_1, \ldots, x_n) \in \mathbf{R}^n$, 将 $\boldsymbol{x}$ 的分量排成递减的次序后, 记作 $x_{[1]} \geqslant \ldots \geqslant x_{[n]}$.

定义 1[4,5] 设 $\boldsymbol{x}, \boldsymbol{y}$ 满足

(1) $\sum\limits_{i=1}^k x_{[i]} \leqslant \sum\limits_{i=1}^k y_{[i]}, k = 1, 2, \ldots, n-1$; (2) $\sum\limits_{i=1}^n x_i = \sum\limits_{i=1}^n y_i$. 则称 $\boldsymbol{x}$ 被 $\boldsymbol{y}$ 所控制, 记作 $\boldsymbol{x} \prec \boldsymbol{y}$.

定义 2[4-5] 设 $\Omega \subset \mathbf{R}^n, \phi : \Omega \to \mathbf{R}$, 若在 Ω 上 $\boldsymbol{x} \prec \boldsymbol{y} \Rightarrow \phi(\boldsymbol{x}) \leqslant \phi(\boldsymbol{y})$, 则称 ϕ 为 Ω 上的Schur凸函数.

引理 1[4-5] 设 $\boldsymbol{x} = (x_1, \ldots, x_n) \in \mathbf{R}^n, \overline{x} = \frac{1}{n} \sum\limits_{i=1}^n x_i$ 则

$$(\overline{x}, \overline{x}, \ldots, \overline{x}) \prec (x_1, x_2, \ldots, x_n)$$

引理 2[5] 初等对称函数 $E_k(\boldsymbol{x}), k = 1, \ldots, n$ 是 $\mathbf{R}_+^n$ 上递增的Schur凹函数.

定理1 的证明 记 $m = \sum\limits_{i=1}^n a_i, s = \sum\limits_{i=1}^n x_i$, 则式(1) 化为

$$\left(\frac{s}{m} \right)^m \geqslant \left(\prod_{i=1}^n \frac{x_i}{a_i} \right)^{a_i} \tag{3}$$

先考虑$a_1, \ldots, a_n$ 为正整数$k_1, \ldots, k_n$的情形, 由引理1 得

$$\boldsymbol{u} = \Bigg(\underbrace{\frac{s}{m}, \ldots, \frac{s}{m}}_{m} \Bigg)$$

$$\prec \Bigg(\underbrace{\frac{x_1}{k_1}, \ldots, \frac{x_1}{k_1}}_{k_1}, \underbrace{\frac{x_2}{k_2}, \ldots, \frac{x_2}{k_2}}_{k_2}, \ldots, \underbrace{\frac{x_n}{k_n}, \ldots, \frac{x_n}{k_n}}_{k_n} \Bigg) = \boldsymbol{v}$$

由引理2 知$E_n(\boldsymbol{x})$ 是$\mathbf{R}_{++}^n$ 上的Schur凹函数, 故$E_n(\boldsymbol{u}) \geqslant E_n(\boldsymbol{v})$, 即式(3) 成立.

再考虑$a_1, \ldots, a_n$ 为正有理数的情形. 设$a_i = \frac{p_i}{q_i}, p_i, q_i, i = 1, \ldots, n$ 为正整数. 不妨设$a_i = \frac{p_i}{q}$, 其中q 是$p_1, \ldots, p_n$ 的公分母, 则$m = \sum\limits_{i=1}^{n} a_i = \frac{1}{q} \sum\limits_{i=1}^{n} p_i$, 此时

$$\left(\frac{s}{m} \right)^m \geqslant \left(\prod_{i=1}^{n} \frac{x_i}{a_i} \right)^{a_i} \Leftrightarrow \left(\frac{s}{\frac{1}{q} \sum\limits_{i=1}^{n} p_i} \right)^{\frac{1}{q} \sum\limits_{i=1}^{n} p_i} \geqslant \prod_{i=1}^{n} \left(\frac{x_i}{\frac{p_i}{q}} \right)^{\frac{p_i}{q}}$$

$$\Leftrightarrow \left(\frac{qs}{\sum\limits_{i=1}^{n} p_i} \right)^{\sum\limits_{i=1}^{n} p_i} \geqslant \prod_{i=1}^{n} \left(\frac{qx_i}{p_i} \right)^{p_i} \Leftrightarrow \left(\frac{s}{\sum\limits_{i=1}^{n} p_i} \right)^{\sum\limits_{i=1}^{n} p_i} \geqslant \prod_{i=1}^{n} \left(\frac{x_i}{p_i} \right)^{p_i}$$

注意$p_i, i = 1, \ldots, n$ 为正整数, 由式(1) 知上面最后一个不等式成立, 故$a_1, \ldots, a_n$ 为有理数时式(3) 成立.

最后考虑 $a_1, \ldots, a_n$ 为正无理数的情形. 此时存在n 个正有理数列$\{r_{i_k}\}, k = 1, 2, \ldots, n$ 且$r_{i_k} \to a_i (k \to \infty)$. 对于有理数$r_{i_k}, i = 1, \ldots, n$, 由式(2) 得

$$\left(\frac{s}{\sum\limits_{i=1}^{n} r_{i_k}} \right)^{\sum\limits_{i=1}^{n} r_{i_k}} \geqslant \prod_{i=1}^{n} \left(\frac{x_i}{r_{i_k}} \right)^{r_{i_k}}$$

令$k \to \infty$, 即知$a_1, \ldots, a_n$ 为正无理数时式(3)成立, 至此, 对于任意正实数$a_1, \ldots, a_n$, 式(3) 成立, 证毕.

1999 年, Paul 和Jack Abad 在美国数学协会(MAA) 上推荐了 "100 个最伟大的定理" (The Hundred Greatest Theorems) , 这些定理的排名是基

于这样的准则: 定理在文献里的地位, 有高质量的证明, 以及突破性的结果. 算术-几何平均值不等式位列 "100 个最伟大的定理" 的第38 位, 足见它在数学及其应用中拥有的崇高地位. 寻求对算术-几何平均值不等式的不同证法, 一直是研究的热点, 至今已有上百种不同的证明方法. 本文利用控制不等式的方法, 结合分析技巧所给出的证明独树一帜.

参考文献

[1] 钱照平. 一个有趣的不等式[J]. 高等数学研究，2007, 10(2) : 33 - 34.

[2] 赵德勤, 殷明. 一个有趣不等式的新证明方法及推论[J]. 大学数学, 2010, 26 (1) : 201 - 202.

[3] 张小明, 褚玉明. 解析不等式新论[M]. 哈尔滨: 哈尔滨工业大学出版社，2009.

[4] MARSHALL A M, OLKIN I. Inequalities: Theory of majorization and its application[M]. New York: Academies Press, 1979.

[5] 王伯英. 控制不等式基础[M]. 北京: 北京师范大学出版社, 1990.

第48篇　关于三个对称函数的
Schur-凸性

(石焕南. 河西学院学报, 2011, 28 (2)：13-17)

摘　要： 本文给出了三个对称函数Schur-凸性的新的证明, 改善了已有的结论.

关键词： 对称函数; Schur-凸性; 不等式

48.1　引言

在本文中, $\mathbf{R}^n$ 和$\mathbf{R}^n_{++}$ 分别表示n 维实数集和n 维正实数集, 并记$\mathbf{R}^1 = \mathbf{R}$, $\mathbf{R}^1_{++} = \mathbf{R}_{++}$.

对于$\boldsymbol{x} = (x_1, ..., x_n) \in \mathbf{R}^n$, 将$\boldsymbol{x}$ 的分量排成递减的次序后, 记作$x_{[1]} \geqslant x_2 \geqslant ... \geqslant x_{[n]}$.

定义 1[1-2]　设$\boldsymbol{x}, \boldsymbol{y} \in \mathbf{R}^n$ 满足

(1) $\sum\limits_{i=1}^k x_{[i]} \leqslant \sum\limits_{i=1}^k y_{[i]}$, $k = 1, 2, ..., n-1$;

(2) $\sum\limits_{i=1}^n x_i = \sum\limits_{i=1}^n y_i$, 则称$\boldsymbol{x}$ 被$\boldsymbol{y}$ 所控制, 记作$\boldsymbol{x} \prec \boldsymbol{y}$.

定义 2[1-2]　设$\Omega \subset \mathbf{R}^n$, $\varphi : \Omega \to \mathbf{R}$, 若在$\Omega$ 上$\boldsymbol{x} \prec \boldsymbol{y} \Rightarrow \varphi(\boldsymbol{x}) \leqslant \varphi(\boldsymbol{y})$, 则称$\varphi$ 为Ω 上的Schur凸函数(简称S-凸函数); 若$-\varphi$ 是Ω 上的S-凸函数, 则称φ 为Ω 上的S- 凹函数.

定理 A[1]　设$\Omega \subset \mathbf{R}^n$ 是有内点的对称凸集, $\varphi : \Omega \to \mathbf{R}$ 在Ω 上连续, 在Ω 的内部Ω^0 可微, 则在Ω 上S-凸(S-凹)的充要条件是φ 在Ω 上对称且$\forall \boldsymbol{x} \in \Omega$, 有

$$(x_1 - x_2)\left(\frac{\partial \varphi}{\partial x_1} - \frac{\partial \varphi}{\partial x_2}\right) \geqslant 0 (\leqslant 0) \tag{1}$$

Schur凸函数是受控理论(Theory of Majorization)最重要的概念, 定理A 是判断Schur凸函数最有力的工具, 故有些文献称其为Schur凸函数基本定

343

理.

应用定理A, 文[1], [3], [4]和[5]分别考察了如下三个对称函数

$$E_k\left(\frac{\boldsymbol{x}}{1-\boldsymbol{x}}\right) = \sum_{1\leqslant i_1<\ldots<i_k\leqslant n}\prod_{j=1}^{k}\frac{x_{i_j}}{1-x_{i_j}}, \quad k=1,2,\ldots,n \qquad (2)$$

$$E_k\left(\frac{1+\boldsymbol{x}}{1-\boldsymbol{x}}\right) = \sum_{1\leqslant i_1<\ldots<i_k\leqslant n}\prod_{j=1}^{k}\frac{1+x_{i_j}}{1-x_{i_j}}, \quad k=1,2,\ldots,n \qquad (3)$$

$$E_k\left(\frac{1-\boldsymbol{x}}{\boldsymbol{x}}\right) = \sum_{1\leqslant i_1<\ldots<i_k\leqslant n}\prod_{j=1}^{k}\frac{1-x_{i_j}}{x_{i_j}}, \quad k=1,2,\ldots,n \qquad (4)$$

的Schur凹凸性, 得到如下结果.

定理 B[3] 设$n\geqslant 2, 2\leqslant k\leqslant n$, 则$E_k(\frac{\boldsymbol{x}}{1-\boldsymbol{x}})$ 在$\left[\frac{k-1}{2(n-1)},1\right]^n$ 上S- 凸, 在$\left[0,\frac{k-1}{2(n-1)}\right]^n$ 上S-凹.

定理 C[4] $E_k(\frac{1+\boldsymbol{x}}{1-\boldsymbol{x}}), k=1,2,\ldots,n$ 在$(0,1)^n$ 上S-凸.

定理 D[1] (1)$E_k(\frac{1-\boldsymbol{x}}{\boldsymbol{x}})$ 在$\mathbf{B} = \{\boldsymbol{x}: \boldsymbol{x}\in(0,1)^n, x_i+x_j\leqslant 1, i\neq j\}$ 上S-凸.

(2)$E_k(\frac{1-\boldsymbol{x}}{\boldsymbol{x}})$ 在$\bar{\mathbf{B}} = \{\boldsymbol{x}: \boldsymbol{x}\in(0,1)^n, x_i+x_j\geqslant 1, i\neq j\}$ 上S- 凹.

定理 E[5] 对于$2\leqslant k\leqslant n$, $E_k(\frac{1-\boldsymbol{x}}{\boldsymbol{x}})$ 在$\left(0,\frac{2n-k-1}{2n-2}\right]^n$ 上S-凸, 在$\left[\frac{2n-k-1}{2n-2},1\right]^n$ 上S-凹.

设$\pi_i(1),\ldots,\pi_i(n)$ 是$1,\ldots,n$ 的任意置换$(i=1,2,\ldots,n!)$, 记$\pi_i=(\pi_i(1),\ldots,\pi_i(n))$, $1,2,\ldots,n!$ 又记$S_n=\{\pi_1,\ldots,\pi_{n!}\}$ 为$\{1,\ldots,n\}$ 的置换群, 且令$\pi_i(\boldsymbol{x})=(x_{\pi_i(1)},\ldots,(x_{\pi_i(n}$ 文[1]证明了如下结论.

定理 F[1] 设$\mathbf{A}\subset\mathbf{R}^k$ 是一个对称凸集, φ 是$\mathbf{A}$ 上的S-凸函数, 具有性质: 对每一个固定的$x_2,\ldots,x_k, \varphi(z,x_2,\ldots,x_k)$ 关于z 在$\{z:(z,x_2,\cdots,x_k)\in\boldsymbol{A}\}$ 上凸, 则对于任何$n>k$

$$\psi(x_1,\cdots,x_n) = \sum_{\pi}\varphi(x_{\pi(1)},\cdots,x_{\pi(k)})$$

在$\mathbf{B} = \{(x_1,\cdots,x_n):(x_{\pi(1)},\cdots,x_{\pi(k)})\in\mathbf{A}$, 对于所有的排列$\pi\}$ 上S-凸.

在大多数应用中, $\mathbf{A}$ 为$\mathbf{I}^k$ 的形式, 其中区间$\mathbf{I}\subset\mathbf{R}$, 且在这种情况下, $\mathbf{B}=\mathbf{I}^n$. 注意固定其他变量, φ 在第一个自变量上的凸性也蕴涵着在其他变量上的凸性, 因为φ 是对称的.

推论 令

$$\bar{\psi}(\boldsymbol{x}) = \frac{\psi(\boldsymbol{x})}{k!(n-k)!}$$

其中ψ 如定理F 所定义, 则

$$\overline{\psi}\,(x_1,...,x_n) = \sum_{1\leqslant i_1 < ... < i_k \leqslant n} \varphi\,(x_{i_1},\cdots,x_{i_k}) \tag{5}$$

且若ψ 是S-凸, 则$\overline{\psi}$ 也是S-凸的.

本文将应用定理F 及其推论研究这三个对称函数的Sehur-凸性, 我们的证明或推广了已有的结论, 或简化了已有的证明, 或纠正了已有结论的疏漏.

48.2　主要结果及其证明

定理 1　设$n \geqslant 2, 2 \leqslant k \leqslant n$, 则$E_k(\frac{x}{1-x})$ 在$\Omega = \{x : x \in (0,1)^n, x_i + x_j \geqslant 1, i \neq j\}$ 上S-凸.

证明　令$\varphi(z) = \prod_{i=1}^{k} \frac{z_i}{1-z_i}$, 则$\ln \varphi(z) = \sum_{i=1}^{k}[\ln z_i - \ln(1-z_i)]$, 于是

$$\frac{\partial \varphi(z)}{\partial z_1} = \varphi(z)\left(\frac{1}{z_1} + \frac{1}{1-z_1}\right), \quad \frac{\partial \varphi(z)}{\partial z_2} = \varphi(z)\left(\frac{1}{z_2} + \frac{1}{1-z_2}\right)$$

$$\begin{aligned}
\Delta :&= (z_1 - z_2)\left(\frac{\partial \varphi(z)}{\partial z_1} - \frac{\partial \varphi(z)}{\partial z_2}\right) \\
&= (z_1 - z_2)\varphi(z)\left(\frac{1}{z_1} - \frac{1}{z_2} + \frac{1}{1-z_1} - \frac{1}{1-z_2}\right) \\
&= (z_1 - z_2)^2 \varphi(z)\frac{z_1 + z_2 - 1}{z_2 z_1(1-z_2)(1-z_1)}
\end{aligned}$$

由此可见, 当$z_i + z_j \geqslant 1, i \neq j$ 且$0 < z_i < 1, i = 1,...,k$ 时, $\Delta \geqslant 0$, 即ψ 在$A = \{z : z \in (0,1)^k, z_i + z_j \geqslant 1, i \neq j\}$ 上S-凸. 又令$g(t) = \frac{t}{1-t}$, 当$t \in (0,1)$ 时, $g''(t) = \frac{2}{(1-t)^3} > 0$, 这意味着$\psi$ 对于单个变量在$(0,1)$ 凸, 故据定理F 和推论知$E_k(\frac{x}{1-x})$ 在Ω 上S-凸.

注 1　定理1 和定理B 的S-凸区域互不包含, 因此我们扩展了定理B 的结论.

定理 C 的证明　令$\varphi(z) = \prod_{i=1}^{k} \frac{1+z_i}{1-z_i}$, 则$\ln \varphi(z) = \sum_{i=1}^{k}[\ln(1+z_i) - \ln(1-z_i)]$ 于是

$$\frac{\partial \varphi(z)}{\partial z_1} = \varphi(z)\left(\frac{1}{1+z_1} + \frac{1}{1-z_1}\right), \quad \frac{\partial \varphi(z)}{\partial z_2} = \varphi(z)\left(\frac{1}{1+z_2} + \frac{1}{1-z_2}\right)$$

$$\Delta := (z_1 - z_2)\left(\frac{\partial \varphi(\boldsymbol{z})}{\partial z_1} - \frac{\partial \varphi(\boldsymbol{z})}{\partial z_2}\right)$$

$$= (z_1 - z_2)\varphi(\boldsymbol{z})\left(\frac{1}{1+z_1} - \frac{1}{1+z_2} + \frac{1}{1-z_1} - \frac{1}{1-z_2}\right)$$

$$= (z_1 - z_2)^2 \varphi(\boldsymbol{z})\frac{2(z_1+z_2)}{(1-z_2^2)(1-z_1^2)}$$

当 $0 < z_i < 1, i = 1, ..., k$ 时, $\Delta \geqslant 0$, 即 φ 在 $(0,1)^k$ 上S-凸. 又令 $g(t) = \frac{1+t}{1-t}$, 当 $t \in (0,1)$ 时, $g''(t) = \frac{4}{(1-t)^3} > 0$, 这意味着 φ 对于单个变量在 $(0,1)$ 凸. 故据定理F 和推论知 $E_k(\frac{1+\boldsymbol{x}}{1-\boldsymbol{x}})$ 在 $(0,1)^n$ 上S-凸.

注 2　上述应用定理F及其推论的证明比之文[4]的证明要简洁许多.

定理 D (1) 的证明　令 $\varphi(\boldsymbol{z}) = \prod\limits_{i=1}^{k} \frac{1-z_i}{z_i}$, 则 $\ln \varphi(z) = \sum\limits_{i=1}^{k}[\ln(1-z_i) - \ln z_i]$, 于是

$$\frac{\partial \varphi(\boldsymbol{z})}{\partial z_1} = \varphi(\boldsymbol{z})\left(\frac{-1}{1-z_1} - \frac{1}{z_1}\right), \quad \frac{\partial \varphi(\boldsymbol{z})}{\partial z_2} = \varphi(\boldsymbol{z})\left(\frac{-1}{1-z_2} - \frac{1}{z_2}\right)$$

$$\Delta := (z_1 - z_2)\left(\frac{\partial \varphi(\boldsymbol{z})}{\partial z_1} - \frac{\partial \varphi(\boldsymbol{z})}{\partial z_2}\right)$$

$$= (z_1 - z_2)\varphi(\boldsymbol{z})\left(\frac{1}{1-z_2} - \frac{1}{1-z_1} + \frac{1}{z_2} - \frac{1}{z_1}\right)$$

$$= (z_1 - z_2)^2 \varphi(\boldsymbol{z})\frac{1-z_1-z_2}{z_1 z_2 (1-z_2)(1-z_1)}$$

由此可见, 当 $z_i + z_j \geqslant 1, i \neq j$ 且 $0 < z_i < 1, i = 1, ..., k$ 时, $\Delta \geqslant 0$, 即 ψ 在 $\mathbf{A} = \{\boldsymbol{z} : \boldsymbol{z} \in (0,1)^k, z_i + z_j \geqslant 1, i \neq j\}$ 上S-凸. 又令 $g(t) = \frac{1-t}{t}$, 当 $t \in (0,1)$ 时, $g''(t) = \frac{2}{t^3} > 0$, 这意味着 ψ 对于单个变量在 $(0,1)$ 凸. 故据定理F 和推论知 $E_k(\frac{1-\boldsymbol{x}}{\boldsymbol{x}})$ 在 $\mathbf{B} = \{\boldsymbol{z} : \boldsymbol{z} \in \mathbf{R}_{++}^n, x_i + x_j \leqslant 1, i \neq j\}$ 上S-凸.

注 3　定理D(2) 的结论有误. 尽管 φ 在 $\bar{\mathbf{A}}$ 上S-凹, 但 φ 对于单个变量在 $(0,1)$ 不凹, 因此不能断言 $E_k(\frac{1-\boldsymbol{x}}{\boldsymbol{x}})$ 在 $\bar{\mathbf{B}} = \{\boldsymbol{x} : \boldsymbol{x} \in \mathbf{R}_{++}^n, x_i + x_j \geqslant 1, i \neq j\}$ 上S-凹. 定理E 的结果也印证了定理D(2) 的结论有误. 事实上, 显然

$$U := \left[\frac{2n-k-2}{2n-2}, \frac{2n-k-1}{2n-2}\right]^n \subset \left(0, \frac{2n-k-1}{2n-2}\right]^n$$

又当 $k \leqslant n-1$ 时, 对于任意 $\boldsymbol{x} \in U$, 有

$$x_i + x_j \geqslant 2 \cdot \frac{2n-k-2}{2n-2} = \frac{2n-k-2}{n-1} \geqslant 1$$

即$U \subset \bar{\mathbf{B}}$ 这样当$k \leqslant n-1$时, 就导出$E_k(\frac{1-x}{x})$ 在U上即S- 凸又S-凹的矛盾.

参考文献

[1]　MARSHALL A M , OLKIN I . Inequalities:theory of majorization and its application[M]. New York :Academies Press, 1979.

[2]　王伯英, 控制不等式基础[M]. 北京:北京师范大学出版社, 1990.

[3]　褚玉明, 夏卫锋, 赵铁洪. 一类对称函数的Schur凸性[J]. 中国科学(A辑: 数学), 2009, 39(11): 1267-1277.

[4]　WEI-FENG XIA, YU-MING CHU. On Schur−Convexity of Some Symmetric Functions[J]. Journal of Inequalities and Applications, Volume 2010, Article ID 543250, 12 pages doi: 10. 1155/2010/543250.

[5]　WEI-FENG XIA, YU-MING CHU. The Schur convexity and Schur multiplicative convexity for a class of symmetric functions with applications[J]. Ukrainian Mathematical Joumal, 2009, 61(10): 1306-1318.

第49篇　A Weak Constrained Majorization with Applications

(WU SHAN-HE, SHI HUAN-NAN. Mathematica Slovaca, 2011, 61

(4): 561-570)

Abstract: For positive real numbers $x_1, \ldots, x_n$ with $x_1 \ldots x_n \geqslant 1$, a weak majorization relation is established, and combining with the Schur-concavity of the elementary symmetric function and its dual form, some inequalities involving the power mean, the arithmetic mean and the geometric mean are obtained.

Keywords: weak majorization; inequality; Schur-concavity; elementary symmetric function; dual form; power mean; arithmetic mean; geometric mean

49.1　Definitions and Lemmas

Throughout the paper we denote the set of n-dimensional row vector on real number field by $\mathbf{R}^n$.

$$\mathbf{R}_+^n = \{\boldsymbol{x} = (x_1, \ldots, x_n) \in \mathbf{R}^n : x_i \geqslant 0, i = 1, \ldots, n\}$$

$$\mathbf{R}_{++}^n = \{\boldsymbol{x} = (x_1, \ldots, x_n) \in \mathbf{R}^n : x_i > 0, i = 1, \ldots, n\}$$

Let $\boldsymbol{x} = (x_1, \ldots, x_n) \in \mathbf{R}^n$. Its elementary symmetric functions are

$$E_k(\boldsymbol{x}) = E_k(x_1, \ldots, x_n) = \sum_{1 \leqslant i_1 < \ldots < i_k \leqslant n} \prod_{j=1}^{k} x_{i_j}, \quad k = 1, \ldots, n$$

The dual form of the elementary symmetric functions are

$$E_k^*(\boldsymbol{x}) = E_k^*(x_1, \ldots, x_n) = \prod_{1 \leqslant i_1 < \ldots < i_k \leqslant n} \sum_{j=1}^{k} x_{i_j}, \quad k = 1, \ldots, n$$

and defined $E_0^*(\boldsymbol{x}) = E_0(\boldsymbol{x}) = 1$, $E_k^*(\boldsymbol{x}) = E_0(\boldsymbol{x}) = 0$ for $k < 0$ or $k > n$.

We denote

$$E_k(\boldsymbol{x}^\alpha) = E_k(x_1^\alpha, \ldots, x_n^\alpha) = \sum_{1 \leqslant i_1 < \ldots < i_k \leqslant n} \prod_{j=1}^{k} x_{i_j}^\alpha$$

$$E_k^*(\boldsymbol{x}^\alpha) = E_k^*(x_1^\alpha, \ldots, x_n^\alpha) = \prod_{1 \leqslant i_1 < \ldots < i_k \leqslant n} \sum_{j=1}^{k} x_{i_j}^\alpha$$

$A = \frac{1}{n} \sum_{i=1}^{n} x_i$ and $G = \sqrt[n]{\prod_{i=1}^{n} x_i}$ denote the arithmetic mean and the geometric mean of the positive real numbers $x_1, \ldots, x_n$ respectively.

In this paper, a weak majorization relation is established for positive real numbers $x_1, \ldots, x_n$ with $x_1 \ldots x_n \geqslant 1$, and combining with the Schur-concavity of the elementary symmetric function and its dual form, some inequalities involving the power mean, the arithmetic mean and the geometric meanare obtained.

We need the following definitions and lemmas.

Definition 1[1-2] Let $\boldsymbol{x} = (x_1, \ldots, x_n)$ and $\boldsymbol{y} = (y_1, \ldots, y_n) \in \mathbf{R}^n$.

(i) $\boldsymbol{x}$ is said to be majorized by $\boldsymbol{y}$ (in symbols $\boldsymbol{x} \prec \boldsymbol{y}$) if $\sum_{i=1}^{k} x_{[i]} \leqslant \sum_{i=1}^{k} y_{[i]}$ for $k = 1, 2, \ldots, n-1$ and $\sum_{i=1}^{n} x_i = \sum_{i=1}^{n} y_i$; $\boldsymbol{x}$ is said to be weakly submajorized by $\boldsymbol{y}$ (in symbols $\boldsymbol{x} \prec_w \boldsymbol{y}$) if $\sum_{i=1}^{k} x_{[i]} \leqslant \sum_{i=1}^{k} y_{[i]}$ for $k = 1, 2, \ldots, n$, where $x_{[1]} \geqslant \cdots \geqslant x_{[n]}$ and $y_{[1]} \geqslant \cdots \geqslant y_{[n]}$ are rearrangements of $\boldsymbol{x}$ and $\boldsymbol{y}$ in a descending order.

(ii) $\boldsymbol{x} \geqslant \boldsymbol{y}$ means $x_i \geqslant y_i$ for all $i = 1, 2, \ldots, n$. Let $\Omega \subset \mathbf{R}^n$, $\varphi: \Omega \to \mathbf{R}$ is said to be increasing if $\boldsymbol{x} \geqslant \boldsymbol{y}$ implies $\varphi(\boldsymbol{x}) \geqslant \varphi(\boldsymbol{y})$. φ is said to be decreasing if and only if $-\varphi$ is increasing.

(iii) let $\Omega \subset \mathbf{R}^n$, $\varphi: \Omega \to \mathbf{R}$ be said to be a Schur-convex function on Ω if $\boldsymbol{x} \prec \boldsymbol{y}$ on Ω implies $\varphi(\boldsymbol{x}) \leq \varphi(\boldsymbol{y})$. φ is said to be the Schur-concave function on Ω if and only if $-\varphi$ is Schur-convex function.

Lemma 1[1] Let $x, y \in \mathbf{R}^n$ and $\delta = \sum_{i=1}^{n}(y_i - x_i)$. If $x \prec_w y$, then

$$\left(x, \underbrace{\frac{\delta}{n}, \ldots, \frac{\delta}{n}}_{n}\right) \prec \left(y, \underbrace{0, \ldots, 0}_{n}\right) \tag{1}$$

Lemma 2[2] Let $x, y \in \mathbf{R}^n$, if $x \prec_w y$, then

$$(x, x_{n+1}) \prec (y, y_{n+1}) \tag{2}$$

where $x_{n+1} = \min\{x_1, \cdots, x_n, y_1, \cdots, y_n\}$, $y_{n+1} = \sum_{i=1}^{n+1} x_i - \sum_{i=1}^{n} y_i$.

Lemma 3[1] Let $x, y \in \mathbf{R}^n$, $I \subset \mathbf{R}$ be an interval and $g: I \to \mathbf{R}$.

(i) $x \prec y$ if and only if

$$\sum_{i=1}^{n} g(x_i) \leqslant \sum_{i=1}^{n} g(y_i) \tag{3}$$

holds for all convex functions g,

(ii) $x \prec y$ if and only if (3) reverses for all concave functions g.

Lemma 4[1-2] Let $A \subset \mathbf{R}^n$, $\phi_i: A \to \mathbf{R}$ for $1 \leqslant i \leq k$, $h: \mathbf{R}^k \to \mathbf{R}$ and $\psi(x) = h(\phi_1(x), \ldots, \phi_k(x))$.

1. If each of ϕ_i for $1 \leqslant i \leqslant k$ is Schur-convex and h is increasing (decreasing), then ψ is Schur-convex (Schur-concave);

2. If each of ϕ_i for $1 \leqslant i \leqslant k$ is Schur-concave and h is increasing (decreasing), then ψ is Schur-concave (Schur-convex).

Lemma 5[1]

$E_k(x)$ be increasing and Schur-concave on $\mathbf{R}_+^n$.

Lemma 6[2]

$E_k^*(x)$ be increasing and Schur-concave on $\mathbf{R}_+^n$.

49.2　Main results and their proofs

In the following, we are in a position to state our main results and give proofs of them.

Theorem 1 Let $x = (x_1, \ldots, x_n) \in \mathbf{R}_{++}^n, n \geqslant 2$ and $\prod\limits_{i=1}^{n} x_i \geqslant 1$. Then

$$\Big(\underbrace{1, \ldots, 1}_{n}\Big) \prec_w (x_1, \ldots, x_n) \tag{4}$$

Proof We prove that (4) holds by induction on n. When $n = 2$, without loss of generality, we may assume that $x_1 \geqslant x_2$. From $x_1, x_2 > 0$ and $x_1 x_2 \geqslant 1$, it is deduced that $x_1 \geqslant 1$ and $x_1 + x_2 \geqslant 2\sqrt{x_1 x_2} \geqslant 2 = 1 + 1$. This means that $(1, 1) \prec_w (x_1, x_2)$.

Suppose when $n = k$, (4) holds. Now we consider the case $n = k + 1$.

Let $x = (x_1, \ldots, x_{k+1}) \in \mathbf{R}_{++}^{k+1}$ and $\prod\limits_{i=1}^{k+1} x_i \geqslant 1$. Without loss of generality, we may assume that $x_1 \geqslant x_2 \geqslant \ldots \geqslant x_{k+1} > 0$. If $x_1 > 1$, then $x_i > 1, i = 1, \ldots, k + 1$, therefore it is clear that

$$\Big(\underbrace{1, \ldots, 1}_{k+1}\Big) \prec_w (x_1, \ldots, x_{k+1})$$

If $x_1 \leqslant 1$, then $x_1 \geqslant x_2 \geqslant \ldots \geqslant x_{k-1} \geqslant x_k x_{k+1}$. By assumption, we have

$$\Big(\underbrace{1, \ldots, 1}_{k}\Big) \prec_w (x_1, \ldots, x_{k-1}, x_k x_{k+1})$$

This means that

$$\sum_{i=1}^{t} x_i \geqslant t, t = 1, \ldots k - 1$$

and

$$\sum_{i=1}^{k-1} x_i + x_k x_{k+1} \geqslant k$$

therefore

$$\sum_{i=1}^{k} x_i \geqslant \sum_{i=1}^{k-1} x_i + x_k x_{k+1} \geqslant k$$

And by geometric-arithmetic mean inequality, we have

$$\sum_{i=1}^{k+1} x_i \geqslant (k + 1) \sqrt[k+1]{x_1 \ldots x_{k+1}} \geqslant k + 1$$

hence (4) holds for $n = k + 1$.

The proof of Theorem 1 is be completed.

Remark 1 By Theorem 1, we obtain easy some weak majorization relations. For example, for positive numbers x_1, x_2, x_3, we have

$$(1,1,1) \prec_w \left(\frac{x_2+x_3}{x_3+x_1}, \frac{x_3+x_1}{x_1+x_2}, \frac{x_2+x_3}{x_1+x_2} \right) \tag{5}$$

$$(1,1,1) \prec_w \left(\frac{x_1}{\sqrt{x_2 x_3}}, \frac{x_2}{\sqrt{x_3 x_1}}, \frac{x_3}{\sqrt{x_1 x_2}} \right) \tag{6}$$

$$(1,1,1) \prec_w \left(\frac{\sqrt{x_2 x_3}}{x_1}, \frac{\sqrt{x_3 x_1}}{x_2}, \frac{\sqrt{x_1 x_2}}{x_3} \right) \tag{7}$$

(6) and (7)are know (see [3]).

Corollary 1 Let $x = (x_1, \ldots, x_n) \in \mathbf{R}_{++}^n, n \geqslant 2$ and $\prod_{i=1}^n x_i \geqslant 1$. Then

$$\left(\underbrace{1, \ldots, 1}_{n}, \underbrace{A-1, \ldots, A-1}_{n} \right) \prec_w \left(x_1, \ldots, x_n, \underbrace{0, \ldots, 0}_{n} \right) \tag{8}$$

$$\left(\underbrace{1, \ldots, 1}_{n}, a \right) \prec_w (x_1, \ldots, x_n, x_{n+1}) \tag{9}$$

where $a = \min\{x_1, \ldots, x_n, 1\}$, $x_{n+1} = n + a - \sum_{i=1}^n x_i$.

Proof By Lemma 1 and Lemma 2, from (4), it is deduced that (8) and (9) are hold respectively.

49.3 Some Applications

Theorem 2 Let $x = (x_1, \ldots, x_n) \in \mathbf{R}_{++}^n, n \geqslant 2$ and $\prod_{i=1}^n x_i \geqslant 1$. If $\alpha \geqslant 1$, then

$$M_\alpha \geqslant \left(1 + \left(\frac{1}{n} \sum_{i=1}^n (x_i - 1) \right)^\alpha \right)^{1/\alpha} \tag{10}$$

further, if $n + a - \sum_{i=1}^n x_i \geqslant 0$, then

$$M_\alpha \geqslant \left(1 + \frac{a^\alpha - (n + a - \sum_{i=1}^n x_i)^\alpha}{n} \right)^{1/\alpha} \tag{11}$$

where $W_\alpha = \left(\frac{1}{n} \sum\limits_{i=1}^{n} x_i^\alpha \right)^{1/\alpha}$, $a = \min\{x_1, \ldots, x_n, 1\}$. If $0 < \alpha < 1$, the inequalities in (10) and (11) are all reversed.

Proof When $\alpha \geqslant 1$, the power function $f(x) = x^\alpha$ is convex on $(0, +\infty)$. by Lemma 3, from (8) and (9), it follows

$$\sum_{i=1}^{n} f(x_i) + n f(0) \geqslant f(1) + n f(A - 1) \tag{12}$$

and

$$\sum_{i=1}^{n} f(x_i) + f\left(n + a - \sum_{i=1}^{n} x_i \right) \geqslant n f(1) + f(a) \tag{13}$$

respectively, making a little deformation, (12) and (13) are deduced (10) and (11) respectively.

When $0 < \alpha < 1$, the power function $f(x) = x^\alpha$ is concave on $(0, +\infty)$, by Lemma 3, it follows the inequalities in (10) and (11) are all reversed.

The proof of Theorem 2 is be completed.

Corollary 2 Let $x = (x_1, \ldots, x_n) \in \mathbf{R}_{++}^n, n \geqslant 2$. If $\alpha \geqslant 1$, then

$$M_\alpha \geqslant (G^\alpha + (A - G)^\alpha)^{1/\alpha} \geqslant G \tag{14}$$

further, if $b - n(A - G) \geqslant 0$, then

$$M_\alpha \geqslant \left(G^\alpha + \frac{b^\alpha - (b - n(A - G))^\alpha}{n} \right)^{1/\alpha} \geqslant G \tag{15}$$

where $b = \min\{x_1, \ldots, x_n, G\}$.

Proof For positive numbers $\frac{x_1}{G}, \ldots, \frac{x_n}{G}$, we have $b = \min\{\frac{x_1}{G}, \ldots, \frac{x_n}{G}, 1\} = \frac{b}{G}$ and

$$\prod_{i=1}^{n} \frac{x_i}{G} = 1, \quad \frac{1}{n} \sum_{i=1}^{n} \frac{x_i}{G} = \frac{A}{G}, \quad \left(\frac{1}{n} \sum_{i=1}^{n} (\frac{x_i}{G})^\alpha \right)^{\frac{1}{\alpha}} = \frac{M_\alpha}{G}$$

So from (10) and (11), it follows

$$\frac{M_\alpha}{G} \geqslant \left(1 + \left(\frac{A}{G} - 1 \right)^\alpha \right)^{1/\alpha} \tag{16}$$

and

$$\frac{M_\alpha}{G} \geqslant \left(1 + \frac{(\frac{b}{G})^\alpha - (n + \frac{b}{G} - \sum_{i=1}^{n} \frac{x_i}{G})^\alpha}{n} \right)^{1/\alpha} \tag{17}$$

respectively, making a little deformation, (16) and (17) are deduced the left inequalities in (14) and (15) respectively. By $A \geqslant G$, it follows that the right inequalities in (14) and (15) are hold.

Theorem 3 Let $x = (x_1, \ldots, x_n) \in \mathbf{R}_{++}^n, n \geqslant 2$ and $\prod\limits_{i=1}^{n} x_i \geqslant 1$, and let $0 \leqslant \alpha \leqslant 1$. If $1 \leqslant k \leqslant n$, then

$$E_k(x^\alpha) \leqslant \sum_{i=0}^{k} C_n^i C_n^{k-i} (A-1)^{(k-i)\alpha} \tag{18}$$

if $n + 1 \leqslant k \leqslant 2n$, then

$$\prod_{l=k-n}^{n} (E_l^*(x^\alpha))^{C_n^{k-l}} \leqslant \prod_{l=k-n}^{n} (l + (k-l)(A-1)^\alpha)^{C_n^l C_n^{k-l}} \tag{19}$$

Proof By Lemma 5 and Lemma 6, from (8), it follows that (18) and (19) are hold respectively.

The proof of Theorem 3 is be completed.

Corollary 3 Let $x = (x_1, \ldots, x_n) \in \mathbf{R}_{++}^n, n \geqslant 2$, and let $0 \leqslant \alpha \leqslant 1$. If $1 \leqslant k \leqslant n$, then

$$E_k(x^\alpha) \leqslant G^{\alpha C_n^k} \sum_{i=0}^{k} C_n^i C_n^{k-i} \left(\frac{A-G}{G} \right)^{(k-i)\alpha} \tag{20}$$

if $n + 1 \leqslant k \leqslant 2n$, then

$$\prod_{l=k-n}^{n} (E_l^*(x^\alpha))^{C_n^{k-l}} \leqslant \prod_{l=k-n}^{n} (lG^\alpha + (k-l)(A-G)^\alpha)^{C_n^l C_n^{k-l}} \tag{21}$$

Proof For positive numbers $\frac{x_1}{G}, \ldots, \frac{x_n}{G}$, we have

$$\prod_{i=1}^{n} \frac{x_i}{G} = 1, \quad \frac{1}{n} \sum_{i=1}^{n} \frac{x_i}{G} = \frac{A}{G}, \quad b = \min \left\{ \frac{x_1}{G}, \ldots, \frac{x_n}{G}, 1 \right\} = \frac{b}{G}$$

So from (18) and (19), it follows

$$\sum_{1 \leqslant i_1 < \ldots < i_k \leqslant n} \prod_{j=1}^{k} \left(\frac{x_{i_j}}{G} \right)^\alpha \leqslant \sum_{i=0}^{k} C_n^i C_n^{k-i} \left(\frac{A}{G} - 1 \right)^{(k-i)\alpha} \tag{22}$$

and

$$\prod_{l=k-n}^{n}\left(E_l^*\left(\frac{x}{G}\right)^\alpha\right)^{C_n^{k-l}} \leqslant \prod_{l=k-n}^{n}\left(l+(k-l)\left(\frac{A-G}{G}\right)^\alpha\right)^{C_n^l C_n^{k-l}} \tag{23}$$

respectively, making a little deformation, (22) and (23) be deduced to (20) and (21) respectively.

Theorem 4 Let $x = (x_1,\ldots,x_n) \in \mathbf{R}_{++}^n, n \geqslant 2$ and $\prod_{i=1}^{n} x_i \geqslant 1$, $n+a-\sum_{i=1}^{n} x_i \geqslant 0$ and let $0 \leqslant \alpha \leqslant 1$. If $1 \leqslant k \leqslant n$, then

$$E_k(x^\alpha) + (n+a-\sum_{i=1}^{n} x_i)^\alpha E_{k-1}(x^\alpha) \leqslant C_n^k + C_n^{k-1} a^\alpha \tag{24}$$

and

$$E_k(x^\alpha) \prod_{1\leqslant i_1<\ldots<i_k\leqslant n}\left(\sum_{j=1}^{k-1} x_{i_j}^\alpha + (n+a-\sum_{i=1}^{n} x_i)^\alpha\right) \leqslant k^{C_n^k}\left(a^\alpha+k-1\right)^{C_n^{k-1}} \tag{25}$$

where $a = \min\{x_1,\ldots,x_n,1\}$.

Proof By Lemma 5 and Lemma 6, from (9), it follows that (20) and (21) are hold respectively.

The proof of Theorem 4 is be completed.

Corollary 4 Let $x = (x_1,\ldots,x_n) \in \mathbf{R}_{++}^n, n \geqslant 2$ and $n+a-\sum_{i=1}^{n} x_i \geqslant 0$, and let $0 \leqslant \alpha \leqslant 1$. If $1 \leqslant k \leqslant n$, then

$$E_k(x^\alpha) + (b-n(A-G))^\alpha E_{k-1}(x^\alpha) \leqslant C_n^k G^{k\alpha} + C_n^{k-1} b^\alpha G^{(k-l)\alpha} \tag{26}$$

and

$$E_k(x^\alpha) \prod_{1\leqslant i_1<\ldots<i_k\leqslant n}\left(\sum_{j=1}^{k-1} x_{i_j}^\alpha + (b-n(A-G))^\alpha\right)$$
$$\leqslant k^{C_n^k} G^{\alpha C_n^k}\left(b^\alpha + (k-1)G^\alpha\right)^{C_n^{k-1}} \tag{27}$$

where $b = \min\{x_1,\ldots,x_n,G\}$.

Proof For positive numbers $\frac{x_1}{G},\ldots,\frac{x_n}{G}$, we have

$$\prod_{i=1}^{n}\frac{x_i}{G} = 1, \frac{1}{n}\sum_{i=1}^{n}\frac{x_i}{G} = \frac{A}{G}, \min\left\{\frac{x_1}{G},\ldots,\frac{x_n}{G},1\right\} = \frac{b}{G}$$

So from (24) and (25), it follows

$$E_k\left(\left(\frac{x}{G}\right)^\alpha\right) + \left(\frac{b}{G} + n - \sum_{j=1}^n \frac{x_j}{G}\right)^\alpha E_{k-1}\left(\left(\frac{x}{G}\right)^\alpha\right)$$

$$\leqslant C_n^k G^{k\alpha} + C_n^{k-1}\left(\frac{b}{G}\right)^\alpha \tag{28}$$

and

$$E_k\left(\left(\frac{x}{G}\right)^\alpha\right) \cdot \prod_{1\leqslant i_1 < \ldots < i_k \leqslant n}\left(\sum_{j=1}^{k-1}\left(\frac{x_{i_j}}{G}\right)^\alpha + \left(n + a - \sum_{i=1}^n \frac{x_i}{G}\right)^\alpha\right)$$

$$\leqslant k^{C_n^k}\left(\left(\frac{b}{G}\right)^\alpha + k - 1\right)^{C_n^{k-1}} \tag{29}$$

respectively, making a little deformation, (28) and (29) be deduced to (26) and (27) respectively.

By taking different n, k, from (20), (21), (26) and (27) we can obtain some interesting inequalities. For example, taking $n = 3, k = 2$ on (20) and taking $n = 3, k = 5$ on (21), we get Corollary 5.

Corollary 5 Let $x_i > 0, i = 1, 2, 3$ and $0 \leqslant \alpha \leqslant 1$. Then

$$(x_1^\alpha x_2^\alpha + x_2^\alpha x_3^\alpha + x_3^\alpha x_1^\alpha)/3 \leqslant G^\alpha(A - G)^{2\alpha} + 3G^{2\alpha}(A - G)^\alpha + G^{3\alpha} \tag{30}$$

$$(x_1^\alpha + x_2^\alpha + x_3^\alpha)\sqrt[3]{(x_1^\alpha + x_2^\alpha) + (x_2^\alpha + x_3^\alpha) + (x_3^\alpha + x_1^\alpha)}$$
$$\leqslant (2G^\alpha + 3(A - G)^\alpha)(3G^\alpha + 2(A - G)^\alpha) \tag{31}$$

where A and G are the arithmetic mean and the geometric mean of x_1, x_2, x_3 respectively.

In particular, taking $\alpha = 1$, from (30) and (31) we get

$$(x_1 x_2 + x_2 x_3 + x_3 x_1)/3 \leqslant G(A^2 + AG + G^2) \tag{32}$$

$$(x_1 + x_2 + x_3)\sqrt[3]{(x_1 + x_2) + (x_2 + x_3) + (x_3 + x_1)} \leqslant (3A - G)(G + 2A) \tag{33}$$

Acknowledgements

The present investigation was supported, in part, by the Scientific Research Common Program of Beijing Municipal Commission of Education under Grant KM200611417009, in part, by the Natural Science Foundation of Fujian province of China under Grant S0850023, and, in part, by the Science Foundation of Project of Fujian Province Education Department of China under Grant JA08231.

References

[1]　B-Y WANG. Foundations of Majorization Inequalities[M]. Beijing: Beijing Normal Univ. Press, 1990.

[2]　A M MARSHALL, I OLKIN. Inequalities:theory of majorization and its application[M]. New York :Academies Press, 1979.

[3]　HUAN-NAN SHI. A simple proposition on the theory of majorization with geometry applications[M]// Xue-zhi Yang. Researches on Inequalities. People' Press of Tibet, 2000.(Chinese)

[4]　HUAN-NAN SHI, SHAN-HE WU, FENG QI. An alternative note on the Schur-convexity of the extended mean values[J]. Mathematical Inequalities and Applications, 2006, 9(2): 219-224.(MR 2007 a: 26016), (Zbl 1104.26014)

[5]　HUAN-NAN SHI, TIE-QUAN XU, FENG QI. Monotonicity Results for Arithmetic Means of Concave and Convex Functions[J]. RGMIA Research Report Collection, Volum 9, Number 3, 2006.

[6]　HUAN-NAN SHI, CHUN GU, DA-MAO LI. Schur-convexity of a mean of convex function[J]. Applied Mathematics Letters, 2009, 22(6): 932-937.

[7]　DA-MAO LI, HUAN-NAN SHI. Schur-Convexity and Schur-Geometric Convexity for a Class of the Means[J]. RGMIA Research Report Collection, 2006, 9, (4).

[8]　HUAN-NAN SHI,SHAN-HE WU. Majorized proof and refinement of the discrete Steffensen's inequality[J]. Taiwanese Journal of Mathematics, 2007, 11(4): 1203-1208.

[9]　HUAN-NAN SHI. Solution of an Open Problem Proposed by Feng Qi[J]. RGMIA Research Report Collection, 2007, 10(4).

[10]　HUAN-NAN SHI. Schur-Convex Functions relate to Hadamard-type inequalities[J]. Journal Mathematical Inequalities, 2007, 1(1): 127-136.

[11] HUAN-NAN SHI. Sharpening of Zhong Kai-lai's Inequality[J]. RGMIA Research Report Collection, 2007, 10 (1).

[12] HUAN-NAN SHI. Generalizations of Bernoulli's Inequality with Applications[J]. Journal Mathematical Inequalities,2007, 2 (1): 101-107.

[13] DA-MAO LI, HUAN-NAN SHI, JIAN ZHANG. Schur-Convexity and Schur-Geometrically Concavity of Seiffert's Mean[J]. RGMIA Research Report Collection, 2008, 11(3).

[14] HUAN-NAN SHI, YONG-MING JIANG, WEI-DONG JIANG.Schur-Convexity and Schur-Geometrically Concavity of Gini Mean[J]. Computers and Mathematics with Applications, 2009, 7(2): 266-274.

[15] HUAN-NAN SHI, CHUN GU, DA-MAO LI. Schur-Convexity of a Mean of Convex Function[J]. RGMIA Research Report Collection, 2006, 9(4).

[16] HUAN-NAN SHI, MIHALY BENCZE, ShAN-HE WU, DA-MAO LI. Schur Convexity of generalized Heronian Means involving two parameters[J]. Journal of Inequalities and Applications,Volume 2008, Article ID 879273, 9 pages doi:10.1155/2008/879273

[17] HUAN-NAN SHI, ShAN-HE WU. Refinement of an Inequality for Generalized Logarithmic Mean[J]. Chinese Quarterly Journal of Mathematics, 2008, 23(4): 594-599.

第50篇　New Proofs of Schur-Concavity for a Class of Symmetric Functions

(SHI HUAN-NAN, ZHANG JIAN, GU CHUN. J. Inequal.　Appl.,

2012:12 doi:10.1186/1029-242X-2012-12)

Abstract: By properties of the Schur-convex function, Schur-concavity for a class of symmetric functions is simply proved uniform.

Keywords: majorization; Schur-concavity; inequality; symmetric functions; concave functions

50.1　Introduction

Throughout the article, $\mathbf{R}$ denotes the set of real numbers, $\boldsymbol{x} = (x_1, x_2, \ldots, x_n)$ denotes n-tuple(n-dimensional real vectors), the set of vectors can be written as

$$\mathbf{R}^n = \{\boldsymbol{x} = (x_1, \ldots, x_n) : x_i \in \mathbf{R}, i = 1, \ldots, n\}$$

$$\mathbf{R}_+^n = \{\boldsymbol{x} = (x_1, \ldots, x_n) : x_i > 0, i = 1, \ldots, n\}$$

In particular, the notations $\mathbf{R}$ and $\mathbf{R}_+$ denote $\mathbf{R}^1$ and $\mathbf{R}_+^1$ respectively. For convenience, we introduce some definitions as follows.

Definition 1[1-2] Let $\boldsymbol{x} = (x_1, \ldots, x_n)$ and $\boldsymbol{y} = (y_1, \ldots, y_n) \in \mathbf{R}^n$.

(i) $\boldsymbol{x} \geqslant \boldsymbol{y}$ means $x_i \geqslant y_i$ for all $i = 1, 2, \ldots, n$.

(ii) Let $\Omega \subset \mathbf{R}^n$, φ: $\Omega \to \mathbf{R}$ is said to be increasing if $\boldsymbol{x} \geqslant \boldsymbol{y}$ implies $\varphi(\boldsymbol{x}) \geqslant \varphi(\boldsymbol{y})$. φ is said to be decreasing if and only if $-\varphi$ is increasing.

Definition 2[1-2] Let $\boldsymbol{x} = (x_1, \ldots, x_n)$ and $\boldsymbol{y} = (y_1, \ldots, y_n) \in \mathbf{R}^n$.

359

(i) $\boldsymbol{x}$ is said to be majorized by $\boldsymbol{y}$ (in symbols $\boldsymbol{x} \prec \boldsymbol{y}$) if $\sum_{i=1}^{k} x_{[i]} \leqslant \sum_{i=1}^{k} y_{[i]}$ for $k = 1, 2, \ldots, n-1$ and $\sum_{i=1}^{n} x_i = \sum_{i=1}^{n} y_i$, where $x_{[1]} \geqslant \cdots \geqslant x_{[n]}$ and $y_{[1]} \geqslant \cdots \geqslant y_{[n]}$ are rearrangements of $\boldsymbol{x}$ and $\boldsymbol{y}$ in a descending order.

(ii) Let $\Omega \subset \mathbf{R}^n$, $\varphi \colon \Omega \to \mathbf{R}$ is said to be a Schur-convex function on Ω if $\boldsymbol{x} \prec \boldsymbol{y}$ on Ω implies $\varphi(x) \leq \varphi(\mathbf{y})$. φ is said to be a Schur-concave function on Ω if and only if $-\varphi$ is Schur-convex function on Ω.

Definition 3[1-2] Let $\boldsymbol{x} = (x_1, \ldots, x_n)$ and $\boldsymbol{y} = (y_1, \ldots, y_n) \in \mathbf{R}^n$.

(i) $\Omega \subseteq \mathbf{R}^n$ is said to be a convex set if $\boldsymbol{x}, \boldsymbol{y} \in \Omega, 0 \leqslant \alpha \leqslant 1$ implies $\alpha \boldsymbol{x} + (1-\alpha)\boldsymbol{y} = (\alpha x_1 + (1-\alpha)y_1, \ldots, \alpha x_n + (1-\alpha)y_n) \in \Omega$.

(ii) Let $\Omega \subset \mathbf{R}^n$ be convex set. A function $\varphi \colon \Omega \to \mathbf{R}$ is said to be a convex function on Ω if

$$\varphi(\alpha \boldsymbol{x} + (1-\alpha)\boldsymbol{y}) \leqslant \alpha \varphi(\boldsymbol{x}) + (1-\alpha)\varphi(\boldsymbol{y})$$

for all $\boldsymbol{x}, \boldsymbol{y} \in \Omega$, and all $\alpha \in [0, 1]$. φ is said to be a concave function on Ω if and only if $-\varphi$ is convex function on Ω.

Recall that the following so-called Schur's condition is very useful for determining whether or not a given function is Schur-convex or Schur-concave.

Theorem A[1] Let $\Omega \subset \mathbf{R}^n$ is symmetric and has a nonempty interior convex set. Ω^0 is the interior of Ω. $\varphi \colon \Omega \to \mathbf{R}$ is continuous on Ω and differentiable in Ω^0. Then φ is the *Schur-convex (Schur-concave) function*, if and only if φ is symmetric on Ω and

$$(x_1 - x_2) \left(\frac{\partial \varphi}{\partial x_1} - \frac{\partial \varphi}{\partial x_2} \right) \geqslant 0 (\leqslant 0) \tag{1}$$

holds for any $\boldsymbol{x} \in \Omega^0$.

In recent years, by using Theorem A, many researchers have studied the Schur-convexity of some of symmetric functions[3-9].

Chu et al.[3] defined the following symmetric functions

$$F_n(\boldsymbol{x}, k) = \prod_{1 \leqslant i_1 < \ldots < i_k \leqslant n} \frac{\sum_{j=1}^{k} x_{i_j}}{\sum_{j=1}^{k} (1 + x_{i_j})}, \quad k = 1, \ldots, n \tag{2}$$

and established the following results by using Theorem A.

Theorem B For $k = 1, \ldots, n$, $F_n(\boldsymbol{x}, k)$ is an Schur-concave function on $\mathbf{R}_+^n$.

Jiang[4] are discussed the following symmetric functions

$$H_k^*(\boldsymbol{x}) = \prod_{1 \leqslant i_1 < \ldots < i_k \leqslant n} \sum_{j=1}^{k} x_{i_j}^{1/k}, \quad k = 1, \ldots, n \tag{3}$$

and established the following results by using Theorem A.

Theorem C For $k = 1, \ldots, n$, $H_k^*(\boldsymbol{x})$ is an Schur-concave function on $\mathbf{R}_+^n$.

Xia and Chu[5] investigated the following symmetric functions

$$\phi_n(\boldsymbol{x}, k) = \prod_{1 \leqslant i_1 < \ldots < i_k \leqslant n} \sum_{j=1}^{k} \frac{x_{i_j}}{1 + x_{i_j}}, \quad k = 1, \ldots, n \tag{4}$$

and established the following results by using Theorem A.

Theorem D For $k = 1, \ldots, n$, $F_n(\boldsymbol{x}, k)$ is an Schur-concave function on $\mathbf{R}_+^n$.

In this note, by properties of the Schur-convex function, we simply prove Theorems B, C and D uniform.

50.2　New proofs three theorems

To prove the above three theorems, we need the following lemmas.

Lemma 1[1-2] If φ is symmetric and convex (concave) on symmetric convex set Ω, then φ is Schur-convex (Schur-concave) on Ω.

Lemma 2[1-2] Let $\Omega \subset \mathbf{R}^n$, $\varphi: \Omega \to \mathbf{R}_+$. Then $\ln \varphi$ is Schur-convex (Schur-concave) if and only if φ is Schur-convex (Schur-concave).

Lemma 3[1-2] Let $\Omega \subset \mathbf{R}^n$ be open convex set, $\varphi: \Omega \to \mathbf{R}$. For $\boldsymbol{x}, \boldsymbol{y} \in \Omega$, defined one variable function $g(t) = \varphi(t\boldsymbol{x} + (1-t)\boldsymbol{y})$ on interval $(0, 1)$. Then φ is convex (concave) on Ω if and only if g is convex (concave) on $(0, 1)$ for all $\boldsymbol{x}, \boldsymbol{y} \in \Omega$.

Lemma 4 Let $\boldsymbol{x} = (x_1, \ldots, x_m)$ and $\boldsymbol{y} = (y_1, \ldots, y_m) \in \mathbf{R}^m$. Then the following functions are concave on $(0, 1)$.

(i) $f(t) = \ln \sum_{j=1}^{m} (tx_j + (1-t)y_j) - \ln \sum_{j=1}^{m} (1 + tx_j + (1-t)y_j)$

(ii) $g(t) = \ln \sum\limits_{j=1}^{m} (tx_j + (1-t)y_j)^{1/m}$

(iii) $h(t) = \frac{1}{m} \ln \psi(t)$, where

$$\psi(t) = \sum_{j=1}^{m} \frac{tx_j + (1-t)y_j}{1 + tx_j + (1-t)y_j}$$

Proof (i) Directly calculating yields

$$f'(t) = \sum_{j=1}^{m} (x_j - y_j) \left[\frac{1}{tx_j + (1-t)y_j} - \frac{1}{1 + tx_j + (1-t)y_j} \right]$$

and

$$f''(t) = -\sum_{j=1}^{m} (x_j - y_j)^2 \left[\frac{1}{(tx_j + (1-t)y_j)^2} - \frac{1}{(1 + tx_j + (1-t)y_j)^2} \right]$$
$$= -\sum_{j=1}^{m} (x_j - y_j)^2 \frac{1 + 2tx_j + 2(1-t)y_j}{(tx_j + (1-t)y_j)^2 (1 + tx_j + (1-t)y_j)^2}$$

Since $f''(t) \leqslant 0$, $f(t)$ is concave on $(0,1)$.

(ii) Directly calculating yields

$$g'(t) = \frac{\frac{1}{m} \sum\limits_{j=1}^{m} (x_j - y_j)^{\frac{1}{m}-1}}{\sum\limits_{j=1}^{m} (tx_j + (1-t)y_j)^{1/m}}$$

and

$$g''(t) = -\frac{\left[\frac{1}{m} \sum\limits_{j=1}^{m} (x_j - y_j)^{\frac{1}{m}-1} \right]^2}{\sum\limits_{j=1}^{m} (tx_j + (1-t)y_j)^{2/m}}$$

Since $g''(t) \leqslant 0$, $f(t)$ is concave on $(0,1)$.

(iii) By computing

$$h'(t) = \frac{1}{m} \frac{\psi'(t)}{\psi(t)},$$

$$h''(t) = \frac{1}{m} \frac{\psi''(t)\psi(t) - (\psi'(t))^2}{\psi^2(t)}$$

where

$$\psi'(t) = \sum_{j=1}^{m} \frac{x_j - y_j}{(1 + tx_j + (1-t)y_j)^2}$$

and

$$\psi''(t) = -\sum_{j=1}^{m} \frac{2(x_j - y_j)^2}{(1 + tx_j + (1-t)y_j)^3}$$

Thus

$$\psi''(t)\psi(t) - (\psi'(t))^2 = -\sum_{j=1}^{m} \frac{2(x_j - y_j)^2}{(1 + tx_j + (1-t)y_j)^3} \sum_{j=1}^{m} \frac{tx_j + (1-t)y_j}{1 + tx_j + (1-t)y_j} -$$

$$\left[\sum_{j=1}^{m} \frac{x_j - y_j}{(1 + tx_j + (1-t)y_j)^2}\right]^2 \leqslant 0$$

and then $h''(t) \leqslant 0$, so $f(t)$ is concave on $(0, 1)$.

The proof of Lemma 4 is completed.

Proof of Theorem A　For any $1 \leqslant i_1 < \cdots < i_k \leqslant n$, by Lemma 3 and Lemma 4(i), it follows that $\ln \sum_{j=1}^{k} x_{i_j} - \ln \sum_{j=1}^{k}(1 + x_{i_j})$ is concave on $\mathbf{R}_+^n$, and then $\ln F_n(\boldsymbol{x}, k) = \prod_{1 \leqslant i_1 < \cdots < i_k \leqslant n} \left(\ln \sum_{j=1}^{k} x_{i_j} - \ln \sum_{j=1}^{k}(1 + x_{i_j})\right)$ is concave on $\mathbf{R}_+^n$. Furthermore, it is clear that $\ln F_n(\boldsymbol{x}, k)$ is symmetric on $\mathbf{R}_+^n$, by Lemma 1, it follows that $\ln F_n(\boldsymbol{x}, k)$ is concave on $\mathbf{R}_+^n$, and then from Lemma 2 we conclude that $F_n(\boldsymbol{x}, k)$ is also concave on $\mathbf{R}_+^n$.

The proof of Theorem A is completed.

Similar to the proof of Theorem A, by Lemma 4 (ii) and Lemma 4 (iii), we can prove Theorems B and C, respectively. Omitted detailed process.

Competing interests

The authors declare that they have no competing interests.

Authors' contributions

All authors read and approved the final manuscript.

Acknowledgment

Shi was supported in part by the Scientific Research Common Program of Beijing Municipal Commission of Education (KM201111417006). This article was typeset by using AmS-LaTeX

References

[1] A M MARSHALL, I OLKIN. Inequalities:theory of majorization and its application[M]. New York :Academies Press, 1979.

[2] B-Y WANG. Foundations of Majorization Inequalities[M]. Beijing: Beijing Normal Univ. Press, 1990.

[3] CHU Y-M, XIA W-F, ZHAO T-H. Some properties for a class of symmetric functions and applications[J]. J. Math. Inequal., 2011, 5 (1): 1–11.

[4] JIANG W-D. Some properties of dual form of the Hamy's symmetric function[J]. J. Math. Inequal., 2007, 1 (1): 117–125.

[5] XIA W-F, CHU Y-M. Schur-convexity for a class of symmetric functions and its applications[J]. J. Inequal. Appl., vol. 2009, Article ID 493759, 15 (2009) doi:10.1155/2009/493759.

[6] XIA W-F, CHU Y-M. On Schur-convexity of some symmetric functions[J]. J. Inequal. Appl., Volume 2010, Article ID 543250, 12 pages doi:10.1155/2010/543250.

[7] QIAN W-M. Schur convexity for the ratios of the Hamy and generalized Hamy symmetric functions[J]. J. Inequal. Appl., 2011:131 doi:10.1186/1029-242X-2011-131.

[8] GUAN K-Z. Schur-convexity of the complete elementary symmetric function[J]. J. Inequal. Appl., 2006 (2006), Article ID 67624, 9 pages doi: 10.1155/ JIA/2006/67624.

[9] CHU Y-M, LV Y-P. The Schur harmonic convexity of the Hamy symmetric function and its applications[J]. J. Inequal. Appl., Volume 2009, Article ID 838529, 10 pages doi:10.1155/2009/838529.

第51篇 一个Schur凸性
判定定理的应用

(石焕南,顾春,张鉴. 四川师范大学学报(自然科学版), 2012, 35 (3):345-348)

摘　要: 本文给出了控制理论中非对称凸集$D = \{\boldsymbol{x} \in \mathbf{R}^n : x_1 \geqslant ... \geqslant x_n\}$上的Schur凸函数判定定理的四个应用: (1) 证明了一个代数不等式; (2) 推广了一个平均值不等式; (3) 确定了一类加权平均的Schur凹性; (4) 验证了一个积函数Schur凸的充分必要条件.

关键词: 受控; Schur凸性; 不等式

51.1　引言

在本文中, $\mathbf{R}^n$ 和$\mathbf{R}_+^n$ 分别表示n 维实数集和n 维正实数集, 并记$\mathbf{R}^1 = \mathbf{R}, \mathbf{R}_+^1 = \mathbf{R}_+$.

定义1[1-2]　设$\boldsymbol{x} = (x_1, ..., x_n)$ 和$\boldsymbol{y} = (y_1, ..., y_n)$ 满足

(1) $\sum\limits_{i=1}^{k} x_{[i]} \leqslant \sum\limits_{i=1}^{k} y_{[i]}$, $k = 1, 2, ..., n - 1$;

(2) $\sum\limits_{i=1}^{n} x_i = \sum\limits_{i=1}^{n} y_i$.

其中$x_{[1]} \geqslant ... \geqslant x_{[n]}$ 和$y_{[1]} \geqslant ... \geqslant y_{[n]}$ 是$\boldsymbol{x}$ 和$\boldsymbol{y}$ 的递减重排, 则称$\boldsymbol{x}$ 被$\boldsymbol{y}$ 所控制, 记作$\boldsymbol{x} \prec \boldsymbol{y}$.

定义2[1-2]　设$\Omega \subset \mathbf{R}^n, \varphi : \Omega \to \mathbf{R}$, 若在$\Omega$上$\boldsymbol{x} \prec \boldsymbol{y} \Rightarrow \varphi(\boldsymbol{x}) \leqslant \varphi(\boldsymbol{y})$, 则称$\varphi$ 为Ω 上的Schur凸函数(简称S-凸函数); 若$-\varphi$ 是Ω 上的S-凸函数, 则称φ 为Ω 上的S-凹函数.

定理A[1]　设$\Omega \subset \mathbf{R}^n$ 是有内点的对称凸集, $\varphi : \Omega \to \mathbf{R}$ 在Ω 上连续, 在Ω 的内部Ω^0 可微, 则在Ω 上S-凸(S-凹)的充要条件是φ 在Ω 上对称且$\forall \boldsymbol{x} \in \Omega$, 有

$$(x_1 - x_2) \left(\frac{\partial \varphi}{\partial x_1} - \frac{\partial \varphi}{\partial x_2} \right) \geqslant 0 (\leqslant 0) \tag{1}$$

S-凸函数是受控理论(Theory of Majorization)最重要的概念, 定理A是

判断S-凸函数最有力的工具, 故有些文献称其为S-凸函数基本定理. 但该定理只适用于对称凸集上的对称函数S-凸性的判定. 对于非对称凸集 $D = \{\boldsymbol{x} \in \mathbf{R}^n : x_1 \geqslant ... \geqslant x_n\}$, 有如下判定定理.

定理B[1]　设 $\varphi : D \to \mathbf{R}$ 在 D 上连续, 在 D 上的内部 D^0 可微, 则 φ 在 D 上S-凸(凹) 的充要条件是 $\forall \boldsymbol{x} \in D^0$, 有

$$\frac{\partial \varphi}{\partial x_i} \geqslant (\leqslant) \frac{\partial \varphi}{\partial x_{i+1}}, i = 1, ..., n-1 \tag{2}$$

注1　若 φ 是 D 上的S-凸函数, 我们容易扩充它到 $\mathbf{R}^n$ 而成为 $\mathbf{R}^n$ 上的S-凸函数: 即对 $\boldsymbol{x} \in \mathbf{R}^n$, 令 $\varphi(\boldsymbol{x}) = \varphi\left(x_{[1]}, ..., x_{[n]}\right)$, 显然这种S-凸扩张是唯一的, 同时 φ 也成为任何集合 $A \subset \mathbf{R}^n$ 上的S-凸函数.

有关定理B的应用的文献较少, 本文将给出该定理的四例应用.

51.2　关于一个代数不等式

定理1[1]　设 $x \geqslant y > 0, \varphi(x,y) = x^\alpha - \alpha xy^{\alpha-1} + (\alpha-1)y^\alpha$, 则当 $\alpha > 1$ 或 $\alpha < 0$ 时, $\varphi(x,y) \geqslant 0$, 当 $0 \leqslant \alpha \leqslant 1$ 时, $\varphi(x,y) \leqslant 0$.

证明　记 $D_+ = \{(x,y) : x \geqslant y > 0\}$.

$$\frac{\partial \varphi}{\partial x} = \alpha x^{\alpha-1} - \alpha y^{\alpha-1}, \quad \frac{\partial \varphi}{\partial y} = -\alpha(\alpha-1)xy^{\alpha-2} + \alpha(\alpha-1)y^{\alpha-1}$$

若 $\alpha > 1$, 则 $\forall (x,y) \in D^0_{++}$, 有

$$\frac{\partial \varphi}{\partial x} - \frac{\partial \varphi}{\partial y} = \alpha x^{\alpha-1} + \alpha(\alpha-1)xy^{\alpha-2} - \alpha^2 y^{\alpha-1}$$

$$\geqslant \alpha y^{\alpha-1} + \alpha(\alpha-1)y \cdot y^{\alpha-2} - \alpha^2 y^{\alpha-1} = 0$$

若 $\alpha < 0$, 则

$$\frac{\partial \varphi}{\partial x} - \frac{\partial \varphi}{\partial y} = -(-\alpha)x^{\alpha-1} + (-\alpha)(1-\alpha)xy^{\alpha-2} - \alpha^2 y^{\alpha-1}$$

$$\geqslant (-\alpha)(1-\alpha) + (-\alpha)(1-\alpha)y \cdot y^{\alpha-2} - \alpha^2 y^{\alpha-1} = 0$$

若 $0 < \alpha < 1$, 则

$$\frac{\partial \varphi}{\partial x} - \frac{\partial \varphi}{\partial y} = \alpha x^{\alpha-1} + \alpha(1-\alpha)xy^{\alpha-2} - \alpha^2 y^{\alpha-1}$$

$$\geqslant \alpha y^{\alpha-1} + \alpha(1-\alpha)y \cdot y^{\alpha-2} - \alpha^2 y^{\alpha-1} = 0$$

据定理B, 由上述讨论知当$\alpha > 1$或$\alpha < 0$时. $\varphi(x, y)$在D_+上S-凸, 当$0 < \alpha < 1$时, 在D_+上S-凹. 因$\left(\frac{x+y}{2}, \frac{x+y}{2}\right) \prec (x, y)$, 故$\alpha > 1$或$\alpha < 0$时, 在$D_+$上有$0 = \varphi\left(\frac{x+y}{2}, \frac{x+y}{2}\right) \leqslant \varphi(x, y)$, 而当$0 < \alpha < 1$时, 在$D_+$上有$0 = \varphi\left(\frac{x+y}{2}, \frac{x+y}{2}\right) \geqslant \varphi(x, y)$, 证毕.

注2　定理1中的$\varphi(x, y)$是非对称函数, 且D_+也不是对称集, 故此例不能利用定理A证明, 而应当用定理B证明.

51.3　关于一个平均值不等式

由Juan Bosco Romero Marquez提供的美国数学月刊第10529号问题[美国数学月刊, 1996, 103(6):509]是: 设$\lambda \geqslant 0, 0 < a \leqslant b, n \in \mathbf{Z}, n > 1$, 证明

$$\sqrt{ab} \leqslant \sqrt[n]{\frac{a^n + b^n + \lambda((a+b)^n - a^n - b^n)}{2 + \lambda(2^n - 2)}} \leqslant \frac{a + b}{2} \tag{3}$$

前不久, 我发现此问题的条件$\lambda \geqslant 0$有误, 应改为$\lambda \geqslant 1$. 否则, 若取$\lambda = 1/2, x = 1, y = 2, n = 2$, 代入(3)右边不等式, 得$\sqrt{\frac{7}{3}} \leqslant \frac{3}{2}$, 即$27 \leqslant 28$, 矛盾. 后来从姜卫东寄来的资料中获知: 1998年, Robin J. Chapman已纠正了此错误, 并就$\lambda \geqslant 1$的情形给予证明. 现应用定理B证明如下结果.

定理2　设$0 < a \leqslant b, \alpha > 0$. 若$\lambda \leqslant 1, 0 < \alpha \leqslant 1$或$\lambda \geqslant 1\alpha \geqslant 1$, 则

$$\frac{a + b}{(2 + \lambda(2^\alpha - 2))^{\frac{1}{\alpha}}} \leqslant \left(\frac{a^\alpha + b^\alpha + \lambda((a+b)^\alpha - a^\alpha - b^\alpha)}{2 + \lambda(2^\alpha - 2)}\right)^{\frac{1}{\alpha}} \leqslant \frac{a + b}{2} \tag{4}$$

若$\lambda \leqslant 1, \alpha \geqslant 1$或$\lambda \geqslant 1, 0 < \alpha \leqslant 1$, 则(4)中的不等式均反向.

证明　令$\varphi(x, y) = (\psi(x, y))^{\frac{1}{\alpha}}$, 其中

$$\psi(a, b) = \frac{a^\alpha + b^\alpha + \lambda((a+b)^\alpha - a^\alpha - b^\alpha)}{2 + \lambda(2^\alpha - 2)}$$

则

$$\frac{\partial \varphi}{\partial a} = \psi^{\frac{1}{\alpha} - 1} \cdot \frac{\alpha a^{\alpha-1} + \lambda\alpha\left((a+b)^{\alpha-1} - a^{\alpha-1}\right)}{2 + \lambda(2^\alpha - 2)}$$

$$\frac{\partial \varphi}{\partial b} = \psi^{\frac{1}{\alpha} - 1} \cdot \frac{\alpha b^{\alpha-1} + \lambda\alpha\left((a+b)^{\alpha-1} - b^{\alpha-1}\right)}{2 + \lambda(2^\alpha - 2)}$$

易见$\frac{\partial \varphi}{\partial a} \geqslant \frac{\partial \varphi}{\partial b}$等价于$\alpha(b^{\alpha-1} - a^{\alpha-1})(1 - \alpha) \leqslant 0$. 据定理B, 当$\lambda \leqslant 1, 0 < \alpha \leqslant 1$时, 或$\lambda \geqslant 1, 0 < \alpha \leqslant 1$时, $\psi(x, y)$在$D_2 = \{(a, b) : b \geqslant a\}$上S-凹. 于

是, 由

$$\left(\frac{a+b}{2}, \frac{a+b}{2}\right) \prec (a,b) \prec ((a+b), 0) \tag{5}$$

有

$$\varphi\left(\frac{a+b}{2}, \frac{a+b}{2}\right) \geqslant \varphi(a,b) \geqslant \varphi((a+b), 0)$$

即(4)成立.

$\frac{\partial \varphi}{\partial a} \leqslant \frac{\partial \varphi}{\partial b}$ 等价于 $\alpha(b^{\alpha-1} - a^{\alpha-1})(1-\alpha) \geqslant 0$. 据定理B, 当 $\lambda \leqslant 1, \alpha \geqslant 1$ 时, 或 $\lambda \geqslant 1, 0 < \alpha \leqslant 1$ 时, $\psi(x,y)$ 在 D_2 上S-凸. 于是, 由(5)有

$$\varphi\left(\frac{a+b}{2}, \frac{a+b}{2}\right) \leqslant \varphi(a,b) \leqslant \varphi((a+b), 0)$$

即(4)反向成立.

注3 与前述第10529号问题比较, 这里将 $n \in \mathbf{Z}, n > 1$ 放宽为 $\alpha > 0$, 改变了左端, 并考虑了 $0 \leqslant \lambda < 1$ 的情形.

51.4 关于一个加权平均

对于 $(a,b) \in \mathbf{R}_+^2$, 专著[3]介绍了加权平均

$$M(\omega; a, b) = \frac{a + \omega b}{1 + \omega}, \quad \omega > 0 \tag{6}$$

利用定理B 易证

定理3 $M(\omega; a, b)$ 在 $D = \{(x,y) : x \geqslant y\}$ 上S-凸(凹)的充要条件是 $1 \geqslant (\leqslant)\omega$.

2009年, 对于 $(a,b) \in \mathbf{R}_+^2$, 匡继昌定义了一类平均[3]

$$K(m,n) = \frac{mA + nG}{m+n} = \frac{m(a+b) + 2n\sqrt{ab}}{2(m+n)} \tag{7}$$

并指出在(6)中取

$$\omega = \frac{(m+2n)a - mb - 2n\sqrt{ab}}{ma - (m+2n)b + 2n\sqrt{ab}}$$

即得 $K(m,n)$.

现利用定理B(当然也可以利用定理A)确定 $K(m,n)$ 的S-凸性.

定理4 $K(m,n)$ 在 $\mathbf{R}_+^2$ 上S-凹.

证明 在$E_2 = \{(a,b) : a \geqslant b\}$上, 令$t = \frac{a}{b}$, 则$t \geqslant 1$, 且

$$\omega = \frac{(m+2n)(\frac{a}{b}) - m - 2n\sqrt{\frac{a}{b}}}{m(\frac{a}{b}) - (m+2n) + 2n\sqrt{\frac{a}{b}}} = \frac{(m+2n)t - m - 2n\sqrt{t}}{mt - (m+2n) + 2n\sqrt{t}}$$

于是$K(m,n)$ 在E_2 上S-凹等价于$\omega \geqslant 1$, 即

$$(m+2n)t - m - 2n\sqrt{t} \geqslant mt - (m+2n) + 2n\sqrt{t}$$

而上式等价于$t - 2\sqrt{t} + 1 = \left(\sqrt{t} - 1\right)^2 \geqslant 0$. 故$K(m,n)$ 在E_2 上S-凹, 再注意$K(m,n)$ 在$\mathbf{R}_+^2$ 上对称, 由注1 知$K(m,n)$ 亦在$\mathbf{R}_+^2$ 上S-凹.

51.5 关于一个积函数Schur凸的充要条件

定理5[1] 设$\boldsymbol{x} \in \mathbf{R}_+^n, \boldsymbol{p} \in \mathbf{R}_+^n$, 则

$$\varphi(x) = \prod_{i=0}^{n} p_i^{x_i - x_{i+1}} \tag{8}$$

是

$$D_+ = \{\boldsymbol{x} : \boldsymbol{x} \in \mathbf{R}_+^n, x_1 \geqslant \cdots \geqslant x_n > 0\}$$

上的S-凸函数的充要条件是

$$\frac{p_1}{p_0} \geqslant \frac{p_2}{p_1} \geqslant \cdots \geqslant \frac{p_n}{p_{n-1}} \tag{9}$$

其中$x_0 = x_{n+1} = 0$.

证明 考虑$\ln \varphi(\boldsymbol{x}) = \sum_{i=0}^{n}(x_i - x_{i+1})\ln p_i$. 由文[2] 推论6.14 (b) 知$\varphi$ 为S-凸$\Leftrightarrow \ln \varphi$ 为S-凸. 注意这里D_+ 是非对称凸集, 由定理B, $\ln \varphi$ 为S-凸$\Leftrightarrow \frac{\partial \ln \varphi(\boldsymbol{x})}{\partial x_i} - \frac{\partial \ln \varphi(\boldsymbol{x})}{\partial x_{i+1}} \geqslant 0, i = 1,...,n-1 \Leftrightarrow \ln \frac{p_i^2}{p_{i-1}p_{i+1}} \geqslant 0, i = 1,...,n-1$, 即(9)成立.

受控理论(Theory of Majorization), 亦称控制不等式理论是一门有着广泛应用并日趋兴旺的数学学科. 截至目前, 我国学者在国内外已发表了二百余篇有关该领域的研究论文(文[4-18]为部分代表作), 已形成了一支在国际上具有一定影响的研究队伍.

参考文献

[1] MARSHALL A M , OLKIN I . Inequalities:theory of majorization and its application[M]. New York :Academies Press, 1979.

[2] 王伯英. 控制不等式基础[M]. 北京: 北京师范大学出版社, 1990.

[3] 匡继昌. 常用不等式[M]. 4版. 济南: 山东科技出版社, 2010.

[4] SHI HUAN-NAN, WU SHAN-HE, QI FENG. An alternative note on the Schur-convexity of the extended mean values[J]. Math. Inequal. Appl., 2006, 9 (2): 219-224.

[5] SHI HUAN-NAN, JIANG YONG-MING, JIANG WEI-DONG. Schur-convexity and Schur-geometrically concavity of Gini mean[J]. Comput. Math. Appl., 2009, 57 (2): 266-274.

[6] SHI HUAN-NAN, LI DA-MAO, GU CHUN. Schur-convexity of a mean of convex function[J]. Appl. Math. Lett., 2009, 22 (6): 932-937.

[7] SHI HUAN-NAN, MIHALY BENCZE, WU SHAN-HE, LI DA-MAO. Schur convexity of generalized Heronian means involving two parameters[J]. J. Inequal. Appl., Volume 2008, Article ID 879273, 9 pages doi:10.1155/2008/879273.

[8] SHI HUAN-NAN, WU SHAN-HE. Majorized proof and refinement of the discrete Steffensen's inequality[J]. Taiwanese Journal of Mathematics, 2007, 11 (4): 1203-1208.

[9] 褚玉明, 夏卫锋. Gini平均值公开问题的解[J]. 中国科学(A辑: 数学), 2009, 39 (8): 996-1002.

[10] 褚玉明, 夏卫锋, 赵铁洪. 一类对称函数的Schur凸性[J]. 中国科学(A辑: 数学), 2009, 39 (11): 1267-1277.

[11] CHU YU-MING, ZHANG XIAO-MING. Necessary and sufficient conditions such that extended mean values are Schur-convex or Schur-concave[J]. Journal of Mathematics of Kyoto University, 2008, 48(1): 229-238.

[12] GUAN KAI-ZHONG. Some properties of a class of symmetric functions[J]. J. Math. Anal. Appl., 2007, 336 (1) : 70-80.

[13] WEN JIA-JIN, YUAN JUN, YUAN SHU-FENG. An optimal version of an inequality involving the third symmetric means[J]. Proc. Indian Acad. Sci. (Math. Sci.), 2008, 118 (4): 505-516.

[14] WU SHAN-HE. Generalization and sharpness of the power means inequality and their applications[J]. J. Math. Anal. Appl., 2005, 312 (2): 637-652.

[15] WU SHAN-HE, DEBNATH LOKENATH. Inequalities for convex sequences and their applications[J]. Comput. Math. Appl., 2007, 54 (4): 525-534.

[16] ZHANG XIAO-MING, CHU YU-MING. Convexity of the integral arithmetic mean of a convex function[J]. Rocky Mountain J. Math., 2010, 40 (3): 1061-1068.

[17] XIA WEI-FENG, CHU YU-MING. The Schur multiplicative convexity of the generalized Muirhead mean[J]. International Journal of Functional Analysis, Operator Theory and Applications, 2009, 1 (1): 1-8.

[18] 张小明, 李世杰. 两个与初等对称函数有关的S-几何凸函数[J]. 四川师范大学学报(自然科学版), 2007, 30 (2): 188-190.

第52篇　Refinements of Inequalities among Difference of Means

(SHI HUAN-NAN, LI DA-MAO, ZHANG JIAN. International Journal

of Mathematics and Mathematical Sciences Volume 2012, Article ID

315697, 15 pages doi:10.1155/2012/315697)

Abstract: In this paper, for the difference of famous means discussed in [1], we study the Schur-geometric convexity in $(0, +\infty) \times (0, +\infty)$ of the difference between these differences. Moreover some inequalities related to the difference of those means are obtained.

Keywords: arithmetic mean; geometric mean; harmonic mean; root-square mean; square-root mean; Heron mean; Schur-geometric convexity; inequality

52.1　Introduction

In 2005, Taneja[1] proved the following chain of inequalities for the binary means for $(a, b) \in \mathbf{R}_+^2 = (0, +\infty) \times (0, +\infty)$

$$H(a, b) \leqslant G(a, b) \leqslant N_1(a, b) \leqslant N_3(a, b) \leqslant N_2(a, b) \leqslant A(a, b) \leqslant S(a, b) \quad (1)$$

where

$$A(a, b) = \frac{a + b}{2}$$

$$G(a, b) = \sqrt{ab}$$

$$H(a, b) = \frac{2ab}{a + b} \quad (2)$$

$$N_1(a, b) = \left(\frac{\sqrt{a} + \sqrt{b}}{2} \right)^2 = \frac{A(a, b) + G(a, b)}{2}$$

$$N_3(a, b) = \frac{a + \sqrt{ab} + b}{3} = \frac{2A(a, b) + G(a, b)}{3}$$

$$N_2(a, b) = \left(\frac{\sqrt{a} + \sqrt{b}}{2} \right) \left(\sqrt{\frac{a + b}{2}} \right) \tag{3}$$

$$S(a, b) = \sqrt{\frac{a^2 + b^2}{2}}$$

The means, A, G, H, S, N_1 and N_3 are called respectively the arithmetic mean, the geometric mean, the harmonic mean, the root-square mean, the square-root mean and Heron's mean. The N_2 one can found in Taneja[2-3].

Furthermore Taneja considered the following difference of means

$$M_{SA}(a, b) = S(a, b) - A(a, b)$$

$$M_{SN_2}(a, b) = S(a, b) - N_2(a, b)$$

$$M_{SN_3}(a, b) = S(a, b) - N_3(a, b)$$

$$M_{SN_1}(a, b) = S(a, b) - N_1(a, b)$$

$$M_{SG}(a, b) = S(a, b) - G(a, b)$$

$$M_{SH}(a, b) = S(a, b) - H(a, b) \tag{4}$$

$$M_{AN_2}(a, b) = A(a, b) - N_2(a, b)$$

$$M_{AG}(a, b) = A(a, b) - G(a, b)$$

$$M_{AH}(a, b) = A(a, b) - H(a, b)$$

$$M_{N_2 N_1}(a, b) = N_2(a, b) - N_1(a, b)$$

$$M_{N_2 G}(a, b) = N_2(a, b) - G(a, b)$$

and established the following:

Theorem A The difference of means given by (4) are nonnegative and convex in $\mathbf{R}_+^2 = (0, +\infty) \times (0, +\infty)$.

Further, using Theorem A, Taneja proved several chains of inequalities, they are refinements of inequalities in (1).

Theorem B The following inequalities among the mean differences hold

$$M_{SA}(a,b) \leqslant \frac{1}{3}M_{SH}(a,b) \leqslant \frac{1}{2}M_{AH}(a,b) \leqslant \frac{1}{2}M_{SG}(a,b) \leqslant M_{AG}(a,b) \quad (5)$$

$$\frac{1}{8}M_{AH}(a,b) \leqslant M_{N_2N_1}(a,b) \leqslant \frac{1}{3}M_{N_2G}(a,b) \leqslant \frac{1}{4}M_{AG}(a,b) \leqslant M_{AN_2}(a,b)$$
$$(6)$$

$$M_{SA}(a,b) \leqslant \frac{4}{5}M_{SN_2}(a,b) \leqslant 4M_{AN_2}(a,b) \quad (7)$$

$$M_{SH}(a,b) \leqslant 2M_{SN_1}(a,b) \leqslant \frac{3}{2}M_{SG}(a,b) \quad (8)$$

$$M_{SA}(a,b) \leqslant \frac{3}{4}M_{SN_3}(a,b) \leqslant \frac{2}{3}M_{SN_1}(a,b) \quad (9)$$

For the difference of means given by (4), we study the Schur-geometric convexity of difference between these differences in order to further improve the inequalities in (1). The main result of this paper reads as follow.

Theorem 1 The following differences are Schur-geometrically convex in $\mathbf{R}_+^2 = (0,+\infty) \times (0,+\infty)$

$$D_{SH-SA}(a,b) = \frac{1}{3}M_{SH}(a,b) - M_{SA}(a,b)$$

$$D_{AH-SH}(a,b) = \frac{1}{2}M_{AH}(a,b) - \frac{1}{3}M_{SH}(a,b)$$

$$D_{SG-AH}(a,b) = M_{SG}(a,b) - M_{AH}(a,b)$$

$$D_{AG-SG}(a,b) = M_{AG}(a,b) - \frac{1}{2}M_{SG}(a,b)$$

$$D_{N_2N_1-AH}(a,b) = M_{N_2N_1}(a,b) - \frac{1}{8}M_{AH}(a,b)$$

$$D_{N_2G-N_2N_1}(a,b) = \frac{1}{3}M_{N_2G}(a,b) - M_{N_2N_1}(a,b) \quad (10)$$

$$D_{AG-N_2G}(a,b) = \frac{1}{4}M_{AG}(a,b) - \frac{1}{3}M_{N_2G}(a,b)$$

$$D_{AN_2-AG}(a,b) = M_{AN_2}(a,b) - \frac{1}{4}M_{AG}(a,b)$$

$$D_{SN_2-SA}(a,b) = \frac{4}{5}M_{SN_2}(a,b) - M_{SA}(a,b)$$

$$D_{AN_2-SN_2}(a,b) = 4M_{AN_2}(a,b) - \frac{4}{5}M_{SN_2}(a,b)$$

$$D_{SN_1-SH}(a,b) = 2M_{SN_1}(a,b) - M_{SH}(a,b)$$

$$D_{SG-SN_1}(a,b) = \frac{3}{2}M_{SG}(a,b) - 2M_{SN_1}(a,b)$$

$$D_{SN_3-SA}(a,b) = \frac{3}{4}M_{SN_3}(a,b) - M_{SA}(a,b) \tag{11}$$

$$D_{SN_1-SN_3}(a,b) = \frac{2}{3}M_{SN_1}(a,b) - \frac{3}{4}M_{SN_3}(a,b)$$

The proof of this theorem will be given in section 3. Applying this result, in section 4, we prove some inequalities related to the considered differences of means. Obtained inequalities are refinements of inequalities (5)-(9).

52.2　Definitions and Auxiliary Lemmas

The Schur-convex function was introduced by I. Schur in 1923, and it has many important applications in analytic inequalities, linear regression, graphs and matrices, combinatorial optimization, information-theoretic topics, Gamma functions, stochastic orderings, reliability, and other related fields(cf. [4 - 14])

In 2003, Xiao-ming Zhang first propose concepts of "Schur-geometrically convex function" which is extension of "Schur-convex function" and establish corresponding decision theorem [15]. Since then, Schur-geometric convexity has evoked the interest of many researchers and numerous applications and extensions have appeared in the literature (cf. [16-19]).

In order to prove the main result of this paper we need the following definitions and auxiliary lemmas.

Definition 1[4,20] Let $\boldsymbol{x} = (x_1, \ldots, x_n) \in \mathbf{R}^n$ and $\boldsymbol{y} = (y_1, \ldots, y_n) \in \mathbf{R}^n$.

(i) $\boldsymbol{x}$ is said to be majorized by $\boldsymbol{y}$ (in symbols $\boldsymbol{x} \prec \boldsymbol{y}$) if $\sum_{i=1}^{k} x_{[i]} \leqslant \sum_{i=1}^{k} y_{[i]}$ for $k = 1, 2, \ldots, n-1$ and $\sum_{i=1}^{n} x_i = \sum_{i=1}^{n} y_i$, where $x_{[1]} \geqslant \cdots \geqslant x_{[n]}$ and $y_{[1]} \geqslant \cdots \geqslant y_{[n]}$ are rearrangements of $\boldsymbol{x}$ and $\boldsymbol{y}$ in a descending order.

(ii) $\Omega \subseteq \mathbf{R}^n$ is called a convex set if $(\alpha x_1 + \beta y_1, \ldots, \alpha x_n + \beta y_n) \in \Omega$ for every $\boldsymbol{x}$ and $\boldsymbol{y} \in \Omega$, where α and $\beta \in [0,1]$ with $\alpha + \beta = 1$.

(iii) Let $\Omega \subseteq \mathbf{R}^n$. The function $\varphi \colon \Omega \to \mathbf{R}$ be said to be a Schur-convex function on Ω if $\boldsymbol{x} \prec \boldsymbol{y}$ on Ω implies $\varphi(\boldsymbol{x}) \leq \varphi(\boldsymbol{y})$. φ is said to be a Schur-concave function on Ω if and only if $-\varphi$ is Schur-convex.

Definition 2[15] Let $\boldsymbol{x} = (x_1, \ldots, x_n) \in \mathbf{R}^n$ and $\boldsymbol{y} = (y_1, \ldots, y_n) \in \mathbf{R}_+^n$.

(i) $\Omega \subseteq \mathbf{R}_+^n$ is called a geometrically convex set if $(x_1^\alpha y_1^\beta, \ldots, x_n^\alpha y_n^\beta) \in \Omega$ for all $\boldsymbol{x}, \boldsymbol{y} \in \Omega$ and $\alpha, \beta \in [0, 1]$ such that $\alpha + \beta = 1$.

(ii) Let $\Omega \subseteq \mathbf{R}_+^n$. The function $\varphi \colon \Omega \to \mathbf{R}_+$ is said to be Schur-geometrically convex function on Ω if $(\ln x_1, \ldots, \ln x_n) \prec (\ln y_1, \ldots, \ln y_n)$ on Ω implies $\varphi(\boldsymbol{x}) \leqslant \varphi(\boldsymbol{y})$. The function φ is said to be a Schur-geometrically concave on Ω if and only if $-\varphi$ is Schur-geometrically convex.

Definition 3[4,20]

(i) The set $\Omega \subseteq \mathbf{R}^n$ is called symmetric set, if $\boldsymbol{x} \in \Omega$ implies $\boldsymbol{P}\boldsymbol{x} \in \Omega$ for every $n \times n$ permutation matrix $\boldsymbol{P}$.

(ii) The function $\varphi : \Omega \to \mathbf{R}$ is called symmetric if for every permutation matrix $\boldsymbol{P}$, $\varphi(\boldsymbol{P}\boldsymbol{x}) = \varphi(\boldsymbol{x})$ for all $\boldsymbol{x} \in \Omega$.

Lemma 1[15] Let $\Omega \subseteq \mathbf{R}_+^n$ be a symmetric and geometrically convex set with a nonempty interior Ω^0. Let $\varphi : \Omega \to \mathbf{R}_+$ be continuous on Ω and differentiable in Ω^0. If φ is symmetric on Ω and

$$(\ln x_1 - \ln x_2)\left(x_1 \frac{\partial \varphi}{\partial x_1} - x_2 \frac{\partial \varphi}{\partial x_2}\right) \geqslant 0 \quad (\leqslant 0) \tag{12}$$

holds for any $\boldsymbol{x} = (x_1, \cdots, x_n) \in \Omega^0$, then φ is a Schur-geometrically convex (Schur-geometrically concave) function.

Lemma 2 For $(a, b) \in \mathbf{R}_+^2 = (0, +\infty) \times (0, +\infty)$ we have

$$1 \geqslant \frac{a+b}{\sqrt{2(a^2+b^2)}} \geqslant \frac{1}{2} + \frac{2ab}{(a+b)^2} \tag{13}$$

$$\frac{a+b}{\sqrt{2(a^2+b^2)}} - \frac{ab}{(a+b)^2} \leqslant \frac{3}{4} \tag{14}$$

and

$$\frac{3}{2} \geqslant \frac{\sqrt{a+b}}{\sqrt{2}(\sqrt{a}+\sqrt{b})} + \frac{\sqrt{a}+\sqrt{b}}{\sqrt{2}\sqrt{a+b}} \geqslant \frac{5}{4} + \frac{ab}{(a+b)^2} \tag{15}$$

Proof It is easy to see that the left-hand inequality in (13) equivalent to $(a-b)^2 \geqslant 0$. And the right-hand inequality in (13) equivalent to

$$\frac{\sqrt{2(a^2+b^2)} - (a+b)}{\sqrt{2(a^2+b^2)}} \leqslant \frac{(a+b)^2 - 4ab}{2(a+b)^2} \tag{16}$$

that is

$$\frac{(a-b)^2}{2(a^2+b^2)+\sqrt{2(a^2+b^2)}(a+b)} \leqslant \frac{(a-b)^2}{2(a+b)^2} \tag{17}$$

Indeed, from the left-hand inequality in (13) we have

$$2(a^2+b^2)+\sqrt{2(a^2+b^2)}(a+b) \geqslant 2(a^2+b^2)+(a+b)^2 \geqslant 2(a+b)^2 \tag{18}$$

so the right-hand inequality in (13) is hold.

The inequality in (14) equivalent to

$$\frac{\sqrt{2(a^2+b^2)}-(a+b)}{\sqrt{2(a^2+b^2)}} \geqslant \frac{(a-b)^2}{4(a+b)^2} \tag{19}$$

Since

$$\frac{\sqrt{2(a^2+b^2)}-(a+b)}{\sqrt{2(a^2+b^2)}} = \frac{2(a^2+b^2)-(a+b)^2}{\sqrt{2(a^2+b^2)}(\sqrt{2(a^2+b^2)}+(a+b))}$$

$$= \frac{(a-b)^2}{2(a^2+b^2)+(a+b)\sqrt{2(a^2+b^2)}} \tag{20}$$

so it is sufficient prove that

$$2(a^2+b^2)+(a+b)\sqrt{2(a^2+b^2)} \leqslant 4(a+b)^2 \tag{21}$$

this is

$$(a+b)\sqrt{2(a^2+b^2)} \leqslant 2(a^2+b^2+4ab) \tag{22}$$

and from of the left-hand inequalities in (13), we have

$$(a+b)\sqrt{2(a^2+b^2)} \leqslant 2(a^2+b^2) \leqslant 2(a^2+b^2+4ab) \tag{23}$$

so the inequality in (14) is hold.

Notice that the functions in the inequalities (15) are homogeneous. So, without loss of generality, we may assume $\sqrt{a}+\sqrt{b}=1$, and set $t=\sqrt{ab}$. Then $0 < t \leqslant 1/4$ and (15) is reduces to

$$\frac{3}{2} \geqslant \frac{\sqrt{1-2t}}{\sqrt{2}} + \frac{1}{\sqrt{2}\sqrt{1-2t}} \geqslant \frac{5}{4} + \frac{t^2}{(1-2t)^2} \tag{24}$$

Squaring every sides in the about inequalities yields

$$\frac{9}{4} \geqslant \frac{1-2t}{2} + \frac{1}{2-4t} + 1 \geqslant \frac{25}{16} + \frac{t^4}{(1-2t)^4} + \frac{5t^2}{2(1-2t)^2} \tag{25}$$

Reducing to common denominator and rearranging, the right-hand inequality in (25) is reduces to

$$\frac{(1-2t)\left(16t^2(2t-1)^2 + \frac{1}{8}(16t-7)^2 + \frac{7}{8}\right)}{16(2t-1)^4} \geqslant 0 \tag{26}$$

and the left-hand inequality in (16) is reduces to

$$\frac{2(1-2t)^2 + 2 - 5(1-2t)}{2(1-2t)} = -\frac{1+2t}{2} \leqslant 0 \tag{27}$$

so two inequalities in (15) are hold.

Lemma 3[16] Let $a \leqslant b, u(t) = ta + (1-t)b, v(t) = tb + (1-t)a$. If $1/2 \leqslant t_2 \leqslant t_1 \leqslant 1$ or $0 \leqslant t_1 \leqslant t_2 \leqslant 1/2$, then

$$\left(\frac{a+b}{2}, \frac{a+b}{2}\right) \prec (u(t_2), v(t_2)) \prec (u(t_1), v(t_1)) \prec (a, b) \tag{28}$$

52.3　Proof of main result

Proof of Theorem 1: Let $(a, b) \in \mathbf{R}_+^2$.

(1) For

$$D_{SH-SA}(a, b) = \frac{1}{3}M_{SH}(a, b) - M_{SA}(a, b)$$

$$= \frac{a+b}{2} - \frac{2ab}{3(a+b)} - \frac{2}{3}\sqrt{\frac{a^2+b^2}{2}} \tag{29}$$

we have

$$\frac{\partial D_{SH-SA}(a, b)}{\partial a} = \frac{1}{2} - \frac{2b^2}{3(a+b)^2} - \frac{2}{3}\frac{a}{\sqrt{2(a^2+b^2)}} \tag{30}$$

$$\frac{\partial D_{SH-SA}(a, b)}{\partial b} = \frac{1}{2} - \frac{2a^2}{3(a+b)^2} - \frac{2}{3}\frac{b}{\sqrt{2(a^2+b^2)}}$$

whence

$$\Lambda := (\ln a - \ln b)\left(a\frac{\partial D_{SH-SA}(a, b)}{\partial a} - b\frac{\partial D_{SH-SA}(a, b)}{\partial b}\right)$$

$$= (a - b)(\ln a - \ln b)\left(\frac{1}{2} + \frac{2ab}{3(a + b)^2} - \frac{2}{3}\frac{a + b}{\sqrt{2(a^2 + b^2)}}\right) \tag{31}$$

From (14) we have

$$\frac{1}{2} + \frac{2ab}{3(a + b)^2} - \frac{2}{3}\frac{a + b}{\sqrt{2(a^2 + b^2)}} \geqslant 0, \tag{32}$$

which implies $\Lambda \geqslant 0$ and, by Lemma 1, it follows that D_{SH-SA} is Schur-geometrically convex in $\mathbf{R}_+^2$.

(2) For

$$D_{AH-SH}(a, b) = \frac{1}{2}M_{AH}(a, b) - \frac{1}{3}M_{SH}(a, b) = \frac{a + b}{4} - \frac{ab}{3(a + b)} - \frac{1}{3}\sqrt{\frac{a^2 + b^2}{2}} \tag{33}$$

To prove that the function D_{AH-SH} is Schur-geometrically convex in $\mathbf{R}_+^2$, it is enough to notice that $D_{AH-SH}(a, b) = \frac{1}{2}D_{SH-SA}(a, b)$.

(3) For

$$D_{SG-AH}(a, b) = M_{SG}(a, b) - M_{AH}(a, b)$$

$$= \sqrt{\frac{a^2 + b^2}{2}} - \sqrt{ab} - \frac{a + b}{2} + \frac{2ab}{a + b} \tag{34}$$

we have

$$\frac{\partial D_{SG-AH}(a, b)}{\partial a} = \frac{a}{\sqrt{2(a^2 + b^2)}} - \frac{b}{2\sqrt{ab}} - \frac{1}{2} + \frac{2b^2}{(a + b)^2}$$

$$\frac{\partial D_{SG-AH}(a, b)}{\partial b} = \frac{b}{\sqrt{2(a^2 + b^2)}} - \frac{a}{2\sqrt{ab}} - \frac{1}{2} + \frac{2a^2}{(a + b)^2} \tag{35}$$

and then

$$\Lambda := (\ln a - \ln b)\left(a\frac{\partial D_{SH-SA}(a, b)}{\partial a} - b\frac{\partial D_{SH-SA}(a, b)}{\partial b}\right)$$

$$= (a - b)(\ln a - \ln b)\left(\frac{a + b}{\sqrt{2(a^2 + b^2)}} - \frac{1}{2} - \frac{2ab}{(a + b)^2}\right) \tag{36}$$

From (13) we have $\Lambda \geqslant 0$, so by the Lemma 1, it follows that D_{SH-SA} is Schur-geometrically convex in $\mathbf{R}_+^2$.

(4) For

$$D_{AG-SG}(a,b) = M_{AG}(a,b) - \frac{1}{2}M_{SG}(a,b) = \frac{1}{2}\left(a+b-\sqrt{ab}-\sqrt{\frac{a^2+b^2}{2}}\right)$$
(37)

we have

$$\frac{\partial D_{AG-SG}(a,b)}{\partial a} = \frac{1}{2}\left(1 - \frac{b}{2\sqrt{ab}} - \frac{a}{\sqrt{2(a^2+b^2)}}\right)$$

$$\frac{\partial D_{AG-SG}(a,b)}{\partial b} = \frac{1}{2}\left(1 - \frac{a}{2\sqrt{ab}} - \frac{b}{\sqrt{2(a^2+b^2)}}\right)$$
(38)

and then

$$\Lambda := (\ln a - \ln b)\left(a\frac{\partial D_{SH-SA}(a,b)}{\partial a} - b\frac{\partial D_{SH-SA}(a,b)}{\partial b}\right)$$

$$= (a-b)(\ln a - \ln b)\left(1 - \frac{a+b}{\sqrt{2(a^2+b^2)}}\right)$$
(39)

By (13) we infer that

$$1 - \frac{a+b}{\sqrt{2(a^2+b^2)}} \geqslant 0$$
(40)

so $\Lambda \geqslant 0$. By the Lemma 1, we get that D_{AG-SG} is Schur-geometrically convex in $\mathbf{R}_+^2$.

(5) For

$$D_{N_2N_1-AH}(a,b) = M_{N_2N_1}(a,b) - \frac{1}{8}M_{AH}(a,b)$$

$$= \left(\frac{\sqrt{a}+\sqrt{b}}{2}\right)\left(\sqrt{\frac{a+b}{2}}\right) - \frac{1}{4}(a+b) - \frac{1}{2}\sqrt{ab} - \frac{1}{8}\left(\frac{a+b}{2} - \frac{2ab}{a+b}\right)$$
(41)

we have

$$\frac{\partial D_{N_2N_1-AH}(a,b)}{\partial a} = \frac{1}{4\sqrt{a}}\sqrt{\frac{a+b}{2}} + \frac{1}{4}\left(\frac{\sqrt{a}+\sqrt{b}}{2}\right)\left(\frac{a+b}{2}\right)^{-1/2} -$$

$$\frac{1}{4} - \frac{b}{4\sqrt{ab}} - \frac{1}{8}\left(\frac{1}{2} - \frac{2b^2}{(a+b)^2}\right)$$

$$\frac{\partial D_{N_2 N_1 - AH}(a, b)}{\partial b} = \frac{1}{4\sqrt{b}}\sqrt{\frac{a+b}{2}} + \frac{1}{4}\left(\frac{\sqrt{a} + \sqrt{b}}{2}\right)\left(\frac{a+b}{2}\right)^{-1/2} -$$
$$\frac{1}{4} - \frac{a}{4\sqrt{ab}} - \frac{1}{8}\left(\frac{1}{2} - \frac{2a^2}{(a+b)^2}\right) \tag{42}$$

and then, from (15) we have

$$\frac{\sqrt{a+b}}{\sqrt{2}(\sqrt{a} + \sqrt{b})} + \frac{\sqrt{a} + \sqrt{b}}{\sqrt{2}\sqrt{a+b}} - \frac{5}{4} - \frac{ab}{(a+b)^2} \geqslant 0 \tag{43}$$

so $\Lambda \geqslant 0$, it follows that $D_{N_2 N_1 - AH}$ is Schur-geometrically convex in $\mathbf{R}_+^2$.

(6) For

$$D_{N_2 G - N_2 N_1}(a, b) = \frac{1}{3}M_{N_2 G}(a, b) - M_{N_2 N_1}(a, b)$$
$$= \frac{a+b}{4} + \frac{\sqrt{ab}}{6} - \frac{2}{3}\left(\frac{\sqrt{a} + \sqrt{b}}{2}\right)\left(\sqrt{\frac{a+b}{2}}\right) \tag{44}$$

we have

$$\frac{\partial D_{N_2 G - N_2 N_1}(a, b)}{\partial a} = \frac{1}{4} + \frac{b}{12\sqrt{ab}} -$$
$$\frac{1}{6\sqrt{a}}\sqrt{\frac{a+b}{2}} - \frac{1}{6}\left(\frac{\sqrt{a} + \sqrt{b}}{2}\right)\left(\frac{a+b}{2}\right)^{-1/2}$$

$$\frac{\partial D_{N_2 G - N_2 N_1}(a, b)}{\partial b} = \frac{1}{4} + \frac{a}{12\sqrt{ab}} -$$
$$\frac{1}{6\sqrt{b}}\sqrt{\frac{a+b}{2}} - \frac{1}{6}\left(\frac{\sqrt{a} + \sqrt{b}}{2}\right)\left(\frac{a+b}{2}\right)^{-1/2} \tag{45}$$

and then

$$\Lambda = (\ln a - \ln b)\left(a\frac{\partial D_{N_2 G - N_2 N_1}(a, b)}{\partial a} - b\frac{\partial D_{N_2 G - N_2 N_1}(a, b)}{\partial b}\right)$$
$$= (\ln a - \ln b)\left(\frac{1}{4}(a - b) - \frac{\sqrt{a} - \sqrt{b}}{6}\sqrt{\frac{a+b}{2}} - \frac{(a-b)(\sqrt{a} + \sqrt{b})}{12}\left(\frac{a+b}{2}\right)^{-1/2}\right)$$
$$= \frac{1}{6}(a - b)(\ln a - \ln b)\left(\frac{3}{2} - \frac{\sqrt{a+b}}{\sqrt{2}(\sqrt{a} + \sqrt{b})} - \frac{\sqrt{a} + \sqrt{b}}{\sqrt{2}\sqrt{a+b}}\right) \tag{46}$$

By (15) we infer that $\Lambda \geqslant 0$, which proves that $D_{N_2G-N_2N_1}$ is Schur-geometrically convex in $\mathbf{R}_+^2$.

(7) For

$$D_{AG-N_2G}(a,b) = \frac{1}{4}M_{AG}(a,b) - \frac{1}{3}M_{N_2G}(a,b)$$

$$= \frac{a+b}{8} + \frac{1}{12}\sqrt{ab} - \frac{1}{3}\left(\frac{\sqrt{a}+\sqrt{b}}{2}\right)\left(\sqrt{\frac{a+b}{2}}\right) \qquad (47)$$

we have

$$\frac{\partial D_{AG-N_2G}(a,b)}{\partial a} = \frac{1}{8} + \frac{b}{24\sqrt{ab}} - \frac{\sqrt{a+b}}{12\sqrt{2a}} - \frac{\sqrt{a}+\sqrt{b}}{12\sqrt{2(a+b)}}$$

(48)

$$\frac{\partial D_{AG-N_2G}(a,b)}{\partial b} = \frac{1}{8} + \frac{a}{24\sqrt{ab}} - \frac{\sqrt{a+b}}{12\sqrt{2b}} - \frac{\sqrt{a}+\sqrt{b}}{12\sqrt{2(a+b)}} \qquad (48)$$

and then

$$\Lambda = (\ln a - \ln b)\left(a\frac{\partial D_{AG-N_2G}(a,b)}{\partial a} - b\frac{\partial D_{AG-N_2G}(a,b)}{\partial b}\right)$$

$$= (\ln a - \ln b)\left(\frac{a-b}{8} - \frac{\sqrt{a+b}(\sqrt{a}-\sqrt{b})}{12\sqrt{2}} - \frac{(a-b)(\sqrt{a}+\sqrt{b})}{12\sqrt{2(a+b)}}\right)$$

$$= \frac{(a-b)(\ln a - \ln b)}{8}\left(1 - \frac{2}{3}\left(\frac{\sqrt{a+b}}{\sqrt{2}(\sqrt{a}+\sqrt{b})} + \frac{\sqrt{a}+\sqrt{b}}{\sqrt{2}\sqrt{a+b}}\right)\right) \qquad (49)$$

From (15) we have $\Lambda \geqslant 0$, and consequently, by Lemma 1, we obtain that D_{AG-N_2G} is Schur-geometrically convex in $\mathbf{R}_+^2$.

(8) In order to prove that the function $D_{AN_2-AG}(a,b)$ is Schur-geometrically convex in $\mathbf{R}_+^2$ it is enough to notice that

$$D_{AN_2-AG}(a,b) = M_{AN_2}(a,b) - \frac{1}{4}M_{AG}(a,b) = 3D_{AG-N_2G}(a,b) \qquad (50)$$

(9) For

$$D_{SN_2-SA}(a,b) = \frac{4}{5}M_{SN_2}(a,b) - M_{SA}(a,b)$$

$$= \frac{a+b}{2} - \frac{1}{5}\sqrt{\frac{a^2+b^2}{2}} - \frac{1}{5}(\sqrt{a}+\sqrt{b})\sqrt{2(a+b)} \qquad (51)$$

we have

$$\frac{\partial D_{SN_2-SA}(a,b)}{\partial a} = \frac{1}{2} - \frac{a}{5\sqrt{2(a^2+b^2)}} - \frac{1}{5}\sqrt{\frac{a+b}{2a}} - \frac{\sqrt{a}+\sqrt{b}}{5\sqrt{2(a+b)}}$$

$$\frac{\partial D_{SN_2-SA}(a,b)}{\partial b} = \frac{1}{2} - \frac{b}{5\sqrt{2(a^2+b^2)}} - \frac{1}{5}\sqrt{\frac{a+b}{2b}} - \frac{\sqrt{a}+\sqrt{b}}{5\sqrt{2(a+b)}} \qquad (52)$$

and then

$$\Lambda = (\ln a - \ln b)\left(\frac{\partial D_{SN_2-SA}(a,b)}{\partial a} - \frac{\partial D_{SN_2-SA}(a,b)}{\partial b}\right)$$

$$= (\ln a - \ln b)(\frac{a-b}{2} - \frac{a^2-b^2}{5\sqrt{2(a^2+b^2)}}$$

$$- \frac{1}{5}(\sqrt{\frac{a(a+b)}{2}} - \sqrt{\frac{b(a+b)}{2}}) - \frac{(\sqrt{a}+\sqrt{b})(a-b)}{5\sqrt{2(a+b)}})$$

$$= \frac{(a-b)(\ln a - \ln b)}{5\sqrt{2}}\left(\frac{5}{\sqrt{2}} - \frac{a+b}{\sqrt{a^2+b^2}} - \frac{\sqrt{a+b}}{\sqrt{a}+\sqrt{b}} - \frac{\sqrt{a}+\sqrt{b}}{\sqrt{a+b}}\right) \qquad (53)$$

From (13) and (15) we obtain that

$$\frac{5}{\sqrt{2}} - \frac{a+b}{\sqrt{a^2+b^2}} - \frac{\sqrt{a+b}}{\sqrt{a}+\sqrt{b}} - \frac{\sqrt{a}+\sqrt{b}}{\sqrt{a+b}} \geqslant \frac{5}{\sqrt{2}} - \sqrt{2} - \frac{3}{\sqrt{2}} = 0 \qquad (54)$$

so $\Lambda \geqslant 0$, which proves that the function $D_{SN_2-SA}(a,b)$ is Schur-geometrically convex in $\mathbf{R}_+^2$.

(10) One can easily check that

$$D_{AN_AN_2-SN_2}(a,b) = 4D_{SN_2-SA}(a,b) \qquad (55)$$

and, consequently, the function that $D_{AN_2-SN_2}$ is Schur-geometrically convex in $\mathbf{R}_+^2$.

(11) To prove that the function

$$D_{SN_1-SH}(a,b) = 2M_{SN_1}(a,b) - M_{SH}(a,b)$$

$$= \sqrt{\frac{a^2+b^2}{2}} - \frac{a+b}{2} - \sqrt{ab} + \frac{2ab}{a+b} \qquad (56)$$

is Schur-geometrically convex in $\mathbf{R}_+^2$, it is enough to notice that

$$D_{SN_1-SH}(a,b) = D_{SG-AH}(a,b) \tag{57}$$

(12) For

$$\begin{aligned}
D_{SG-SN_1}(a,b) &= \frac{3}{2}M_{SG}(a,b) - 2M_{SN_1}(a,b) \\
&= \frac{1}{2}\left(a + b - \sqrt{ab} - \sqrt{\frac{a^2+b^2}{2}}\right)
\end{aligned} \tag{58}$$

we have

$$\frac{\partial D_{SG-SN_1}(a,b)}{\partial a} = \frac{1}{2}\left(1 - \frac{b}{2\sqrt{ab}} - \frac{a}{\sqrt{2(a^2+b^2)}}\right)$$

$$\frac{\partial D_{SG-SN_1}(a,b)}{\partial b} = \frac{1}{2}\left(1 - \frac{a}{2\sqrt{ab}} - \frac{b}{\sqrt{2(a^2+b^2)}}\right) \tag{59}$$

and then

$$\begin{aligned}
\Lambda &= (\ln a - \ln b)\left(a\frac{\partial D_{SG-SN_1}(a,b)}{\partial a} - b\frac{\partial D_{SG-SN_1}(a,b)}{\partial b}\right) \\
&= \frac{(a-b)(\ln a - \ln b)}{2}\left(1 - \frac{a+b}{\sqrt{2(a^2+b^2)}}\right)
\end{aligned} \tag{60}$$

By the inequality (13) we get that $\Lambda \geqslant 0$, which proves that D_{SG-SN_1} is Schur-geometrically convex in $\mathbf{R}_+^2$.

(13) It is easy to check that

$$D_{SN_3-SA}(a,b) = \frac{1}{2}D_{AG-SG}(a,b) \tag{61}$$

which means that the function D_{SN_3-SA} is Schur-geometrically convex in $\mathbf{R}_+^2$.

(14) To prove that the function $D_{SN_1-SN_3}$ is Schur-geometrically convex in $\mathbf{R}_+^2$ it is enough to notice that

$$D_{SN_1-SN_3}(a,b) = \frac{1}{6}D_{AG-SG}(a,b) \tag{62}$$

The proof of Theorem I is complete.

52.4　Applications

Applying Theorem 1, Lemma 3 and Definition 2 one can easily prove the following

Theorem 2 Let $0 < a \leqslant b$. $1/2 \leqslant t \leqslant 1$ or $0 \leqslant t \leqslant 1/2$, $u = a^t b^{1-t}$ and $v = b^t a^{1-t}$. Then

$$M_{SA}(a,b) \leqslant \frac{1}{3}M_{SH}(a,b) - \left(\frac{1}{3}M_{SH}(u,v) - M_{SA}(u,v)\right) \leqslant \frac{1}{3}M_{SH}(a,b)$$

$$\leqslant \frac{1}{2}M_{AH}(a,b) - \left(\frac{1}{2}M_{AH}(u,v) - \frac{1}{3}M_{SH}(u,v)\right) \leqslant \frac{1}{2}M_{AH}(a,b)$$

$$\leqslant \frac{1}{2}M_{SG}(a,b) - \left(\frac{1}{2}M_{SG}(u,v) - \frac{1}{2}M_{AH}(u,v)\right) \leqslant \frac{1}{2}M_{SG}(a,b)$$

$$\leqslant M_{AG}(a,b) - \left(M_{AG}(u,v) - \frac{1}{2}M_{SG}(u,v)\right) \leqslant M_{AG}(a,b) \quad (63)$$

$$\frac{1}{8}M_{AH}(a,b) \leqslant M_{N_2 N_1}(a,b) - \left(M_{N_2 N_1}(u,v) - \frac{1}{8}M_{AH}(u,v)\right) \leqslant M_{N_2 N_1}(a,b)$$

$$\leqslant \frac{1}{3}M_{N_2 G}(a,b) - \left(\frac{1}{3}M_{N_2 G}(u,v) - M_{N_2 N_1}(u,v)\right) \leqslant \frac{1}{3}M_{N_2 G}(a,b)$$

$$\leqslant \frac{1}{4}M_{AG}(a,b) - \left(\frac{1}{4}M_{AG}(u,v) - \frac{1}{3}M_{N_2 G}(u,v)\right) \leqslant \frac{1}{4}M_{AG}(a,b)$$

$$\leqslant M_{AN_2}(a,b) - \left(M_{AN_2}(u,v) - \frac{1}{4}M_{AG}(u,v)\right) \leqslant M_{AN_2}(a,b)$$

$$(64)$$

$$M_{SA}(a,b) \leqslant \frac{4}{5}M_{SN_2}(a,b) - \left(\frac{4}{5}M_{SN_2}(u,v) - \frac{4}{5}M_{SN_2}(u,v)\right) \leqslant \frac{4}{5}M_{SN_2}(a,b)$$

$$\leqslant 4M_{AN_2}(a,b) - \left(4M_{AN_2}(u,v) - \frac{4}{5}M_{SN_2}(u,v)\right) \leqslant 4M_{AN_2}(a,b)$$

$$(65)$$

$$M_{SH}(a,b) \leqslant 2M_{SN_1}(a,b) - (2M_{SN_1}(u,v) - M_{SH}(u,v)) \leqslant 2M_{SN_1}(a,b)$$

$$\leqslant \frac{3}{2}M_{SG}(a,b) - \left(\frac{3}{2}M_{SG}(u,v) - \frac{3}{2}M_{SG}(u,v)\right) \leqslant \frac{3}{2}M_{SG}(a,b)$$

$$(66)$$

$$M_{SA}(a,b) \leqslant \frac{3}{4}M_{SN_3}(a,b) - \left(\frac{3}{4}M_{SN_3}(u,v) - M_{SA}(u,v)\right) \leqslant \frac{3}{4}M_{SN_3}(a,b)$$

$$\leqslant \frac{2}{3}M_{SN_1}(a,b) - \left(\frac{2}{3}M_{SN_1}(u,v) - \frac{3}{4}M_{SN_3}(u,v)\right) \leqslant \frac{2}{3}M_{SN_1}(a,b)$$

$$(67)$$

Remark (63), (64), (65), (66) and (67) is a refinement of (5), (6), (7), (8) and (9) respectively.

Acknowledgements

The authors are grateful to the referees for their helpful comments and suggestions.

References

[1] I J TANEJA. Refinement of inequalities among means[J]. Journal of Combinatorics, Information & System Sciences,2006, Volume 31, ISSUE 1-4, 343-364, arXiv:math/0505192v2 [math.GM] 12 Jul 2005.

[2] I J TANEJA. On a Difference of Jensen Inequality and its Applications to Mean Divergence Measures[J]. RGMIA Research Report Collection, http://rgmia.vu.edu.au, 7(4)(2004), Art. 16. Also in:arXiv:math.PR/0501302 v1 19 Jan 2005.

[3] I J TANEJA. On Symmetric and Non-Symmetric Divergence Measures and Their Generalizations[J]. Advances in Imaging and Electron Physics, 2005, 138: 177-250.

[4] A M MARSHALL, I. OLKINIKIN. Inequalities: Theory of Majorization and Its Application[M]. New York : Academies Press, 1979.

[5] BO-YING WANG. Foundations of Majorization Inequalities[M]. Beijing: Beijing Normal Univ. Press, 1990.

[6] XIAO-MING ZHANG. Geometrically Convex Functions[M].Hefei: An'hui University Press, 2004.

[7] HUAN-NAN SHI, YONG-MING JIANG, WEI-DONG JIANG. Schur-Convexity and Schur-Geometrically Concavity of Gini Mean[J]. Computers and Mathematics with Applications, 2009, 57: 266-274.

[8] YU-MING CHU, XIAO-MING ZHANG. The Schur geometrical convexity of the extended mean values[J]. Journal of Convex Analysis, 2008, 15(4): 869-890.

[9]　KAI-ZHONG GUAN. A class of symmetric functions for multiplicatively convex function[J]. Math. Inequal. Appl., 2007, 10 (4): 745-753.

[10]　HUAN-NAN SHI, MIHALY BENCZE, SHAN-HE WU, DA-MAO LI. Schur Convexity of generalized Heronian Means involving two parameters[J]. J. Inequal. Appl., Volume 2008, Article ID 879273, 9 pages doi:10.1155/2008/879273.

[11]　XIAO-MING ZHANG. The Schur geometrical convexity of integral arithmetic mean[J]. International Journal of Pure and Applied Mathematics, 2007, 41(7): 919-925.

[12]　KAI-ZHONG GUAN. Schur-convexity of the complete symmetric function[J]. Math. Inequal. Appl. 2006, 9(4): 567-576.

[13]　KAI-ZHONG GUAN. Some properties of a class of symmetric functions[J]. J. Math. Anal. Appl., 2007, 336(1): 70-80.

[14]　C STEPNIAK. An effective characterization of Schur-convex functions with applications[J]. J. Convex Anal. 2007, 14(1): 103-108.

[15]　HUAN-NAN SHI. Schur-Convex Functions relate to Hadamard-type inequalities[J]. J. Math. Inequal., 2007, 1(1): 127-136.

[16]　HUAN-NAN SHI, DA-MAO LI, CHUN GU. Schur-Convexity of a Mean of Convex Function[J]. Applied Mathematics Letters, 2009, 22 : 932-937.

[17]　YU-MING CHU, XIAO-MING ZHANG. Necessary and sufficient conditions such that extended mean values are Schur-convex or Schur-concave[J]. Journal of Mathematics of Kyoto University, 2008, 48 (1): 229–238.

[18]　ELEZOVIC N, PECARIC J. Note on Schur-convex functions[J]. Rocky Mountain J. Math., 1998, 29: 853–856.

[19]　JÓZSEF SÁNDOR. The Schur-convexity of Stolarsky and Gini means[J]. Banach J. Math. Anal., 2007, 1(2): 212-215.

[20]　HUAN-NAN ShI, SHAN-HE WU, FENG QI. An alternative note on the Schur-convexity of the extended mean values[J]. Math. Inequal. Appl., 2006, 9(2): 219–224.

第53篇 一个不等式命题的概率证明 及推广

(石焕南. 湖南理工学院学报(自然科学版), 2012, 25 (3): 1-2,72)

摘 要: 本文借助于概率的简单性质简洁地证明了一个不等式命题, 并将其推广到一般初等对称函数的情形, 显示了概率方法的巧妙性与优越性及其应用上的广泛性. 从一个侧面揭示了数学不同学科之间的内在联系.

关键词: 不等式; 概率证明; 初等对称函数

文[1]提出如下

命题 当 $0 < x_1, x_2, \ldots, x_n < 1$ 时, 有不等式

$$\sum_{i=1}^{n} \frac{x_1 x_2 \cdots x_n}{x_i} - (n-1) \prod_{i=1}^{n} x_i < 1 \tag{1}$$

文[2] 用数学归纳法证明了该命题. 本文利用概率方法给出该命题一个简单的证明, 并将其推广到一般初等对称函数的情形.

证明 由概率的一般加法公式

$$\sum_{i=1}^{n} \frac{x_1 x_2 \cdots x_n}{x_i} - n \prod_{i=1}^{n} x_i < 1 - \prod_{i=1}^{n} x_i \tag{2}$$

设 $A_1, A_2, \ldots, A_n$ 是同一随机试验中的 n 个相互独立的随机事件, 且 $P(A_i) = x_i, i = 1, \ldots, n.$ 又记

$$B_1 = \bar{A}_1 A_2 \cdots A_n, B_2 = A_1 \bar{A}_2 \cdots A_n, \ldots, B_{n-1} = A_1 \cdots \bar{A}_{n-1} A_n, B_n = A_1 \cdots A_{n-1} \bar{A}_n$$

不难验证 $B_1, B_2, \ldots, B_n$ 互不相容且 $B_i \subset \bar{A}_i, i = 1, \ldots, n$. 于是

$$\sum_{i=1}^{n} \frac{x_1 x_2 \cdots x_n}{x_i} - n \prod_{i=1}^{n} x_i$$

$$= P(A_2 A_3 \cdots A_n) + P(A_1 A_3 \cdots A_n) + \cdots +$$

$$\quad P(A_1 \cdots A_{n-2} A_n) + P(A_1 \cdots A_{n-1}) - nP(A_1 \cdots A_n)$$

$$= [P(A_2 A_3 \cdots A_n) - P(A_1 \cdots A_n)] + [P(A_1 A_3 \cdots A_n) - P(A_1 \cdots A_n)] + \cdots +$$

$$\quad [P(A_1 \cdots A_{n-2} A_n) - P(A_1 \cdots A_n)] + [P(A_1 \cdots A_{n-1}) - P(A_1 \cdots A_n)]$$

$$= P(\bar{A}_1 A_2 \cdots A_n) + P(A_1 \bar{A}_2 \cdots A_n) + \cdots + P(A_1 \cdots A_{n-1} \bar{A}_n)$$

$$= P(B_1 \cup B_2 \cup \cdots \cup B_n) \quad (\text{注意} B_1, B_2, \ldots, B_n \text{互不相容})$$

$$\leqslant P(\bar{A}_1 \cup \bar{A}_2 \cdots \cup \bar{A}_n) \quad (\text{注意} B_i \subset \bar{A}_i, i = 1, \ldots, n)$$

$$= 1 - P(A_1 \cdots A_n) = 1 - \prod_{i=1}^{n} x_i \quad (\text{注意} A_1, \cdots, A_n \text{ 相互独立})$$

$$\leqslant 1$$

式(2)得证, 从而式(1)成立, 证毕.

为了将命题推广为一般初等对称函数的情形, 我们先考察一个简单的情形.

定理 1　设 $0 < x_1, x_2, x_3 x_4 < 1$, 则

$$2 \sum_{1 \leqslant i_1 < i_2 \leqslant 4} x_{i_1} x_{i_2} - 3 \sum_{1 \leqslant i_1 < i_2 \leqslant 4} x_{i_1} x_{i_2} x_{i_3} < 4 \tag{3}$$

证明　设 A_1, A_2, A_3, A_4 是同一随机试验中的 4 个相互独立的随机事件, 且 $P(A_i) = x_i, 1, 2, 3, 4$, 则

$$4 > 4 - \sum_{1 \leqslant i_1 < i_2 \leqslant 4} x_{i_1} x_{i_2} x_{i_3}$$

$$= 1 - P(A_1 A_2 A_3) + 1 - P(A_1 A_2 A_4) + 1 - P(A_1 A_3 A_4) + 1 - P(A_2 A_3 A_4)$$

$$= P(\bar{A}_1 \cup \bar{A}_2 \cup \bar{A}_3) + P(\bar{A}_1 \cup \bar{A}_2 \cup \bar{A}_4) + P(\bar{A}_1 \cup \bar{A}_3 \cup \bar{A}_4) + P(\bar{A}_2 \cup \bar{A}_3 \cup \bar{A}_4)$$

$$\geqslant P(\bar{A}_1 A_2 A_3) + P(A_1 \bar{A}_2 A_3) + P(A_1 A_2 \bar{A}_3) + P(\bar{A}_1 A_2 A_4) +$$

$$\quad P(A_1 \bar{A}_2 A_4) + P(A_1 A_2 \bar{A}_4) + P(\bar{A}_1 A_3 A_4) + P(A_1 \bar{A}_2 A_4) +$$

$$\quad P(A_1 A_3 \bar{A}_4) + P(\bar{A}_2 A_3 A_4) + P(A_2 \bar{A}_3 A_4) + P(A_2 A_3 \bar{A}_4)$$

$$= [P(A_2 A_3) - P(A_1 A_2 A_3)] + [P(A_1 A_3) - P(A_1 A_2 A_3)] +$$

$$\quad [P(A_1 A_2) - P(A_1 A_2 A_3)] + [P(A_2 A_4) - P(A_1 A_2 A_4)] +$$

$$[P(A_1A_4) - P(A_1A_2A_4)] + [P(A_1A_2) - P(A_1A_2A_4)]+$$

$$[P(A_3A_4) - P(A_1A_3A_4)] + [P(A_1A_4) - P(A_1A_3A_4)]+$$

$$[P(A_1A_3) - P(A_1A_3A_4)] + [P(A_3A_4) - P(A_2A_3A_4)]+$$

$$[P(A_2A_4) - P(A_2A_3A_4)] + [P(A_2A_3) - P(A_2A_3A_4)]$$

$$=2\sum_{1\leqslant i_1 < i_2 \leqslant 4} x_{i_1}x_{i_2} - 3\sum_{1\leqslant i_1 < i_2 \leqslant 4} x_{i_1}x_{i_2}x_{i_3}$$

一般地, 我们有

定理 2 设 $0 < x_1, x_2, \ldots, x_n < 1$, 则

$$(n-k+1)\sum_{1\leqslant i_1 < \cdots < i_{k-1} \leqslant n}\prod_{j=1}^{k-1}x_{i_j} - k\sum_{1\leqslant i_1 < \cdots < i_k \leqslant n}\prod_{j=1}^{k}x_{i_j} < \mathrm{C}_n^k \qquad (4)$$

证明 设 $A_1, A_2, \ldots, A_n$ 是同一随机试验中的 n 个相互独立的随机事件, 且 $P(A_i) = x_i, 1, 2, \ldots, n$, 则

$$\mathrm{C}_n^k > \mathrm{C}_n^k - \sum_{1\leqslant i_1 < \cdots < i_k \leqslant n}\prod_{j=1}^{k}x_{i_j} = \sum_{1\leqslant i_1 < \cdots < i_k \leqslant n}\left(1 - \prod_{j=1}^{k}x_{i_j}\right)$$

$$= \sum_{1\leqslant i_1 < \cdots < i_k \leqslant n}\left(1 - \prod_{j=1}^{k}P(A_{i_j})\right) = \sum_{1\leqslant i_1 < \cdots < i_k \leqslant n}\left[1 - P\left(\prod_{j=1}^{k}A_{i_j}\right)\right]$$

$$= \sum_{1\leqslant i_1 < \cdots < i_k \leqslant n} P\left(\bar{A}_{i_1} \cup \bar{A}_{i_2} \cdots \cup \bar{A}_{i_k}\right)$$

$$\geqslant \sum_{1\leqslant i_1 < \cdots < i_k \leqslant n}\sum_{j=1}^{k} P(A_{i_1}\cdots A_{i_{j-1}}\bar{A}_{i_j}A_{i_{j+1}}\cdots A_{i_k})$$

$$= \sum_{1\leqslant i_1 < \cdots < i_k \leqslant n}\sum_{j=1}^{k}\left[P(A_{i_1}\cdots A_{i_{j-1}}A_{i_{j+1}}\cdots A_{i_k}) - P(A_{i_1}A_{i_2}\cdots A_{i_k})\right]$$

$$=(n-k+1)\sum_{1\leqslant i_1 < \cdots < i_k \leqslant n} P(A_{i_1}A_{i_2}\cdots A_{i_{k-1}}) - k\sum_{1\leqslant i_1 < \cdots < i_k \leqslant n} P(A_{i_1}A_{i_2}\cdots A_{i_k})$$

$$=(n-k+1)\sum_{1\leqslant i_1 < \cdots < i_{k-1} \leqslant n}\prod_{j=1}^{k-1}x_{i_j} - k\sum_{1\leqslant i_1 < \cdots < i_k \leqslant n}\prod_{j=1}^{k}x_{i_j}$$

由此即得到式(4), 证毕.

著名数学家王梓坤院士在文[3]指出:"用概率的方法来证明一些关系式或解决其他数学分析中的问题, 是概率论的重要研究方向之一." 用概率的方法解决一些问题, 思路别开生面, 过程常常简洁直观.

参考文献

[1]　阳凌云. 数学素质教育导论[M]. 长沙：湖南科学技术出版社, 2005.

[2]　符云锦. 一个不等式的证明与应用[J]. 中国初等数学研究, 2012(4)：80-83.

[3]　王梓坤. 概率论基础及其应用[M]. 北京：科学出版社, 1979：24.

第54篇 Sharpening of Kai-lai Zhong's Inequality

(SHI HUAN-NAN, GU CHUN. Journal of Latex Class Files, Vol. 6,

No. 1, January 2007:1-4)

Abstract: By using means of the theory of majorization, Kai-lai Zhong's Inequality is sharpened. As an application, some triangular inequalities are sharpened.

Keywords: Kai-lai Zhong's inequality; majorization; triangular inequalities

54.1 Introduction

Let $a_1 \geqslant a_2 \geqslant \cdots \geqslant a_n \geqslant 0$. If $\sum_{j=1}^{k} a_j \leqslant \sum_{j=1}^{k} b_j, k = 1, \ldots, n$, then

$$\sum_{j=1}^{n} a_j^2 \leqslant \sum_{j=1}^{n} b_j^2$$

with the equality holding only if $a_k = b_k, k = 1, \ldots, n$.

It is known as the Kai-lai Zhong's inequality [1]. In 1989, Ji Chen [2] obtained the following exponential generalization of this inequality:

Let $a_1 \geqslant a_2 \geqslant \ldots \geqslant a_n \geqslant 0, b_1 \geqslant b_2 \geqslant \cdots \geqslant b_n \geqslant 0$. If $\sum_{j=1}^{k} a_j \leqslant \sum_{j=1}^{k} b_j, k = 1, \ldots, n$, then for $p > 1$, we have

$$\sum_{j=1}^{n} a_j^p \leqslant \sum_{j=1}^{n} b_j^p \qquad (1)$$

with the equality holding only if $a_k = b_k, k = 1, \ldots, n$.

In 1996, Ke Hu[3-4] given the following sharpening of the inequality in (1)

$$\sum_{j=1}^{n} a_j^p \leqslant \sum_{j=1}^{n} |b_j|^p \cdot \left[1 - \frac{A^2}{\left(\sum_{i=1}^{n} a_i^p \sum_{j=1}^{n} |b_j|^p \right)^2} \right]^{\frac{\theta(p)}{2}} \tag{2}$$

where

$$A = \sum_{i=1}^{n} a_1^p e_i \sum_{j=1}^{n} |b_j|^p - \sum_{i=1}^{n} a_i^p \sum_{j=1}^{n} |b_j|^p e_j$$

and $1 - e_k - e_m \geqslant 0$, for $k, m = 1, 2, \ldots, n$. $\theta(p) = p - 1$ for $p > 2$ and $\theta(p) = 1$ for $p < 2$.

In recent years, some further generalizations and applications about the Kai-lai Zhong's inequality have been obtained in[5-7] and the references therein. The purpose of this note is to establish a sharped Kai-lai Zhong's inequality which is very simple and clear by means of the theory of majorization. As an application, some triangular inequalities are sharpened.

Definition[8] Let $a = (a_1, \ldots, a_n)$ and $b = (b_1, \ldots, b_n) \in \mathbf{R}^n$. Then a is said to be majorized by b (in symbols $a \prec b$) if

(i)

$$\sum_{i=1}^{k} a_{[i]} \leqslant \sum_{i=1}^{k} b_{[i]} \quad \text{for} \quad k = 1, 2, \ldots, n - 1$$

(ii)

$$\sum_{i=1}^{n} a_i = \sum_{i=1}^{n} b_i$$

where $a_{[1]} \geqslant a_{[2]} \geqslant \cdots \geqslant a_{[n]}$ and $b_{[1]} \geqslant b_{[2]} \geqslant \cdots \geqslant b_{[n]}$ are components of a and b rearranged in descending order, and a is said to strictly majors by b (written $a \prec\prec b$) if a is not permutation of b. And a is said to be weakly submajorized by b (written $a \prec_w b$) if

$$\sum_{i=1}^{k} a_{[i]} \leqslant \sum_{i=1}^{k} b_{[i]}, \quad k = 1, 2, \ldots, n$$

Lemma 1 [10] Let $a \in \mathbf{R}_+^n, b \in \mathbf{R}^n$ and $\delta = \sum_{i=1}^{n} (b_i - a_i)$. If $a \prec_w b$, then

$$\left(a, \underbrace{\frac{\delta}{k}, \cdots, \frac{\delta}{k}}_{k}\right) \prec \left(b, \underbrace{0, \ldots, 0}_{k}\right) \tag{3}$$

Lemma 2[10] Let a, $b \in \mathbf{R}^n$, if $a \prec_w b$, then

$$(a, \ a_{n+1}) \prec (b, \ b_{n+1}) \tag{4}$$

where

$$a_{n+1} = \min\{a_1, \cdots, a_n, \ b_1, \cdots, b_n\} \quad b_{n+1} = \sum_{i=1}^{n+1} a_i - \sum_{i=1}^{n} b_i$$

Lemma 3[8] Let a, $b \in \mathbf{R}^n$, $b_1 \leqslant \cdots \leqslant b_n$, if $a \prec_w b$, then

$$(a_1, \cdots, a_{n-1}, a_n) \prec (b_1, \cdots, b_{n-1}, c) \tag{5}$$

where $c = b_n - (\sum\limits_{i=1}^{n} a_i - \sum\limits_{i=1}^{n} b_i)$.

Lemma 4[8] Let $I \subset \mathbf{R}$ be an interval, $a, b \in I^n \subset \mathbf{R}^n$, and $g : I \to \mathbf{R}$. Then

(i) $a \prec b$ if and only if

$$\sum_{i=1}^{n} g(a_i) \leqslant (\geqslant) \sum_{i=1}^{n} g(b_i) \tag{6}$$

holds for all convex(concave) functions g.

(ii) $a \prec\prec b$ if and only if

$$\sum_{i=1}^{n} g(a_i) < (>) \sum_{i=1}^{n} g(b_i) \tag{7}$$

holds for all strictly convex(concave) functions g.

Lemma 5[8] Let $I \subset \mathbf{R}$ be an interval, $a, b \in I^n \subset \mathbf{R}^n$, and $g : I \to \mathbf{R}$. If $a \prec_w b$, then

$$(g(a_1), g(a_2), \ldots, g(a_n)) \prec_w (g(b_1), g(b_2), \ldots, g(b_n)) \tag{8}$$

holds for all increasing convex functions g.

54.2　Main results and Proofs

Theorem 1　Let $a_1 \geqslant a_2 \geqslant \cdots \geqslant a_n \geqslant 0$, $b_1 \geqslant b_2 \geqslant \cdots \geqslant b_n \geqslant 0$, $\sum\limits_{j=1}^{k} a_j \leqslant \sum\limits_{j=1}^{k} b_j$, $k = 1, \ldots, n$, i.e. $\boldsymbol{a} \prec_w \boldsymbol{b}$, and let $\delta = \sum\limits_{j=1}^{n}(b_j - a_j)$. If $p > 1$. Then

$$\sum_{j=1}^{n} a_j^p \leqslant \sum_{j=1}^{n} b_j^p - \frac{\delta^p}{k^{p-1}} \tag{9}$$

if $0 < p \leqslant 1$, then (9) reverses, with the equality holding only if $a_j = b_j, j = 1, \ldots, n$.

Proof　According to Lemma 1 and Lemma 4, it follows that Theorem 1 is holds.

Theorem 2　Let $a_1 \geqslant a_2 \geqslant \cdots \geqslant a_n$, $b_1 \geqslant b_2 \geqslant \cdots \geqslant b_n$, $\sum\limits_{j=1}^{k} a_j \leqslant \sum\limits_{j=1}^{k} b_j$, $k = 1, \ldots, n$, i.e. $\boldsymbol{a} \prec_w \boldsymbol{b}$, and let $a_{n+1} = \min\{a_1, \cdots, a_n, b_1, \cdots, b_n\}$, $b_{n+1} = \sum\limits_{i=1}^{n+1} a_i - \sum\limits_{i=1}^{n} b_i$. Then

$$\sum_{j=1}^{n+1} a_j^p \leqslant \sum_{j=1}^{n+1} b_j^p \tag{10}$$

if $0 < p \leqslant 1$, then (10) reverses, with the equality holding only if $a_j = b_j, j = 1, \ldots, n$.

Proof　According to Lemma 2 and Lemma 4, it follows that Theorem 2 is holds.

Theorem 3　Let $\boldsymbol{a}, \boldsymbol{b} \in \mathbf{R}^n$, $b_1 \leqslant \cdots \leqslant b_n$, if $\boldsymbol{a} \prec_w \boldsymbol{b}$, and let $c = b_n - (\sum\limits_{i=1}^{n} a_i - \sum\limits_{i=1}^{n} b_i)$. Then

$$\sum_{j=1}^{n} a_j^p \leqslant \sum_{j=1}^{n-1} b_j^p + c^p \tag{11}$$

if $0 < p \leqslant 1$, then (11) reverses, with the equality holding only if $a_j = b_j, j = 1, \ldots, n$.

Proof　According to Lemma 3 and Lemma 4, it follows that Theorem 3 is holds.

54.3 Geometrical Application

Let $\triangle A_1 A_2 A_3$ be a triangle with vertices A_1, A_2, A_3, sides a_1, a_2, a_3 (with a_j opposite A_j), altitudes h_1, h_2, h_3 (with h_j from A_j), medians m_1, m_2, m_3 (with m_j from A_j), angle-bisectors w_1, w_2, w_3 (with w_j from A_j) exradii r_1, r_2, r_3 (with r_j tangent to a_j), radius of circumcircle R, radius of circle r and semi-perimeter s. And let P be an interior point of $\triangle A_1 A_2 A_3$ or point on sides of $\triangle A_1 A_2 A_3$, R_j be distance from P to the vertex A_j, j=1, 2, 3. The symbol $\sum$ denote the cyclic sum.

Lemma 6

$$(\ln h_2 h_3, \ln h_3 h_1, \ln h_1 h_2)$$

$$\prec_w (\ln h_1 r_1, \ln h_2 r_2, \ln h_3 r_3) \tag{12}$$

$$(\ln w_2 w_3, \ln w_3 w_1, \ln w_1 w_2)$$

$$\prec_w (\ln w_1 r_1, \ln w_2 r_2, \ln w_3 r_3) \tag{13}$$

Proof We prove only (13). (12) can be proved similarly. Without loss of generality, we may assume $a_1 \geqslant a_2 \geqslant a_3$, and then $w_3 \geqslant w_2 \geqslant w_1$. It follows that $w_2 w_3 \geqslant w_3 w_1 \geqslant w_1 w_2$. In order to prove (13), we need to prove

$$w_1 r_1 \geqslant w_2 r_2 \geqslant w_3 r_3 \tag{14}$$

$$w_2 w_3 \leqslant w_1 r_1 \tag{15}$$

$$(w_2 w_3)(w_3 w_1) \leqslant (w_1 r_1)(w_2 r_2) \tag{16}$$

$$(w_2 w_3)(w_3 w_1)(w_1 w_2) \leqslant (w_1 r_1)(w_2 r_2)(w_3 r_3) \tag{17}$$

From

$$w_1 = \frac{2\sqrt{a_2 a_3 s(s - a_1)}}{a_2 + a_3}, r_1 = \sqrt{\frac{s(s - a_2)(s - a_3)}{s - a_1}} \tag{18}$$

it is easy to see that the first inequality in (14) equivalent to

$$\sqrt{\frac{a_2(s - a_2)}{a_1(s - a_1)}} \geqslant \frac{a_2 + a_3}{a_1 + a_3} \tag{19}$$

Since $a_2(s - a_2) \geqslant a_1(s - a_1), a_2 + a_3 \geqslant a_1 + a_3$, (19) holds, the first inequality in (14) follows immediately. The second inequality in (14) can be proved similarly. From (17), it is easy to see that (15) equivalent to

$(a_3 + a_1)(a_1 + a_2) \geqslant 2a_1(a_2 + a_3)$, i.e. $(a_1 - a_3)(a_1 + a_2) \geqslant 0$, so (15) holds. And by $m_1^2 \leqslant r_1 r_2, m_2^2 \leqslant r_2 r_3, m_3^2 \leqslant r_3 r_1$, (16) and (17) can be deduced.

The proof of Lemma 6 is now completed. (This proof Due to Jian Liu)

Theorem 4　For $\triangle A_1 A_2 A_3$, if $p > 1$, then

$$\sum m_j^p \leqslant \sum a_j^p - \frac{(\sum a_j - \sum m_j)^p}{k^{p-1}} \tag{20}$$

$$\sum R_j^p \leqslant \sum a_j^p - \frac{(\sum a_j - \sum R_j)^p}{k^{p-1}} \tag{21}$$

$$\sum m_j^p \leqslant \sum r_j^p - \frac{(\sum r_j - \sum m_j)^p}{k^{p-1}} \tag{22}$$

$$\left(\frac{\sqrt{3}}{2}\right)^p \sum a_j^p \leqslant \sum r_j^p - \frac{(\sum r_j - \frac{\sqrt{3}}{2}\sum a_j)^p}{k^{p-1}} \tag{23}$$

where $k = 1, 2, 3$. If $0 < p \leqslant 1$, then inequalities in (20)-(23) are all reverses.

Proof　Notice that

$$m_1 < \frac{1}{2}(a_1 + a_2) \leqslant a_1$$

$$m_2 < \frac{1}{2}(a_2 + a_3) \leqslant a_2$$

$$m_3 < \frac{1}{2}(a_3 + a_1) \leqslant a_3$$

it is easy to check that $(m_1, m_2, m_3) \prec_w (a_1, a_2, a_3)$, and then by Theorem 1, (20) is proved. It is easy to check that $(R_1, R_2, R_3) \prec_w (a_1, a_2, a_3)$, and then by Theorem 1, (21) is proved.

The inequalities(22) and (23) can be proved, by the following majorization in [10] respectively

$$(m_1, m_2, m_3) \prec_w (r_1, r_2, r_3)$$

and

$$\left(\frac{\sqrt{3}}{2}a_1, \frac{\sqrt{3}}{2}a_2, \frac{\sqrt{3}}{2}a_3\right) \prec_w (r_1, r_2, r_3)$$

Remark 1 (21) is sharpening of a result due to Zhen-ping An[9], (22) is sharpening of a result due to Ji Chen[1], and (23) is also sharpening of a known result (see [1]).

Theorem 5 For $\triangle A_1 A_2 A_3$, if $p > 1$, then

$$\sum h_2^p h_3^p \leqslant \sum h_1^p r_1^p - \frac{(\sum h_1 r_1 - \sum h_2 h_3)^p}{k^{p-1}} \tag{24}$$

$$\sum w_2^p w_3^p \leqslant \sum w_1^p r_1^p - \frac{(\sum w_1 r_1 - \sum w_2 w_3)^p}{k^{p-1}} \tag{25}$$

where $k = 1, 2, 3$. If $< p \leqslant 1$, then inequalities in (24) and (25) are all reverses.

Proof Notice that $g(x) = e^x$ be increasing convex function, by Lemma 5, from (12) and (13) it follows

$$(h_2 h_3, h_3 h_1, h_1 h_2) \prec_w (h_1 r_1, h_2 r_2, h_3 r_3)$$

and

$$(w_2 w_3, w_3 w_1, w_1 w_2) \prec_w (w_1 r_1, w_2 r_2, w_3 r_3)$$

respectively, and then by Theorem 1, (24) and (25) are proved.

Acknowledgments The author is indebted to Dr. Jian Liu for his valuable comments and suggestions.

References

[1] J -CH KUANG.Cháng yòng bù děng shì (Applied Inequalities) [M]. 3rd ed. Jinan: Shandong Press of science and technology,2004.

[2] J CHEN. Solution collecting[J]. shùxúetongxùn (Mathematics of Communications), 1989(12) : 3.

[3] K HU. On pseudo-wean value inequalities[J]. J. Fuzhou Tcachers College, 1996, 48(1):1-3.

[4] K HU. Several problems in analysis inequalities[M]. Wuhan: Wuhan University Press, 2003.

[5] W R LU. Discovery of a new inequality[J]. J. Bingzhou Tcachers College, 1996, 12 (2) : 31-34.

[6] T H LI. Some generalizations of an algebraic inequality[J]. J. Sichuan Three-gorges University,2000, 16 (3) : 80-84.

[7] Q SH YANG, X M YIN. On an Extension Zhong Kai-lai's Inequality[J]. J. Huan City University (Natural Science), 2004, 13 (3) : 48-50.

[8] B Y WANG. Kòngzhì Bùděngshì Jíchǔ, (Foundations of Majorization Inequalities)[M]. Beijing: Beijing Normal University Press, 1990.

[9] ZH P AN. Extensions of several triangular inequalities[J]. In Researches on China elementary mathematics, edited by SH.-G. Yang, Henan Education Press, Xinziang, China, 1992:241

[10] A M MARSHALL, I OLKIN, B C ARNOLD. Inequalities: Theory of majorization and its application[M]. 2nd ed. New York: Springer, 2011.

[11] L -B SITU. Proof of CWX-335[J]. Research Communication on Inequalities, 2004, 11(3) : 396-399.

第55篇　Schur-Convexity of Dual Form of Some Symmetric Functions

(Huan-Nan Shi and Jing Zhang. J. Inequal. Appl. , 2013, 2013:295)

Abstract:　By properties of Schur-convex function, Schur-convexity of the dual form of some symmetric functions is simply proved.

Keywords:　majorization; Schur-convexity; inequality; symmetric functions; dual form; convex function

55.1　Introduction

Throughout the article, $\mathbf{R}$ denotes the set of real numbers, $\boldsymbol{x} = (x_1, x_2, \ldots, x_n)$ denotes n-tuple (n-dimensional real vectors), the set of vectors can be written as

$$\mathbf{R}^n = \{\boldsymbol{x} = (x_1, \ldots, x_n) : x_i \in \mathbf{R}, i = 1, \ldots, n\}$$

$$\mathbf{R}_+^n = \{\boldsymbol{x} = (x_1, \ldots, x_n) : x_i > 0, i = 1, \ldots, n\}$$

In particular, the notations $\mathbf{R}$ and $\mathbf{R}_+$ denote $\mathbf{R}^1$ and $\mathbf{R}_+^1$ respectively. For convenience, we introduce some definitions as follows.

Definition 1[1-2] Let $\boldsymbol{x} = (x_1, \ldots, x_n)$ and $\boldsymbol{y} = (y_1, \ldots, y_n) \in \mathbf{R}^n$.

(i) $\boldsymbol{x} \geq \boldsymbol{y}$ means $x_i \geq y_i$ for all $i = 1, 2, \ldots, n$.

(ii) Let $\Omega \subset \mathbf{R}^n$, $\varphi\colon \Omega \to \mathbf{R}$ is said to be increasing if $\boldsymbol{x} \geq \boldsymbol{y}$ implies $\varphi(\boldsymbol{x}) \geq \varphi(\boldsymbol{y})$. φ is said to be decreasing if and only if $-\varphi$ is increasing.

Definition 2[1-2] Let $\boldsymbol{x} = (x_1, \ldots, x_n)$ and $\boldsymbol{y} = (y_1, \ldots, y_n) \in \mathbf{R}^n$.

(i) $\boldsymbol{x}$ is said to be majorized by $\boldsymbol{y}$ (in symbols $\boldsymbol{x} \prec \boldsymbol{y}$) if $\sum_{i=1}^{k} x_{[i]} \leq \sum_{i=1}^{k} y_{[i]}$

for $k = 1, 2, \ldots, n - 1$ and $\sum\limits_{i=1}^{n} x_i = \sum\limits_{i=1}^{n} y_i$, where $x_{[1]} \geqslant \cdots \geqslant x_{[n]}$ and $y_{[1]} \geqslant \cdots \geqslant y_{[n]}$ are rearrangements of x and y in a descending order.

(ii) Let $\Omega \subset \mathbf{R}^n$, $\varphi \colon \Omega \to \mathbf{R}$ is said to be a Schur-convex function on Ω if $x \prec y$ on Ω implies $\varphi(x) \leq \varphi(y)$. φ is said to be a Schur-concave function on Ω if and only if $-\varphi$ is Schur-convex function on Ω.

Definition 3[1-2] Let $x = (x_1, \ldots, x_n)$ and $y = (y_1, \ldots, y_n) \in \mathbf{R}^n$.

(i) $\Omega \subset \mathbf{R}^n$ is said to be a convex set if $x, y \in \Omega, 0 \leqslant \alpha \leqslant 1$ implies $\alpha x + (1 - \alpha)y = (\alpha x_1 + (1 - \alpha)y_1, \ldots, \alpha x_n + (1 - \alpha)y_n) \in \Omega$.

(ii) Let $\Omega \subset \mathbf{R}^n$ be convex set. A function $\varphi \colon \Omega \to \mathbf{R}$ is said to be a convex function on Ω, if

$$\varphi(\alpha x + (1 - \alpha)y) \leqslant \alpha\varphi(x) + (1 - \alpha)\varphi(y)$$

for all $x, y \in \Omega$, and all $\alpha \in [0, 1]$. φ is said to be a concave function on Ω if and only if $-\varphi$ is convex function on Ω.

(iii) Let $\Omega \subset \mathbf{R}^n$. A function $\varphi \colon \Omega \to \mathbf{R}$ is said to be a log-convex function on Ω if function $\ln\varphi$ is convex.

Definition 4[1]

(i) $\Omega \subset \mathbf{R}^n$ is called symmetric set, if $x \in \Omega$ implies $Px \in \Omega$ for every $n \times n$ permutation matrix P.

(ii) The function $\varphi \colon \Omega \to \mathbf{R}$ is called symmetric if for every permutation matrix P, $\varphi(Px) = \varphi(x)$ for all $x \in \Omega$.

Theorem A(Schur-Convex Function Decision Theorem)[1] Let $\Omega \subset \mathbf{R}^n$ is symmetric and has a nonempty interior convex set. Ω^0 is the interior of Ω. $\varphi \colon \Omega \to \mathbf{R}$ is continuous on Ω and differentiable in Ω^0. Then φ is the Schur-convex(Schur-concave)function, if and only if φ is symmetric on Ω and

$$(x_1 - x_2)\left(\frac{\partial\varphi}{\partial x_1} - \frac{\partial\varphi}{\partial x_2}\right) \geqslant 0(\leqslant 0) \tag{1}$$

holds for any $x \in \Omega^0$.

The Schur-convex functions were introduced by I. Schur in 1923 and have important applications in analytic inequalities, elementary quantum mechanics and quantum information theory. See [1].

In recent years, many scholars use the Schur-convex function decision theorem to determine the Schur-convexity of many symmetric function.

Wei-feng Xia et. al.[3] proved that the symmetric function

$$E_k\left(\frac{x}{1+x}\right) = \sum_{1\leqslant i_1<...<i_k\leqslant n}\prod_{j=1}^{k}\frac{x_{i_j}}{1+x_{i_j}}, \quad k=1,\ldots,n \tag{2}$$

is Schur-convex on $\mathbf{R}_+^n$.

Yu-ming Chu et. al.[4] proved that the symmetric function

$$E_k\left(\frac{x}{1-x}\right) = \sum_{1\leqslant i_1<...<i_k\leqslant n}\prod_{j=1}^{k}\frac{x_{i_j}}{1-x_{i_j}}, \quad k=1,\ldots,n \tag{3}$$

is Schur-convex on $[\frac{k-1}{2(n-1)},1)^n$ and Schur-concave on $[0,\frac{k-1}{2(n-1)}]^n$.

Wei-feng Xia and Yu-ming Chu[5] proved that the symmetric function

$$E_k\left(\frac{1-x}{x}\right) = \sum_{1\leqslant i_1<...<i_k\leqslant n}\prod_{j=1}^{k}\frac{1-x_{i_j}}{x_{i_j}}, \quad k=1,\ldots,n \tag{4}$$

is Schur-convex on $(0,\frac{2n-k-1}{2(n-1)}]^n$ and Schur-concave on $[\frac{2n-k-1}{2(n-1)},1]^n$.

Wei-feng Xia and Yu-ming Chu[6] also proved that the symmetric function

$$E_k\left(\frac{1+x}{1-x}\right) = \sum_{1\leqslant i_1<...<i_k\leqslant n}\prod_{j=1}^{k}\frac{1+x_{i_j}}{1-x_{i_j}}, \quad k=1,\ldots,n \tag{5}$$

is Schur-convex on $(0,1)^n$.

Hua Mei et. al.[7] proved that the symmetric function

$$E_k\left(\frac{1}{x}-x\right) = \sum_{1\leqslant i_1<...<i_k\leqslant n}\prod_{j=1}^{k}\left(\frac{1}{x_{i_j}}-x_{i_j}\right), \quad k=1,\ldots,n \tag{6}$$

is Schur-convex on $(0,1)^n$.

In this paper, by properties of Schur-convex function, we study Schur-convexity of the dual form of the above symmetric functions, and obtained the following results.

Theorem 1 The symmetric function

$$E_k^*\left(\frac{x}{1+x}\right) = \prod_{1\leqslant i_1<...<i_k\leqslant n}\sum_{j=1}^{k}\frac{x_{i_j}}{1+x_{i_j}}, \quad k=1,\ldots,n \tag{7}$$

is a Schur-concave function on $\mathbf{R}_+^n$.

Theorem 2 The symmetric function

$$E_k^* \left(\frac{x}{1-x} \right) = \prod_{1 \leqslant i_1 < \dots < i_k \leqslant n} \sum_{j=1}^k \frac{x_{i_j}}{1-x_{i_j}}, \quad k = 1, \dots, n \tag{8}$$

is a Schur-convex function on $[\frac{1}{2}, 1)^n$.

Theorem 3 The symmetric function

$$E_k^* \left(\frac{1-x}{x} \right) = \prod_{1 \leqslant i_1 < \dots < i_k \leqslant n} \sum_{j=1}^k \frac{1-x_{i_j}}{x_{i_j}}, \quad k = 1, \dots, n \tag{9}$$

is a Schur-convex function on $(0, \frac{1}{2}]^n$.

Theorem 4 The symmetric function

$$E_k^* \left(\frac{1+x}{1-x} \right) = \prod_{1 \leqslant i_1 < \dots < i_k \leqslant n} \sum_{j=1}^k \frac{1+x_{i_j}}{1-x_{i_j}}, \quad k = 1, \dots, n \tag{10}$$

is a Schur-convex function on $(0, 1)^n$.

Theorem 5 The symmetric function

$$E_k^* \left(\frac{1}{x} - x \right) = \prod_{1 \leqslant i_1 < \dots < i_k \leqslant n} \sum_{j=1}^k \left(\frac{1}{x_{i_j}} - x_{i_j} \right), \quad k = 1, \dots, n \tag{11}$$

is a Schur-convex function on $\left(0, \sqrt{\sqrt{5} - 2} \right)^n$.

55.2　Lemmas

To prove the above three theorems, we need the following lemmas.

Lemma 1[1-2] If φ is symmetric and convex (concave) on symmetric convex set Ω, then φ is Schur-convex (Schur-concave) on Ω.

Lemma 2[2] Let $\Omega \subset \mathbf{R}^n$, $\varphi: \Omega \to \mathbf{R}_+$. Then $\log \varphi$ is Schur-convex (Schur-concave) if and only if φ is Schur-convex (Schur-concave).

Lemma 3[1-2] Let $\Omega \subset \mathbf{R}^n$ be open convex set, $\varphi: \Omega \to \mathbf{R}$. For $x, y \in \Omega$, defined one variable function $g(t) = \varphi(tx + (1-t)y)$ on interval $(0, 1)$. Then φ is convex (concave) on Ω if and only if g is convex (concave) on $[0, 1]$ for all $x, y \in \Omega$.

Lemma 4 Let $x = (x_1, \dots, x_m)$ and $y = (y_1, \dots, y_m) \in \mathbf{R}_+^m$. Then the function $p(t) = \log g(t)$ is concave on $[0, 1]$, where

$$g(t) = \sum_{j=1}^{m} \frac{tx_j + (1 - t)y_j}{1 + tx_j + (1 - t)y_j}$$

Proof

$$p'(t) = \frac{g'(t)}{g(t)}$$

where

$$g'(t) = \sum_{j=1}^{m} \frac{x_j - y_j}{(1 + tx_j + (1 - t)y_j)^2}$$

$$p''(t) = \frac{g''(t)g(t) - (g'(t))^2}{g^2(t)}$$

where

$$g''(t) = -\sum_{j=1}^{m} \frac{2(x_j - y_j)^2}{(1 + tx_j + (1 - t)y_j)^3}$$

Thus

$$
g''(t)g(t) - (g'(t))^2
$$
$$
= \left(-\sum_{j=1}^{m} \frac{2(x_j - y_j)^2}{(1 + tx_j + (1 - t)y_j)^3} \right) \left(\sum_{j=1}^{m} \frac{tx_j + (1 - t)y_j}{1 + tx_j + (1 - t)y_j} \right) -
$$
$$
\left(\sum_{j=1}^{m} \frac{x_j - y_j}{(1 + tx_j + (1 - t)y_j)^2} \right)^2 \leqslant 0
$$

and then $p''(t) \leqslant 0$, that is $p(t)$ is concave on $[0, 1]$.

The proof of Lemma 4 is completed.

Lemma 5 Let $\boldsymbol{x} = (x_1, \ldots, x_m)$ and $\boldsymbol{y} = (y_1, \ldots, y_m) \in \left[\frac{1}{2}, 1\right)^m$. Then the function $q(t) = \log \psi(t)$ is convex on $[0, 1]$, where

$$\psi(t) = \sum_{j=1}^{m} \frac{tx_j + (1 - t)y_j}{1 - tx_j - (1 - t)y_j}$$

Proof

$$q'(t) = \frac{\psi'(t)}{\psi(t)}$$

where

$$\psi'(t) = \sum_{j=1}^{m} \frac{x_j - y_j}{(1 - tx_j - (1 - t)y_j)^2}$$

$$q''(t) = \frac{\psi''(t)\psi(t) - (\psi'(t))^2}{\psi^2(t)}$$

where

$$\psi''(t) = \sum_{j=1}^{m} \frac{2(x_j - y_j)^2}{(1 - tx_j - (1-t)y_j)^3}$$

By the Cauchy inequality, we have

$$\psi''(t)\psi(t) - (\psi'(t))^2$$

$$= \left(\sum_{j=1}^{m} \frac{2(x_j - y_j)^2}{(1 - tx_j - (1-t)y_j)^3} \right) \left(\sum_{j=1}^{m} \frac{tx_j + (1-t)y_j}{1 - tx_j - (1-t)y_j} \right) -$$

$$\left(\sum_{j=1}^{m} \frac{x_j - y_j}{(1 - tx_j - (1-t)y_j)^2} \right)^2$$

$$\geqslant \left(\sum_{j=1}^{m} \frac{\sqrt{2}|x_j - y_j|}{(1 - tx_j - (1-t)y_j)^{\frac{3}{2}}} \frac{\sqrt{tx_j + (1-t)y_j}}{\sqrt{1 - tx_j - (1-t)y_j}} \right)^2 -$$

$$\left(\sum_{j=1}^{m} \frac{x_j - y_j}{(1 - tx_j - (1-t)y_j)^2} \right)^2$$

$$= \left(\sum_{j=1}^{m} \frac{\sqrt{2}|x_j - y_j|\sqrt{tx_j + (1-t)y_j}}{(1 - tx_j - (1-t)y_j)^2} \right)^2 - \left(\sum_{j=1}^{m} \frac{x_j - y_j}{(1 - tx_j - (1-t)y_j)^2} \right)^2$$

From $x_j, y_j \in [\frac{1}{2}, 1)$, it follows that $\sqrt{2}\sqrt{tx_j + (1-t)y_j} \geqslant 1$, hence $\psi''(t)\psi(t) - (\psi'(t))^2 \geqslant 0$, and then $q''(t) \geqslant 0$, that is $q(t)$ is convex on $[0, 1]$.

The proof of Lemma 5 is completed.

Lemma 6 Let $x = (x_1, \ldots, x_m)$ and $y = (y_1, \ldots, y_m) \in (0, \frac{1}{2}]^m$. Then the function $r(t) = \log \varphi(t)$ is convex on $[0, 1]$, where

$$\varphi(t) = \sum_{j=1}^{m} \frac{1 - tx_j - (1-t)y_j}{tx_j + (1-t)y_j}$$

Proof

$$r'(t) = \frac{\varphi'(t)}{\varphi(t)}$$

where

$$\varphi'(t) = -\sum_{j=1}^{m} \frac{x_j - y_j}{(tx_j + (1-t)y_j)^2}$$

$$r''(t) = \frac{\varphi''(t)\varphi(t) - (\varphi'(t))^2}{\varphi^2(t)}$$

where

$$\varphi''(t) = \sum_{j=1}^{m} \frac{2(x_j - y_j)^2}{(tx_j + (1-t)y_j)^3}$$

by the Cauchy inequality, we have

$$\varphi''(t)\varphi(t) - (\varphi'(t))^2$$

$$= \left(\sum_{j=1}^{m} \frac{2(x_j - y_j)^2}{(tx_j + (1-t)y_j)^3} \right) \left(\sum_{j=1}^{m} \frac{1 - tx_j - (1-t)y_j}{tx_j + (1-t)y_j} \right) -$$

$$\left(-\sum_{j=1}^{m} \frac{x_j - y_j}{(tx_j + (1-t)y_j)^2} \right)^2$$

$$\geqslant \left(\sum_{j=1}^{m} \frac{\sqrt{2}|x_j - y_j|}{(tx_j + (1-t)y_j)^{\frac{3}{2}}} \frac{\sqrt{1 - tx_j - (1-t)y_j}}{\sqrt{tx_j + (1-t)y_j}} \right)^2 -$$

$$\left(\sum_{j=1}^{m} \frac{x_j - y_j}{(tx_j + (1-t)y_j)^2} \right)^2$$

$$= \left(\sum_{j=1}^{m} \frac{\sqrt{2}|x_j - y_j|\sqrt{1 - tx_j - (1-t)y_j}}{(tx_j + (1-t)y_j)^2} \right)^2 - \left(\sum_{j=1}^{m} \frac{x_j - y_j}{(tx_j + (1-t)y_j)^2} \right)^2$$

From $x_j, y_j \in (0, \frac{1}{2}]$, it follows that $\sqrt{2}\sqrt{1 - tx_j - (1-t)y_j} \geqslant 1$, hence $\varphi''(t)\varphi(t) - (\varphi'(t))^2 \geqslant 0$, and then $r''(t) \geqslant 0$, that is $r(t)$ is convex on $[0, 1]$.

The proof of Lemma 6 is completed.

Lemma 7 Let $\boldsymbol{x} = (x_1, \ldots, x_m)$ and $\boldsymbol{y} = (y_1, \ldots, y_m) \in (0, 1)^m$. Then the function $h(t) = \log f(t)$ is convex on $[0, 1]$, where

$$f(t) = \sum_{j=1}^{m} \frac{1 + tx_j + (1-t)y_j}{1 - tx_j - (1-t)y_j}$$

Proof

$$h'(t) = \frac{f'(t)}{f(t)}$$

where

$$f'(t) = \sum_{j=1}^{m} \frac{2(x_j - y_j)}{(1 - tx_j - (1-t)y_j)^2}$$

$$h''(t) = \frac{f''(t)f(t) - (f'(t))^2}{f^2(t)}$$

where
$$f''(t) = \sum_{j=1}^{m} \frac{4(x_j - y_j)^2}{(1 - tx_j - (1-t)y_j)^3}$$

by the Cauchy inequality, we have

$$f''(t)f(t) - (f'(t))^2$$

$$= \left(\sum_{j=1}^{m} \frac{4(x_j - y_j)^2}{(1 - tx_j - (1-t)y_j)^3} \right) \left(\sum_{j=1}^{m} \frac{1 + tx_j + (1-t)y_j}{1 - tx_j - (1-t)y_j} \right) -$$

$$\left(\sum_{j=1}^{m} \frac{2(x_j - y_j)}{(1 - tx_j - (1-t)y_j)^2} \right)^2$$

$$\geqslant \left(\sum_{j=1}^{m} \frac{2|x_j - y_j|}{(1 - tx_j - (1-t)y_j)^{\frac{3}{2}}} \frac{\sqrt{1 + tx_j + (1-t)y_j}}{\sqrt{1 - tx_j - (1-t)y_j}} \right)^2 -$$

$$\left(\sum_{j=1}^{m} \frac{2(x_j - y_j)}{(1 - tx_j - (1-t)y_j)^2} \right)^2$$

$$= \left(\sum_{j=1}^{m} \frac{2|x_j - y_j|\sqrt{1 + tx_j + (1-t)y_j}}{(1 - tx_j - (1-t)y_j)^2} \right)^2 - \left(\sum_{j=1}^{m} \frac{2(x_j - y_j)}{(1 - tx_j - (1-t)y_j)^2} \right)^2$$

From $x_j, y_j \in (0, 1)$ it follows that $\sqrt{2}\sqrt{1 + tx_j + (1-t)y_j} \geqslant 1$, hence $f''(t)f(t) - (f'(t))^2 \geqslant 0$, and then $h''(t) \geqslant 0$, that is $h(t)$ is convex on $[0, 1]$.

The proof of Lemma 7 is completed.

Lemma 8 Let $\boldsymbol{x} = (x_1, \ldots, x_m)$ and $\boldsymbol{y} = (y_1, \ldots, y_m) \in \left(0, \sqrt{\sqrt{5} - 2} \right)^m$. Then the function $s(t) = \log w(t)$ is convex on $[0, 1]$, where

$$w(t) = \sum_{j=1}^{m} \left(\frac{1}{tx_j + (1-t)y_j} - (tx_j + (1-t)y_j) \right)$$

Proof
$$s'(t) = \frac{w'(t)}{w(t)}$$

where
$$w'(t) = -\sum_{j=1}^{m} (x_j - y_j) \left(\frac{1}{(tx_j + (1-t)y_j)^2} + 1 \right)$$

$$s''(t) = \frac{w''(t)w(t) - (w'(t))^2}{w^2(t)}$$

where

$$w''(t) = \sum_{j=1}^{m} \frac{2(x_j - y_j)^2}{(tx_j + (1-t)y_j)^3}$$

by the Cauchy inequality, we have

$$w''(t)w(t) - (w'(t))^2$$

$$= \left(\sum_{j=1}^{m} \frac{2(x_j - y_j)^2}{(tx_j + (1-t)y_j)^3} \right) \left(\sum_{j=1}^{m} \left(\frac{1}{tx_j + (1-t)y_j} - (tx_j + (1-t)y_j) \right) \right) -$$

$$\left(-\sum_{j=1}^{m} (x_j - y_j) \left(\frac{1}{(tx_j + (1-t)y_j)^2} + 1 \right) \right)^2$$

$$\geq \left(\sum_{j=1}^{m} \frac{\sqrt{2}|x_j - y_j|}{(tx_j + (1-t)y_j)^{\frac{3}{2}}} \sqrt{\frac{1}{tx_j + (1-t)y_j} - (tx_j + (1-t)y_j)} \right)^2 -$$

$$\left(\sum_{j=1}^{m} (x_j - y_j) \left(\frac{1}{(tx_j + (1-t)y_j)^2} + 1 \right) \right)^2$$

$$= \left(\sum_{j=1}^{m} \frac{\sqrt{2}|x_j - y_j| \sqrt{1 - (tx_j + (1-t)y_j)^2}}{(tx_j + (1-t)y_j)^2} \right)^2 -$$

$$\left(\sum_{j=1}^{m} (x_j - y_j) \frac{1 + (tx_j + (1-t)y_j)^2}{(tx_j + (1-t)y_j)^2} \right)^2$$

Let $u_j := tx_j + (1-t)y_j$. From $x_j, y_j \in \left(0, \sqrt{\sqrt{5} - 2} \right)$ it follows that $u_j^2 \leqslant \sqrt{5} - 2$. Since

$$u_j^2 \leqslant \sqrt{5} - 2 \Leftrightarrow (u_j^2 + 2)^2 \leqslant 5 \Leftrightarrow u_j^4 + 4u_j^2 - 1 \leqslant 0$$

$$\Leftrightarrow 2(1 - u_j^2) \geqslant (1 + u_j^2)^2 \Leftrightarrow \sqrt{2} \sqrt{1 - u_j^2} \geqslant 1 + u_j^2$$

so $w''(t)w(t) - (w'(t))^2 \geqslant 0$, and then $s''(t) \geqslant 0$, that is $s(t)$ is convex on $[0, 1]$.

The proof of Lemma 8 is completed.

55.3 Proof of Main Results

Proof of Theorem 4 For any $1 \leqslant i_1 < \cdots < i_k \leqslant n$, by Lemma 3 and Lemma 7, it follows that $\log \sum_{j=1}^{k} \frac{1+x_{i_j}}{1-x_{i_j}}$ is convex on $(0,1)^k$. Obviously, $\log \sum_{j=1}^{k} \frac{1+x_{i_j}}{1-x_{i_j}}$ is also convex on $(0,1)^n$, and then $\log E_k^* \left(\frac{1+x}{1-x} \right) = \sum_{1 \leqslant i_1 < \ldots < i_k \leqslant n} \log \sum_{j=1}^{k} \frac{1+x_{i_j}}{1-x_{i_j}}$ is convex on $(0,1)^n$. Furthermore, it is clear that $\log E_k^* \left(\frac{1+x}{1-x} \right)$ is symmetric on $(0,1)^n$, by Lemma 1, it follows that $\log E_k^* \left(\frac{1+x}{1-x} \right)$ is Schur-convex on $(0,1)^n$, and then from Lemma 2 we conclude that $E_k^* \left(\frac{1+x}{1-x} \right)$ is also Schur-convex on $(0,1)^n$.

The proof of Theorem 4 is completed.

Similar to the proof of Theorem 4, we can use Lemma 4, Lemma 5, Lemma 6 and Lemma 8, respectively prove Theorem 1, Theorem 2, Theorem 3 and Theorem 5, therefore omitted details of the proof.

Remark 1　Using the Schur-convex function decision theorem, Qinghua Liu et al.[8] have proved Theorem 3. Weifeng Xia and Yuming Chu[9] has proved that the symmetric function

$$E_k^* \left(\frac{1+x}{x} \right) = \prod_{1 \leqslant i_1 < \ldots < i_k \leqslant n} \sum_{j=1}^{k} \frac{1+x_{i_j}}{x_{i_j}}, \quad k = 1, \ldots, n \qquad (12)$$

is a Schur-convex function on $\mathbf{R}_+^n$.

The reader may wish to be proved the inequality (12) by properties of Schur-convex function.

Acknowledgment

The work was supported by Funding Project for Academic Human Resources Development in Institutions of Higher Learning under the Jurisdiction of Beijing Municipality (PHR (IHLB)) (PHR201108407). Thanks for the help. This article was typeset by using $\mathcal{AMS}$-LaTeX.

References

[1]　A M MARSHALL, I OLKIN, B C ARNOLD. Inequalities: Theory of majorization and its application[M]. 2nd ed. New York: Springer , 2011.

[2]　B-Y WANG. Kòngzhì Bùděngshì Jíchǔ (Foundations of Majorization Inequalities)[M]. Beijing: Beijing Normal University Press, 1990.

[3]　W F XIA, G D WANG, Y M CHU. Schur convexity and inequalities for a class of symmetric functions[J]. International Journal of Pure and Applied Mathematics, 2010, 58 (4): 435-452.

[4]   Y M CHU, W F XIA, T H ZHAO. Schur convexity for a class of symmetric functions [J]. Sci. China Math, 2010, 53(2): 465-474.

[5]   W F XIA, Y M CHU. Schur convexity and Schur multiplicative convexity for a class of symmetric functions with applications[J]. Ukrainian Mathematical Journal,2009, 61(10):1541-1555.

[6]   WEI-FENG XIA, YU-MING CHU. On Schur-convexity of some symmetric functions[J]. J. Inequal. Appl., Volume 2010, Article ID 543250, 12 pages doi:10.1155/2010/543250.

[7]   H MEI, C L BAI, H MAN. Extension of an Inequality Guess[J]. Journal of Inner Mongolia University for Nationalities, 2006, 21(2):127-129.

[8]   H Q LIU, Q YU, Y ZHANG. Some properties of a class of symmetric functions and its applications[J]. Journal of Hengyang Normal University, 2012, 33 (6):167-171.

[9]   WEIFENG XIA, YUMING CHU. Schur convexity with respect to a class of symmetric functions and their applications[J]. Bulletin of Mathematical Analysis and Applications, 2011, 3 (3): 84-96.

[10] H-N SHI. Theory of majorization and analytic Inequalities[M]. Harbin: Harbin Institute of Technology Press, 2012.

第56篇　Schur Convexity of Dual Form of the Complete Symmetric Function

(ZHANG KONGSHENG, SHI HUANNAN. Mathematical Inequalities

and Applications, Volume 16, Number 4 (2013): 963 - 970)

Abstract: The complete symmetric function is defined as follows

$$c_r(\boldsymbol{x}) = \sum_{i_1+i_2+\ldots+i_n=r} x_1^{i_1} \ldots x_n^{i_n} = \sum_{1\leqslant i_1\leqslant i_2\leqslant\ldots\leqslant i_r\leqslant n} \prod_{j=1}^{r} x_{i_j}$$

where $c_0(\boldsymbol{x}) = 1$, $r \in \{1, 2, \ldots, n\}$, $i_1, i_2, \ldots, i_n$ are non-negative integers.

Its dual form is

$$f(\boldsymbol{x}, r) = \prod_{i_1+i_2+\ldots+i_n=r} \sum_{j=1}^{n} i_j x_j = \prod_{1\leqslant i_1\leqslant i_2\leqslant\ldots\leqslant i_r\leqslant n} \sum_{j=1}^{r} x_{i_j}$$

where $i_j, j = 1, 2, \cdots, n$ can take min value 0 in last equality.

In this paper, the Schur-convexity, the Schur-geometrically-convexity and the Schur-harmonically-convexity of dual form of the complete symmetric function are investigated. As consequences, some new inequalities are established via majorilization theory.

Keywords:　complete symmetric function; dual form; Schur-convexity; Schur-geometrically-convexity; Schur-harmonically convexity

56.1 Introduction

Throughout this paper, $\mathbf{R}_+^n = \{x | x = (x_1, x_2, \ldots, x_n), x_i > 0, i = 1, 2 \ldots, n\}$ and

$$x = (x_1, x_2, \ldots, x_n) \in \mathbf{R}_+^n, \quad y = (y_1, y_2, \ldots, y_n) \in \mathbf{R}_+^n$$

For simplicity, $\mathbf{R}_+$ stands for $\mathbf{R}_+^1$.

Whiteley[12] introduced the Whitely's symmetric function $T_n^{[r,s]}(x)$ through the following equation

$$\sum_{r=0}^{\infty} T_n^{[r,s]}(x) t^r = \begin{cases} \prod_{i=1}^n (1 + x_i t)^s, & s > 0 \\ \\ \prod_{i=1}^n (1 - x_i t)^s, & s < 0 \end{cases}$$

The complete symmetric function[1,8] which is Whiteley's symmetric function as $s = -1$ reads as follows

$$c_r(x) = T_n^{[r,-1]}(x) = \sum_{i_1 + i_2 + \ldots + i_n = r} x_1^{i_1} \ldots x_n^{i_n} \tag{1}$$

where $c_0(x) = 1$, $r \in \{1, 2, \ldots, n\}$, $i_1, i_2, \ldots, i_n$ are non-negative integers.

K. Z. Guan[4] discussed the Schur convexity of $c_r(x)$ and proved that $c_r(x)$ is Schur convex in $\mathbf{R}_+^n$. Subsequently, Y. M. Chu et al.[17] proved that $c_r(x)$ is Schur multiplicatively convex and harmonic convex in $\mathbf{R}_+^n$.

It is not difficult to show that the equivalent definition of the complete symmetric function is

$$c_r(x) = \sum_{1 \leqslant i_1 \leqslant i_2 \leqslant \ldots \leqslant i_r \leqslant n} \prod_{j=1}^r x_{i_j} \tag{2}$$

In fact, for any $x = (x_1, x_2, \ldots, x_n) \in \mathbf{R}_+^n$, let

$$A = \{x_1^{i_1} \cdots x_n^{i_n} : i_1 + \ldots + i_n = r\}$$

$$B = \{x_{j_1} \cdots x_{j_r} : 1 \leqslant j_1 \leqslant \ldots \leqslant j_r \leqslant n\}$$

It suffices to show that $A = B$ in order to prove that (1) is equivalent to (2). For any $x_1^{i_1} \cdots x_n^{i_n} \in A$, without loss of generality, we may assume $i_1 \cdots i_n$ are not equal to zero. And then we can choose appropriate $j_1, \ldots, j_r$

such that $j_1 = \ldots = j_{i_1} = 1$, $j_{i_1+1} = \cdots = j_{i_1+i_2} = 2, \ldots, j_{i_1+\ldots+i_{n-1}+1} =$
$\cdots = j_{i_1+\ldots+i_n} = n$, this indicates that $x_1^{i_1} \cdots x_n^{i_n} = x_{j_1} \cdots x_{j_r}$ which implies
$A \subseteq B$. On the other hand, $B \subseteq A$ is obviously true. Together with $A \subseteq B$
we have that $A = B$.

The main purpose of this paper is to discuss the Schur-convexity, Schur-
geometrically-convexity and Schur-harmonically-convexity of the dual form
of the complete symmetric function, that is, we shall study the function
given by

$$f(\boldsymbol{x}, r) = \prod_{i_1+i_2+\ldots+i_n=r} \sum_{j=1}^{n} i_j x_j = \prod_{1 \leqslant i_1 \leqslant i_2 \leqslant \ldots \leqslant i_r \leqslant n} \sum_{j=1}^{r} x_{i_j} \qquad (3)$$

In (3), $i_j, j = 1, 2, \cdots, n$ can take min value 0 in last equality. As applications, some new inequalities are established.

56.2　Definitions and Lemmas

Schur-convexity of symmetric functions introduced by Schur[11] plays
an important role in establishing new inequalities, see, for example, [3-
9,11,13,14]. The Schur-geometrically-convexity and Schur-harmonically-
convexity involving some special functions have been investigated, see, for
example, [1,2,10,15,16]. We recall some well known definitions and lemmas
in order to carry out our work.

Let $x_{[1]} \geqslant x_{[2]} \geqslant \cdots \geqslant x_{[n]}$, $y_{[1]} \geqslant y_{[2]} \geqslant \cdots \geqslant y_{[n]}$ be their ordered
components.

Definition 1[9] The n-tuple $\boldsymbol{x}$ is said to be majorized by $\boldsymbol{y}$ (in symbols
$\boldsymbol{x} \prec \boldsymbol{y}$), if

$$\sum_{i=1}^{m} x_{[i]} \leqslant \sum_{i=1}^{m} y_{[i]}, \quad m = 1, 2, \ldots, n-1$$

and

$$\sum_{i=1}^{n} x_{[i]} = \sum_{i=1}^{n} y_{[i]}$$

Definition 2[15] The n-tuple $\boldsymbol{x}$ is said to be logarithmically majorized by y (in symbols $\log x \prec \log y$) if

$$\prod_{i=1}^{m} x_{[i]} \leqslant \prod_{i=1}^{m} y_{[i]}, \quad m = 1, 2, \ldots, n-1$$

and

$$\prod_{i=1}^{n} x_{[i]} = \prod_{i=1}^{n} y_{[i]}$$

Definition 3[15] A function $f : \mathbf{R}_+^n \to \mathbf{R}_+$ is called Schur geometrically convex (or Schur multiplicatively convex) if

$$\log x \prec \log y \quad \Rightarrow f(x) \leqslant f(y)$$

Lemma 1[15] Suppose that $f : \mathbf{R}_+^n \to \mathbf{R}_+$ is symmetric and differentiable, if

$$(x_1 - x_2)\left(x_1 \frac{\partial f}{\partial x_1} - x_2 \frac{\partial f}{\partial x_2}\right) \geqslant 0 \tag{4}$$

for all $x_1 \neq x_2$, then f is Schur geometrically convex.

Remark 1 In order to show that f is Schur geometrically convex, it suffices to show that inequality (4) holds for $x_1 > x_2$, because when $x_1 < x_2$,

$$(x_1 - x_2)\left(x_1 \frac{\partial f}{\partial x_1} - x_2 \frac{\partial f}{\partial x_2}\right) = (x_2 - x_1)\left(x_2 \frac{\partial f}{\partial x_2} - x_1 \frac{\partial f}{\partial x_1}\right) \geqslant 0$$

Definition 4[1] A function $f : \mathbf{R}_+^n \to \mathbf{R}_+$ is called Schur harmonically convex if

$$x \prec y \Rightarrow f\left(\frac{1}{x}\right) \leqslant f\left(\frac{1}{y}\right)$$

where

$$\frac{1}{x} = \left(\frac{1}{x_1}, \frac{1}{x_2}, \ldots, \frac{1}{x_n}\right), \frac{1}{y} = \left(\frac{1}{y_1}, \frac{1}{y_2}, \ldots, \frac{1}{y_n}\right)$$

Lemma 2[1] Suppose that $f : \mathbf{R}_+^n \to \mathbf{R}_+$ is symmetric and differentiable, if

$$(x_1 - x_2)\left(x_1^2 \frac{\partial f}{\partial x_1} - x_2^2 \frac{\partial f}{\partial x_2}\right) \geqslant 0 \tag{5}$$

for all $x_1 \neq x_2$, then f is Schur harmonically convex.

Lemma 3[7] Let $x = (x_1, x_2, \ldots, x_n) \in \mathbf{R}_+^n$ and $\sum\limits_{i=1}^{n} x_i = s$. If $c \geqslant s$,
then

$$\frac{c - x}{\frac{nc}{s} - 1} = \left(\frac{c - x_1}{\frac{nc}{s} - 1}, \frac{c - x_2}{\frac{nc}{s} - 1}, \ldots, \frac{c - x_n}{\frac{nc}{s} - 1} \right) \prec (x_1, x_2, \ldots, x_n) = x$$

Lemma 4[4] Let $x = (x_1, x_2, \ldots, x_n) \in \mathbf{R}_+^n$ and $\sum\limits_{i=1}^{n} x_i = s$. If $c \geqslant 0$,
then

$$\frac{c + x}{\frac{nc}{s} + 1} = \left(\frac{c + x_1}{\frac{nc}{s} + 1}, \frac{c + x_2}{\frac{nc}{s} + 1}, \ldots, \frac{c + x_n}{\frac{nc}{s} + 1} \right) \prec (x_1, x_2, \ldots, x_n) = x$$

56.3 Main results

In this section, we mainly investigate the Schur geometrically convexity,
the Schur convexity and the Schur harmonically convexity with respect to
$f(x, r)$.

Theorem 1 The dual form of the complete symmetric function given
by

$$f(x, r) = \prod_{i_1 + i_2 + \ldots + i_n = r} \sum_{j=1}^{n} i_j x_j$$

is Schur geometrically convex in $\mathbf{R}_+^n$.

Proof The cases $r = 1$ and $r = 2$ can be easily proved.

Next, we consider the case $r \geqslant 3$ and $x_1 > x_2$.

$$f(x, r) = \prod_{\substack{i_1 + i_2 + \ldots + i_n = r \\ i_1 \neq 0, i_2 = 0}} \sum_{j=1}^{n} i_j x_j \cdot \prod_{\substack{i_1 + i_2 + \ldots + i_n = r \\ i_1 = 0, i_2 \neq 0}} \sum_{j=1}^{n} i_j x_j$$

$$\cdot \prod_{\substack{i_1 + i_2 + \ldots + i_n = r \\ i_1 \neq 0, i_2 \neq 0}} \sum_{j=1}^{n} i_j x_j \cdot \prod_{\substack{i_1 + i_2 + \ldots + i_n = r \\ i_1 = 0, i_2 = 0}} \sum_{j=1}^{n} i_j x_j$$

It is not difficult to show that

$$x_1 \frac{\partial f(x, r)}{\partial x_1} f(x, r) \left(\sum_{\substack{i_1 + i_2 + \ldots + i_n = r \\ i_1 \neq 0, i_2 = 0}} \frac{i_1 x_1}{\sum\limits_{j=1}^{n} i_j x_j} + \sum_{\substack{i_1 + i_2 + \ldots + i_n = r \\ i_1 \neq 0, i_2 \neq 0}} \frac{i_1 x_1}{\sum\limits_{j=1}^{n} i_j x_j} \right)$$

$$= f(x,r)\Big(\sum_{\substack{k+k_3+\dots+k_n=r\\k\neq0}}\frac{kx_1}{kx_1+\sum\limits_{j=3}^{n}k_jx_j}+\sum_{\substack{k+m+i_3+\dots+i_n=r\\k\neq0,m\neq0}}\frac{kx_1}{kx_1+mx_2+\sum\limits_{j=3}^{n}i_jx_j}\Big)$$

and

$$x_2\frac{\partial f(\boldsymbol{x},r)}{\partial x_2}=f(x,r)\Big(\sum_{\substack{i_1+i_2+\dots+i_n=r\\i_1=0,i_2\neq0}}\frac{i_2x_2}{\sum\limits_{j=1}^{n}i_jx_j}+\sum_{\substack{i_1+i_2+\dots+i_n=r\\i_1\neq0,i_2\neq0}}\frac{i_2x_2}{\sum\limits_{j=1}^{n}i_jx_j}\Big)$$

$$= f(\boldsymbol{x},r)\Big(\sum_{\substack{k+k_3+\dots+k_n=r\\k\neq0}}\frac{kx_2}{kx_2+\sum\limits_{j=3}^{n}k_jx_j}+\sum_{\substack{k+m+i_3+\dots+i_n=r\\k\neq0,m\neq0}}\frac{kx_2}{kx_2+mx_1+\sum\limits_{j=3}^{n}i_jx_j}\Big)$$

Thus

$$x_1\frac{\partial f(\boldsymbol{x},r)}{\partial x_1}-x_2\frac{\partial f(x,r)}{\partial x_2}=f(x,r)(\Lambda_1+\Lambda_2)$$

where

$$\Lambda_1=\sum_{\substack{k+k_3+\dots+k_n=r\\k\neq0}}\Big(\frac{kx_1}{kx_1+\sum\limits_{j=3}^{n}k_jx_j}-\frac{kx_2}{kx_2+\sum\limits_{j=3}^{n}k_jx_j}\Big)$$

and

$$\Lambda_2=\sum_{\substack{k+m+i_3+\dots+i_n=r\\k\neq0,m\neq0}}\Big(\frac{kx_1}{kx_1+mx_2+\sum\limits_{j=3}^{n}i_jx_j}-\frac{kx_2}{kx_2+mx_1+\sum\limits_{j=3}^{n}i_jx_j}\Big).$$

Notice that $h(t)=\frac{t}{c+t}=1-\frac{c}{c+t}$ $(c>0)$ is increasing in $(0,+\infty)$, this yields $\Lambda_1\geqslant0$. Meanwhile, let $a=\sum\limits_{j=3}^{n}i_jx_j$. Then

$$\frac{kx_1}{kx_1+mx_2+\sum\limits_{j=3}^{n}i_jx_j}-\frac{kx_2}{kx_2+mx_1+\sum\limits_{j=3}^{n}i_jx_j}$$

$$=\frac{kx_1}{kx_1+mx_2+a}-\frac{kx_2}{kx_2+mx_1+a}$$

$$=\frac{mk(x_1^2-x_2^2)+ak(x_1-x_2)}{(kx_1+mx_2+a)(kx_2+mx_1+a)}$$

$$\geqslant0$$

Thus $\Lambda_2\geqslant0$. Consequently

$$x_1\frac{\partial f(\boldsymbol{x},r)}{\partial x_1}-x_2\frac{\partial f(\boldsymbol{x},r)}{\partial x_2}\geqslant0$$

which completes the proof.

Theorem 3 The dual form of the complete symmetric function $f(\boldsymbol{x},r)$ is increasing and Schur concave for each $r \in \{1,2,...,n\}$ in $\mathbf{R}_+^n$.

Proof It is not difficult to show that $\ln f(\boldsymbol{x},r)$ is increasing and concave in $\mathbf{R}_+^n$, and notice that $\ln f(\boldsymbol{x},r)$ is symmetric, this leads to $\ln f(\boldsymbol{x},r)$ is Schur concave by the proposition C.2 in [9]. And then $f(\boldsymbol{x},r)$ is also increasing and Schur concave for each $r \in \{1,2,...,n\}$ in $\mathbf{R}_+^n$.

Theorem 3　The dual form of the complete symmetric function is Schur harmonically convex in $\mathbf{R}_+^n$.

Proof In view of

$$x_1 \frac{\partial f(\boldsymbol{x},r)}{\partial x_1} - x_2 \frac{\partial f(\boldsymbol{x},r)}{\partial x_2} \geqslant 0$$

$\frac{\partial f(\boldsymbol{x},r)}{\partial x_1} > 0, \frac{\partial f(\boldsymbol{x},r)}{\partial x_2} > 0$ and $x_1 > x_2$, one has $x_1^2 \frac{\partial f(\boldsymbol{x},r)}{\partial x_1} - x_2^2 \frac{\partial f(\boldsymbol{x},r)}{\partial x_2} \geqslant 0$, which completes the proof.

56.4　Some applications

In this section, some new inequalities involving the dual form of the complete symmetric function shall be established by utilizing the previous results and majorilization theory.

Corollary 1 Let $\boldsymbol{x} = (x_1, x_2, \cdots, x_n) \in \mathbf{R}_+^n$, $\sum_{i=1}^{n} x_i = s$.

(1) If $c \geqslant s$, then $f(\frac{1}{x},r) \geqslant f(\frac{\frac{nc}{s}-1}{c-x},r)$.

(2) If $c \geqslant 0$, then $f(\frac{1}{x},r) \geqslant f(\frac{\frac{nc}{s}+1}{c+x},r)$.

Proof By the Schur harmonically convexity of $f(x,r)$, parts (1) and (2) follow from Lemma 3 and Lemma 4 respectively.

Corollary 2 If $\boldsymbol{x} = (x_1, x_2, \cdots, x_n) \in \mathbf{R}_+^n$, then we have

$$\prod_{1 \leqslant i_1 \leqslant i_2 \leqslant ... \leqslant i_r \leqslant n} \sum_{j=1}^{r} x_{i_j} \geqslant \prod_{1 \leqslant i_1 \leqslant i_2 \leqslant ... \leqslant i_r \leqslant n} (r \sqrt[n]{x_1 x_2 \cdots x_n}) \tag{6}$$

Proof Corollary 2 follows from the Schur geometrically convexity of

$f(\boldsymbol{x}, r)$ and the fact $\log \boldsymbol{x} \succ \log \sqrt[n]{\boldsymbol{x}}$, where

$$\sqrt[n]{\boldsymbol{x}} = \left(\sqrt[n]{x_1 x_2 \cdots x_n}, \sqrt[n]{x_1 x_2 \cdots x_n}, \dots, \sqrt[n]{x_1 x_2 \cdots x_n} \right)$$

Corollary 3 Let $\boldsymbol{A} = (a_{ij})_{n \times n} (n \geqslant 3)$ be a positive definite Hermitian matrix, and $\boldsymbol{I}$ denote $n \times n$ unit matrix. Then we have

$$\prod_{1 \leqslant i_1 \leqslant i_2 \leqslant \dots \leqslant i_r \leqslant n} \sum_{j=1}^{r} \frac{\lambda_{i_j}}{\operatorname{tr}(\boldsymbol{A})} \geqslant \prod_{1 \leqslant i_1 \leqslant i_2 \leqslant \dots \leqslant i_r \leqslant n} \left(r \sqrt[n]{\det(\boldsymbol{A})} / \operatorname{tr}(\boldsymbol{A}) \right) \quad (7)$$

and

$$\prod_{1 \leqslant i_1 \leqslant i_2 \leqslant \dots \leqslant i_r \leqslant n} \sum_{j=1}^{r} \frac{\lambda_{i_j}}{1 + \lambda_{i_j}} \geqslant \prod_{1 \leqslant i_1 \leqslant i_2 \leqslant \dots \leqslant i_r \leqslant n} \left(r \sqrt[n]{\det(\boldsymbol{A}) / \det(\boldsymbol{I} + \boldsymbol{A})} \right),$$

$$(8)$$

where $\lambda_i (1 \leqslant i \leqslant n)$ is eigenvalue of matrix $\boldsymbol{A}$, $\operatorname{tr}(\boldsymbol{A}) = \sum_{i=1}^{n} \lambda_i$ and $\det(\boldsymbol{A}) = \prod_{i=1}^{n} \lambda_i$.

Proof As mentioned in [5], we know that

$$\log \left(\frac{\sqrt[n]{\det(\boldsymbol{A})}}{\operatorname{tr}(\boldsymbol{A})}, \dots, \frac{\sqrt[n]{\det(\boldsymbol{A})}}{\operatorname{tr}(\boldsymbol{A})} \right) \prec \log \left(\frac{\lambda_1}{\operatorname{tr}(\boldsymbol{A})}, \dots, \frac{\lambda_n}{\operatorname{tr}(\boldsymbol{A})} \right)$$

and

$$\log \left(\sqrt[n]{\det(\boldsymbol{A}) / \det(\boldsymbol{I} + \boldsymbol{A})}, \dots, \sqrt[n]{\det(\boldsymbol{A}) / \det(\boldsymbol{I} + \boldsymbol{A})} \right)$$
$$\prec \log \left(\frac{\lambda_1}{1 + \lambda_1}, \dots, \frac{\lambda_n}{1 + \lambda_n} \right)$$

By using Theorem 1, we assert that both (7) and (8) hold.

Acknowledgements This work was supported by the NSF of Anhui Province(KJ2010B001, KJ2012Z009) and the Scientific Research Common Program of Beijing Municipal Commission of Education (KM201111417006).

References

[1] Y M CHU, Y P LV. The Schur harmonic convexity of the Hamy symmetric function and its applications[J]. J. Inequal. Appl., 2009, Art. ID. 838529, 10 pages.

[2]　Y M Chu, X M ZHANG, G D WANG. The Schur geometrical convexity of the extended mean values[J]. J. Conv. Anal., 2008, 15: 707-718.

[3]　Y M CHU, X M ZHANG. Necessary and sufficient conditions such that extended mean values are Schur-convex or Schur-concave[J]. J. Math. Kyoto Univ.,2008, 48: 229-238.

[4]　K Z GUAN. Schur-convexity of the complete symmetric function[J]. Math. Inequal. Appl., 2006, 9: 567-576.

[5]　K Z GUAN. Some properties of a class of symmetric functions[J]. J. Math. Anal. Appl., 2007, 33: 670-680.

[6]　K Z GUAN. The Hamy symmetric function and its generalization[J]. Math. Inequal. Appl., 2006, 9: 797-805.

[7]　K Z GUAN, J H SHEN. Schur-convexity for a class of symmetric function and its applications[J]. Math. Inequal. Appl.,2006, 9: 199-210.

[8]　W D JIANG. Some properties of dual form of the Hamy's symmetric function[J]. J. Math. Inequal., 2007, 1: 117-125.

[9]　A W MARSHALL, I OLKIN. Inequalities: Theory of Majorization and Its Applications[M]. New York: Academic Press, 1979.

[10]　J X MENG, Y M CHU, X M TANG. The Schur-harmonic-convexity of dual form of the Hamy symmetric function[J]. Matematicki Vesnik, 2010, 62: 37-46.

[11]　I SCHUR. Über eine Klasse von Mittelbildungen mit Anwendungen auf die Determinantentheorie[J]. Sitzunsber Berlin Math Ges,1923, 22: 9-20.

[12]　J N WHITELEY. Some inequalities concerning symmetric forms[J]. Mathematika, 1958, 5: 47-49.

[13]　W F XIA, Y M CHU. On Schur-convexity of some symmetric functions[J]. J. Inequal. Appl., 2010, Art. ID. 543250, 12 pages.

[14]　W F XIA, Y M CHU. Schur-convexity for a class of symmetric functions and its applications[J]. J. Inequal. Appl., 2009, Art. ID. 493759, 15 pages.

[15]　X M ZHANG. Geometrically Convex Functions[M]. Hefei: An hui University Press, 2004.

[16] X M ZHANG. S-Geometric convexity of a function involving Maclaurin's elementary symmetric mean[J]. J. Inequal. Pure. Appl. Math., Vol. 8, 2007, Article 51, 6 pages.

[17] Y M CHU, G D WANG, X H ZHANG. The Schur multiplicative and harmonic convexities of the complete symmetric function[J]. Math. Nachr., 1-11(2011)/DOI 10.1002/mana.200810197.

第57篇　关于一类对称函数的Schur凸性

(张静，石焕南. 数学的实践与认识，2013,43 (19):292-296)

摘　要:　利用Schur凸函数、Schur几何凸函数和Schur调和凸函数的有关性质简化证明了一类与对数凸函数有关的对称函数的Schur凸性、Schur几何凸性和Schur 调和凸性.

关键词:　受控; Schur凸性; Schur几何凸性; Schur调和凸性; 不等式; 凸函数; 对数凸函数

57.1　引言

受控(Majorization) 理论中函数保序性质的系统研究最先是由Issai Schur 在1923 年进行的, 为纪念其贡献, 也因此将他所研究的这类函数命名为Schur 凸函数[1]. Schur 凸函数是新兴数学学科受控理论的核心概念之一. 受控理论在分析不等式、广义平均值、统计试验、图和矩阵、组合优化、可靠性、信息安全和其他相关领域均有重要应用[1-2]. 2004 年张小明[3-4] 提出Schur 几何凸函数的概念, 2009 年褚玉明等提出Schur 调和凸函数的概念. 这些概念丰富了受控理论, 并引起国内外学者们的关注.

各种对称函数的Schur凸性、Schur几何凸性和Schur 调和凸性的研究是近年国内外解析不等式研究领域的热点[8-17]. 2010 年, 克罗地亚学者I. Roventa[9]定义了如下一类对称函数. 设区间$I \subset \mathbf{R}, f : I \to \mathbf{R}_+$为对数凸函数

$$F_k(\boldsymbol{x}) = \sum_{1 \leqslant i_1 < \cdots < i_k \leqslant n} \prod_{j=1}^{k} f(x_{i_j}), \, k = 1, 2, \ldots, n$$

对于$k = 1, 2, n-1$, Roveta 证得$F_k(\boldsymbol{x})$ 是I^n 上的Schur 凸函数, 而对于$2 < k < n-1$ 的情形未加以讨论. 2011 年, 王淑红等[10] 不仅全面考察了$F_k(\boldsymbol{x})$ 在I^n 上的Schur 凸性, 并且还研究了$F_k(\boldsymbol{x})$ 在I^n 上的Schur 几何凸性和Schur 调和凸性, 得到如下结果.

定理 A 设 $I \subset \mathbf{R}$ 是一具有非空内部的对称凸集, 函数 $f : I \to \mathbf{R}$ 为在 I 上连续、在 I 的内部可微的对数凸函数, 则对任意的 $k = l, 2 \ldots, n$, $F_k(\boldsymbol{x})$ 为 I^n 上的 Schur 凸函数.

定理 B 设 $I \subset \mathbf{R}_+$ 是一具有非空内部的对称凸集, 函数 $f : I \to \mathbf{R}_+$ 在 I 上连续, 在 I 的内部可微, 且 f 为 I 上递增的对数凸函数, 则对任意的 $k = l, 2 \ldots, n$, $F_k(\boldsymbol{x})$ 为 I^n 上的 Schur 几何凸函数.

定理 C 设 $I \subset \mathbf{R}_+$ 是一具有非空内部的对称凸集, 函数 $f : I \to \mathbf{R}_+$ 在 I 上连续, 在 I 的内部可微, 且 f 为 I 上递增的对数凸函数, 则对任意的 $k = l, 2 \ldots, n$, $F_k(\boldsymbol{x})$ 为在 I^n 上的 Schur 调和凸函数.

对于各类对称函数的 Schur 凸性、Schur 几何凸性和 Schur 调和凸性的研究, 经典的方法是, 依据 Schur 凸函数、Schur 几何凸函数和 Schur 调和凸函数的判定定理[2,4-5] 来判断, 即通过计算判别式, 再利用各种不等式的放缩技巧确定其符号来进行判断. 这一过程往往是繁琐的, 特别是对于一些表达式复杂的函数未必有效, 例如 Stolarsky - Gini 均值函数的 Schur 凸性、Schur 调和凸性的研究[11]. 王淑红等即采用经典的方法. 本文另辟新径, 利用 Muirhead 定理的一个相关结果[1], 提出了一种新方法, 有助于该类问题的解决, 并且可以给出非常简洁的证明. 它为研究对称函数各种 Schur 凸性提供了新的思路.

57.2 基本概念

在本文中, $\mathbf{R}^n$ 和 $\mathbf{R}_+^n$ 分别表示 n 维实数集和 n 维正实数集, 并记 $\mathbf{R}^1 = \mathbf{R}, \mathbf{R}_+^1 = \mathbf{R}_+$.

定义 1[1-2] 设 $\boldsymbol{x} = (x_1, ..., x_n), \boldsymbol{y} = (y_1, ..., y_n) \in \mathbf{R}^n$.

(1) 若 $\sum_{i=1}^{k} x_{[i]} \leqslant \sum_{i=1}^{k} y_{[i]}$, $k = 1, 2, ..., n-1$, 且 $\sum_{i=1}^{n} x_i = \sum_{i=1}^{n} y_i$, 则称 $\boldsymbol{x}$ 被 $\boldsymbol{y}$ 所控制, 记作 $\boldsymbol{x} \prec \boldsymbol{y}$, 其中 $x_{[1]} \geqslant \cdots \geqslant x_{[n]}$ 和 $y_{[1]} \geqslant \cdots \geqslant y_{[n]}$ 分别是 $\boldsymbol{x}$ 和 $\boldsymbol{y}$ 的递减重排.

(2) 设 $\Omega \subset \mathbf{R}^n, \varphi : \Omega \to \mathbf{R}$. 若在 Ω 上 $\boldsymbol{x} \prec \boldsymbol{y} \Rightarrow \varphi(\boldsymbol{x}) \leqslant \varphi(\boldsymbol{y})$, 则称 φ 为 Ω 上的 Schur 凸函数; 若 $-\varphi$ 是 Ω 上的 Schur 凸函数, 则称 φ 为 Ω 上的 Schur 凹函数.

(3) $\boldsymbol{x} \leqslant \boldsymbol{y}$ 表示 $x_i \leqslant y_i, i = 1, ..., n$. 设 $\Omega \subset \mathbf{R}^n, \varphi : \Omega \to \mathbf{R}$. 若 $\forall \boldsymbol{x}, \boldsymbol{y} \in \Omega, \boldsymbol{x} \leqslant \boldsymbol{y} \Rightarrow \varphi(\boldsymbol{x}) \leqslant \varphi(\boldsymbol{y})$, 则称 φ 为 Ω 上的增函数; 若 $-\varphi$ 是 Ω 上的增函数, 则称 φ 为 Ω 上的减函数.

定义 2[1-2] 设 $\Omega \subset \mathbf{R}^n, \alpha, \beta \in [0,1]$ 且 $\alpha + \beta = 1$.

(1) 若对于任何 $\boldsymbol{x}, \boldsymbol{y} \in \Omega$ 总有 $\alpha \boldsymbol{x} + \beta \boldsymbol{y} = (\alpha x_1 + \beta y_1, ..., \alpha x_n + \beta y_n) \in \Omega$, 则称 Ω 称为凸集.

(2) 设 Ω 称为凸集, $\varphi : \Omega \to \mathbf{R}$. 若对于任何 $\boldsymbol{x}, \boldsymbol{y} \in \Omega$ 总有

$$\varphi(\alpha \boldsymbol{x} + \beta \boldsymbol{y}) \leqslant \alpha \varphi(\boldsymbol{x}) + \beta \varphi(\boldsymbol{y})$$

则称 φ 为 Ω 上的凸函数. 若 $-\varphi$ 是 Ω 上的凸函数, 则称 φ 为 Ω 上的凹函数.

(3) 设 Ω 称为凸集, $\varphi : \Omega \to \mathbf{R}_+$. 若 $\ln \varphi$ 为 Ω 上的凸函数,则称 φ 为 Ω 上的对数凸函数.

定义 3[1-2]　设 $\Omega \subset \mathbf{R}^n$.

(1) 若 $\boldsymbol{x} \in \Omega \Rightarrow \boldsymbol{Px} \in \Omega$, $\forall n \times n$ 置换矩阵 P 则称 Ω 为对称集,

(2) 设 Ω 为对称集, $\varphi : \Omega \to \mathbf{R}$. 若对于任何 $\boldsymbol{x} \in \Omega$ 和对于任意 $n \times n$ 置换矩阵 $\boldsymbol{P}$ 均有 $\varphi(\boldsymbol{Px}) = \varphi(\boldsymbol{x})$, 则称 φ 为 Ω 上的对称函数.

引理 1[3]　设 φ 在开凸集 $\Omega \subset \mathbf{R}^n$ 上可微, 则 φ 是 Ω 上的增函数的充要条件是 $\nabla \varphi(\boldsymbol{x}) \geqslant 0, \forall \boldsymbol{x} \in \Omega$, 其中

$$\nabla \varphi(\boldsymbol{x}) = \left(\frac{\partial \varphi(\boldsymbol{x})}{\partial x_1}, \ldots, \frac{\partial \varphi(\boldsymbol{x})}{\partial x_n} \right) \in \mathbf{R}^n$$

Schur 凸函数判定定理 [2]　设 $\Omega \subset \mathbf{R}^n$ 是有内点的对称凸集, $\varphi : \Omega \to \mathbf{R}$ 在 Ω 上连续, 在 Ω 的内部 Ω^0 可微, 则在 Ω 上Schur-凸(Schur-凹)的充要条件是 φ 在 Ω 上对称且 $\forall \boldsymbol{x} \in \Omega$, 有

$$(x_1 - x_2) \left(\frac{\partial \varphi}{\partial x_1} - \frac{\partial \varphi}{\partial x_2} \right) \geqslant 0 (\leqslant 0) \tag{1}$$

定义 4[4]　设 $\Omega \subset \mathbf{R}^n_{++}$, $\varphi : \Omega \to \mathbf{R}_+$.

(i) 若 $\forall \boldsymbol{x}, \boldsymbol{y} \in \Omega$ 总有 $(x_1^\alpha y_1^\beta, \ldots, x_n^\alpha y_n^\beta) \in \Omega$, 则称 Ω 为几何凸集, 其中 $\alpha, \beta \in [0, 1]$, 且 $\alpha + \beta = 1$.

(ii) $\forall \boldsymbol{x}, \boldsymbol{y} \in \Omega$, 若 $(\ln x_1, \ldots, \ln x_n) \prec (\ln y_1, \ldots, \ln y_n)$ 有 $\varphi(\boldsymbol{x}) \leqslant \varphi(\boldsymbol{y})$, 则称为 φ 为 Ω 上的Schur几何凸函数.

Schur 几何凸函数判定定理 [4]　设 $\Omega \subset \mathbf{R}^n_{++}$ 是有内点的对称几何凸集, $\varphi : \Omega \to \mathbf{R}$ 在 Ω 上连续, 在 Ω 的内部 Ω^0 可微, 则在 Ω 上Schur 几何凸(Schur-凹)的充要条件是 φ 在 Ω 上对称且 $\forall \boldsymbol{x} \in \Omega$, 有

$$(\ln x_1 - \ln x_2) \left(x_1 \frac{\partial \varphi}{\partial x_1} - x_2 \frac{\partial \varphi}{\partial x_2} \right) \geqslant 0 (\leqslant 0) \tag{2}$$

定义 5[5-6,12]　设 $\Omega \subset \mathbf{R}^n_+$, $\varphi : \Omega \to \mathbf{R}_+$. $\forall \boldsymbol{x} = (x_1, \cdots, x_n), \boldsymbol{y} = (y_1, \cdots, y_n) \in \Omega$.

(1) 若 $\left(\dfrac{x_1 y_1}{\alpha x_1 + \beta y_1}, \cdots, \dfrac{x_n y_n}{\alpha x_n + \beta y_n} \right) \in \Omega$, 则称 Ω 是调和凸集, 其中 $\alpha, \beta \in$

$[0,1]$ 且 $\alpha + \beta = 1$.

(2) 若

$$\left(\frac{1}{x_1}, \cdots, \frac{1}{x_n}\right) \prec \left(\frac{1}{y_1}, \cdots, \frac{1}{y_n}\right) \Rightarrow \varphi(\boldsymbol{x}) \leq \varphi(\boldsymbol{y})$$

或等价地

$$\boldsymbol{x} \prec \boldsymbol{y} \Rightarrow \varphi\left(\frac{1}{x_1}, \cdots, \frac{1}{x_n}\right) \leqslant \varphi\left(\frac{1}{y_1}, \cdots, \frac{1}{y_n}\right)$$

则称 φ 为 Ω 上的 Schur 调和凸函数; 若 $-\varphi$ 是 Ω 上 Schur 调和凸函数, 则称 φ 为 Ω 上的 Schur 调和凹函数.

Schur 调和凸函数判定定理[5,12]　设 $\Omega \subset \mathbf{R}^n$ 是有内点的对称调和凸集, $\varphi : \Omega \to \mathbf{R}_+$ 于 Ω 上连续, 在 Ω 的内部 Ω^0 一阶可微. 若 φ 在 Ω 上对称, 且对于任意 $\boldsymbol{x} = (x_1, \cdots, x_n) \in \Omega^0$, 有

$$(x_1 - x_2)\left(x_1^2 \frac{\partial \varphi}{\partial x_1} - x_2^2 \frac{\partial \varphi}{\partial x_2}\right) \geqslant 0 \, (\leqslant 0) \tag{3}$$

则 φ 是 Ω 上 Schur 调和凸(凹)函数.

57.3　引理

为给出我们的证明, 需要以下引理.

设 $\pi = (\pi(1), ..., \pi(n))$ 是 $1, ..., n$ 的任一置换, 共 $n!$ 个. 专著[1]证明了如下结论.

引理 1[1]　设 $\mathbf{A} \subset \mathbf{R}^k$ 是一个对称凸集, φ 是 $\mathbf{A}$ 上的 S-凸函数, 具有性质: 对每一个固定的 $x_2, ..., x_k$, $\varphi(z, x_2, ..., x_k)$ 关于 z 在 $\{z : (z, x_2, \cdots, x_k) \in A\}$ 上凸, 则对于任何 $n > k$,

$$\psi(x_1, \cdots, x_n) = \sum_\pi \varphi\left(x_{\pi(1)}, \cdots, x_{\pi(k)}\right)$$

在 $\mathbf{B} = \{(x_1, \cdots, x_n) : (x_{\pi(1)}, \cdots, x_{\pi(k)}) \in \mathbf{A}$, 对于所有的排列 $\pi\}$ 上 S-凸.

在大多数应用中, $\mathbf{A}$ 为 $\mathbf{I}^k$ 的形式, 其中区间 $\mathbf{I} \subset \mathbf{R}$, 且在这种情况下, $\mathbf{B} = \mathbf{I}^n$. 注意固定其他变量, φ 在第一个自变量上的凸性也蕴涵着在其他变量上的凸性, 因为 φ 是对称的.

推论[1]　令

$$\bar{\psi}(\boldsymbol{x}) = \frac{\psi(\boldsymbol{x})}{k!(n-k)!}$$

其中ψ 如引理1 所定义, 则

$$\overline{\psi}(x_1,...,x_n) = \sum_{1 \leqslant i_1 < ... < i_k \leqslant n} \varphi(x_{i_1}, \cdots, x_{i_k}) \tag{4}$$

且若ψ 是S-凸, 则$\overline{\psi}$ 也是S-凸的.

引理 2[1]   设$g : I \to \mathbf{R}_+$ 连续, $\varphi(\boldsymbol{x}) = \prod\limits_{i=1}^{n} g(x_i)$, 则$\varphi$ 在I^n 上Schur 凸(Schur 凹) 的充要条件是$\ln g$ 在I 上凸(凹).

引理 3 [1]   设$\Omega \subset \mathbf{R}^n, g : \Omega \to \mathbf{R}_+$. 若$\ln g$ 凸, 则g 亦凸.

引理 4 [12]   设$\varphi(\boldsymbol{x})$ 在$\Omega \subset \mathbf{R}_+^n$ 上非负可微. 若$\varphi(\boldsymbol{x})$ 是递增的Schur 凸函数, 则$\varphi(\boldsymbol{x})$ 一定是Schur 几何凸函数.

引理 5[12]   设$\varphi(\boldsymbol{x})$ 在$\Omega \subset \mathbf{R}_+^n$ 上非负可微. 若$\varphi(\boldsymbol{x})$ 是递增的Schur 凸函数或Schur 几何凸函数, 则$\varphi(\boldsymbol{x})$ 一定是Schur 调和凸函数.

57.4   定理的证明

定理 A 的证明   令$\varphi(\boldsymbol{x}) = \prod\limits_{i=1}^{k} f(x_k)$ 由引理2 知φ 在I^k 上Schur 凸. 又因f 是I 上的对数凸函数, 据引理3,f 亦是I 上的凸函数, 这样对每一个固定的$x_2, \ldots, x_k$

$$\varphi(z, x_2, \ldots, x_k) = f(z) \prod_{i=2}^{k} f(x_i)$$

关于z 在$\left\{z : (z, x_2, \ldots, x_k) \in I^k\right\}$上凸. 于是由引理1 的推论可知, 对任意的$k = 1, 2, \ldots, n$, $F_k(\boldsymbol{x})$ 为I^n 上的Schur 凸函数.

定理 B 的证明   因f 为I 上递增的对数凸函数, 故f 在I 上非负递增, 于是

$$F_k(\boldsymbol{x}) = \sum_{1 \leqslant i_1 < \cdots < i_k \leqslant n} \prod_{j=1}^{k} f(x_{i_j})$$

亦在I^n 上非负递增. 又由定理A 知$F_k(\boldsymbol{x})$ 是Schur 凸函数, 再据引理4 即得证.

定理 C 的证明   仿照定理B 的证明, 应用引理5 即得证.

致谢   作者对审稿人所提的修改建议表示衷心感谢.

参考文献

[1]   A M MARSHALL, I OLKIN, B C ARNOLD. Inequalities: Theory of majorization and its application[M]. 2nd ed. New York: Springer, 2011.

[2] 王伯英. 控制不等式基础[M]. 北京: 北京师范大学出版社, 1990.

[3] 张小明. 几何凸函数的几个定理及其应用[M]. 首都师范大学学报(自然科学版)[J]. 2004, 25(2): 11-13,18.

[4] 张小明. 几何凸函数[M]. 合肥: 安徽大学出版社, 2004.

[5] CHU Y-M, SUN T-C. The Schur harmonic convexity for a class of symmetric functions [J]. Acta Mathematica Scientia, 2010, 30B (5): 1501 - 1506.

[6] CHU Y-M WANG G-D, ZHANG X-H. Schur muftiplicative and harmonic convexities of the complete symmetric fanction[J]. Mathematische Nachrichten, 2011, 284 (5-6): 53-66.

[7] CHU YU-MING, LV YU-PEI. The Schur harmonic convexity of the Hamy symmetric function and its applications [J]. J. Inequal. Appl., Volume 2009, Article ID 838529, 10 pages, doi:10.1155/2009/838529.

[8] ČULJAK V, PEČARIĆ J. Schur-convexity of ČEBIŠEV functional [J]. Mathematical Inequalities and Applications, 2011, 14(4): 911-916.

[9] IONEL ROVENTA. Schur convexity of a class of symmetric functions [J]. Annals of the University of Craiova, Mathematics and Computer Science Series. 2010, 37(1): 12-18.

[10] 王淑红，张天宇, 华志强. 一类对称函数的Schur-几何凸性及Schur-调和凸性[J]. 内蒙古民族大学学报(自然科学版), 2011, 26 (4): 387-390.

[11] WITKOWSKI ALFRED. On Schur-convexity and Schur-geometric convexity of four-parameter family of means [J]. Mathematical Inequalities and Applications, 2011, 14(4): 897 - 903.

[12] 石焕南. 受控理论与解析不等式[M]. 哈尔滨: 哈尔滨工业大学出版社, 2012.

[13] ČULJAK V, FRANJIČ I, GHULAM R,PEČARIĆ J. Schur-convexity of averages of convex functions[J]. Journal of Inequalities and Applications, Volume 2011, Article ID 581918,25 pages, doi: 10.1155/2011/581918.

[14] SHI H-N, ZHANG J. Schur-convexity of dual form of some symmetrc functions[J]. Journal of Inequalities and Applications, Volume 2013, Article ID 295,9 pages, doi: 10. 1186/1029 - 242 X - 2013 - 295.

[15] SHI H N. Two Schur-convex functions related to Hadamard-type integral inequalities[J]. Publicationes Mathematicae Debrecen, 2011, 78(2): 393-403.

[16] GUAN K, GUAN R. Some properties of a generalized Hamy symmetric function and its applications[J]. Journal of Mathematical Analysis and Applications, 2011, 376(2): 494-505.

[17] WU Y, QI F. Schur-harmonic convexity for differences of some means [J]. Analysis, 2012, 32(4): 263-270.

第58篇　一类条件不等式的
控制证明与应用

(石焕南, 张静. 纯粹数学与应用数学, 2013, 29 (5)：441-449)

摘　要： 本文通过判断相关函数的 Schur 凸性、Schur 几何凸性和 Schur 调和凸性, 证明并推广了一类条件不等式, 并据此建立了某些单形不等式.

关键词： Schur 凸性; Schur 调和凸性; Schur 几何凸性; 条件不等式; 单形

在本文中, $\mathbf{R}$ 和 $\mathbf{R}_+^n$ 分别表示 n 维实数集和 n 维正实数集, 并记 $\mathbf{R}^1 = \mathbf{R}$, $\mathbf{R}_+^1 = \mathbf{R}_+$. 又记集合

$$U = \left\{ (x_1, \cdots, x_n) : x_k > 0, k = 1, \cdots, n, \sum_{k=1}^{n} \frac{1}{x_k} \leqslant 1 \right\}$$

1923年数学家 Issai Schur 率先系统研究了受控理论中函数的保序性, 为纪念其贡献, 将他所研究的这类函数命名为 Schur 凸函数. 随着受控理论研究的深入, Schur 凸函数已经在解析不等式、矩阵论、广义平均值、概率统计、图论、数值分析、可靠性、信息安全和其他相关领域发挥着愈来愈重要的作用[1-2]. 下面给出 Schur 凸函数的定义、判定及其推广.

定义 1[1-2]　设 $x = (x_1, \cdots, x_n), y = (y_1, \cdots, y_n) \in \mathbf{R}^n$.

(1) 若 $\sum_{i=1}^{k} x_{[i]} \leqslant \sum_{i=1}^{k} y_{[i]}, k = 1, 2, \cdots, n-1$, 且 $\sum_{i=1}^{n} x_i = \sum_{i=1}^{n} y_i$, 则称 x 被 y 所控制, 记作 $x \prec y$, 其中 $x_{[1]} \geqslant \cdots \geqslant x_{[n]}$ 和 $y_{[1]} \geqslant \cdots \geqslant y_{[n]}$ 分别是 x 和 y 的分量的递减重排.

(2) 设 $\Omega \subset \mathbf{R}^n, \varphi : \Omega \to \mathbf{R}$. 若在 Ω 上 $x \prec y \Rightarrow \varphi(x) \leqslant \varphi(y)$, 则称 φ 为 Ω 上的 Schur 凸函数. 若 $-\varphi$ 是 Ω 上的 Schur 凸函数, 则称 φ 为 Ω 上的 Schur 凹函数.

引理 1[1-2]　设 $\boldsymbol{x} = (x_1, \cdots, x_n) \in \mathbf{R}^n$, 则

$$(\bar{\boldsymbol{x}}, \cdots, \bar{\boldsymbol{x}}) \prec (x_1, \cdots, x_n) \tag{1}$$

其中 $\bar{x} = \dfrac{1}{n} \displaystyle\sum_{i=1}^{n} x_i$.

定义 2[1-2]　设 $\Omega \subset \mathbf{R}^n$.

(1) 若 $\boldsymbol{x} \in \Omega \Rightarrow \boldsymbol{x}\boldsymbol{P} \in \Omega, \forall n \times n$ 置换矩阵 $\boldsymbol{P}$, 则称 Ω 为对称集.

(2) 设 Ω 为对称集, $\varphi : \Omega \to \mathbf{R}$. 若对于任何 $\boldsymbol{x} \in \Omega$ 和任意 $n \times n$ 置换矩阵 $\boldsymbol{P}$, 都有 $\varphi(\boldsymbol{x}\boldsymbol{P}) = \varphi(\boldsymbol{x})$, 则称 φ 为 Ω 上的对称函数.

引理 2[2]　设 $\Omega \subset \mathbf{R}^n$ 是有内点的对称凸集, $\varphi : \Omega \to \mathbf{R}$ 在 Ω 上连续, 在 Ω 的内部 Ω^0 可微, 则 φ 在 Ω 上 Schur 凸(凹)$\Leftrightarrow$ φ 在 Ω 上对称且 $\forall \boldsymbol{x} = (x_1, \cdots, x_n) \in \Omega^0$, 有

$$(x_1 - x_2)\left(\frac{\partial \varphi}{\partial x_1} - \frac{\partial \varphi}{\partial x_2}\right) \geqslant 0 (\leqslant 0) \tag{2}$$

定义 3[3]　设 $\Omega \subset \mathbf{R}_+^n$, $\varphi : \Omega \to \mathbf{R}_+$. $\forall \boldsymbol{x} = (x_1, \cdots, x_n), \boldsymbol{y} = (y_1, \cdots, y_n) \in \Omega$.

(1) 若 $\left(x_1^\alpha y_1^\beta, \cdots, x_n^\alpha y_n^\beta\right) \in \Omega$, 则称 Ω 是几何凸集, 其中 $\alpha, \beta \in [0, 1]$ 且 $\alpha + \beta = 1$.

(2) 若 $(\ln x_1, \cdots, \ln x_n) \prec (\ln y_1, \cdots, \ln y_n) \Rightarrow \varphi(\boldsymbol{x}) \leqslant \varphi(\boldsymbol{y})$, 则称 φ 为 Ω 上的 Schur 几何凸函数; 若 $-\varphi$ 是 Ω 上 Schur 几何凸函数, 则称 φ 为 Ω 上的 Schur 几何凹函数.

我国学者张小明建立了下述 Schur 几何凸函数判定定理.

引理 3[3]　设 $\Omega \subset \mathbf{R}^n$ 是有内点的对称几何凸集, $\varphi : \Omega \to \mathbf{R}_+$ 于 Ω 上连续, 在 Ω 的内部 Ω^0 一阶可微. 若 φ 在 Ω 上对称, 且对于任意 $\boldsymbol{x} = (x_1, \cdots, x_n) \in \Omega^0$, 有

$$(\ln x_1 - \ln x_2)\left(x_1 \frac{\partial \varphi}{\partial x_1} - x_2 \frac{\partial \varphi}{\partial x_2}\right) \geqslant 0 (\leqslant 0) \tag{3}$$

则 φ 是 Ω 上 Schur 几何凸(凹)函数.

2009年, 我国学者褚玉明、吕瑜佩[4]提出 Schur 调和凸函数的概念, 它同 Schur 凸函数、Schur 几何凸函数等概念一起受到国内外学者们的注意[5-18].

定义 4[5-6]　设 $\Omega \subset \mathbf{R}_+^n$, $\varphi : \Omega \to \mathbf{R}_+$. $\forall \boldsymbol{x} = (x_1, \cdots, x_n), \boldsymbol{y} = (y_1, \cdots, y_n) \in \Omega$.

(1) 若 $\left(\dfrac{x_1 y_1}{\alpha x_1 + \beta y_1}, \cdots, \dfrac{x_n y_n}{\alpha x_n + \beta y_n}\right) \in \Omega$, 则称 Ω 是调和凸集, 其中 $\alpha, \beta \in [0, 1]$ 且 $\alpha + \beta = 1$.

(2) 若

$$\left(\frac{1}{x_1},\cdots,\frac{1}{x_n}\right) \prec \left(\frac{1}{y_1},\cdots,\frac{1}{y_n}\right) \Rightarrow \varphi(\boldsymbol{x}) \leqslant \varphi(\boldsymbol{y})$$

或等价地

$$\boldsymbol{x} \prec \boldsymbol{y} \Rightarrow \varphi\left(\frac{1}{x_1},\cdots,\frac{1}{x_n}\right) \leqslant \varphi\left(\frac{1}{y_1},\cdots,\frac{1}{y_n}\right)$$

则称 φ 为 Ω 上的 Schur 调和凸函数; 若 $-\varphi$ 是 Ω 上 Schur 调和凸函数, 则称 φ 为 Ω 上的 Schur 调和凹函数.

引理 4[5-6] 设 $\Omega \subset \mathbf{R}^n$ 是有内点的对称调和凸集, $\varphi : \Omega \to \mathbf{R}_+$ 于 Ω 上连续, 在 Ω 的内部 Ω^0 一阶可微. 若 φ 在 Ω 上对称, 且对于任意 $\boldsymbol{x} = (x_1,\cdots,x_n) \in \Omega^0$, 有

$$(x_1 - x_2)\left(x_1^2 \frac{\partial \varphi}{\partial x_1} - x_2^2 \frac{\partial \varphi}{\partial x_2}\right) \geqslant 0\,(\leqslant 0) \tag{4}$$

则 φ 是 Ω 上 Schur 调和凸(凹)函数.

本文将通过判断相关函数的 Schur 凸性、Schur 几何凸性和 Schur 调和凸性, 证明一类条件不等式并据此建立某些单形不等式.

定理 1 设 $\boldsymbol{x} = (x_1,\cdots,x_n) \in \mathbf{R}_+^n$ 且 $\sum_{k=1}^n \frac{1}{x_k} = \lambda$, 则 $\forall m \in \mathbf{N}$ 成立

$$n\left(n^{m-1}-1\right)G^{-m} \leqslant \left(\sum_{k=1}^n x_k^{-1}\right)^m - \left(\sum_{k=1}^n x_k^{-m}\right) \leqslant \left(1 - \frac{1}{n^{m-1}}\right)\lambda^m \tag{5}$$

$$\left(\sum_{k=1}^n x_k\right)^m - \left(\sum_{k=1}^n x_k^m\right) \geqslant \frac{n^{2m}-n^{m+1}}{\lambda^m} \tag{6}$$

其中 $G = \sqrt[n]{\prod_{i=1}^n x_i}$.

注 1 当 $\lambda = 1$ 时, 式(5)中右边不等式见文献[19].

证明 令 $\varphi(x_1,\cdots,x_n) = \left(\sum_{k=1}^n x_k\right)^m - \left(\sum_{k=1}^n x_k^m\right)$. 显然 φ 在 $\mathbf{R}_+^n$ 上对称, 且有

$$\frac{\partial \varphi}{\partial x_i} = m\left(\sum_{k=1}^n x_k\right)^{m-1} - m x_i^{m-1} \geqslant 0, i = 1,\cdots,n$$

不妨设 $x_1 \neq x_2$, 于是

$$\Delta_1 := (x_1 - x_2)\left(\frac{\partial \varphi}{\partial x_1} - \frac{\partial \varphi}{\partial x_2}\right) = -m(x_1 - x_2)\left(x_1^{m-1} - x_2^{m-1}\right) \leqslant 0$$

据引理2, $\varphi(x_1, \cdots, x_n)$ 在 $\mathbf{R}_+^n$ 上 Schur 凹. 由引理1 有

$$\left(\frac{\lambda}{n}, \cdots, \frac{\lambda}{n}\right) \prec \left(\frac{1}{x_1}, \cdots, \frac{1}{x_n}\right)$$

故

$$\varphi\left(\frac{\lambda}{n}, \cdots, \frac{\lambda}{n}\right) \geqslant \varphi\left(\frac{1}{x_1}, \cdots, \frac{1}{x_n}\right)$$

即式(5)中右边不等式成立.

$$\begin{aligned}
\Delta_2 := &(x_1 - x_2)\left(x_1 \frac{\partial \varphi}{\partial x_1} - x_2 \frac{\partial \varphi}{\partial x_2}\right) \\
= &m(x_1 - x_2)\left[(x_1 - x_2)\left(\sum_{k=1}^{n} x_k\right)^{m-1} - (x_1^m - x_2^m)\right] \\
= &m(x_1 - x_2)^2\left[\left(\sum_{k=1}^{n} x_k\right)^{m-1} - \frac{x_1^m - x_2^m}{x_1 - x_2}\right] \\
\geqslant &m(x_1 - x_2)^2\left[(x_1 + x_2)^{m-1} - \frac{x_1^m - x_2^m}{x_1 - x_2}\right] \\
= &m(x_1 - x_2)^2\left[\left(x_1^{m-1} + (m-1)x_1^{m-2}x_2 + \cdots + (m-1)x_1 x_2^{m-2} + x_2^{m-1}\right) - \right. \\
&\left.\left(x_1^{m-1} + x_1^{m-2}x_2 + \cdots + x_1 x_2^{m-2} + x_2^{m-1}\right)\right] \geqslant 0
\end{aligned}$$

注意到, 总有 $\dfrac{\ln x_1 - \ln x_2}{x_1 - x_2} > 0$, 故据引理3, $\varphi(x_1, \cdots, x_n)$ 在 $\mathbf{R}_+^n$ 上 Schur 几何凸. 由引理1 有

$$\left(\ln G^{-1}, \cdots, \ln G^{-1}\right) \prec \left(\ln x_1^{-1}, \cdots, \ln x_n^{-1}\right)$$

故

$$\varphi\left(G^{-1}, \cdots, G^{-1}\right) \leqslant \varphi\left(x_1^{-1}, \cdots, x_n^{-1}\right)$$

即式(5)中左边不等式成立.

$$\begin{aligned}
\Delta_3 := &(x_1 - x_2)\left(x_1^2 \frac{\partial \varphi}{\partial x_1} - x_2^2 \frac{\partial \varphi}{\partial x_2}\right) \\
= &m(x_1 - x_2)\left[(x_1^2 - x_2^2)\left(\sum_{k=1}^{n} x_k\right)^{m-1} - (x_1^{m+1} - x_2^{m+1})\right] \\
= &m(x_1 - x_2)^2\left[(x_1 + x_2)\left(\sum_{k=1}^{n} x_k\right)^{m-1} - \frac{x_1^{m+1} - x_2^{m+1}}{x_1 - x_2}\right]
\end{aligned}$$

$$\geqslant m(x_1 - x_2)^2 \left[(x_1 + x_2)^m - \frac{x_1^{m+1} - x_2^{m+1}}{x_1 - x_2} \right]$$

$$= m(x_1 - x_2)^2 \left[\left(x_1^m + m x_1^{m-1} x_2 + \cdots + m x_1 x_2^{m-1} + x_2^m \right) - \right.$$
$$\left. \left(x_1^m + x_1^{m-1} x_2 + \cdots + x_1 x_2^{m-1} + x_2^m \right) \right] \geqslant 0$$

据引理4, $\varphi(x_1, \cdots, x_n)$在$\mathbf{R}_+^n$上 Schur 调和凸. 由引理1 有

$$\left(\frac{\lambda}{n}, \cdots, \frac{\lambda}{n} \right) \prec \left(\frac{1}{x_1}, \cdots, \frac{1}{x_n} \right)$$

故

$$\varphi\left(\frac{n}{\lambda}, \cdots, \frac{n}{\lambda} \right) \leqslant \varphi(x_1, \cdots, x_n)$$

即式(6)成立, 证毕.

定理 2 设$x_k > 0, k = 1, \cdots, n, n \geqslant 2$且$\sum_{k=1}^{n} \frac{1}{x_k} = \lambda \leqslant 1$, 则

$$\prod_{k=1}^{n} \left(x_k^{-1} - 1 \right) \leqslant \left(G^{-1} - 1 \right)^n \tag{7}$$

$$\prod_{k=1}^{n} \left(x_k - 1 \right) \geqslant \left(\frac{n}{\lambda} - 1 \right)^n \tag{8}$$

其中 $G = \sqrt[n]{\prod_{i=1}^{n} x_i}$.

证明 根据定理条件, 可知$x_k > 1, k = 1, \cdots, n$. 令$\psi(x_1, \cdots, x_n) = \prod_{k=1}^{n} (x_k - 1)$. 易见$U$是调和凸集, 显然$\psi$在$U$上对称, 且有

$$\frac{\partial \psi}{\partial x_1} = \frac{\psi(x_1, \cdots, x_n)}{x_1 - 1}, \frac{\partial \psi}{\partial x_2} = \frac{\psi(x_1, \cdots, x_n)}{x_2 - 1}$$

于是

$$\Delta_2 = (x_1 - x_2) \left(x_1 \frac{\partial \psi}{\partial x_1} - x_2 \frac{\partial \psi}{\partial x_2} \right)$$

$$= (x_1 - x_2) \psi(x_1, \cdots, x_n) \left(\frac{x_1}{x_1 - 1} - \frac{x_2}{x_2 - 1} \right)$$

$$= -\frac{(x_1 - x_2)^2 \psi(x_1, \cdots, x_n)}{(x_1 - 1)(x_2 - 1)} \leqslant 0$$

注意到, 总有$\dfrac{\ln x_1 - \ln x_2}{x_1 - x_2} > 0$, 故据引理3, $\psi(x_1, \cdots, x_n)$在$\mathbf{R}_+^n$上 Schur 几

何凹. 由

$$\left(\ln G^{-1}, \cdots, \ln G^{-1}\right) \prec \left(\ln x_1^{-1}, \cdots, \ln x_n^{-1}\right)$$

有

$$\psi\left(G^{-1}, \cdots, G^{-1}\right) \geqslant \psi\left(x_1^{-1}, \cdots, x_n^{-1}\right)$$

即式(7)成立.

$$\Delta_3 = (x_1 - x_2)\left(x_1^2 \frac{\partial \psi}{\partial x_1} - x_2^2 \frac{\partial \psi}{\partial x_2}\right)$$

$$= (x_1 - x_2)\psi(x_1, \cdots, x_n)\left(\frac{x_1^2}{x_1 - 1} - \frac{x_2^2}{x_2 - 1}\right)$$

$$= (x_1 - x_2)^2 \psi(x_1, \cdots, x_n) \frac{x_1 x_2 - (x_1 + x_2)}{(x_1 - 1)(x_2 - 1)}$$

因 $\sum_{k=1}^{n} \frac{1}{x_k} = \lambda \leqslant 1$, 有 $1 \geqslant \frac{1}{x_1} + \frac{1}{x_2}$, 即 $x_1 x_2 - (x_1 + x_2) \geqslant 0$, 于是 $\Delta_3 \geqslant 0$. 据引理4, $\psi(x_1, \cdots, x_n)$ 在 U 上 Schur 调和凸. 由

$$\left(\frac{\lambda}{n}, \cdots, \frac{\lambda}{n}\right) \prec \left(\frac{1}{x_1}, \cdots, \frac{1}{x_n}\right)$$

有

$$\psi\left(\frac{n}{\lambda}, \cdots, \frac{n}{\lambda}\right) \leqslant \psi(x_1, \cdots, x_n)$$

即式(8)成立, 证毕.

推论 1[20]　设 $x_k > 0, k = 1, \cdots, n, n \geqslant 2$ 且 $\sum_{k=1}^{n} \frac{1}{1 + x_k} = \lambda \leqslant 1$, 则

$$\prod_{k=1}^{n} x_k \geqslant \left(\frac{n}{\lambda} - 1\right)^n \tag{9}$$

证明　作置换 $x_k \to x_k - 1, k = 1, \cdots, n$, 则推论1化为定理2的式(8).

推论 2　$x_k > 1, k = 1, \cdots, n, n \geqslant 2$ 且 $\sum_{k=1}^{n} \frac{x_k^2}{1 + x_k^2} = 1$, 则

$$\prod_{k=1}^{n} x_k \leqslant (n - 1)^{-\frac{n}{2}} \tag{10}$$

证明　作置换 $x_k \to \frac{1}{\sqrt{x_k - 1}}, k = 1, \cdots, n$, 则推论2 化为定理2式(8)中 $\lambda = 1$ 的情形.

推论 3　设 $x_k > 1, k = 1, \cdots, n, n \geqslant 2$ 且 $\sum\limits_{k=1}^{n} \dfrac{1}{1 + x_k^n} = 1$, 则

$$\prod_{k=1}^{n} x_k \geqslant n - 1 \tag{11}$$

证明　作置换 $x_k \to \sqrt[n]{x_k - 1}, k = 1, \cdots, n$, 则推论3化为定理2式(8)中 $\lambda = 1$ 的情形.

定理 3　设 $x_k > 0$, $k = 1, \cdots, n, n \geqslant 2$ 且 $\sum\limits_{k=1}^{n} \dfrac{1}{x_k} = \lambda \leqslant 1$, 则

$$\sum_{k=1}^{n} \frac{x_k}{1 - x_k} \geqslant \frac{nG}{1 - G} \tag{12}$$

$$\sum_{k=1}^{n} \frac{1}{x_k - 1} \geqslant \frac{n\lambda}{n - \lambda} \tag{13}$$

其中 $G = \sqrt[n]{\prod\limits_{i=1}^{n} x_i}$.

证明　令 $\xi(x_1, \cdots, x_n) = \sum\limits_{k=1}^{n} \dfrac{1}{x_k - 1}$, 显然 ξ 在 U 上对称, 且有

$$\frac{\partial \xi}{\partial x_1} = -\frac{1}{(x_1 - 1)^2}, \ \frac{\partial \xi}{\partial x_2} = -\frac{1}{(x_2 - 1)^2}$$

于是

$$\begin{aligned}
\Delta_2 &= (x_1 - x_2)\left(x_1 \frac{\partial \xi}{\partial x_1} - x_2 \frac{\partial \xi}{\partial x_2}\right) \\
&= (x_1 - x_2)\left[\frac{x_2}{(x_2 - 1)^2} - \frac{x_1}{(x_1 - 1)^2}\right] \\
&= (x_1 - x_2)\left[\frac{x_2(x_1 - 1)^2 - x_1(x_2 - 1)^2}{(x_1 - 1)^2(x_2 - 1)^2}\right] \\
&= (x_1 - x_2)^2 \frac{x_1 x_2 - 1}{(x_1 - 1)^2(x_2 - 1)^2}
\end{aligned}$$

由定理3 的条件可断定 $x_k > 1, k = 1, \cdots, n$, 于是 $x_1 x_2 \geqslant 1$, 故 $\Delta_2 \geqslant 0$, 据引理3, $\xi(x_1, \cdots, x_n)$ 在 U 上 Schur 几何凸. 由 $(\ln G^{-1}, \cdots, \ln G^{-1}) \prec (\ln x_1^{-1}, \cdots, \ln x_n^{-1})$, 有

$$\xi\left(G^{-1}, \cdots, G^{-1}\right) \leqslant \xi\left(x_1^{-1}, \cdots, x_n^{-1}\right)$$

即式(12)成立.

$$\Delta_3 = (x_1 - x_2)\left(x_1^2 \frac{\partial \xi}{\partial x_1} - x_2^2 \frac{\partial \xi}{\partial x_2}\right)$$

$$= (x_1 - x_2)\left[\frac{x_2^2}{(x_2-1)^2} - \frac{x_1^2}{(x_1-1)^2}\right]$$

$$= (x_1 - x_2)\left[\frac{x_2^2(x_1-1)^2 - x_1^2(x_2-1)^2}{(x_1-1)^2(x_2-1)^2}\right]$$

$$= (x_1 - x_2)^2 \frac{(x_1-1)x_2 + x_1(x_2-1)}{(x_1-1)^2(x_2-1)^2}$$

因 $x_k > 1, k = 1, \cdots, n$, 故 $\Delta_3 \geqslant 0$. 据引理4, $\xi(x_1, \cdots, x_n)$ 在 U 上 Schur 调和凸.

由 $\left(\frac{\lambda}{n}, \cdots, \frac{\lambda}{n}\right) \prec \left(\frac{1}{x_1}, \cdots, \frac{1}{x_n}\right)$ 有 $\xi\left(\frac{n}{\lambda}, \cdots, \frac{n}{\lambda}\right) \leqslant \xi(x_1, \cdots, x_n)$, 即式(13)成立, 证毕.

推论 4　设 $x_k > 0, k = 1, \cdots, n, n \geqslant 2$ 且 $\sum_{k=1}^{n} \frac{x_k}{1+x_k} = \lambda \leqslant 1$, 则

$$\sum_{k=1}^{n} x_k \geqslant \frac{n\lambda}{n-\lambda} \tag{14}$$

证明　作置换 $x_k \to \frac{1}{x_k - 1}, k = 1, \cdots, n$, 则推论4化为定理3中式(13)的情形.

利用上文所建立的代数不等式给出几个单形不等式. 设 A 表示 $\mathbf{R}^n(n \geqslant 3)$ 中以 $A_1, A_2, \cdots, A_{n+1}$ 为顶点的 n 维单形, 顶点 A_k 所对 $n-1$ 维界面上的高为 h_k, 该界面外的旁切球半径为 $r_k, k = 1, 2, \cdots, n+1$, r 为单形的内切球半径, 则成立如下等式[19,21]

$$\sum_{k=1}^{n+1} \frac{1}{h_k} = \frac{1}{r} \tag{15}$$

$$\sum_{k=1}^{n+1} \frac{1}{r_k} = \frac{n-1}{r} \tag{16}$$

定理 4　设 $A = A_1 A_2 \cdots A_{n+1}$ 是 $\mathbf{R}^n(n \geqslant 3)$ 中的一个单形, 则 $\forall m \in \mathbf{N}$ 成立

$$(n+1)\left[(n+1)^{m-1} - 1\right]\frac{1}{G^m} \leqslant \left(\sum_{k=1}^{n+1} \frac{1}{h_k}\right)^m - \left(\sum_{k=1}^{n+1} \frac{1}{h_k^m}\right)$$

$$\leqslant \left[1 - \frac{1}{(n+1)^{m-1}}\right]\frac{1}{r^m} \tag{17}$$

$$\left(\sum_{k=1}^{n+1} h_k\right)^m - \left(\sum_{k=1}^{n+1} h_k^m\right) \geqslant r^m \left[(n+1)^{2m} - (n+1)^{m+1}\right] \tag{18}$$

其中 $G = \sqrt[n+1]{\prod_{i=1}^{n+1} h_i}$.

证明 据式(15)条件, 由定理1 得证.

定理 5 设 $A = A_1 A_2 \cdots A_{n+1}$ 是 $\mathbf{R}^n (n \geqslant 3)$ 中的一个单形, 则 $\forall m \in \mathbf{N}$ 成立

$$(n+1)\left[(n+1)^{m-1} - 1\right]\frac{1}{G^m} \leqslant \left(\sum_{k=1}^{n+1} \frac{1}{r_k}\right)^m - \left(\sum_{k=1}^{n+1} \frac{1}{r_k^m}\right)$$

$$\leqslant \left[1 - \frac{1}{(n+1)^{m-1}}\right]\frac{(n-1)^m}{r^m} \tag{19}$$

$$\left(\sum_{k=1}^{n+1} r_k\right)^m - \left(\sum_{k=1}^{n+1} r_k^m\right) \geqslant \frac{r^m\left[(n+1)^{2m} - (n+1)^{m+1}\right]}{(n-1)^m} \tag{20}$$

其中 $G = \sqrt[n+1]{\prod_{i=1}^{n+1} r_i}$.

证明 据式(6)条件, 由定理1 得证.

定理 6 设 $A = A_1 A_2 \cdots A_{n+1}$ 是 $\mathbf{R}^n (n \geqslant 3)$ 中的一个单形. 若 $r \geqslant n-1$, 则

$$\prod_{k=1}^{n+1} (h_k - 1) \geqslant \left[r(n+1) - 1\right]^{n+1} \tag{21}$$

$$\prod_{k=1}^{n} (r_k - 1) \geqslant \left[\frac{r(n+1)}{n-1} - 1\right]^{n+1} \tag{22}$$

证明 由定理2 的式(8)可得证.

定理 7 设 $A = A_1 A_2 \cdots A_{n+1}$ 是 $\mathbf{R}^n (n \geqslant 3)$ 中的一个单形. 若 $r \geqslant n-1$, 则

$$\sum_{k=1}^{n+1} \frac{1}{h_k - 1} \geqslant \frac{r(n+1)}{r(n+1) - 1} \tag{23}$$

$$\sum_{k=1}^{n+1} \frac{1}{r_k - 1} \geqslant \frac{(n+1)(n-1)}{r(n+1) - (n-1)} \tag{24}$$

证明 由定理3 的式(13)可得证.

致谢 作者感谢张晗方教授给予本文的热情帮助.

参考文献

[1]　MARSHALL A M, OLKIN I, ARNOLD B C. Inequalities: Theory of Majorization and Its Application [M]. 2nd ed. New York: Springer Press, 2011.

[2]　王伯英. 控制不等式基础 [M]. 北京: 北京师范大学出版社, 1990.

[3]　张小明. 几何凸函数 [M]. 合肥: 安徽大学出版社, 2004.

[4]　CHU YUMING, Lü YUPEI. The Schur harmonic convexity of the Hamy symmetric function and its applications [J]. Journal of Inequalities and Applications, 2009, Volume 2009, Article ID 838529, 10 pages.

[5]　石焕南. 受控理论与解析不等式 [M]. 哈尔滨: 哈尔滨工业大学出版社, 2012.

[6]　CHU YUMING, SUN TIANCHUAN. The Schur harmonic convexity for a class of symmetric functions [J]. Acta Mathematica Scientia, 2010, 30B(5): 1501-1506.

[7]　CHU Y M, WANG G D, ZHANG X H. The Schur multiplicative and harmonic convexities of the complete symmetric function [J]. Mathematische Nachrichten, 2011, 284(5, 6): 653-663.

[8]　GUAN KAIZHONG, GUAN RUKE. Some properties of a generalized Hamy symmetric function and its applications [J]. Journal of Mathematical Analysis and Applications, 2011, 376(2): 494-505.

[9]　SHI HUANNAN. Two Schur-convex functions related to Hadamard-type integral inequalities [J]. Publicationes Mathematicae Debrecen, 2011, 78(2): 393-403.

[10]　ČULJAK V. Schur-convexity of the weighted Čebišev functional [J]. Journal of Mathematical Inequalities, 2011, 5(2): 213-217.

[11]　XIA WEIFENG, CHU YUMING. The Schur convexity of Gini mean values in the sense of harmonic mean [J]. Acta Mathematica Scientia, 2011, 31B(3): 1103-1112.

[12]　YANG ZHENHANG. Schur harmonic convexity of Gini means [J]. International Mathematical Forum, 2011, 6(16): 747-762.

[13]　CHU YUMING, XIA WEIFENG. Necessary and sufficient conditions for the Schur harmonic convexity of the Generalized Muirhead Mean [J]. Proceedings of A. Razmadze Mathematical Institute, 2010, 152: 19-27.

[14] WU YING, QI FENG. Schur-harmonic convexity for differences of some means [J]. Analysis, 2012, 32(4): 263-270.

[15] CHU YUMING, XIA WEIFENG, ZHANG XIAOHUI. The Schur concavity, Schur multiplicative and harmonic convexities of the second dual form of the Hamy symmetric function with applications [J]. Journal of Multivariate Analysis, 2012, 105(1): 412-421.

[16] XIA WEIFENG, CHU YUMING, WANG GENDI. Necessary and sufficient conditions for the Schur harmonic convexity or concavity of the extended mean values [J]. Revista De La Uniòn Matemática Argentina, 2010, 51(2): 121-132.

[17] 夏卫锋, 褚玉明. 一类对称函数的Schur凸性与应用 [J]. 数学进展, 2012, 41(4): 436-446.

[18] 邵志华. 一类对称函数的Schur-几何凸性Schur-调和凸性 [J]. 数学的实践与认识, 2012, 42(16): 199-206.

[19] 匡继昌. 常用不等式 [M]. 4版. 济南: 山东科学技术出版社, 2010.

[20] 杨学枝. 数学奥林匹克不等式研究 [M]. 哈尔滨: 哈尔滨工业大学出版社, 2009.

[21] MITRRINOVIĆ D S, PEČRIĆ J E, VOLENEC V. Recent Advances in Geometric Inequalities [M]. Dordrecht: Kluwer Academic Publishers, 1989.

[22] 石焕南. 一个有理分式不等式的加细 [J]. 纯粹数学与应用数学, 2006, 22(2): 256-262.

[23] 张晗方. 几何不等式导引 [M]. 北京: 中国科学文化出版社, 2003.

第59篇　Judgement Theorems of Schur Geometric and Harmonic Convexities

(SHI HUAN-NAN, ZHANG JING. J. Inequal. Appl., 2013, 2013:527

doi:10.1186/ 1029-242X-2013-527)

Abstract: The judgement theorems of Schur geometric and Schur harmonic convexities for a class of symmetric functions are given. As its application, some analytic inequalities are established.

Keywords: Schur geometric convexity; Schur harmonic convexity; inequality; symmetric function

59.1　Introduction

Throughout this paper, $\mathbf{R}$ denotes the set of real numbers, $\boldsymbol{x} = (x_1, x_2, \cdots, x_n)$ denotes n-tuple (n-dimensional real vectors), the set of vectors can be written as

$$\mathbf{R}^n = \{\boldsymbol{x} = (x_1, \cdots, x_n) : x_i \in \mathbf{R}, i = 1, \cdots, n\}$$

$$\mathbf{R}^n_+ = \{\boldsymbol{x} = (x_1, \cdots, x_n) : x_i > 0, i = 1, \cdots, n\}$$

In particular, the notations $\mathbf{R}$ and $\mathbf{R}_+$ denote $\mathbf{R}^1$ and $\mathbf{R}^1_+$, respectively.

Let $\pi = (\pi(1), \cdots, \pi(n))$ be a permutation of $(1, \cdots, n)$, all permutations is totally $n!$. The following conclusion is proved in [1].

Theorem A Let $A \subset \mathbf{R}^k$ be a symmetric convex set and let φ be a Schur-convex function defined on A with the property that for each fixed $x_2, \cdots, x_k$, $\varphi(z, x_2, \cdots, x_k)$ is convex in z on $\{z : (z, x_2, \cdots, x_k) \in A\}$. Then

439

for any $n > k$

$$\psi(x_1, \cdots, x_n) = \sum_\pi \varphi\left(x_{\pi(1)}, \cdots, x_{\pi(k)}\right) \tag{1}$$

is Schur-convex on

$$B = \left\{(x_1, \cdots, x_n) : \left(x_{\pi(1)}, \cdots, x_{\pi(k)}\right) \in A, \text{for all permutations } \pi\right\}.$$

Furthermore, the symmetric function

$$\overline{\psi}(\boldsymbol{x}) = \sum_{1 \leqslant i_1 < \cdots < i_k \leqslant n} \varphi(x_{i_1}, \cdots, x_{i_k}) \tag{2}$$

is also Schur-convex on B.

Theorem A is very effective for judgement of the Schur-convexity of the symmetric functions of the form (2), see the references [1] and [9].

The Schur geometrically convex functions were proposed by Zhang[2] in 2004. Further, the Schur harmonically convex functions were proposed by Chu and Lü [3] in 2009. The theory of majorization was enriched and expanded by using these concepts [10−20]. Regarding Schur geometrically convex functions and Schur harmonically convex functions, the aim of this paper is to establish the following judgement theorems which are similar to Theorem A.

Theorem 1 Let $A \subset \mathbf{R}^k$ be a symmetric geometrically convex set and let φ be a Schur geometrically convex (concave) function defined on A with the property that for each fixed $x_2, \cdots, x_k$, $\varphi(z, x_2, \cdots, x_k)$ is GA convex (concave) in z on $\{z : (z, x_2, \cdots, x_k) \in A\}$. Then for any $n > k$

$$\psi(x_1, \cdots, x_n) = \sum_\pi \varphi\left(x_{\pi(1)}, \cdots, x_{\pi(k)}\right)$$

is Schur geometrically convex (concave) on

$$B = \left\{(x_1, \cdots, x_n) : \left(x_{\pi(1)}, \cdots, x_{\pi(k)}\right) \in A, \text{for all permutations } \pi\right\}$$

Furthermore, the symmetric function

$$\overline{\psi}(\boldsymbol{x}) = \sum_{1 \leqslant i_1 < \cdots < i_k \leqslant n} \varphi(x_{i_1}, \cdots, x_{i_k})$$

is also Schur geometrically convex (concave) on B.

Theorem 2 Let $A \subset \mathbf{R}^k$ be a symmetric harmonically convex set and let φ be a Schur harmonically convex (concave) function defined on A with the property that for each fixed $x_2, \cdots, x_k$, $\varphi(z, x_2, \cdots, x_k)$ is HA convex (concave) in z on $\{z : (z, x_2, \cdots, x_k) \in A\}$. Then for any $n > k$

$$\psi(x_1, \cdots, x_n) = \sum_\pi \varphi\left(x_{\pi(1)}, \cdots, x_{\pi(k)}\right)$$

is Schur harmonically convex (concave) on

$$B = \left\{(x_1, \cdots, x_n) : \left(x_{\pi(1)}, \cdots, x_{\pi(k)}\right) \in A, \text{ for all permutations } \pi\right\}$$

Furthermore, the symmetric function

$$\overline{\psi}(\boldsymbol{x}) = \sum_{1 \leqslant i_1 < \cdots < i_k \leqslant n} \varphi\left(x_{i_1}, \cdots, x_{i_k}\right)$$

is also Schur harmonically convex (concave) on B.

59.2　Definitions and lemmas

In order to prove some further results, in this section we will recall usefull definitions and lemmas.

Definition 1 [1,4] Let $\boldsymbol{x} = (x_1, \cdots, x_n)$ and $\boldsymbol{y} = (y_1, \cdots, y_n) \in \mathbf{R}^n$.

(i) We say $\boldsymbol{y}$ majorizes $\boldsymbol{x}$ ($\boldsymbol{x}$ is said to be majorized by $\boldsymbol{y}$), denoted by $\boldsymbol{x} \prec \boldsymbol{y}$, if $\sum_{i=1}^k x_{[i]} \leqslant \sum_{i=1}^k y_{[i]}$ for $k = 1, 2, \cdots, n-1$ and $\sum_{i=1}^n x_i = \sum_{i=1}^n y_i$, where $x_{[1]} \geqslant \cdots \geqslant x_{[n]}$ and $y_{[1]} \geqslant \cdots \geqslant y_{[n]}$ are rearrangements of $\boldsymbol{x}$ and $\boldsymbol{y}$ in a descending order.

(ii) Let $\Omega \subset \mathbf{R}^n$, a function $\varphi\colon \Omega \to \mathbf{R}$ is said to be a Schur-convex function on Ω if $\boldsymbol{x} \prec \boldsymbol{y}$ on Ω implies $\varphi(\boldsymbol{x}) \leqslant \varphi(\boldsymbol{y})$. A function φ is said to be a Schur-concave function on Ω if and only if $-\varphi$ is Schur-convex function on Ω.

Definition 2 [1,4] Let $\boldsymbol{x} = (x_1, \cdots, x_n)$ and $\boldsymbol{y} = (y_1, \cdots, y_n) \in \mathbf{R}^n, 0 \leqslant \alpha \leqslant 1$. A set $\Omega \subset \mathbf{R}^n$ is said to be a convex set if $\boldsymbol{x}, \boldsymbol{y} \in \Omega$ implies $\alpha \boldsymbol{x} + (1-\alpha)\boldsymbol{y} = (\alpha x_1 + (1-\alpha)y_1, \cdots, \alpha x_n + (1-\alpha)y_n) \in \Omega$.

Definition 1[1,4]

(i) A set $\Omega \subset \mathbf{R}^n$ is called a symmetric set, if $\boldsymbol{x} \in \Omega$ implies $\boldsymbol{x}P \in \Omega$ for every $n \times n$ permutation matrix $\boldsymbol{P}$.

(ii) A function $\varphi : \Omega \to \mathbf{R}$ is called symmetric if for every permutation matrix P, $\varphi(\boldsymbol{x}P) = \varphi(\boldsymbol{x})$ for all $\boldsymbol{x} \in \Omega$.

Definition 4 Let $\Omega \subset \mathbf{R}_+^n$, $\boldsymbol{x} = (x_1, \cdots, x_n) \in \Omega$ and $\boldsymbol{y} = (y_1, \cdots, y_n) \in \Omega$.

(i) [2] A set Ω is called a geometrically convex set if $(x_1^\alpha y_1^\beta, \cdots, x_n^\alpha y_n^\beta) \in \Omega$ for all $\boldsymbol{x}, \boldsymbol{y} \in \Omega$ and $\alpha, \beta \in [0, 1]$ such that $\alpha + \beta = 1$.

(ii) [2] A function $\varphi \colon \Omega \to \mathbf{R}_+$ is said to be a Schur geometrically convex function on Ω, if $(\log x_1, \cdots, \log x_n) \prec (\log y_1, \cdots, \log y_n)$ on Ω implies $\varphi(\boldsymbol{x}) \leqslant \varphi(\boldsymbol{y})$. A function φ is said to be a Schur geometrically concave function on Ω if and only if $-\varphi$ is a Schur geometrically convex function.

Definition 5 [5] Let $\Omega \subset \mathbf{R}_+^n$.

(i) A set Ω is said to be a harmonically convex set if $\frac{\boldsymbol{x}\boldsymbol{y}}{\lambda \boldsymbol{x} + (1-\lambda)\boldsymbol{y}} \in \Omega$ for every $\boldsymbol{x}, \boldsymbol{y} \in \Omega$ and $\lambda \in [0, 1]$, where $\boldsymbol{x}\boldsymbol{y} = \sum\limits_{i=1}^{n} x_i y_i$ and $\frac{1}{\boldsymbol{x}} = \left(\frac{1}{x_1}, \cdots, \frac{1}{x_n}\right)$.

(ii) A function $\varphi : \Omega \to \mathbf{R}_+$ is said to be a Schur harmonically convex function on Ω if $\frac{1}{\boldsymbol{x}} \prec \frac{1}{\boldsymbol{y}}$ implies $\varphi(\boldsymbol{x}) \leqslant \varphi(\boldsymbol{y})$. A function φ is said to be a Schur harmonically concave function on Ω if and only if $-\varphi$ is a Schur harmonically convex function.

Definition 6[6] Let $I \subset \mathbf{R}_+$, $\varphi : I \to \mathbf{R}_+$ be continuous.

(i) A function φ is said to be a GA convex (concave) function on I if

$$\varphi(\sqrt{xy}) \leqslant (\geqslant) \frac{\varphi(x) + \varphi(y)}{2}$$

for all $x, y \in I$.

(ii) A function φ is said to be a HA convex (concave) function on I if

$$\varphi\left(\frac{2xy}{x+y}\right) \leqslant (\geqslant)\frac{\varphi(x)+\varphi(y)}{2}$$

for all $x, y \in I$.

Lemma 1[4] Let $\Omega \subset \mathbf{R}^n$ be a symmetric convex set with a nonempty interior Ω^0. $\varphi : \Omega \to \mathbf{R}$ is continuous on Ω and differentiable on Ω^0. Then φ is a Schur-convex (Schur-concave) function if and only if φ is symmetric on Ω and

$$(x_1 - x_2)\left(\frac{\partial\varphi}{\partial x_1} - \frac{\partial\varphi}{\partial x_2}\right) \geqslant 0(\leqslant 0) \tag{3}$$

holds for any $x = (x_1, \cdots, x_n) \in \Omega^0$.

Lemma 2[2] Let $\Omega \subset \mathbf{R}_+^n$ be a symmetric geometrically convex set with a nonempty interior Ω^0. Let $\varphi : \Omega \to \mathbf{R}_+$ be continuous on Ω and differentiable on Ω^0. Then φ is a Schur geometrically convex (Schur geometrically concave) function if and only if φ is symmetric on Ω and

$$(x_1 - x_2)\left(x_1\frac{\partial\varphi}{\partial x_1} - x_2\frac{\partial\varphi}{\partial x_2}\right) \geqslant 0 \quad (\leqslant 0) \tag{4}$$

holds for any $x = (x_1, \cdots, x_n) \in \Omega^0$.

Lemma 3[5,7] Let $\Omega \subset \mathbf{R}_+^n$ be a symmetric harmonically convex set with a nonempty interior Ω^0. Let $\varphi : \Omega \to \mathbf{R}_+$ be continuous on Ω and differentiable on Ω^0. Then φ is a Schur harmonically convex (Schur harmonically concave) function if and only if φ is symmetric on Ω and

$$(x_1 - x_2)\left(x_1^2\frac{\partial\varphi}{\partial x_1} - x_2^2\frac{\partial\varphi}{\partial x_2}\right) \geqslant 0 \quad (\leqslant 0) \tag{5}$$

holds for any $x = (x_1, \cdots, x_n) \in \Omega^0$.

Lemma 4[6] Let $I \subset \mathbf{R}_+$ be an open subinterval and let $\varphi : I \to \mathbf{R}_+$ be differentiable.

(i) φ is GA-convex (concave) if and only if $x\varphi'(x)$ is increasing (decreasing)

(ii) φ is HA-convex (concave) if and only if $x^2\varphi'(x)$ is increasing (decreasing)

59.3 Proofs of main results

Proof of Theorem 1 To verify condition (4) of Lemma 2, denote by $\sum\limits_{\pi(i,j)}$ the summation over all permutations π such that $\pi(i) = 1, \pi(j) = 2$. Because φ is symmetric

$$
\begin{aligned}
&\psi(x_1, \cdots, x_n) \\
&= \sum_{\substack{i,j \leqslant k \\ i \neq j}} \sum_{\pi(i,j)} \varphi\left(x_1, x_2, x_{\pi(1)}, \cdots, x_{\pi(i-1)}, x_{\pi(i+1)}, \cdots, x_{\pi(j-1)}, x_{\pi(j+1)}, \cdots, x_{\pi(k)}\right) + \\
&\quad \sum_{i \leqslant k < j} \sum_{\pi(i,j)} \varphi\left(x_1, x_{\pi(1)}, \cdots, x_{\pi(i-1)}, x_{\pi(i+1)}, \cdots, x_{\pi(k)}\right) + \\
&\quad \sum_{j \leqslant k < i} \sum_{\pi(i,j)} \varphi\left(x_2, x_{\pi(1)}, \cdots, x_{\pi(j-1)}, x_{\pi(j+1)}, \cdots, x_{\pi(k)}\right) + \\
&\quad \sum_{\substack{k < i,j \\ i \neq j}} \sum_{\pi(i,j)} \varphi\left(x_{\pi(1)}, \cdots, x_{\pi(k)}\right)
\end{aligned}
$$

Then

$$
\begin{aligned}
\Delta_1 :=& \left(x_1 \frac{\partial \psi}{\partial x_1} - x_2 \frac{\partial \psi}{\partial x_2}\right)(x_1 - x_2) \\
=& \sum_{\substack{i,j \leqslant k \\ i \neq j}} \sum_{\pi(i,j)} \left[x_1 \varphi_{(1)}\left(x_1, x_2, x_{\pi(1)}, \cdots, x_{\pi(i-1)}, x_{\pi(i+1)}, \cdots, x_{\pi(j-1)}, x_{\pi(j+1)}, \cdots, x_{\pi(k)}\right) - \right. \\
& \left. x_2 \varphi_{(2)}\left(x_1, x_2, x_{\pi(1)}, \cdots, x_{\pi(i-1)}, x_{\pi(i+1)}, \cdots, x_{\pi(j-1)}, x_{\pi(j+1)}, \cdots, x_{\pi(k)}\right)\right](x_1 - x_2) + \\
& \sum_{i \leqslant k < j} \sum_{\pi(i,j)} \left[x_1 \varphi_{(1)}\left(x_1, x_{\pi(1)}, \cdots, x_{\pi(i-1)}, x_{\pi(i+1)}, \cdots, x_{\pi(k)}\right) - \right. \\
& \left. x_2 \varphi_{(1)}\left(x_2, x_{\pi(1)}, \cdots, x_{\pi(i-1)}, x_{\pi(i+1)}, \cdots, x_{\pi(k)}\right)\right](x_1 - x_2)
\end{aligned}
$$

Here

$$
\left(x_1 \varphi_{(1)} - x_2 \varphi_{(2)}\right)(x_1 - x_2) \geqslant 0 \ (\leqslant 0)
$$

because φ is Schur geometrically convex (concave), and

$$
\left[x_1 \varphi_{(1)}(x_1, z) - x_2 \varphi_{(1)}(x_2, z)\right](x_1 - x_2) \geqslant 0 \ (\leqslant 0)
$$

because $\varphi(z, x_2, \cdots, x_k)$ is GA convex (concave) in its first argument on $\{z : (z, x_2, \cdots, x_k) \in A\}$. Accordingly, $\Delta_1 \geqslant 0 \ (\leqslant 0)$. This shows that ψ is

Schur geometrically convex (concave) on

$$B = \left\{ (x_1, \cdots, x_n) : \left(x_{\pi(1)}, \cdots, x_{\pi(k)} \right) \in A, \text{for all permutations } \pi \right\}$$

Notice that

$$\overline{\psi}(\boldsymbol{x}) = \psi(\boldsymbol{x}) / k! (n-k)!$$

Of course, $\overline{\psi}$ is Schur geometrically convex (concave) whenever ψ is Schur geometrically convex (concave).

The proof of Theorem 1 is completed.

Proof of Theorem 2 We only need to verify condition (5) of Lemma 3, the proof is similar with Theorem 1 and is omitted.

Remark 1 In most applications, A has the form I^k for some interval $I \subset \mathbf{R}$ and in this case $B = I^n$. Notice that the convexity of φ in its first argument also implies that φ is convex in each argument, the other arguments being fixed, because φ is symmetric.

59.4　Applications

Let

$$E_k \left(\frac{\boldsymbol{x}}{1-\boldsymbol{x}} \right) = \sum_{1 \leq i_1 < \cdots < i_k \leq n} \prod_{j=1}^{k} \frac{x_{i_j}}{1 - x_{i_j}} \tag{6}$$

In 2011, Guan and Guan[8] proved the following Theorem 3 through Lemma 2.

Theorem 3 The symmetric function $E_k \left(\frac{\boldsymbol{x}}{1-\boldsymbol{x}} \right)$, $k = 1, \cdots, n$, is Schur geometrically convex on $(0, 1)^n$.

Now, we give a new proof of Theorem 3 by using the Theorem 1. Furthermore, we prove the following Theorem 4 through Theorem 2.

Theorem 4 The symmetric function $E_k \left(\frac{\boldsymbol{x}}{1-\boldsymbol{x}} \right)$, $k = 1, \cdots, n$, is Schur harmonically convex on $(0, 1)^n$.

Proof of Theorem 3 Let $\varphi(\boldsymbol{z}) = \prod_{i=1}^{k} [z_i / (1 - z_i)]$. Then

$$\log \varphi(\boldsymbol{z}) = \sum_{i=1}^{k} [\log z_i - \log (1 - z_i)]$$

and

$$\frac{\partial \varphi(z)}{\partial z_1} = \varphi(z) \left(\frac{1}{z_1} + \frac{1}{1 - z_1} \right), \quad \frac{\partial \varphi(z)}{\partial z_2} = \varphi(z) \left(\frac{1}{z_2} + \frac{1}{1 - z_2} \right) \tag{7}$$

$$\begin{aligned}
\Delta :&= (z_1 - z_2) \left(z_1 \frac{\partial \varphi(z)}{\partial z_1} - z_2 \frac{\partial \varphi(z)}{\partial z_2} \right) \\
&= (z_1 - z_2) \varphi(z) \left(\frac{z_1}{1 - z_1} - \frac{z_2}{1 - z_2} \right) \\
&= (z_1 - z_2)^2 \varphi(z) \frac{1}{(1 - z_2)(1 - z_1)}
\end{aligned}$$

This shows that $\Delta \geqslant 0$ when $0 < z_i < 1$, $i = 1, \cdots, k$. According to Lemma 2, φ is Schur geometrically convex on $A = \{z : z \in (0,1)^k\}$. Let $g(t) = \frac{t}{1-t}$, then $h(t) := tg'(t) = \frac{t}{(1-t)^2}$. From $t \in (0,1)$, it follows that $h'(t) = \frac{1+t}{(1-t)^3} \geqslant 0$. According to Lemma 4 (i), φ is GA convex in its single variable on $(0,1)$. So $E_k \left(\frac{x}{1-x} \right)$ is Schur geometrically convex on $(0,1)^n$ from Theorem 1. The proof of Theorem 3 is completed.

Proof of Theorem 4 Let $\varphi(z) = \prod_{i=1}^{k} (z_i / 1 - z_i)$, then

$$\log \varphi(z) = \sum_{i=1}^{k} [\log z_i - \log(1 - z_i)]$$

From (7), we get

$$\begin{aligned}
\Delta_1 :&= (z_1 - z_2) \left(z_1^2 \frac{\partial \varphi(z)}{\partial z_1} - z_2^2 \frac{\partial \varphi(z)}{\partial z_2} \right) \\
&= (z_1 - z_2) \varphi(z) \left(z_1 - z_2 + \frac{z_1^2}{1 - z_1} - \frac{z_2^2}{1 - z_2} \right) \\
&= (z_1 - z_2)^2 \varphi(z) \left[1 + \frac{z_1 + z_2 - z_1 z_2}{(1 - z_2)(1 - z_1)} \right]
\end{aligned}$$

This shows that $\Delta_1 \geqslant 0$ when $0 < z_i < 1$, $i = 1, \cdots, k$. According to Lemma 3, φ is Schur harmonically convex on $A = \{z : z \in (0,1)^k\}$. Let $g(t) = \frac{t}{1-t}$, then $p(t) := t^2 g'(t) = \frac{t^2}{(1-t)^2}$. From $t \in (0,1)$, it follows that $p'(t) = \frac{2t}{(1-t)^3} \geqslant 0$. According to Lemma 4 (ii), φ is HA convex in its single variable on $(0,1)$. So $E_k \left(\frac{x}{1-x} \right)$ is Schur harmonically convex on $(0,1)^n$ from Theorem 2. The proof of Theorem 4 is completed.

By using the Theorem A, the following conclusion is proved in [1].

The symmetric function

$$\overline{\psi}\left(\boldsymbol{x}\right) = \sum_{1 \leqslant i_1 < \cdots < i_k \leqslant n} \frac{x_{i_1} + \cdots + x_{i_k}}{x_{i_1} \cdots x_{i_k}} \tag{8}$$

is Schur-convex on $\mathbf{R}_+^n$.

Now we use Theorem 1 and Theorem 2 respectively to study Schur geometric convexity and Schur harmonic convexity of $\overline{\psi}\left(\boldsymbol{x}\right)$.

Theorem 5 The symmetric function $\overline{\psi}\left(\boldsymbol{x}\right)$ is Schur geometrically convex and Schur harmonically concave on $\mathbf{R}_+^n$.

Proof Let $\varphi\left(\boldsymbol{y}\right) = \sum_{i=1}^{k} y_i \left/ \prod_{i=1}^{k} y_i \right.$, then $\log\varphi(\boldsymbol{y}) = \log(\sum_{i=1}^{k} y_i) - \sum_{i=1}^{k} \log y_i$.
Thus

$$\frac{\partial\varphi(\boldsymbol{y})}{\partial y_1} = \varphi(\boldsymbol{y}) \left(\frac{1}{\sum_{i=1}^{k} y_i} - \frac{1}{y_1} \right), \quad \frac{\partial\varphi(\boldsymbol{y})}{\partial y_2} = \varphi(\boldsymbol{y}) \left(\frac{1}{\sum_{i=1}^{k} y_i} - \frac{1}{y_2} \right)$$

$$\Delta := (y_1 - y_2) \left(y_1 \frac{\partial\varphi\left(\boldsymbol{y}\right)}{\partial y_1} - y_2 \frac{\partial\varphi\left(\boldsymbol{y}\right)}{\partial y_2} \right)$$

$$= (y_1 - y_2) \varphi\left(\boldsymbol{y}\right) \left(\frac{y_1 - y_2}{\sum_{i=1}^{k} y_i} \right)$$

$$= \frac{(y_1 - y_2)^2}{\prod_{i=1}^{k} y_i} \geqslant 0$$

According to Lemma 2, $\varphi\left(\boldsymbol{y}\right)$ is Schur geometrically convex on $\mathbf{R}_+^k$. Let $g(z) = \varphi\left(z, x_2, \cdots, x_k\right) = \frac{z+a}{bz} = \frac{1}{b} + \frac{a}{bz}$, where $a = \sum_{i=2}^{k} x_i, b = \prod_{i=2}^{k} x_i$, then $h(z) := zg'\left(z\right) = -\frac{a}{bz}$. From $z \in \mathbf{R}_+$, it follows that $h'(z) = \frac{a}{bz^2} \geqslant 0$. According to Lemma 4 (i), φ is GA convex in its single variable on $\mathbf{R}_+$. So $\overline{\psi}\left(\boldsymbol{x}\right)$ is Schur geometrically convex on $\mathbf{R}_+$ from Theorem 1.

It is easy to check that

$$\Delta_1 := (y_1 - y_2) \left(y_1^2 \frac{\partial\varphi\left(\boldsymbol{y}\right)}{\partial y_1} - y_2^2 \frac{\partial\varphi\left(\boldsymbol{y}\right)}{\partial y_2} \right)$$

$$= \frac{(y_1 - y_2)^2 \left(y_1 + y_2 - \sum\limits_{i=1}^{k} y_i\right)}{\prod\limits_{i=1}^{k} y_i} \leqslant 0$$

According to Lemma 3, $\varphi(y)$ is Schur harmonically concave on $\mathbf{R}_+^k$. Let $h(z) := z^2 g'(z) = -\frac{a}{b}$. $h'(z) = 0$ when $z \in \mathbf{R}_+$. According to Lemma 4 (ii), φ is HA concave in its single variable on $\mathbf{R}_+$. So $\overline{\psi}(x)$ is Schur harmonically concave on $\mathbf{R}_+^n$ from Theorem 2.

Remark 2 Let

$$H = \frac{n}{\sum\limits_{i=1}^{n} \frac{1}{x_i}}, \quad G = \left(\prod\limits_{i=1}^{n} x_i\right)^{\frac{1}{n}}$$

where $x_i > 0$, $i = 1, \cdots, n$. Then

$$(\log G, \cdots, \log G) \prec (\log x_1, \cdots, \log x_n) \tag{9}$$

$$\left(\frac{1}{H}, \cdots, \frac{1}{H}\right) \prec \left(\frac{1}{x_1}, \cdots, \frac{1}{x_n}\right) \tag{10}$$

From Theorem 5, it follows that

$$\frac{k C_n^k}{H^{k-1}} \geqslant \sum_{1 \leqslant i_1 < \cdots < i_k \leqslant n} \frac{x_{i_1} + \cdots + x_{i_k}}{x_{i_1} \cdots x_{i_k}} \geqslant \frac{k C_n^k}{G^{k-1}} \tag{11}$$

By using Theorem A, the following conclusion is proved in [1].

The symmetric function

$$\psi(x) = \sum_{1 \leqslant i_1 < \cdots < i_k \leqslant n} \frac{x_{i_1} \cdots x_{i_k}}{x_{i_1} + \cdots + x_{i_k}}$$

is Schur-concave on $\mathbf{R}_+^n$.

By applying the Theorem 2, we further obtain the following result.

Theorem 6 The symmetric function $\psi(x)$ is Schur harmonically convex on $\mathbf{R}_+^n$.

Proof Let $\lambda(y) = \prod\limits_{i=1}^{k} y_i \Big/ \sum\limits_{i=1}^{k} y_i$. According to the proof of Theorem 5, $\varphi(y)$ is Schur harmonically concave on $\mathbf{R}_+^k$. Let $\lambda(y) = \frac{1}{\varphi(y)}$. From the definition of Schur harmonically convex, it follows that $\lambda(y)$ is Schur harmonically convex on $\mathbf{R}_+^k$. Let $g(z) = \lambda(z, x_2, \cdots, x_k) = \frac{bz}{z+a}$, where

$a = \sum_{i=2}^{k} x_i, b = \prod_{i=2}^{k} x_i$. Then $h(z) := z^2 g'(z) = \frac{z^2 ab}{(z+a)^2}$. With the fact that $h'(z) = \frac{2za^2 b}{(z+a)^3} \geqslant 0$ for $z \in \mathbf{R}_+$, it follows that φ is HA convex in its single variable on $\mathbf{R}_+$. So from Theorem 2, $\psi(\boldsymbol{x})$ is Schur harmonically convex on $\mathbf{R}_+^n$.

Remark 3 From Theorem 6 and (10), it follows that

$$\sum_{1 \leqslant i_1 < \cdots < i_k \leqslant n} \frac{x_{i_1} \cdots x_{i_k}}{x_{i_1} + \cdots + x_{i_k}} \geqslant \frac{H^{k-1} C_n^k}{k} \qquad (12)$$

where $x_i > 0$, $i = 1, \cdots, n$.

Remark 4 It needs further discussion that $\psi(\boldsymbol{x})$ is Schur geometrically convex on $\mathbf{R}_+^n$.

Acknowledgments

The work was supported by the Funding Project for Academic Human Resources Development in Institutions of Higher Learning under the Jurisdiction of Beijing Municipality (PHR (IHLB)) (PHR201108407) and the National Natural Science Foundation of China (Grant No. 11101034).

References

[1]　MARSHALL A M, OLKIN I, ARNOLD B C. Inequalities: Theory of Majorization and Its Application [M]. 2nd ed. New York: Springer Press, 2011.

[2]　X M ZHANG. Geometrically Convex Functions[M]. Hefei: An hui University Press, 2004.

[3]　CHU YUMING, LÜ YUPEI. The Schur harmonic convexity of the Hamy symmetric function and its applications [J]. Journal of Inequalities and Applications, 2009, Volume 2009, Article ID 838529, 10 pages.

[4]　B.-Y.WANG. Kòngzhì Bùděngshì Jīchǔ (Foundations of Majorization Inequalities)[M]. Beijing: Beijing Normal University Press, 1990.

[5]　H-N SHI. Theory of majorization and analytic Inequalities[J]. Harbin: Harbin Institute of Technology Press, 2012.

[6]　G. D. ANDERSON, M K VAMANAMUTHY, M VUORINEN. Generalized convexity and inequalities[J]. Journal of Mathematical Analysis and Applications, 2007, 335 (2): 1294-1308.

[7] Y-M CHU, TIAN-CHUAN SUN. The Schur harmonic convexity for a class of symmetric functions[J]. Acta Mathematica Scientia, 2010, 30B (5): 1501-1506.

[8] K GUAN, R GUAN. Some properties of a generalized Hamy symmetric function and its applications[J]. Journal of Mathematical Analysis and Applications, 2011, 376 (2): 494-505.

[9] H-N SHI. Schur convexity of three symmetric functions[J]. Journal of Hexi University, 2011, 27 (2): 13-17.

[10] WEI-FENG XIA, YU-MING CHU. Schur-convexity for a class of symmetric functions and its applications[J]. Journal of Inequalities and Applications, Volume 2009, Article ID 493759, 15 pages, doi: 10.1155/2009/493759.

[11] IONEL ROVENŢA. Schur convexity of a class of symmetric functions[J]. Annals of the University of Craiova, Mathematics and Computer Science Series, 2010, 37 (1): 12-18.

[12] WEI-FENG XIA, YU-MING CHU. On Schur convexity of some symmetric functions[J]. Journal of Inequalities and Applications, Volume 2010, Article ID 543250, 12 pages, doi: 10.1155/2010/543250.

[13] JUNXIA MENG, YUMING CHU, XIAOMIN TANG. The Schur-harmonic-convexity of dual form of the Hamy symmetric function[J]. Matematički Vesnik, 2010, 62 (1): 37-46.

[14] Y-M CHU, G-D WANG, X-H ZHANG. The Schur multiplicative and harmonic convexities of the complete symmetric function[J]. Mathematische Nachrichten, 2011, 284 (5-6): 653-663.

[15] YU-MING CHU, WEI-FENG XIA, TIE-HONG ZHAO. Some properties for a class of symmetric functions and applications[J]. Journal of Mathematical Inequalities, 2011, 5 (1): 1-11.

[16] WEI-MAO QIAN. Schur convexity for the ratios of the Hamy and generalized Hamy symmetric functions[J]. Journal of Inequalities and Applications, Volume 2011, Article 131, 8 pages, doi: 10.1186/1029-242X-2011-131.

[17] YU-MING CHU, WEI-FENG XIA, XIAO-HUI ZHANG. The Schur concavity, Schur multiplicative and harmonic convexities of the second dual form of the Hamy symmetric function with applications[J]. Journal of Multivariate Analysis, 2012, 105 (1): 412-421.

[18] IONEL ROVENȚA. A note on Schur-concave functions[J]. Journal of Inequalities and Applications, Volume 2012, Article 159, 9 pages, doi: 10.1186/1029-242X-2012-159.

[19] WEI-FENG XIA, XIAO-HUI ZHAN, GEN-DI WANG, YU-MING CHU. Some properties for a class of symmetric functions with applications[J]. Indian Journal of Pure and Applied Mathematics, 2012, 43 (3): 227-249.

[20] HUAN-NAN SHI, JING ZHANG. Schur-convexity of dual form of some symmetric functions[J]. Journal of Inequalities and Applications, Volume 2013, Article 295, 9 pages, doi: 10.1186/1029-242X-2013-295.

第60篇　A Reverse Analytic Inequality for the Elementary Symmetric Function

(SHI HUAN-NAN, ZHANG JING. Journal of Applied Mathematics,

vol. 2013, Article ID 674567, 5 pages, 2013. doi:10.1155/2013/674567)

Abstract: In the article, we give a reverse inequality involving the elementary symmetric function by use of the Schur harmonic convexity theory. As applications, several new analytic inequalities for the n-dimensional simplex are established.

Keywords: majorization; Schur harmonic convexity; analytic inequality; elementary symmetric function; simplex

60.1　Introduction

Let $\boldsymbol{x} = (x_1, x_2, \cdots, x_n) \in \mathbf{R}^n = (-\infty, +\infty)^n$. The elementary symmetric functions are defined by

$$E_k(\boldsymbol{x}) = E_k(x_1, \cdots, x_n) = \sum_{1 \leqslant i_1 < \cdots < i_k \leqslant n} \prod_{j=1}^{k} x_{i_j}, \quad k = 1, \cdots, n$$

$E_0(\boldsymbol{x}) = 1$ and $E_k(\boldsymbol{x}) = 0$ for $k < 0$ or $k > n$.

In 1997, Leng and Tang[1] in order to achieve higher dimensional generalization of the famous Pedoe inequality[2-5], presented the following analytic inequality.

Let $x \in \left(0, \dfrac{1}{2}\right)^n$, and $E_1(x) = 1$. Then

$$E_{n-1}\left(\frac{1}{x} - 2\right) = E_{n-1}\left(\frac{1}{x_1} - 2, \cdots, \frac{1}{x_n} - 2\right) \geqslant n(n-2)^{n-1} \qquad (1)$$

Chen et al.[6] gave an analytic proof which is skillful. By the method of successive adjustment, Fan[7] gave a simple elementary proof for inequality under conditions $x \in (0,1)^n$, $n \geqslant 2$ and $E_1(x) = 1$. Using the same method, Shi[8] extended inequality (1) to the cases of $E_k\left(\dfrac{1}{x} - 2\right)$ for $k = 1, 2, \cdots, n-1$, where $x \in \mathbf{R}_+^n$.

Theorem A Let $x \in \mathbf{R}_+^n = (0, +\infty)^n$, $n \geqslant 2$ and $E_1(x) = 1$. Then for $k = 1, 2, \cdots, n-1$, we have

$$E_k\left(\frac{1}{x} - 2\right) = E_k\left(\frac{1}{x_1} - 2, \cdots, \frac{1}{x_n} - 2\right) \geqslant \binom{n}{k}(n-2)^k \qquad (2)$$

where $\dbinom{n}{k} = \dfrac{n!}{k!(n-k)!}$.

And Shi also pointed out that the inequality (2) does not hold for $k = n$.

In 2002, using analytical methods, Ma and Pu[9] gave parameter extension of the inequality (2).

Theorem B Let $x \in \mathbf{R}_+^n, n, k \in \mathbf{N} = \{1, 2, \cdots\}, \mu > 0, n \geqslant 2$ and $E_1(x) = 1$. Then for $n \geqslant k + \mu - 1$, we have

$$E_k\left(\frac{1}{x} - \mu\right) = E_k\left(\frac{1}{x_1} - \mu, \cdots, \frac{1}{x_n} - \mu\right) \geqslant \binom{n}{k}(n-\mu)^k \qquad (3)$$

In 2005, Li et al.[10] gave an elementary proof for Theorem B when $k \leqslant n - 1$.

In this paper, by studying the Schur harmonic convexity of $E_k\left(\dfrac{1}{x} - \mu\right)$, we obtain a reverse of the inequality (3) and accordingly establish the reverse of the corresponding simplex inequalities.

Our main result is the following theorem.

Theorem 1 Let $x \in \mathbf{R}_+^n, n, k \in \mathbf{N}, \mu > 0, n \geqslant 2$. Then for $n \geqslant k+\mu-1$, $E_k\left(\dfrac{1}{x} - \mu\right)$ is Schur harmonically concave on $\Omega = \{x : x \in \mathbf{R}_+^n, E_1(x) \leqslant$

1}. When $E_1(\boldsymbol{x}) \leqslant 1$, we have

$$E_k\left(\frac{1}{\boldsymbol{x}} - \mu\right) = E_k\left(\frac{1}{x_1} - \mu, \cdots, \frac{1}{x_n} - \mu\right) \leqslant \binom{n}{k}\left(\frac{1}{H} - \mu\right)^k \quad (4)$$

where $H = n(\sum\limits_{i=1}^{n} x_i^{-1})^{-1}$ is the harmonic mean of $\boldsymbol{x}$.

Taking $\mu = 2$ in Theorem 1, we infer the following corollary.

Corollary 1 Let $\boldsymbol{x} \in \mathbf{R}_+^n, n, k \in \mathbf{N}$, and $E_1(\boldsymbol{x}) \leqslant 1$. Then for $n \geqslant k+1$, we have

$$E_k\left(\frac{1}{\boldsymbol{x}} - 2\right) = E_k\left(\frac{1}{x_1} - 2, \cdots, \frac{1}{x_n} - 2\right) \leqslant \binom{n}{k}\left(\frac{1}{H} - 2\right)^k \quad (5)$$

where $H = n(\sum\limits_{i=1}^{n} x_i^{-1})^{-1}$ is the harmonic mean of $\boldsymbol{x}$.

Taking $k = n - 1$ in Corollary 1, we obtain the following inequality.

Corollary 2 Let $\boldsymbol{x} \in \mathbf{R}_+^n, n \in \mathbf{N}$, and $E_1(\boldsymbol{x}) \leqslant 1$. Then for $n \geqslant 2$, we have

$$\sum_{i=1}^{n} \prod_{j=1, j \neq i}^{n} \left(\frac{1}{x_j} - 2\right) \leqslant n\left(\frac{1}{H} - 2\right)^{n-1} \quad (6)$$

where $H = n(\sum\limits_{i=1}^{n} x_i^{-1})^{-1}$ is the harmonic mean of $\boldsymbol{x}$.

60.2 Definitions and Lemmas

Definition 1[11–12] Let $\boldsymbol{x} = (x_1, \cdots, x_n)$ and $\boldsymbol{y} = (y_1, \cdots, y_n) \in \mathbf{R}^n$.

(i) $\boldsymbol{x}$ is said to be majorized by $\boldsymbol{y}$ (in symbols $\boldsymbol{x} \prec \boldsymbol{y}$) if $\sum\limits_{i=1}^{k} x_{[i]} \leqslant \sum\limits_{i=1}^{k} y_{[i]}$ for $k = 1, 2, \cdots, n - 1$ and $\sum\limits_{i=1}^{n} x_i = \sum\limits_{i=1}^{n} y_i$, where $x_{[1]} \geqslant \cdots \geqslant x_{[n]}$ and $y_{[1]} \geqslant \cdots \geqslant y_{[n]}$ are rearrangements of $\boldsymbol{x}$ and $\boldsymbol{y}$ in a descending order.

(ii) $\Omega \subseteq \mathbf{R}^n$ is called a convex set if $(\alpha x_1 + \beta y_1, \cdots, \alpha x_n + \beta y_n) \in \Omega$ for every $\boldsymbol{x}$ and $\boldsymbol{y} \in \Omega$, where α and $\beta \in [0, 1]$ with $\alpha + \beta = 1$.

(iii) Let $\Omega \subseteq \mathbf{R}^n$. The function $\varphi \colon \Omega \to \mathbf{R}$ is said to be a Schur convex function on Ω if $\boldsymbol{x} \prec \boldsymbol{y}$ on Ω implies $\varphi(\boldsymbol{x}) \leqslant \varphi(\boldsymbol{y})$. φ is said to be a Schur concave function on Ω if and only if $-\varphi$ is Schur convex.

Lemma 1[11−12] Let $x = (x_1, x_2, \cdots, x_n) \in \mathbf{R}^n$. Then

$$(A, \cdots, A) \prec (x_1, x_2, \cdots, x_n) \tag{7}$$

where $A = \dfrac{1}{n}\displaystyle\sum_{i=1}^{n} x_i$ is the arithmetic mean of x.

Definition 2[11−12] Let $\Omega \subseteq \mathbf{R}^n$.

(i) Ω is called a symmetric set, if $x \in \Omega$ implies $xP \in \Omega$ for every $n \times n$ permutation matrix P.

(ii) The function $\varphi : \Omega \to \mathbf{R}$ is called symmetric if for every permutation matrix P, $\varphi(xP) = \varphi(x)$ for all $x \in \Omega$.

Lemma 2[11] Let $\Omega \subset \mathbf{R}^n$ be symmetric and have a nonempty interior Ω^0 convex set, and let $\varphi : \Omega \to \mathbf{R}$ be continuous on Ω and differentiable in Ω^0. Then φ is the Schur-convex (Schur-concave) function if and only if φ is symmetric on Ω and

$$(x_1 - x_2)\left(\frac{\partial \varphi(x)}{\partial x_1} - \frac{\partial \varphi(x)}{\partial x_2}\right) \geqslant 0 (\leqslant 0) \tag{8}$$

holds for any $x \in \Omega^0$.

The Schur-convexity described the ordering of majorization, the order-preserving functions were first comprehensively studied by Issai Schur in 1923. It has important applications in analytic inequalities, combinatorial optimization, quantum physics, information theory, and so on. See [11]

In 2009, Chu and Lv[13] introduced the notion of harmonically Schur convex function and some interesting inequalities were obtained. The Schur harmonic convexity involving some special functions has been investigated, see e.g. [14-20].

Definition 3[13] Let $\Omega \subset \mathbf{R}_+^n$.

(i) A set Ω is said to be harmonically convex if $\dfrac{xy}{\lambda x + (1-\lambda)y} \in \Omega$ for every $x, y \in \Omega$ and $\lambda \in [0, 1]$, where $xy = \displaystyle\sum_{i=1}^{n} x_i y_i$ and $\dfrac{1}{x} = \left(\dfrac{1}{x_1}, \cdots, \dfrac{1}{x_n}\right)$.

(ii) A function $\varphi : \Omega \to \mathbf{R}_+$ is said to be Schur harmonically convex on Ω if $\dfrac{1}{x} \prec \dfrac{1}{y}$ implies $\varphi(x) \leq \varphi(y)$. A function φ is said to be a

Schur harmonically concave function on Ω if and only if $-\varphi$ is a Schur harmonically convex function.

Lemma 3 Let $\Omega \subset \mathbf{R}_+^n$ be a symmetric and harmonically convex set with inner points and let $\varphi : \Omega \to \mathbf{R}_+$ be a continuously symmetric function which is differentiable on Ω^0. Then φ is Schur harmonically convex (Schur harmonically concave) on Ω if and only if

$$(x_1 - x_2)\left(x_1^2 \frac{\partial \varphi(x)}{\partial x_1} - x_2^2 \frac{\partial \varphi(x)}{\partial x_2}\right) \geqslant 0 \quad (\leqslant 0), \quad x \in \Omega^0 \tag{9}$$

Lemma 4 Let $x \in \mathbf{R}_+^n, n, k \in \mathbf{N}, \mu > 0, n \geqslant 2$ and $E_1(x) \leqslant 1$. Then for $n \geqslant k + \mu - 1$, we have

$$E_k\left(\frac{1}{x} - \mu\right) \geqslant 0 \tag{10}$$

Proof Let $E_1(x) = \sigma \leqslant 1$. Then $E_1\left(\frac{x}{\sigma}\right) = \frac{x_1}{\sigma} + \cdots + \frac{x_n}{\sigma} = 1$, and $n \geqslant k + \mu - 1 \geqslant k + \sigma\mu - 1$, from Theorem B, we have

$$E_k\left(\frac{\sigma}{x} - \sigma\mu\right) = E_k\left(\frac{\sigma}{x_1} - \sigma\mu, \cdots, \frac{\sigma}{x_n} - \sigma\mu\right) \geqslant \binom{n}{k}(n - \sigma\mu)^k$$

and then

$$E_k\left(\frac{1}{x} - \mu\right) = E_k\left(\frac{1}{x_1} - \mu, \cdots, \frac{1}{x_n} - \mu\right) \geqslant \binom{n}{k}\left(\frac{n - \sigma\mu}{\sigma}\right)^k \geqslant 0$$

Lemma 4 is proved.

60.3   Proof of Main Result

Proof of Theorem 1 When $n \geqslant 2$, $k = 1$, obviously, the inequality (4) holds. When $n \geqslant 2$, $k \geqslant 2$, note that

$$\varphi(x) := E_k\left(\frac{1}{x} - \mu\right) = \left(\frac{1}{x_1} - \mu\right)\left(\frac{1}{x_2} - \mu\right) E_{k-2}\left(\frac{1}{x_3} - \mu, \cdots, \frac{1}{x_n} - \mu\right) +$$
$$\left(\frac{1}{x_1} - \mu + \frac{1}{x_2} - \mu\right) E_{k-1}\left(\frac{1}{x_3} - \mu, \cdots, \frac{1}{x_n} - \mu\right) +$$
$$E_k\left(\frac{1}{x_3} - \mu, \cdots, \frac{1}{x_n} - \mu\right)$$

we have

$$\frac{\partial \varphi(\boldsymbol{x})}{\partial x_1} = -\frac{1}{x_1^2}\left(\frac{1}{x_2} - \mu\right)E_{k-2}\left(\frac{1}{x_3} - \mu, \cdots, \frac{1}{x_n} - \mu\right) - $$
$$\frac{1}{x_1^2}E_{k-1}\left(\frac{1}{x_3} - \mu, \cdots, \frac{1}{x_n} - \mu\right)$$

and

$$\frac{\partial \varphi(\boldsymbol{x})}{\partial x_2} = -\frac{1}{x_2^2}\left(\frac{1}{x_1} - \mu\right)E_{k-2}\left(\frac{1}{x_3} - \mu, \cdots, \frac{1}{x_n} - \mu\right) - $$
$$\frac{1}{x_2^2}E_{k-1}\left(\frac{1}{x_3} - \mu, \cdots, \frac{1}{x_n} - \mu\right)$$

And then

$$\Delta := (x_1 - x_2)\left(x_1^2 \frac{\partial \varphi(\boldsymbol{x})}{\partial x_1} - x_2^2 \frac{\partial \varphi(\boldsymbol{x})}{\partial x_2}\right)$$
$$= -\frac{(x_1 - x_2)^2}{x_1 x_2}E_{k-2}\left(\frac{1}{x_3} - \mu, \cdots, \frac{1}{x_n} - \mu\right)$$

Now we distinguish two cases to prove $\Delta \leqslant 0$.

Case 1 If $n \geqslant 2$, $k = 2$, then from the definition of elementary symmetric functions, it follows that $\Delta \leqslant 0$.

If $n = 3$, $k = 3$, then from the condition $n \geqslant k + \mu - 1$, it follows that $\mu \leqslant n - k + 1 \leqslant 3 - 3 + 1 = 1$, and by $x_i \leqslant E_1(\boldsymbol{x}) \leqslant 1$, it follows that $\frac{1}{x_i} \geqslant 1, i = 3, \cdots, n$, therefore

$$E_{k-2}\left(\frac{1}{x_3} - \mu, \cdots, \frac{1}{x_n} - \mu\right)$$
$$= E_1\left(\frac{1}{x_3} - \mu, \cdots, \frac{1}{x_n} - \mu\right)$$
$$= \sum_{i=3}^{n}\frac{1}{x_i} - (n - 2)\mu \geqslant 0$$

Case 2 If $n \geqslant 4$. Then $n - 2 \geqslant 2$, from $n \geqslant k + \mu - 1$, it follows that

$$(n - 2) \geqslant (k - 2) + \mu - 1$$

and $\sum_{i=3}^{n} x_i \leqslant 1$. So by Lemma 4, it follows that $E_{k-2}\left(\frac{1}{x_3} - \mu, \cdots, \frac{1}{x_n} - \mu\right) \geqslant 0$.

Thus, for all $n \geqslant 2$, $\Delta \leqslant 0$. By Lemma 3, we can derive that $E_k\left(\frac{1}{\boldsymbol{x}} - \mu\right)$

is Schur harmonically concave on Ω.

From Lemma 1, it is seen that

$$\left(\frac{1}{H},\cdots,\frac{1}{H}\right) \prec \left(\frac{1}{x_1},\cdots,\frac{1}{x_n}\right) \tag{11}$$

According to definition 3 (ii), inequality (4) follows.

The proof of Theorem 1 is completed.

60.4 Applications in geometry

Let Ω_n be an n-dimensional simplex in the n-dimensional Euclidean space $\mathbf{R}^n$ ($n \geqslant 2$) with vertices $A_0, A_1, \cdots, A_n$ whose volume is V. For $i = 0, 1, \cdots, n$, let r_i be the radius of i-th escribed sphere of Ω_n, F_i be the area of the i-th face $f_i = \{A_0, A_1, \cdots, A_{i-1}, A_{i+1}, \cdots, A_n\}$ of Ω_n, $F = \sum_{i=0}^{n} F_i$. R and r is circumradius and inradius of Ω_n, respectively. Let P be an arbitrary interior point of the simplex Ω_n, d_i be the distance from the point P to the i-th face f_i of Ω_n , h_i be the altitude of Ω_n from vertex A_i for $i = 0, 1, \cdots, n$.

In [5], by the Theorem B, Ma and Pu obtained the following two theorems.

Theorem C In the simplex Ω_n ($n \geqslant 2$), for $k = 1, 2, \cdots, n$, we have

$$E_k\left(\frac{h_0}{r_0}, \frac{h_1}{r_1}, \cdots, \frac{h_n}{r_n}\right) \geqslant \binom{n+1}{k}(n-1)^k \tag{12}$$

with equality if Ω_n is an orthocentric simplex.

Theorem D Let P be an arbitrary point in the interior of n-dimensional simplex Ω_n. Let B_i be the intersection of the line A_iP with the hyperplane f_i, further let $R_i = |A_iP|, d_i' = |PB_i|, M_i = |A_iB_i|, i = 0, 1, \cdots, n$. Then for $k = 1, 2, \cdots, n+1$, we have

$$E_k\left(\frac{R_0}{d_0'}, \frac{R_1}{d_1'}, \cdots, \frac{R_n}{d_n'}\right) \geqslant \binom{n+1}{k}n^k \tag{13}$$

with equality if Ω_n is an orthocentric simplex, and P is its orthocentre.

Now, by Theorem 1, we give the reverse of the inequalities (12) and (13).

Theorem 2　In the simplex Ω_n, for $k = 1, 2, \cdots, n$, we have

$$E_k\left(\frac{h_0}{r_0}, \frac{h_1}{r_1}, \cdots, \frac{h_n}{r_n}\right) \leqslant \binom{n+1}{k}\left(\frac{1}{H} - 2\right)^k \tag{14}$$

where $H = (n+1)\left(\displaystyle\sum_{i=0}^{n}\left(\frac{h_i}{r_i} + 2\right)\right)^{-1}$.

Proof　In the n-dimensional simplex, for $k = 0, 1, \cdots, n$, it is well-known that

$$r_i = \frac{nV}{\displaystyle\sum_{j=0}^{n} F_j - 2F_i} \tag{15}$$

and

$$h_i = \frac{nV}{F_i} \tag{16}$$

Let $x_i = \dfrac{F_i}{\displaystyle\sum_{j=0}^{n} F_j}, i = 0, 1, \cdots, n$. Then $E_1(\boldsymbol{x}) = 1$. From equalities (15) and (16), we have

$$\frac{1}{x_i} - 2 = \frac{\displaystyle\sum_{j=0}^{n} F_j - 2F_i}{F_i} = \frac{h_i}{r_i}, i = 0, 1, \cdots, n$$

Thus from the inequality (5), it is deduced that the inequality(14) holds.

The proof of Theorem 2 is completed.

Theorem 3　Let P be an arbitrary point in the interior of n-dimensional simplex Ω_n. Let B_i be the intersection of the line A_iP with the hyperplane f_i, further let $R_i = |A_iP|, d_i' = |PB_i|, M_i = |A_iB_i|, i = 0, 1, \cdots, n$. Then for $k = 1, 2, \cdots, n+1$

$$E_k\left(\frac{R_0}{d_0'}, \frac{R_1}{d_1'}, \cdots, \frac{R_n}{d_n'}\right) \leqslant \binom{n+1}{k}\left(\frac{1}{H} - 1\right)^k \tag{17}$$

where $H = (n+1)\left(\displaystyle\sum_{i=0}^{n} \frac{M_i}{d_i'}\right)^{-1}$.

Proof　It is easy to see that

$$\sum_{i=0}^{n} \frac{d_i'}{M_i} = 1, \frac{R_i}{M_i} = 1 - \frac{d_i'}{M_i}, i = 0, 1, \cdots, n$$

In the inequality (4), let $\mu = 1$, $x_i = \dfrac{d_i'}{M_i}$. Then $1 \leqslant k \leqslant n+1$, $\dfrac{1}{x_i} - 1 = \dfrac{R_i}{d_i'}$, $i = 0, 1, \cdots, n$. Thus the inequality(17) holds.

The proof of Theorem 3 is completed.

When $k = n + 1$, the inequality (17) is reduced to

Corollary 3 Under the conditions of Theorem 3, we have

$$\prod_{i=0}^{n} R_i \leqslant \left(\frac{1}{H} - 1\right)^{n+1} \prod_{i=0}^{n} d_i' \tag{18}$$

Theorem 4 In the simplex Ω_n, let

$$\lambda_i = \frac{F - 2F_i}{F_i}, \quad \mu_i = \frac{F_0^2 + \cdots + F_n^2 - 2F_i^2}{F_i^2} \quad (i = 0, 1, \cdots, n)$$

Then

$$\sum_{i=0}^{n} \prod_{j=0, j \neq i}^{n} \lambda_j \leqslant (n+1) \left(\frac{1}{H} - 2\right)^n \tag{19}$$

where $H = n(\sum_{i=0}^{n} (\lambda_i + 2))^{-1}$;

$$\sum_{i=0}^{n} \prod_{j=0, j \neq i}^{n} \mu_j \leqslant (n+1) \left(\frac{1}{H} - 2\right)^n \tag{20}$$

where $H = n(\sum_{i=0}^{n} (\mu_i + 2))^{-1}$.

Proof Taking $x_i = \dfrac{F_i}{F}(i = 0, 1, \cdots, n)$ and applying Corollary 2 for these $n+1$ positive real numbers $x_0, x_1, \cdots, x_n$, the inequality (19) follows.

Taking $x_i = \dfrac{F_i^2}{F_0^2 + F_1^2 + \cdots + F_n^2}$ and applying Corollary 2, the desired inequality (20) follows.

The proof of Theorem 4 is completed.

Remark 1 The inequalities (19) and (20) give the reverse of the inequalities (2.10) and (2.11) in [10], respectively.

Conflict of Interests

The authors declare that there is no conflict of interests regarding the publication of this article.

Acknowledgments

The work was supported by Funding Project for Academic Human Resources Development in Institutions of Higher Learning under the Jurisdiction of Beijing Municipality (PHR (IHLB)) (PHR201108407). We are grateful to Professor Fernando Simões and the four anonymous reviewers for helpful advices that improved its presentation.

References

[1]　G S LENG, L H TANG. Some generalizations to several dimensional of the Pedoe inequality with applications[J]. Acta Mathematica Sinica, 1997, 40 (1): 14–21.

[2]　L YANG, J Z ZHANG. Generalization to higher dimensions of the Neuberg-Pedoe inequqlity with applitions[J]. Acta Mathematica Sinica,1981, 24(3): 401–408.

[3]　L YANG, J Z ZHANG. A generalization to several dimensions of the Neuberg-Pedoe inequality, with applications[J]. Bulletin of the Australian Mathematical Society, 1983, 27: 203–214.

[4]　D S MITRNOVIĆ, J E PEČARIĆ. About the Neuberg-Pedoe and the Oppenheim inequalities[J]. Journal of mathematical analysis and applications, 1988, 129 (1): 196–210.

[5]　D S MITRINOVIĆ, J E PEčARIĆ, V VOLENEC. Recent advances in geometric inequalities[M]. Dordrecht: Kluwer Academic Publishers, 1989.

[6]　J CHEN, Y HUANG, S H XIA. A note on generalizations of Neuberg-Pedoe inequality to higher dimensional case[J]. Journal of Sichuan University (Natural Science Edition), 1999, 36(2): 197–200.

[7]　Y W FAN. An generalization of Neuberg-Pedoe inequality for simplies[J]. Journal of Yanan University (Natural Science Edition), 1999, 18(3): 19–22.

[8]　H N SHI. The generalization of a analytic inequality[J]. Journal of Chengdu University (Natural Science), 2001, 20(1): 5–7.

[9]　T Y MA, Z N PU. Some generalizations of an analytical ineguality with applications[J]. Journal of Sichuan University (Natural Science Edition), 2002, 39(1): 1–6.

[10] X Y LI, G S LENG, L H TANG. Inequalities for a simplex and any point[J]. Mathematical Inequalities & Applications, 2005, 8(3): 547–557.

[11] MARSHALL A M, OLKIN I, ARNOLD B C. Inequalities: Theory of Majorization and Its Application [M]. 2nd ed. New York: Springer Press, 2011.

[12] B.-Y.WANG. Kòngzhì Bùděngshì Jīchǔ (Foundations of Majorization Inequalities[M]. Beijing: Beijing Normal University Press, 1990.

[13] Y M CHU, Y P LV. The Schur harmonic convexity of the Hamy symmetric function and its applications[J]. Journal of Inequalities and Applications, vol. 2009, Article ID 838529, 10 pages, 2009.

[14] W F XIA, Y M CHU. The Schur harmonic convexity of Lehmer means [J]. International Mathematical Forum, 2009, 4(41): 2009–2015.

[15] Y M CHU, T C SUN. The Schur harmonic convexity for a class of symmetric functions[J]. Acta Mathematica Scientia, 2010, 30B(5): 1501–1506.

[16] Y M CHU, W F XIA. Necessary and sufficient conditions for the Schur harmonic convexity of the generalized Muirhead mean[J]. Proceedings of A. Razmadze Mathematical Institute, 2010, 152: 19–27.

[17] Z H YANG. Schur harmonic convexity of Gini means[J]. International Mathematical Forum, 2011, 6(16): 747–762.

[18] W F XIA, Y M CHU, G D WANG. Necessary and sufficient conditions for the Schur harmonic convexity or concavity of the extended mean values[J]. Revista De La Unión Matemática Argentina, 2010, 51(2): 121–132.

[19] H N SHI. Theory of majorization and analytic Inequalities[M]. Harbin: Harbin Institute of Technology Press, 2012.

[20] H N SHI, J ZHANG. Some new judgement theorems of Schur geometric and Schur harmonic convexities for a class of symmetric functions[J]. Journal of Inequalities and Applications, vol. 2013, Article ID 527, 9 pages, 2013.

第61篇　Two Double Inequalities for k-Gamma and k-Riemann Zeta Functions

(ZHANG JING, SHI HUAN-NAN. J. Inequal. Appl., 2014, 2014:191)

Abstract: By using methods in the theory of majorization, a double inequality for the gamma function is extended to the k-gamma function and the k-Riemann zeta function.

Keywords: majorization; Schur convexity; k-gamma function; k-Riemann zeta function; Apéry' s constant; log-convexity

61.1　Introduction

The Euler gamma function $\Gamma(x)$ is defined for $x > 0$ by

$$\Gamma(x) = \int_0^\infty \mathrm{e}^{-t} t^{x-1} \mathrm{d}t \tag{1}$$

In 2005, by using a geometrical method, C. Alsina and M. S. Toms[2] have proved the following double inequality

$$\frac{1}{n!} \leqslant \frac{\Gamma(1+x)^n}{\Gamma(1+nx)} \leqslant 1 \tag{2}$$

In 2009, Nguyen V. Vinh and Ngo P.N. Ngoc[3] obtained the following generalization of the double inequality (2)

$$\frac{\prod\limits_{i=1}^n \Gamma(1+\alpha_i)}{\Gamma(\beta + \sum\limits_{i=1}^n \alpha_i)} \leqslant \frac{\prod\limits_{i=1}^n \Gamma(1+\alpha_i x)}{\Gamma(\beta + (\sum\limits_{i=1}^n \alpha_i)x)} \leqslant \frac{1}{\Gamma(\beta)} \tag{3}$$

where $x \in [0,1], \beta \geqslant 1, \alpha_i > 0, n \in \mathbf{N}$.

For $k > 0$, the Γ_k function [4] is defined by

$$\Gamma_k(x) = \lim_{n \to \infty} \frac{n!k^n(nk)^{\frac{x}{k}-1}}{(x)_{n,k}}, \quad x \in \mathbf{C} \backslash k \in \mathbf{Z}^- \tag{4}$$

where $(x)_{n,k} = x(x+k(x+2k)\cdots(x+(n-1)k)$.

The above definition is a generalization of the definition of $\Gamma(x)$ function. For $x \in \mathbf{C}$ with $R(x) > 0$; the function $\Gamma_k(x)$ is given by the integral[4]

$$\Gamma_k(x) = \int_0^\infty \mathrm{e}^{-\frac{t^k}{k}} t^{x-1} \mathrm{d}t \tag{5}$$

In this note, by using methods on the theory of majorization, we extended the double inequality (3) to the $\Gamma_k(x)$ function, namely establish the following theorem.

Theorem 1

$$\frac{\prod\limits_{i=1}^{n} \Gamma_k(1+\alpha_i)}{\Gamma_k(\beta + \sum\limits_{i=1}^{n} \alpha_i)} \leqslant \frac{\prod\limits_{i=1}^{n} \Gamma_k(1+\alpha_i x)}{\Gamma_k(\beta + (\sum\limits_{i=1}^{n} \alpha_i)x)} \leqslant \frac{1}{\Gamma_k(\beta)} \tag{6}$$

where $x \in [0,1], \beta \geqslant 1, \alpha_i > 0, n \in \mathbf{N}$.

61.2　Definitions and Lemmas

We need the following definitions and auxiliary lemmas.

Definition 1 [6-7] Let $\boldsymbol{x} = (x_1, \cdots, x_n)$ and $\boldsymbol{y} = (y_1, \cdots, y_n) \in \mathbf{R}^n$.

(i) $\boldsymbol{x}$ is said to be majorized by $\boldsymbol{y}$ (in symbols $\boldsymbol{x} \prec \boldsymbol{y}$) if $\sum\limits_{i=1}^{k} x_{[i]} \leqslant \sum\limits_{i=1}^{k} y_{[i]}$ for $k = 1, 2, \cdots, n-1$ and $\sum\limits_{i=1}^{n} x_i = \sum\limits_{i=1}^{n} y_i$, where $x_{[1]} \geqslant \cdots \geqslant x_{[n]}$ and $y_{[1]} \geqslant \cdots \geqslant y_{[n]}$ are rearrangements of $\boldsymbol{x}$ and $\boldsymbol{y}$ in a descending order.

(ii) $\Omega \subseteq \mathbf{R}^n$ is called a convex set if $(\alpha x_1 + \beta y_1, \cdots, \alpha x_n + \beta y_n) \in \Omega$ for every $\boldsymbol{x}$ and $\boldsymbol{y} \in \Omega$, where α and $\beta \in [0,1]$ with $\alpha + \beta = 1$.

(iii) Let $\Omega \subseteq \mathbf{R}^n$. The function $\varphi \colon \Omega \to \mathbf{R}$ is said to be a Schur convex function on Ω if $\boldsymbol{x} \prec \boldsymbol{y}$ on Ω implies $\varphi(\boldsymbol{x}) \leqslant \varphi(\boldsymbol{y})$. φ is said to be a Schur concave function on Ω if and only if $-\varphi$ is Schur convex.

Lemma 1[6] Let $\boldsymbol{x}, \boldsymbol{y} \in \Omega$, $x_1 \geqslant x_2 \geqslant \cdots \geqslant x_n$, and $\sum\limits_{i=1}^{n} x_i = \sum\limits_{i=1}^{n} y_i$. If for some $k, 1 \leqslant k < n, x_i \leqslant y_i, i = 1, ..., k, x_i \geqslant y_i$ for $i = k+1, ..., n$, then $\boldsymbol{x} \prec \boldsymbol{y}$.

Lemma 2 Let f, g be a continuous nonnegative functions defined on an interval $[a, b] \subset \mathbf{R}$. Then $I(x) = \int_a^b g(t)(f(t))^x \mathrm{d}t$ is log-convex on $[0, +\infty)$.

Proof Let $\alpha, \beta \geqslant 0, 0 < s < 1$ by Hölder integral inequality, we have

$$
\begin{aligned}
I\left(s\alpha + (1-s)\beta\right) &= \int_a^b g(t)(f(t))^{s\alpha + (1-s)\beta} \mathrm{d}t \\
&= \int_a^b \left(g(t)(f(t))^\alpha\right)^s \left(g(t)(f(t))^\beta\right)^{1-s} \mathrm{d}t \\
&\leqslant \left(\int_a^b g(t)(f(t))^\alpha \mathrm{d}t\right)^s \left(\int_a^b g(t)(f(t))^\beta \mathrm{d}t\right)^{1-s} \\
&= (I(\alpha))^s (I(\beta))^{1-s}
\end{aligned}
$$

i.e.

$$
\log I\left(s\alpha + (1-s)\beta\right) \leqslant s \log I(\alpha) + (1-s) \log I(\beta)
$$

this means that $I(x)$ is log-convex on $[0, +\infty)$.

Lemma 3[6] Let g be a continuous nonnegative function defined on an interval $[a, b] \subset \mathbf{R}$. Then

$$
\varphi(\boldsymbol{x}) = \prod_{i=1}^{n} g(x_i), \quad \boldsymbol{x} \in [a, b]^n
$$

is Schur-convex on $[a, b]^n$ if and only if $\log g$ is convex on $[a, b]$.

The Schur-convexity described the ordering of majorization, the order-preserving functions were first comprehensively studied by Issai Schur in 1923. It has important applications in analytic inequalities, combinatorial optimization, quantum physics, information theory, and so on. See [6,8-10].

61.3　Proof of Main Result

Proof of Theorem 1 Let

$$
\boldsymbol{u} = (u_1, ..., u_n, u_{n+1}) = \left(\beta + (\sum_{i=1}^{n} \alpha_i)x - 1, \alpha_1, ..., \alpha_n\right)
$$

and

$$v = (v_1, ..., v_n, v_{n+1}) = \left(\beta + \sum_{i=1}^{n} \alpha_i - 1, \alpha_1 x, ..., \alpha_n x \right)$$

It is clear that $\sum_{i=1}^{n+1} u_i = \sum_{i=1}^{n+1} v_i$.

Without loss of generality, we may assume that $\alpha_1 \geqslant \alpha_2 \geqslant \cdots \geqslant \alpha_n$. The following discussion is divided into two cases:

Case 1 $\beta + (\sum_{i=1}^{n} \alpha_i) x - 1 \geqslant \alpha_1$. Notice that $x \in [0, 1]$, we have $u_1 = \beta + (\sum_{i=1}^{n} \alpha_i) x - 1 \leqslant \beta + \sum_{i=1}^{n} \alpha_i - 1 = v_1$, and $u_i = \alpha_{i-1} \geqslant \alpha_{i-1} x = v_i, i = 2, ..., n+1$. Hence from the Lemma 1, it follows that $\boldsymbol{u} \prec \boldsymbol{v}$.

Case 2 $\beta + (\sum_{i=1}^{n} \alpha_i) x - 1 < \alpha_1$. There exist $k(k = 2, ..., n)$ such that

$$\alpha_1 \geqslant ... \geqslant \alpha_{k-1} \geqslant \beta + (\Sigma_{i=1}^{n} \alpha_i) x - 1 \geqslant \alpha_{k+1} \geqslant ... \geqslant \alpha_n$$

Now notice that $\beta - 1 \geqslant 0$, we have $u_1 = \alpha_1 \leqslant \beta + \sum_{i=1}^{n} \alpha_i - 1 = v_1$, and $u_i = \alpha_i \geqslant \alpha_{i-1} x = v_i, i = 2, ..., k-1$. $u_k = \beta + (\sum_{i=1}^{n} \alpha_i) x - 1 \geqslant \alpha_k x = v_k$. $u_i = \alpha_{i-1} \geqslant \alpha_{i-1} x = v_i, i = k+1, ..., n+1$. Hence from the Lemma 1, it follows that $\boldsymbol{u} \prec \boldsymbol{v}$.

Let

$$\boldsymbol{w} = (w_1, ..., w_n, w_{n+1}) = (\beta - 1, \alpha_1 x, ..., \alpha_n x)$$

$$\boldsymbol{z} = (z_1, ..., z_n, z_{n+1}) = \left(\beta + (\sum_{i=1}^{n} \alpha_i) x - 1, \underbrace{0, ..., 0}_{n} \right)$$

It is clear that $\sum_{i=1}^{n+1} w_i = \sum_{i=1}^{n+1} z_i$.

The following discussion is divided into two cases:

Case 1 $\beta - 1 \geqslant \alpha_1 x$. Notice that $x \in [0, 1]$, we have $w_1 = \beta - 1 \leqslant \beta + (\sum_{i=1}^{n} \alpha_i) x - 1 = z_1$, and $w_i = \alpha_{i-1} x \geqslant 0 = z_i, i = 2, ..., n+1$. Hence from the Lemma 1, it follows that $\boldsymbol{w} \prec \boldsymbol{z}$.

Case 2 $\beta - 1 < \alpha_1 x$. There exist $k(k = 2, ..., n)$ such that

$$\alpha_1 x \geqslant \cdots \geqslant \alpha_{k-1} x \geqslant \beta - 1 \geqslant \alpha_{k+1} x \geqslant \cdots \geqslant \alpha_n x$$

Now notice that $\beta - 1 \geqslant 0$, we have $w_1 = \alpha_1 x \leqslant \beta + (\sum_{i=1}^{n} \alpha_i) x - 1 = z_1$, and $w_i = \alpha_i x \geqslant 0 = z_i, i = 2, ..., k-1$. $w_k = \beta - 1 \geqslant 0 = z_k$. $w_i = \alpha_{i-1} x \geqslant$

$0 = z_i, i = k+1, ..., n+1$. Hence from the Lemma 1, it follows that $\boldsymbol{w} \prec \boldsymbol{z}$.

Taking $g(t) = \mathrm{e}^{-\frac{t^k}{k}}, f(t) = t, a = 0, b = +\infty$, then

$$I(x) = \int_a^b g(t)(f(t))^x \mathrm{d}t = \int_0^\infty \mathrm{e}^{-\frac{t^k}{k}} t^x \mathrm{d}t = \Gamma_k(x+1) \qquad (7)$$

By the Lemma 2, $I(x)$ is log-convex on $[0, +\infty)$, and then from the Lemma 3, $\varphi(\boldsymbol{x}) = \prod_{i=1}^n I(x_i)$ is Schur-convex on $[0, +\infty)^n$. Then and from $\boldsymbol{u} \prec \boldsymbol{v}$ and $\boldsymbol{w} \prec \boldsymbol{z}$, we have

$$\varphi(\boldsymbol{u}) \leqslant \varphi(\boldsymbol{v})$$

and

$$\varphi(\boldsymbol{w}) \leqslant \varphi(\boldsymbol{z})$$

i.e.

$$\Gamma_k(\beta + (\sum_{i=1}^n \alpha_i)x) \prod_{i=1}^n \Gamma_k(\beta + \sum_{i=1}^n \alpha_i) \leqslant \Gamma_k(1 + \alpha_i) \prod_{i=1}^n \Gamma_k(1 + \alpha_i x) \qquad (8)$$

and

$$\Gamma_k(\beta) \prod_{i=1}^n \Gamma_k(1 + \alpha_i x) \leqslant \Gamma_k(\beta + (\sum_{i=1}^n \alpha_i)x) \qquad (9)$$

Thus, we have proved the double inequality (6).

The proof of Theorem 1 is completed.

Acknowledgments

The work was supported by Funding Project for Academic Human Resources Development in Institutions of Higher Learning under the Jurisdiction of Beijing Municipality (PHR (IHLB)) (PHR201108407). Thanks for the help. This article was typeset by using AmS-LaTeX

References

[1]　GEORGE E ANDREWS, RICHARD ASKEY, RANJAN ROY. Special Functions[M]. London: Cambridge Uni Press. 1999.

[2]　ALSINA C , TOMS M S . A geometrical proof of a new inequality for the gamma function[J]. J. Ineq. Pure Appl. Math., 2005(62), Art.48.

[3] NGUYEN V VINH, NGO P N NGOC. An Inequality for the Gamma Function[J]. International Mathematical Forum,2009, 4(28): 1379 - 1382.

[4] R DIAZ - E PARIGUAN. On hypergeometric functions and k-Pochhammer symbol[J]. Divulgaciones Matematicas, 2007, 15 (2):179 - 192.

[5] CHRYSI G KOKOLOGIANNAKI-VALMIR KRASNIQI. Some Properties Of The K-Gamma Function[J]. Le Matematiche,2013, LXVIII - Fasc. I, : 13 - 22.

[6] MARSHALL A M, OLKIN I, ARNOLD B C. Inequalities: Theory of Majorization and Its Application [M]. 2nd ed. New York: Springer Press, 2011.

[7] B.-Y.WANG. Kòngzhì Bùdĕngshì Jīchŭ (Foundations of Majorization Inequalities)[M]. Beijing: Beijing Normal University Press, 1990.

[8] MARSHALL A M, OLKIN I. Schur-Convexity, Gamma Functions,and Moments[J]. International Series of Numerical Mathematics, 2008, 157: 245 - 250.

[9] MILAN MERKLE. On log-convexity of a ratio of gamma functions[J]. Univ. Beograd. Publ. Elektrotehn. Fak. Ser. Mat., 1997, 8 : 114-119.

[10] MILAN MERKLE. Conditions for convexity of a derivative and some applications to the Gamma function[J]. Aequationes Math., 1998, 55: 273-280.

第62篇　Schur-Convexity of Dual Form of a Class Symmetric Function

(SHI HUAN-NAN, ZHANG JING. J.Math. Inequal., Volume 8, Number 2 (2014):349 - 358)

Abstract: By properties of Schur-convex function, Schur-geometrically convex function and Schur-harmonically convex function, Schur-convexity, Schur-geometric and harmonic convexities of the dual form for a class of symmetric functions are simply proved. As an application, several inequalities are obtained, some of which extend the known ones.

Keywords: Schur-convexity; Schur-geometric convexity; Schur-harmonic convexity; inequality; log-convex function; symmetric functions; dual form

62.1　Introduction

Throughout the article, $\mathbf{R}$ denotes the set of real numbers, $\boldsymbol{x} = (x_1, x_2, \ldots, x_n)$ denotes n-tuple (n-dimensional real vectors), the set of vectors can be written as

$$\mathbf{R}^n = \{\boldsymbol{x} = (x_1, \ldots, x_n) : x_i \in \mathbf{R}, i = 1, \ldots, n\}$$

$$\mathbf{R}_+^n = \{\boldsymbol{x} = (x_1, \ldots, x_n) : x_i > 0, i = 1, \ldots, n\}$$

In particular, the notations $\mathbf{R}$ and $\mathbf{R}_+$ denote $\mathbf{R}^1$ and $\mathbf{R}_+^1$ respectively. For convenience, we introduce some definitions as follows.

Definition 1 [1-2] Let $\boldsymbol{x} = (x_1, \ldots, x_n)$ and $\boldsymbol{y} = (y_1, \ldots, y_n) \in \mathbf{R}^n$.

(i) $\boldsymbol{x} \geqslant \boldsymbol{y}$ means $x_i \geqslant y_i$ for all $i = 1, 2, \ldots, n$.

(ii) Let $\Omega \subset \mathbf{R}^n$, $\varphi \colon \Omega \to \mathbf{R}$ is said to be increasing if $\boldsymbol{x} \geqslant \boldsymbol{y}$ implies $\varphi(\boldsymbol{x}) \geqslant \varphi(\boldsymbol{y})$. φ is said to be decreasing if and only if $-\varphi$ is increasing.

Definition 2 [1-2] Let $\boldsymbol{x} = (x_1, \ldots, x_n)$ and $\boldsymbol{y} = (y_1, \ldots, y_n) \in \mathbf{R}^n$.

(i) $\boldsymbol{x}$ is said to be majorized by $\boldsymbol{y}$ (in symbols $\boldsymbol{x} \prec \boldsymbol{y}$) if $\sum_{i=1}^{k} x_{[i]} \leqslant \sum_{i=1}^{k} y_{[i]}$ for $k = 1, 2, \ldots, n-1$ and $\sum_{i=1}^{n} x_i = \sum_{i=1}^{n} y_i$, where $x_{[1]} \geqslant \cdots \geqslant x_{[n]}$ and $y_{[1]} \geqslant \cdots \geqslant y_{[n]}$ are rearrangements of $\boldsymbol{x}$ and $\boldsymbol{y}$ in a descending order.

(ii) Let $\Omega \subset \mathbf{R}^n$, $\varphi \colon \Omega \to \mathbf{R}$ is said to be a Schur-convex function on Ω if $\boldsymbol{x} \prec \boldsymbol{y}$ on Ω implies $\varphi(\boldsymbol{x}) \leqslant \varphi(\boldsymbol{y})$. φ is said to be a Schur-concave function on Ω if and only if $-\varphi$ is Schur-convex function on Ω.

Definition 3 [1-2] Let $\boldsymbol{x} = (x_1, \ldots, x_n)$ and $\boldsymbol{y} = (y_1, \ldots, y_n) \in \mathbf{R}^n$.

(i) $\Omega \subset \mathbf{R}^n$ is said to be a convex set if $\boldsymbol{x}, \boldsymbol{y} \in \Omega, 0 \leqslant \alpha \leqslant 1$ implies $\alpha \boldsymbol{x} + (1-\alpha)\boldsymbol{y} = (\alpha x_1 + (1-\alpha)y_1, \ldots, \alpha x_n + (1-\alpha)y_n) \in \Omega$.

(ii) Let $\Omega \subset \mathbf{R}^n$ be convex set. A function $\varphi \colon \Omega \to \mathbf{R}$ is said to be a convex function on Ω if

$$\varphi(\alpha \boldsymbol{x} + (1-\alpha)\boldsymbol{y}) \leqslant \alpha \varphi(\boldsymbol{x}) + (1-\alpha)\varphi(\boldsymbol{y})$$

for all $\boldsymbol{x}, \boldsymbol{y} \in \Omega$, and all $\alpha \in [0,1]$. φ is said to be a concave function on Ω if and only if $-\varphi$ is convex function on Ω.

(iii) Let $\Omega \subset \mathbf{R}^n$. A function $\varphi \colon \Omega \to \mathbf{R}$ is said to be a log-convex function on Ω if function $\log \varphi$ is convex.

Theorem A (Schur-Convex Function Decision Theorem)[1] Let $\Omega \subset \mathbf{R}^n$ is symmetric and has a nonempty interior convex set. Ω^0 is the interior of Ω. $\varphi \colon \Omega \to \mathbf{R}$ is continuous on Ω and differentiable in Ω^0. Then φ is the Schur-convex (Schur-concave) function, if and only if φ is symmetric on Ω and

$$(x_1 - x_2)\left(\frac{\partial \varphi}{\partial x_1} - \frac{\partial \varphi}{\partial x_2}\right) \geqslant 0(\leqslant 0) \tag{1}$$

holds for any $\boldsymbol{x} \in \Omega^0$.

The Schur-convex function was introduced by Issai Schur in 1923 and has important applications in analytic inequalities, combinatorial optimization, quantum physics, information theory, and so on. See[1].

Definition 4 [3] Let $\boldsymbol{x} = (x_1, \ldots, x_n) \in \mathbf{R}_+^n$ and $\boldsymbol{y} = (y_1, \ldots, y_n) \in \mathbf{R}_+^n$.

(i) $\Omega \subset \mathbf{R}_{+}^{n}$ is called a geometrically convex set if $(x_1^{\alpha} y_1^{\beta}, \ldots, x_n^{\alpha} y_n^{\beta}) \in \Omega$ for all $\boldsymbol{x}, \boldsymbol{y} \in \Omega$ and $\alpha, \beta \in [0, 1]$ such that $\alpha + \beta = 1$.

(ii) Let $\Omega \subset \mathbf{R}_{+}^{n}$. The function $\varphi : \Omega \to \mathbf{R}_{+}$ is said to be Schur-geometrically convex function on Ω if $(\log x_1, \ldots, \log x_n) \prec (\log y_1, \ldots, \log y_n)$ on Ω implies $\varphi(\boldsymbol{x}) \leqslant \varphi(\boldsymbol{y})$. The function φ is said to be a Schur-geometrically concave function on Ω if and only if $-\varphi$ is Schur-geometrically convex function.

Theorem B (Schur-Geometrically Convex Function Decision Theorem)[3]
Let $\Omega \subset \mathbf{R}_{+}^{n}$ be a symmetric and geometrically convex set with a nonempty interior Ω^{0}. Let $\varphi : \Omega \to \mathbf{R}_{+}$ be continuous on Ω and differentiable in Ω^{0}. If φ is symmetric on Ω and

$$(\log x_1 - \log x_2) \left(x_1 \frac{\partial \varphi}{\partial x_1} - x_2 \frac{\partial \varphi}{\partial x_2} \right) \geqslant 0 \quad (\leqslant 0) \tag{2}$$

holds for any $\boldsymbol{x} = (x_1, \cdots, x_n) \in \Omega^{0}$, then φ is a Schur-geometrically convex (Schur-geometrically concave) function.

The geometric Schur-convexity was investigated by Chu et al.[9], Guan[10], and Niculescu [11]. We also note that the authors use the term "Schur-multiplicative (geometric) convexity".

Recently, Xia et al.[12] introduced the notion of harmonically Schur-convex function and some interesting inequalities were obtained.

Definition 5[4] Let $\Omega \subset \mathbf{R}_{+}^{n}$.

(i) A set Ω is said to be harmonically convex if $\frac{\boldsymbol{x}\boldsymbol{y}}{\lambda \boldsymbol{x} + (1-\lambda)\boldsymbol{y}} \in \Omega$ for every $\boldsymbol{x}, \boldsymbol{y} \in \Omega$ and $\lambda \in [0, 1]$, where $\boldsymbol{x}\boldsymbol{y} = \sum\limits_{i=1}^{n} x_i y_i$ and $\frac{1}{\boldsymbol{x}} = \left(\frac{1}{x_1}, \cdots, \frac{1}{x_n} \right)$.

(ii) A function $\varphi : \Omega \to \mathbf{R}_{+}$ is said to be Schur-harmonically convex on Ω if $\frac{1}{\boldsymbol{x}} \prec \frac{1}{\boldsymbol{y}}$ implies $\varphi(\boldsymbol{x}) \leqslant \varphi(\boldsymbol{y})$.

Theorem C (Schur-Harmonically Convex Function Decision Theorem)[4]
Let $\Omega \subset \mathbf{R}_{+}^{n}$ be a symmetric and harmonically convex set with inner points and let $\varphi : \Omega \to \mathbf{R}_{+}$ be a continuously symmetric function which is differentiable on Ω^{0}. Then φ is Schur-harmonically convex(Schur-harmonically concave) on Ω if and only if

$$(x_1 - x_2) \left(x_1^{2} \frac{\partial \varphi(\boldsymbol{x})}{\partial x_1} - x_2^{2} \frac{\partial \varphi(\boldsymbol{x})}{\partial x_2} \right) \geqslant 0 \quad (\leqslant 0), \quad \boldsymbol{x} \in \Omega^{0} \tag{3}$$

Let interval $I \subset \mathbf{R}$ and let $f : I \to \mathbf{R}_+$ be a log-convex function. Define the symmetric function F_k by

$$F_k(\boldsymbol{x}) = \sum_{1 \leqslant i_1 < \ldots < i_k \leqslant n} \prod_{j=1}^{k} f(x_{i_j}), \quad k = 1, \ldots, n \tag{4}$$

In 2010, for 1,2 and $n-1$, I. Roventa [5] proved that $F_k(\boldsymbol{x})$ is a Schur-convex function on I^n, but without discuss the case of $2 < k < n-1$. In 2011, Shu-hong Wang et al.[6] studied completely Schur convexity, Schur geometric and harmonic convexities of $F_k(\boldsymbol{x})$ on I^n, using the above decision theorems, i.e. Theorem A, Theorem B and Theorem C respectively proved the following three theorems.

Theorem D Let $I \subset \mathbf{R}$ is a symmetric convex set with non-empty interior and let $f : I \to \mathbf{R}$ be continuous on I and differentiable in the interior of I. If f is a log-convex function, then for any $k = 1, 2, ..., n$, $F_k(\boldsymbol{x})$ is a Schur-convex function on I^n

Theorem E Let $I \subset \mathbf{R}_+$ is a symmetric convex set with non-empty interior and let $f : I \to \mathbf{R}_+$ be continuous on I and differentiable in the interior of I. If f is an increasing log-convex function, then for any $k = 1, 2, ..., n$, $F_k(\boldsymbol{x})$ is a Schur-geometrically convex function on I^n.

Theorem F Let $I \subset \mathbf{R}_+$ is a symmetric convex set with non-empty interior and let $f : I \to \mathbf{R}_+$ be continuous on I and differentiable in the interior of I. If f is an increasing log-convex function, then for any $k = 1, 2, ..., n$, $F_k(\boldsymbol{x})$ is a Schur-harmonically convex function on I^n.

In this paper, we study the dual form of $F_k(\boldsymbol{x})$

$$F_k^*(\boldsymbol{x}) = \prod_{1 \leqslant i_1 < \ldots < i_k \leqslant n} \sum_{j=1}^{k} f(x_{i_j}), \quad k = 1, \ldots, n \tag{5}$$

By properties of Schur-convex function, Schur-geometrically convex function and Schur-harmonically convex function, we obtained the following results:

Theorem 1 Let $I \subset \mathbf{R}$ is a symmetric convex set with non-empty interior and let $f : I \to \mathbf{R}$ be continuous on I and differentiable in the interior of I. If f is a log-convex function, then for any $k = 1, 2, ..., n$, $F_k^*(\boldsymbol{x})$ is a Schur-convex function on I^n

Theorem 2 Let $I \subset \mathbf{R}_+$ is a symmetric convex set with non-empty interior and let $f : I \to \mathbf{R}_+$ be continuous on I and differentiable in the interior of I. If f is an increasing log-convex function, then for any $k = 1, 2, ..., n$, $F_k^*(x)$ is a Schur-geometrically convex function on I^n.

Theorem 3 Let $I \subset \mathbf{R}_+$ is a symmetric convex set with non-empty interior and let $f : I \to \mathbf{R}_+$ be continuous on I and differentiable in the interior of I. If f is an increasing log-convex function, then for any $k = 1, 2, ..., n$, $F_k^*(x)$ is a Schur-harmonically convex function on I^n.

62.2　Lemmas

To prove the above three theorems, we need the following lemmas.

Lemma 1[1-2] If φ is symmetric and convex (concave) on symmetric convex set Ω, then φ is Schur-convex (Schur-concave) on Ω.

Lemma 2[2] Let $\Omega \subset \mathbf{R}^n$, $\varphi\colon \Omega \to \mathbf{R}_+$. Then $\log \varphi$ is Schur-convex (Schur-concave) if and only if φ is Schur-convex (Schur-concave).

Lemma 3[1-2] Let $\Omega \subset \mathbf{R}^n$ be open convex set, $\varphi : \Omega \to \mathbf{R}$. For $x, y \in \Omega$, defined one variable function $g(t) = \varphi(tx + (1-t)y)$ on interval $(0, 1)$. Then φ is convex (concave) on Ω if and only if g is convex (concave) on $[0, 1]$ for all $x, y \in \Omega$.

Lemma 4 Let $x = (x_1, \ldots, x_m)$ and $y = (y_1, \ldots, y_m) \in \mathbf{R}^m$. If f is a log-convex function, then the functions $p(t) = \log g(t)$ is convex on $[0, 1]$, where

$$g(t) = \sum_{j=1}^{m} f(tx_j + (1-t)y_j)$$

Proof

$$p'(t) = \frac{g'(t)}{g(t)}$$

where

$$g'(t) = \sum_{j=1}^{m}(x_j - y_j)f'(tx_j + (1-t)y_j)$$

$$p''(t) = \frac{g''(t)g(t) - (g'(t))^2}{g^2(t)}$$

where

$$g''(t) = \sum_{j=1}^{m}(x_j - y_j)^2 f''(tx_j + (1-t)y_j)$$

by the Cauchy inequality, we have

$$g''(t)g(t) - (g'(t))^2$$

$$= \left(\sum_{j=1}^{m} (x_j - y_j)^2 f''(tx_j + (1-t)y_j) \right) \left(\sum_{j=1}^{m} f(tx_j + (1-t)y_j) \right) -$$

$$\left(\sum_{j=1}^{m} (x_j - y_j) f'(tx_j + (1-t)y_j) \right)^2$$

$$\geqslant \left(\sum_{j=1}^{m} |x_j - y_j| \sqrt{f''(tx_j + (1-t)y_j)} \cdot \sqrt{f(tx_j + (1-t)y_j)} \right)^2 -$$

$$\left(\sum_{j=1}^{m} (x_j - y_j) f'(tx_j + (1-t)y_j) \right)^2$$

From the log-convexity of f it follows that $(\log f(u))'' = \frac{f''(u)f(u) - (f'(u))^2}{f^2(u)} \geqslant 0$, hence

$$\sqrt{f''(tx_j + (1-t)y_j)} \cdot \sqrt{f(tx_j + (1-t)y_j)} \geqslant f'(tx_j + (1-t)y_j)$$

and then $g''(t)g(t) - (g'(t))^2 \geqslant 0$, i.e. $p''(t) \geqslant 0$, that is $p(t) = \log g(t)$ is convex on $[0,1]$.

The proof of Lemma 4 is completed.

Lemma 5 Let

$$f(t) = \frac{x^t - 1}{t}$$

If $x > 1$, then $f(t)$ is a log-convex function on $\mathbf{R}_+$.

Proof By computing, we have

$$(\log f(t))'' = -\frac{x^t (\log x)^2}{(x^t - 1)^2} + \frac{1}{t^2} \tag{6}$$

We need only prove $(\log f(t))'' \geqslant 0$. It equivalent to

$$t^2 x^t (\log x)^2 \leqslant (x^t - 1)^2$$

In both sides the inequality (6), dividing by x^t and extracting the square root, then the inequality (6) equivalent to

$$g(t) := x^{\frac{t}{2}} - x^{-\frac{t}{2}} - t \log x \geqslant 0$$

When $x > 1$, $g'(t) = \frac{1}{2} \log x \left(x^{\frac{t}{2}} + x^{-\frac{t}{2}} - 2 \right) = \frac{1}{2} \log(x) \left(x^{\frac{t}{4}} - 1 \right)^2 x^{-\frac{t}{2}} \geqslant 0$, hence $g(t)$ is increasing on $\mathbf{R}_+$, and then $g(t) \geqslant g(0) = 0$, that is $(\log f(t))'' \geqslant 0$.

The proof of Lemma 5 is completed.

62.3　Proof of Main Results

Proof of Theorem 1　For any $1 \leqslant i_1 < \cdots < i_k \leqslant n$, by Lemma 3 and Lemma 4, it follows that $\log \sum\limits_{j=1}^{k} f(x_{i_j})$ is convex on I^k. Obviously, $\log \sum\limits_{j=1}^{k} f(x_{i_j})$ is also convex on I^n, and then

$$\log F_k^*(\boldsymbol{x}) = \sum_{1 \leqslant i_1 < \cdots < i_k \leqslant n} \log \sum_{j=1}^{k} f(x_{i_j})$$

is convex on I^n. Furthermore, it is clear that $\log F_k^*(\boldsymbol{x})$ is symmetric on I^n, by Lemma 1, it follows that $\log F_k^*(\boldsymbol{x})$ is Schur-convex on I^n, and then from Lemma 2 we conclude that $F_k^*(\boldsymbol{x})$ is also Schur-convex on I^n.

The proof of Theorem 1 is completed.

Proof of Theorem 2　For $\boldsymbol{x} \in I \subset \mathbf{R}_+$ and $x_1 \neq x_2$, we have

$$\Delta = (\log x_1 - \log x_2) \left(x_1 \frac{\partial F_k^*}{\partial x_1} - x_2 \frac{\partial F_k^*}{\partial x_2} \right)$$

$$= (\log x_1 - \log x_2) \left(x_1 \frac{\partial F_k^*}{\partial x_1} - x_1 \frac{\partial F_k^*}{\partial x_2} + x_1 \frac{\partial F_k^*}{\partial x_2} - x_2 \frac{\partial F_k^*}{\partial x_2} \right)$$

$$= x_1 \frac{\log x_1 - \log x_2}{x_1 - x_2} (x_1 - x_2) \left(\frac{\partial F_k^*}{\partial x_1} - \frac{\partial F_k^*}{\partial x_2} \right) + \frac{\partial F_k^*}{\partial x_2} (x_1 - x_2)(\log x_1 - \log x_2)$$

Since $F_k^*(\boldsymbol{x})$ is Schur-convex on I^n, by Theorem A, we have

$$(x_1 - x_2) \left(\frac{\partial F_k^*}{\partial x_1} - \frac{\partial F_k^*}{\partial x_2} \right) \geqslant 0$$

Notice that f and $\log t$ is increasing, we have $\frac{\partial F_k^*}{\partial x_2} \geqslant 0$, $\frac{\log x_1 - \log x_2}{x_1 - x_2} \geqslant 0$ and $(x_1 - x_2)(\log x_1 - \log x_2) \geqslant 0$, so that $\Delta \geqslant 0$, by Theorem B, it follows that $F_k^*(\boldsymbol{x})$ is Schur-geometric convex on I^n.

Proof of Theorem 3　The proof of Theorem 3 similar to Theorem 2, the detailed proof is left to the reader.

Remark 1 If using the decision theorems, i.e. Theorem A, Theorem B and Theorem C respectively direct to prove Theorem 1, Theorem 2 and Theorem 3, I am afraid not above proofs are simple, interested readers may wish to try.

62.4 Applications

Theorem 4 The symmetric function

$$Q_k(\boldsymbol{x}) = \prod_{1 \leqslant i_1 < \cdots < i_k \leqslant n} \sum_{j=1}^{k} \frac{1 + x_{i_j}}{1 - x_{i_j}}, \quad k = 1, \ldots, n \tag{7}$$

is Schur-convex function, Schur-geometrically and harmonically convex function on $(0,1)^n$. And for $\boldsymbol{x} \in (0,1)^n$, we have

$$\prod_{1 \leqslant i_1 < \cdots < i_k \leqslant n} \sum_{j=1}^{k} \frac{1 + x_{i_j}}{1 - x_{i_j}} \geqslant \left(\frac{k(n+s)}{n-s} \right)^{C_n^k}, \quad k = 1, \ldots, n \tag{8}$$

where $s = \sum_{i=1}^{n} x_i$ and $C_n^k = \frac{n!}{k!(n-k)!}$.

Proof Let $f(x) = \frac{1+x}{1-x}, x \in (0,1)$. By computing, we have $f'(x) = \frac{2}{(1-x)^2} > 0$ and $\log(f(x))'' = \frac{4x}{(1+x)^2(1-x)^2} \geqslant 0$, that is f is an increasing log-convex function. By Theorem 1, Theorem 2 and Theorem 3, it follows that $Q_k(\boldsymbol{x})$ is respectively Schur-convex function, Schur-geometrically and harmonically convex function on $(0,1)^n$.

Since $\boldsymbol{y} = \left(\frac{s}{n}, \frac{s}{n}, \ldots, \frac{s}{n} \right) \prec \boldsymbol{x} = (x_1, x_2, \ldots, x_n)$, from Schur-convexity of $Q_k(\boldsymbol{x})$, it follows that $Q_k(\boldsymbol{y}) \leqslant Q_k(\boldsymbol{x})$, i.e. inequality (8) holds.

The proof of Theorem 4 is completed.

Specially, taking $k = 1, s = 1$, from the inequality (8) we can get the known Klamkin inequality

$$\prod_{i=1}^{n} \frac{1 + x_i}{1 - x_i} \geqslant \left(\frac{n+1}{n-1} \right)^n \tag{9}$$

By analogous proof with Theorem 4, we can obtain the following theorem.

Theorem 5 The symmetric function

$$R_k(\boldsymbol{x}) = \prod_{1 \leqslant i_1 < \cdots < i_k \leqslant n} \sum_{j=1}^{k} \frac{x_{i_j}}{1 - x_{i_j}}, \quad k = 1, \ldots, n \tag{10}$$

is Schur-convex function, Schur-geometrically and harmonically convex function on $[\frac{1}{2}, 1)^n$. And for $\boldsymbol{x} \in [\frac{1}{2}, 1)^n$, we have

$$\prod_{1 \leqslant i_1 < \cdots < i_k \leqslant n} \sum_{j=1}^{k} \frac{x_{i_j}}{1 - x_{i_j}} \geqslant \left(\frac{ks}{n - s} \right)^{C_n^k}, \quad k = 1, \ldots, n \tag{11}$$

where $s = \sum\limits_{i=1}^{n} x_i$ and $C_n^k = \frac{n!}{k!(n-k)!}$.

Theorem 6 The symmetric function

$$D_k(\boldsymbol{x}) = \prod_{1 \leqslant i_1 < \cdots < i_k \leqslant n} \sum_{j=1}^{k} x_{i_j}^{x_{i_j}}, \quad k = 1, \ldots, n \tag{12}$$

is Schur-convex on $\mathbf{R}_+^n$ and Schur-geometric and harmonic convex on $[e^{-1}, +\infty)^n$. And for $\boldsymbol{x} \in \mathbf{R}_+^n$, we have

$$\prod_{1 \leqslant i_1 < \cdots < i_k \leqslant n} \sum_{j=1}^{k} x_{i_j}^{x_{i_j}} \geqslant \left(k[A(\boldsymbol{x})]^{A(\boldsymbol{x})} \right)^{C_n^k}, \quad k = 1, \ldots, n \tag{13}$$

where $A(\boldsymbol{x}) = \frac{1}{n} \sum\limits_{i=1}^{n} x_i$ and $C_n^k = \frac{n!}{k!(n-k)!}$.

Proof It is not difficult to verify that x^x is log-convex function on $(0, +\infty)$ and increasing on $[e^{-1}, +\infty)$. By Theorem 1, Theorem 2 and Theorem 3, it follows that $D_k(\boldsymbol{x})$ is Schur-convex on $\mathbf{R}_+^n$, is Schur-geometric and harmonic convex on $[e^{-1}, +\infty)^n$.

Since $\boldsymbol{y} = (A(\boldsymbol{x}), A(\boldsymbol{x}), \ldots, A(\boldsymbol{x})) \prec \boldsymbol{x} = (x_1, x_2, \ldots, x_n)$, from Schur-convexity of $D_k(\boldsymbol{x})$, it follows that $D_k(\boldsymbol{y}) \leqslant D_k(\boldsymbol{x})$, i.e. inequality (13) holds.

The proof of Theorem 6 is completed.

From Lemma 5 and Theorem 1, we can obtain the following Theorem 7.

Theorem 7 Let $x > 1$.

$$P_k(\boldsymbol{t}) = \prod_{1 \leqslant i_1 < \cdots < i_k \leqslant n} \sum_{j=1}^{k} \frac{x^{t_{i_j}} - 1}{t_{i_j}}, \quad k = 1, \ldots, n \tag{14}$$

is Schur-convex on $\mathbf{R}_+^n$.

And for $p, q \in \mathbf{R}_+^n$ and $p \prec q$, we have

$$\prod_{1 \leqslant i_1 < \cdots < i_k \leqslant n} \sum_{j=1}^{k} \frac{x^{p_{i_j}} - 1}{p_{i_j}} \leqslant \prod_{1 \leqslant i_1 < \cdots < i_k \leqslant n} \sum_{j=1}^{k} \frac{x^{q_{i_j}} - 1}{q_{i_j}}, \quad k = 1, \dots, n \quad (15)$$

Taking $n = 2, k = 1$ and $p = (m, m), q = (m + r, m - r)$, from the inequality (15) we can get the known inequality

$$(x^{m-r} - 1)(x^{m+r} - 1) \geqslant \left(1 - \frac{r^2}{m^2}\right)(x^m - 1)^2 \quad (16)$$

where $r \in \mathbf{N}, m \geqslant 2, r < m$.

Taking $k = 1$, from the inequality (15) we can get the inequality (3) in [7].

$$\prod_{j=1}^{n} q_j \left(x^{p_j} - 1\right) \leqslant \prod_{j=1}^{n} p_j \left(x^{q_j} - 1\right) \quad (17)$$

Taking $k = n$, from the inequality (15) we can get the inequality

$$\sum_{i=1}^{n} \frac{x^{p_i} - 1}{p_i} \leqslant \sum_{i=1}^{n} \frac{x^{q_i} - 1}{q_i} \quad (18)$$

Acknowledgment

The work was supported by Funding Project for Academic Human Resources Development in Institutions of Higher Learning under the Jurisdiction of Beijing Municipality (PHR (IHLB)) (PHR201108407). Thanks for the help. This article was typeset by using AmS-LaTeX

References

[1] MARSHALL A M, OLKIN I, ARNOLD B C. Inequalities: Theory of Majorization and Its Application [M]. 2nd ed. New York: Springer Press, 2011.

[2] B.-Y.WANG. Kòngzhì Bùděngshì Jīchǔ (Foundations of Majorization Inequalities)[M]. Beijing: Beijing Normal University Press, 1990.

[3] X M ZHANG. Geometrically Convex Functions[M]. Hefei: An hui University Press, 2004.

[4]　Y M CHU, T C SUN. The Schur harmonic convexity for a class of symmetric functions[J]. Acta Mathematica Scientia 2010, 30B (5):1501-1506.

[5]　IONEL ROVENŢA. Schur convexity of a class of symmetric functions[J]. Annals of the University of Craiova, Mathematics and Computer Science Series, 2010, 37 (1): 12-18.

[6]　S H WANG, T Y ZHANG, Z Q HUA. Schur Convexity and Schur Multiplicatively Convexity and Schur Harmonic Convexity for a Class of Symmetric Functions[J]. Journal of Inner Mongolia University for the Nationalities (Natural Sciences), 2011, 26(4): 387-390.

[7]　KEIICHI WATANABE. On relation between a Schur, Hardy-Littlewood-Pólya and Karamata's theorem and an inequality of some products of $x^p - 1$ derived from the Furuta inequality[J]. J. Inequal. Appl., 2013, 2013:137 doi:10.1186/1029-242X-2013-137.

[8]　H N Shi. Theory of majorization and analytic Inequalities[M]. Harbin: Harbin Institute of Technology Press, 2012.

[9]　Y CHU, X M ZHANG, G WANG. The Schur geometrical convexity of the extended mean values[J]. J. Convex Anal., 2008, 15(4): 707-718.

[10]　K Z GUAN. A class of symmetric functions for multiplicatively convex function[J]. Math. Inequal. Appl., 2007, 10(4): 745-753.

[11]　C P NICULESCU. Convexity according to the geometric mean[J]. Math. Inequal. Appl., 2000, 3(2): 155-167.

[12]　W F XIA, Y M CHU. Schur-convexity for a class of symmetric functions and its applications[J]. J. Inequal. Appl. 2009 (2009), Article ID 493759, 15 pages.

第63篇　等差数列的凸性和对数凸性

(石焕南, 李明. 湖南理工学院学报(自然科学版),2014,27 (3):1-6)

摘　要:　本文研究了等差数列的凸性和对数凸性. 进而利用受控理论证明了一些等差数列不等式.

关键词:　等差数列; 凸性; 对数凸性; 不等式; 受控

本文研究等差数列的凸性和对数凸性并利用受控理论证明一些等差数列不等式.

设 $\{a_i\}$ 是公差为 d 的等差数列, 则其通项 $a_i = a_1 + (i-1)d$, 前 n 项之和

$$S_n = \sum_{i=1}^{n} a_i = \frac{n(a_1 + a_n)}{2} = na_1 + \frac{n(n-1)d}{2}$$

在本文中, $\mathbf{R}^n$ 和 $\mathbf{R}_+^n$ 分别表示 n 维实数集和 n 维正实数集, 并记 $\mathbf{R}^1 = \mathbf{R}, \mathbf{R}_+^1 = \mathbf{R}_+, \mathbf{Z}_+^n$ 表示非负整数集. 先回忆一下数列的凸性和对数凸性的定义概念.

定义 1[1-2]　若实数列 $\{a_i\}$ (有限的 $\{a_i\}_{i=1}^n$ 或无限的 $\{a_i\}_{i=1}^\infty$) 满足条件

$$a_{i-1} + a_{i+1} \geqslant 2a_i \ (i = 2, \ldots, n-1 \text{或 } i \geqslant 2) \tag{1}$$

则称 $\{a_i\}$ 是一个凸数列. 若不等式(1)反向, 则称数列 $\{a_i\}$ 是一个凹数列.

显然, 等差数列既是凸数列, 也是凹数列.

定义 2[1-2]　若非负实数列 $\{a_i\}$ (有限的 $\{a_i\}_{i=1}^n$ 或无限的 $\{a_i\}_{i=1}^\infty$) 满足条件

$$a_{i-1}a_{i+1} \geqslant a_i^2 \ (i = 2, \ldots, n-1 \text{或 } i \geqslant 2) \tag{2}$$

则称 $\{a_i\}$ 是一个对数凸数列. 若不等式(2)反向, 则称数列 $\{a_i\}$ 是一个对数凹数列.

63.1　主要结果

我们的主要结果是下述六个定理.

定理1 若 $\{a_i\}$ 是非负等差数列, 则 $\{a_i\}$ 是对数凹数列.

定理2 若 $\{a_i\}$ 是正项等差数列, 公差 $d \geqslant 0$, 则 $\left\{\dfrac{a_i}{a_{i-1}}\right\}$ 既是凸数列, 也是对数凸数列.

定理3 若 $\{a_i\}$ 是非负等差数列, 公差 $d \geqslant 0$, 则 $\{ia_i\}$ 既是凸数列, 也是对数凹数列.

定理4 若 $\{a_i\}$ 为正项等差数列, 则数列 $\left\{\dfrac{1}{a_i}\right\}$ 是凸数列.

定理5 若非负等差数列 $\{a_i\}$ 的公差 $d \geqslant 0$, 则其前 n 项和数列 $\{S_i\}$ 既是凸数列, 也是对数凹数列.

定理6 若 $\{a_i\}$ 为非负等差数列, 公差 $d \geqslant 0$, $T_i = \prod_{j=1}^{i} a_j$, 则 $\{T_i\}$ 既是凸数列, 也是对数凸数列.

63.2　定义和引理

定义 3[3-4]　设 $x = (x_1, ..., x_n), y = (y_1, ..., y_n) \in \mathbf{R}^n$.

(i) 若 $\sum_{i=1}^{k} x_{[i]} \leqslant \sum_{i=1}^{k} y_{[i]}$, $k = 1, 2, ..., n-1$, 且 $\sum_{i=1}^{n} x_i = \sum_{i=1}^{n} y_i$, 则称 x 被 y 所控制, 记作 $x \prec y$. 其中 $x_{[1]} \geqslant ... \geqslant x_{[n]}$ 和 $y_{[1]} \geqslant ... \geqslant y_{[n]}$ 分别是 x 和 y 的递减重排.

(ii) $x \leqslant y$ 表示对所有的 $i = 1, \dots, n$, $x_i \leqslant y_i$.

定义 4[3-4]　设 $\Omega \subset \mathbf{R}^n, \varphi : \Omega \to \mathbf{R}$, 若 $\forall x, y \in \Omega$, $x \leqslant y$, 恒有 $\varphi(x) \leqslant \varphi(y)$, 则称 φ 为 Ω 上的增函数; 若 $-\varphi$ 是 Ω 上的增函数, 则称 φ 为 Ω 上的减函数.

引理 1[3-4]　设 $x = (x_1, ..., x_n) \in \mathbf{R}^n$, $\bar{x} = \dfrac{1}{n} \sum_{i=1}^{n} x_i$, 则

$$(\bar{x}, \bar{x}, \dots, \bar{x}) \prec (x_1, x_2, \dots, x_n) \tag{3}$$

引理 2[1,5]　数列 $\{a_i\}$ 是凸数列的充要条件为: $\forall p, q \in \mathbf{Z}_+^n$, 当 $p \prec q$ 时, 恒有

$$a_{q_1} + \cdots + a_{q_n} \geqslant a_{p_1} + \cdots + a_{p_n} \tag{4}$$

引理 3[1,6]　设 $f : I \subset \mathbf{R} \to \mathbf{R}$ 是递增的凸函数, $\{a_n\}$ 是凸集 I 上的凸数列, $p, q \in \mathbf{Z}_+^n$, 若 $p \prec q$, 则

$$\sum_{i=1}^{n} f(a_{q_i}) \geqslant \sum_{i=1}^{n} f(a_{p_i}) \tag{5}$$

引理 4[5-6]   非负数列 $\{a_n\}$ 是对数凸数列的充要条件为: 对任意 $\boldsymbol{p}, \boldsymbol{q} \in \mathbf{Z}_+^n$, 若 $\boldsymbol{p} \prec \boldsymbol{q}$, 则

$$\prod_{i=1}^n a_{q_i} \geqslant \prod_{i=1}^n a_{p_i} \tag{6}$$

引理 5   若 $\{a_n\}$ 是对数凸数列, 则 $\{a_n\}$ 是凸数列.

证明   因 $\{a_n\}$ 是对数凸数列, 即 $a_{i-1}a_{i+1} \geqslant a_i^2$, 于是, $a_{i-1} + a_{i+1} \geqslant 2\sqrt{a_{i-1}a_{i+1}} \geqslant 2a_i$, 故 $\{a_n\}$ 是凸数列.

引理 6   若 $\{a_n\}$ 既是凸数列也是对数凹数列, 则 $\{\frac{1}{a_n}\}$ 是凸数列.

证明   若 $\{a_n\}$ 既是凸数列也是对数凹数列, 则 $(a_{i-1} + a_{i+1})a_i = 2a_i^2 \geqslant 2a_{i-1}a_{i+1}$, 即 $\frac{1}{a_{i+1}} + \frac{1}{a_{i-1}} \geqslant \frac{2}{a_i}$, 故 $\{\frac{1}{a_n}\}$ 是凸数列.

引理 7   设 $\{a_n\}$ 是单调递增的非负数列. 若 $\{a_n\}$ 是凸数列, 则 $\{ia_n\}$ 也是凸数列; 若 $\{a_n\}$ 是对数凹数列, 则 $\{ia_n\}$ 也是对数凹数列.

证明   若 $\{a_n\}$ 是凸数列, 则 $a_{i-1} + a_{i+1} \geqslant 2a_i$; 又因 $\{a_n\}$ 单调递增, 有 $a_{i+1} - a_{i-1} \geqslant 0$, 于是

$$(i+1)a_{i+1} + (i-1)a_{i-1} = i(a_{i-1} + a_{i+1}) + (a_{i+1} - a_{i-1}) \geqslant 2ia_i$$

故 $\{ia_n\}$ 是凸数列.

若 $\{a_n\}$ 是对数凹数列, 则 $a_{i-1}a_{i+1} \leqslant a_i^2$, 于是

$$(i+1)a_{i+1} \cdot (i-1)a_{i-1} = (i^2-1)a_{i-1}a_{i+1} \leqslant i^2 a_i^2$$

故 $\{ia_n\}$ 是对数凹数列.

引理 8[3-4]   设 $\boldsymbol{x}, \boldsymbol{y} \in \mathbf{R}^n, x_1 \geqslant x_2 \geqslant \cdots \geqslant x_n, \sum_{i=1}^n x_i = \sum_{i=1}^n y_i$, 若存在 $k, 1 \leqslant k \leqslant n$, 使得 $x_i \leqslant y_i, i = 1, ..., k, x_i \geqslant y_i, i = k+1, ..., n$, 则 $\boldsymbol{x} \prec \boldsymbol{y}$.

引理 9[3-4]   设 $I \subset \mathbf{R}$ 为一个区间, $\boldsymbol{x}, \boldsymbol{y} \in I^n \subset \mathbf{R}^n$, 则 $\boldsymbol{x} \prec \boldsymbol{y} \Leftrightarrow \forall$ 凸(凹)函数 $g: I \to \mathbf{R}$, 有

$$\sum_{i=1}^n g(x_i) \leqslant (\geqslant) \sum_{i=1}^n g(y_i)$$

63.3   主要结果的证明

定理1 的证明   因 $a_{i-1}a_{i+1} = (a_i - d)(a_i + d) = a_i^2 - d^2 \leqslant a_i^2$, 故 $\{a_n\}$ 是对数凹数列.

定理2 的证明　由定理1, $a_{i-2}a_i \leqslant a_{i-1}^2$, 于是

$$\frac{a_{i-1}}{a_{i-2}} \cdot \frac{a_{i+1}}{a_i} = \left(1 + \frac{d}{a_{i-2}}\right)\left(1 + \frac{d}{a_i}\right) = d\left(\frac{1}{a_{i-2}} + \frac{1}{a_i}\right) + \frac{d}{a_{i-2}} \cdot \frac{d}{a_i}$$

$$\geqslant 1 + \frac{2d}{\sqrt{a_{i-2}a_i}} + \left(\frac{d}{a_{i-1}}\right)^2 \geqslant 1 + \frac{2d}{a_{i-1}} + \left(\frac{d}{a_{i-1}}\right)^2 = \left(1 + \frac{d}{a_{i-1}}\right)^2 = \left(\frac{a_i}{a_{i-1}}\right)^2$$

故 $\left\{\frac{a_i}{a_{i-1}}\right\}$ 是对数凸数列. 由引理5 知 $\left\{\frac{a_i}{a_{i-1}}\right\}$ 也是凸数列.

定理3 的证明　由定理1 知, $\{a_i\}$ 是对数凹数列; 又 $d \geqslant 0$, $\{a_i\}$ 是递增的, 于是由引理7 知 $\{ia_i\}$ 既是凸数列, 也是对数凹数列.

定理4 的证明　由定理1 知, 等差数列 $\{a_i\}$ 是对数凹数列, 由引理6 知 $\{\frac{1}{a_i}\}$ 是凸数列.

定理5 的证明　因公差 $d \geqslant 0$, 所以 $a_{i+1} \geqslant a_i$, 于是

$$S_{i+1} + S_{i-1} = 2S_i + a_{i+1} - a_i \geqslant 2S_i$$

故 S_i 是凸数列.

因 $\{a_i\}$ 是对数凹数列, 所以

$$
\begin{aligned}
S_{i+1}S_{i-1} &= \frac{(i-1)(a_1 + a_{i-1})}{2} \cdot \frac{(i-1)(a_1 + a_{i+1})}{2} \\
&= \frac{1}{4}(i^2 - 1)\left[a_i^2 + a_1(a_{i+1} + a_{i-1}) + a_{i+1}a_{i-1}\right] \\
&\leqslant \frac{1}{4}(i^2 - 1)\left[a_1^2 + 2a_1 a_i + a_i^2\right] \\
&= \frac{1}{4}(i^2 - 1)(a_1 + a_i)^2 \\
&\leqslant \frac{1}{4}i^2(a_1 + a_i)^2 = S_i^2
\end{aligned}
$$

故 $\{S_i\}$ 也是对数凹数列.

定理6 的证明　因公差 $d \geqslant 0$, 所以 $a_{i+1} \geqslant a_i$, 于是

$$T_{i-1}T_{i+1} = \prod_{j=1}^{i-1} a_j \prod_{j=1}^{i+1} a_j \geqslant \left(\prod_{j=1}^{i} a_j\right)^2 = T_i^2$$

所以 $\{T_i\}$ 是对数凸数列. 由引理5 知 $\{T_i\}$ 也是凸数列.

63.4 应用

命题 1[7]   设 a_i 是正项等差数列, 则当 $r \geqslant 1$ 或 $r \leqslant 0$ 时, 有

$$n\left(\frac{a_1 + a_n}{2}\right)^r \leqslant \sum_{i=1}^{n} a_i^r \leqslant \frac{n(a_1^r + a_n^r)}{2} \tag{7}$$

当 $0 < r \leqslant 1$ 时, 不等式(7)均反向.

证明   不妨设公差 $d \geqslant 0$, 则 $a_1 \leqslant a_2 \leqslant \cdots \leqslant a_n$, 于是由引理1 与引理8, 有

$$\left(\underbrace{\frac{a_1 + a_n}{2}, \cdots, \frac{a_1 + a_n}{2}}_{2n}\right) \prec (a_n, a_n, a_{n-1}, a_{n-1}, \ldots, a_2, a_2, a_1, a_1)$$

$$\prec \left(\underbrace{a_n, \cdots, a_n}_{n}, \underbrace{a_1, \cdots, a_1}_{n}\right)$$

当 $r \geqslant 1$ 或 $r \leqslant 0$ 时, x^r 是凸函数, 由引理9, 有

$$2n\left(\frac{a_1 + a_n}{2}\right)^r \leqslant 2\sum_{i=1}^{n} a_i^r \leqslant n(a_1^r + a_n^r)$$

由此即得式(7). 当 $0 < r \leqslant 1$ 时, x^r 是凹函数, 由引理9 知不等式(7)均反向.

命题 2[8]   设 $\{a_i\}$ 是正项等差数列, 公差 $d \geqslant 0$, n 为自然数, 则

$$\frac{a_2}{a_1}C_n^0 + \frac{a_3}{a_2}C_n^1 + \cdots + \frac{a_{n+2}}{a_{n+1}}C_n^n \leqslant 2^{n-1}\left(\frac{a_2}{a_1} + \frac{a_{n+2}}{a_{n+1}}\right) \tag{8}$$

证明   由定理2 知 $\left\{\frac{a_i}{a_{i-1}}\right\}$ 是凸数列. 由引理8, 不难验证

$$\boldsymbol{x} = \left(\underbrace{n+2, \cdots, n+2}_{C_n^n}, \underbrace{n+1, \cdots, n+1}_{C_n^1}, \ldots, \underbrace{2, \cdots, 2}_{C_n^0}\right)$$

$$\prec \left(\underbrace{n+2, \cdots, n+2}_{2^{n-1}}, \underbrace{2, \cdots, 2}_{2^{n-1}}\right) = \boldsymbol{y} \tag{9}$$

事实上, 显然 $x_1 \geqslant x_2 \geqslant \cdots \geqslant x_{2^n}$. 又

$$\sum_{i=1}^{2^n} x_i = \sum_{i=0}^{n} \mathrm{C}_n^i (i+2) = \sum_{i=0}^{n} i\mathrm{C}_n^i + 2\sum_{i=0}^{n} \mathrm{C}_n^i = n2^{n-1} + 2^{n+1}$$

$$\sum_{i=1}^{2^n} y_i = (n+2)2^{n-1} + 2 \cdot 2^{n-1} = n2^{n-1} + 2^{n+1} = \sum_{i=1}^{2^n} x_i$$

易见对于 $k = 2^{n-1}$, 当 $i = 1, 2, \ldots, k$ 时, $x_i \leqslant y_i$, 当 $i = k+1, k+2, \ldots, n$ 时, $x_i \geqslant y_i$. 这样 $\boldsymbol{x}$ 和 $\boldsymbol{y}$ 满足引理8 的条件, 故有 $\boldsymbol{x} \prec \boldsymbol{y}$, 从而由引理2 知(8) 成立.

命题 3[9]   设 $\{a_i\}$ 是正项等差数列, n 为自然数, 则

$$a_1\mathrm{C}_n^0 + a_2\mathrm{C}_n^1 + \cdots + a_{n+1}\mathrm{C}_n^n = 2^{n-1}(a_1 + a_{n+1}) \tag{10}$$

$$\frac{\mathrm{C}_n^0}{a_1} + \frac{\mathrm{C}_n^1}{a_2} + \cdots + \frac{\mathrm{C}_n^n}{a_{n+1}} \leqslant 2^{n-1}\left(\frac{1}{a_1} + \frac{1}{a_{n+1}}\right) \tag{11}$$

证明   类似于(8), 可得

$$\boldsymbol{x} = \left(\underbrace{n+1, \cdots, n+1}_{\mathrm{C}_n^n}, \underbrace{n, \cdots, n}_{\mathrm{C}_n^1}, \ldots, \underbrace{1, \cdots, 1}_{\mathrm{C}_n^0}\right)$$

$$\prec \left(\underbrace{n+1, \cdots, n+1}_{2^{n-1}}, \underbrace{1, \cdots, 1}_{2^{n-1}}\right) = \boldsymbol{y} \tag{12}$$

因等差数列 $\{a_i\}$ 是凸数列, 从而据引理2, 由(12), 有

$$a_1\mathrm{C}_n^0 + a_2\mathrm{C}_n^1 + \cdots + a_{n+1}\mathrm{C}_n^n \leqslant 2^{n-1}(a_1 + a_{n+1}) \tag{13}$$

因等差数列 $\{-a_i\}$ 是凸数列, 从而据引理2, 由(12), 有

$$-a_1\mathrm{C}_n^0 - a_2\mathrm{C}_n^1 - \cdots - a_{n+1}\mathrm{C}_n^n \leqslant 2^{n-1}(-a_1 - a_{n+1})$$

即

$$a_1\mathrm{C}_n^0 + a_2\mathrm{C}_n^1 + \cdots + a_{n+1}\mathrm{C}_n^n \geqslant 2^{n-1}(a_1 + a_{n+1}) \tag{14}$$

结合(13)和(14)得(10).

据定理4, 数列 $\{\frac{1}{a_i}\}$ 是凸数列, 结合(12)可得(11).

命题 4   设 $\{a_i\}$ 是正项等差数列, 若 $n < p < m, n < q < m$, 且 $n + m = $

$p+q, n, m, p, q, k$ 均为正整数, 则

$$(ma_m)^k + (na_n)^k \geqslant (pa_p)^k + (qa_q)^k \tag{15}$$

$$(ma_m)^k \cdot (na_n)^k \leqslant (pa_p)^k \cdot (qa_q)^k \tag{16}$$

$$\frac{1}{(ma_m)^k} + \frac{1}{(na_n)^k} \geqslant \frac{1}{(pa_p)^k} + \frac{1}{(qa_q)^k} \tag{17}$$

证明 条件 $n < p < m, n < q < m$, 且 $n+m = p+q$ 意味着 $(p,q) \prec (n,m)$, 由定理3 知 $\{ia_i\}$ 既是凸数列也是对数凹数列, 注意对于 $k \geqslant 1$, t^k 是递增的凸函数, 由引理3 知(15)成立. 而由引理4, $(ma_m) \cdot (na_n) \leqslant (pa_p) \cdot (qa_q)$, 进而(16)成立. 由引理6 知 $\{\frac{1}{ia_i}\}$ 是凸数列, 进而由引理3 知(17)成立.

命题 5[10] 设 $\{a_i\}$ 是正项等差数列, 公差 $d \geqslant 0$, S_n 是前 n 项和. 若 $n < p < m, n < q < m$, 且 $n+m = p+q$, 则

$$(mS_m)^k + (nS_n)^k \geqslant (pS_p)^k + (qS_q)^k \tag{18}$$

$$(mS_m)^k \cdot (nS_n)^k \leqslant (pS_p)^k \cdot (qS_q)^k \tag{19}$$

$$\frac{1}{(mS_m)^k} + \frac{1}{(nS_n)^k} \geqslant \frac{1}{(pS_p)^k} + \frac{1}{(qS_q)^k} \tag{20}$$

证明 由定理5, $\{S_i\}$ 既是凸数列, 也是对数凹数列. 由引理7 知 $\{iS_i\}$ 既是凸数列也是对数凹数列, 余下证明类似于命题4, 从略.

命题 6[11] 设 $n > 1$, 则有下述阶乘不等式

$$(n!)^{m-1} \leqslant (m!)^{n-1}, \ m > n \tag{21}$$

$$((n+1)!)^n \leqslant \prod_{k=1}^{n}(2k)! \tag{22}$$

$$(n!)^n \leqslant \prod_{k=1}^{n}(2k-1)! \tag{23}$$

$$2^n \cdot n! \leqslant (2n)! \tag{24}$$

$$(2n)!! \leqslant (n+1)^n \tag{25}$$

证明 由定理6 知正项等差数列 $\{i\}$ 的前 i 项乘积数列 $\{i!\}$ 是对数凸数

列. 利用引理8 不难验证

$$\left(\underbrace{n,\cdots,n}_{m-1}\right) \prec \left(\underbrace{m,\cdots,m}_{n-1},\underbrace{1,\cdots,1}_{m-n}\right)$$

据引理4, 由上式即得式(21).

由引理1 有

$$\left(\underbrace{n+1,\cdots,n+1}_{n}\right) \prec (2,4,\ldots,2n) \tag{26}$$

和

$$\left(\underbrace{n,\cdots,n}_{n}\right) \prec (1,3,\ldots,2n-1) \tag{27}$$

易见

$$\left(n,\underbrace{2,\cdots,2}_{n}\right) \prec \left(2n,\underbrace{1,\cdots,1}_{n}\right) \tag{28}$$

据引理4, 由(26), (27)和(28)分别可得(22), (23)和(24).

利用引理8 不难验证

$$\left(n,n,n,\underbrace{2,\cdots,2}_{2n}\right) \prec \left(2n,n+1,\underbrace{1,\cdots,1}_{4n-1}\right)$$

据引理4, 由上式可得(25).

命题 7(Khinchin 不等式)[11]　设 n_k 为非负整数, 且 $\sum\limits_{i=1}^{k} n_i = n$, 则

$$\prod_{j=1}^{k} n_j! \leqslant \left(\frac{1}{2}\right)^n \prod_{j=1}^{k}(2n_j)! \tag{29}$$

证明　由定理6 知正项等差数列 $\{i\}$ 的前 i 项乘积数列 $\{i!\}$ 是对数凸数列. 利用引理8 不难验证

$$\left(n_1,\ldots,n_k,\underbrace{2,\cdots,2}_{n}\right) \prec \left(2n_1,\ldots,2n_k,\underbrace{1,\cdots,1}_{n}\right) \tag{30}$$

据引理4, 由(30)可得

$$2^n \prod_{j=1}^{k} n_j! \leqslant \prod_{j=1}^{k} (2n_j)!$$

即(29)成立.

命题 8　设$n \in \mathbf{Z}_+$, 则

$$\frac{1}{n} + \frac{1}{n+1} + \frac{1}{n+2} + \cdots + \frac{1}{n^2} \geqslant \frac{2(n^2-n+1)}{n(n+1)} > 1,\ n > 1 \quad (31)$$

$$\frac{1}{n} + \frac{1}{n+1} + \frac{1}{n+2} + \cdots + \frac{1}{2n} \leqslant \frac{3(n+1)}{4n} \leqslant \frac{3}{2} \quad (32)$$

证明　由定理4 知正项等差数列$\{i\}$ 的倒数数列$\{\frac{1}{i}\}$ 是凸数列. 由引理1 有

$$\left(\underbrace{\frac{n(n+1)}{2}, \cdots, \frac{n(n+1)}{2}}_{n^2-n+1} \right) \prec \left(n, n+1, \ldots, n^2 \right) \quad (33)$$

据引理2, 由(33)可得

$$\frac{1}{n} + \frac{1}{n+1} + \frac{1}{n+2} + \cdots + \frac{1}{n^2} \leqslant (n^2-n+1)\frac{1}{\frac{n(n+1)}{2}} = \frac{2(n^2-n+1)}{n(n+1)}$$

当$n > 2$ 时, 因

$$\frac{2(n^2-n+1)}{n(n+1)} > 1 \Leftrightarrow 2(n^2-n+1) > n(n+1) \Leftrightarrow (n-2)(n-1) > 0$$

成立. 而当$n=2$ 时, $\frac{1}{2} + \frac{1}{3} + \frac{1}{4} = \frac{13}{12} > 1$, 故(31)成立.

利用引理8 不难验证

$$(2n, 2n, 2n-1, 2n-1, \ldots, n+1, n+1, n, n) \prec \left(\underbrace{2n, \cdots, 2n}_{n+1}, \underbrace{n, \cdots, n}_{n+1} \right) \quad (34)$$

因$\{\frac{1}{i}\}$ 是凸数列. 由引理2 有

$$2\left(\frac{1}{n} + \frac{1}{n+1} + \frac{1}{n+2} + \cdots + \frac{1}{2n} \right) \leqslant (n+1)\left(\frac{1}{2n} + \frac{1}{n} \right) = \frac{3(n+1)}{2n}$$

于是

$$\frac{1}{n} + \frac{1}{n+1} + \frac{1}{n+2} + \cdots + \frac{1}{2n} \leqslant \frac{3(n+1)}{4n}$$

而 $\frac{3(n+1)}{4n} \leqslant \frac{3}{2} \Leftrightarrow 2 \leqslant 2n$, 故(32)成立.

(31)和(32)两式分别加细了文献[12] 中第185 页的两个不等式.

参考文献

[1]　石焕南. 受控理论与解析不等式[M]. 哈尔滨: 哈尔滨工业大学出版社, 2012.

[2]　PEČARIĆ J E, FRANK PROSCHAN, TONG Y L. Convex functions, partial orderings, and statistical applications [M]. New York: Academic Press.Inc., 1992.

[3]　王伯英. 控制不等式基础[M]. 北京: 北京师范大学出版社, 1990.

[4]　MARSHALL A W, OLKIN I, ARNOLD B C. Inequalities: theory of majorization and its application[M]. 2nd ed. New York: Springer Press, 2011.

[5]　石焕南, 李大矛. 凸数列的一个等价条件及其应用[J]. 曲阜师范大学学报(自然科学版), 2001, 27(4): 4-6.

[6]　石焕南. 凸数列的一个等价条件及其应用 II [J]. 数学杂志, 2004, 24(4): 390-394

[7]　李明. 关于正项等差数列幂和式的双边不等式[J]. 中国初等数学研究, 2014,5:41-42.

[8]　盛宏礼. 正项等差数列一类分式不等式[J]. 中国初等数学研究, 2014, 5: 58-61.

[9]　盛宏礼. 正项等差数列一类新不等式[J]. 数学通讯, 2011,11(下半月): 34 35.

[10]　李玉群, 李永利. 关于正项等差数列方幂的若干不等式研究[J]. 济源职业技术学院学报, 2005, 4(3): 21-23.

[11]　匡继昌. 常用不等式[M]. 第4 版. 济南: 山东科学技术出版社, 2010.

[12]　甘志国. 数列与不等式[M]. 哈尔滨: 哈尔滨工业大学出版社, 2014.

第64篇　等比数列的凸性和对数凸性

(石焕南. 广东第二师范学院学报,2015,35 (3):9-15)

摘　要:　本文研究等比数列的凸性和对数凸性, 进而利用受控理论证明一些等比数列不等式.

关键词:　等比数列; 凸性; 对数凸性; 不等式; 受控

在本文中, $\mathbf{R}^n$ 和$\mathbf{R}^n_+$ 分别表示n 维实数集和n 维正实数集, 并记$\mathbf{R}^1 = \mathbf{R}, \mathbf{R}^1_+ = \mathbf{R}_+, \mathbf{Z}^n_+$ 表示非负整数集.

设$\{b_i\}$ 是公比为q 的等比数列, 则其通项

$$b_i = b_1 q^{i-1} \tag{1}$$

前n 项之和

$$S_n = \sum_{i=1}^{n} b_i = \frac{b_1(1-q^n)}{1-q} = \frac{b_1 - b_n q}{1-q} \tag{2}$$

其前n 项积

$$T_n = \prod_{i=1}^{n} b_i = b_1^n q^{\frac{n(n-1)}{2}} \tag{3}$$

本文研究等比数列的凸性和对数凸性并利用受控理论证明一些等比数列不等式. 我们先回忆一下数列的凸性和对数凸性的定义.

定义 1[1-2]　若实数列$\{a_i\}$ (有限的$\{a_i\}_{i=1}^n$ 或无限的$\{a_i\}_{i=1}^{\infty}$) 满足条件

$$a_{i-1} + a_{i+1} \geqslant 2a_i \ (i = 2,\dots,n-1\text{或} i \geqslant 2) \tag{4}$$

则称$\{a_i\}$ 是一个凸数列. 若上述不等式反向, 则称数列$\{a_i\}$ 是一个凹数列.

定义 2[1-2]　若非负实数列$\{a_i\}$ (有限的$\{a_i\}_{i=1}^n$ 或无限的$\{a_i\}_{i=1}^{\infty}$) 满足条件

$$a_{i-1} a_{i+1} \geqslant a_i^2 \ (i = 2,\dots,n-1\text{或} i \geqslant 2) \tag{5}$$

则称$\{a_i\}$ 是一个对数凸数列. 若上述不等式(反向, 则称数列$\{a_i\}$ 是一个对

数凹数列.

我们的主要结论是下述12个定理.

定理 1　等比数列 $\{b_i\}$ 满足

$$b_{i-1}b_{i+1} = b_i^2,\ i \geqslant 2 \tag{6}$$

也就是说, $\{b_i\}$ 既是对数凸数列也是对数凹数列.

定理 2　若 $\{b_i\}$ 是非负等比数列, 则 $\{b_i\}$ 是凸数列.

定理 3　若 $\{b_i\}$ 是正项等比数列, 则 $\{b_i - b_{i-1}\}$ 既是对数凸数列也是对数凹数列, 还是凸数列.

定理 4　若 $\{b_i\}$ 是非负等比数列, 公比 $q \geqslant 1$, 则 $\{ib_i\}$ 既是凸数列也是对数凹数列.

定理 5　若 $\{b_i\}$ 是非负等比数列, 公比 $q \geqslant 1$, 则 $\{b_i^i\}$ 既是对数凸数列也是凸数列. 公比 $q \leqslant 1$, 则 $\{b_i^i\}$ 是对数凹数列.

定理 6　若 $\{b_i\}$ 是非负等比数列, 公比 $q > 0, b_1 \leqslant q$, 则 $\{b_i^{\frac{1}{i}}\}$ 是对数凹数列.

定理 7　若 $\{b_i\}$ 是非负等比数列, 则 $\{\frac{b_i}{i}\}$ 既是对数凸数列也是凸数列.

定理 8　设 $\{b_i\}$ 为正项等比数列, 则下述命题成立:

(1) $\{\frac{1}{b_i}\}$ 既是对数凸数列也是对数凹数列, 还是凸数列;

(2) 若 c 是非负常数, 则数列 $\{\frac{1}{b_i+c}\}$ 是对数凹数列;

(3) 若公比 $q \geqslant 1, c < b_1$, 则数列 $\{\frac{1}{b_i-c}\}$ 是对数凸数列也是凸数列.

定理 9　非负等比数列 $\{b_i\}$ 的前 n 项和数列 $\{S_i\}$ 是对数凹数列. 若公比 $q \geqslant 1$, 则 $\{S_i\}$ 也是凸数列.

定理 10　设 $\{b_i\}$ 是非负等比数列, c 是正常数, 若公比 $q \geqslant 1$ 且 $c < b_1$, 则 $\{S_i - c\}$ 既是凸数列也是对数凹数列.

定理 11　若 $\{b_i\}$ 是非负等比数列, 则数列 $\{\frac{S_i}{i}\}$ 是凸数列.

定理 12　设 $\{b_i\}$ 为非负等比数列, 公比 $q \geqslant 1$, $T_i = \prod_{j=1}^{i} b_j$ 为前 i 项乘积, 则 $\{T_i\}$ 既是凸数列也是对数凸数列.

定义 3[3-4]　设 $\boldsymbol{x} = (x_1, ..., x_n), \boldsymbol{y} = (y_1, ..., y_n) \in \mathbf{R}^n$.

(i) 若 $\sum\limits_{i=1}^{k} x_{[i]} \leqslant \sum\limits_{i=1}^{k} y_{[i]}, k = 1, 2, ..., n-1$, 且 $\sum\limits_{i=1}^{n} x_i = \sum\limits_{i=1}^{n} y_i$, 则称 $\boldsymbol{x}$ 被 $\boldsymbol{y}$ 所控制, 记作 $\boldsymbol{x} \prec \boldsymbol{y}$. 其中 $x_{[1]} \geqslant \cdots \geqslant x_{[n]}$ 和 $y_{[1]} \geqslant \cdots \geqslant y_{[n]}$ 分别是 $\boldsymbol{x}$ 和 $\boldsymbol{y}$ 的递减重排.

(ii) $\boldsymbol{x} \leqslant \boldsymbol{y}$ 表示对所有的 $i = 1, \ldots, n, x_i \leqslant y_i$.

定义 4[3-4]　设 $\Omega \subset \mathbf{R}^n, \varphi : \Omega \to \mathbf{R}$, 若 $\forall \boldsymbol{x}, \boldsymbol{y} \in \Omega, \boldsymbol{x} \leqslant \boldsymbol{y}$, 恒有 $\varphi(\boldsymbol{x}) \leqslant \varphi(\boldsymbol{y})$, 则称 φ 为 Ω 上的增函数; 若 $-\varphi$ 是 Ω 上的增函数, 则称 φ 为 Ω 上的减函数.

引理 1[3-4]   设 $\boldsymbol{x} = (x_1, ..., x_n) \in \mathbf{R}^n$, $\bar{x} = \frac{1}{n}\sum_{i=1}^{n} x_i$, 则

$$(\bar{x}, \bar{x}, \ldots, \bar{x}) \prec (x_1, x_2, \ldots, x_n) \tag{7}$$

引理 2[1,5]   设 $n \geqslant 2$, 数列 $\{a_i\}$ 是凸数列的充要条件为: $\forall \boldsymbol{p}, \boldsymbol{q} \in \mathbf{Z}_+^n$, 若 $\boldsymbol{p} \prec \boldsymbol{q}$, 恒有

$$a_{q_1} + \cdots + a_{q_n} \geqslant a_{p_1} + \cdots + a_{p_n} \tag{8}$$

引理 3[1,6]   设 $f: I \subset \mathbf{R} \to \mathbf{R}$ 是增的凸函数, $\{a_n\}$ 是凸集 I 上的凸数列, $\boldsymbol{p}, \boldsymbol{q} \in \mathbf{Z}_+^n$, 若 $\boldsymbol{p} \prec \boldsymbol{q}$, 则

$$\sum_{i=1}^{n} f(a_{q_i}) \geqslant \sum_{i=1}^{n} f(a_{p_i}) \tag{9}$$

引理 4[5-6]   非负数列 $\{a_n\}$ 是对数凸(凹)数列的充要条件为: 对任意 $\boldsymbol{p}, \boldsymbol{q} \in \mathbf{Z}_+^n$, 若 $\boldsymbol{p} \prec \boldsymbol{q}$, 则

$$\prod_{i=1}^{n} a_{q_i} \geqslant (\leqslant) \prod_{i=1}^{n} a_{p_i} \tag{10}$$

引理 5   若 $\{a_n\}$ 是对数凸数列, 则 $\{a_n\}$ 是凸数列.

证明   因 $\{a_n\}$ 是对数凸数列, 即 $a_{i-1}a_{i+1} \geqslant a_i^2$, 于是

$$a_{i-1} + a_{i+1} \geqslant 2\sqrt{a_{i-1}a_{i+1}} \geqslant 2a_i$$

即 $\{a_n\}$ 是凸数列.

引理 6   若 $\{a_n\}$ 既是凸数列也是对数凹数列, 则 $\left\{\frac{1}{a_n}\right\}$ 是凸数列.

证明   若 $\{a_n\}$ 既是凸数列也是对数凹数列, 则

$$(a_{i-1} + a_{i+1})a_i = 2a_i^2 \geqslant 2a_{i-1}a_{i+1}$$

即

$$\frac{1}{a_{i+1}} + \frac{1}{a_{i-1}} \geqslant \frac{2}{a_i}$$

由此得证.

引理 7   设 $\{a_n\}$ 是单调递增的非负数列. 若 $\{a_n\}$ 是凸数列(对数凹数列), 则 $\{ia_n\}$ 也是凸数列(对数凹数列).

证明   若 $\{a_n\}$ 是凸数列, 则 $a_{i-1} + a_{i+1} \geqslant 2a_i$; 又因 $\{a_n\}$ 单调递增,

有 $a_{i+1} - a_{i-1} \geqslant 0$, 于是

$$(i+1)a_{i+1} + (i-1)a_{i-1} = i(a_{i-1} + a_{i+1}) + (a_{i+1} - a_{i-1}) \geqslant 2ia_i$$

故 $\{ia_n\}$ 是凸数列. 若 $\{a_n\}$ 是对数凹数列, 则 $a_{i-1}a_{i+1} \leqslant a_i^2$, 于是

$$(i+1)a_{i+1} \cdot (i-1)a_{i-1} = (i^2 - 1)a_{i-1}a_{i+1} \leqslant i^2 a_i^2$$

故 $\{ia_n\}$ 是对数凹数列.

引理 8[3-4]　设 $\boldsymbol{x}, \boldsymbol{y} \in \mathbf{R}^n, x_1 \geqslant x_2 \geqslant \cdots \geqslant x_n, \sum\limits_{i=1}^{n} x_i = \sum\limits_{i=1}^{n} y_i$, 若存在 $k, 1 \leqslant k \leqslant n$, 使得 $x_i \leqslant y_i, i = 1, ..., k, x_i \geqslant y_i, i = k+1, ..., n$, 则 $\boldsymbol{x} \prec \boldsymbol{y}$.

引理 9[3-4]　设 $I \subset \mathbf{R}$ 为一个区间, $\boldsymbol{x}, \boldsymbol{y} \in I^n \subset \mathbf{R}^n$, 则 $\boldsymbol{x} \prec \boldsymbol{y} \Leftrightarrow \forall$ 凸 (凹) 函数 $g : I \to \mathbf{R}$, 有

$$\sum_{i=1}^{n} g(x_i) \leqslant (\geqslant) \sum_{i=1}^{n} g(y_i)$$

引理 10[1]　若 $\{a_n\}$ 是一个凸序列, 则 $\{A_k\}$ 也是一个凸序列, 其中 $A_k = \frac{1}{k}\sum\limits_{i=1}^{k} a_i$.

定理 1 的证明　因为

$$b_{i-1}b_{i+1} = (b_i q) \cdot (\frac{b_i}{q}) = b_i^2$$

所以据定义 2, $\{b_i\}$ 既是对数凸数列也是对数凹数列.

定理 2 的证明　由定理 1 和引理 5 即得证.

定理 3 的证明　由定理 1 和式 (1) 有

$$
\begin{aligned}
&(b_{i-1} - b_{i-2})(b_{i-1} - b_i) - (b_i - b_{i-1})^2 \\
=&b_{i-1}b_{i+1} - b_{i-1}b_i - b_{i-2}b_{i+1} + b_{i-2}b_i - (b_i^2 - 2b_{i-1}b_i + b_{i-1}^2) \\
=&b_i^2 - b_{i-1}b_i - b_{i-2}b_{i+1} + b_{i-1}^2 - (b_i^2 - 2b_{i-1}b_i + b_{i-1}^2) \\
=&b_{i-1}b_i - b_{i-2}b_{i+1} = b_1 q^{2i-3} - b_1 q^{2i-3} = 0
\end{aligned}
$$

故 $\{b_i - b_{i-1}\}$ 既是对数凸数列也是对数凹数列, 由引理 5 知 $\{b_i - b_{i-1}\}$ 也是凸数列.

定理 4 的证明　由定理 1 和定理 2 知 $\{b_i\}$ 是凸数列和对数凹数列, 又 $q \geqslant 1, \{b_i\}$ 递增. 于是由引理 7 知 $\{ib_i\}$ 是凸数列和对数凹数列.

定理 5 的证明　因 $q \geqslant (\leqslant)1$, 则 $b_{i+1} \geqslant (\leqslant)b_i$, 于是

$$b_{i-1}^{i} b_{i+1}^{i+1} = (b_{i-1}b_{i+1})^{i-1}b_{i+1}^2 = (b_i^2)^{i-1}b_{i+1}^2 \geqslant (\leqslant)(b_i^2)^{i-1}b_i^2 = (b_i^i)^2$$

故 $\{b_i^i\}$ 是对数凸(凹)数列, 进而由引理5 知 $q \geqslant 1$ 时, $\{b_i^i\}$ 也是凸数列.

定理 6 的证明　由于

$$b_{i-1}^{\frac{1}{i-1}} b_{i+1}^{\frac{1}{i+1}} = (b_1 q^{i-2})^{\frac{1}{i-1}} (b_1 q^i)^{\frac{1}{i+1}} = \left(q_{i-1}^{\frac{i^2-i-1}{i^2-1}} b_1^{\frac{i}{i^2-1}}\right)^2$$

$$\left(b_i^{\frac{1}{i}}\right)^2 = \left((b_1 q^{i-1})^{\frac{1}{i}}\right)^2 = \left(b_1^{\frac{1}{i}} q^{\frac{i-1}{i}}\right)^2$$

欲证

$$b_{i-1}^{\frac{1}{i-1}} b_{i+1}^{\frac{1}{i+1}} \leqslant \left(b_i^{\frac{1}{i}}\right)^2$$

只需证

$$q_{i-1}^{\frac{i^2-i-1}{i^2-1}} b_1^{\frac{i}{i^2-1}} \leqslant b_1^{\frac{1}{i}} q^{\frac{i-1}{i}}$$

此式等价于

$$\left(\frac{b_1}{q}\right)^{\frac{1}{i(i^2-1)}} \leqslant 1$$

因 $b_1 \leqslant q$, 上式成立.

定理 7 的证明　因

$$\frac{b_{i+1}}{i+1}\frac{b_{i-1}}{i-1} = \frac{b_i^2}{i^2-1} \geqslant \frac{b_i^2}{i^2}$$

故 $\left\{\frac{b_i}{i}\right\}$ 是对数凸数列, 进而由引理 5 知 $\left\{\frac{b_i}{i}\right\}$ 也是凸数列.

定理 8 的证明　(1) 因 $\frac{1}{b_{i-1}} \cdot \frac{1}{b_{i+1}} = \left(\frac{1}{b_i}\right)^2$, 故 $\left\{\frac{1}{b_i}\right\}$ 既是对数凸数列也是对数凹数列. 由引理 5 知 $\left\{\frac{1}{b_i}\right\}$ 还是凸数列.

(2) 若 c 是非负常数, 由定理 1 和定理2, 有

$$\frac{1}{b_{i-1}+c} \cdot \frac{1}{b_{i+1}+c} = \frac{1}{b_{i-1}b_{i+1}+c(b_{i-1}+b_{i+1})+c^2} \leqslant \frac{1}{b_i^2+2b_ic+c^2} = \left(\frac{1}{b_i+c}\right)^2$$

故 $\left\{\frac{1}{b_i+c}\right\}$ 是对数凹数列.

(3) 若公比 $q \geqslant 1, c < b_1$, 由定理 1 和定理2, 有

$$\frac{1}{b_{i-1}-c} \cdot \frac{1}{b_{i+1}-c} = \frac{1}{b_{i-1}b_{i+1}-c(b_{i-1}+b_{i+1})+c^2} \geqslant \frac{1}{b_i^2-2b_ic+c^2} = \left(\frac{1}{b_i-c}\right)^2$$

故 $\left\{\frac{1}{b_i-c}\right\}$ 是对数凸数列.

定理 9 的证明　因 $\{b_i\}$ 是凸数列, 则 $b_{i+1}+b_{i-1} \geqslant 2b_i$, 又 $b_{i+1}b_{i-1}=b_i^2$, 有

$$
\begin{aligned}
S_{i+1}S_{i-1} &= \frac{b_1+b_{i+1}q}{1-q} \cdot \frac{b_1+b_{i-1}q}{1-q} \\
&= \frac{b_1^2 - b_1 q(b_{i+1}+b_{i-1}) + b_{i+1}b_{i-1}q^2}{(1-q)^2} \\
&\leqslant \frac{b_1^2 - 2b_1 q b_i + b_i^2 q^2}{(1-q)^2} = \frac{(b_1-b_i q)^2}{(1-q)^2} = S_i^2
\end{aligned}
$$

故 $\{S_n\}$ 是对数凹数列. 若公比 $q \geqslant 1$, 有 $b_{i+1} \geqslant b_i$, 于是

$$
S_{i+1}+S_{i-1} = 2S_i + b_{i+1} - b_i \geqslant 2S_i
$$

故 $\{S_n\}$ 也是凸数列.

定理 10 的证明　据定理 9, $\{S_i\}$ 是凸数列, 有

$$
(S_{i-1}-c) + (S_{i+1}-c) = S_{i-1}+S_{i+1}-2c \geqslant 2S_i - 2c = 2(S_i-c)
$$

和

$$
(S_{i-1}-c)\cdot(S_{i+1}-c) = S_{i-1}S_{i+1}-c(S_{i-1}+S_{i+1})+c^2 \leqslant S_i^2 - 2cS_i + c^2 = (S_i-c)^2
$$

故 $\{S_i - c\}$ 既是凸数列也是对数凹数列.

定理 11 的证明　由定理 2 知 $\{b_i\}$ 是凸数列, 进而由引理 10 知 $\{\frac{S_i}{i}\}$ 是凸数列.

定理 12 的证明　因公差 $q \geqslant 1$, 有 $b_{i+1} \geqslant b_i$, 于是

$$
T_{i+1}T_{i-1} = \prod_{j=1}^{i+1} b_j \prod_{j=1}^{i-1} b_j \geqslant \left(\prod_{j=1}^{i} b_j\right)^2 = T_i^2
$$

所以 $\{T_i\}$ 是对数凸数列, 进而由引理 5 知 $\{T_i\}$ 也是凸数列.

命题 1　设 $\{b_i\}$ 是正项等差比数列, 则当 $r \geqslant 1$ 或 $r \leqslant 0$ 时, 有

$$
\left(\frac{b_1-b_n q}{n(1-q)}\right)^r \leqslant \frac{1}{n}\sum_{i=1}^{n} b_i^r \leqslant \frac{1}{n}\left(\frac{b_1-b_n q}{1-q}\right)^r \tag{11}
$$

当 $0 < r \leqslant 1$ 时, 上述不等式均反向.

证明　不妨设公比$q \geqslant 1$, 据引理 1 和引理8, 有

$$\left(\underbrace{\frac{b_1 - b_n q}{n(1-q)}, \cdots, \frac{b_1 - b_n q}{n(1-q)}}_{n}\right) \prec (b_n, b_{n-1}, \ldots, b_1) \prec \left(\frac{b_1 - b_n q}{1-q}, \underbrace{0, \cdots, 0}_{n-1}\right)$$

(12)

当$r \geqslant 1$ 或 $r \leqslant 0$ 时, x^r 是凸函数, 由引理9, 有

$$n\left(\frac{b_1 - b_n q}{n(1-q)}\right)^r \leqslant \sum_{i=1}^{n} b_i^r \leqslant \left(\frac{b_1 - b_n q}{1-q}\right)^r$$

由此即得式(11). 当$0 < r \leqslant 1$ 时, x^r 是凹函数, 由引理9 知上述不等式均反向.

命题 2[8]　设$\{a_i\}$ 是正项等比数列, n 为自然数, 则

$$(b_2 - b_1)^{C_n^0} (b_3 - b_2)^{C_n^1} \cdots (b_{n+2} - b_{n+1})^{C_n^n} = (b_2 - b_1)^{2^{n-1}} (b_{n+2} - b_{n+1})^{2^{n-1}}$$

(13)

证明　由定理1.2 知$\{b_i - b_{i-1}\}$ 既是对数凸数列也是对数凹数列. 由引理 8, 不难验证

$$\boldsymbol{x} = \left(\underbrace{n+2, \cdots, n+2}_{C_n^n}, \underbrace{n+1, \cdots, n+1}_{C_n^1}, \ldots, \underbrace{2, \cdots, 2}_{C_n^0}\right)$$

$$\prec \left(\underbrace{n+2, \cdots, n+2}_{2^{n-1}}, \underbrace{2, \cdots, 2}_{2^{n-1}}\right) = \boldsymbol{y}$$

(14)

事实上, 显然$x_1 \geqslant x_2 \geqslant \cdots \geqslant x_{2^n}$. 又

$$\sum_{i=1}^{2^n} x_i = \sum_{i=0}^{n} C_n^i (i+2) = \sum_{i=0}^{n} i C_n^i + 2 \sum_{i=0}^{n} C_n^i = n2^{n-1} + 2^{n+1}$$

$$\sum_{i=1}^{2^n} y_i = (n+2)2^{n-1} + 2 \cdot 2^{n-1} = n2^{n-1} + 2^{n+1} = \sum_{i=1}^{2^n} x_i$$

易见对于$k = 2^{n-1}$, 当$i = 1, 2, \ldots, k$ 时, $x_i \leqslant y_i$, 当$i = k+1, k+2, \ldots, n$ 时, $x_i \geqslant y_i$. 这样$\boldsymbol{x}$ 和 $\boldsymbol{y}$ 满足引理8 的条件, 故有$\boldsymbol{x} \prec \boldsymbol{y}$, 从而由引理4 知(13) 成立.

命题 3[7]　设 $\{b_i\}$ 是正项等比数列, n 为自然数, 则

$$\frac{C_n^0}{b_1} + \frac{C_n^1}{b_2} + \cdots + \frac{C_n^n}{b_{n+1}} \leqslant 2^{n-1}\left(\frac{1}{b_1} + \frac{1}{b_{n+1}}\right) \tag{15}$$

证明　类似于(14), 可得

$$\boldsymbol{x} = \left(\underbrace{n+1,\cdots,n+1}_{C_n^n}, \underbrace{n,\cdots,n}_{C_n^1}, \ldots, \underbrace{1,\cdots,1}_{C_n^0}\right)$$

$$\prec \left(\underbrace{n+1,\cdots,n+1}_{2^{n-1}}, \underbrace{1,\cdots,1}_{2^{n-1}}\right) = \boldsymbol{y} \tag{16}$$

据定理8(1), 数列 $\{\frac{1}{b_i}\}$ 是凸数列, 结合(16) 可得(15).

命题 4[8]　设 $\{b_i\}$ 是正项等比数列, 若 $n < p < m, n < q < m$, 且 $n + m = p + q, n, m, p, q, k$ 均为正整数, 则

$$(mb_m)^k + (nb_n)^k \geqslant (pb_p)^k + (qb_q)^k \tag{17}$$

$$(mb_m)^k \cdot (nb_n)^k \leqslant (pb_p)^k \cdot (qb_q)^k \tag{18}$$

$$\frac{1}{(mb_m)^k} + \frac{1}{(nb_n)^k} \geqslant \frac{1}{(pb_p)^k} + \frac{1}{(qb_q)^k} \tag{19}$$

证明　条件 $n < p < m, n < q < m$, 且 $n + m = p + q$ 意味着 $(p,q) \prec (n,m)$, 由定理4 知 $\{ib_i\}$ 是凸数列也是对数凹数列, 注意对于 $k \geqslant 1$, t^k 是递增的凸函数, 由引理3 知(17) 成立. 而由引理4, $(mb_m) \cdot (nb_n) \leqslant (pb_p) \cdot (qb_q)$, 进而(18)成立. 由引理6 知 $\{\frac{1}{ib_i}\}$ 是凸数列, 进而由引理3 知(19)成立.

命题 5[8]　设 $\{b_i\}$ 是正项等比数列, 公比 $q \geqslant 1$, S_n 是前 n 项和. 若 $n < p < m, n < q < m$, 且 $n + m = p + q$, 则

$$(mS_m)^k + (nS_n)^k \geqslant (pS_p)^k + (qS_q)^k \tag{20}$$

$$(mS_m)^k \cdot (nS_n)^k \leqslant (pS_p)^k \cdot (qS_q)^k \tag{21}$$

$$\frac{1}{(mS_m)^k} + \frac{1}{(nS_n)^k} \geqslant \frac{1}{(pS_p)^k} + \frac{1}{(qS_q)^k} \tag{22}$$

证明　由定理7, $\{S_i\}$ 既是凸数列, 也是对数凹数列. 由引理7 知 $\{iS_i\}$ 既是凸数列也是对数凹数列, 余下证明类似于命题4, 从略.

命题 6[9]   设 $P_n(x) = \sum\limits_{i=0}^{n} x^i, n \geqslant 2$, 则当 $x > 0$ 时,

$$\frac{P_n(x)}{P_n(x) - 1 - x^n} \geqslant \frac{n+1}{n-1} \tag{23}$$

$$P_n(x) \geqslant (2n+1)x^n \tag{24}$$

证明   不难验证(23)等价于

$$2P(x) \leqslant (n+1)(1+x^n) \tag{25}$$

利用引理 8 不难验证

$$(n, n, n-1, n-1, \ldots, 1, 1, 0, 0) \prec \left(\underbrace{n, \cdots, n}_{n+1}, \underbrace{0, \cdots, 0}_{n+1} \right) \tag{26}$$

据引理2, 由(26) 即得(25).

由引理 1 有

$$\left(\underbrace{n, \cdots, n}_{2n+1} \right) \prec (2n, 2n-1, \ldots, 1, 0) \tag{27}$$

据引理2, 由(27)即得(24).

命题 7[10]   设 $x > 0$, 且 $x \neq 1, n \in \mathbf{N}$, 则

$$x + x^{-n} \geqslant 2n \cdot \frac{x-1}{x^n - 1} \tag{28}$$

证明   不难验证(28)等价于

$$\frac{(x^{n+1}+1)(x^n-1)}{x-1} = (x^{n+1}+1)(x^{n-1}+x^{n-2}+\cdots+x+1) \geqslant 2nx^n$$

即

$$x^{2n} + x^{2n-1} + \cdots + x^{n+1} + x^{n-1} + x^{n-2} + \cdots + x + 1 \geqslant 2nx^n \tag{29}$$

据引理 1 知

$$\left(\underbrace{n, \cdots, n}_{2n} \right) \prec (2n, 2n-1, \ldots, n+1, n-1, \ldots, 1, 0) \tag{30}$$

从而据引理2, 由(30)即得(29).

命题 8　设 $\{b_i\}$ 是正项等比数列, 其公比小于 1 , n 为大于 1 的自然数, 求证

$$b_1^1 b_2^2 b_3^3 \cdots b_{2n-1}^{2n-1} \leqslant b_n^{(2n-1)n} \tag{31}$$

证明　由定理 5 知 $\{b_i^i\}$ 是对数凹数列. 由引理 1 有

$$\left(\underbrace{n, \cdots, n}_{2n-1} \right) \prec (1, 2, \ldots, 2n-1) \tag{32}$$

据引理4,由(32)即得(31).

命题 9[11]　设 $\{b_i\}$ 是首项不大于公比的正项等比数列, n 为正整数, 求证:

$$b_1^1 b_2^{\frac{1}{2}} b_3^{\frac{1}{3}} \cdots b_n^{\frac{1}{n}} \geqslant \sqrt{b_1^n b_n} \tag{33}$$

证明　由定理 6 知 $\{b_i^{\frac{1}{i}}\}$ 是对数凹数列. 利用引理 8 不难验证

$$(n, n, n-1, n-1, \ldots, 2, 2, 1, 1) \prec \left(\underbrace{n, \cdots, n}_{n}, \underbrace{0, \cdots, 0}_{n} \right) \tag{34}$$

据引理4, 由(34)可得

$$\left(b_1^1 b_2^{\frac{1}{2}} b_3^{\frac{1}{3}} \cdots b_n^{\frac{1}{n}} \right)^2 \geqslant b_1^n \left(b_n^{\frac{1}{n}} \right)^n = b_1^n b_n \tag{35}$$

对(35)两边开方即得(33).

命题 10　设正项等比数列 $\{b_i\}$ 的公比 $q \geqslant 1$, 其前 n 项的积为 T_n, 求证

$$T_1 T_2 \cdots T_{n-1} \geqslant T_n^{2n-1} (n > 1) \tag{36}$$

证明　由定理12 知 $\{T_i\}$ 是对数凸数列,据引理 1 有

$$\left(\underbrace{n, \cdots, n}_{2n-1} \right) \prec (1, 2, \ldots, 2n-1) \tag{37}$$

据引理4, 由(37)可得(36).

命题 11[12]　若 $x > 0$, 证明

$$2x^3 + 3x^2 - 12x + 7 \geqslant 0 \tag{38}$$

证明　因$x > 0$, 则$\{x^k\}$是公比为x的非负等比数列, 由定理2 知$\{x^k\}$是凸数列, 又由引理1 有

$$\left(\underbrace{1, \cdots, 1}_{12}\right) \prec \left(3, 3, 2, 2, 2, \underbrace{0, \cdots, 0}_{7}\right) \tag{39}$$

据引理2, 由上式有$2x^3 + 3x^2 + 7 \geqslant 12x$, 即(38)成立.

命题 12[12]　若$x > 0$, 证明

$$x^{n-1} + \frac{1}{x^{n-1}} \leqslant x^n + \frac{1}{x^n} \tag{40}$$

证明　因$x > 0$, 则$\{x^k\}$是公比为x的非负等比数列, 由定理 2 知$\{x^k\}$是凸数列, 又显然$(2n - 1, 1) \prec (2n, 0)$, 据引理2 有$x^{2n-1} + x \leqslant x^{2n} + 1$, 即(40)成立.

命题 13[13]　$x > 0$, p, q 是非负整数, $p \leqslant q$, 证明

$$x^p - x^q \geqslant (q - p)(x^{q-1} - x^q) \tag{41}$$

证明　不妨设$p < q$, 易见(41)等价于

$$x_q + qx^{q-1} + px^q \leqslant x^p + qx^q + px^{q-1} \tag{42}$$

利用引理 8 不难验证

$$\left(\underbrace{q, \cdots, q}_{p+1}, \underbrace{q - 1, \cdots, q - 1}_{q}\right) \prec \left(\underbrace{q, \cdots, q}_{q}, \underbrace{q - 1, \cdots, q - 1}_{p}, p\right) \tag{43}$$

因$x > 0$, 则$\{x^k\}$是公比为x的非负等比数列, 由定理2 知$\{x^k\}$是凸数列, 据引理2, 由(43)即得(44).

命题 14[12]　证明

$$x + x^2 + \cdots + x^{2n} \leqslant n(x^{2n+1} + 1)(x \geqslant 0) \tag{44}$$

证明　若$x > 0$, 则$\{x^k\}$是公比为x的非负等比数列, 由定理 2 知$\{x^k\}$是凸数列, 利用引理 8 不难验证

$$(2n, 2n - 1, \ldots, 2, 1) \prec \left(\underbrace{2n + 1, \cdots, 2n + 1}_{n}, \underbrace{0, \cdots, 0}_{n}\right) \tag{45}$$

据引理2 由(45)即得(44).

命题 15 (Nanson 不等式)[9] 设 $x > 0$, 证明对任意数 n 都有

$$\frac{1 + x^2 + x^4 + \cdots + x^{2n}}{1 + x^3 + x^5 + \cdots + x^{2n-1}} \geqslant \frac{n+1}{n} \tag{46}$$

证明 因 $x > 0$ 则 $\{x^k\}$ 是公比为 x 的正数等比数列, 由定理2 知 $\{x^k\}$ 是凸数列, 又

$$\left(\underbrace{1, \cdots, 1}_{n+1}, \underbrace{3, \cdots, 3}_{n+1}, \underbrace{5, \cdots, 5}_{n+1}, ..., \underbrace{2n-1, \cdots, 2n-1}_{n+1}\right)$$

$$\prec \left(\underbrace{0, \cdots, 0}_{n}, \underbrace{2, \cdots, 2}_{n}, \underbrace{4, \cdots, 4}_{n}, ..., \underbrace{2n, \cdots, 2n}_{n}\right) \tag{47}$$

(参见[1]), 由引理2 及式(47) 有

$$n(1 + x^2 + x^4 + \cdots + x^{2n}) \geqslant (n+1)(1 + x^3 + x^5 + \cdots + x^{2n-1})$$

由此得证.

参考文献

[1] 石焕南. 受控理论与解析不等式[M]. 哈尔滨: 哈尔滨工业大学出版社, 2012.

[2] PEČARIĆJ E, FRANK PROSCHAN, TONG Y L. Convex functions, partial orderings, and statistical applications [M]. New York: Academic Press.Inc., 1992.

[3] 王伯英. 控制不等式基础[M]. 北京: 北京师范大学出版社, 1990.

[4] MARSHALL A W, OLKIN I, ARNOLD B C. Inequalities: theory of majorization and its application [M]. 2nd ed. New York: Springer Press, 2011.

[5] 石焕南, 李大矛.凸数列的一个等价条件及其应用[J]. 曲阜师范大学学报(自然科学版), 2001, 27(4): 4-6.

[6] 石焕南. 凸数列的一个等价条件及其应用, II [J]. 数学杂志, 2004, 24(4): 390 - 394.

[7] 盛宏礼. 一类正项等比数列的新不等式[J]. 数学通讯, 2010, 6(下半月): 17 - 18.

[8] 李永利, 孙秀亭. 关于正项等比数列方幂的不等式[J]. 数学通讯,2004, 11: 27 - 28.

[9]   匡继昌. 常用不等式[M]. 第4版. 济南: 山东科学技术出版社, 2010.

[10] 何志民. 等比数列的前n 项和公式在不等式中的应用[J]. 内蒙古科技与经济, 2002,12:330.

[11] 盛宏礼. 有关正项等比数列的不等式[J]. 数学通讯, 2007(19): 15 - 17.

[12] MITRINOVIĆ D S, BARNES E S, MARSH D C B, RADOK J R M. Elementary inequalities [M]. The Hetherlands: P. Noordhoff, Ltd. Groningen, 1964.

[13] 甘志国. 数列与不等式[M]. 哈尔滨: 哈尔滨工业大学出版社, 2014.

第65篇　Compositions Involving Schur-Geometrically Convex Functions

(SHI HUAN-NAN, ZHANG JING. J. Inequal. Appl., (2015) 2015:320

DOI 10.1186/s13660-015-0842-x)

Abstract: The decision theorem of the Schur-geometric convexity for the compositions involving Schur-geometrically convex functions is established and used to determine the Schur-geometric convexity of some symmetric functions.

Keywords: Schur-geometrically convex function; geometrically convex function; complex function; symmetric function

65.1　Introduction

Throughout the article, $\mathbf{R}$ denotes the set of real numbers, $\boldsymbol{x} = (x_1, x_2, \ldots, x_n)$ denotes n-tuple (n-dimensional real vectors), the set of vectors can be written as

$$\mathbf{R}^n = \{\boldsymbol{x} = (x_1, \ldots, x_n) : x_i \in \mathbf{R}, i = 1, \ldots, n\}$$

$$\mathbf{R}^n_{++} = \{\boldsymbol{x} = (x_1, \ldots, x_n) : x_i > 0, i = 1, \ldots, n\}$$

$$\mathbf{R}^n_+ = \{\boldsymbol{x} = (x_1, \ldots, x_n) : x_i \geqslant 0, i = 1, \ldots, n\}$$

In particular, the notations $\mathbf{R}$, $\mathbf{R}_{++}$ and $\mathbf{R}_+$ denote $\mathbf{R}^1$, $\mathbf{R}^1_{++}$ and $\mathbf{R}^1_+$, respectively.

The following conclusion is proved in reference [1-2].

Theorem A Let the interval $[a, b] \subset \mathbf{R}$, $\varphi : \mathbf{R}^n \to \mathbf{R}$, $f : [a, b] \to \mathbf{R}$ and $\psi(x_1, \cdots, x_n) = \varphi(f(x_1), \cdots, f(x_n)) : [a, b]^n \to \mathbf{R}$.

(i) If φ is increasing and Schur-convex and f is convex, then ψ is Schur-convex.

(ii) If φ is increasing and Schur-concave and f is concave, then ψ is Schur-concave.

(iii) If φ is decreasing and Schur-convex and f is concave, then ψ is Schur-convex.

(iv) If φ is increasing and Schur-convex and f is increasing and convex, then ψ is increasing and Schur-convex.

(v) If φ is decreasing and Schur-convex and f is decreasing and concave, then ψ is increasing and Schur-convex.

(vi) If φ is increasing and Schur-convex and f is decreasing and convex, then ψ is decreasing and Schur-convex.

(vii) If φ is decreasing and Schur-convex and f is increasing and concave, then ψ is decreasing and Schur-convex.

(viii) If φ is decreasing and Schur-concave and f is decreasing and concave, then ψ is increasing and Schur-concave.

Theorem A is very effective for determine of the Schur-convexity of the complex functions.

The Schur geometrically convex functions were proposed by Zhang [3] in 2003. The theory of majorization was enriched and expanded by using these concepts. Regarding the Schur geometrically convex functions, the aim of this paper is to establish the following theorem which is similar to Theorem A.

Theorem 1 Let the interval $[a, b] \subset \mathbf{R}_{++}$, $\varphi : \mathbf{R}^n \to \mathbf{R}$, $f : [a, b] \to \mathbf{R}$ and $\psi(x_1, \cdots, x_n) = \varphi(f(x_1), \cdots, f(x_n)) : [a, b]^n \to \mathbf{R}$.

(i) If φ is increasing and Schur-geometrically convex and f is geometrically convex, then ψ is Schur-geometrically convex.

(ii) If φ is increasing and Schur-geometrically concave and f is geometrically concave, then ψ is Schur-geometrically concave.

(iii) If φ is decreasing and Schur-geometrically convex and f is geometrically concave, then ψ is Schur-geometrically convex.

(iv) If φ is increasing and Schur-geometrically convex and f is increasing and geometrically convex, then ψ is increasing and Schur-geometrically convex.

(v) If φ is decreasing and Schur-geometrically convex and f is decreasing and geometrically concave, then ψ is increasing and Schur-geometrically convex.

(vi) If φ is increasing and Schur-geometrically convex and f is decreasing and geometrically convex, then ψ is decreasing and Schur-geometrically convex.

(vii) If φ is decreasing and Schur-geometrically convex and f is increasing and geometrically concave, then ψ is decreasing and Schur-geometrically convex.

(viii) If φ is decreasing and Schur-geometrically concave and f is decreasing and geometrically concave, then ψ is increasing and Schur-geometrically concave.

65.2　Definitions and lemmas

In order to prove our results, in this section we will recall usefull definitions and lemmas.

Definition 1 [1-2] Let $x = (x_1, \ldots, x_n)$ and $y = (y_1, \ldots, y_n) \in \mathbf{R}^n$.

We say y majorizes x (x is said to be majorized by y), denoted by $x \prec y$, if $\sum_{i=1}^{k} x_{[i]} \leqslant \sum_{i=1}^{k} y_{[i]}$ for $k = 1, 2, \ldots, n-1$ and $\sum_{i=1}^{n} x_i = \sum_{i=1}^{n} y_i$, where $x_{[1]} \geqslant \cdots \geqslant x_{[n]}$ and $y_{[1]} \geqslant \cdots \geqslant y_{[n]}$ are rearrangements of x and x in a descending order.

Definition 2 [1-2] Let $x = (x_1, \ldots, x_n)$ and $y = (y_1, \ldots, y_n) \in \mathbf{R}^n$.

(i) A set $\Omega \subset \mathbf{R}^n$ is said to be a convex set if

$$\alpha x + (1-\alpha)y = (\alpha x_1 + (1-\alpha)y_1, \ldots, \alpha x_n + (1-\alpha)y_n) \in \Omega$$

for all $x, y \in \Omega$, and all $\alpha \in [0, 1]$.

(ii) Let $\Omega \subset \mathbf{R}^n$ be convex set. A function $\varphi \colon \Omega \to \mathbf{R}$ is said to be a convex function on Ω if

$$\varphi\left(\alpha\boldsymbol{x} + (1-\alpha)\boldsymbol{y}\right) \leqslant \alpha\varphi(\boldsymbol{x}) + (1-\alpha)\varphi(\boldsymbol{y})$$

holds for all $\boldsymbol{x}, \boldsymbol{y} \in \Omega$, and all $\alpha \in [0,1]$. φ is said to be a concave function on Ω if and only if $-\varphi$ is convex function on Ω.

(iii) Let $\Omega \subset \mathbf{R}^n$. A function $\varphi: \Omega \to \mathbf{R}$ is said to be a Schur-convex function on Ω if $\boldsymbol{x} \prec \boldsymbol{y}$ on Ω implies $\varphi(\boldsymbol{x}) \leqslant \varphi(\boldsymbol{y})$. A function φ is said to be a Schur-concave function on Ω if and only if $-\varphi$ is Schur-convex function on Ω.

Lemma 1 (Schur-convex function decision theorem)[1-2] Let $\Omega \subset \mathbf{R}^n$ be symmetric and have a nonempty interior convex set. Ω^0 is the interior of Ω. $\varphi : \Omega \to \mathbf{R}$ is continuous on Ω and differentiable in Ω^0. Then φ is the *Schur − convex (Schur − concave) function* if and only if φ is symmetric on Ω and

$$(x_1 - x_2)\left(\frac{\partial\varphi}{\partial x_1} - \frac{\partial\varphi}{\partial x_2}\right) \geqslant 0 (\leqslant 0) \tag{1}$$

holds for any $\boldsymbol{x} \in \Omega^0$.

Definition 3 [3] Let $\boldsymbol{x} = (x_1, \ldots, x_n) \in \mathbf{R}^n_{++}$ and $\boldsymbol{y} = (y_1, \ldots, y_n) \in \mathbf{R}^n_{++}$.

(i) A set $\Omega \subset \mathbf{R}^n_{++}$ is called a geometrically convex set if

$$\boldsymbol{x}^\alpha \boldsymbol{y}^{1-\alpha} = (x_1^\alpha y_1^{1-\alpha}, \ldots, x_n^\alpha y_n^{1-\alpha}) \in \Omega$$

for all $\boldsymbol{x}, \boldsymbol{y} \in \Omega$ and $\alpha \in [0,1]$.

(ii) Let $\Omega \subset \mathbf{R}_{++}$ be geometrically convex set. A function $\varphi : I \to \mathbf{R}_{++}$ is called a geometrically convex(concave) function, if

$$\varphi\left(\boldsymbol{x}^\alpha \boldsymbol{y}^{1-\alpha}\right) \leqslant (\geqslant)\varphi^\alpha(\boldsymbol{x})\varphi^{1-\alpha}(\boldsymbol{y})$$

holds for all $\boldsymbol{x}, \boldsymbol{y} \in \Omega$ and all $\alpha \in [0,1]$.

(iii) Let $\Omega \subset \mathbf{R}^n_{++}$. A function $\varphi : \Omega \to \mathbf{R}_{++}$ is said to be a Schur-geometrically convex (or concave, respectively) function on Ω if

$$(\log x_1, \ldots, \log x_n) \prec (\log y_1, \ldots, \log y_n)$$

on Ω implies $\varphi(\boldsymbol{x}) \leqslant (or \geqslant, respectively) \varphi(\boldsymbol{y})$.

By Definitions 3 (iii), the following are obvious.

Corollary 1 Let $\Omega \subset \mathbf{R}_{++}^n$ be a set, and let $\log \Omega = \{(\log x_1, \cdots, \log x_n) : (x_1, \cdots, x_n) \in \Omega\}$. Then $\varphi : \Omega \to \mathbf{R}_{++}$ is a Schur-geometrically convex (or concave, respectively) function on Ω if and only if $\varphi(e^{x_1}, ..., e^{x_n})$ is a Schur-convex (or concave, respectively) function on $\log \Omega$.

Lemma 2 (Schur-geometrically convex function decision theorem) [3] Let $\Omega \subset \mathbf{R}_{++}^n$ be a symmetric and geometrically convex set with a nonempty interior Ω^0. Let $\varphi : \Omega \to \mathbf{R}_{++}$ be continuous on Ω and differentiable in Ω^0. If φ is symmetric on Ω and

$$(\log x_1 - \log x_2)\left(x_1 \frac{\partial \varphi}{\partial x_1} - x_2 \frac{\partial \varphi}{\partial x_2}\right) \geqslant 0 \quad (\leqslant 0) \tag{2}$$

holds for any $\boldsymbol{x} = (x_1, \cdots, x_n) \in \Omega^0$, then φ is a Schur geometrically convex (Schur geometrically concave) function.

The Schur-geometric convexity was proposed by Zhang[3] in 2004, and was investigated by Chu et al.[4], Guan[5], Sun et al.[6], and so on. We also note that some authors use the term "Schur-multiplicative convexity".

Lemma 3 [5] If $f : [a, b] \subset \mathbf{R}_{++} \to \mathbf{R}_{++}$ is geometrically convex (or concave, respectively) if and only if $\log f(e^x)$ is convex (or concave, respectively) on $[\log a, \log b]$.

Lemma 4 [5] If $f : [a, b] \subset \mathbf{R}_{++} \to \mathbf{R}_{++}$ is a twice differentiable function, then f is a geometrically convex (or concave, respectively) if and only if

$$x\left[f''(x)f(x) - (f'(x))^2\right] + f(x)f'(x) \geqslant 0 (or \leqslant 0, respectively) \tag{3}$$

65.3　Proof of main results

Proof of Theorem 1 We only give the proof of Theorem 1 (iv) in detail. Similar argument leads to the proof of the rest part.

If φ is increasing and Schur-geometrically convex and f is increasing and geometrically convex, then by Proposition 1, $\varphi(e^{x_1}, ..., e^{x_n})$ is increasing and Schur convex and by Lemma 3, $g(x) = \log f(e^x)$ is increasing and convex on $[\log a, \log b]$. And then from Theorem A (iv), it follows that $\varphi(e^{\log f(e^{x_1})}, ..., e^{\log f(e^{x_n})}) = \varphi(f(e^{x_1}), ..., f(e^{x_n}))$ is increasing and Schur-convex. Again by Proposition 1, it follows that $\psi(x_1, \cdots, x_n) = $

$\varphi(f(x_1), \cdots, f(x_n))$ is increasing and Schur-geometrically convex.

The proof of Theorem 1 is completed.

65.4   Applications

Let $\boldsymbol{x} = (x_1, \ldots, a_n) \in \mathbf{R}^n$. Its elementary symmetric functions are

$$E_r(\boldsymbol{x}) = E_r(x_1, \ldots, x_n) = \sum_{1 \leqslant i_1 < \ldots < i_r \leqslant n} \prod_{j=1}^{r} x_{i_j}, \quad r = 1, \ldots, n$$

The dual form of the elementary symmetric functions are

$$E_r^*(\boldsymbol{x}) = E_r^*(x_1, \ldots, x_n) = \prod_{1 \leqslant i_1 < \ldots < i_r \leqslant n} \sum_{j=1}^{r} x_{i_j}, \quad r = 1, \ldots, n$$

and defined $E_0(\boldsymbol{x}) = E_0^*(\boldsymbol{x}) = 1$, and $E_r(\boldsymbol{x}) = E_r^*(\boldsymbol{x}) = 0$ for $r < 0$ or $r > n$.

It is well-known that $E_r(\boldsymbol{x})$ is a increasing and Schur-concave function on $\mathbf{R}_+^n$ [1]. By Lemma 2, it is easy to prove that $E_r(\boldsymbol{x})$ is a Schur-geometrically convex function on $\mathbf{R}_{++}$. In fact

$$(\log x_1 - \log x_2)\left(x_1 \frac{\partial E_r(\boldsymbol{x})}{\partial x_1} - x_2 \frac{\partial E_r(\boldsymbol{x})}{\partial x_2}\right)$$

$$= x_1 x_2 (x_1 - x_2)(\log x_1 - \log x_2) E_{r-2}(x_3, ..., x_n) +$$
$$(x_1^2 - x_2^2)(\log x_1 - \log x_2) E_{r-1}(x_3, ..., x_n) \geqslant 0$$

In [7-8], Shi proved that $E_r^*(\boldsymbol{x})$ is a increasing and Schur-concave function and Schur-geometrically convex function on $\mathbf{R}_{++}$.

For $\boldsymbol{x} = (x_1, \cdots, x_n) \in \mathbf{R}^n$, the complete symmetric function $c_n(\boldsymbol{x}, r)$ is defined as

$$c_n(\boldsymbol{x}, r) = \sum_{i_1+i_2+\cdots+i_n=r} \prod_{j=1}^{n} x_j^{i_j}$$

where $c_0(\boldsymbol{x}, r) = 1$, $r \in \{1, 2, \cdots, n\}$, $i_1, i_2, \cdots, i_n$ are non-negative integers.

The dual form of the complete symmetric function are

$$c_n^*(\boldsymbol{x}, r) = \prod_{i_1+i_2+\cdots+i_n=r} \sum_{j=1}^{n} i_j x_j$$

where $i_j (j = 1, 2, \cdots, n)$ can take min value 0 in last equality.

Guan[9] discussed the Schur-convexity of $c_n(\boldsymbol{x}, r)$ and proved that $c_n(\boldsymbol{x}, r)$

is increasing and Schur-convex on $\mathbf{R}_{++}^n$. Subsequently, Chu et al. [10] proved that $c_n(\boldsymbol{x}, r)$ is Schur-geometrically convex on $\mathbf{R}_{++}^n$.

Zhang and Shi [11] proved that $c_n^*(\boldsymbol{x}, r)$ is increasing, Schur-concave and Schur-geometrically convex on $\mathbf{R}_{++}^n$.

In the following, we prove that the Schur-geometric convexity of the complex functions involving the above symmetric functions and their dual form by using Theorem 1.

Let $f(x) = \frac{1+x}{1-x}, x \neq 1$. Directly calculating yields

$$x \left[f''(x)f(x) - (f'(x))^2 \right] + f(x)f'(x) = \frac{2(x^2+1)}{(1-x)^4} \geqslant 0$$

That is, f is increasing and geometrically convex on $\mathbf{R}\backslash\{1\}$. Since $E_r(\boldsymbol{x})$, $E_r^*(\boldsymbol{x})$, $c_n(\boldsymbol{x}, r)$ and $c_n^*(\boldsymbol{x}, r)$ are all increasing and Schur-geometrically convex function on $\mathbf{R}_{++}$, and notice that $f(x) = \frac{1+x}{1-x} > 0$, for $-1 < x < 1$, by Theorem 1 (iv), it follows

Theorem 2 The following symmetric functions are increasing and Schur-geometrically convex on $(-1, 1)^n$

$$E_r \left(\frac{1+\boldsymbol{x}}{1-\boldsymbol{x}} \right) = \sum_{1 \leqslant i_1 < \cdots < i_r \leqslant n} \prod_{j=1}^{r} \frac{1 + x_{i_j}}{1 - x_{i_j}} \tag{4}$$

$$E_r^* \left(\frac{1+\boldsymbol{x}}{1-\boldsymbol{x}} \right) = \prod_{1 \leqslant i_1 < \cdots < i_r \leqslant n} \sum_{j=1}^{r} \frac{1 + x_{i_j}}{1 - x_{i_j}} \tag{5}$$

$$c_n \left(\frac{1+\boldsymbol{x}}{1-\boldsymbol{x}}, r \right) = \sum_{i_1 + i_2 + \cdots + i_n = r} \prod_{j=1}^{n} \left(\frac{1 + x_j}{1 - x_j} \right)^{i_j} \tag{6}$$

and

$$c_n^* \left(\frac{1+\boldsymbol{x}}{1-\boldsymbol{x}}, r \right) = \prod_{i_1 + i_2 + \cdots + i_n = r} \sum_{j=1}^{n} i_j \left(\frac{1 + x_j}{1 - x_j} \right) \tag{7}$$

Remark 1 By Lemma 2, Xia et al. [12] proved that $E_r \left(\frac{1+\boldsymbol{x}}{1-\boldsymbol{x}} \right)$ is Schur-geometrically convex on $(0, 1)^n$. By the properties of Schur-geometrically convex function, Shi and Zhang [13] proved that $E_r^* \left(\frac{1+\boldsymbol{x}}{1-\boldsymbol{x}} \right)$ is Schur-geometrically convex on $(0, 1)^n$. By Theorem 1, this conclusion is extended to the collection $(-1, 1)^n$.

For $r \geqslant 1$, let $g(x) = x^{\frac{1}{r}}, x \in \mathbf{R}_{++}$. Directly calculating yields

$$x \left[g''(x)g(x) - (g'(x))^2 \right] + g(x)g'(x) = 0$$

That is, g is increasing and geometrically convex(concave) on $\mathbf{R}_{++}^n$. Since $E_r(\boldsymbol{x})$, $E_r^*(\boldsymbol{x})$, $c_n(\boldsymbol{x}, r)$ and $c_n^*(\boldsymbol{x}, r)$ are all increasing and Schur-geometrically convex function on $\mathbf{R}_{++}$, by Theorem 1 (iv), it follows

Theorem 3 The following symmetric functions are increasing and Schur-geometrically convex on $\mathbf{R}_{++}^n$

$$E_r\left(\boldsymbol{x}^{\frac{1}{r}}\right) = \sum_{1 \leqslant i_1 < \cdots < i_r \leqslant n} \prod_{j=1}^{r} x_{i_j}^{\frac{1}{r}} \tag{8}$$

$$E_r^*\left(\boldsymbol{x}^{\frac{1}{r}}\right) = \prod_{1 \leqslant i_1 < \cdots < i_r \leqslant n} \sum_{j=1}^{r} x_{i_j}^{\frac{1}{r}} \tag{9}$$

$$c_n\left(\boldsymbol{x}^{\frac{1}{r}}, r\right) = \sum_{i_1+i_2+\cdots+i_n=r} \prod_{j=1}^{n} x_j^{\frac{i_j}{r}} \tag{10}$$

and

$$c_n^*\left(\boldsymbol{x}^{\frac{1}{r}}, r\right) = \prod_{i_1+i_2+\cdots+i_n=r} \sum_{j=1}^{n} i_j x_j^{\frac{1}{r}} \tag{11}$$

Remark 2 (i) By Lemma 1, Guan[14] and Jiang [15] proved respectively that the Hamy's symmetric function $E_r\left(\boldsymbol{x}^{\frac{1}{r}}\right)$ and its dual form $E_r^*\left(\boldsymbol{x}^{\frac{1}{r}}\right)$ is Schur-geometrically convex on $\mathbf{R}_{++}^n$. In contrast, Our proof is very simple by Theorem 1.

(ii) Here we prove Schur-geometric convexity of $c_n\left(\boldsymbol{x}^{\frac{1}{r}}, r\right)$ on $\mathbf{R}_{++}^n$ by the Theorem 1. Guan [16] proved that $c_n\left(\boldsymbol{x}^{\frac{1}{r}}, r\right)$ is Schur-concave on $\mathbf{R}_{++}^n$ by the Lemma 1.

Since $f(x) = \frac{1+x}{1-x}$ is increasing and geometrically convex on $(-1, 1)^n$, from Theorem 1 and Theorem 3, it follows

Theorem 4 The following symmetric functions are increasing and Schur-geometrically convex on $(-1, 1)^n$

$$E_r\left(\left(\frac{1+\boldsymbol{x}}{1-\boldsymbol{x}}\right)^{\frac{1}{r}}\right) = \sum_{1 \leqslant i_1 < \cdots < i_r \leqslant n} \prod_{j=1}^{r} \left(\frac{1+x_{i_j}}{1-x_{i_j}}\right)^{\frac{1}{r}} \tag{12}$$

$$E_r^*\left(\left(\frac{1+\boldsymbol{x}}{1-\boldsymbol{x}}\right)^{\frac{1}{r}}\right) = \prod_{1 \leqslant i_1 < \cdots < i_r \leqslant n} \sum_{j=1}^{r} \left(\frac{1+x_{i_j}}{1-x_{i_j}}\right)^{\frac{1}{r}} \tag{13}$$

$$c_n\left(\left(\frac{1+\boldsymbol{x}}{1-\boldsymbol{x}}\right)^{\frac{1}{r}}, r\right) = \sum_{i_1+i_2+\cdots+i_n=r} \prod_{j=1}^{n} \left(\frac{1+x_j}{1-x_j}\right)^{\frac{i_j}{r}} \tag{14}$$

and

$$c_n^* \left(\left(\frac{1+x}{1-x} \right)^{\frac{1}{r}}, r \right) = \prod_{i_1+i_2+\cdots+i_n=r} \sum_{j=1}^n i_j \left(\frac{1+x_j}{1-x_j} \right)^{\frac{1}{r}} \tag{15}$$

Remark 3 By Lemma 2, Long and Chu [17] proved that $E_r^* \left(\left(\frac{1+x}{1-x} \right)^{\frac{1}{r}} \right)$ is Schur-geometrically convex on $(0,1)^n$. By Theorem 1, this conclusion is extended to the collection $(-1,1)^n$.

Let $h(x) = \frac{x}{1-x}, x \neq 1$. Then $h'(x) = \frac{1}{(1-x)^2} > 0$, $h''(x) = \frac{2}{(1-x)^3}$, and

$$x \left[h''(x)h(x) - (h'(x))^2 \right] + h(x)h'(x) = \frac{x^2}{(1-x)^4} > 0$$

That is, h is increasing and geometrically convex on $\mathbf{R} \backslash \{1\}$. Since $E_r(x)$, $E_r^*(x)$, $c_n(x,r)$ and $c_n^*(x,r)$ are all increasing and Schur-geometrically convex function on $\mathbf{R}_{++}$, and notice that $h(x) = \frac{x}{1-x} > 0$, for $0 \leqslant x < 1$, by Theorem 1 (iv), it follows

Theorem 5 The following symmetric functions are increasing and Schur-geometrically convex on $[0,1)^n$

$$E_r \left(\frac{x}{1-x} \right) = \sum_{1 \leqslant i_1 < \cdots < i_r \leqslant n} \prod_{j=1}^r \frac{x_{i_j}}{1-x_{i_j}} \tag{16}$$

$$E_r^* \left(\frac{x}{1-x} \right) = \prod_{1 \leqslant i_1 < \cdots < i_r \leqslant n} \sum_{j=1}^r \frac{x_{i_j}}{1-x_{i_j}} \tag{17}$$

$$c_n \left(\frac{x}{1-x}, r \right) = \sum_{i_1+i_2+\cdots+i_n=r} \prod_{j=1}^n \left(\frac{x_j}{1-x_j} \right)^{i_j} \tag{18}$$

and

$$c_n^* \left(\frac{x}{1-x}, r \right) = \prod_{i_1+i_2+\cdots+i_n=r} \sum_{j=1}^n i_j \left(\frac{x_j}{1-x_j} \right) \tag{19}$$

Remark 4 By Lemma 2, Guan [18] proved that $E_r \left(\frac{x}{1-x} \right)$ is Schur-geometrically convex on $[0,1)^n$. By the judgement theorems of Schur-geometric convexity for a class of symmetric functions, Shi and Zhang [19] give another proof. Here by Theorem 1, we give a new proof.

By the properties of Schur-geometrically convex function, Shi and Zhang [13] proved that $E_r^* \left(\frac{x}{1-x} \right)$ is Schur-geometrically convex on $[\frac{1}{2}, 1)^n$. By Theorem 1, this conclusion is extended to the collection $[0,1)^n$.

By Lemma 2, Sun et al. [20] proved that $c_n \left(\frac{x}{1-x}, r \right)$ is Schur-geometrically convex on $[0,1)^n$, here by Theorem 1, we give a new proof.

Since $f(x) = \frac{x}{1-x}$ is increasing and geometrically convex on $[0,1)$, from Theorem 1 and Theorem 3, it follows

Theorem 6 The following symmetric functions are increasing and Schur-geometrically convex on $[0,1)^n$

$$E_r\left(\left(\frac{\boldsymbol{x}}{1-\boldsymbol{x}}\right)^{\frac{1}{r}}\right) = \sum_{1\leqslant i_1<\cdots<i_r\leqslant n}\prod_{j=1}^{r}\left(\frac{x_{i_j}}{1-x_{i_j}}\right)^{\frac{1}{r}} \tag{20}$$

$$E_r^*\left(\left(\frac{\boldsymbol{x}}{1-\boldsymbol{x}}\right)^{\frac{1}{r}}\right) = \prod_{1\leqslant i_1<\cdots<i_r\leqslant n}\sum_{j=1}^{r}\left(\frac{x_{i_j}}{1-x_{i_j}}\right)^{\frac{1}{r}} \tag{21}$$

$$c_n\left(\left(\frac{\boldsymbol{x}}{1-\boldsymbol{x}}\right)^{\frac{1}{r}},r\right) = \sum_{i_1+i_2+\cdots+i_n=r}\prod_{j=1}^{n}\left(\frac{x_j}{1-x_j}\right)^{\frac{i_j}{r}} \tag{22}$$

and

$$c_n^*\left(\left(\frac{\boldsymbol{x}}{1-\boldsymbol{x}}\right)^{\frac{1}{r}},r\right) = \prod_{i_1+i_2+\cdots+i_n=r}\sum_{j=1}^{n}i_j\left(\frac{x_j}{1-x_j}\right)^{\frac{1}{r}} \tag{23}$$

Acknowledgments

The work was supported by Funding Project for Academic Human Resources Development in Institutions of Higher Learning under the Jurisdiction of Beijing Municipality (PHR (IHLB)) (PHR201108407). Thanks for the help. This article was typeset by using $\mathcal{AMS}$-LaTeX.

References

[1] MARSHALL A W, OLKIN I, ARNOLD B C. Inequalities: theory of majorization and its application[M].2nd ed. New York: Springer Press, 2011.

[2] B.-Y.WANG. Kòngzhì Bùděngshì Jīchǔ(Foundations of Majorization Inequalities[M]. Beijing: Beijing Normal University Press, 1990.

[3] X M ZHANG. Geometrically Convex Functions[M]. Hefei: An hui University Press, 2004.

[4] Y M CHU, X M ZHANG, G D WANG. The Schur geometrical convexity of the extended mean values[J]. Journal of Convex Analysis, 2008, 15(4): 707-718.

[5] CONSTANTIN P NICULESCU. Convexity according to the geometric mean[J]. Math. Inequal. Appl. 2000, 3(2):155-167.

[6] Y-M CHU, G-D WANG, X -H ZHANG. The Schur multiplicative and harmonic convexities of the complete symmetric function[J]. Mathematische Nachrichten, 2011, (5-6): 653 – 663.

[7] H-N SHI. Schur-Concavity and Schur-Geometrically Convexity of Dual Form for Elementary Symmetric Function with Applications[J]. RGMIA Research Report Collection, 2007, 10(2).

[8] H -N SHI. Theory of majorization and analytic Inequalities[M]. Harbin: Harbin Institute of Technology Press, 2012.

[9] K -Z GUAN. Schur-convexity of the complete symmetric function[J]. Mathematical Inequalities & Applications, 2006, 9(4): 567-576.

[10] Y-M CHU, G-D WANG, X-H ZHANG. The Schur multiplicative and harmonic convexities of the complete symmetric function[J]. Mathematische Nachrichten, 2011, 284(5-6): 653-663.

[11] KONGSHENG ZHANG, HUANNAN SHI. Schur convexity of dual form of the complete symmetric function[J]. Mathematical Inequalities & Applications, 2013, 16(4): 963 – 970.

[12] W-F XIA, Y-M CHU. On Schur convexity of some symmetric functions [J]. Journal of Inequalities and Applications, 2010, Article ID 543250, 12 pages.

[13] HUAN-NAN SHI, JING ZHANG. Schur-convexity, Schur-geometric and harmonic convexities of dual form of a class symmetric functions[J]. J.Math. Inequal, 2014, 8 (2): 349 – 358.

[14] K-Z GUAN. A class of symmetric functions for multiplicatively convex function[J]. Math. Inequal. Appl., 2007, 10 (4): 745-753.

[15] W-D JIANG. Some properties of dual form of the Hamy' s symmetric function[J]. J. Math. Inequal., 2007, 1(1): 117-125.

[16] K-Z GUAN. The Hamy symmetric function and its generalization[J]. Math. Inequal. Appl., 2006, 9 (4): 797-805.

[17] B Y LONG, Y M CHU. The Schur convexity and inequalities for a class of symmetric functions[J]. Acta Mathematica Scientia, 2012, 32 A (1):80-89.

[18] K-Z GUAN. Some properties of a class of symmetric functions[J]. J. Math. Anal. Appl., 2007, 336: 70-80.

[19] H-N SHI, J ZHANG. Some new judgment theorems of Schur geometric and Schur harmonic convexities for a class of symmetric functions[J]. J. Inequal. Appl., 2013, 2013:527 doi:10.1186/ 1029-242X-2013-527.

[20] M B SUN, N B CHEN, S H LI. Some properties of a class of symmetric functions and its applications[J]. Mathematische Nachrichten, 2014, doi: 10.1002/mana.201300073.

第66篇　Hermite-Hadamard type Inequalities for Functions with a Bounded Second Derivative

(SHI HUAN-NAN. Proceedings of the Jangjeon Mathematical Society,

19 (2016), No. 1. pp. 135-144)

Abstract: By applying results from the theory of majorization, some new inequalities of Hermite-Hadamard type for functions with a bounded second derivative are established.

Keywords: Hermite-Hadamard type inequality; majorization; second derivative

66.1　Introduction

Throughout the article, for $a \leqslant b$ and $0 \leqslant p \leqslant \frac{1}{2}$, write

$$u := u(p) = pa + (1-p)b, v := v(p) = pb + (1-p)a$$

In [1-2], the following double integral inequalities were obtained.

Theorem A[1−2] Let $f : [a, b] \to \mathbf{R}$ be a twice differentiable mapping on (a, b) and suppose that $\gamma \leqslant f''(t) \leqslant \Gamma$ for all $t \in (a, b)$. Then we have

$$\frac{\gamma(b-a)^2}{24} \leqslant \frac{1}{b-a} \int_a^b f(t)\mathrm{d}t - f\left(\frac{a+b}{2}\right) \leqslant \frac{\Gamma(b-a)^2}{24} \qquad (1)$$

$$\frac{\gamma(b-a)^2}{12} \leqslant \frac{f(a)+f(b)}{2} - \frac{1}{b-a} \int_a^b f(t)\mathrm{d}t \leqslant \frac{\Gamma(b-a)^2}{12} \qquad (2)$$

In [3], the above inequalities were refined as follows.

Theorem B[3] Let $f : [a, b] \to \mathbf{R}$ be a twice differentiable mapping on (a, b) and suppose that $\gamma \leqslant f''(t) \leqslant \Gamma$ for all $t \in (a, b)$. Then we have the double inequality

$$\frac{3S - 2\Gamma}{24}(b - a)^2 \leqslant \frac{1}{b-a}\int_a^b f(t)\mathrm{d}t - f\left(\frac{a+b}{2}\right) \leqslant \frac{3S - 2\gamma}{24}(b-a)^2 \quad (3)$$

$$\frac{3S - 2\Gamma}{24}(b - a)^2 \leqslant \frac{f(a) + f(b)}{2} - \frac{1}{b-a}\int_a^b f(t)\mathrm{d}t \leqslant \frac{3S - 2\gamma}{24}(b-a)^2 \quad (4)$$

where $S = \frac{f'(b)-f'(a)}{b-a}$.

In this paper, by applying results from the theory of majorization, we prove and generalize the double inequalities (1) and (2). Our main results are as follows:

Theorem 1 Let $f : [a, b] \to \mathbf{R}$ be a twice differentiable mapping on (a, b) and suppose that $\gamma \leqslant f''(t) \leqslant \Gamma$ for all $t \in (a, b)$. If $0 \leqslant p \leqslant \frac{1}{2}$, then for $a < b$, we have

$$\frac{\gamma p(1 - p)(b - a)^2}{6} \leqslant \frac{1}{b-a}\int_a^b f(t)\mathrm{d}t - \frac{1}{v-u}\int_u^v f(t)\mathrm{d}t$$
$$\leqslant \frac{\Gamma p(1 - p)(b - a)^2}{6} \quad (5)$$

Taking $p = \frac{1}{4}$ in inequalities (5), we have the following corollary:

Corollary 1 Let $f : [a, b] \to \mathbf{R}$ be a twice differentiable mapping on (a, b) and suppose that $\gamma \leqslant f''(t) \leqslant \Gamma$ for all $t \in (a, b)$. Then

$$\frac{\Gamma(b - a)^2}{32} \leqslant \frac{1}{b-a}\int_a^b f(t)\mathrm{d}t - \frac{2}{b-a}\int_{\frac{3a+b}{4}}^{\frac{a+3b}{4}} f(t)\mathrm{d}t \leqslant \frac{\gamma(b - a)^2}{32} \quad (6)$$

Theorem 2 Let $f : [a, b] \to \mathbf{R}$ be a twice differentiable mapping on (a, b) and suppose that $\gamma \leqslant f''(t) \leqslant \Gamma$ for all $t \in (a, b)$. If $0 \leqslant p \leqslant \frac{1}{2}$, then for $a < b$, we have

$$\frac{\gamma p(1 - p)(b - a)^2}{3} \leqslant \left(\frac{f(a) + f(b)}{2} - \frac{1}{b-a}\int_a^b f(t)\mathrm{d}t\right)$$
$$-\left(\frac{f(u) + f(v)}{2} - \frac{1}{v-u}\int_u^v f(t)\mathrm{d}t\right)$$
$$\leqslant \frac{\Gamma p(1 - p)(b - a)^2}{3} \quad (7)$$

Remark 1 By L'Hospital Rule, it is not difficult to verify that

$$\lim_{p \to \frac{1}{2}} \frac{1}{v-u} \int_u^v f(t)\mathrm{d}t = f\left(\frac{a+b}{2}\right) \tag{8}$$

and then making $p \to \frac{1}{2}$, inequalities (5) and (7) is respectively reduces to (1) and (2).

Taking $p = \frac{1}{4}$ in inequality (7), we have the following corollary:

Corollary 2 Let $f : [a,b] \to \mathbf{R}$ be a twice differentiable mapping on (a,b) and suppose that $\gamma \leqslant f''(t) \leqslant \Gamma$ for all $t \in (a,b)$. Then

$$\frac{\gamma(b-a)^2}{16} \leqslant \left(\frac{f(a)+f(b)}{2} - \frac{1}{b-a}\int_a^b f(t)\mathrm{d}t\right) -$$
$$\left(\frac{f(\frac{a+3b}{4})+f(\frac{3a+b}{4})}{2} - \frac{2}{b-a}\int_{\frac{3a+b}{4}}^{\frac{a+3b}{4}} f(t)\mathrm{d}t\right) \leqslant \frac{\Gamma(b-a)^2}{16} \tag{9}$$

Theorem 3 Let $f : [a,b] \to \mathbf{R}$ be a twice differentiable mapping on (a,b) and suppose that $\gamma \leqslant f''(t) \leqslant \Gamma$ for all $t \in (a,b)$. If $0 \leqslant p \leqslant \frac{1}{2}$, then for $a < b$, we have

$$\frac{\gamma p(1-p)(b-a)^2}{12} \leqslant \left(\frac{f(a)+f(b)}{4} - \frac{1}{b-a}\int_a^b f(t)\mathrm{d}t\right) -$$
$$\left(\frac{f(u)+f(v)}{4} - \frac{1}{v-u}\int_u^v f(t)\mathrm{d}t\right)$$
$$\leqslant \frac{\Gamma p(1-p)(b-a)^2}{12} \tag{10}$$

Making $p \to \frac{1}{2}$, from (8) and (10), we have the following corollary:

Corollary 3 Let $f : [a,b] \to \mathbf{R}$ be a twice differentiable mapping on (a,b) and suppose that $\gamma \leqslant f''(t) \leqslant \Gamma$ for all $t \in (a,b)$. Then for $a < b$, we have

$$\frac{\gamma(b-a)^2}{24} \leqslant \frac{f(a)+f(b)}{2} - \frac{2}{b-a}\int_a^b f(t)\mathrm{d}t - f\left(\frac{a+b}{2}\right) \leqslant \frac{\Gamma(b-a)^2}{24} \tag{11}$$

Taking $p = \frac{1}{4}$ in inequality (10), we have the following corollary:

Corollary 4 Let $f : [a,b] \to \mathbf{R}$ be a twice differentiable mapping on

(a, b) and suppose that $\gamma \leqslant f''(t) \leqslant \Gamma$ for all $t \in (a, b)$. Then

$$\frac{\gamma(b-a)^2}{64} \leqslant \left(\frac{f(a) + f(b)}{4} - \frac{1}{b-a} \int_a^b f(t)\mathrm{d}t \right) -$$

$$\left(\frac{f(\frac{a+3b}{4}) + f(\frac{3a+b}{4})}{4} - \frac{2}{b-a} \int_{\frac{3a+b}{4}}^{\frac{a+3b}{4}} f(t)\mathrm{d}t \right) \leqslant \frac{\Gamma(b-a)^2}{64} \qquad (12)$$

Theorem 4 Let $f : [a, b] \to \mathbf{R}$ such that $f^{(4)}$ is continuous on $[a, b]$, and suppose that $\gamma \leqslant f^{(4)}(t) \leqslant \Gamma$ for all $t \in (a, b)$. If $0 \leqslant p \leqslant \frac{1}{2}$, then for $a < b$, we have

$$\frac{\gamma p(1-p)(p^2 + (1-p)^2)(b-a)^4}{360} \leqslant \left(\frac{f(a) + f(b)}{6} - \frac{1}{b-a} \int_a^b f(t)\mathrm{d}t \right) -$$

$$\left(\frac{f(u) + f(v)}{6} - \frac{1}{v-u} \int_u^v f(t)\mathrm{d}t \right)$$

$$\leqslant \frac{\Gamma p(1-p)(p^2 + (1-p)^2)(b-a)^4}{360} \qquad (13)$$

Taking $p \to \frac{1}{2}$, from(8) and (13), we have the following corollary.

Corollary 5 Let $f : [a, b] \to \mathbf{R}$ such that $f^{(4)}$ is continuous on $[a, b]$, and suppose that $\gamma \leqslant f^{(4)}(t) \leqslant \Gamma$ for all $t \in (a, b)$. Then for $a < b$, we have

$$\frac{\gamma(b-a)^4}{2880} \leqslant \frac{1}{6} \left(f(a) + 4f\left(\frac{a+b}{2} \right) + f(b) \right) - \frac{1}{b-a} \int_a^b f(t)\mathrm{d}t \leqslant \frac{\Gamma(b-a)^4}{2880} \qquad (14)$$

Taking $p = \frac{1}{4}$ in inequalities (13), we have the following corollary:

Corollary 6 Let $f : [a, b] \to \mathbf{R}$ such that $f^{(4)}$ is continuous on $[a, b]$, and suppose that $\gamma \leqslant f^{(4)}(t) \leqslant \Gamma$ for all $t \in (a, b)$. Then for $a < b$, we have

$$\frac{\gamma(b-a)^4}{3072} \leqslant \left(\frac{f(a) + f(b)}{6} - \frac{1}{b-a} \int_a^b f(t)\mathrm{d}t \right) -$$

$$\left(\frac{f(\frac{3a+b}{4}) + f(\frac{a+3b}{4})}{6} - \frac{2}{b-a} \int_{\frac{3a+b}{4}}^{\frac{a+3b}{4}} f(t)\mathrm{d}t \right) \leqslant \frac{\Gamma(b-a)^4}{3072} \qquad (15)$$

66.2 Definitions and Lemmas

We need the following definitions and lemmas.

Definition 1 [4-5] Let $\boldsymbol{x} = (x_1, \ldots, x_n)$ and $\boldsymbol{y} = (y_1, \ldots, y_n) \in \mathbf{R}^n$.

(i) $\boldsymbol{x}$ is said to be majorized by $\boldsymbol{y}$ (in symbols $\boldsymbol{x} \prec \boldsymbol{y}$) if $\displaystyle\sum_{i=1}^k x_{[i]} \leqslant \sum_{i=1}^k y_{[i]}$

for $k = 1, 2, \ldots, n-1$ and $\sum\limits_{i=1}^{n} x_i = \sum\limits_{i=1}^{n} y_i$, where $x_{[1]} \geqslant \cdots \geqslant x_{[n]}$ and $y_{[1]} \geqslant \cdots \geqslant y_{[n]}$ are of x and y in a descending order.

(ii) Let $\Omega \subseteq \mathbf{R}^n$. The function $\varphi: \Omega \to \mathbf{R}$ be said to be a Schur-convex function on Ω if $x \prec y$ on Ω implies $\varphi(x) \leq \varphi(y)$. φ is said to be a Schur-concave function on Ω if and only if $-\varphi$ is Schur-convex.

Lemma 1[6] If $\frac{1}{2} \leqslant p \leqslant 1$ or $0 \leqslant p \leqslant \frac{1}{2}$, then

$$\left(\frac{a+b}{2}, \frac{a+b}{2}\right) \prec (u(p), v(p)) \prec (a, b) \tag{16}$$

Lemma 2 [7] Suppose that I is an open interval and $f : I \to \mathbf{R}$ is a continuous function. If

$$F(x, y) = \begin{cases} \frac{1}{y-x} \int_x^y f(t)dt - f(\frac{x+y}{2}), & x, y \in I, \ x \neq y \\ 0, \ x = y \in I \end{cases} \tag{17}$$

then $F(x, y)$ is Schur convex (concave) on I^2 if and only if f is convex (concave) on I.

Lemma 3 [7] Suppose that I is an open interval and $f : I \to \mathbf{R}$ is a continuous function. If

$$G(x, y) = \begin{cases} \frac{f(x)+f(y)}{2} - \frac{1}{y-x} \int_x^y f(t)dt, & x, y \in I, \ x \neq y \\ 0, \ x = y \in I \end{cases} \tag{18}$$

then $G(x, y)$ is Schur convex (concave) on I^2 if and only if f is convex (concave) on I.

Lemma 4 [8] Suppose that I is an open interval and $f : I \to \mathbf{R}$ is a continuous function. If

$$P(x, y) = \begin{cases} \frac{f(x)+f(y)}{4} + \frac{1}{2}f\left(\frac{x+y}{2}\right) - \frac{1}{y-x} \int_x^y f(t)dt, & x, y \in I, \ x \neq y \\ 0, \ x = y \in I \end{cases} \tag{19}$$

then $P(x, y)$ is Schur convex (concave) on I^2 if and only if f is convex (concave) on I.

A function $f(x)$ is said to be r-convex on $[a, b]$ with $r \geqslant 2$ if and only if $f^{(r)}(x)$ exists and $f^{(r)}(x) \geqslant 0$.

Lemma 5 [9] Suppose that I is an open interval and $f : I \to \mathbf{R}$ such that $f^{(4)}$ is continuous on I, If

$$S(x,y) = \begin{cases} \frac{f(x)+f(y)}{6} + \frac{2}{3}f\left(\frac{x+y}{2}\right) - \frac{1}{y-x}\int_x^y f(t)\mathrm{d}t, & x,y \in I,\ x \neq y \\ 0, & x = y \in I \end{cases} \tag{20}$$

then $S(x,y)$ is Schur convex (concave) on I^2 if and only if f is 4-convex (concave) on I.

66.3   Proofs of main results

Proof of Theorem 1 Let

$$h(t) = f(t) - \frac{\alpha t^2}{2}$$

and let

$$F_\alpha(x,y) = \begin{cases} \frac{1}{y-x}\int_x^y h(t)\mathrm{d}t - h\left(\frac{x+y}{2}\right), & x,y \in (a,b),\ x \neq y \\ 0, & x = y \in (a,b) \end{cases} \tag{21}$$

For $x \neq y$, we have

$$\begin{aligned} F_\alpha(x,y) &= \frac{1}{y-x}\int_x^y \left(f(t) - \frac{\alpha t^2}{2}\right)\mathrm{d}t - \left(f\left(\frac{x+y}{2}\right) - \frac{\alpha\left(\frac{x+y}{2}\right)^2}{2}\right) \\ &= \frac{1}{y-x}\int_x^y f(t)\mathrm{d}t - \frac{\alpha(y^3-x^3)}{6(y-x)} - f\left(\frac{x+y}{2}\right) + \frac{\alpha(x+y)^2}{8} \\ &= \frac{1}{y-x}\int_x^y f(t)\mathrm{d}t - f\left(\frac{x+y}{2}\right) - \frac{\alpha(y-x)^2}{24} \end{aligned}$$

that is

$$F_\alpha(x,y) = \frac{1}{y-x}\int_x^y f(t)\mathrm{d}t - f\left(\frac{x+y}{2}\right) - \frac{\alpha(y-x)^2}{24} \tag{22}$$

Since $h''(t) = f''(t) - \gamma \geqslant 0$, by Lemma 2, it follows that $F_\gamma(x,y)$ is Schur convex on $(a,b) \times (a,b)$. Combining (16), we have

$$F_\gamma\left(\frac{a+b}{2}, \frac{a+b}{2}\right) \leqslant F_\gamma(u,v) \leqslant F_\gamma(a,b)$$

namely

$$0 \leqslant \frac{1}{v-u} \int_u^v f(t)\mathrm{d}t - f\left(\frac{a+b}{2}\right) - \frac{\gamma((b-a)(2p-1))^2}{24}$$

$$\leqslant \frac{1}{b-a} \int_a^b f(t)\mathrm{d}t - f\left(\frac{a+b}{2}\right) - \frac{\gamma(b-a)^2}{24} \tag{23}$$

Making a little rearrangement, from the second inequality in (23) we obtain the first inequality in (5).

Since $h''(t) = f''(t) - \Gamma \leqslant 0$, by Lemma 2, it follows that $F_\Gamma(x,y)$ is Schur concave on $(a,b) \times (a,b)$, combining (16), we have

$$F_\Gamma\left(\frac{a+b}{2}, \frac{a+b}{2}\right) \geqslant F_\Gamma(u,v) \geqslant F_\Gamma(a,b)$$

that is

$$0 \geqslant \frac{1}{v-u} \int_u^v f(t)\mathrm{d}t - f\left(\frac{a+b}{2}\right) - \frac{\Gamma((b-a)(2p-1))^2}{24}$$

$$\geqslant \frac{1}{b-a} \int_a^b f(t)\mathrm{d}t - f\left(\frac{a+b}{2}\right) - \frac{\Gamma(b-a)^2}{24} \tag{24}$$

Making a little rearrangement, from the second inequality in (24) we obtain the second inequality in (5).

The proof of Theorem 1 is completed.

Proof of Theorem 2 Let

$$G_\alpha(x,y) = \begin{cases} \frac{h(x)+h(y)}{2} - \frac{1}{y-x} \int_x^y h(t)\mathrm{d}t, & x, y \in (a,b), \ x \neq y \\ 0, & x = y \in (a,b) \end{cases} \tag{25}$$

where $h(t) = f(t) - \frac{\alpha t^2}{2}$.

For $x \neq y$, by computing, we have

$$G_\alpha(x,y) = \frac{f(x)+f(y)}{2} - \frac{1}{y-x} \int_x^y f(t)\mathrm{d}t - \frac{\alpha(y-x)^2}{12} \tag{26}$$

The remaining part of the proof is similar to theorem 1, so omitted.

The proof of Theorem 2 is completed.

Proof of Theorem 3 Let

$$P_\alpha(x,y) = \begin{cases} \frac{h(x)+h(y)}{4} + \frac{1}{2}h\left(\frac{x+y}{2}\right) - \frac{1}{y-x} \int_x^y h(t)\mathrm{d}t, & x, y \in I, \ x \neq y \\ 0, & x = y \in I \end{cases} \tag{27}$$

where $h(t) = f(t) - \frac{\alpha t^2}{2}$.

For $x \neq y$, by computing, we have

$$P_\alpha(x,y) = \frac{f(x) + f(y)}{4} + \frac{1}{2}f\left(\frac{x+y}{2}\right) - \frac{1}{y-x}\int_x^y f(t)\mathrm{d}t - \frac{\alpha(y-x)^2}{48} \quad (28)$$

The remaining part of the proof is similar to theorem 1, so omitted.
The proof of Theorem 3 is completed.

Proof of Theorem 4 Let

$$Q_\alpha(x,y) = \begin{cases} \frac{w(x)+w(y)}{6} + \frac{2}{3}w\left(\frac{x+y}{2}\right) - \frac{1}{y-x}\int_x^y w(t)\mathrm{d}t, & x,y \in I, \ x \neq y \\ 0, & x = y \in I \end{cases}$$

$$(29)$$

where $w(t) = f(t) - \frac{\alpha t^4}{24}$.

For $x \neq y$, by computing, we have

$$Q_\alpha(x,y) = \frac{f(x) + f(y)}{6} + \frac{2}{3}f\left(\frac{x+y}{2}\right) - \frac{1}{y-x}\int_x^y f(t)\mathrm{d}t - \frac{\alpha(y-x)^4}{2880} \quad (30)$$

Since $w^{(4)}(t) = f^{(4)}(t) - \gamma \geqslant 0$, by Lemma 5, it follows that $Q_\gamma(x,y)$ is Schur convex on $(a,b) \times (a,b)$, combining (16), we have

$$Q_\gamma\left(\frac{a+b}{2}, \frac{a+b}{2}\right) \leqslant Q_\gamma(u,v) \leqslant Q_\gamma(a,b)$$

that is

$$0 \leqslant \frac{f(u) + f(v)}{6} + \frac{2}{3}f\left(\frac{a+b}{2}\right) - \frac{1}{v-u}\int_u^v f(t)\mathrm{d}t - \frac{\gamma((b-a)(2p-1))^4}{2880}$$

$$\leqslant \frac{f(a) + f(b)}{6} + \frac{2}{3}f\left(\frac{a+b}{2}\right) - \frac{1}{b-a}\int_a^b f(t)\mathrm{d}t - \frac{\gamma(b-a)^4}{2880} \quad (31)$$

Making a little rearrangement, from the second inequality in (31) we obtain the first inequality in (13).

Since $w^{(4)}(t) = f^{(4)}(t) - \Gamma \leqslant 0$, by Lemma 5, it follows that $Q_\Gamma(x,y)$ is Schur concave on $(a,b) \times (a,b)$, combining (16), we have

$$Q_\Gamma\left(\frac{a+b}{2}, \frac{a+b}{2}\right) \geqslant Q_\Gamma(u,v) \geqslant Q_\Gamma(a,b)$$

making a little rearrangement, we obtain the second inequality in (13).

The proof of Theorem 4 is completed.

66.4　A remark on Theorem B

In [10], the following identity was derived: if $f : [a, b] \to \mathbf{R}$ is twice differentiable, then

$$\frac{1}{b-a} \int_a^b f(t)dt - \frac{f(a)+f(b)}{2} + \frac{b-a}{8}[f'(b) - f'(a)]$$

$$= \frac{1}{2(b-a)} \int_a^b \left(t - \frac{a+b}{2}\right)^2 f''(t)dt \qquad (32)$$

By this identity, we can get the following the double inequality

$$\frac{3S-\Gamma}{24}(b-a)^2 \leqslant \frac{f(a)+f(b)}{2} - \frac{1}{b-a} \int_a^b f(t)dt \leqslant \frac{3S-\gamma}{24}(b-a)^2 \quad (33)$$

In fact, since $f''(t) \leqslant \Gamma$, from (44), we have

$$\frac{f(a)+f(b)}{2} - \frac{1}{b-a} \int_a^b f(t)dt$$

$$= \frac{b-a}{8}[f'(b) - f'(a)] - \frac{1}{2(b-a)} \int_a^b \left(t - \frac{a+b}{2}\right)^2 f''(t)dt$$

$$\geqslant \frac{b-a}{8}[f'(b) - f'(a)] - \frac{\Gamma}{2(b-a)} \int_a^b \left(t - \frac{a+b}{2}\right)^2 dt$$

$$= \frac{b-a}{8}[f'(b) - f'(a)] - \frac{\Gamma(b-a)^2}{24}$$

$$= \frac{3S-\Gamma}{24}(b-a)^2$$

Similarly, from $f''(t) \geqslant \gamma$, it follows that the second inequality in (45) holds.

If $\Gamma > 0$, then the first inequality in (45) stronger than the first inequality in (4), and if $\gamma < 0$, then the second inequality in (45) stronger than the second inequality in (4).

References

[1]　P CERONE, S S DRAGOMIR. Midpoint-type rules from an inequality point of view [M] //ANASTASSIOV, ed. Handbook of analytic-computational methods in applied mathematics. New York: CRC Press, 2000.

[2]　P CERONE, S S DRAGOMIR. Trapezoidal-type rules from an in-

equality point of view[M]//ANASTASSIOU,ed. Handbook of analytic-computational methods in applied mathematics[M]. New York: CRC Press, 2000.

[3] N UJEVIj′ C. Some double integral inequalities and applications[J]. Acta Math. Univ. Comenian. (N.S.), 2002, 71 : 189–199.

[4] A M MARSHALL, I OLKIN. Inequalities:theory of majorization and its application[M]. New York : Academies Press, 1979.

[5] B-Y WANG. Foundations of Majorization Inequalities (Kong zhi bu deng shi ji chu)[M]. Beijing: Beijing Normal Univ. Press, 1990 .

[6] H -N SHI, Y-M JIANG, W-D JIANG. Schur-Convexity and Schur-Geometrically Concavity of Gini Mean[J]. Comput. Math. Appl., 2009, 57: 266-274.

[7] YU-MING CHU, GEN-DI WANG, XIAO-HUI ZHANG. Schur convexity and Hadmard′ s inequality[J]. Math. Ineq. & Appl., 2010,13 (4): 725-731.

[8] V ĆULJAK, I FRANJIĆ, R GHULAM, J PEČARIĆ. Schur-convexity of averages of convex functions[J]. J. Inequal. Appl., Volume, 2011, Article ID 581918, 25 pages doi:10.1155/2011/581918.

[9] IVA FRANJIĆ, JOSIP PEČARIĆ. Schur-convexity and the Simpson formula[J]. Applied Mathematics Letters 2011, 24 : 1565-1568.

[10] X-L CHENG, J SUN. A note on the perturbed trapezoid inequality[J]. Journal of Inequalities in Pure and Applied Mathematics, 2002(3):1-7.

第67篇　Multi-Parameter Generalization of Rado-Popoviciu inequalities

(ZHANG JING, SHI HUAN-NAN. J.Math. Inequal., Volume 10, Number 2 (2016), 577 ‑ 582)

Abstract: By using methods on the theory of majorization, new generalizations of Rado's inequality and Popoviciu's inequality which involves multi-parameter are established.

Keywords: Rado's inequality; Popoviciu's inequality; majorization; Schur-concavity; symmetric function; dual form; convex function

67.1　Introduction

Throughout the article, $\mathbf{R}$ denotes the set of real numbers, $\boldsymbol{x} = (x_1, \cdots, x_n)$ denotes n-tuple (n-dimensional real vectors), the set of vectors can be written as

$$\mathbf{R}^n = \{\boldsymbol{x} = (x_1, \cdots, x_n) : x_i \in \mathbf{R}, i = 1, \cdots, n\}$$

$$\mathbf{R}_+^n = \{\boldsymbol{x} = (x_1, \cdots, x_n) : x_i \geqslant 0, i = 1, \cdots, n\}$$

$$\mathbf{R}_{++}^n = \{\boldsymbol{x} = (x_1, \cdots, x_n) : x_i > 0, i = 1, \cdots, n\}$$

In particular, the notations $\mathbf{R}$, $\mathbf{R}_+$ and $\mathbf{R}_{++}$ denote $\mathbf{R}^1$, $\mathbf{R}_+^1$ and $\mathbf{R}_{++}^1$, respectively.

Let $\boldsymbol{x} = (x_1, \cdots, x_n) \in \mathbf{R}^n$. The elementary symmetric functions are defined by

$$E_k(\boldsymbol{x}) = E_k(x_1, \cdots, x_n) = \sum_{1 \leqslant i_1 < \cdots < i_k \leqslant n} \prod_{j=1}^{k} x_{i_j}, \quad k = 1, \cdots, n$$

$E_0(\boldsymbol{x}) = 1$, and $E_k(\boldsymbol{x}) = 0$ for $k < 0$ or $k > n$. The dual form of the elementary symmetric functions are

$$E_k^*(\boldsymbol{x}) = E_k^*(x_1, \cdots, x_n) = \prod_{1 \leqslant i_1 < \cdots < i_k \leqslant n} \sum_{j=1}^{k} x_{i_j}, \quad k = 1, \cdots, n$$

$E_0^*(\boldsymbol{x}) = 1$, and $E_k^*(\boldsymbol{x}) = 0$ for $k < 0$ or $k > n$.

For $\boldsymbol{x} \in \mathbf{R}_{++}^n$, let

$$A_n(\boldsymbol{x}) = \frac{\sum_{i=1}^{n} x_i}{n}, G_n(\boldsymbol{x}) = \sqrt[n]{\prod_{i=1}^{n} x_i}$$

The following inequalities

$$n(A_n(\boldsymbol{x}) - G_n(\boldsymbol{x})) \geqslant (n-1)(A_{n-1}(\boldsymbol{x}) - G_{n-1}(\boldsymbol{x})) \tag{1}$$

$$\left(\frac{A_n(\boldsymbol{x})}{G_n(\boldsymbol{x})}\right)^n \geqslant \left(\frac{A_{n-1}(\boldsymbol{x})}{G_{n-1}(\boldsymbol{x})}\right)^{n-1} \tag{2}$$

are known in the bibliography as Rado's inequality and Popoviciu's inequality respectively (see Mitrinović and Vasić [1]. Inequalities (1) and (2) furnish a good route for joining arithmetic means and geometric means of positive numbers. This pair of inequalities has attracted considerable attention by many mathematicians, and has actuated a quantity of research articles giving their simple proofs providing diverse improvements, generalizations and analogs (see [1-8] and references therein).

The aim of this paper is to establish a new generalization of Rado's inequality and Popoviciu's inequality which involves multi-parameter, by Schur-concavity of the elementary symmetric function and its dual formula, as well as a simple majorization relation.

Our main results are the following.

Theorem 1 Let $\boldsymbol{x} \in \mathbf{R}_{++}^n$, $n \geqslant 2, 0 < \alpha \leqslant 1, \lambda > 0$. Then for

$k = 1, \cdots, n$, we have

$$\left(A_n(x) + \frac{(\lambda - 1)x_n}{n}\right)^{\alpha}$$

$$\geqslant \left(\left(1 - \frac{1}{k}\right)(A_{n-1}(x))^{\alpha} + \frac{1}{k}\lambda^{\alpha}(x_n)^{\alpha}\right)^{\frac{k}{n}} \cdot (A_{n-1}(x))^{(1-\frac{k}{n})\alpha} \qquad (3)$$

Taking $\alpha = \lambda = 1$ and $k = n$, from the inequality (3), it follows that

Corollary 1 Let $x \in \mathbf{R}_{++}^n$, $n \geqslant 2$, we have

$$nA_n(x) - (n - 1)A_{n-1}(x) \geqslant x_n \qquad (4)$$

Remark 1 By the arithmetic-geometric mean inequality, it follows that

$$\frac{x_n + G_{n-1}(x) + \cdots + G_{n-1}(x)}{n} \geqslant (x_n G_{n-1}(x) \cdots G_{n-1}(x))^{\frac{1}{n}}$$

i.e.

$$x_n + (n - 1)G_{n-1}(x) \geqslant n \left(x_n G_{n-1}^{n-1}(x)\right)^{\frac{1}{n}} = nG_n(x)$$

i.e.

$$x_n \geqslant nG_n(x) - (n - 1)G_{n-1}(x) \qquad (5)$$

(4) and (5) together give

$$nA_n(x) - (n - 1)A_{n-1}(x) \geqslant x_n \geqslant nG_n(x) - (n - 1)G_{n-1}(x) \qquad (6)$$

The inequality (6) refine the equivalent form of Rado's inequality (1).

Taking $\alpha = \lambda = 1$ and $k = 1$, from the inequality (3), it follows that

Corollary 2 Let $x \in \mathbf{R}_{++}^n$, $n \geqslant 2$, we have

$$A_n(x) \geqslant (x_n)^{\frac{1}{n}}(A_{n-1}(x))^{1-\frac{1}{n}} \qquad (7)$$

Remark 2 It is clear that inequality (7) is equivalent to

$$(A_n(x))^n \geqslant x_n(A_{n-1}(x))^{n-1}$$

and then

$$\left(\frac{A_n(\boldsymbol{x})}{G_n(\boldsymbol{x})}\right)^n \geqslant \frac{x_n(A_{n-1}(\boldsymbol{x}))^{n-1}}{(G_n(\boldsymbol{x}))^n} = \left(\frac{A_{n-1}(\boldsymbol{x})}{G_{n-1}(\boldsymbol{x})}\right)^{n-1}$$

This shows that the inequality (7) is equivalent to Popoviciu's inequality (2).

Theorem 1 Let $\boldsymbol{x} \in \mathbf{R}_{++}^n$, $n \geqslant 2, 0 < \alpha \leqslant 1, \lambda > 0$. Then for $k = 1, \cdots, n$, we have

$$\frac{\left(A_n(\boldsymbol{x}) + \dfrac{(\lambda - 1)x_n}{n}\right)^{k\alpha}}{(G_n(\boldsymbol{x}))^{n\alpha}}$$
$$\geqslant \lambda^\alpha \cdot \frac{k}{n} \cdot \frac{(A_{n-1}(\boldsymbol{x}))^{(k-1)\alpha}}{(G_{n-1}(\boldsymbol{x}))^{(n-1)\alpha}} + \left(1 - \frac{k}{n}\right)\frac{(A_{n-1}(\boldsymbol{x}))^{k\alpha}}{(G_n(\boldsymbol{x}))^{n\alpha}} \qquad (8)$$

Remark 2 When $\alpha = \lambda = 1$ and $k = n$, the inequality (8) is reduces to Popoviciu's inequality (2), and when $\alpha = \lambda = 1$ and $k = 1$, the inequality (8) is reduces (4).

67.2 Definitions and Lemmas

We need the following definitions and auxiliary lemmas.

Definition 1[9-10] Let $\boldsymbol{x} = (x_1, \cdots, x_n)$ and $\boldsymbol{y} = (y_1, \cdots, y_n) \in \mathbf{R}^n$.

(i) $\boldsymbol{x} \geqslant \boldsymbol{y}$ means $x_i \geqslant y_i$ for all $i = 1, 2, \cdots, n$.

(ii) Let $\Omega \subset \mathbf{R}^n$, φ: $\Omega \to \mathbf{R}$ is said to be increasing if $\boldsymbol{x} \geqslant \boldsymbol{y}$ implies $\varphi(\boldsymbol{x}) \geqslant \varphi(\boldsymbol{y})$. φ is said to be decreasing if and only if $-\varphi$ is increasing.

Definition 2[9-10] Let $\boldsymbol{x} = (x_1, \cdots, x_n)$ and $\boldsymbol{y} = (y_1, \cdots, y_n) \in \mathbf{R}^n$.

(i) $\boldsymbol{x}$ is said to be majorized by $\boldsymbol{y}$ (in symbols $\boldsymbol{x} \prec \boldsymbol{y}$) if $\sum_{i=1}^k x_{[i]} \leqslant \sum_{i=1}^k y_{[i]}$ for $k = 1, 2, \cdots, n - 1$ and $\sum_{i=1}^n x_i = \sum_{i=1}^n y_i$, where $x_{[i]}$ denotes the i th largest component of $\boldsymbol{x}$, $x_{[1]} \geqslant \cdots \geqslant x_{[n]}$ and $y_{[1]} \geqslant \cdots \geqslant y_{[n]}$ are rearrangements of $\boldsymbol{x}$ and $\boldsymbol{y}$ in a descending order.

(ii) Let $\Omega \subset \mathbf{R}^n$, φ: $\Omega \to \mathbf{R}$ is said to be a Schur-convex function on Ω if $\boldsymbol{x} \prec \boldsymbol{y}$ on Ω implies $\varphi(\boldsymbol{x}) \leq \varphi(\boldsymbol{y})$. φ is said to be a Schur-concave function on Ω if and only if $-\varphi$ is Schur-convex function on Ω.

The Schur-convexity described the ordering of majorization, the order-preserving functions were first comprehensively studied by I. Schur in 1923. It has important applications in combinatorial analysis, geometric inequalities, matrix theory, numerical analysis, etc. See [9].

Lemma 1[9-11] Let $x = (x_1, \cdots, x_n) \in \mathbf{R}^n$ and $\bar{x} = \dfrac{1}{n} \sum\limits_{i=1}^{n} x_i$. Then

$$(\bar{x}, \cdots, \bar{x}) \prec (x_1, \cdots, x_n)$$

Lemma 2[9] The elementary symmetric functions $E_k(x)$ are increasing and Schur-concave on $\mathbf{R}_+^n$.

Lemma 3[9] The dual form of the elementary symmetric functions $E_k^*(x)$ are increasing and Schur-concave on $\mathbf{R}_+^n$.

Lemma 4[10] Let the set $\mathbf{A} \subset \mathbf{R}$. The function $\varphi : \mathbf{R}_+^n \to \mathbf{R}$ is increasing and Schur-concave, and the function $g : \mathbf{A} \to \mathbf{R}_+$ is concave. Then the composite function $\psi(x_1, \cdots, x_n) = \varphi(g(x_1), \cdots, g(x_n)) : \mathbf{A}^n \to \mathbf{R}$ is Schur-concave.

Lemma 5 Let $n \geqslant 2, t, \lambda > 0$, $g : \mathbf{R}_{++} \to \mathbf{R}_+$ is concave. Then for $k = 1, \cdots, n$, we have

$$g\left(\frac{t+\lambda}{n}\right) \geqslant \left(\left(1 - \frac{1}{k}\right) g\left(\frac{t}{n-1}\right) + \frac{1}{k} \cdot g(\lambda)\right)^{\frac{k}{n}} \cdot \left(g\left(\frac{t}{n-1}\right)\right)^{1-\frac{k}{n}} \tag{9}$$

Proof By Lemma 1, it follows that

$$u = \Big(\underbrace{\frac{t+\lambda}{n}, \cdots, \frac{t+\lambda}{n}}_{n}\Big) \prec \Big(\underbrace{\frac{t}{n-1}, \cdots, \frac{t}{n-1}}_{n-1}, \lambda\Big) = v \tag{10}$$

and from Lemma 3 and Lemma 4, it follows that $E_k^*(g(x_1), \cdots, g(x_n))$ is Schur-concave on $\mathbf{R}_{++}^n$, combining (10) we get

$$E_k^*\left(g\left(\frac{t+\lambda}{n}\right), \cdots, g\left(\frac{t+\lambda}{n}\right)\right) \geqslant E_k^*\left(g\left(\frac{t}{n-1}\right), \cdots, g\left(\frac{t}{n-1}\right), g(\lambda)\right)$$

i.e.

$$\left(kg\left(\frac{t+\lambda}{n}\right)\right)^{\binom{n}{k}} \geqslant \left((k-1)g\left(\frac{t}{n-1}\right) + g(\lambda)\right)^{\binom{n-1}{k-1}} \left(kg\left(\frac{t}{n-1}\right)\right)^{\binom{n-1}{k}} \tag{11}$$

Extracting root of $\binom{n}{k}$ the both sides in the inequality (11), we get

$$k\left(g\left(\frac{t+\lambda}{n}\right)\right) \geqslant \left((k-1)g\left(\frac{t}{n-1}\right)+g(\lambda)\right)^{\frac{k}{n}} \cdot \left(kg\left(\frac{t}{n-1}\right)\right)^{1-\frac{k}{n}} \tag{12}$$

Dividing both sides in the inequality (12) by $k = k^{\frac{k}{n}} \cdot k^{(1-\frac{k}{n})}$, we obtain the inequality (9).

Lemma 6 Let $n \geqslant 2, t, \lambda > 0$, $g : \mathbf{R}_{++} \to \mathbf{R}_+$ is concave. Then for $k = 1, \cdots, n$, we have

$$\left(g\left(\frac{t+\lambda}{n}\right)\right)^k \geqslant \frac{k}{n} \cdot g(\lambda)\left(g\left(\frac{t}{n-1}\right)\right)^{k-1} + \left(1-\frac{k}{n}\right)\left(g\left(\frac{t}{n-1}\right)\right)^k \tag{13}$$

Proof From Lemma 2 and Lemma 4, it follows that $E_k(g(x_1), \cdots, g(x_n))$ is Schur-concave on $\mathbf{R}_{++}^n$, combining (10) we get

$$E_k\left(g\left(\frac{t+\lambda}{n}\right), \cdots, g\left(\frac{t+\lambda}{n}\right)\right) \geqslant E_k\left(g\left(\frac{t}{n-1}\right), \cdots, g\left(\frac{t}{n-1}\right), g(\lambda)\right)$$

i.e.

$$\binom{n}{k}\left(g\left(\frac{t+\lambda}{n}\right)\right)^k$$
$$\geqslant \binom{n-1}{k-1}g(\lambda)\left(g\left(\frac{t}{n-1}\right)\right)^{k-1} + \binom{n-1}{k}\left(g\left(\frac{t}{n-1}\right)\right)^k \tag{14}$$

Dividing both sides in the inequality (14) by $\binom{n}{k}$, we obtain the inequality (13).

67.3 Proof of Main Results

Proof of Theorem 1 Taking $g(y) = y^\alpha, 0 < \alpha \leqslant 1$ and $t = \dfrac{\sum_{i=1}^{n-1} x_i}{x_n}$, from (9) it is deduced that

$$\left(\frac{A_n(\boldsymbol{x}) + \frac{(\lambda-1)x_n}{n}}{x_n}\right)^\alpha$$
$$\geqslant \left(\left(1-\frac{1}{k}\right)\left(\frac{A_{n-1}(\boldsymbol{x})}{x_n}\right)^\alpha + \frac{1}{k}\lambda^\alpha\right)^{\frac{k}{n}} \cdot \left(\frac{A_{n-1}(\boldsymbol{x})}{x_n}\right)^{(1-\frac{k}{n})\alpha} \tag{15}$$

Multiplying both sides in the inequality (15) by $(x_n)^\alpha = (x_n)^{\frac{k}{n}\alpha}(x_n)^{(1-\frac{k}{n})\alpha}$, we obtain the inequality (3).

The proof of Theorem 1 is completed.

Proof of Theorem 2　Taking $g(y) = y^\alpha, 0 < \alpha \leqslant 1$ and $t = \dfrac{\sum\limits_{i=1}^{n-1} x_i}{x_n}$, from (13) it is deduced that

$$
\left(\frac{A_n(\boldsymbol{x}) + \dfrac{(\lambda - 1)x_n}{n}}{x_n} \right)^{k\alpha}
$$

$$
\geqslant \lambda^\alpha \frac{k}{n} \left(\frac{A_{n-1}(\boldsymbol{x})}{x_n} \right)^{(k-1)\alpha} + \left(1 - \frac{k}{n} \right) \left(\frac{A_{n-1}(\boldsymbol{x})}{x_n} \right)^{k\alpha} \tag{16}
$$

Multiplying both sides in the inequality (16) by $(x_n)^{k\alpha} \left(\prod\limits_{i=1}^{n} x_i^\alpha \right)^{-1}$, we obtain the inequality (8).

The proof of Theorem 2 is completed.

Conflict of Interests

The authors declare that there is no conflict of interests regarding the publication of this article.

Acknowledgment

The work was supported by Funding Project for Academic Human Resources Development in Institutions of Higher Learning under the Jurisdiction of Beijing Municipality (PHR (IHLB)) (PHR201108407).

References

[1]　D S MITRINOVIĆ, P M VASIĆ. Analytic Inequalities[M]. Berlin: Springer-Verlag, 1970: 74-94.

[2]　D S MITRINOVIĆ, J E PEČARIĆ, A M FINK. Classical and New Inequalities in Analysis[M]. Dordrecht: Kluwer Academic Publishers, 1993: 21-56.

[3]　CHUNG-LIE WANG. Inequalities of the Rado-Popoviciu type for functions and their applications[J]. J. Math. Anal. Appl., 1984, 100 (2) : 436-446.

[4] HUAN-NAN SHI. A new generalization of the Popoviciu's inequality[J]. J. Sichuan Norm. Univ. Nat. Sci. Ed., 2002, 25 (5): 510-511.

[5] VASILE MIHEŞAN. Rado and Popoviciu type inequalities for pseudo arithmetic and geometric means[J]. Int. J. Pure Appl. Math., 2005, 23 (3) : 293-297.

[6] M BENCZE. A generalization of T. Popoviciu's inequality[J]. Gazeta Mat., 1991, 96 : 159-161.

[7] H ALZER. Rado-type inequalities for geometric and harmonic means [J]. J. Pure Appl. Math. Sci., 1989, 24 : 125-130.

[8] S WU, L DEBNATH. Weighted generalization of Rado's inequality and Popoviciu's inequality[J]. Appl. Math. Lett., 2008, 21 : 313-319.

[9] MARSHALL A M, OLKIN I, ARNOLD B C. Inequalities: Theory of Majorization and Its Application [M]. 2nd ed. New York: Springer Press, 2011.

[10] B-Y WANG. Foundations of Majorization Inequalities (Kong zhi bu deng shi ji chu)[M]. Beijing: Beijing Normal Univ. Press, 1990.

[11] H-N SHI. Theory of Majorization and Analytic Inequalities[M]. Harbin: Harbin Institute of Technology Press, 2012.

第68篇　Schur-m Power Convexity for a Mean of Two Variables with Three Parameters

(WANG DONGSHENG ,FU CUN-RU, SHI HUANNAN. Journal of

Nonlinear Science and Applications, J. Nonlinear Sci. Appl. 9 (2016),

2298-2304)

Abstract: The Schur-m power convexity of a mean for two variables with three parameters is investigated, a judging condition about the Schur-m power convexity of a mean for two variables with three parameters is given.

Keywords: Schur convexity; Schur geometric convexity; Schur harmonic convexity; Schur-m power convexity; majorization

68.1　Introduction

Throughout the paper we assume that the set of n-dimensional row vector on the real number field by $\mathbf{R}^n$.

$$\mathbf{R}^n_+ = \{x = (x_1, \ldots, x_n) \in \mathbf{R}^n : x_i > 0, i = 1, \ldots, n\}$$

In particular, $\mathbf{R}^1$ and $\mathbf{R}^1_+$ denoted by $\mathbf{R}$ and $\mathbf{R}_+$ respectively.

In 2009, Kuang[1] defined a mean of two variables with three parameters as follows

$$K(\omega_1, \omega_2, p; a, b) = \left[\frac{\omega_1 A(a^p, b^p) + \omega_2 G(a^p, b^p)}{\omega_1 + \omega_2} \right]^{\frac{1}{p}}$$

where $A(a, b) = \frac{a+b}{2}$ and $G(a, b) = \sqrt{ab}$ respectively is the arithmetic mean and geometric mean of two positive numbers a and b, parameters $p \neq 0$,

$\omega_1, \omega_2 \geqslant 0$ with $\omega_1 + \omega_2 \neq 0$.

In particular

$$K\left(1, \frac{\omega}{2}, 1; a, b\right) = \frac{a + \omega\sqrt{ab} + b}{\omega + 2}$$

is the generalized Heron mean, which was introduced by Janous [2] in 2001.

$$K\left(1, \frac{\omega}{2}, p, a, b\right) = \frac{a^p + \omega(ab)^{p/2} + b^p}{\omega + 2}$$

is the generalized Heron mean with parameter.

For simplicity, sometimes we will be $K(\omega_1, \omega_2, p; a, b)$ for $K(\omega_1, \omega_2, p)$ or $K(a, b)$.

In recent years, the study on the properties of the mean with two variables by using theory of majorization is unusually active.

Yang[3-5] generalized the notion of Schur convexity to Schur f-convexity, which contains the Schur geometrical convexity, Schur harmonic convexity and so on. Moreover, he discussed Schur m-power convexity of Stolarsky means [3], Gini means [4] and Daróczy means [5]. Subsequently, many scholars have aroused the interest of Schur m-power convexity(see references [6-10]).

In this paper, the Schur-m power convexity of the mean $K(\omega_1, \omega_2, p)$ is discussed, a judging condition about the Schur-m power convexity of the mean $K(\omega_1, \omega_2, p)$ is given.

Our main result is as follows:

Theorem 1 (I) For $m > 0$.

(i) If $p \geqslant \max\{(1 + \frac{\omega_2}{\omega_1})m, 2m\}$, then $K(\omega_1, \omega_2, p)$ is Schur-m power convex with $(a, b) \in \mathbf{R}_+^2$.

(ii) If $m \leqslant p \leqslant \min\{(1 + \frac{\omega_2}{\omega_1})m, 2m\}$, then $K(\omega_1, \omega_2, p)$ is Schur-m power concave with $(a, b) \in \mathbf{R}_+^2$.

(iii) If $0 \leqslant p < m$, then $K(\omega_1, \omega_2, p)$ is Schur-m power concave with $(a, b) \in \mathbf{R}_+^2$.

(iv) If $p < 0$, then $K(\omega_1, \omega_2, p)$ is Schur-m power concave with $(a, b) \in \mathbf{R}_+^2$.

(II) For $m < 0$.

(i) If $p \geqslant 0$, then $K(\omega_1, \omega_2, p)$ is Schur-m power convex with $(a, b) \in \mathbf{R}_+^2$.

(ii) If $m \leqslant p < 0$, then $K(\omega_1, \omega_2, p)$ is Schur-m power convex with $(a, b) \in \mathbf{R}_+^2$.

(iii) If $2m \leqslant p < m$, and $p = (1 + \frac{\omega_2}{\omega_1})m$, $(0 < \frac{\omega_2}{\omega_1} < 1)$, then $K(\omega_1, \omega_2, p)$ is Schur-m power convex with $(a, b) \in \mathbf{R}_+^2$.

(iv) If $p < 2m$, and $p = (1 + \frac{\omega_1}{\omega_2})m$, $(\frac{\omega_2}{\omega_1} > 1)$, then $K(\omega_1, \omega_2, p)$ is Schur-m power concave with $(a, b) \in \mathbf{R}_+^2$.

68.2　Definitions and Lemmas

We need the following definitions and lemmas.

Definition 1 [11–12] Let $\boldsymbol{x} = (x_1, \ldots, x_n)$ and $\boldsymbol{y} = (y_1, \ldots, y_n) \in \mathbf{R}^n$.

(i) $\boldsymbol{x}$ is said to be majorized by $\boldsymbol{y}$ (in symbols $\boldsymbol{x} \prec \boldsymbol{y}$) if $\sum\limits_{i=1}^{k} x_{[i]} \leqslant \sum\limits_{i=1}^{k} y_{[i]}$ for $k = 1, 2, \ldots, n - 1$ and $\sum\limits_{i=1}^{n} x_i = \sum\limits_{i=1}^{n} y_i$, where $x_{[1]} \geqslant \cdots \geqslant x_{[n]}$ and $y_{[1]} \geqslant \cdots \geqslant y_{[n]}$ are rearrangements of $\boldsymbol{x}$ and $\boldsymbol{y}$ in a descending order.

(ii) $\Omega \subset \mathbf{R}^n$ is called a convex set if $(\alpha x_1 + \beta y_1, \ldots, \alpha x_n + \beta y_n) \in \Omega$ for any $\boldsymbol{x}$ and $\boldsymbol{y} \in \Omega$, where α and $\beta \in [0, 1]$ with $\alpha + \beta = 1$.

(iii) Let $\Omega \subset \mathbf{R}^n$, $\varphi \colon \Omega \to \mathbf{R}$ is said to be a Schur-convex function on Ω if $\boldsymbol{x} \prec \boldsymbol{y}$ on Ω implies $\varphi(\boldsymbol{x}) \leq \varphi(\boldsymbol{y})$. φ is said to be a Schur-concave function on Ω if and only if $-\varphi$ is Schur-convex function.

Definition 2 [13–14] Let $\boldsymbol{x} = (x_1, \ldots, x_n)$ and $\boldsymbol{y} = (y_1, \ldots, y_n) \in \mathbf{R}_+^n$.

(i) $\Omega \subset \mathbf{R}_+^n$ is called a geometrically convex set if $(x_1^\alpha y_1^\beta, \ldots, x_n^\alpha y_n^\beta) \in \Omega$ for any $\boldsymbol{x}$ and $\boldsymbol{y} \in \Omega$, where α and $\beta \in [0, 1]$ with $\alpha + \beta = 1$.

(ii) Let $\Omega \subset \mathbf{R}_+^n$, $\varphi \colon \Omega \to \mathbf{R}_+$ is said to be a Schur-geometrically convex function on Ω if $(\log x_1, \ldots, \log x_n) \prec (\log y_1, \ldots, \log y_n)$ on Ω implies $\varphi(\boldsymbol{x}) \leqslant \varphi(\boldsymbol{y})$. φ is said to be a Schur-geometrically concave function on Ω if and only if $-\varphi$ is Schur-geometrically convex function.

Definition 3 [15–16] Let $\Omega \subset \mathbf{R}_+^n$.

(i) A set Ω is said to be a harmonically convex set if $\frac{\boldsymbol{x}\boldsymbol{y}}{\lambda\boldsymbol{x} + (1-\lambda)\boldsymbol{y}} \in \Omega$ for every $\boldsymbol{x}, \boldsymbol{y} \in \Omega$ and $\lambda \in [0, 1]$, where $\boldsymbol{x}\boldsymbol{y} = \sum\limits_{i=1}^{n} x_i y_i$ and $\frac{1}{\boldsymbol{x}} = \left(\frac{1}{x_1}, \cdots, \frac{1}{x_n}\right)$.

(ii) A function $\varphi : \Omega \to \mathbf{R}_+$ is said to be a Schur harmonically convex function on Ω if $\frac{1}{\boldsymbol{x}} \prec \frac{1}{\boldsymbol{y}}$ implies $\varphi(\boldsymbol{x}) \le \varphi(\boldsymbol{y})$. A function φ is said to be a Schur harmonically concave function on Ω if and only if $-\varphi$ is a Schur harmonically convex function.

Definition 4 [3] Let $f : \mathbf{R}_+ \to \mathbf{R}$ be defined by

$$f(x) = \begin{cases} \dfrac{x^m - 1}{m}, & m \ne 0 \\ \ln x, & m = 0 \end{cases} \tag{2}$$

Then a function $\phi : \Omega \subset \mathbf{R}_+^n \to \mathbf{R}$ is said to be Schur m-power convex on Ω if

$$(f(x_1), f(x_2), \ldots, f(x_n)) \prec (f(y_1), f(y_2), \ldots, f(y_n))$$

for all $(x_1, x_2, \ldots, x_n) \in \Omega$ and $(y_1, y_2, \ldots, y_n) \in \Omega$ implies $\phi(\boldsymbol{x}) \le \phi(\boldsymbol{y})$.

If $-\phi$ is Schur m-power convex, then we say that ϕ is Schur m-power concave.

If putting $f(x) = x, \ln x, \frac{1}{x}$ in Definition 4, then definitions of the Schur-convex, Schur-geometrically convex, and Schur-harmonically convex functions can be deduced respectively.

Lemma 1[12] Let $\Omega \subset \mathbf{R}^n$ is convex set, and has a nonempty interior set Ω^0. Let $\varphi : \Omega \to \mathbf{R}$ is continuous on Ω and differentiable in Ω^0. Then φ is the Schur-convex(Schur-concave)function, if and only if it is symmetric on Ω and if

$$(x_1 - x_2)\left(\frac{\partial \varphi}{\partial x_1} - \frac{\partial \varphi}{\partial x_2}\right) \ge 0 (\le 0)$$

holds for any $\boldsymbol{x} = (x_1, x_2, \cdots, x_n) \in \Omega^0$.

Lemma 2[13-14] Let $\Omega \subset \mathbf{R}_+^n$ be a symmetric geometrically convex set with a nonempty interior Ω^0. Let $\varphi : \Omega \to \mathbf{R}_+$ be continuous on Ω and differentiable on Ω^0. Then φ is a Schur geometrically convex (Schur geometrically concave) function if and only if φ is symmetric on Ω and

$$(x_1 - x_2)\left(x_1 \frac{\partial \varphi}{\partial x_1} - x_2 \frac{\partial \varphi}{\partial x_2}\right) \ge 0 \quad (\le 0) \tag{3}$$

holds for any $\boldsymbol{x} = (x_1, \cdots, x_n) \in \Omega^0$.

Lemma 3[15-16] Let $\Omega \subset \mathbf{R}_+^n$ be a symmetric harmonically convex set with a nonempty interior Ω^0. Let $\varphi : \Omega \to \mathbf{R}_+$ be continuous on Ω and differentiable on Ω^0. Then φ is a Schur harmonically convex (Schur harmonically concave) function if and only if φ is symmetric on Ω and

$$(x_1 - x_2)\left(x_1^2 \frac{\partial\varphi}{\partial x_1} - x_2^2 \frac{\partial\varphi}{\partial x_2}\right) \geqslant 0 \quad (\leqslant 0) \tag{4}$$

holds for any $\boldsymbol{x} = (x_1, \cdots, x_n) \in \Omega^0$.

Lemma 4[3] Let $\Omega \subset \mathbf{R}_+^n$ be a symmetric set with nonempty interior Ω^0 and $\varphi : \Omega \to \mathbf{R}_+$ be continuous on Ω and differentiable in Ω^0. Then φ is Schur m-power convex on Ω if and only if φ is symmetric on Ω and

$$\frac{x_1^m - x_2^m}{m}\left[x_1^{1-m} \frac{\partial\varphi(\boldsymbol{x})}{\partial x_1} - x_2^{1-m} \frac{\partial\varphi(\boldsymbol{x})}{\partial x_2}\right] \geqslant 0, \quad \text{if } m \neq 0 \tag{5}$$

and

$$(\log x_1 - \log x_2)\left[x_1 \frac{\partial\varphi(\boldsymbol{x})}{\partial x_1} - x_2 \frac{\partial\varphi(\boldsymbol{x})}{\partial x_2}\right] \geqslant 0, \quad \text{if } m = 0 \tag{6}$$

for all $\boldsymbol{x} \in \Omega^0$.

Lemma 5 Let

$$g(x) = \omega_1(p - m)x^{\frac{p}{2}} - \omega_2 \frac{p}{2} x^m + \omega_2\left(\frac{p}{2} - m\right), \quad x \in [1, \infty) \tag{7}$$

where $\omega_1 \geqslant 0, \omega_2 \geqslant 0, m \in \mathbf{R}$ and $m \neq 0$.

(I) For $m > 0$.

(i) If $p \geqslant \max\{(1 + \frac{\omega_2}{\omega_1})m, 2m\}$, then $g(x) \geqslant 0$.

(ii) If $m \leqslant p \leqslant \min\{(1 + \frac{\omega_2}{\omega_1})m, 2m\}$, then $g(x) \leqslant 0$.

(iii) If $0 \leqslant p < m$, then $g(x) \leqslant 0$.

(iv) If $p < 0$, then the symbol of $g(x)$ is not fixed (from negative to positive).

(II) For $m < 0$.

(i) If $p \geqslant 0$, then $g(x) > 0$.

(ii) If $m \leqslant p < 0$, then $g(x) \geqslant 0$.

(iii) If $2m \leqslant p < m$, and $p = (1 + \frac{\omega_2}{\omega_1})m(0 < \frac{\omega_2}{\omega_1} < 1)$, then $g(x) \geqslant 0$.

(iv) If $p < 2m$, and $p = (1 + \frac{\omega_1}{\omega_2})m(\frac{\omega_2}{\omega_1} > 1)$, then $g(x) \leqslant 0$.

Proof From (7), we have

$$g(1) = \omega_1(p - m) - \omega_2 \frac{p}{2} + \omega_2\left(\frac{p}{2} - m\right) = \omega_1[p - (1 + \frac{\omega_2}{\omega_1})m] \tag{8}$$

and

$$g^{'}(x) = \omega_1(p-m)\frac{p}{2}x^{\frac{p}{2}-1} - \omega_2 m\frac{p}{2}x^{m-1} = \frac{p}{2}x^{m-1}h(x) \tag{9}$$

where

$$h(x) = \omega_1(p-m)x^{\frac{p}{2}-m} - \omega_2 m \tag{10}$$

(I) For $m > 0$.

(i) If $p \geqslant \max\{(1+\frac{\omega_2}{\omega_1})m, 2m\}$, then $p > 0$ and

$$h(x) \geqslant \omega_1(p-m) - \omega_2 m = g(1) \geqslant 0 \tag{11}$$

so for $x \in [1, \infty)$, we have $g^{'}(x) \geqslant 0$, and then, $g(x) \geqslant g(1) \geqslant 0$.

(ii) If $m \leqslant p \leqslant \min\{(1+\frac{\omega_2}{\omega_1})m, 2m\}$, then it is easy to see that the inequality (12) is reversed, so for $x \in [1, +\infty)$, we have $g^{'}(x) \leqslant 0$, and then, $g(x) \leqslant g(1) \leqslant 0$.

(iii) If $0 \leqslant p < m$, then $h(x) \leqslant 0$ and $g(1) \leqslant 0$, so for $x \in [1, +\infty)$, we have $g^{'}(x) \leqslant 0$, and then, $g(x) \leqslant g(1) \leqslant 0$.

(iv) If $p < 0$, then $h(x) \leqslant 0$, so for $x \in [1, +\infty)$, we have $g^{'}(x) \geqslant 0$, this means that $g(x)$ is increasing on $[1, +\infty)$. For $p < 0$, notice that $\lim\limits_{x \to +\infty} x^{\frac{p}{2}} = 0$, from (7), it is easy to see that $\lim\limits_{x \to +\infty} g(x) = +\infty$, but $g(1) < 0$, so the symbol of $g(x)$ is not fixed (from negative to positive) on $[1, +\infty)$.

(II) For $m < 0$.

(i) If $p \geqslant 0$, then $h(x) \geqslant 0$, so for $x \in [1, +\infty)$, we have $g^{'}(x) \geqslant 0$, and then, $g(x) \geqslant g(1) > 0$.

(ii) If $m \leqslant p < 0$, then $h(x) \geqslant 0$, so for $x \in [1, +\infty)$, we have $g^{'}(x) \leqslant 0$, and then, $g(x) \geqslant \lim\limits_{x \to +\infty} g(x) = \omega_2(\frac{p}{2} - m) \geqslant 0$.

(iii) If $2m \leqslant p < m$, and $p = (1 + \frac{\omega_2}{\omega_1})m$, $(0 < \frac{\omega_2}{\omega_1} < 1)$, then

$$h(x) \leqslant \omega_1(p-m) - \omega_2 m = g(1) = 0 \tag{12}$$

so for $x \in [1, +\infty)$, we have $g^{'}(x) \geqslant 0$, and then, $g(x) \geqslant g(1) = 0$.

(iv) If $p < 2m$, and $p = (1 + \frac{\omega_2}{\omega_1})m$, $(\frac{\omega_2}{\omega_1} > 1)$, then

$$h(x) \geqslant \omega_1(p-m) - \omega_2 m = g(1) = 0 \tag{13}$$

so for $x \in [1, +\infty)$, we have $g^{'}(x) \leqslant 0$, and then, $g(x) \leqslant g(1) = 0$.

68.3　Proofs of Theorem

Proof of Theorem 1

From the definition of $K(\omega_1, \omega_2, p)$, we have

$$K(\omega_1, \omega_2, p) = \left(\frac{\omega_1 \frac{a^p + a^p}{2} + \omega_2 a^{\frac{p}{2}} b^{\frac{p}{2}}}{\omega_1 + \omega_2} \right)^{\frac{1}{p}}$$

It is clear that $K(\omega_1, \omega_2, p)$ is symmetric with $(a, b) \in \mathbf{R}_+^2$.

Write

$$s(a, b) := \left[\frac{\omega_1 (a^p + a^p) + 2\omega_2 a^{\frac{p}{2}} b^{\frac{p}{2}}}{2(\omega_1 + \omega_2)} \right]^{\frac{1}{p} - 1}$$

$$\frac{\partial K}{\partial a} = s(a, b) \left(\frac{\omega_1 a^{p-1} + \omega_2 a^{\frac{p}{2} - 1} b^{\frac{p}{2}}}{\omega_1 + \omega_2} \right)$$

$$\frac{\partial K}{\partial b} = s(a, b) \left(\frac{\omega_1 b^{p-1} + \omega_2 a^{\frac{p}{2}} b^{\frac{p}{2} - 1}}{\omega_1 + \omega_2} \right)$$

and then

$$\Delta := \frac{a^m - b^m}{m} \left(a^{1-m} \frac{\partial K}{\partial a} - b^{1-m} \frac{\partial K}{\partial b} \right) = \frac{s(a, b)}{2(\omega_1 + \omega_2)} f(a, b)$$

where

$$f(a, b) := \frac{a^m - b^m}{m} [\omega_1 (a^{p-m} - b^{p-m}) + \omega_2 (a^{\frac{p}{2} - m} b^{\frac{p}{2}} - a^{\frac{p}{2}} b^{\frac{p}{2} - m})]$$

Without loss of generality, we may assume that $a \geqslant b$, then $z := \frac{a}{b} \geqslant 1$, and then

$$\Delta = \frac{s(a, b) b^p}{2(\omega_1 + \omega_2)} \cdot \frac{z^m - 1}{m} q(z) \tag{14}$$

where

$$q(z) = \omega_1 (z^{p-m} - 1) + \omega_2 (z^{\frac{p}{2} - m} - z^{\frac{p}{2}}) = \omega_1 z^{p-m} + \omega_2 z^{\frac{p}{2} - m} - \omega_2 z^{\frac{p}{2}} - \omega_1$$

$$q'(z) = \omega_1 (p - m) z^{p-m-1} + \omega_2 (\frac{p}{2} - m) z^{\frac{p}{2} - m - 1} - \omega_2 \frac{p}{2} z^{\frac{p}{2} - 1} = z^{\frac{p}{2} - m - 1} g(z)$$

(I) For $m > 0$.

(i) If $p \geqslant \max\{(1 + \frac{\omega_2}{\omega_1})m, 2m\}$, $(I)(i)$ in from Lemma 5, it follows that

$q^{'}(z) \geqslant 0$, so $q(z) \geqslant q(1) = 0$. Notice that

$$\frac{s(a,b)b^p}{2(\omega_1 + \omega_2)} > 0, \quad \frac{z^m - 1}{m} \geqslant 0$$

from (14), we have $\Delta \geqslant 0$, and by Lemma 4, it follows that $K(\omega_1, \omega_2, p)$ is Schur-m power convex with $(a,b) \in \mathbf{R}_+^2$.

By the same arguments, from (I)(ii) and (I)(iii) in Lemma 5 we can prove (I)(ii) and (I)(iii) in Theorem 1, respectively.

(iv) If $p < 0$, then from Lemma 5 (I)(iv), it follows that the symbol of $q^{'}(z)$ is not fixed (from negative to positive). This means that $q(x)$ first decreases and then increases, but $q(1) = 0$ and

$$\lim_{z \to +\infty} q(z) = \lim_{z \to +\infty} (\omega_1 z^{p-m} + \omega_2 z^{\frac{p}{2}-m} - \omega_2 z^{\frac{p}{2}} - \omega_1) = -\omega_1 < 0,$$

hence $q(z) \leqslant 0$, and then $\Delta \leqslant 0$, by Lemma 4, it follows that $K(\omega_1, \omega_2, p)$ is Schur-m power concave with $(a,b) \in \mathbf{R}_+^2$.

By analogous discussing with case (I) , from (II)(i),(II)(ii), (II)(iii) and (II)(iv) in Lemma 5, we can prove (II)(i), (II)(ii), (II)(iii) and (II)(iv) in Theorem 1, respectively. The detailed proofs are left to the reader.

The proof of Theorem 1 is complete.

68.4   Acknowledgements

The authors are indebted to the referees for their helpful suggestions.

References

[1]   J-C KUANG. Applied Inequalities (Chang yong bu deng shi)[M]. 4th ed. Jinan: Shandong Press of Science and Technology, 2010.

[2]   W JANOUS. A note on generalized Heronian means[J]. Mathematical Inequalities & Applications, 2001, 4 (3): 369-375.

[3]   Z-H YANG. Schur power convexity of Stolarsky means[J]. Publ. Math. Debrecen, 2012,80(1-2): 43-66.

[4]   Z-H YANG. Schur power convexity of Gini means[J]. Bull. Korean Math. Soc., 2013, 50 (2): 485-498.

[5]   Z-H YANG. Schur power comvexity of the daróczy means[J]. Math. Inequal. Appl., 2013, 16 (3): 751-762.

[6]　W WANG, S-G YANG. Schur m-power convexity of a class of multi-plicatively convex functions and applications[J]. Abstract and Applied Analysis Volume 2014 (2014), Article ID 258108, 12 pages.

[7]　H-P YIN, H-N SHI, F QI. On Schur m-power convexity for ratios of some means[J]. J.Math. Inequal., 2015, 9 (1): 145-153.

[8]　Q XU. Reserch on Schur p power-convexity of the quotient of arithmetic mean and geometric mean[J]. Journal of Fudan University(Natural Science), 2015, 54 (3): 299-295.

[9]　Y-P DENG, S-H WU, D HE. The Schur Power convexity for the gen-eralized Muirhead mean[J]. Mathematics in practice and theory, 2014, 44 (5):255-268.

[10]　H-P YIN, H-N SHI, F QI. On Schur m-power convexity for ratios of some means[J]. J. Math. Inequal., 2015, 9 (1): 145-153.

[11]　B-Y WANG. Foundations of majorization inequalities[M]. Beijing: Bei-jing Normal Univ. Press, 1990.

[12]　A W MARSHALL, I OLKIN. Inequalities: Theory of Majorization and Its Applications[M]. New York: Academic Press, 1979.

[13]　X M ZHANG. Geometrically Convex Functions[M]. Hefei: An hui University Press, 2004.

[14]　C P NICULESCU. Convexity according to the geometric mean[J]. Mathematical Inequalities & Applications, 2000, 3(2): 155-167.

[15]　Y-M CHU, G-D WANG, X-H ZHANG. The Schur multiplicative and harmonic convexities of the complete symmetric function[J]. Mathe-matische Nachrichten, 2011, 284 (5-6): 653-663.

[16]　J-X MENG, Y M CHU, X-M TANG. The Schur-harmonic-convexity of dual form of the Hamy symmetric function[J]. Matematički Vesnik, 2010, 62 (1): 37-46.

[17]　H-N SHI, Y-M JIANG, W-D JIANG. Schur-convexity and Schur-geometrically concavity of Gini mean[J]. Computers and Mathematics with Applications,2009 57 : 266-274.

[18]　A WITKOWSKI. On Schur convexity and Schur-geometrical convexity of four-parameter family of means[J]. Math. Inequal. Appl., 2011, 14 (4)：897-903.

[19]　J SÁNDOR. The Schur-convexity of Stolarsky and Gini means[J]. Ba-nach J. Math. Anal., 2007, 1(2): 212-215.

[20] Y-M CHU, X-M ZHANG. Necessary and sufficient conditions such that extended mean values are Schur-convex or Schur-concave[J]. Journal of Mathematics of Kyoto University, 2008, 48(1): 229-238.

[21] Y-M CHU, X-M ZHANG. The Schur geometrical convexity of the extended mean values[J]. Journal of Convex Analysis, 2008, 15 (4): 869-890.

[22] W-F XIA, Y-M CHU. The Schur convexity of Gini mean values in the sense of harmonic mean[J]. Acta Mathematica Scientia 2011, 31B(3) : 1103-1112.

[23] H-N SHI, B MIHALY, S-H WU, D-M LI. Schur convexity of generalized Heronian means involving two parameters[J]. J. Inequal. Appl., vol.2008, Article ID 879273, 9 pages.

[24] W-F XIA, Y-M CHU. The Schur multiplicative convexity of the generalized Muirhead mean[J]. International Journal of Functional Analysis, Operator Theory and Applications, 2009, 1(1): 1-8.

[25] Y-M CHU, W-F XIA. Necessary and sufficient conditions for the Schur harmonic convexity of the generalized Muirhead Mean[J]. Proceedings of A. Razmadze Mathematical Institute, 2010, 152 : 19-27.

[26] Z-H YANG. Necessary and sufficient conditions for Schur geometrical convexity of the four-parameter homogeneous means[J]. Abstr. Appl. Anal., Volume 2010, Article ID 830163, 16 pages doi: 10.1155/2010/830163.

[27] W-F XIA, Y-M CHU. The Schur convexity of the weighted generalized logarithmic mean values according to harmonic mean[J]. International Journal of Modern Mathematics, 2009, 4(3) : 225-233.

[28] A WITKOWSKI. On Schur-convexity and Schur-geometric convexity of four-parameter family of means[J]. Math. Inequal. Appl., 2011, 14 (4): 897-903.

[29] Z-H YANG. Schur harmonic convexity of Gini means[J]. International Mathematical Forum, 2011, 6(16): 747-762.

[30] F QI, J SÁNDOR, S S DRAGOMIR, A SOFO. Notes on the Schur-convexity of the extended mean values[J]. Taiwanese J. Math., 2005, 9(3): 411-420.

[31] L -L FU, B-Y Xi, H M SRIVASTAVA. Schur-convexity of the generalized Heronian means involving two positive numbers[J]. Taiwanese Journal of Mathematics, 2011, 15(6): 2721-2731.

[32] T-Y ZHANG, A-P JI. Schur-convexity of generalized Heronian mean[J]. Communications in Computer and Information Science, 1, Volume 244, Information Computing and Applications, Part 1, Pages 25-33.

[33] W-F XIA, Y-M CHU, G-D WANG. Necessary and sufficient conditions for the Schur harmonic convexity or concavity of the extended mean values[J]. Revista De La Uniòn Matemática Argentina, 2010, 51(2) : 121-132.

[34] Y WU, F QI. Schur-harmonic convexity for differences of some means [J]. Analysis, 2012, 32: 1001‐1008.

[35] V LOKESHA, K M NAGARAJA, B NAVEEN KUMAR, Y-D WU. Schur convexity of Gnan mean for two variables[J]. NNTDM, 2011, 17 (4): 37-41.

[36] Y WU, F QI, H-N SHI. Schur-harmonic convexity for differences of some special means in two variables[J]. J. Math. Inequal., J.Math. Inequal., 2014, 8 (2): 321-330.

[37] W-M GONG, X-H SHEN, Y-M CHU. The Schur convexity for the generalized Muirhead mean[J]. J. Math. Inequal., 2014, 8 (4): 855-862.

[38] K M NAGARAJA, SUDHIR KUMAR SAHU. Schur harmonic convexity of Stolarsky extended mean values[J]. Scientia Magna, 2013, 9 (2): 18-29.

[39] V LOKESHA, B NAVEEN KUMAR, K M NAGARAJA, S PADMANADHAN. Schur geometric convexity for ratio of difference of means[J]. Journal of Scientific Research & Reports, 2014 3(9): 1211-1219; Article no. JSRR.2014.9.008.

[40] Y-P DENG, S-H WU, Y-M CHU, D HE. The Schur convexity of the generalized Muirhead-Heronian means[J]. Abstract and Applied Analysis, Volume 2014 (2014), Article ID 706518, 11 pages.

[41] W-M GONG, H SUN, Y-M CHU. The Schur convexity for the generalized muirhead mean[J]. J. Math. Inequal., 2014, 8 (4): 855-862.

第69篇　Schur-Convexity, Schur-Geometric and Schur-Harmonic Convexity for a Composite Function of Complete Symmetric Function

(SHI HUAN-NAN, ZHANG JING; Qing-hua Ma. SpringerPlus, (2016)

5:296. DOI 10.1186/s40064-016-1940-z)

Abstract: In this paper, using the properties of Schur-convex function, Schur-geometrically convex function and Schur-harmonically convex function, we provide much simpler proofs of the Schur-convexity, Schur-geometric convexity and Schur-harmonic convexity for a composite function of the complete symmetric function.

Keywords: Schur-convexity; Schur-geometric convexity; Schur-harmonic convexity; complete symmetric function

69.1　Introduction

Throughout the article, $\mathbf{R}$ denotes the set of real numbers, $\boldsymbol{x} = (x_1, x_2, \ldots, x_n)$ denotes n-tuple (n-dimensional real vectors), the set of vectors can be written as

$$\mathbf{R}^n = \{\boldsymbol{x} = (x_1, x_2, \cdots, x_n) : x_i \in \mathbf{R}, i = 1, 2, \ldots, n\}$$

$$\mathbf{R}^n_+ = \{\boldsymbol{x} = (x_1, x_2, \ldots, x_n) : x_i > 0, i = 1, 2, \ldots, n\}$$

$$\mathbf{R}^n_- = \{\boldsymbol{x} = (x_1, x_2, \ldots, x_n) : x_i < 0, i = 1, 2, \ldots, n\}$$

In particular, the notations $\mathbf{R}$ and $\mathbf{R}_+$ denote $\mathbf{R}^1$ and $\mathbf{R}^1_+$, respectively.

The following complete symmetric function is an important class of symmetric functions.

For $x = (x_1, x_2, \ldots, x_n) \in \mathbf{R}^n$, the complete symmetric function $c_n(x, r)$ is defined as

$$c_n(x, r) = \sum_{i_1 + i_2 + \cdots + i_n = r} x_1^{i_1} x_2^{i_2} \cdots x_n^{i_n} \tag{1}$$

where $c_0(x, r) = 1$, $r \in \{1, 2, \ldots, n\}$, $i_1, i_2, \ldots, i_n$ are non-negative integers.

It has been investigated by many mathematicians and there are many interesting results in the literature.

Guan[1] discussed the Schur-convexity of $c_n(x, r)$ and proved that $c_n(x, r)$ is increasing and Schur-convex on $\mathbf{R}_+^n$. Subsequently, Chu et al.[2] proved that $c_n(x, r)$ is Schur-geometrically convex and harmonically convex on $\mathbf{R}_+^n$.

Recently, Sun et al. [3] studied the Schur-convexity, Schur-geometric and harmonic convexities of the following composite function of $c_n(x, r)$

$$F_n(x, r) = \sum_{i_1 + i_2 + \cdots + i_n = r} \left(\frac{x_1}{1 - x_1}\right)^{i_1} \left(\frac{x_2}{1 - x_2}\right)^{i_2} \cdots \left(\frac{x_n}{1 - x_n}\right)^{i_n} \tag{2}$$

Using the Lemma 1, Lemma 2 and Lemma 3 in second section, they proved as follows: Theorem A, Theorem B and Theorem C, respectively.

Theorem A For $x = (x_1, x_2, \ldots, x_n) \in [0, 1)^n \cup (1, +\infty)^n$ and $r \in \mathbf{N}$

(i) $F_n(x, r)$ is increasing in x_i for all $i \in \{1, 2, \ldots, n\}$ and Schur-convex on $[0, 1)^n$ for each r fixed;

(ii) If r is even integer (or odd integer, respectively), then $F_n(x, r)$ is Schur-convex (or Schur-concave, respectively) on $(1, +\infty)^n$, and it is decreasing (or increasing, respectively) in x_i for all $i \in \{1, 2, \ldots, n\}$.

Theorem B For $x = (x_1, x_2, \ldots, x_n) \in [0, 1)^n \cup (1, +\infty)^n$ and $r \in \mathbf{N}$.

(i) $F_n(x, r)$ is Schur-geometrically convex on $[0, 1)^n$;

(ii) If r is even integer (or odd integer, respectively), then $F_n(x, r)$ is Schur-geometrically convex (or Schur-geometrically concave, respectively) on $(1, +\infty)^n$.

Theorem C For $x = (x_1, x_2, \ldots, x_n) \in [0, 1)^n \cup (1, +\infty)^n$ and $r \in \mathbf{N}$.

(i) $F_n(x, r)$ is Schur-harmonically convex on $[0, 1)^n$;

(ii) If r is even integer (or odd integer, respectively), then $F_n(\boldsymbol{x}, r)$ is Schur-harmonically convex (or Schur-harmonically concave, respectively) on $(1, +\infty)^n$.

In this paper, using the properties of Schur-convex function, Schur-geometrically convex function and Schur-harmonically convex function, we will provide much simpler proofs of the above results.

69.2　Definitions and Lemmas

For convenience, we recall some definitions as follows.

Definition 1 Let $\boldsymbol{x} = (x_1, x_2, \ldots, x_n)$ and $\boldsymbol{y} = (y_1, y_2, \ldots, y_n) \in \mathbf{R}^n$.

(i) $\boldsymbol{x} \geq \boldsymbol{y}$ means $x_i \geqslant y_i$ for all $i = 1, 2, \ldots, n$.

(ii) Let $\Omega \subset \mathbf{R}^n$, $\varphi\colon \Omega \to \mathbf{R}$ is said to be increasing if $\boldsymbol{x} \geq \boldsymbol{y}$ implies $\varphi(\boldsymbol{x}) \geqslant \varphi(\boldsymbol{y})$. φ is said to be decreasing if and only if $-\varphi$ is increasing.

Definition 2 Let $\boldsymbol{x} = (x_1, x_2, \ldots, x_n)$ and $\boldsymbol{y} = (y_1, y_2, \ldots, y_n) \in \mathbf{R}^n$.

(i) $\boldsymbol{x}$ is said to be majorized by $\boldsymbol{y}$ (in symbols $\boldsymbol{x} \prec \boldsymbol{y}$) if $\sum\limits_{i=1}^{k} x_{[i]} \leqslant \sum\limits_{i=1}^{k} y_{[i]}$ for $k = 1, 2, \ldots, n-1$ and $\sum\limits_{i=1}^{n} x_i = \sum\limits_{i=1}^{n} y_i$, where $x_{[1]} \geqslant x_{[2]} \geqslant \cdots \geqslant x_{[n]}$ and $y_{[1]} \geqslant y_{[2]} \geqslant \cdots \geqslant y_{[n]}$ are rearrangements of $\boldsymbol{x}$ and $\boldsymbol{y}$ in a descending order.

(ii) Let $\Omega \subset \mathbf{R}^n$, $\varphi\colon \Omega \to \mathbf{R}$ is said to be a Schur-convex function on Ω if $\boldsymbol{x} \prec \boldsymbol{y}$ on Ω implies $\varphi(\boldsymbol{x}) \leq \varphi(\boldsymbol{y})$. The function φ is said to be Schur-concave on Ω if and only if $-\varphi$ is a Schur-convex function on Ω.

Definition 3 Let $\boldsymbol{x} = (x_1, x_2, \ldots, x_n)$ and $\boldsymbol{y} = (y_1, y_2, \ldots, y_n) \in \mathbf{R}^n$.

(i) $\Omega \subset \mathbf{R}^n$ is said to be a convex set if $\boldsymbol{x}, \boldsymbol{y} \in \Omega, 0 \leqslant \alpha \leqslant 1$, implies $\alpha\boldsymbol{x} + (1-\alpha)\boldsymbol{y} = (\alpha x_1 + (1-\alpha)y_1, \alpha x_2 + (1-\alpha)y_2, \ldots, \alpha x_n + (1-\alpha)y_n) \in \Omega$.

(ii) Let $\Omega \subset \mathbf{R}^n$ be a convex set. A function $\varphi\colon \Omega \to \mathbf{R}$ is said to be convex on Ω if

$$\varphi(\alpha\boldsymbol{x} + (1-\alpha)\boldsymbol{y}) \leqslant \alpha\varphi(\boldsymbol{x}) + (1-\alpha)\varphi(\boldsymbol{y})$$

for all $\boldsymbol{x}, \boldsymbol{y} \in \Omega$, and all $\alpha \in [0, 1]$. The function φ is said to be concave on Ω if and only if $-\varphi$ is a convex function on Ω.

Definition 4

(i) A set $\Omega \subset \mathbf{R}^n$ is called symmetric, if $\boldsymbol{x} \in \Omega$ implies $\boldsymbol{x}\boldsymbol{P} \in \Omega$ for every $n \times n$ permutation matrix $\boldsymbol{P}$.

(ii) A function $\varphi : \Omega \to \mathbf{R}$ is called symmetric if for every permutation matrix $\boldsymbol{P}$, $\varphi(\boldsymbol{x}\boldsymbol{P}) = \varphi(\boldsymbol{x})$ for all $\boldsymbol{x} \in \Omega$.

Lemma 1(Schur-convex function decision theorem) [4] Let $\Omega \subset \mathbf{R}^n$ be symmetric convex set with nonempty interior. Ω^0 is the interior of Ω. The function $\varphi : \Omega \to \mathbf{R}$ is continuous on Ω and continuously differentiable on Ω^0. Then φ is a $Schur-convex$ $(or$ $Schur-concave,$ $respectively)$ $function$ if and only if φ is symmetric on Ω and

$$(x_1 - x_2)\left(\frac{\partial \varphi}{\partial x_1} - \frac{\partial \varphi}{\partial x_2}\right) \geqslant 0 (\text{or} \leqslant 0, \text{ respectively}) \tag{3}$$

holds for any $\boldsymbol{x} \in \Omega^0$.

The first systematical study of the functions preserving the ordering of majorization was made by Issai Schur in 1923. In Schur's honor, such functions are said to be "Schur-convex". It has many important applications in analytic inequalities, combinatorial optimization, quantum physics, information theory, and other related fields. See [4,12-14].

Definition 5 Let $\Omega \subset \mathbf{R}^n_+$, $\boldsymbol{x} = (x_1, x_2, \ldots, x_n)$ and $\boldsymbol{y} = (y_1, y_2, \ldots, y_n) \in \mathbf{R}^n_+$.

(i) [5] Ω is called a geometrically convex set if $(x_1^\alpha y_1^\beta, x_2^\alpha y_2^\beta, \ldots, x_n^\alpha y_n^\beta) \in \Omega$ for all $\boldsymbol{x}, \boldsymbol{y} \in \Omega$ and $\alpha, \beta \in [0, 1]$ such that $\alpha + \beta = 1$.

(ii) [5] The function $\varphi: \Omega \to \mathbf{R}_+$ is said to be a Schur-geometrically convex function on Ω, for any $\boldsymbol{x}, \boldsymbol{y} \in \Omega$, if

$$(\log x_1, \log x_2, \ldots, \log x_n) \prec (\log y_1, \log y_2, \ldots, \log y_n)$$

implies $\varphi(\boldsymbol{x}) \leqslant \varphi(\boldsymbol{y})$. The function φ is said to be a Schur-geometrically concave function on Ω if and only if $-\varphi$ is a Schur-geometrically convex function on Ω.

By Definition 5, the following is obvious.

Proposition 1 Let $\Omega \subset \mathbf{R}^n_+$, and let

$$\log \Omega = \{(\log x_1, \log x_2, \ldots, \log x_n) : (x_1, x_2, \ldots, x_n) \in \Omega\}$$

Then $\varphi : \Omega \to \mathbf{R}_+$ is a Schur-geometrically convex (or Schur-geometrically concave, respectively) function on Ω if and only if $\varphi(e^{x_1}, e^{x_2}, \ldots, e^{x_n})$ is a Schur-convex (or Schur-concave, respectively) function on $\log \Omega$.

Lemma 2(Schur-geometrically convex function decision theorem) [5] Let $\Omega \subset \mathbf{R}_+^n$ be a symmetric and geometrically convex set with a nonempty interior Ω^0. Let $\varphi : \Omega \to \mathbf{R}_+$ be continuous on Ω and differentiable in Ω^0. If φ is symmetric on Ω and

$$(\log x_1 - \log x_2) \left(x_1 \frac{\partial \varphi}{\partial x_1} - x_2 \frac{\partial \varphi}{\partial x_2} \right) \geqslant 0 \quad (\text{or} \leqslant 0, \text{respectively}) \quad (4)$$

holds for any $\boldsymbol{x} = (x_1, x_2, \ldots, x_n) \in \Omega^0$, then φ is a Schur-geometrically convex (or Schur-geometrically concave, respectively) function.

The Schur-geometric convexity was proposed by Zhang[5] in 2004, and was investigated by Chu et al.[6], Guan [7], Sun et al. [8], and so on. We also note that some authors use the term "Schur multiplicative convexity".

In 2009, Chu[2,9−10] introduced the notion of Schur-harmonically convex function.

Definition 6[9] Let $\Omega \subset \mathbf{R}_+^n$, $\boldsymbol{x} = (x_1, x_2, \ldots, x_n)$ and $\boldsymbol{y} = (y_1, y_2, \ldots, y_n) \in \mathbf{R}_+^n$.

(i) A set Ω is said to be harmonically convex if $\left(\dfrac{2x_1 y_1}{x_1 + y_1}, \dfrac{2x_2 y_2}{x_2 + y_2}, \ldots, \dfrac{2x_n y_n}{x_n + y_n} \right) \in \Omega$ for every $\boldsymbol{x}, \boldsymbol{y} \in \Omega$.

(ii) A function $\varphi : \Omega \to \mathbf{R}_+$ is said to be Schur-harmonically convex on Ω, for any $\boldsymbol{x}, \boldsymbol{y} \in \Omega$, if $\left(\dfrac{1}{x_1}, \dfrac{1}{x_2}, \ldots, \dfrac{1}{x_n} \right) \prec \left(\dfrac{1}{y_1}, \dfrac{1}{y_2}, \ldots, \dfrac{1}{y_n} \right)$ implies $\varphi(\boldsymbol{x}) \leq \varphi(\boldsymbol{y})$. A function φ is said to be a Schur-harmonically concave function on Ω if and only if $-\varphi$ is a Schur-harmonically convex function on Ω.

By Definition 6, the following is obvious.

Proposition 2 Let $\Omega \subset \mathbf{R}_+^n$ be a set, and let $\dfrac{1}{\Omega} = \{ \left(\dfrac{1}{x_1}, \dfrac{1}{x_2}, \ldots, \dfrac{1}{x_n} \right) : (x_1, x_2, \ldots, x_n) \in \Omega \}$. Then $\varphi : \Omega \to \mathbf{R}_+$ is a Schur-harmonically convex (or Schur-harmonically concave, respectively) function on Ω if and only if $\varphi(\dfrac{1}{x_1}, \dfrac{1}{x_2}, \ldots, \dfrac{1}{x_n})$ is a Schur-convex (or Schur-concave, respectively) function on $\dfrac{1}{\Omega}$.

Lemma 3(Schur-harmonically convex function decision theorem)[9] Let $\Omega \subset \mathbf{R}_+^n$ be a symmetric and harmonically convex set with inner points and let $\varphi : \Omega \to \mathbf{R}_+$ be a continuous symmetric function which is differentiable on Ω^0. Then φ is Schur-harmonically convex (or Schur-harmonically concave, respectively) on Ω if and only if

$$(x_1 - x_2)\left(x_1^2 \frac{\partial \varphi}{\partial x_1} - x_2^2 \frac{\partial \varphi}{\partial x_2}\right) \geqslant 0 \quad (\text{or} \leqslant 0, \text{respectively}), \quad \boldsymbol{x} \in \Omega^0 \quad (5)$$

Lemma 4 If r is even integer (or odd integer, respectively), then $c_n(\boldsymbol{x}, r)$ is decreasing and Schur-convex (or increasing and Schur-concave, respectively) on $\mathbf{R}_-^n$.

Proof Notice that

$$c_n(-\boldsymbol{x}, r)$$
$$= \sum_{i_1+i_2+\cdots+i_n=r} (-x_1)^{i_1}(-x_2)^{i_2}\cdots(-x_n)^{i_n}$$
$$= (-1)^{i_1+i_2+\cdots+i_n} \sum_{i_1+i_2+\cdots+i_n=r} x_1^{i_1} x_2^{i_2} \cdots x_n^{i_n}$$
$$= (-1)^r c_n(\boldsymbol{x}, r),$$

i.e.

$$c_n(-\boldsymbol{x}, r) = (-1)^r c_n(\boldsymbol{x}, r)$$

If r is even integer, then $c_n(\boldsymbol{x}, r) = c_n(-\boldsymbol{x}, r)$. For $\boldsymbol{x}, \boldsymbol{y} \in \mathbf{R}_-^n$, if $\boldsymbol{x} \prec \boldsymbol{y}$, then $-\boldsymbol{x} \prec -\boldsymbol{y}$ and $-\boldsymbol{x}, -\boldsymbol{y} \in \mathbf{R}_+^n$, but $c_n(\boldsymbol{x}, r)$ is Schur-convex in $\mathbf{R}_+^n$, so that $c_n(-\boldsymbol{x}, r) \leqslant c_n(-\boldsymbol{y}, r)$, i.e. $c_n(\boldsymbol{x}, r) \leqslant c_n(\boldsymbol{y}, r)$, this shows that $c_n(\boldsymbol{x}, r)$ is Schur-convex in $\mathbf{R}_-^n$. If $\boldsymbol{x} \leqslant \boldsymbol{y}$, then $-\boldsymbol{x} \geqslant -\boldsymbol{y}$, but $c_n(\boldsymbol{x}, r)$ is increasing in $\mathbf{R}_+^n$, so that $c_n(-\boldsymbol{x}, r) \geqslant c_n(-\boldsymbol{y}, r)$, i.e. $c_n(\boldsymbol{x}, r) \geqslant c_n(\boldsymbol{y}, r)$, this shows that $c_n(\boldsymbol{x}, r)$ is decreasing in $\mathbf{R}_-^n$.

If r is odd integer, then $c_n(\boldsymbol{x}, r) = -c_n(-\boldsymbol{x}, r)$. For $\boldsymbol{x}, \boldsymbol{y} \in \mathbf{R}_-^n$, if $\boldsymbol{x} \prec \boldsymbol{y}$, then $-\boldsymbol{x} \prec -\boldsymbol{y}$ and $-\boldsymbol{x}, -\boldsymbol{y} \in \mathbf{R}_+^n$, but $c_n(\boldsymbol{x}, r)$ is Schur-convex in $\mathbf{R}_+^n$, so that $c_n(-\boldsymbol{x}, r) \leqslant c_n(-\boldsymbol{y}, r)$, i.e. $c_n(\boldsymbol{x}, r) \geqslant c_n(\boldsymbol{y}, r)$, this shows that $c_n(\boldsymbol{x}, r)$ is Schur-concave in $\mathbf{R}_-^n$. If $\boldsymbol{x} \leqslant \boldsymbol{y}$, then $-\boldsymbol{x} \geqslant -\boldsymbol{y}$, but $c_n(\boldsymbol{x}, r)$ is increasing in $\mathbf{R}_+^n$, so that $c_n(-\boldsymbol{x}, r) \geqslant c_n(-\boldsymbol{y}, r)$, i.e. $c_n(\boldsymbol{x}, r) \leqslant c_n(\boldsymbol{y}, r)$, this shows that $c_n(\boldsymbol{x}, r)$ is increasing in $\mathbf{R}_-^n$.

Lemma 5[4,11] Let the set $\mathbf{A}, \mathbf{B} \subset \mathbf{R}$, $\varphi : \mathbf{B}^n \to \mathbf{R}$, $f : \mathbf{A} \to \mathbf{B}$ and $\psi(x_1, x_2, \ldots, x_n) = \varphi(f(x_1), f(x_2), \ldots, f(x_n)) : \mathbf{A}^n \to \mathbf{R}$.

(i) If φ is increasing and Schur-convex and f is increasing and convex, then ψ is increasing and Schur-convex.

(ii) If φ is decreasing and Schur-convex and f is increasing and concave, then ψ is decreasing and Schur-convex.

(iii) If φ is increasing and Schur-concave and f is increasing and concave, then ψ is increasing and Schur-concave.

(iv) If φ is decreasing and Schur-convex and f is decreasing and concave, then ψ is increasing and Schur-convex.

(v) If φ is increasing and Schur-concave and f is decreasing and concave, then ψ is decreasing and Schur-concave.

Lemma 6 Let the set $\Omega \subset \mathbf{R}_+^n$. The function $\varphi : \Omega \to \mathbf{R}_+$ is differentiable.

(i) If φ is increasing and Schur-convex, then φ is Schur-geometrically convex.

(ii) If φ is decreasing and Schur-concave, then φ is Schur-geometrically concave.

Proof We only give the proof of Lemma 6 (i) in detail. Similar argument leads to the proof of Lemma 6 (ii).

For $x \in I \subset \mathbf{R}_+$ and $x_1 \neq x_2$, we have

$$\Delta = (\log x_1 - \log x_2)\left(x_1 \frac{\partial \varphi}{\partial x_1} - x_2 \frac{\partial \varphi}{\partial x_2}\right)$$

$$= (\log x_1 - \log x_2)\left(x_1 \frac{\partial \varphi}{\partial x_1} - x_1 \frac{\partial \varphi}{\partial x_2} + x_1 \frac{\partial \varphi}{\partial x_2} - x_2 \frac{\partial \varphi}{\partial x_2}\right)$$

$$= x_1 \frac{\log x_1 - \log x_2}{x_1 - x_2}(x_1 - x_2)\left(\frac{\partial \varphi}{\partial x_1} - \frac{\partial \varphi}{\partial x_2}\right) + \frac{\partial \varphi}{\partial x_2}(x_1 - x_2)(\log x_1 - \log x_2)$$

Since φ is Schur-convex on Ω, by Lemma 1, we have

$$(x_1 - x_2)\left(\frac{\partial \varphi}{\partial x_1} - \frac{\partial \varphi}{\partial x_2}\right) \geqslant 0$$

Notice that φ and $y = \log x$ is increasing, we have $\dfrac{\partial \varphi}{\partial x_2} \geqslant 0$, $\dfrac{\log x_1 - \log x_2}{x_1 - x_2} \geqslant 0$ and $(x_1 - x_2)(\log x_1 - \log x_2) \geqslant 0$, so that $\Delta \geqslant 0$, by Lemma 2, it follows that φ is Schur-geometrically convex on Ω.

Lemma 7 Let the set $\Omega \subset \mathbf{R}_+^n$. The function $\varphi : \Omega \to \mathbf{R}_+$ is differentiable.

(i) If φ is increasing and Schur-convex, then φ is Schur-harmonically convex.

(ii) If φ is decreasing and Schur-concave, then φ is Schur-harmonically concave.

Proof　We only give the proof of Lemma 7 (ii) in detail. Similar argument leads to the proof of Lemma 7 (i).

For $\boldsymbol{x} \in I \subset \mathbf{R}_+$ and $x_1 \neq x_2$, we have

$$\begin{aligned}
\Lambda &= (x_1 - x_2)\left(x_1^2 \frac{\partial \varphi}{\partial x_1} - x_2^2 \frac{\partial \varphi}{\partial x_2}\right) \\
&= (x_1 - x_2)\left(x_1^2 \frac{\partial \varphi}{\partial x_1} - x_1^2 \frac{\partial \varphi}{\partial x_2} + x_1^2 \frac{\partial \varphi}{\partial x_2} - x_2^2 \frac{\partial \varphi}{\partial x_2}\right) \\
&= x_1^2 (x_1 - x_2)\left(\frac{\partial \varphi}{\partial x_1} - \frac{\partial \varphi}{\partial x_2}\right) + \frac{\partial \varphi}{\partial x_2}(x_1 - x_2)\left(x_1^2 - x_2^2\right)
\end{aligned}$$

Since φ is Schur-concave on Ω, by Lemma 1, we have

$$(x_1 - x_2)\left(\frac{\partial \varphi}{\partial x_1} - \frac{\partial \varphi}{\partial x_2}\right) \leqslant 0$$

Notice that φ is decreasing and $y = x^2 (x > 0)$ is increasing, we have $\dfrac{\partial \varphi}{\partial x_2} \leqslant 0$ and $(x_1 - x_2)\left(x_1^2 - x_2^2\right) \geqslant 0$, so that $\Lambda \leqslant 0$, by Lemma 3, it follows that φ is Schur-harmonically concave on Ω.

69.3　Simple Proof of Theorems

Proof of Theorem A　Let $g(t) = \dfrac{t}{1-t}$. Directly calculating yields $g'(t) = \dfrac{1}{(1-t)^2}$ and $g''(t) = \dfrac{2}{(1-t)^3}$, it is to see that g is increasing and convex on $(0, 1)$ and g is increasing and concave on $(1, +\infty)$.

Since $c_n(\boldsymbol{x}, r)$ is increasing and Schur-convex in $\mathbf{R}_+^n$, from Lemma 5 (i) it follows that $F_n(\boldsymbol{x}, r)$ is increasing and Schur-convex in $(0, 1)^n$, and then by continuity of $F_n(\boldsymbol{x}, r)$ on $[0, 1)^n$, it follows that $F_n(\boldsymbol{x}, r)$ is increasing and Schur-convex on $[0, 1)^n$.

If r is even integer, then from Lemma 4, we known that $c_n(\boldsymbol{x}, r)$ is decreasing and Schur-convex, moreover g is increasing and concave on $(1, +\infty)$. By Lemma 5 (ii), it follows that $F_n(\boldsymbol{x}, r)$ is decreasing and Schur-convex.

551

If r is odd integer, then from Lemma 4, we known that $c_n(\boldsymbol{x}, r)$ is increasing and Schur-concave, moreover g is increasing and concave on $(1, +\infty)$. By Lemma 5 (iii), it follows that $F_n(\boldsymbol{x}, r)$ is increasing and Schur-concave.

The proof of Theorem A is completed.

Proof of Theorem B From Theorem A (i) and Lemma 6 (i), it follows that Theorem B (i) holds.

Considering

$$F_n(\mathrm{e}^{\boldsymbol{x}}, r) = \sum_{i_1+i_2+\cdots+i_n=r} \left(\frac{\mathrm{e}^{x_1}}{1-\mathrm{e}^{x_1}}\right)^{i_1} \left(\frac{\mathrm{e}^{x_2}}{1-\mathrm{e}^{x_2}}\right)^{i_2} \cdots \left(\frac{\mathrm{e}^{x_n}}{1-\mathrm{e}^{x_n}}\right)^{i_n} \quad (6)$$

Let $h(t) = \dfrac{\mathrm{e}^t}{1-\mathrm{e}^t}$. Then $h < 0$ on $(0, +\infty)$. Directly calculating yields $h'(t) = \dfrac{\mathrm{e}^t}{(1-\mathrm{e}^t)^2}$ and $h''(t) = \dfrac{\mathrm{e}^t(1+\mathrm{e}^t)}{(1-\mathrm{e}^t)^3}$, it is to see that h is increasing and concave on $(0, +\infty)$. From Lemma 4 and Lemma 5 (ii) (or (iii), respectively), it follows that if r is even integer (or odd integer, respectively), then $F_n(\mathrm{e}^{\boldsymbol{x}}, r)$ is Schur-convex (or Schur-concave, respectively) on $(0, +\infty)$. And then, by Proposition 1, Theorem B (ii) holds.

The proof of Theorem B is completed.

Proof of Theorem C From Theorem A (i) and Lemma 7 (i), it follows that Theorem C (i) holds.

Considering

$$F_n\left(\frac{1}{\boldsymbol{x}}, r\right) = \sum_{i_1+i_2+\cdots+i_n=r} \left(\frac{1}{x_1-1}\right)^{i_1} \left(\frac{1}{x_2-1}\right)^{i_2} \cdots \left(\frac{1}{x_n-1}\right)^{i_n} \quad (7)$$

Let $p(t) = \dfrac{1}{t-1}$. Then $p < 0$ on $(0, 1)$. Directly calculating yields $p'(t) = -\dfrac{1}{(t-1)^2}$ and $p''(t) = \dfrac{2}{(t-1)^3}$, it is to see that p is decreasing and concave on $(0, 1)$. From Lemma 4 and Lemma 5 (iv) (or (v), respectively), it follows that if r is even integer (or odd integer, respectively), then $F_n\left(\dfrac{1}{\boldsymbol{x}}, r\right)$ is Schur-convex (or Schur-concave, respectively) on $(0, 1)$. And then, by Proposition 2, Theorem C (ii) holds.

The proof of Theorem C is completed.

69.4　Conclusions

In this paper, by the properties of Schur-convex function, Schur-geometrically convex function and Schur-harmonically convex function, we provide much simpler proofs of Theorem A, B, C.

Authors' contributions

The main idea of this paper was proposed by H-NS. This work was carried out in collaboration between all authors. They read and approved the final manuscript.

Acknowledgment

The work was supported by the Importation and Development of High-Caliber Talents Project of Beijing Municipal Institutions (Grant No. ID-HT201304089) and the National Natural Science Foundation of China (Grant No. 11501030). The authors are indebted to the referees for their helpful suggestions.

Competing interests

The authors declare that there is no conflict of interests regarding the publication of this article.

References

[1]　K-Z GUAN. Schur-convexity of the complete symmetric function[J]. Mathematical Inequalities & Applications, 2006, 9(4): 567-576.

[2]　Y-M CHU, G-D WANG, X-H ZHANG. The Schur multiplicative and harmonic convexities of the complete symmetric function[J]. Mathematische Nachrichten, 2011, 284 (5-6): 653-663.

[3]　M B SUN, N B CHEN, S H LI. Some properties of a class of symmetric functions and its applications[J]. Mathematische Nachrichten, 2014, doi: 10.1002/mana.201300073.

[4]　MARSHALL A W, OLKIN I, ARNOLD B C. Inequalities: theory of majorization and its application[M]. 2nd ed. New York: Springer Press, 2011.

[5] XIAO-MING ZHANG. Geometrically Convex Functions[M]. Hefei: An hui University Press, 2004.

[6] Y M CHU, X M ZHANG, G D WANG. The Schur geometrical convexity of the extended mean values[J]. Journal of Convex Analysis, 2008, 15(4): 707-718.

[7] K Z GUAN. A class of symmetric functions for multiplicatively convex function[J]. Mathematical Inequalities & Applications, 2007, 10(4): 745-753.

[8] T-C SUN, Y-P LV, Y-M CHU. Schur multiplicative and harmonic convexities of generalized Heronian mean in n variables and their applications[J]. International Journal of Pure and Applied Mathematics, 2009, 55(1): 25-33.

[9] Y M CHU, T C SUN. The Schur harmonic convexity for a class of symmetric functions[J]. Acta Mathematica Scientia, 2010, 30B(5): 1501-1506.

[10] Y-M CHU, Y-P LV. The Schur harmonic convexity of the Hamy symmetric function and its applications[J]. Journal of Inequalities and Applications, 2009, Article ID 838529, 10 pages.

[11] BO-YING WANG. Foundations of Majorization Inequalities[M]. Beijing: Beijing Normal Univ. Press, 1990.

[12] I ROVENŢA. Schur convexity of a class of symmetric functions[J]. Annals of the University of Craiova, Mathematics and Computer Science Series, 2010, 37(1): 12-18.

[13] V ČULJAK, I FRANJIĆ, R GHULAM, J PEČARIĆ. Schur-convexity of averages of convex functions[J]. Journal of Inequalities and Applications, Volume 2011, Article ID 581918, 25 pages, doi: 10.1155/2011/581918.

[14] ZHANG J, SHI H-N. Two double inequalities for k-gamma and k-Riemann zeta functions[J]. J Inequal Appl 2014:191. doi:10.1186/1029-242X-2014-191

第70篇 Majorized Proof of Arithmetic-Geometric-Harmonic Means Inequality

(SHI HUAN-NAN. Advanced studies in contemporary mathematics, 2016, 26 (4): 681 - 684)

Abstract: As we all know, arithmetic -geometric mean inequalities are the most basic and important inequalities. To seek different method to prove them has been one of the study focuses and they have been proven by more than a hundred ways. By using methods based on the theory of majorization, the arithmetic-geometric-harmonic means inequality is proved in a new way.

Keywords: arithmetic mean; geometric mean; harmonic mean; Schur convexity; Schur geometric convexity; majorization

70.1 Introduction

At a mathematics conference in July, 1999, Paul and Jack Abad presented their list of "The Hundred Greatest Theorems." Their ranking is based on the following criteria: "the place the theorem holds in the literature, the quality of the proof, and the unexpectedness of the result."

The arithmetic geometric mean inequality ranking thirty-eighth in the "100 theorems", which shows its lofty position in mathematics and its application. Looking for the different proof method of the arithmetic geometric mean inequality, has been a hot research, so far, there are hundreds of different methods of proof(see [4-13] and references therein).

In this note, a new proof of the arithmetic-geometric-harmonic mean

555

inequality is given, by using methods based on the theory of majorization.

Throughout the paper we assume that the set of n-dimensional row vector on the real number field by $\mathbf{R}^n$.

$$\mathbf{R}^n_+ = \{\boldsymbol{x} = (x_1, \ldots, x_n) \in \mathbf{R}^n : x_i > 0, i = 1, \ldots, n\}$$

In particular, $\mathbf{R}^1$ and $\mathbf{R}^1_+$ denoted by $\mathbf{R}$ and $\mathbf{R}_+$ respectively.

Recall the arithmetic mean, geometric mean, and harmonic mean

$$A_n(\boldsymbol{x}) = \frac{\sum\limits_{i=1}^{n} x_i}{n}, \ G_n(\boldsymbol{x}) = \sqrt[n]{\prod_{i=1}^{n} x_i}, \ H_n(\boldsymbol{x}) = \frac{n}{\sum\limits_{i=1}^{n} \frac{1}{x_i}}$$

where $n \in \mathbf{N}$ and $\boldsymbol{x} \in \mathbf{R}^n_+$.

The order relation among these means is well-known:

Theorem 1

$$A_n(\boldsymbol{x}) \geqslant G_n(\boldsymbol{x}) \geqslant H_n(\boldsymbol{x}) \tag{1}$$

70.2 Definitions and Lemmas

We need the following definitions and lemmas.

Definition 1 [1-2] Let $\boldsymbol{x} = (x_1, \ldots, x_n)$ and $\boldsymbol{y} = (y_1, \ldots, y_n) \in \mathbf{R}^n$.

(i) $\boldsymbol{x}$ is said to be majorized by $\boldsymbol{y}$ (in symbols $\boldsymbol{x} \prec \boldsymbol{y}$) if $\sum\limits_{i=1}^{k} x_{[i]} \leqslant \sum\limits_{i=1}^{k} y_{[i]}$ for $k = 1, 2, \ldots, n-1$ and $\sum\limits_{i=1}^{n} x_i = \sum\limits_{i=1}^{n} y_i$, where $x_{[1]} \geqslant \cdots \geqslant x_{[n]}$ and $y_{[1]} \geqslant \cdots \geqslant y_{[n]}$ are rearrangements of $\boldsymbol{x}$ and $\boldsymbol{y}$ in a descending order.

(ii) $\Omega \subset \mathbf{R}^n$ is called a convex set if $(\alpha x_1 + \beta y_1, \ldots, \alpha x_n + \beta y_n) \in \Omega$ for any $\boldsymbol{x}$ and $\boldsymbol{y} \in \Omega$, where α and $\beta \in [0, 1]$ with $\alpha + \beta = 1$.

(iii) let $\Omega \subset \mathbf{R}^n$, $\varphi\colon \Omega \to \mathbf{R}$ is said to be a Schur-convex function on Ω if $\boldsymbol{x} \prec \boldsymbol{y}$ on Ω implies $\varphi(\boldsymbol{x}) \leq \varphi(\boldsymbol{y})$. φ is said to be a Schur-concave function on Ω if and only if $-\varphi$ is Schur-convex function.

Definition 2 [3-4] Let $\boldsymbol{x} = (x_1, \ldots, x_n)$ and $\boldsymbol{y} = (y_1, \ldots, y_n) \in \mathbf{R}^n_+$.

(i) $\Omega \subset \mathbf{R}^n_+$ is called a geometrically convex set if $(x_1^\alpha y_1^\beta, \ldots, x_n^\alpha y_n^\beta) \in \Omega$ for any $\boldsymbol{x}$ and $\boldsymbol{y} \in \Omega$, where α and $\beta \in [0, 1]$ with $\alpha + \beta = 1$.

(ii) Let $\Omega \subset \mathbf{R}_+^n$, $\varphi: \Omega \to \mathbf{R}_+$ is said to be a Schur-geometrically convex function on Ω if $(\log x_1, \ldots, \log x_n) \prec (\log y_1, \ldots, \log y_n)$ on Ω implies $\varphi(\boldsymbol{x}) \leqslant \varphi(\boldsymbol{y})$. φ is said to be a Schur-geometrically concave function on Ω if and only if $-\varphi$ is Schur-geometrically convex function.

Lemma 1[1-2] Let $\Omega \subset \mathbf{R}^n$ is convex set, and has a nonempty interior set Ω^0. Let $\varphi : \Omega \to \mathbf{R}$ is continuous on Ω and differentiable in Ω^0. Then φ is the $Schur-convex(Schur-concave)function$, if and only if it is symmetric on Ω and if

$$(x_1 - x_2) \left(\frac{\partial \varphi}{\partial x_1} - \frac{\partial \varphi}{\partial x_2} \right) \geqslant 0 (\leqslant 0)$$

holds for any $\boldsymbol{x} = (x_1, x_2, \cdots, x_n) \in \Omega^0$.

Lemma 2[3-4] Let $\Omega \subset \mathbf{R}_+^n$ be a symmetric geometrically convex set with a nonempty interior Ω^0. Let $\varphi : \Omega \to \mathbf{R}_+$ be continuous on Ω and differentiable on Ω^0. Then φ is a Schur geometrically convex (Schur geometrically concave) function if and only if φ is symmetric on Ω and

$$(x_1 - x_2) \left(x_1 \frac{\partial \varphi}{\partial x_1} - x_2 \frac{\partial \varphi}{\partial x_2} \right) \geqslant 0 \quad (\leqslant 0) \tag{2}$$

holds for any $\boldsymbol{x} = (x_1, \cdots, x_n) \in \Omega^0$.

Lemma 3 [1] Let $\boldsymbol{x} = (x_1, \cdots, x_n) \in \mathbf{R}^n$ and $A_n(\boldsymbol{x}) = \frac{1}{n} \sum_{i=1}^{n} x_i$. Then

$$(A_n(\boldsymbol{x}), \cdots, A_n(\boldsymbol{x})) \prec (x_1, \cdots, x_n)$$

70.3　Proofs of Theorems

Proof of Theorem 1

For $\boldsymbol{x} \in \mathbf{R}_+^1$, let

$$\varphi(\boldsymbol{x}) = nA_n(\boldsymbol{x}) + \frac{1}{G_n^n(\boldsymbol{x})} = \sum_{i=1}^{n} x_i + \frac{1}{\prod\limits_{i=1}^{n} x_i}$$

Then

$$\frac{\partial \varphi(\boldsymbol{x})}{\partial x_1} = 1 - \frac{1}{x_1^2 x_2 \prod\limits_{i=3}^{n} x_i}$$

and

$$\frac{\partial \varphi(\boldsymbol{x})}{\partial x_2} = 1 - \frac{1}{x_1 x_2^2 \prod\limits_{i=3}^{n} x_i}$$

and then

$$(x_1 - x_2)\left(\frac{\partial \varphi(\boldsymbol{x})}{\partial x_1} - \frac{\partial \varphi(\boldsymbol{x})}{\partial x_2}\right) = \frac{(x_1 - x_2)^2}{x_1^2 x_2^2 x_3 \cdots x_n} \geqslant 0$$

by Lemma 1, it follows that $\varphi(\boldsymbol{x})$ is the Schur-convex with $\boldsymbol{x} \in \mathbf{R}_+^n$, combining Lemma 3, we have

$$\varphi\left(A_n(\boldsymbol{x}), \cdots, A_n(\boldsymbol{x})\right) \leqslant \varphi\left(x_1, \cdots, x_n\right)$$

that is

$$n A_n(\boldsymbol{x}) + \frac{1}{A_n^n(\boldsymbol{x})} \leqslant n A_n(\boldsymbol{x}) + \frac{1}{G_n^n(\boldsymbol{x})}$$

namely

$$A_n(\boldsymbol{x}) \geqslant G_n(\boldsymbol{x}) \tag{3}$$

$$(x_1 - x_2)\left(x_1 \frac{\partial \varphi(\boldsymbol{x})}{\partial x_1} - x_2 \frac{\partial \varphi(\boldsymbol{x})}{\partial x_2}\right) = (x_1 - x_2)^2 \geqslant 0$$

By Lemma 2, it follows that $\varphi(\boldsymbol{x})$ is the Schur-geometrically convex with $\boldsymbol{x} \in \mathbf{R}_+^n$. From Lemma 3, we have

$$\left(\log \frac{1}{G_n(\boldsymbol{x})}, \cdots, \log \frac{1}{G_n(\boldsymbol{x})}\right) \prec \left(\log \frac{1}{x_1}, \cdots, \log \frac{1}{x_n}\right)$$

hence

$$\varphi\left(\frac{1}{G_n(\boldsymbol{x})}, \cdots, \frac{1}{G_n(\boldsymbol{x})}\right) \leqslant \varphi\left(\frac{1}{x_1}, \cdots, \frac{1}{x_n}\right)$$

that is

$$\frac{n}{G_n(\boldsymbol{x})} + G_n^n(\boldsymbol{x}) \leqslant \sum_{i=1}^{n} \frac{1}{x_i} + G_n^n(\boldsymbol{x})$$

namely

$$G_n(\boldsymbol{x}) \geqslant H_n(\boldsymbol{x}) \tag{4}$$

The proof of Theorem 1 is complete.

Remark 1 If taking $\psi(\boldsymbol{x}) = A_n(\boldsymbol{x}) + G_n(\boldsymbol{x})$, Theorem 1 can be proved in the same way as shown before, detailed proofs are left to the reader.

References

[1]　A M MARSHALL, I OLKIN. Inequalities: Theory of Majorization and Its Application[M]. New York : Academies Press, 1979.

[2]　BO-YING WANG. Foundations of Majorization Inequalities[M]. Beijing: Beijing Normal Univ. Press, 1990.

[3]　XIAO-MING ZHANG. Geometrically Convex Functions[M].Hefei: An hui University Press, 2004.

[4]　C P NICULESCU. Convexity According to the Geometric Mean[J]. Mathematical Inequalities & Applications, 2000, 3(2):155-167.

[5]　P S BULLEN, D S MITRINOVIC, P M VASIC. Means and their inequalities[M]. Dordrecht: Reidel publishing Co., 1988.

[6]　J-C KUANG. Applied Inequalities (Chang yong bu deng shi)[M]. 4th ed. Ji'nan: Shandong Press of Science and Technology, 2010.

[7]　YASUHARU UCHIDA. Geometric-arithmetic mean inequality[J]. J. Inequal. Pure and Appl. Math, 2008, 9(2), art. 56.

[8]　JOSIP PĚCARIĆ. A new proof of the arithmetic mean-the geometric mean inequality[J]. J. Math. Anal. Appl., 1997, 215 : 577-578.

[9]　HAO ZHI CHUAN. Note on the inequality of the arithmetic and geometric means[J]. Pacific Journal of Mathematics, 1990, 143(1): 43-46.

[10]　CHUNG LIE WANG. A generalization of the HGA inequalities[J]. Soochow J. Math. Natur. Sci. 1980, 6 : 149-152.

[11]　HANSHENG YANG, HENG YANG. The arithmetic-geometric mean inequality and the constant[J]. Mathematics Magazine, 2001, 74(4): 321-323.

[12]　WAN-LAN WANG. Some Inequalities Involving Means and Their Converses[J]. J. Math. Anal. Appl., 1999, 238(2): 567-579.

[13]　CHUNG-LIE WANG. Convexity and inequalities[J]. J. Math. Anal. Appl., 1979, 72(1) : 355-361.

第71篇　凸函数的两个性质的控制证明

(张鉴, 石焕南, 顾春. 高等数学研究, 2016, 19 (4):32-33.)

摘　要:　本文利用受控理论给出有关凸函数的两个性质的新颖而简洁的证明.

关键词:　凸函数; 不等式; 受控

71.1　引言

在本文中, $\mathbf{R}^n$ 和 $\mathbf{R}^n_+$ 分别表示 n 维实数集和 n 维正实数集, 并记 $\mathbf{R}^1 = \mathbf{R}, \mathbf{R}^1_+ = \mathbf{R}_+$.

文[1]根据凸函数的已有性质, 运用数学归纳法证得关于凸函数的如下两个定理.

定理 1　设 $F : \mathbf{R} \to \mathbf{R}$ 是凸函数, 则对任意的 $x, y \in \mathbf{R}$ 和任意的自然数 n, 有

$$F\left(\frac{x}{2^{2n-1}}\right) \leqslant \frac{1}{2^{2n-1}} F(x+y) + \sum_{p=1}^{2n-1} \frac{1}{2^p} F\left(\frac{(-1)^p y}{2^{2n-1-p}}\right) \qquad (1)$$

定理 2　设 $F : \mathbf{R} \to \mathbf{R}$ 是凸函数, 则对于任意的自然数 m 和 $v_1, v_2, v_3 \in \mathbf{R}$, 有

$$-F(v_1+v_2+v_3) \leqslant -2^{2m} F\left(\frac{v_1}{2^{2m}}\right) + 2^{2m-1} F\left(\frac{-v_2}{2^{2m}}\right) + \sum_{p=1}^{2m-1} p^{2^{2m-1-p}} F\left(\frac{(-1)^p v_3}{2^{2m-1-p}}\right) \qquad (2)$$

本文利用受控理论给出这两个定理的新颖而简单的证明.

71.2　定义和引理

我们需要如下定义和引理.

定义 1[1-4]　设 $x = (x_1, ..., x_n), y = (y_1, ..., y_n) \in \mathbf{R}^n$.

(1) 若 $\sum_{i=1}^{k} x_{[i]} \leqslant \sum_{i=1}^{k} y_{[i]}$, $k = 1, 2, ..., n-1$, 且 $\sum_{i=1}^{n} x_i = \sum_{i=1}^{n} y_i$, 则称 x 被 y 所控制, 记作 $x \prec y$. 其中 $x_{[1]} \geqslant ... \geqslant x_{[n]}$ 和 $y_{[1]} \geqslant ... \geqslant y_{[n]}$ 分别是 x 和 y 的递减重排.

(2) $x \leqslant y$ 表示对所有的 $i = 1, ..., n$, $x_i \leqslant y_i$.

引理 1　设 $\{b_i\}$ 是公比为 q 的等比数列, 则其前 n 项之和

$$S_n = \sum_{i=1}^{n} b_i = \frac{b_1(1 - q^n)}{1 - q} = \frac{b_1 - b_n q}{1 - q} \tag{3}$$

引理 2[3-4]　设 $x = (x_1, ..., x_n) \in \mathbf{R}^n$, $\bar{x} = \frac{1}{n} \sum_{i=1}^{n} x_i$, 则

$$(\bar{x}, \bar{x}, ..., \bar{x}) \prec (x_1, x_2, ..., x_n) \tag{4}$$

引理 3[3-4]　设 $I \subset \mathbf{R}$ 为一个区间, $x, y \in I^n \subset \mathbf{R}^n$, 则 $x \prec y \Leftrightarrow \forall$ 凸(凹)函数 $g : I \to \mathbf{R}$, 有

$$\sum_{i=1}^{n} g(x_i) \leqslant (\geqslant) \sum_{i=1}^{n} g(y_i)$$

71.3　定理的证明

定理 1 的证明　式(1)等价于

$$2^{2n-1} F\left(\frac{x}{2^{2n-1}}\right) \leqslant F(x+y) + \sum_{p=1}^{2n-1} 2^{2n-1-p} F\left(\frac{(-1)^p y}{2^{2n-1-p}}\right) \tag{5}$$

利用引理1 可得

$$2^{2n-1} = 1 + \sum_{p=1}^{2n-1} 2^{2n-1-p}$$

记

$$\boldsymbol{u} = (u_1, ..., u_{2^{2n-1}})$$
$$= (x+y, \underbrace{\frac{(-1)^1 y}{2^{2n-1-1}}, \cdots, \frac{(-1)^1 y}{2^{2n-1-1}}}_{2^{2n-1-1}}, \underbrace{\frac{(-1)^2 y}{2^{2n-1-2}}, \cdots, \frac{(-1)^2 y}{2^{2n-1-2}}}_{2^{2n-1-2}}, \cdots,$$

$$\underbrace{\frac{(-1)^{2n-1}y}{2^{2n-1-(2n-1)}}, \cdots, \frac{(-1)^{2n-1}y}{2^{2n-1-(2n-1)}}}_{2^{2n-1-(2n-1)}})$$

易见 $\sum\limits_{i=1}^{2^{2n-1}} u_i = x$, 于是 $\frac{1}{2^{2n-1}}\sum\limits_{i=1}^{2^{2n-1}} u_i = \frac{x}{2^{2n-1}}$, 据引理2 知

$$\overline{u} = \left(\underbrace{\frac{x}{2^{2n-1}}, \cdots, \frac{x}{2^{2n-1}}}_{2^{2n-1}} \right) \prec u$$

因 $F : \mathbf{R} \to \mathbf{R}$ 是凸函数, 由引理3 即得式(5), 由此得证.

定理2 的证明　式(2)即为

$$2^{2m}F\left(\frac{v_1}{2^{2m}}\right) \leqslant F(v_1+v_2+v_3)+2^{2m-1}F\left(\frac{-v_2}{2^{2m}}\right)+\sum_{p=1}^{2m-1} p^{2^{2m-1-p}}F\left(\frac{(-1)^p v_3}{2^{2m-1-p}}\right)$$

(6)

利用引理1 可得 $2^{2m} = 1 + 2^{2m-1} + \sum\limits_{p=1}^{2m-1} 2^{2m-1-p}$, 记

$$w = (w_1, \ldots, w_{2^{2m}})$$
$$= (v_1+v_2+v_3, \underbrace{\frac{-v_2}{2^{2m}}, \ldots, \frac{-v_2}{2^{2m}}}_{2^{2m-1}}, \underbrace{\frac{(-1)^1 v_3}{2^{2m-1-1}}, \cdots, \frac{(-1)^1 v_3}{2^{2m-1-1}}}_{2^{2m-1-1}}, \cdots,$$
$$\underbrace{\frac{(-1)^{2m-1}v_3}{2^{2m-1-(2n-1)}}, \cdots, \frac{(-1)^{2m-1}v_3}{2^{2m-1-(2n-1)}}}_{2^{2m-1-(2n-1)}})$$

易见 $\sum\limits_{i=1}^{2^{2m}} w_i = v_1$, 于是 $\frac{1}{2^{2m}}\sum\limits_{i=1}^{2^{2m}} w_i = \frac{v_1}{2^{2m}}$, 据引理2 知

$$\overline{w} = \left(\underbrace{\frac{v_1}{2^{2m}}, \cdots, \frac{v_1}{2^{2m}}}_{2^{2m}} \right) \prec w$$

因 $F : \mathbf{R} \to \mathbf{R}$ 是凸函数, 由引理3 即得式(6), 由此得证.

参考文献

[1]　吴燕, 李武. 凸函数的一些新性质[J]. 高等数学研究, 2014, 17(4): 16 – 18, 22.

[2]　王伯英. 控制不等式基础[M]. 北京:北京师范大学出版社, 1990.

[3]　MARSHALL A W, OLKIN I, ARNOLD B C. Inequalities: theory of majorization and its application[M]. 2nd ed. New York: Springer Press, 2011.

[4]　石焕南. 受控理论与解析不等式[M]. 哈尔滨: 哈尔滨工业大学出版社, 2012.

第72篇　关于"一类对称函数的Schur- m 指数凸性"的注记

(张鑑, 顾春, 石焕南. 系统科学与数学(核心期刊),2016, 36 (10):1779-1782)

　　摘　要:　文[2]采用通常的手段, 即利用Schur - m 指数凸函数判定定理研究了一类对称函数的Schur -m 指数凸性. 本文修正了文[2]中的主要定理的条件, 并且利用Schur - m 指数凸函数的一个已知性质, 非常简洁地证明了这个定理.

　　关键词:　Schur 指数凸性; Schur凸性; Schur几何凸性; 不等式; 乘凸

　　在本文中, $\mathbf{R}^n$ 和 $\mathbf{R}_+^n$ 分别表示 n 维实数集和 n 维正实数集, 并记 $\mathbf{R}^1 = \mathbf{R}, \mathbf{R}_+^1 = \mathbf{R}_+$.

　　2007年, 关开中[1] 研究了Hamy对称函数的推广形式

$$\sum_r^n (f(\boldsymbol{x})) = \sum_{1 \leqslant i_1 < \ldots < i_r \leqslant n} f\left(\prod_{j=1}^r x_{i_j}^{\frac{1}{r}}\right), \quad r = 1, \ldots, n \tag{1}$$

的Schur-几何凸性, 获得了如下结果.

　　定理A　函数 $f(x)$ 是定义在区间 $I \subset \mathbf{R}_+$ 上的非负函数, 且有连续的偏导数, 若函数 $f(x)$ 是单调的且是乘凸函数, 则 $\sum_r^n (f(\boldsymbol{x}))$ 是Schur-几何凸函数.

　　最近, 王文和杨世国[2] 进一步研究了 $\sum_r^n (f(\boldsymbol{x}))$ 的Schur m - 指数凸性. 文[2] 定理3.1 (i) 的证明中写到

$$\Delta_1 = \cdots = U(m; x_1, x_2)(x_1 - x_2)\left(x_1^{1-m} f'(x_1) - x_2^{1-m} f'(x_2)\right)$$

应用引理3.2和引理3.3, 则有:当 $m \leqslant 1$ 时, $\Delta_1 \geqslant 0$.

　　此处有误. 引理3.2 的式(3.1): $(x_1 - x_2)\left(x_1^k f'(x_1) - x_2^k f'(x_2)\right) \geqslant 0$ 要求 $k \geqslant 1$, 因此这里应要求 $1 - m \geqslant 1$, 即 $m \leqslant 0$. 故定理(i) 的条件" $m \leqslant 1$ "应改为" $m \leqslant 0$". 这样一来, 文[2] 中的定理3.1 可简述为

定理1　设$I \subset \mathbf{R}_+, f: I \to \mathbf{R}_+$ 在I 内有二阶连续的偏导数. 若f 在I 上是单调递增的且是乘凸的, 则

(1) 对于$m \leqslant 0$, $\sum\limits_{r}^{n}(f(\boldsymbol{x}))$ 在I^n上Schur m -指数凸;

(2) 当$r = 2, n = 2$ 时, 对于$m > 0$, $\sum\limits_{r}^{n}(f(\boldsymbol{x}))$ 在I^n 上Schur m - 指数凹.

由文[2]中的定理3.1 得到的两个推论, 即文[2]的推论3.4 和推论3.5 也应作相应的调整.

文[2] 是采用通常的方法,即利用本文引理1 所给出的Schur m - 指数凸性判定定理证明其主要结果定理1(1)的. 本文利用Schur m - 指数凸函数的一个已知性质, 即本文的引理2 给出定理1 (1) 的非常简单的证明.

我们需要如下定义和引理.

定义1[3-4]　设区间$I \subset \mathbf{R}_+$, 若函数$f: I \to \mathbf{R}_+$ 满足, $\forall \boldsymbol{x}, \boldsymbol{y} \in \mathbf{R}_+^n, \alpha \in [0,1]$, 有

$$f\left(x_1^\alpha y_1^{1-\alpha}, \ldots, x_n^\alpha y_n^{1-\alpha}\right) \leqslant f^\alpha(x_1, \ldots, x_n) f^{1-\alpha}(y_1, \ldots, y_n)$$

则称f为I上的乘凸函数(或几何凸函数).

定义2[5-6]　设$\boldsymbol{x} = (x_1, \cdots, x_n), \boldsymbol{y} = (y_1, \cdots, y_n) \in \mathbf{R}^n$.

(i) $\boldsymbol{x} \geqslant \boldsymbol{y}$ 表示$x_i \geqslant y_i, i = 1, 2, \cdots, n$;

(ii) 设$\Omega \subset \mathbf{R}^n, \varphi: \Omega \to \mathbf{R}$. 若在$\Omega$ 上$\boldsymbol{x} \geqslant \boldsymbol{y} \Rightarrow \varphi(\boldsymbol{x}) \geqslant \varphi(\boldsymbol{y})$, 则称$\varphi$ 为Ω 上的增函数, 若$-\varphi$ 是Ω 上的增函数, 则称φ 为Ω 上的减函数.

定义3[5-6]　设$\boldsymbol{x} = (x_1, \ldots, x_n)$ 和$\boldsymbol{y} = (y_1, \ldots, y_n)$ 满足

(1) $\sum\limits_{i=1}^{k} x_{[i]} \leqslant \sum\limits_{i=1}^{k} y_{[i]}, k = 1, 2, \ldots, n-1$;

(2) $\sum\limits_{i=1}^{n} x_i = \sum\limits_{i=1}^{n} y_i$.

其中$x_{[1]} \geqslant \ldots \geqslant x_{[n]}$ 和$y_{[1]} \geqslant \cdots \geqslant y_{[n]}$ 是$\boldsymbol{x}$ 和$\boldsymbol{y}$ 的递减重排, 则称$\boldsymbol{x}$ 被$\boldsymbol{y}$ 所控制, 记作$\boldsymbol{x} \prec \boldsymbol{y}$.

定义4[5-6]　设$\Omega \subset \mathbf{R}^n, \varphi: \Omega \to \mathbf{R}$, 若在$\Omega$ 上$\boldsymbol{x} \prec \boldsymbol{y} \Rightarrow \varphi(\boldsymbol{x}) \leqslant \varphi(\boldsymbol{y})$, 则称$\varphi$ 为Ω 上的Schur凸函数; 若$-\varphi$ 是Ω 上的Schur凸函数, 则称φ 为Ω 上的Schur凹函数.

定义5[7]　设$\Omega \subset \mathbf{R}_+^n, \varphi: \Omega \to \mathbf{R}_+$. 对于任意$\boldsymbol{x}, \boldsymbol{y} \in \Omega$, 若

$$(\ln x_1, \ldots, \ln x_n) \prec (\ln y_1, \ldots, \ln y_n) \Rightarrow \varphi(\boldsymbol{x}) \leqslant (\geqslant)\varphi(\boldsymbol{y})$$

则称φ 为Ω 上的Schur几何凸(凹)函数.

定义6[8-10]　设$\Omega \subset \mathbf{R}^n$ 且Ω的内部非空, $f: \Omega \to \mathbf{R}$ 是严格单调函数.

$\varphi:\Omega\to\mathbf{R}$, 若对于任意$\boldsymbol{x},\boldsymbol{y}\in\Omega$, 总有

$$(f(x_1),\ldots,f(x_n))\prec(f(y_1),\ldots,f(y_n))\Rightarrow\varphi(\boldsymbol{x})\leqslant(\geqslant)\varphi(\boldsymbol{y})$$

则称φ 为Ω 上的Schur-f 凸函数；若$-\varphi$ 是Ω 上Schur-f 凸函数, 则称φ 为Ω 上Schur-f 凹函数.

定义7[8-10]   在定义7 中若取

$$f(x)=\begin{cases}\dfrac{x^m-1}{m}, & m\neq 0\\ \ln x, & m=0\end{cases}\tag{2}$$

则称φ为Ω上的m 阶Schur-幂凸函数(或Schur m-指数凸函数);若$-\varphi$ 为Ω 上的m 阶Schur-幂凸函数, 则称φ 为Ω上的m 阶Schur-幂凹函数.

注1   在定义7 中取$f(x)=x$ 和$\ln x$ 可分别得Schur-凸函数和Schur-几何凸函数的定义.

引理1[8-10]   设$f:\mathbf{R}\to\mathbf{R}$是严格单调的可微函数, $I^n\subset\mathbf{R}^n_+$ 是有内点的对称集, $\varphi:I^n\to\mathbf{R}^n_+$ 于I^n 上连续, 在I^n 的内部可微, φ 是I^n 上Schur f-凸（Schur$f$- 凹）的充要条件是φ 在I^n 上对称, 且对于一切I^n的内点x, 有

$$(f(x_1)-f(x_2))\left(\frac{1}{f'(x_1)}\frac{\partial\varphi(\boldsymbol{x})}{\partial x_1}-\frac{1}{f'(x_2)}\frac{\partial\varphi(\boldsymbol{x})}{\partial x_2}\right)\geqslant 0(\leqslant 0)\tag{3}$$

特别, 对于m 阶Schur-幂凸函数, 若$m=0$, 式(3)化为

$$(\ln x_1-\ln x_2)\left(x_1\frac{\partial\varphi(\boldsymbol{x})}{\partial x_1}-x_2\frac{\partial\varphi(\boldsymbol{x})}{\partial x_2}\right)\geqslant 0(\leqslant 0)\tag{4}$$

若$m\neq 0$, 式(3)化为

$$\frac{x_1^m-x_2^m}{m}\left(x_1^{1-m}\frac{\partial\varphi(\boldsymbol{x})}{\partial x_1}-x_2^{1-m}\frac{\partial\varphi(\boldsymbol{x})}{\partial x_2}\right)\geqslant 0(\leqslant 0)\tag{5}$$

关于不同阶的Schur-幂凸函数之间的关系，张小明证得

引理2[11-12]   设$p>q$, 区间$I\subset\mathbf{R}_+$, $\varphi:I^n\to\mathbf{R}$ 为对称可微函数. 若φ 为递增的p 阶Schur-幂凸函数, 则φ 为q 阶Schur-幂凸函数.

定理1 (1) 的证明   由定理A 知$\sum_r^n(f(\boldsymbol{x}))$ 是Schur-几何凸的, 即当$m=0$ 时, $\sum_r^n(f(\boldsymbol{x}))$ 是Schur 0 阶指数凸的. 因f 在I 上单调递增, 显然$\sum_r^n(f(\boldsymbol{x}))$ 在I^n 上单调递增. 若$m<0$, 由引理2 即可断定$\sum_r^n(f(\boldsymbol{x}))$ 是I^n 上的Schur-m阶指数凸的. 定理1 (1) 得证.

致谢：　作者衷心感谢审稿专家所提供的宝贵的修改意见!

参考文献

[1] GUAN KAI-ZHONG. A class of symmetric functions for multiplicatively convex function [J]. Math. Inequal. Appl., 2007, 10(4): 745-753.

[2] 王文，杨世国. 一类对称函数的Schur -指数凸性[J]. 系统科学与数学, 2014, 34 (3):367-375.

[3] 张小明, 褚玉明. 解析不等式新论[M]. 哈尔滨：哈尔滨工业大学出版社, 2009.

[4] CONSTANTIN P. NICULESCU. Convexity according to the geometric mean[J]. Math. Inequal. Appl., 2000, 3 (2):155-167.

[5] MARSHALL A W, OLKIN I, ARNOLD B C. Inequalities: theory of majorization and its application[M]. 2nd ed. New York: Springer Press, 2011.

[6] 王伯英. 控制不等式基础[M]. 北京: 北京师范大学出版社, 1990.

[7] 张小明. 几何凸函数[M]. 合肥: 安徽大学出版社, 2004.

[8] YANG ZHEN-HANG. Schur power convexity of Stolarsky means[J]. Publ. Math. Debrecen, 2012,80(1-2): 43-66.

[9] YANG ZHEN-HANG. Schur power convexity of Gini means[J]. Bull. Korean Math. Soc., 2013,50 (2):485-498.

[10] YANG ZHEN-HANG. Schur power convexity of the Daróczy means[J]. Math. Inequal. Appl., 2013, 16(3): 751‒762.

[11] 张小明. 几个N元平均的积的Schur-p阶幂凸性[J]. 湖南理工学院学报（自然科学版）, 2011, 24 (2):1-6,13.

[12] 石焕南. 受控理论与解析不等式[M]. 哈尔滨: 哈尔滨工业大学出版社, 2012.

第73篇 凸数列的几个加权和性质的控制证明

(石焕南, 张鉴,顾春, 四川师范大学学报(自然版), 2016, 39 (3):373-376)

摘　要: 本文利用受控理论并结合概率方法给出凸数列的几个加权和性质的新证明.

关键词: 凸数列; 加权和; 受控; 概率方法; 不等式

在本文中, $\mathbf{R}^n$ 和 $\mathbf{Z}_+^n$ 分别表示 n 维实数集和 n 维非负整数集, 并记 $\mathbf{R}^1 = \mathbf{R}, \mathbf{Z}_+^1 = \mathbf{Z}_+$. $\mathbf{N}^*$ 表示正整数集.

若实数列 $\{a_i\}$ (有限的 $\{a_i\}_{i=1}^n$ 或无限的 $\{a_i\}_{i=1}^\infty$) 满足条件

$$a_{i-1} + a_{i+1} \geqslant 2a_i \tag{1}$$

其中 $i = 2, ..., n-1$ 或 $i \geqslant 2$, 则称 $\{a_i\}$ 是一个凸数列. 若上述不等式反向, 则称数列 $\{a_i\}$ 是一个凹数列.

文[1]和[2]利用数学归纳法和Abel 变换给出凸数列的几个有趣的加权和性质, 即下述五个定理. 本文利用受控理论并结合概率方法给出这些结果的新的证明.

定理1 [1] 若 $\{a_i\}_{i=0}^\infty$ 是一个凸数列, $p \in \mathbf{Z}_+$, 则对任意的 $n \in \mathbf{N}^*$, 有

$$\sum_{i=0}^n a_i C_{p+i}^p \geqslant \frac{1}{p+2} C_{p+n+1}^{p+1} [na_{n-1} + (p+2-n)a_n] \tag{2}$$

$$\sum_{i=0}^n a_i C_{p+n-i}^p \geqslant \frac{1}{p+2} C_{p+n+1}^{p+1} [(p+2-n)a_0 + na_1] \tag{3}$$

定理2 [1] 若 $\{a_i\}_{i=0}^\infty$ 是一个凸数列, $p \in \mathbf{Z}_+$, 则对任意的 $n \in \mathbf{N}^*$, 有

$$\sum_{i=1}^n i^2 a_i \geqslant \frac{1}{12} n(n+1)[n(n-1)a_{n-1} - (n^2 - 5n - 2)a_n] \tag{4}$$

定理3 [2] 若$\{a_i\}_{i=0}^{\infty}$ 是一个凸数列, 则对任意的$n \in \mathbf{N}^*, n \geqslant 3$, 有

$$\sum_{i=1}^{n} i^2 a_i \geqslant \frac{1}{12} n(n+1)[na_1 + (3n+2)a_n] \tag{5}$$

定理4 [2] 若$\{a_i\}_{i=0}^{\infty}$ 是一个凸数列, 则对任意的$n \in \mathbf{N}^*, n \geqslant 3$, 有

$$2^{n-2} n(a_1 + a_n) \geqslant \sum_{i=1}^{n} i a_i \mathrm{C}_n^i \tag{6}$$

$$2^{n-3} n[na_1 + (n+2)a_n] \geqslant \sum_{i=1}^{n} i^2 a_i \mathrm{C}_n^i \tag{7}$$

定理5 [2] 若$\{a_i\}_{i=0}^{\infty}$ 是一个凸数列, 则对任意的$n \in \mathbf{N}^*, n \geqslant 2$, 有

$$\sum_{i=0}^{n} a_i \mathrm{C}_n^i \leqslant 2^{n-1}(a_0 + a_n) \tag{8}$$

$$\sum_{i=0}^{n} a_i \mathrm{C}_k^i \mathrm{C}_m^{n-i} \geqslant \frac{ma_0 + ka_n}{m+k} \mathrm{C}_{m+k}^n \tag{9}$$

其中k, m 是不小于n 的正整数.

我们需要如下定义和引理.

定义1 [3-4]　设$\boldsymbol{x} = (x_1, ..., x_n)$ 和$\boldsymbol{y} = (y_1, ..., y_n)$ 满足

(1) $\sum\limits_{i=1}^{k} x_{[i]} \leqslant \sum\limits_{i=1}^{k} y_{[i]}$, $k = 1, 2, ..., n-1$;

(2) $\sum\limits_{i=1}^{n} x_i = \sum\limits_{i=1}^{n} y_i$.

其中$x_{[1]} \geqslant \cdots \geqslant x_{[n]}$ 和$y_{[1]} \geqslant \cdots \geqslant y_{[n]}$ 是$\boldsymbol{x}$ 和$\boldsymbol{y}$ 的递减重排, 则称$\boldsymbol{x}$ 被$\boldsymbol{y}$ 所控制, 记作$\boldsymbol{x} \prec \boldsymbol{y}$.

引理1 [3-4]　设$\boldsymbol{x} = (x_1, \cdots, x_n) \in \mathbf{R}^n$. 则

$$(\overline{x}, \cdots, \overline{x}) \prec (x_1, \cdots, x_n) \tag{10}$$

其中$\overline{x} = \dfrac{1}{n} \sum\limits_{i=1}^{n} x_i$ 是$\boldsymbol{x}$ 的算术平均值.

引理2 [5-6]　设$n \geqslant 2$, 数列$\{a_i\}$ 是凸数列的充要条件为: 对于一切$\boldsymbol{p}, \boldsymbol{q} \in \mathbf{Z}_+^n$, 若$\boldsymbol{p} \prec \boldsymbol{q}$, 恒有

$$a_{p_1} + \cdots + a_{p_n} \leqslant a_{q_1} + \cdots + a_{q_n} \tag{11}$$

引理3 [3−4]　设 $\boldsymbol{x}, \boldsymbol{y} \in \mathbf{R}^n$, $x_1 \geqslant x_2 \geqslant \cdots \geqslant x_n$ 且 $\sum\limits_{i=1}^{n} x_i = \sum\limits_{i=1}^{n} y_i$. 若存在 $k, 1 \leqslant k < n$, 使得 $x_i \leqslant y_i, i = 1, \ldots, k, x_i \geqslant y_i, i = k+1, \ldots, n$, 则 $\boldsymbol{x} \prec \boldsymbol{y}$.

引理4 [2]　对任意 $n \in \mathbf{N}^*$, 有下列等式

$$\sum_{i=1}^{n} i \mathrm{C}_n^i = 2^{n-1} n \tag{12}$$

$$\sum_{i=1}^{n} i^2 \mathrm{C}_n^i = 2^{n-2} n(n+1) \tag{13}$$

$$\sum_{i=1}^{n} i^3 \mathrm{C}_n^i = 2^{n-3} n^2(n+3) \tag{14}$$

下面用概率方法证明几个组合恒等式.

引理5　设 $p \in \mathbf{Z}_+$, 则对任意的 $n \in \mathbf{N}^*$, 有

$$\sum_{i=0}^{n} \mathrm{C}_{p+n-i}^p = \mathrm{C}_{p+n+1}^{p+1} (\text{朱世杰恒等式}) \tag{15}$$

证明　考虑随机试验: 从自然数 1 到 $p+n+1$ 中任取 $p+1$ 个数, 令 A_i 表示取出的 $p+1$ 个数的最大数是 $p+i+1$, 则

$$P(A_i) = \frac{\mathrm{C}_{p+i}^p}{\mathrm{C}_{p+n+1}^{p+1}}, i = 0, 1, \ldots, n$$

显然诸 A_i 互不相容, 且 $\bigcup\limits_{i=0}^{n} A_i = \Omega$, 故 $\sum\limits_{i=0}^{n} P(A_i) = 1$, 由此即得 (15).

引理6　设 $p \in \mathbf{Z}_+$, 则对任意的 $n \in \mathbf{N}^*$, 及任意不小于 n 的正整数 k, m, 有

$$\sum_{i=0}^{n} \mathrm{C}_k^i \mathrm{C}_m^{n-i} = \mathrm{C}_{m+k}^n (\text{范德蒙恒等式}) \tag{16}$$

$$\sum_{i=0}^{n} i \mathrm{C}_k^i \mathrm{C}_m^{n-i} = \frac{kn}{k+m} \mathrm{C}_{m+k}^n \tag{17}$$

证明　考虑随机试验: 袋里装有 $n+k$ 个球, 其中 m 个红球, k 个白球. 现从中任取 $n(n \leqslant k, n \leqslant m)$ 个球, 令 X 表示取出的 n 个球中的白球数, 则

$$P(X = i) = \frac{\mathrm{C}_k^i \mathrm{C}_m^{n-i}}{\mathrm{C}_{m+k}^n}, i = 0, 1, \ldots, n$$

从而 $\sum_{i=0}^{n} P(X=i)=1$, 由此即得(16). 又

$$E(X)=\sum_{i=0}^{n} i \frac{\mathrm{C}_k^i \mathrm{C}_m^{n-i}}{\mathrm{C}_{m+k}^n} \tag{18}$$

若令

$$X_i=\begin{cases}1, & \text{若第}i\text{个白球被取出}\\ 0, & \text{若第}i\text{个白球未被取出}\end{cases}, i=1,...,k \tag{19}$$

则 $X=X_1+\cdots+X_k$, 而

$$E(X_i)=P(X_i=1)=P(\text{若第}i\text{个白球被取出})$$
$$=\frac{\mathrm{C}_{m+k-1}^{n-1}}{\mathrm{C}_{m+k}^n}=\frac{n}{k+m}, i=1,...,k,$$

于是

$$E(X)=E(X_1)+\cdots+E(X_k)=\frac{nk}{k+m} \tag{20}$$

结合(18)和(20)即得(17).

引理7　设 $p \in \mathbf{Z}_+$, 则对任意 $n \in \mathbf{N}^*$, 有

$$\sum_{i=1}^{n} i\mathrm{C}_{p+i}^p=\frac{n(p+1)}{p+2}\mathrm{C}_{p+n+1}^{p+1} \tag{21}$$

$$\sum_{i=0}^{n} i\mathrm{C}_{p+n-i}^p=\frac{n}{p+2}\mathrm{C}_{p+n+1}^{p+1} \tag{22}$$

证明　考虑随机试验: 从自然数1到 $p+n+1$ 中任取 $p+1$ 个数, 随机变量 X 表示取出的最大数与 $p+1$ 的差, 则

$$P(X=i)=\frac{\mathrm{C}_{p+i}^p}{\mathrm{C}_{p+n+1}^{p+1}}, i=1,...,n$$

从而, 我们有

$$P(X=i)=\sum_{i=1}^{n}\frac{\mathrm{C}_{p+i}^p}{\mathrm{C}_{p+n+1}^{p+1}}=1$$

即

$$\sum_{i=1}^{n}\mathrm{C}_{p+i}^p=\mathrm{C}_{p+n+1}^{p+1} \tag{23}$$

又有

$$E(X) = \sum_{i=1}^{n} i \frac{C_{p+i}^{p}}{C_{p+n+1}^{p+1}} \tag{24}$$

另一方面, 据文献[7], 对于整值随机变量 X, 有

$$E(X) = \sum_{i=1}^{n} P(X \geqslant i) = n - \sum_{i=1}^{n} P(X < i)$$

$$= n - \sum_{i=2}^{n} P(X < i) = n - \sum_{i=2}^{n} \frac{C_{p+i}^{p+1}}{C_{p+n+1}^{p+1}} \tag{25}$$

结合(24)和(25), 有

$$\sum_{i=1}^{n} i C_{p+i}^{p} = n C_{p+n+1}^{p+1} - \sum_{i=2}^{n} C_{p+i}^{p+1}$$

$$= n C_{p+n+1}^{p+1} - \sum_{i=1}^{n-1} C_{p+i}^{p+1} = n C_{p+n+1}^{p+1} - C_{p+n+1}^{p+2} (\text{由}(23))$$

$$= n C_{p+n+1}^{p+1} - \frac{2}{p+2} C_{p+n+1}^{p+2} = \frac{n(p+1)}{p+2} C_{p+n+1}^{p+1}$$

(21)得证. 对(21)式作变换 $n - i \to i$, 可知

$$\sum_{i=1}^{n} i C_{p+n-i}^{p+1} = \sum_{i=1}^{n} (n-i) C_{p+i}^{p} = n \sum_{i=1}^{n} C_{p+i}^{p} - \sum_{i=1}^{n} i C_{p+i}^{p}$$

$$= n C_{p+n+1}^{p+1} - \frac{n(p+1)}{p+2} C_{p+n+1}^{p+1} (\text{由}(23)\text{和}(21))$$

$$= \frac{n}{p+2} C_{p+n+1}^{p+1},$$

(22)得证.

定理1 的证明　令

$$\boldsymbol{x} = \left(\underbrace{n, \cdots, n}_{(p+2-n)C_{p+n+1}^{p+1}}, \underbrace{n-1, \cdots, n-1}_{nC_{p+n+1}^{p+1}} \right)$$

$$\boldsymbol{y} = \left(\underbrace{n, \cdots, n}_{(p+2)C_{p+n}^{p}}, \ldots, \underbrace{1, \cdots, 1}_{(p+2)C_{p+1}^{p}}, \underbrace{0, \cdots, 0}_{(p+2)C_{p+0}^{p}} \right)$$

由(23)有

$$(p+2)\sum_{i=0}^{n} C_{p+i}^p = C_{p+n+1}^{p+1}[(p+2-n)+n] = (p+2)C_{p+n+1}^{p+1} =: m$$

又由(21) 有

$$\sum_{i=1}^{m} x_i = C_{p+n+1}^{p+1}[n(p+2-n)+(n-1)n] = n(p+1)C_{p+n+1}^{p+1} = (p+2)\sum_{i=0}^{n} iC_{p+i}^p = \sum_{i=1}^{m} y_i$$

再由$\boldsymbol{x}$ 和$\boldsymbol{y}$ 的结构, 易见存在$k, 1 \leqslant k \leqslant m$, 使得$x_i \leqslant y_i, i = 1, \ldots, k, x_i \geqslant y_i$, $i = k+1, \ldots, m$, 故据引理3 知$\boldsymbol{x} \prec \boldsymbol{y}$, 从而由引理2 可得(2).

令

$$\boldsymbol{u} = \left(\underbrace{1, \cdots, 1,}_{nC_{p+n+1}^{p+1}} \quad \underbrace{0, \cdots, 0}_{(p+2-n)C_{p+n+1}^{p+1}} \right)$$

$$\boldsymbol{v} = \left(\underbrace{n, \cdots, n}_{(p+2)C_{p+n-n}^p}, \ldots, \underbrace{1, \cdots, 1}_{(p+2)C_{p+n-1}^p}, \underbrace{0, \cdots, 0}_{(p+2)C_{p+n-0}^p} \right)$$

由(15)有

$$(p+2)\sum_{i=0}^{n} C_{p+n-i}^p = C_{p+n+1}^{p+1}[(p+2-n)+n] = (p+2)C_{p+n+1}^{p+1} := m$$

又由(22)有

$$\sum_{i=1}^{m} u_i = C_{p+n+1}^{p+1}[0\cdot(p+2-n)+1\cdot n] = nC_{p+n+1}^{p+1} = (p+2)\sum_{i=0}^{n} iC_{p+n-i}^p = \sum_{i=1}^{m} v_i$$

再由$\boldsymbol{u}$ 和$\boldsymbol{v}$ 的结构, 易见存在$k, 1 \leqslant k \leqslant m$, 使得$u_i \leqslant v_i, i = 1, \ldots, k, u_i \geqslant v_i$, $i = k+1, \ldots, m$, 故据引理3 知$\boldsymbol{u} \prec \boldsymbol{v}$, 从而由引理2 可得(3).

定理2 的证明　令

$$\boldsymbol{x} = \left(\underbrace{n-1, \cdots, n-1}_{n^2(n+1)(n-1)} \right)$$

$$y = \left(\underbrace{n, \cdots, n}_{12n^2 + n(n+1)(n^2 - 5n - 2)}, \ldots, \underbrace{2, \cdots, 2}_{12 \cdot 2^2}, \underbrace{1, \cdots, 1}_{12 \cdot 1^2} \right)$$

注意

$$n(n+1)(n^2 - 5n - 2) + 12 \sum_{i=1}^{n} i^2$$

$$= n(n+1)(n^2 - 5n - 2) + 12 \times \frac{1}{6} n(n+1)(2n+1)$$

$$= n^2(n+1)(n-1) := m$$

$$\sum_{i=1}^{m} y_i = n^2(n+1)(n^2 - 5n - 2) + 12 \sum_{i=1}^{n} i^3$$

$$= n^2(n+1)(n^2 - 5n - 2) + 12 \times \left(\frac{1}{2} n(n+1) \right)^2 = n^2(n+1)(n-1)^2$$

我们有 $\frac{1}{m} \sum_{i=1}^{m} y_i = n - 1$, 由引理1 知 $\boldsymbol{x} \prec \boldsymbol{y}$, 从而由引理2 可得(4).

定理3 的证明 令

$$\boldsymbol{x} = \left(\underbrace{n, \cdots, n}_{n(3n+2)(n+1)}, \underbrace{1, \cdots, 1}_{n^2(n+1)} \right)$$

$$\boldsymbol{y} = \left(\underbrace{n, \cdots, n}_{12n^2}, \ldots, \underbrace{2, \cdots, 2}_{12 \cdot 2^2}, \underbrace{1, \cdots, 1}_{12 \cdot 1^2} \right)$$

$$12 \sum_{i=1}^{n} i^2 = 12 \times \frac{1}{6} n(n+1)(2n+1) = n^2(n+1) + n(n+1)(3n+2) := m$$

$$\sum_{i=1}^{m} y_i = 12 \sum_{i=1}^{n} i^3 = 12 \times \left(\frac{1}{2} n(n+1) \right)^2 = n^2(n+1) + n^2(n+1)(3n+2) = \sum_{i=1}^{m} x_i$$

再由 $\boldsymbol{x}$ 和 $\boldsymbol{y}$ 的结构, 易见存在 $k, 1 \leqslant k \leqslant m$, 使得 $x_i \leqslant y_i, i = 1, \ldots, k, x_i \geqslant y_i,$
$i = k+1, \ldots, m$, 故据引理3 知 $\boldsymbol{x} \prec \boldsymbol{y}$, 从而由引理2 可得(5).

定理4 的证明 令

$$\boldsymbol{x} = \left(\underbrace{n, \cdots, n}_{nC_n^n}, \cdots, \underbrace{2, \cdots, 2}_{2C_n^2}, \underbrace{1, \cdots, 1}_{C_n^1} \right)$$

$$\boldsymbol{y} = \left(\underbrace{n, \cdots, n}_{2^{n-2}n}, \underbrace{1, \cdots, 1}_{2^{n-2}n} \right)$$

由引理4 有

$$2 \times 2^{n-2}n = 2^{n-1}n = \sum_{i=1}^{n} i\mathrm{C}_n^i := m$$

$$\sum_{i=1}^{m} y_i = 2^{n-2}n^2 + 2^{n-2}n = 2^{n-2}n(n+1) = \sum_{i=1}^{n} i^2\mathrm{C}_n^i = \sum_{i=1}^{m} x_i$$

注意 $n\mathrm{C}_n^n \leqslant 2^{n-2}n$, 易见存在 $k, 1 \leqslant k \leqslant m$, 使得 $x_i \leqslant y_i, i = 1, \ldots, k, x_i \geqslant y_i$, $i = k+1, \ldots, m$, 故据引理3 知 $\boldsymbol{x} \prec \boldsymbol{y}$, 从而由引理2 可得(6).

令

$$\boldsymbol{u} = \left(\underbrace{n, \cdots, n}_{n^2\mathrm{C}_n^n}, \cdots, \underbrace{2, \cdots, 2}_{2^2\mathrm{C}_n^2}, \underbrace{1, \cdots, 1}_{\mathrm{C}_n^1} \right)$$

$$\boldsymbol{v} = \left(\underbrace{n, \cdots, n}_{2^{n-3}n(n+2)}, \underbrace{1, \cdots, 1}_{2^{n-3}n^2} \right)$$

由引理4 有

$$2^{n-3}n(n+2) + 2^{n-3}n^2 = 2^{n-2}n(n+1) = \sum_{i=1}^{n} i^2\mathrm{C}_n^i := m$$

$$\sum_{i=1}^{m} v_i = 2^{n-3}n^2(n+2) + 2^{n-3}n^2 = 2^{n-3}n^2(n+3) = \sum_{i=1}^{n} i^3\mathrm{C}_n^i = \sum_{i=1}^{m} u_i$$

注意 $n^2\mathrm{C}_n^n \leqslant 2^{n-3}n(n+3)$, 易见存在 $k, 1 \leqslant k \leqslant m$, 使得 $u_i \leqslant v_i, i = 1, \ldots, k, u_i \geqslant v_i$, $i = k+1, \ldots, m$, 故据引理3 知 $\boldsymbol{u} \prec \boldsymbol{v}$, 从而由引理2 可得(7).

定理5 的证明　令

$$\boldsymbol{x} = \left(\underbrace{n, \cdots, n}_{\mathrm{C}_n^n}, \underbrace{n-1, \cdots, n-1}_{\mathrm{C}_n^{n-1}}, \cdots, \underbrace{1, \cdots, 1}_{\mathrm{C}_n^1}, \underbrace{0, \cdots, 0}_{\mathrm{C}_n^0} \right)$$

$$\boldsymbol{y} = \left(\underbrace{n, \cdots, n}_{2^{n-1}}, \underbrace{0, \cdots, 0}_{2^{n-1}} \right)$$

注意$2^n = \sum\limits_{i=1}^{n} C_n^i$, 易见对于$k = 2^{n-1}$, 满足$1 \leqslant k \leqslant 2^n$, 使得$x_i \leqslant y_i, i = 1, \ldots, k, x_i \geqslant y_i$, $i = k+1, \ldots, 2^n$, 故据引理3 知$\boldsymbol{x} \prec \boldsymbol{y}$, 从而由引理2 可得(8).

令

$$\boldsymbol{u} = \left(\underbrace{n, \cdots, n}_{k C_{m+k}^n}, \underbrace{0, \cdots, 0}_{m C_{m+k}^n} \right)$$

$$\boldsymbol{v} = \left(\underbrace{n, \cdots, n}_{(m+k) C_k^n C_m^{n-n}}, \cdots, \underbrace{1, \cdots, 1}_{(m+k) C_k^1 C_m^{n-1}}, \underbrace{0, \cdots, 0}_{(m+k) C_k^0 C_m^{n-0}} \right)$$

由引理7 有

$$(k+m) C_{m+k}^n = (m+k) \sum_{i=1}^{n} C_k^i C_m^{n-i} := s$$

$$\sum_{i=1}^{s} u_i = nk C_{m+k}^n = \sum_{i=0}^{n} i C_k^i C_m^{n-i} = \sum_{i=1}^{s} v_i$$

再由$\boldsymbol{u}$ 和$\boldsymbol{v}$ 的结构, 易见存在$j, 1 \leqslant j \leqslant s$, 使得$u_i \leqslant v_i, i = 1, \ldots, j, u_i \geqslant v_i$, $i = j+1, \ldots, s$, 故据引理3 知$\boldsymbol{u} \prec \boldsymbol{v}$, 从而由引理2 可得(9).

受控理论(Theory of Majorization), 亦称控制不等式理论是不等式研究的有力武器. 近些年我国学者在国内外已发表了众多有关该领域的研究论文, 例如文[8]-[13], 详细的内容请参见专著[14].

参考文献

[1] 卢小宁, 萧振纲. 凸数列的几个加权和性质[J]. 湖南理工学院学报(自然科学版), 2014,27 (4):6-9.

[2] 萧振纲. 凸数列的几个封闭性质与加权和性质[J]. 湖南理工学院学报(自然科学版), 2012, 25 (2):1-6.

[3] 王伯英. 控制不等式基础[M]. 北京: 北京师范大学出版社, 1990.

[4] MARSHALLl A M, OLKIN I. Inequalities: Theory of Majorization and Its Application[M]. New York : Academies Press, 1979.

[5] 石焕南, 李大矛.凸数列的一个等价条件及其应用[J]. 曲阜师范大学学报(自然科学版), 2001, 27(4)： 4-6.

[6] 石焕南. 凸数列的一个等价条件及其应用 II [J]. 数学杂志, 2004, 24(4): 390-394.

[7] 朱秀娟, 洪再吉. 概率统计150 题[M].修订本. 长沙: 湖南科学技术出版社, 1987.

[8] 张小明, 李世杰. 两个与初等对称函数有关的S-几何凸函数[J]. 四川师范大学学报(自然科学版), 2007, 30 (2): 188-190.

[9] 石焕南. Popoviciu不等式的新推广[J]. 四川师范大学学报(自然科学版), 2002, 25(5)：510-511.

[10] 石焕南,顾春,张鉴. 一个Schur 凸性判定定理的应用[J]. 四川师范大学学报(自然科学版), 2012, 35 (3):345-348.

[11] 杨洪, 文家金. 涉及Hardy函数的不等式[J]. 四川师范大学学报(自然科学版), 2007, 30 (5): 374-277.

[12] 谢巍,文家金. Jensen- Pecaric -Svrtan型不等式[J]. 四川师范大学学报(自然科学版), 2009, 32 (5) :621-625.

[13] 张勇, 文家金, 王挽澜. 含幂指数的一个不等式猜想的研究[J]. 四川师范大学学报(自然科学版), 2005, 28(2)：245-249.

[14] 石焕南. 受控理论与解析不等式[M]. 哈尔滨: 哈尔滨工业大学出版社, 2012.

第74篇　Schur-Convexity for Lehmer Mean of n Variables

(FU CUN-RU, WANG DONGSHENG, SHI HUANNAN. J. Nonlinear

Sci. Appl. 9 (2016), 5510-5520)

Abstract: Schur-convexity, Schur-geometric convexity and Schur-harmonic convexity for Lehmer mean of n variables are investigated, and some mean value inequalities of n variables are established.

Keywords: n-variables Lehmer mean; Schur convexity; Schur geometric convexity; Schur harmonic convexity; majorization; inequalities.

74.1　Introduction and Preliminaries

Throughout the paper we assume that the set of n-dimensional row vector on the real number field by $\mathbf{R}^n$.

$$\mathbf{R}_+^n = \{\boldsymbol{x} = (x_1, \ldots, x_n) \in \mathbf{R}^n : x_i > 0, i = 1, \ldots, n\}$$

In particular, $\mathbf{R}^1$ and $\mathbf{R}_+^1$ denoted by $\mathbf{R}$ and $\mathbf{R}_+$ respectively.

For $x, y > 0$ and $p \in \mathbf{R}$, the Lehmer mean values $L_p(x, y)$ were introduced by Lehmer[1] as follows

$$L_p(x, y) = \frac{x^p + y^p}{x^{p-1} + y^{p-1}}$$

Many mean values are special cases of the Lehmer mean values, for example

$$A(x, y) = \frac{x + y}{2} = L_1(x, y)$$

is the arithmetic mean

$$G(x,y) = \sqrt{xy} = L_{\frac{1}{2}}(x,y)$$

is the geometric mean

$$H(x,y) = \frac{2xy}{x+y} = L_0(x,y)$$

is the harmonic mean

$$\tilde{H}(x,y) = \frac{x^2+y^2}{x+y} = L_2(x,y)$$

is the anti-harmonic mean.

Investigation of the elementary properties and inequalities for $L_p(x,y)$ has attracted the attention of a considerable number of mathematicians (see [1-3,10-12,14,21,23,26,28-31]).

In 2009, Gu and Shi[11] discussed the Schur convexity and Schur geometric convexity of the Lehmer means $L_p(x,y)$ with respect to $(x,y) \in \mathbf{R}_+^2$ for fixed p. Subsequently, Xia and Chu[36] researched the Schur harmonic convexity of the Lehmer means $L_p(x,y)$ with respect to $(x,y) \in \mathbf{R}_+^2$ for fixed p.

Let $\boldsymbol{x} = (x_1, x_2, \ldots, x_n) \in \mathbf{R}_+^2$. For Schur-convexity and Schur-geometric convexity of n variables Lehmer mean

$$L_p(\boldsymbol{x}) = L_p(x_1, x_2, \ldots, x_n) = \frac{\displaystyle\sum_{i=1}^{n} x_i^p}{\displaystyle\sum_{i=1}^{n} x_i^{p-1}}$$

Gu and Shi[11] obtained the following results:

Theorem 1 Let $\boldsymbol{x} = (x_1, x_2, \ldots, x_n) \in \mathbf{R}_+^n$ and $p \in \mathbf{R}$.

If $1 \leqslant p \leqslant 2$, then $L_p(\boldsymbol{x})$ is Schur-convex with $\boldsymbol{x} \in \mathbf{R}_+^n$, if $0 \leqslant p \leqslant 1$, then $L_p(\boldsymbol{x})$ is Schur-concave with $\boldsymbol{x} \in \mathbf{R}_+^n$.

Furthermore, Gu and Shi[11] proposed the following conjecture.

Conjecture 1 If $p \geqslant 2$, then $L_p(\boldsymbol{x})$ is Schur-convex with $\boldsymbol{x} \in \mathbf{R}_+^n$, if $p \leqslant 0$, then $L_p(\boldsymbol{x})$ is Schur-concave with $\boldsymbol{x} \in \mathbf{R}_+^n$.

We first point out that this conjecture does not hold.

In fact, for $n = 3, p = 3$, by computing, we have

$$\Delta := (x_1 - x_2) \left(\frac{\partial L_3(x)}{\partial x_1} - \frac{\partial L_3(x)}{\partial x_2} \right) = \frac{(x_1 - x_2)^2 \lambda(x)}{(x_1^2 + x_2^2 + x_3^2)^2}$$

where

$$\lambda(x) = \lambda(x_1, x_2, x_3) = 3(x_1 + x_2)(x_1^2 + x_2^2 + x_3^2) - 2(x_1^3 + x_2^3 + x_3^3)$$

if taking $x = (1, 3, 7)$, then $\lambda(x) = -34$, so that $\Delta < 0$, but taking $y = (1, 2, 3)$, then $\lambda(y) = 54$, so that $\Delta > 0$. According to Lemma 2.4 in second section, we assert that the Schur-convexity of $L_3(x_1, x_2, x_3)$ is not determined on the whole $\mathbf{R}_+^3$.

It can easily be shown that $L_{-2}(x_1, x_2, x_3) = \frac{1}{L_3(\frac{1}{x_1}, \frac{1}{x_2}, \frac{1}{x_3})}$, since the Schur-convexity of $L_3(x_1, x_2, x_3)$ is not determined on the whole $\mathbf{R}_+^3$, $L_{-2}(x_1, x_2, x_3)$ so does.

In this paper, we will study Schur-convexity, Schur-geometric convexity and Schur-harmonic convexity of $L_p(x)$ on certain subsets of $\mathbf{R}_+^n$. As consequences, some interesting inequalities are obtained.

Our main results are as follows:

Theorem 2 Let $x = (x_1, x_2, \ldots, x_n) \in \mathbf{R}_+^n, n \geqslant 2$ and $p \in \mathbf{R}$.

(I) If $p \geqslant 2$, then for any $a > 0$, $L_p(x)$ is Schur-convex with $x \in \left[\frac{(p-2)a}{p}, a \right]^n$;

(II) If $p < 0$, then for any $a > 0$, $L_p(x)$ is Schur-concave with $x \in \left[a, \frac{(p-2)a}{p} \right]^n$.

Theorem 3 Let $x = (x_1, x_2, \ldots, x_n) \in \mathbf{R}_+^n, n \geqslant 2$ and $p \in \mathbf{R}$.

(I) If $p < \frac{1}{2}$ and $p \neq 0$, then for any $a > 0$, $L_p(x)$ is Schur-geometrically concave with $x \in \left[a, (\frac{p-1}{p})^2 a \right]^n$;

(II) If $p > \frac{1}{2}$, then for any $a > 0$, $L_p(x)$ is Schur-geometrically convex with $x \in \left[(\frac{p-1}{p})^2 a, a \right]^n$;

(III) If $p = 0$, then $L_p(x)$ is Schur-geometrically convex with $x \in \mathbf{R}_+^n$.

Theorem 4 Let $x = (x_1, x_2, \ldots, x_n) \in \mathbf{R}_+^n, n \geqslant 2$ and $p \in \mathbf{R}$.

(I) If $0 \leqslant p \leqslant 1$, then $L_p(x)$ is Schur-harmonically convex with $x \in \mathbf{R}_+^n$, if $-1 \leqslant p \leqslant 0$, then $L_p(x)$ is Schur-harmonically concave with $x \in \mathbf{R}_+^n$;

(II) If $p > 1$, then for any $a > 0$, $L_p(\boldsymbol{x})$ is Schur-harmonically convex with $\boldsymbol{x} \in \left[\frac{(p-1)a}{p+1}, a \right]^n$;

(III) If $p < -1$, then for any $a > 0$, $L_p(\boldsymbol{x})$ is Schur-harmonically concave with $\boldsymbol{x} \in \left[a, \frac{(p-1)a}{p+1} \right]^n$.

74.2　Definitions and Lemmas

We need the following definitions and lemmas.

Definition 1 [17,27] Let $\boldsymbol{x} = (x_1, x_2, \ldots, x_n)$ and $\boldsymbol{y} = (y_1, y_2, \ldots, y_n) \in \mathbf{R}^n$.

(i) $\boldsymbol{x}$ is said to be majorized by $\boldsymbol{y}$ (in symbols $\boldsymbol{x} \prec \boldsymbol{y}$) if $\sum_{i=1}^{k} x_{[i]} \leqslant \sum_{i=1}^{k} y_{[i]}$
for $k = 1, 2, \ldots, n-1$ and $\sum_{i=1}^{n} x_i = \sum_{i=1}^{n} y_i$, where $x_{[1]} \geqslant \cdots \geqslant x_{[n]}$ and $y_{[1]} \geqslant \cdots \geqslant y_{[n]}$ are rearrangements of $\boldsymbol{x}$ and $\boldsymbol{y}$ in a descending order.

(ii) $\Omega \subset \mathbf{R}^n$ is called a convex set if $(\alpha x_1 + \beta y_1, \alpha x_2 + \beta y_2, \ldots, \alpha x_n + \beta y_n) \in \Omega$ for any $\boldsymbol{x}$ and $\boldsymbol{y} \in \Omega$, where α and $\beta \in [0, 1]$ with $\alpha + \beta = 1$.

(iii) Let $\Omega \subset \mathbf{R}^n$, $\varphi \colon \Omega \to \mathbf{R}$ is said to be a Schur-convex function on Ω if $\boldsymbol{x} \prec \boldsymbol{y}$ on Ω implies $\varphi(\boldsymbol{x}) \leq \varphi(\boldsymbol{y})$. φ is said to be a Schur-concave function on Ω if and only if $-\varphi$ is Schur-convex function.

Definition 2 [20,44] Let $\boldsymbol{x} = (x_1, x_2, \ldots, x_n)$ and $\boldsymbol{y} = (y_1, y_2, \ldots, y_n) \in \mathbf{R}_+^n$.

(i) $\Omega \subset \mathbf{R}_+^n$ is called a geometrically convex set if $(x_1^\alpha y_1^\beta, x_2^\alpha y_2^\beta, \ldots, x_n^\alpha y_n^\beta) \in \Omega$ for any $\boldsymbol{x}$ and $\boldsymbol{y} \in \Omega$, where α and $\beta \in [0, 1]$ with $\alpha + \beta = 1$.

(ii) Let $\Omega \subset \mathbf{R}_+^n$, $\varphi \colon \Omega \to \mathbf{R}_+$ is said to be a Schur-geometrically convex function on Ω if $(\ln x_1, \ln x_2, \ldots, \ln x_n) \prec (\ln y_1, \ln y_2, \ldots, \ln y_n)$ on Ω implies $\varphi(\boldsymbol{x}) \leqslant \varphi(\boldsymbol{y})$. φ is said to be a Schur-geometrically concave function on Ω if and only if $-\varphi$ is Schur-geometrically convex function.

Definition 3 [4,18] Let $\boldsymbol{x} = (x_1, x_2, \ldots, x_n)$ and $\boldsymbol{y} = (y_1, y_2, \ldots, y_n) \in \mathbf{R}_+^n$.

(i) A set $\Omega \subset \mathbf{R}_+^n$ is said to be a harmonically convex set if

$$\left(\frac{x_1 y_1}{\lambda x_1 + (1-\lambda)y_1}, \frac{x_2 y_2}{\lambda x_2 + (1-\lambda)y_2}, \ldots, \frac{x_n y_n}{\lambda x_n + (1-\lambda)y_n} \right) \in \Omega$$

for every $x, y \in \Omega$ and $\lambda \in [0, 1]$.

(ii) A function $\varphi : \Omega \to \mathbf{R}_+$ is said to be a Schur-harmonically convex function on Ω if $\left(\frac{1}{x_1}, \frac{1}{x_2}, \ldots, \frac{1}{x_n}\right) \prec \left(\frac{1}{y_1}, \frac{1}{y_2}, \ldots, \frac{1}{y_n}\right)$ implies $\varphi(x) \le \varphi(y)$. A function φ is said to be a Schur-harmonically concave function on Ω if and only if $-\varphi$ is a Schur-harmonically convex function.

Lemma 1 [17,27] Let $\Omega \subset \mathbf{R}^n$ is convex set, and has a nonempty interior set Ω^0. Let $\varphi : \Omega \to \mathbf{R}$ is continuous on Ω and differentiable in Ω^0. Then φ is the $Schur - convex(or Schur - concave, resp.) function$, if and only if it is symmetric on Ω and if

$$(x_1 - x_2)\left(\frac{\partial \varphi(x)}{\partial x_1} - \frac{\partial \varphi(x)}{\partial x_2}\right) \ge 0 (\text{or} \le 0, \text{resp.}) \tag{1}$$

holds for any $x = (x_1, x_2, \cdots, x_n) \in \Omega^0$.

Remark 2 [9,19] It is easy to see that the condition (1) is equivalent to

$$\frac{\partial \varphi(x)}{\partial x_i} \le (\text{or} \ge, \text{resp.}) \frac{\partial \varphi(x)}{\partial x_{i+1}}, \quad i = 1, \ldots, n - 1, \quad \text{for all } x \in \mathbf{D} \cap \Omega$$

where $\mathbf{D} = \{x : x_1 \le x_2 \le \cdots \le x_n\}$.

The condition (1) is also equivalent to

$$\frac{\partial \varphi(x)}{\partial x_i} \ge (\text{or} \le, \text{resp.}) \frac{\partial \varphi(x)}{\partial x_{i+1}}, \quad i = 1, \ldots, n - 1, \quad \text{for all } x \in \mathbf{E} \cap \Omega$$

where $\mathbf{E} = \{x : x_1 \ge x_2 \ge \cdots \ge x_n\}$.

Lemma 2 [20,44] Let $\Omega \subset \mathbf{R}_+^n$ be a symmetric geometrically convex set with a nonempty interior Ω^0. Let $\varphi : \Omega \to \mathbf{R}_+$ be continuous on Ω and differentiable on Ω^0. Then φ is a Schur-geometrically convex (or Schur-geometrically concave, resp.) function if and only if φ is symmetric on Ω and

$$(x_1 - x_2)\left(x_1 \frac{\partial \varphi(x)}{\partial x_1} - x_2 \frac{\partial \varphi(x)}{\partial x_2}\right) \ge 0 \quad (\text{or} \le 0, \text{resp.}) \tag{2}$$

holds for any $x = (x_1, x_2, \cdots, x_n) \in \Omega^0$.

Remark 2 It is easy to see that the condition (2) is equivalent to

$$x_i \frac{\partial \varphi(x)}{\partial x_i} \le (\text{or} \ge, \text{resp.}) x_{i+1} \frac{\partial \varphi(x)}{\partial x_{i+1}}, \quad i = 1, \ldots, n - 1, \quad \text{for all } x \in \mathbf{D} \cap \Omega$$

where $\mathbf{D} = \{x : x_1 \le x_2 \le \cdots \le x_n\}$.

The condition (2) is also equivalent to

$$x_i \frac{\partial \varphi(\boldsymbol{x})}{\partial x_i} \geqslant (\text{or} \leqslant, \text{ resp.}) \ x_{i+1} \frac{\partial \varphi(\boldsymbol{x})}{\partial x_{i+1}}, \quad i = 1, \ldots, n-1, \quad \text{for all } \boldsymbol{x} \in \mathbf{D} \cap \Omega$$

where $\mathbf{E} = \{\boldsymbol{x} : x_1 \geqslant x_2 \geqslant \cdots \geqslant x_n\}$.

Lemma 3 [4,18] Let $\Omega \subset \mathbf{R}_+^n$ be a symmetric harmonically convex set with a nonempty interior Ω^0. Let $\varphi : \Omega \to \mathbf{R}_+$ be continuous on Ω and differentiable on Ω^0. Then φ is a Schur- harmonically convex (or Schur-harmonically concave, $resp.$) function if and only if φ is symmetric on Ω and

$$(x_1 - x_2) \left(x_1^2 \frac{\partial \varphi(\boldsymbol{x})}{\partial x_1} - x_2^2 \frac{\partial \varphi(\boldsymbol{x})}{\partial x_2} \right) \geqslant 0 \quad (\text{or} \leqslant 0, \text{resp.}) \tag{3}$$

holds for any $\boldsymbol{x} = (x_1, x_2, \cdots, x_n) \in \Omega^0$.

Remark 2 It is easy to see that the condition (3) is equivalent to

$$x_i^2 \frac{\partial \varphi(\boldsymbol{x})}{\partial x_i} \leqslant (\text{or} \geqslant, \text{ resp.}) \ x_{i+1}^2 \frac{\partial \varphi(\boldsymbol{x})}{\partial x_{i+1}}, \quad i = 1, \ldots, n-1, \quad \text{for all } \boldsymbol{x} \in \mathbf{D} \cap \Omega$$

where $\mathbf{D} = \{\boldsymbol{x} : x_1 \leqslant x_2 \leqslant \cdots \leqslant x_n\}$.

The condition (3) is also equivalent to

$$x_i^2 \frac{\partial \varphi(\boldsymbol{x})}{\partial x_i} \geqslant (\text{or} \leqslant, \text{ resp.}) \ x_{i+1}^2 \frac{\partial \varphi(\boldsymbol{x})}{\partial x_{i+1}}, \quad i = 1, \ldots, n-1, \quad \text{for all } \boldsymbol{x} \in \mathbf{E} \cap \Omega,$$

where $\mathbf{E} = \{\boldsymbol{x} : x_1 \geqslant x_2 \geqslant \cdots \geqslant x_n\}$.

Lemma 3 Let $x_1 \geqslant x_2 \geqslant \cdots \geqslant x_n > 0, m \in \mathbf{R}$. Then

$$x_1 \geqslant \frac{x_1^m + x_2^m + \cdots + x_n^m}{x_1^{m-1} + x_2^{m-1} + \cdots + x_n^{m-1}} \geqslant x_n$$

Proof

$$x_1(x_1^{m-1} + x_2^{m-1} + \cdots + x_n^{m-1}) - (x_1^m + x_2^m + \cdots + x_n^m)$$
$$= x_1^{m-1}(x_1 - x_1) + x_2^{m-1}(x_1 - x_2) + \cdots + x_n^{m-1}(x_1 - x_n) \geqslant 0$$

$$x_n(x_1^{m-1} + x_2^{m-1} + \cdots + x_n^{m-1}) - (x_1^m + x_2^m + \cdots + x_n^m)$$
$$= x_1^{m-1}(x_n - x_1) + x_2^{m-1}(x_n - x_2) + \cdots + x_n^{m-1}(x_n - x_n) \leqslant 0$$

We have thus proved the Lemma 3.

Lemma 4 [17] Let $\boldsymbol{x} = (x_1, x_2, \cdots, x_n) \in \mathbf{R}_+^n$ and $A_n(\boldsymbol{x}) = \frac{1}{n}\sum\limits_{i=1}^{n} x_i$. Then

$$\boldsymbol{u} = \Big(\underbrace{A_n(\boldsymbol{x}), A_n(\boldsymbol{x}), \cdots, A_n(\boldsymbol{x})}_{n} \Big) \prec (x_1, x_2, \cdots, x_n) = \boldsymbol{x}$$

74.3 Proofs of Theorems

Proofs of Theorems 2 Straightforward computation gives

$$\frac{\partial L_p(\boldsymbol{x})}{\partial x_i} = \frac{px_i^{p-1}\sum\limits_{j=1}^{n} x_j^{p-1} - (p-1)x_i^{p-2}\sum\limits_{j=1}^{n} x_j^p}{(\sum\limits_{j=1}^{n} x_j^{p-1})^2}, i = 1, 2, \ldots, n \qquad (4)$$

and then

$$\frac{\partial L_p(\boldsymbol{x})}{\partial x_i} - \frac{\partial L_p(\boldsymbol{x})}{\partial x_{i+1}} = \frac{f_i(\boldsymbol{x})}{(\sum\limits_{i=1}^{n} x_i^{p-1})^2}, i = 1, 2, \ldots, n-1$$

where

$$f_i(\boldsymbol{x}) = p(x_i^{p-1} - x_{i+1}^{p-1})\sum\limits_{j=1}^{n} x_j^{p-1} - (p-1)(x_i^{p-2} - x_{i+1}^{p-2})\sum\limits_{j=1}^{n} x_j^p$$

It is clear that $L_p(\boldsymbol{x})$ is symmetric with $\boldsymbol{x} \in \mathbf{R}_+^n$. Without loss of generality, we may assume that $x_1 \geqslant x_2 \geqslant \cdots \geqslant x_n > 0$.

For any $a > 0$, according to the integral mean value theorem, there is a ξ lies between x_i and x_{i+1}, such that

$$p(x_i^{p-1} - x_{i+1}^{p-1}) - a(p-1)(x_i^{p-2} - x_{i+1}^{p-2})$$

$$= (p-1)p\int_{x_{i+1}}^{x_i} x^{p-2}\mathrm{d}x - a(p-2)(p-1)\int_{x_{i+1}}^{x_i} x^{p-3}\mathrm{d}x$$

$$= (p-1)\int_{x_{i+1}}^{x_i} [px^{p-2} - a(p-2)x^{p-3}]\mathrm{d}x$$

$$= (p-1)[p\xi^{p-2} - a(p-2)\xi^{p-3}](x_i - x_{i+1})$$

$$= (p-1)p\xi^{p-3}\left(\xi - \frac{(p-2)a}{p}\right)(x_i - x_{i+1}) \qquad (5)$$

Proof of (I)　When $p \geqslant 2$ and $a \geqslant x_1 \geqslant x_2 \geqslant \cdots \geqslant x_n \geqslant \frac{(p-2)a}{p} > 0$, from (5), we have

$$p(x_i^{p-1} - x_{i+1}^{p-1}) - a(p-1)(x_i^{p-2} - x_{i+1}^{p-2}) \geqslant 0$$

this is

$$\frac{p(x_i^{p-1} - x_{i+1}^{p-1})}{(p-1)(x_i^{p-2} - x_{i+1}^{p-2})} \geqslant a$$

and then from Lemma 3, it follows that

$$\frac{p(x_i^{p-1} - x_{i+1}^{p-1})}{(p-1)(x_i^{p-2} - x_{i+1}^{p-2})} \geqslant x_1 \geqslant \frac{\sum\limits_{j=1}^{n} x_j^p}{\sum\limits_{j=1}^{n} x_j^{p-1}}$$

namely, $f_i(\boldsymbol{x}) \geqslant 0$, and then $\frac{\partial L_p(\boldsymbol{x})}{\partial x_i} \geqslant \frac{\partial L_p(\boldsymbol{x})}{\partial x_{i+1}}$. By Lemma 1 and Remark 1, it follows that $L_p(\boldsymbol{x})$ is Schur-convex with $\boldsymbol{x} \in \left[\frac{p-2}{p}a, a\right]^n$.

Proof of (II)　When $p < 0$ and $\frac{(p-2)a}{p} \geqslant x_1 \geqslant x_2 \geqslant \cdots \geqslant x_n \geqslant a > 0$ from (5), we have

$$p(x_i^{p-1} - x_{i+1}^{p-1}) - a(p-1)(x_i^{p-2} - x_{i+1}^{p-2}) \leqslant 0$$

this is

$$\frac{p(x_i^{p-1} - x_{i+1}^{p-1})}{(p-1)(x_i^{p-2} - x_{i+1}^{p-2})} \leqslant a$$

and then from Lemma 3, it follows that

$$\frac{p(x_i^{p-1} - x_{i+1}^{p-1})}{(p-1)(x_i^{p-2} - x_{i+1}^{p-2})} \leqslant x_n \leqslant \frac{\sum\limits_{j=1}^{n} x_j^p}{\sum\limits_{j=1}^{n} x_j^{p-1}}$$

namely, $f_i(\boldsymbol{x}) \leqslant 0$, and then $\frac{\partial L_p(\boldsymbol{x})}{\partial x_i} \leqslant \frac{\partial L_p(\boldsymbol{x})}{\partial x_{i+1}}$. By Lemma 1 and Remark 1, it follows that $L_p(\boldsymbol{x})$ is Schur-concave with $\boldsymbol{x} \in \left[a, \frac{p-2}{p}a\right]^n$.

The proof of Theorem 2 is complete.

Proofs of Theorems 3　From (4), we have

$$x_i \frac{\partial L_p(\boldsymbol{x})}{\partial x_i} - x_{i+1} \frac{\partial L_p(\boldsymbol{x})}{\partial x_{i+1}} = \frac{g_i(\boldsymbol{x})}{(\sum\limits_{i=1}^{n} x_i^{p-1})^2}, i = 1, 2, \ldots, n-1$$

where

$$g_i(\boldsymbol{x}) = p(x_i^p - x_{i+1}^p) \sum_{j=1}^{n} x_j^{p-1} - (p-1)(x_i^{p-1} - x_{i+1}^{p-1}) \sum_{j=1}^{n} x_j^p$$

It is clear that $L_p(\boldsymbol{x})$ is symmetric with $\boldsymbol{x} \in \mathbf{R}_+^n$. Without loss of generality, we may assume that $x_1 \geqslant x_2 \geqslant \cdots \geqslant x_n > 0$.

For any $a > 0$, according to the integral mean value theorem, there is a ξ lies between x_i and x_{i+1}, such that

$$p(x_i^p - x_{i+1}^p) - a(p-1)(x_i^{p-1} - x_{i+1}^{p-1})$$
$$= p^2 \int_{x_{i+1}}^{x_i} x^{p-1} dx - a(p-1)^2 \int_{x_{i+1}}^{x_i} x^{p-2} dx$$
$$= \int_{x_{i+1}}^{x_i} [p^2 x^{p-1} - a(p-1)^2 x^{p-2}] dx$$
$$= [p^2 \xi^{p-1} - a(p-1)^2 \xi^{p-2}](x_i - x_{i+1})$$
$$= p^2 \xi^{p-2} \left[\xi - (\frac{p-1}{p})^2 a \right] (x_i - x_{i+1}) \tag{6}$$

Proof of (I)　When $p \geqslant \frac{1}{2}$ and $a \geqslant x_1 \geqslant x_2 \geqslant \cdots \geqslant x_n \geqslant (\frac{p-1}{p})^2 a > 0$, from (6), we have

$$p(x_i^p - x_{i+1}^p) - a(p-1)(x_i^{p-1} - x_{i+1}^{p-1}) \geqslant 0$$

this is

$$\frac{p(x_i^p - x_{i+1}^p)}{(p-1)(x_i^{p-1} - x_{i+1}^{p-1})} \geqslant a$$

and then from Lemma 3, it follows that

$$\frac{p(x_i^p - x_{i+1}^p)}{(p-1)(x_i^{p-1} - x_{i+1}^{p-1})} \geqslant x_1 \geqslant \frac{\sum_{j=1}^{n} x_j^p}{\sum_{j=1}^{n} x_j^{p-1}}$$

namely, $g_i(\boldsymbol{x}) \geqslant 0$, and then $x_i \frac{\partial L_p(\boldsymbol{x})}{\partial x_i} \geqslant x_{i+1} \frac{\partial L_p(\boldsymbol{x})}{\partial x_{i+1}}$. By Lemma 2 and Remark 2, it follows that $L_p(\boldsymbol{x})$ is Schur-geometrically convex with $\boldsymbol{x} \in \left[(\frac{p-1}{p})^2 a, a \right]^n$.

Proof of (II)　When $p < \frac{1}{2}, p \neq 0$ and $(\frac{p-1}{p})^2 a \geqslant x_1 \geqslant x_2 \geqslant \cdots \geqslant x_n \geqslant$

$a > 0$, from (6), we have

$$p(x_i^p - x_{i+1}^p) - a(p-1)(x_i^{p-1} - x_{i+1}^{p-1}) \leqslant 0$$

this is

$$\frac{p(x_i^p - x_{i+1}^p)}{(p-1)(x_i^{p-1} - x_{i+1}^{p-1})} \leqslant a$$

and then from Lemma 3, it follows that

$$\frac{p(x_i^p - x_{i+1}^p)}{(p-1)(x_i^{p-1} - x_{i+1}^{p-1})} \leqslant x_n \leqslant \frac{\sum\limits_{j=1}^{n} x_j^p}{\sum\limits_{j=1}^{n} x_j^{p-1}}$$

namely, $g_i(\boldsymbol{x}) \leqslant 0$, and then $x_i \frac{\partial L_p(\boldsymbol{x})}{\partial x_i} \leqslant x_{i+1} \frac{\partial L_p(\boldsymbol{x})}{\partial x_{i+1}}$. By Lemma 2 and Remark 2, it follows that $L_p(\boldsymbol{x})$ is Schur-geometrically concave with $\boldsymbol{x} \in \left[a, (\frac{p-1}{p})^2 a\right]^n$.

Proof of (III)　When $p = 0$, $g_i(\boldsymbol{x}) \leqslant 0$, it follows that $L_p(\boldsymbol{x})$ is Schur-geometrically concave with $\boldsymbol{x} \in \mathbf{R}_+^n$.

The proof of Theorem 2 is complete.

Proofs of Theorems 4　From (4), we have

$$x_i^2 \frac{\partial L_p(\boldsymbol{x})}{\partial x_i} - x_{i+1}^2 \frac{\partial L_p(\boldsymbol{x})}{\partial x_{i+1}} = \frac{h_i(\boldsymbol{x})}{(\sum\limits_{i=1}^{n} x_i^{p-1})^2}, i = 1, 2, \ldots, n-1 \qquad (7)$$

where

$$h_i(\boldsymbol{x}) = p(x_i^{p+1} - x_{i+1}^{p+1}) \sum_{j=1}^{n} x_j^{p-1} - (p-1)(x_i^p - x_{i+1}^p) \sum_{j=1}^{n} x_j^p$$

It is clear that $L_p(\boldsymbol{x})$ is symmetric with $\boldsymbol{x} \in \mathbf{R}_+^n$. Without loss of generality, we may assume that $x_1 \geqslant x_2 \geqslant \cdots \geqslant x_n > 0$.

Proof of (I)　According to the integral mean value theorem, there is a ξ lies between x_i and x_{i+1}, such that

$$h_i(\boldsymbol{x}) = \sum_{j=1}^{n} x_j^{p-1} \left[p(x_i^{p+1} - x_{i+1}^{p+1}) - (p-1)(x_i^p - x_{i+1}^p) \frac{\sum\limits_{j=1}^{n} x_j^p}{\sum\limits_{j=1}^{n} x_j^{p-1}} \right]$$

$$= \sum_{j=1}^{n} x_j^{p-1} \left[(p+1)p \int_{x_{i+1}}^{x_i} x^p \mathrm{d}x - p(p-1) \frac{\sum\limits_{j=1}^{n} x_j^p}{\sum\limits_{j=1}^{n} x_j^{p-1}} \int_{x_{i+1}}^{x_i} x^{p-1} \mathrm{d}x \right]$$

$$= \sum_{j=1}^{n} x_j^{p-1} p \int_{x_{i+1}}^{x_i} \left[(p+1)x^p - (p-1) \frac{\sum\limits_{j=1}^{n} x_j^p}{\sum\limits_{j=1}^{n} x_j^{p-1}} x^{p-1} \right] \mathrm{d}x$$

$$= \sum_{j=1}^{n} x_j^{p-1} p \left[(p+1)\xi^p - (p-1) \frac{\sum\limits_{j=1}^{n} x_j^p}{\sum\limits_{j=1}^{n} x_j^{p-1}} \xi^{p-1} \right] (x_i - x_{i+1})$$

$$= \sum_{j=1}^{n} x_j^{p-1} (p+1)p\xi^{p-1} \left[\xi - \frac{p-1}{p+1} \frac{\sum\limits_{j=1}^{n} x_j^p}{\sum\limits_{j=1}^{n} x_j^{p-1}} \right] (x_i - x_{i+1}) \qquad (8)$$

Notice that for $-1 < p \leqslant 1$, $\xi - \frac{p-1}{p+1} \frac{\sum\limits_{j=1}^{n} x_j^p}{\sum\limits_{j=1}^{n} x_j^{p-1}} \geqslant 0$.

When $0 < p \leqslant 1$, from (8), we have $h_i(\boldsymbol{x}) \geqslant 0$, and then $x_i^2 \frac{\partial L_p(\boldsymbol{x})}{\partial x_i} \geqslant x_{i+1}^2 \frac{\partial L_p(\boldsymbol{x})}{\partial x_{i+1}}$. By Lemma 3 and Remark 2, it follows that $L_p(\boldsymbol{x})$ is Schur-harmonically convex with $\boldsymbol{x} \in \mathbf{R}_+^n$.

When $-1 < p \leqslant 0$, $h_i(\boldsymbol{x}) \leqslant 0$, and then $x_i^2 \frac{\partial L_p(\boldsymbol{x})}{\partial x_i} \leqslant x_{i+1}^2 \frac{\partial L_p(\boldsymbol{x})}{\partial x_{i+1}}$. By Lemma 3 and Remark 2, it follows that $L_p(\boldsymbol{x})$ is Schur-harmonically concave with $\boldsymbol{x} \in \mathbf{R}_+^n$.

When $p = -1$, $h_i(\boldsymbol{x}) = 2 \sum\limits_{j=1}^{n} x_j^{-1}(x_i^{-1} - x_{i+1}^{-1}) \leqslant 0$, it follows that $L_p(\boldsymbol{x})$ is Schur-harmonically concave with $\boldsymbol{x} \in \mathbf{R}_+^n$.

Proof of (II) For any $a > 0$, according to the integral mean value theorem, there is a ξ lies between x_i and x_{i+1}, such that

$$p(x_i^{p+1} - x_{i+1}^{p+1}) - a(p-1)(x_i^p - x_{i+1}^p)$$

$$= p(p+1) \int_{x_{i+1}}^{x_i} x^p dx - a(p-1)p \int_{x_{i+1}}^{x_i} x^{p-1} dx$$

$$= p \int_{x_{i+1}}^{x_i} [(p+1)x^p - a(p-1)x^{p-1}] dx$$

$$= p[(p+1)\xi^p - a(p-1)\xi^{p-1}](x_i - x_{i+1})$$

$$= p(p+1)\xi^{p-1} \left[\xi - \frac{(p-1)a}{p+1} \right] (x_i - x_{i+1}) \qquad (9)$$

When $p \geqslant 1$ and $a \geqslant x_1 \geqslant x_2 \geqslant \cdots \geqslant x_n \geqslant \frac{p-1}{p+1}a > 0$, from (9), we have

$$p(x_i^{p+1} - x_{i+1}^{p+1}) - a(p-1)(x_i^p - x_{i+1}^p) \geqslant 0$$

this is

$$\frac{p(x_i^{p+1} - x_{i+1}^{p+1})}{(p-1)(x_i^p - x_{i+1}^p)} \geqslant a$$

and then from Lemma 3, it follows that

$$\frac{p(x_i^{p+1} - x_{i+1}^{p+1})}{(p-1)(x_i^p - x_{i+1}^p)} \geqslant x_1 \geqslant \frac{\sum\limits_{j=1}^{n} x_j^p}{\sum\limits_{j=1}^{n} x_j^{p-1}}$$

namely, $h_i(\boldsymbol{x}) \geqslant 0$, and then $x_i^2 \frac{\partial L_p(\boldsymbol{x})}{\partial x_i} \geqslant x_{i+1}^2 \frac{\partial L_p(\boldsymbol{x})}{\partial x_{i+1}}$. By Lemma 3 and Remark 2, it follows that $L_p(\boldsymbol{x})$ is Schur-harmonically convex with $\boldsymbol{x} \in \left[\frac{p-1}{p+1}a, a\right]^n$.

Proof of (III)　When $p < -1$ and $\frac{p-1}{p+1}a \geqslant x_1 \geqslant x_2 \geqslant \cdots \geqslant x_n \geqslant a > 0$, from (9), we have

$$p(x_i^{p+1} - x_{i+1}^{p+1}) - a(p-1)(x_i^p - x_{i+1}^p) \leqslant 0$$

this is

$$\frac{p(x_i^{p+1} - x_{i+1}^{p+1})}{(p-1)(x_i^p - x_{i+1}^p)} \leqslant a$$

and then from Lemma 3, it follows that

$$\frac{p(x_i^{p+1} - x_{i+1}^{p+1})}{(p-1)(x_i^p - x_{i+1}^p)} \leqslant x_n \leqslant \frac{\sum\limits_{j=1}^{n} x_j^p}{\sum\limits_{j=1}^{n} x_j^{p-1}}$$

namely, $h_i(\boldsymbol{x}) \leqslant 0$, and then $x_i^2 \frac{\partial L_p(\boldsymbol{x})}{\partial x_i} \leqslant x_{i+1}^2 \frac{\partial L_p(\boldsymbol{x})}{\partial x_{i+1}}$. By Lemma 3 and Remark 2, it follows that $L_p(\boldsymbol{x})$ is Schur-harmonically concave with $\boldsymbol{x} \in \left[a, \frac{p-1}{p+1}a\right]^n$.

The proof of Theorem 4 is complete.

74.4 Applications

Theorem 5 For any $a > 0$, if $p \geqslant 2$ and $\boldsymbol{x} = (x_1, x_2, \ldots, x_n) \in \left[\frac{p-2}{p}a, a\right]^n$, then we have

$$A_n(\boldsymbol{x}) \geqslant L_p(\boldsymbol{x}) \tag{10}$$

if $p < 0$ and $\boldsymbol{x} \in \left[a, \frac{p-2}{p}a\right]^n$, then the inequality (10) is reversed.

Proof If $p \geqslant 2$ and $\boldsymbol{x} \in \left[\frac{p-2}{p}a, a\right]^n$, then by Theorem 2, from Lemma 4, we have

$$L_p(\boldsymbol{u}) \geqslant L_p(\boldsymbol{x})$$

rearranging gives (10), if $p < 0$ and $\boldsymbol{x} \in \left[a, \frac{p-2}{p}a\right]^n$, then the inequality (10) is reversed.

The proof is complete.

Theorem 6 For any $a > 0$, if $p > \frac{1}{2}$ and $\boldsymbol{x} = (x_1, x_2, \ldots, x_n) \in \left[(\frac{p-1}{p})^2 a, a\right]^n$, then we have

$$G_n(\boldsymbol{x}) \leqslant L_p(\boldsymbol{x}) \tag{11}$$

where $G_n(\boldsymbol{x}) = \sqrt[n]{x_1 x_2 \cdots x_n}$ is the geometric mean of $\boldsymbol{x}$.

If $p < \frac{1}{2}, p \neq 0$ and $\boldsymbol{x} \in \left[a, (\frac{p-1}{p})^2 a\right]^n$, then the inequality (11) is reversed.

Proof By Lemma 4, we have

$$\left(\underbrace{\log G_n(\boldsymbol{x}), \cdots, \log G_n(\boldsymbol{x})}_{n}\right) \prec (\log x_1, \log x_2, \cdots, \log x_n)$$

if $p > \frac{1}{2}$ and $\boldsymbol{x} \in \left[(\frac{p-1}{p})^2 a, a\right]^n$, by Theorem 3, it follows

$$L_p\left(\underbrace{G_n(\boldsymbol{x}), \cdots, G_n(\boldsymbol{x})}_{n}\right) \leqslant L_p(x_1, x_2, \cdots, x_n)$$

rearranging gives (11). If $p < \frac{1}{2}, p \neq 0$ and $\boldsymbol{x} \in \left[a, (\frac{p-1}{p})^2 a\right]^n$, then the inequality (11) is reversed.

The proof is complete.

Theorem 7 For any $a > 0$, if $p > 1$ and $\boldsymbol{x} \in \left[\frac{p-1}{p+1}a, a\right]^n$, then we have

$$H_n(\boldsymbol{x}) \leqslant L_p(\boldsymbol{x}) \tag{12}$$

where $H_n(\boldsymbol{x}) = \frac{n}{\sum\limits_{i=1}^{n}\frac{1}{x_i}}$ is the harmonic mean of $\boldsymbol{x}$. If $p < -1$ and $\boldsymbol{x} \in \left[a, \frac{p-1}{p+1}a\right]^n$, then the inequality (12) is reversed.

Proof By Lemma 4, we have

$$\left(\underbrace{\frac{1}{H_n(\boldsymbol{x})}, \cdots, \frac{1}{H_n(\boldsymbol{x})}}_{n}\right) \prec \left(\frac{1}{x_1}, \frac{1}{x_2}, \cdots, \frac{1}{x_n}\right)$$

if $p > 1$ and $\boldsymbol{x} \in \left[\frac{p-1}{p+1}a, a\right]^n$, by Theorem 4, it follows

$$L_p\left(\underbrace{H_n(\boldsymbol{x}), \cdots, H_n(\boldsymbol{x})}_{n}\right) \leqslant L_p\left(x_1, x_2, \cdots, x_n\right)$$

rearranging gives (12), if $p < -1$ and $\boldsymbol{x} \in \left[a, \frac{p-1}{p+1}a\right]^n$, then the inequality (12) is reversed.

The proof is complete.

In recent years, the study on the properties of the mean by using theory of majorization is unusually active, interested readers may refer to the literature [5-9, 15, 16, 19, 22, 24, 25, 32-35, 37-43, 45].

Acknowledgements

The authors are indebted to the referees for their helpful suggestions.

References

[1] H ALZER, ÜBER. Lehmers Mittelwertfamilie[J]. Elem. Math., 1988, 43(2): 50–54.

[2] H ALZER. Bestmögliche Abschätzungen für spezielle Mittelwerte[J]. Zb. Rad. Prirod. -Mat. Fak. Ser. Mat., 1993, 23(1): 331–346.

[3] E F BECKENBACH. A class of mean value functions[J]. Amer. Math. Monthly, 1950, 57: 1–6.

[4] Y -M CHU, G -D WANG, X-H ZHANG. The Schur multiplicative and harmonic convexities of the complete symmetric function[J]. Mathematische Nachrichten, 2011, 284 (5-6): 653–663.

[5] Y -M CHU, W -F XIA. Necessary and sufficient conditions for the Schur harmonic convexity of the generalized Muirhead mean[J]. Proceedings of A. Razmadze Mathematical Institute, 2010,152: 19–27.

[6] Y -M CHU, X -M ZHANG. Necessary and sufficient conditions such that extended mean values are Schur-convex or Schur-concave[J]. Journal of Mathematics of Kyoto University, 2008, 48(1): 229–238.

[7] Y-M CHU, X M ZHANG, G-D WANG. The Schur geometrical convexity of the extended mean values[J]. J. Convex Anal., 2008, 15: 707-718.

[8] L -L FU, B-Y XI, H M SRIVASTAVA. Schur-convexity of the generalized Heronian means involving two positive numbers[J]. Taiwanese Journal of Mathematics, 2011, 15 (6): 2721–2731.

[9] W-M GONG, X -H SHEN, Y -M CHU. The Schur convexity for the generalized Muirhead mean[J]. J. Math. Inequal., 2014, 8 (4): 855–862.

[10] H W GOULD, M E MAYS. Series expansions of means[J]. J. Math. Anal. Appl.,1984, 101(2): 611–621.

[11] C GU, H -N SHI. Schur-convexity and Schur-geometric concavity of Lehmer means[J]. Mathematics in practice and theory, 2009, 39(12) : 183–188.

[12] Z -J GUO, X -H SHEN, Y -M CHU. The best possible Lehmer mean bounds for a convex combination of logarithmic and harmonic means [J]. International Mathematical Forum, 2013, 8(31): 1539 –1551.

[13] D H Lehmer. On the compounding of certain means[J]. J. Math. Anal. Appl., 1971, 36: 183–200.

[14] Z LIU. Remark on inequalities between Hölder and Lehmer means[J]. J. Math. Anal. Appl., 2000, 247 (1): 309–313.

[15] V LOKESHA, B NAVEEN KUMAR, K M NAGSRAJA, S PADMAN-ABHAN. Schur geometric convexity for ratio of difference of means[J]. Journal of Scientific Research & Reports, 2014, 3(9): 1211–1219.

[16] V LOKESHA, K M NAGSRAJA, NAVEEN, KUMAR B, Y-D WU. Schur convexity of Gnan mean for two variables[J]. Notes Number Theory Discrete Math., 2011, 17(4): 37-41.

[17] A M MARSHALL, I OLKIN. Inequalities:theory of majorization and its application[M]. New York : Academies Press, 1979.

[18] J -X MENG, Y M CHU, X -M Tang. The Schur-harmonic-convexity of dual form of the Hamy symmetric function[J]. Matematički Vesnik, 2010, 62 (1): 37–46.

[19] K M NAGARAJA, SUDHIR KUMAR SAHU. Schur harmonic convexity of Stolarsky extended mean values[J]. Scientia Magna, 2013, 9 (2): 18–29.

[20] C P NICULESCU. Convexity according to the geometric mean[J]. Math. Inequal. Appl., 2000,3 (2): 155–167.

[21] Z PÁLES. Inequalities for sums of powers[J]. J. Math. Anal. Appl., 1988, 131(1): 265–270.

[22] F QI, J SÁNDOR, S S DRAGOMIR, A SOFO. Notes on the Schur-convexity of the extended mean values[J]. Taiwanese J. Math., 2005, 9 (3): 411–420.

[23] Y-F QIU, M -K WANG, Y -M CHU G -D WANG. Two sharp inequalities for Lehmer mean, identric mean and logarithmic mean[J]. J. Math. Inequal., 2011, 5 (3): 301–306.

[24] J SÁNDOR. The Schur-convexity of Stolarsky and Gini means[J]. Banach J. Math. Anal., 2007, 1 (2): 212–215.

[25] H-N SHI, Y -M JIANG, W -D JIANG. Schur-convexity and Schur-geometrically concavity of Gini mean[J]. Computers and Mathematics with Applications, 2009(57): 266–274.

[26] K B STOLARSKY. Hölder means, Lehmer means, and $x^{-1}\log\cosh x$[J]. J. Math. Anal. Appl., 1996, 202 (3): 810–818.

[27] B-Y WANG. Foundations of majorization inequalities[M]. Beijing: Beijing Normal Univ. Press, 1990.

[28] M -K WANG, Y -M CHU, G. -D. WANG. A sharp double inequality between the Lehmer and arithmetic-geometric means[J]. Pac. J. Appl. Math, 2012, 4 (1): 1–25.

[29] M-K WANG, Y -F QIU, Y -M CHU. Sharp bounds for seiffert means in terms of Lehmer means[J]. J. Math. Inequal., 2010, 4(4):581–586.

[30] S R WASSELL. Rediscovering a family of means[J]. Math. Intelligencer, 2002, 24 (2): 58–65.

[31] A WITKOWSKI. Convexity of weighted Stolarsky means[J]. J. Inequal. Pure Appl. Math., 2006, 7(2), Article 73, 6 pp.(electronic).

[32] A WITKOWSKI. On Schur convexity and Schur-geometrical convexity of four-parameter family of means[J]. Math. Inequal. Appl.,2011, 14 (4): 897–903.

[33] Y WU, F QI. Schur-harmonic convexity for differences of some means [J]. Analysis (Munich),2012, 32: 263-270.

[34] Y WU, F QI, H-N SHI. Schur-harmonic convexity for differences of some special means in two variables[J]. J. Math. Inequal., 2014, 8: 321-330.

[35] W -F XIA, Y -M CHU. The Schur convexity of the weighted generalized logarithmic mean values according to harmonic mean[J]. International Journal of Modern Mathematics, 2009, 4(3): 225–233.

[36] W -F XIA, Y -M CHU. The Schur harmonic convexity of Lehmer means[J]. International Mathematical Forum, 2009, (41): 2009–2015.

[37] W -F XIA, Y -M CHU. The Schur multiplicative convexity of the generalized Muirhead mean[J]. International Journal of Functional Analysis, Operator Theory and Applications, 2009(1) : 1–8.

[38] W-F XIA, Y-M CHU. The Schur convexity of Gini mean values in the sense of harmonic mean[J]. Acta Mathematica Scientia, 2011, 31B(3): 1103–1112.

[39] W -F XIA, Y -M CHU, G-D WANG. Necessary and sufficient conditions for the Schur harmonic convexity or concavity of the extended mean values[J]. Revista De La Uniòn Matemática Argentina, 2010, 51 (2): 121–132.

[40] Z -H YANG. Necessary and sufficient conditions for Schur geometrical convexity of the four-parameter homogeneous means[J]. Abstr. Appl. Anal., 2010(2010), Article ID 830163, 16 pages doi: 10.1155/2010/830163.

[41] Z -H YANG. Schur harmonic convexity of Gini means[J]. International Mathematical Forum, 2011,6 (16): 747–762.

[42] Z -H YANG. Schur power convexity of Stolarsky means[J]. Publ. Math. Debrecen, 2012, 80 (1-2): 43–66.

[43] H-P YIN, H -N SHI, F QI. On Schur m-power convexity for ratios of some means[J]. J.Math. Inequal., 2015, 9(1): 145–153.

[44] X -M ZHANG. Geometrically convex functions[M]. Hefei: An hui University Press, 2004.

[**45**] T -Y ZHANG, A -P JI. Schur-convexity of generalized Heronian mean[J]. Communications in computer and information science, 2011, 244, Information Computing and Applications, Part 1, pages 25–33.

第75篇　Compositions Involving Schur Harmonically Convex Functions

(SHI HUAN-NAN, ZHANG JING. Journal of Computational Analysis

and Applications, 2017, 22 (5): 907-922)

Abstract: The decision theorem of the Schur harmonic convexity for the compositions involving Schur harmonically convex functions is established and used to determine the Schur harmonic convexity of some symmetric functions.

Keywords: Schur harmonically convex function; harmonically convex function; composite function; symmetric function

75.1　Introduction

Throughout the article, $\mathbf{R}$ denotes the set of real numbers, $\boldsymbol{x} = (x_1, x_2, \ldots, x_n)$ denotes n-tuple (n-dimensional real vectors), the set of vectors can be written as

$$\mathbf{R}^n = \{\boldsymbol{x} = (x_1, x_2, \ldots, x_n) : x_i, \in \mathbf{R}, i = 1, 2, \ldots, n\}$$

$$\mathbf{R}^n_{++} = \{\boldsymbol{x} = (x_1, x_2, \ldots, x_n) : x_i > 0, i = 1, 2, \ldots, n\}$$

$$\mathbf{R}^n_{+} = \{\boldsymbol{x} = (x_1, x_2, \ldots, x_n) : x_i \geqslant 0, i = 1, 2, \ldots, n\}$$

In particular, the notations $\mathbf{R}$, $\mathbf{R}_{++}$ and $\mathbf{R}_{+}$ denote $\mathbf{R}^1$, $\mathbf{R}^1_{++}$ and $\mathbf{R}^1_{+}$, respectively.

The following conclusion is proved in reference [1 − 2].

Theorem A Let the interval $[a, b] \subset \mathbf{R}$, $\varphi : \mathbf{R}^n \to \mathbf{R}$, $f : [a, b] \to \mathbf{R}$ and $\psi(x_1, x_2, \ldots, x_n) = \varphi(f(x_1), f(x_2), \ldots, f(x_n)) : [a, b]^n \to \mathbf{R}$.

(i) If φ is increasing and Schur-convex and f is convex, then ψ is Schur-

596

convex.

(ii) If φ is increasing and Schur-concave and f is concave, then ψ is Schur-concave.

(iii) If φ is decreasing and Schur-convex and f is concave, then ψ is Schur-convex.

(iv) If φ is increasing and Schur-convex and f is increasing and convex, then ψ is increasing and Schur-convex.

(v) If φ is decreasing and Schur-convex and f is decreasing and concave, then ψ is increasing and Schur-convex.

(vi) If φ is increasing and Schur-convex and f is decreasing and convex, then ψ is decreasing and Schur-convex.

(vii) If φ is decreasing and Schur-convex and f is increasing and concave, then ψ is decreasing and Schur-convex.

(viii) If φ is decreasing and Schur-concave and f is decreasing and convex, then ψ is increasing and Schur-concave.

Theorem A is very effective for determine of the Schur-convexity of the composite functions.

The Schur harmonically convex functions were proposed by Chu et al.[3−5] in 2009. The theory of majorization was enriched and expanded by using this concepts. Regarding the Schur harmonically convex functions, the aim of this paper is to establish the following theorem which is similar to Theorem A.

Theorem 1 Let the interval $[a, b] \subset \mathbf{R}_{++}$, $\varphi : \mathbf{R}_{++}^n \to \mathbf{R}_{++}$, $f : [a, b] \to \mathbf{R}_{++}$ and $\psi(x_1, x_2, \ldots, x_n) = \varphi(f(x_1), f(x_2), \ldots, f(x_n)) : [a, b]^n \to \mathbf{R}_{++}$.

(i) If φ is increasing and Schur harmonically convex and f is harmonically convex, then ψ is Schur harmonically convex.

(ii) If φ is increasing and Schur harmonically concave and f is harmonically concave, then ψ is Schur harmonically concave.

(iii) If φ is decreasing and Schur harmonically convex and f is harmonically concave, then ψ is Schur harmonically convex.

(iv) If φ is increasing and Schur harmonically convex and f is increasing and harmonically convex, then ψ is increasing and Schur harmonically convex.

(v) If φ is decreasing and Schur harmonically convex and f is decreasing and harmonically concave, then ψ is increasing and Schur harmonically convex.

(vi) If φ is increasing and Schur harmonically convex and f is decreasing and harmonically convex, then ψ is decreasing and Schur harmonically convex.

(vii) If φ is decreasing and Schur harmonically convex and f is increasing and harmonically concave, then ψ is decreasing and Schur harmonically convex.

(viii) If φ is decreasing and Schur harmonically concave and f is decreasing and harmonically convex, then ψ is increasing and Schur harmonically concave.

75.2   Definitions and lemmas

In order to prove our results, in this section we will recall useful definitions and lemmas.

Definition 1 [1-2] Let $x = (x_1, x_2, \ldots, x_n)$ and $y = (y_1, y_2, \ldots, y_n) \in \mathbf{R}^n$.

(i) $x \geqslant y$ means $x_i \geqslant y_i$ for all $i = 1, 2, \ldots, n$.

(ii) Let $\Omega \subset \mathbf{R}^n$, $\varphi \colon \Omega \to \mathbf{R}$ is said to be increasing if $x \geqslant y$ implies $\varphi(x) \geqslant \varphi(y)$. φ is said to be decreasing if and only if $-\varphi$ is increasing.

Definition 2 [1-2] Let $x = (x_1, x_2, \ldots, x_n)$ and $y = (y_1, y_2, \ldots, y_n) \in \mathbf{R}^n$.

We say y majorizes x (x is said to be majorized by y), denoted by $x \prec y$, if $\sum\limits_{i=1}^{k} x_{[i]} \leqslant \sum\limits_{i=1}^{k} y_{[i]}$ for $k = 1, 2, \ldots, n-1$ and $\sum\limits_{i=1}^{n} x_i = \sum\limits_{i=1}^{n} y_i$, where $x_{[1]} \geqslant x_{[2]} \geqslant \cdots \geqslant x_{[n]}$ and $y_{[1]} \geqslant y_{[2]} \geqslant \cdots \geqslant y_{[n]}$ are rearrangements of x and y in a descending order.

Definition 3 [1-2] Let $x = (x_1, x_2, \ldots, x_n)$ and $y = (y_1, y_2, \ldots, y_n) \in \mathbf{R}^n$.

(i) A set $\Omega \subset \mathbf{R}^n$ is said to be a convex set if

$$\alpha\boldsymbol{x}+(1-\alpha)\boldsymbol{y} = (\alpha x_1+(1-\alpha)y_1, \alpha x_2+(1-\alpha)y_2, \ldots, \alpha x_n+(1-\alpha)y_n) \in \Omega$$

for all $\boldsymbol{x}, \boldsymbol{y} \in \Omega$, and $\alpha \in [0, 1]$.

(ii) Let $\Omega \subset \mathbf{R}^n$ be convex set. A function $\varphi \colon \Omega \to \mathbf{R}$ is said to be a convex function on Ω if

$$\varphi\left(\alpha\boldsymbol{x} + (1 - \alpha)\boldsymbol{y}\right) \leqslant \alpha\varphi(\boldsymbol{x}) + (1 - \alpha)\varphi(\boldsymbol{y})$$

holds for all $\boldsymbol{x}, \boldsymbol{y} \in \Omega$, and $\alpha \in [0, 1]$. φ is said to be a concave function on Ω if and only if $-\varphi$ is convex function on Ω.

(iii) Let $\Omega \subset \mathbf{R}^n$. A function $\varphi \colon \Omega \to \mathbf{R}$ is said to be a Schur-convex function on Ω if $\boldsymbol{x} \prec \boldsymbol{y}$ on Ω implies $\varphi(\boldsymbol{x}) \leq \varphi(\boldsymbol{y})$. A function φ is said to be a Schur-concave function on Ω if and only if $-\varphi$ is Schur-convex function on Ω.

Lemma 1 (Schur-convex function decision theorem) [1-2]　Let $\Omega \subset \mathbf{R}^n$ be symmetric and have a nonempty interior convex set. Ω^0 is the interior of Ω. $\varphi \colon \Omega \to \mathbf{R}$ is continuous on Ω and differentiable in Ω^0. Then φ is the $Schur - convex$ (or $Schur - concave$, respectively) $function$ if and only if φ is symmetric on Ω and

$$(x_1 - x_2)\left(\frac{\partial\varphi}{\partial x_1} - \frac{\partial\varphi}{\partial x_2}\right) \geqslant 0(\text{or} \leqslant 0, \text{respectively}) \qquad (1)$$

holds for any $\boldsymbol{x} \in \Omega^0$.

Definition 2 [6]　Let $\Omega \subset \mathbf{R}^n_{++}$.

(i) A set Ω is said to be a harmonically convex set if $\dfrac{\boldsymbol{x}\boldsymbol{y}}{\lambda\boldsymbol{x} + (1 - \lambda)\boldsymbol{y}} \in \Omega$ for every $\boldsymbol{x}, \boldsymbol{y} \in \Omega$ and $\lambda \in [0, 1]$, where $\boldsymbol{x}\boldsymbol{y} = \sum\limits_{i=1}^n x_i y_i$ and $\dfrac{1}{\boldsymbol{x}} = \left(\dfrac{1}{x_1}, \dfrac{1}{x_2}, \ldots, \dfrac{1}{x_n}\right)$.

(ii) Let $\Omega \subset \mathbf{R}^n_{++}$ be a harmonically convex set. A function $\varphi \colon \Omega \to \mathbf{R}_{++}$ be a continuous function, then φ is called a harmonically convex (or concave, respectively) function, if

$$\varphi\left(\frac{1}{\frac{\alpha}{x} + \frac{1-\alpha}{y}}\right) \leqslant (\text{or} \geqslant, \text{respectively})\frac{1}{\frac{\alpha}{\varphi(x)} + \frac{1-\alpha}{\varphi(y)}}$$

holds for any $x, y \in \Omega$, and $\alpha \in [0, 1]$.

(iii) A function $\varphi : \Omega \to \mathbf{R}_{++}$ is said to be a Schur harmonically convex (or concave, respectively) function on Ω if $\dfrac{1}{x} \prec \dfrac{1}{y}$ implies $\varphi(x) \leqslant$ (or $\geqslant$, respectively) $\varphi(y)$.

By Definition[4], it is not difficult to prove the following propositions.

Proposition 1 Let $\Omega \subset \mathbf{R}^n_{++}$ be a set, and let $\dfrac{1}{\Omega} = \{(\dfrac{1}{x_1}, \dfrac{1}{x_2}, \ldots, \dfrac{1}{x_n}) :$

$(x_1, x_2, \ldots, x_n) \in \Omega\}$. Then $\varphi : \Omega \to \mathbf{R}_{++}$ is a Schur harmonically convex (or concave, respectively) function on Ω if and only if $\varphi(\dfrac{1}{x})$ is a Schur-convex (or concave, respectively) function on $\dfrac{1}{\Omega}$.

In fact, for any $u, v \in \dfrac{1}{\Omega}$, there exist $x, y \in \Omega$ such that $u = \dfrac{1}{x}, v = \dfrac{1}{y}$. Let $u \prec v$, that is $\dfrac{1}{x} \prec \dfrac{1}{y}$, if $\varphi : \Omega \to \mathbf{R}_{++}$ is a Schur harmonically convex (or concave, respectively) function on Ω, then $\varphi(x) \leqslant$ (or $\geqslant$, respectively)$\varphi(y)$, namely, $\varphi(\dfrac{1}{u}) \leqslant$ (or $\geqslant$, respectively)$\varphi(\dfrac{1}{v})$, this means that $\varphi(\dfrac{1}{x})$ is a Schur-convex (or concave, respectively) function on $\dfrac{1}{\Omega}$. The necessity is proved. The sufficiency can be similar to proof.

Proposition 2 $f : [a, b](\subset \mathbf{R}_{++}) \to \mathbf{R}_{++}$ is harmonically convex (or concave, respectively) if and only if $g(x) = \dfrac{1}{f(\frac{1}{x})}$ is concave (or convex, respectively) on $\left[\dfrac{1}{b}, \dfrac{1}{a}\right]$.

In fact, for any $x, y \in \left[\dfrac{1}{b}, \dfrac{1}{a}\right]$, then $\dfrac{1}{x}, \dfrac{1}{y} \in [a, b]$. If $f : [a, b](\subset \mathbf{R}_{++}) \to \mathbf{R}_{++}$ is harmonically convex (or concave, respectively), then

$$f\left(\dfrac{1}{\alpha x + (1 - \alpha)y}\right) \leqslant \text{(or} \geqslant, \text{respectively)} \dfrac{1}{\frac{\alpha}{f(\frac{1}{x})} + \frac{1-\alpha}{f(\frac{1}{y})}}$$

this is

$$\dfrac{1}{f(\frac{1}{\alpha x + (1-\alpha)y})} \geqslant \text{(or} \leqslant, \text{respectively)} \dfrac{\alpha}{f(\frac{1}{x})} + \dfrac{1-\alpha}{f(\frac{1}{y})}$$

this means that $g(x) = \dfrac{1}{f(\frac{1}{x})}$ is concave (or convex, respectively) on $\left[\dfrac{1}{b}, \dfrac{1}{a}\right]$. The necessity is proved. The sufficiency can be similar to proof.

Lemma 2(Schur harmonically convex function decision theorem)[5] Let $\Omega \subset \mathbf{R}^n_{++}$ be a symmetric and harmonically convex set with inner points, and let $\varphi : \Omega \to \mathbf{R}_{++}$ be a continuously symmetric function which is differentiable on interior Ω^0. Then φ is Schur harmonically convex (or Schur harmonically concave, respectively) on Ω if and only if

$$(x_1 - x_2)\left(x_1^2 \frac{\partial \varphi(\boldsymbol{x})}{\partial x_1} - x_2^2 \frac{\partial \varphi(\boldsymbol{x})}{\partial x_2}\right) \geqslant 0 \quad (\text{or} \leqslant 0, \text{ respectively}), \quad \boldsymbol{x} \in \Omega^0 \ (2)$$

75.3　Proof of main results

Proof of Theorem 1　We only give the proof of Theorem 1 (vi) in detail. Similar argument leads to the proof of the rest part.

If φ is increasing and Schur harmonically convex and f is decreasing and harmonically convex, then by Proposition 1, it follows that $\varphi(\dfrac{1}{x_1}, \dfrac{1}{x_2}, \ldots, \dfrac{1}{x_n})$ is decreasing and Schur convex, and by Proposition 2, it follows that $g(x) = \dfrac{1}{f(\frac{1}{x})}$ is decreasing and concave on $\left[\dfrac{1}{b}, \dfrac{1}{a}\right]$. And then from Theorem A (iii), it follows that

$$\varphi\left(\frac{1}{g(x_1)}, \frac{1}{g(x_2)}, \ldots, \frac{1}{g(x_n)}\right) = \varphi\left(f(\frac{1}{x_1}), f(\frac{1}{x_2}), \ldots, f(\frac{1}{x_n})\right)$$

is increasing and Schur-convex. Again by Proposition 1, it follows that

$$\psi(x_1, x_2, \ldots, x_n) = \varphi(f(x_1), f(x_2), \ldots, f(x_n))$$

is decreasing and Schur harmonically convex.

75.4　Applications

Let $\boldsymbol{x} = (x_1, x_2, \ldots, x_n) \in \mathbf{R}^n$. Its elementary symmetric functions are

$$E_r(\boldsymbol{x}) = E_r(x_1, x_2, \ldots, x_n) = \sum_{1 \leqslant i_1 < i_2 < \cdots < i_r \leqslant n} \prod_{j=1}^{r} x_{i_j}, \quad r = 1, 2, \ldots, n$$

and defined $E_0(\boldsymbol{x}) = 1$, and $E_r(\boldsymbol{x}) = 0$ for $r < 0$ or $r > n$. The dual forms of the elementary symmetric functions are

$$E_r^*(\boldsymbol{x}) = E_r^*(x_1, x_2, \ldots, x_n) = \prod_{1 \leqslant i_1 < i_2 < \cdots < i_r \leqslant n} \sum_{j=1}^{r} x_{i_j}, \quad r = 1, 2, \ldots, n$$

and defined $E_0^*(\boldsymbol{x}) = 1$, and $E_r^*(\boldsymbol{x}) = 0$ for $r < 0$ or $r > n$.

It is well-known that $E_r(\boldsymbol{x})$ is a increasing and Schur-concave function on $\mathbf{R}_+^n$[1].

In [6-7], Shi proved that $E_r^*(\boldsymbol{x})$ is a increasing and Schur-concave function on $\mathbf{R}_+^n$.

Theorem 2 For $r = 1, 2, \ldots, n, n \geqslant 2$, $E_r(\boldsymbol{x})$ and $E_r^*(\boldsymbol{x})$ are Schur harmonically convex function on $\mathbf{R}_{++}^n$.

Proof Noting that

$$
\begin{aligned}
E_r(\boldsymbol{x}) =& x_1 x_2 E_{r-2}(x_3, x_4, \ldots, x_n) + (x_1 + x_2) E_{r-1}(x_3, x_4, \ldots, x_n) + \\
& E_r(x_3, x_4, \ldots, x_n), \qquad r = 1, 2, \ldots, n
\end{aligned}
$$

then

$$
(x_1 - x_2) \left(x_1^2 \frac{\partial E_r(\boldsymbol{x})}{\partial x_1} - x_2^2 \frac{\partial E_r(\boldsymbol{x})}{\partial x_2} \right)
$$
$$
\begin{aligned}
=& (x_1 - x_2)[x_1^2 (x_2 E_{r-2}(x_3, x_4, \ldots, x_n) + E_{r-1}(x_3, x_4, \ldots, x_n)) - \\
& x_2^2 (x_1 E_{r-2}(x_3, x_4, \ldots, x_n) + E_{r-1}(x_3, x_4, \ldots, x_n))] \\
=& (x_1 - x_2)^2 [x_1 x_2 E_{r-2}(x_3, x_4, \ldots, x_n) + (x_1 + x_2) E_{r-1}(x_3, x_4, \ldots, x_n)] \geqslant 0
\end{aligned}
$$

by Lemma 2, it follows that $E_r(\boldsymbol{x})$ is Schur harmonically convex on $\mathbf{R}_{++}^n$.

By a direct, though tedious, calculation, and according to Lemma 2, $E_1^*(\boldsymbol{x})$, $E_2^*(\boldsymbol{x})$ is Schur harmonically convex on $\mathbf{R}_{++}^n$. When $r > 2$, it is easy to see that

$$
E_r^*(\boldsymbol{x}) = E_r^*(x_1, x_2, \ldots, x_n) = E_r^*(x_2, x_3, \ldots, x_n) \times \prod_{2 \leqslant i_1 < i_2 < \cdots < i_{r-1} \leqslant n} \left(x_1 + \sum_{j=1}^{r-1} x_{i_j} \right)
$$

then

$$
\log E_r^*(\boldsymbol{x}) = \log E_r^*(x_2, x_3, \ldots, x_n) + \sum_{2 \leqslant i_1 < i_2 < \cdots < i_{r-1} \leqslant n} \log \left(x_1 + \sum_{j=1}^{r-1} x_{i_j} \right)
$$

Now, it leads to

$$
\frac{1}{E_r^*(\boldsymbol{x})} \frac{\partial E_r^*(\boldsymbol{x})}{\partial x_1} = \sum_{2 \leqslant i_1 < i_2 < \cdots < i_{r-1} \leqslant n} \frac{1}{x_1 + \sum_{j=1}^{r-1} x_{i_j}}
$$

and then

$$\frac{\partial E_r^*(\boldsymbol{x})}{\partial x_1} = E_r^*(\boldsymbol{x}) \times$$

$$\left[\sum_{3 \leqslant i_1 < i_2 < \cdots < i_{r-1} \leqslant n} \frac{1}{x_1 + \sum\limits_{j=1}^{r-1} x_{i_j}} + \sum_{3 \leqslant i_1 < i_2 < \cdots < i_{r-2} \leqslant n} \frac{1}{x_1 + x_2 + \sum\limits_{j=1}^{r-2} x_{i_j}} \right]$$

By the same arguments

$$\frac{\partial E_r^*(\boldsymbol{x})}{\partial x_2} = E_r^*(\boldsymbol{x}) \times$$

$$\left[\sum_{3 \leqslant i_1 < i_2 < \cdots < i_{r-1} \leqslant n} \frac{1}{x_2 + \sum\limits_{j=1}^{r-1} x_{i_j}} + \sum_{3 \leqslant i_1 < i_2 < \cdots < i_{r-2} \leqslant n} \frac{1}{x_1 + x_2 + \sum\limits_{j=1}^{r-2} x_{i_j}} \right]$$

Thus

$$(x_1 - x_2) \left(x_1^2 \frac{\partial E_r^*(\boldsymbol{x})}{\partial x_1} - x_2^2 \frac{\partial E_r^*(\boldsymbol{x})}{\partial x_2} \right)$$

$$= (x_1 - x_2) E_r^*(\boldsymbol{x}) \times \left[\sum_{3 \leqslant i_1 < i_2 < \cdots < i_{r-1} \leqslant n} \left(\frac{x_1^2}{x_1 + \sum\limits_{j=1}^{r-1} x_{i_j}} - \frac{x_2^2}{x_2 + \sum\limits_{j=1}^{r-1} x_{i_j}} \right) + \right.$$

$$\left. (x_1^2 - x_2^2) \cdot \sum_{3 \leqslant i_1 < i_2 < \cdots < i_{r-2} \leqslant n} \frac{1}{x_1 + x_2 + \sum\limits_{j=1}^{r-2} x_{i_j}} \right]$$

$$= (x_1 - x_2)^2 E_r^*(\boldsymbol{x}) \times \left[\sum_{3 \leqslant i_1 < i_2 < \cdots < i_{r-1} \leqslant n} \frac{x_1 x_2 + (x_1 + x_2) \sum\limits_{j=1}^{r-1} x_{i_j}}{\left(x_1 + \sum\limits_{j=1}^{r-1} x_{i_j}\right)\left(x_2 + \sum\limits_{j=1}^{r-1} x_{i_j}\right)} + \right.$$

$$\left. (x_1 + x_2) \cdot \sum_{3 \leqslant i_1 < i_2 < \cdots < i_{r-2} \leqslant n} \frac{1}{x_1 + x_2 + \sum\limits_{j=1}^{r-2} x_{i_j}} \right] \geqslant 0$$

by Lemma 2, it follows that $E_r^*(\boldsymbol{x})$ is Schur harmonically convex on $\mathbf{R}_{++}^n$.

For $\boldsymbol{x} = (x_1, x_2, \ldots, x_n) \in \mathbf{R}^n$, the complete symmetric functions $c_n(\boldsymbol{x}, r)$ are defined as

$$c_n(\boldsymbol{x}, r) = \sum_{i_1+i_2+\cdots+i_n=r} \prod_{j=1}^{n} x_j^{i_j}, \quad r = 1, 2, \ldots, n$$

where $c_0(\boldsymbol{x}, r) = 1$, $r \in \{1, 2, \ldots, n\}$, $i_1, i_2, \ldots, i_n$ are non-negative integers. The dual forms of the complete symmetric functions $c_n^*(\boldsymbol{x}, r)$ are

$$c_n^*(\boldsymbol{x}, r) = \prod_{i_1+i_2+\cdots+i_n=r} \sum_{j=1}^{n} i_j x_j, \quad r = 1, 2, \ldots, n$$

where $i_j(j = 1, 2, \ldots, n)$ are non-negative integers.

Guan[8] discussed the Schur-convexity of $c_n(\boldsymbol{x}, r)$ and proved that $c_n(\boldsymbol{x}, r)$ is increasing and Schur-convex on $\mathbf{R}_{++}^n$. Subsequently, Chu et al. [5] proved that $c_n(\boldsymbol{x}, r)$ is Schur harmonically convex on $\mathbf{R}_{++}^n$.

Zhang and Shi [9] proved that $c_n^*(\boldsymbol{x}, r)$ is increasing, Schur-concave and Schur harmonically convex on $\mathbf{R}_{++}^n$.

In the following, we prove that the Schur harmonic convexity of the composite functions involving the above symmetric functions and their dual form by using Theorem 1.

Theorem 3 The following symmetric functions are increasing and Schur harmonically convex on $(0, 1)^n$, $r = 1, 2, \ldots, n$,

$$E_r\left(\frac{1+\boldsymbol{x}}{1-\boldsymbol{x}}\right) = \sum_{1 \leqslant i_1 < i_2 < \cdots < i_r \leqslant n} \prod_{j=1}^{r} \frac{1+x_{i_j}}{1-x_{i_j}} \tag{3}$$

$$E_r^*\left(\frac{1+\boldsymbol{x}}{1-\boldsymbol{x}}\right) = \prod_{1 \leqslant i_1 < i_2 < \cdots < i_r \leqslant n} \sum_{j=1}^{r} \frac{1+x_{i_j}}{1-x_{i_j}} \tag{4}$$

$$c_n\left(\frac{1+\boldsymbol{x}}{1-\boldsymbol{x}}, r\right) = \sum_{i_1+i_2+\cdots+i_n=r} \prod_{j=1}^{n} \left(\frac{1+x_j}{1-x_j}\right)^{i_j} \tag{5}$$

and

$$c_n^*\left(\frac{1+\boldsymbol{x}}{1-\boldsymbol{x}}, r\right) = \prod_{i_1+i_2+\cdots+i_n=r} \sum_{j=1}^{n} i_j \left(\frac{1+x_j}{1-x_j}\right) \tag{6}$$

Proof Let $f(x) = \dfrac{1+x}{1-x}, x \in (0, 1)$. Then $f(x) > 0$, $f'(x) = \dfrac{2}{(1-x)^2} > 0$, so f is increasing on $(0, 1)$.

And let $g(x) = \dfrac{1}{f(\frac{1}{x})} = \dfrac{x-1}{x+1}$. Then $g''(x) = -\dfrac{4}{(x+1)^3} < 0$, this means that $\dfrac{1}{f(\frac{1}{x})}$ is concave on $(1, \infty)$, by Proposition 2, it follows that f is harmonically convex on $(0, 1)$. Since $E_r(\boldsymbol{x})$, $E_r^*(\boldsymbol{x})$, $c_n(\boldsymbol{x}, r)$ and $c_n^*(\boldsymbol{x}, r)$ are all increasing and Schur harmonically convex function on $\mathbf{R}_{++}^n$, by Theorem 1 (iv), it follows that Theorem 3 holds.

Remark 1 By Lemma 2, Xia and Chu [10] proved that $E_r\left(\dfrac{1+\boldsymbol{x}}{1-\boldsymbol{x}}\right)$ is Schur harmonically convex on $(0, 1)^n$. By the properties of Schur harmonically convex function, Shi and Zhang [11] proved that $E_r^*\left(\dfrac{1+\boldsymbol{x}}{1-\boldsymbol{x}}\right)$ is Schur harmonically convex on $(0, 1)^n$. By Theorem 1, we give a new proof.

Theorem 4 The following symmetric functions are increasing and Schur harmonically convex on $\mathbf{R}_{++}^n, r = 1, 2, \ldots, n$

$$E_r\left(\boldsymbol{x}^{\frac{1}{r}}\right) = \sum_{1 \leqslant i_1 < i_2 < \cdots < i_r \leqslant n} \prod_{j=1}^{r} x_{i_j}^{\frac{1}{r}} \tag{7}$$

$$E_r^*\left(\boldsymbol{x}^{\frac{1}{r}}\right) = \prod_{1 \leqslant i_1 < i_2 < \cdots < i_r \leqslant n} \sum_{j=1}^{r} x_{i_j}^{\frac{1}{r}} \tag{8}$$

$$c_n\left(\boldsymbol{x}^{\frac{1}{r}}, r\right) = \sum_{i_1+i_2+\cdots+i_n=r} \prod_{j=1}^{n} x_j^{\frac{i_j}{r}} \tag{9}$$

and

$$c_n^*\left(\boldsymbol{x}^{\frac{1}{r}}, r\right) = \prod_{i_1+i_2+\cdots+i_n=r} \sum_{j=1}^{n} i_j x_j^{\frac{1}{r}} \tag{10}$$

Proof For $r \geqslant 1$, let $p(x) = x^{\frac{1}{r}}, x \in \mathbf{R}_{++}$. Then $p'(x) = \frac{1}{r}x^{\frac{1}{r}-1} > 0$, so p is increasing on $\mathbf{R}_{++}$.

And let $q(x) = \dfrac{1}{p\left(\frac{1}{x}\right)} = x^{\frac{1}{r}} = p(x)$. Then $q''(x) = \frac{1}{r}(\frac{1}{r} - 1)x^{\frac{1}{r}-2} \leqslant 0$, this means that $\dfrac{1}{p\left(\frac{1}{x}\right)}$ is concave on $\mathbf{R}_{++}$, by Proposition 2, it follows that p is harmonically convex on $\mathbf{R}_{++}$. Since $E_r(\boldsymbol{x})$, $E_r^*(\boldsymbol{x})$, $c_n(\boldsymbol{x}, r)$ and $c_n^*(\boldsymbol{x}, r)$ are all increasing and Schur harmonically convex function on $\mathbf{R}_{++}^n$, by Theorem 1 (iv), it follows that Theorem 4 holds.

Remark 2 By Lemma 2, Chu and Lv[3] proved that the Hamy's symmetric function $E_r\left(\boldsymbol{x}^{\frac{1}{r}}\right)$ is Schur harmonically convex on $\mathbf{R}_{++}^n$. Later, K. Z. Guan and R. K. Guan [12] further studied the harmonic convexity of the generalized Hamy symmetric function.

By Lemma 2, Meng et al.[13] proved that the dual form of the Hamy's symmetric function $E_r^*\left(\boldsymbol{x}^{\frac{1}{r}}\right)$ is Schur harmonically convex on $\mathbf{R}_{++}^n$.

By Lemma 2, Chu and Sun [4] proved that $c_n\left(\boldsymbol{x}^{\frac{1}{r}}, r\right)$ is Schur harmonically convex on $\mathbf{R}_{++}^n$.

By Theorem 1, we give a new proof.

Since $f(x) = \dfrac{1+x}{1-x}$ is increasing and harmonically convex on $(0,1)$, from Theorem 1 (iv) and Theorem 4, it follows

Theorem 5 The following symmetric functions are increasing and Schur harmonically convex on $(0,1)^n, r = 1, 2, \ldots, n$

$$E_r\left(\left(\frac{1+x}{1-x}\right)^{\frac{1}{r}}\right) = \sum_{1 \leqslant i_1 < i_2 < \cdots < i_r \leqslant n} \prod_{j=1}^{r} \left(\frac{1+x_{i_j}}{1-x_{i_j}}\right)^{\frac{1}{r}} \tag{11}$$

$$E_r^*\left(\left(\frac{1+x}{1-x}\right)^{\frac{1}{r}}\right) = \prod_{1 \leqslant i_1 < i_2 < \cdots < i_r \leqslant n} \sum_{j=1}^{r} \left(\frac{1+x_{i_j}}{1-x_{i_j}}\right)^{\frac{1}{r}} \tag{12}$$

$$c_n\left(\left(\frac{1+x}{1-x}\right)^{\frac{1}{r}}, r\right) = \sum_{i_1+i_2+\cdots+i_n=r} \prod_{j=1}^{n} \left(\frac{1+x_j}{1-x_j}\right)^{\frac{i_j}{r}} \tag{13}$$

and

$$c_n^*\left(\left(\frac{1+x}{1-x}\right)^{\frac{1}{r}}, r\right) = \prod_{i_1+i_2+\cdots+i_n=r} \sum_{j=1}^{n} i_j \left(\frac{1+x_j}{1-x_j}\right)^{\frac{1}{r}} \tag{14}$$

Remark 3 By Lemma 2, Long and Chu [14] proved that $E_r^*\left(\left(\frac{1+x}{1-x}\right)^{\frac{1}{r}}\right)$ is Schur harmonically convex on $(0,1)^n$. By Theorem 1, we give a new proof.

Theorem 6 The following symmetric functions are increasing and Schur harmonically convex on $(0,1)^n, r = 1, 2, \ldots, n$

$$E_r\left(\frac{x}{1-x}\right) = \sum_{1 \leqslant i_1 < i_2 < \cdots < i_r \leqslant n} \prod_{j=1}^{r} \frac{x_{i_j}}{1-x_{i_j}} \tag{15}$$

$$E_r^*\left(\frac{x}{1-x}\right) = \prod_{1 \leqslant i_1 < i_2 < \cdots < i_r \leqslant n} \sum_{j=1}^{r} \frac{x_{i_j}}{1-x_{i_j}} \tag{16}$$

$$c_n\left(\frac{x}{1-x}, r\right) = \sum_{i_1+i_2+\cdots+i_n=r} \prod_{j=1}^{n} \left(\frac{x_j}{1-x_j}\right)^{i_j} \tag{17}$$

and

$$c_n^*\left(\frac{x}{1-x}, r\right) = \prod_{i_1+i_2+\cdots+i_n=r} \sum_{j=1}^{n} i_j \left(\frac{x_j}{1-x_j}\right) \tag{18}$$

Proof Let $h(x) = \dfrac{x}{1-x}, x \in (0,1)$. Then $h'(x) = \dfrac{1}{(1-x)^2} > 0$, so h is increasing on $(0,1)$.

And let $k(x) = \dfrac{1}{h(\frac{1}{x})} = x - 1$. Then $k''(x) = 0$, this means that $\dfrac{1}{h(\frac{1}{x})}$ is concave and convex on $(1, \infty)$, by Proposition 2, it follows that h is harmonically convex on $(0, 1)$. Since $E_r(\boldsymbol{x})$, $E_r^*(\boldsymbol{x})$, $c_n(\boldsymbol{x}, r)$ and $c_n^*(\boldsymbol{x}, r)$ are all increasing and Schur harmonically convex function on $\mathbf{R}_{++}^n$, by Theorem 1 (iv), it follows that Theorem 5 holds.

Remark 4 By the judgement theorem of Schur harmonic convexity for a class of symmetric functions, Shi and Zhang [15] proved that $E_r\left(\dfrac{\boldsymbol{x}}{1-\boldsymbol{x}}\right)$ is Schur harmonically convex on $(0, 1)^n$. Here by Theorem 1, we give a new proof.

By the properties of Schur harmonically convex function, Shi and Zhang[11] proved that $E_r^*\left(\dfrac{\boldsymbol{x}}{1-\boldsymbol{x}}\right)$ is Schur harmonically convex on $\left[\dfrac{1}{2}, 1\right)^n$. By Theorem 1, this conclusion is extended to the collection $(0, 1)^n$.

By Lemma 2, Sun et al.[16] proved that $c_n\left(\dfrac{\boldsymbol{x}}{1-\boldsymbol{x}}, r\right)$ is Schur harmonically convex on $[0, 1)^n$, here by Theorem 1, we give a new proof.

Since $f(x) = \dfrac{x}{1-x}$ is increasing and harmonically convex on $(0, 1)$, from Theorem 1 (iv) and Theorem 4, it follows

Theorem 7 The following symmetric functions are increasing and Schur harmonically convex on $(0, 1)^n$, $r = 1, 2, \ldots, n$,

$$E_r\left(\left(\frac{\boldsymbol{x}}{1-\boldsymbol{x}}\right)^{\frac{1}{r}}\right) = \sum_{1 \leqslant i_1 < i_2 < \cdots < i_r \leqslant n} \prod_{j=1}^{r}\left(\frac{x_{i_j}}{1-x_{i_j}}\right)^{\frac{1}{r}} \tag{19}$$

$$E_r^*\left(\left(\frac{\boldsymbol{x}}{1-\boldsymbol{x}}\right)^{\frac{1}{r}}\right) = \prod_{1 \leqslant i_1 < i_2 < \cdots < i_r \leqslant n} \sum_{j=1}^{r}\left(\frac{x_{i_j}}{1-x_{i_j}}\right)^{\frac{1}{r}} \tag{20}$$

$$c_n\left(\left(\frac{\boldsymbol{x}}{1-\boldsymbol{x}}\right)^{\frac{1}{r}}, r\right) = \sum_{i_1+i_2+\cdots+i_n=r} \prod_{j=1}^{n}\left(\frac{x_j}{1-x_j}\right)^{\frac{i_j}{r}} \tag{21}$$

and

$$c_n^*\left(\left(\frac{\boldsymbol{x}}{1-\boldsymbol{x}}\right)^{\frac{1}{r}}, r\right) = \prod_{i_1+i_2+\cdots+i_n=r} \sum_{j=1}^{n} i_j \left(\frac{x_j}{1-x_j}\right)^{\frac{1}{r}} \tag{22}$$

Remark 5 By Lemma 2, Sun[17] proved that $E_r\left(\left(\dfrac{\boldsymbol{x}}{1-\boldsymbol{x}}\right)^{\frac{1}{r}}\right)$ is Schur harmonically convex on $[0, 1)^n$. Here by Theorem 1, we give a new proof.

Conflict of Interests

The authors declare that there is no conflict of interests regarding the publication of this article.

Acknowledgments

The work was supported by the Importation and Development of High-Caliber Talents Project of Beijing Municipal Institutions (Grant No. ID-HT201304089). Thanks for the help.

References

[1]   A W MARSHALL, I OLKIN, B C ARNOLD. Inequalities: Theory of Majorization and Its Application[M]. 2nd ed. New York: Springer, 2011.

[2]   B Y WANG. Foundations of Majorization Inequalities[M]. Beijing: Beijing Normal University Press, 1990.

[3]   Y M CHU, Y P LV. The Schur harmonic convexity of the Hamy symmetric function and its applications[J]. Journal of Inequalities and Applications, 2009, Article ID 838529, 10 pages.

[4]   Y M CHU, T C SUN. The Schur harmonic convexity for a class of symmetric functions[J]. Acta Mathematica Scientia, 2010, 30B(5): 1501-1506.

[5]   Y M CHU, G D WANG, X H ZHANG. The Schur multiplicative and harmonic convexities of the complete symmetric function[J]. Mathematische Nachrichten, 2011, 284(5-6): 653-663.

[6]   H N SHI. Theory of Majorization and Analytic Inequalities[M].Harbin: Harbin Institute of Technology Press, 2012.

[7]   H N SHI. Schur-concavity and Schur-geometrically convexity of dual form for elementary symmetric function with applications[J]. RGMIA Research Report Collection, 2007, 10(2).

[8]   K Z GUAN. Schur-convexity of the complete symmetric function[J]. Mathematical Inequalities & Applications, 2006, 9(4): 567-576.

[9]   K S ZHANG, H N SHI. Schur convexity of dual form of the complete symmetric function[J].Mathematical Inequalities & Applications, 2013, 16(4): 963-970.

[10] W F XIA, Y M CHU. On Schur convexity of some symmetric functions [J]. Journal of Inequalities and Applications, 2010, Article ID 543250, 12 pages.

[11] H N SHI, J ZHANG. Schur-convexity, Schur geometric and Schur harmonic convexities of dual form of a class symmetric functions[J]. Journal of Mathematical Inequalities, 2014, 8(2): 349-358.

[12] K Z GUAN, R K GUAN. Some properties of a generalized Hamy symmetric function and its applications[J]. Journal of Mathematical Analysis and Applications, 2011, 376: 494-505.

[13] JUNXIA MENG, YUMING CHU, XIAOMING TANG. The Schur-harmonic-convexity of dual form of the Hamy symmetric function[J]. Matematički Vesnik, 2010, 62(1): 37-46.

[14] B Y LONG, Y M CHU. The Schur convexity and inequalities for a class of symmetric functions[J]. Acta Mathematica Scientia, 2012, 32A(1): 80-89.

[15] H N SHI, J ZHANG. Some new judgement theorems of Schur geometric and Schur harmonic convexities for a class of symmetric functions[J]. Journal of Inequalities and Applications 2013, 2013: 527, doi: 10.1186/1029-242X-2013-527.

[16] M B SUN, N B CHEN, S H LI, Some properties of a class of symmetric functions and its applications [J]. Mathematische Nachrichten, 2014, 287(13): 1530-1544, doi: 10.1002/mana.201300073.

[17] M B SUN. The Schur convexity for two classes of symmetric functions [J]. Sci. Sin. Math., 2014, 44(6): 633-656, doi: 10.1360/N012013-00157.

第76篇　Two Schur-Convex Functions Related to the Generalized Integral Quasiarithmetic Means

(SUNYI-JIN, WANG DONGSHENG, SHI HUANNAN. Advances in Inequalities and Applications,2017, 2017:7.　Available online at http://scik.org)

Abstract: The Schur-convexity of two functions which related to the generalized integral quasiarithmetic means are researched, and two new inequalities are established. As applications, some refinements of Hadamard-type inequalities for convex functions and log-convex function are obtained.

Keywords: Schur-convex function; inequality; convex function; log-convex function; Hadamard's inequality; quasiarithmetic means

76.1　Introduction

Throughout the paper we assume that the set of n-dimensional row vector on real number field by $\mathbf{R}^n$, and $\mathbf{R}^n_+ = \{\boldsymbol{x} = (x_1, \ldots, x_n) \in \mathbf{R}^n : x_i > 0, i = 1, \ldots, n\}$. In particular, $\mathbf{R}^1$ and $\mathbf{R}^1_+$ denoted by $\mathbf{R}$ and $\mathbf{R}_+$ respectively.

Let f be a convex function defined on the interval $I \subseteq \mathbf{R} \to \mathbf{R}$ and the real numbers $a, b \in I$ with $a < b$. Then

$$f\left(\frac{a+b}{2}\right) \leqslant \frac{1}{b-a} \int_a^b f(x)\,\mathrm{d}x \leqslant \frac{f(a) + f(b)}{2} \tag{1}$$

is known as the Hadamard's inequality for convex function [1]. For some recent results which generalize, improve, and extend this classical inequality,

610

see [2-8].

When $f, -g$ both are convex functions satisfying $\int_a^b g(x)\mathrm{d}x > 0$ and $f\left(\frac{a+b}{2}\right) \geqslant 0$, S.-J. Yang in [5] generalized (1) as

$$\frac{f\left(\frac{a+b}{2}\right)}{g\left(\frac{a+b}{2}\right)} \leqslant \frac{\frac{1}{b-a}\int_a^b f(x)\,\mathrm{d}x}{\frac{1}{b-a}\int_a^b g(x)\,\mathrm{d}x} \tag{2}$$

To go further in exploring (2), Lan He in [8] define two mappings L and F by

$L : [a, b] \times [a, b] \to \mathbf{R}$

$$L(x, y; f, g) = \left[\left|\int_x^y f(t)\,\mathrm{d}t - (y-x)f\left(\frac{x+y}{2}\right)\right]\left[(y-x)g\left(\frac{x+y}{2}\right) - \int_x^y g(t)\,\mathrm{d}t\right]$$

and

$F : [a, b] \times [a, b] \to \mathbf{R}$

$$F(x, y; f, g) = g\left(\frac{x+y}{2}\right)\int_x^y f(t)\,\mathrm{d}t - f\left(\frac{x+y}{2}\right)\int_x^y g(t)\,\mathrm{d}t$$

Huan-nan Shi in [9] studied the Schur-convexity of $L(x, y; f, g)$ and $F(x, y; f, g)$ with variables (x, y) in $[a, b] \times [a, b] \subseteq \mathbf{R}^2$, obtained the following results.

Theorem A Let f and $-g$ both be convex function on $[a, b]$. Then $L(x, y; f, g)$ is Schur-convex on $[a, b] \times [a, b] \subseteq \mathbf{R}^2$.

Theorem B Let f and $-g$ both be nonnegative convex function on $[a, b]$. Then $F(x, y; f, g)$ is Schur-convex on $[a, b] \times [a, b] \subseteq \mathbf{R}^2$.

And then Shi established the refinement of the inequality of (2).

Theorem C Let f and $-g$ both be convex function on $[a, b] \subseteq \mathbf{R}$. If $\int_b^a g(x)\,\mathrm{d}x > 0$ and $f\left(\frac{a+b}{2}\right) \geqslant 0$, then

$$\frac{f\left(\frac{a+b}{2}\right)}{g\left(\frac{a+b}{2}\right)} \leqslant \frac{\int_a^b f(t)\,\mathrm{d}t - \int_{ta+(1-t)b}^{tb+(1-t)a} f(t)\,\mathrm{d}t}{\int_a^b g(t)\,\mathrm{d}t - \int_{ta+(1-t)b}^{tb+(1-t)a} g(t)\,\mathrm{d}t} \leqslant \frac{\int_a^b f(t)\,\mathrm{d}t}{\int_a^b g(t)\,\mathrm{d}t} \tag{3}$$

where $\frac{1}{2} \leqslant t < 1$ or $0 \leqslant t \leqslant \frac{1}{2}$.

Vera Čuljak et al in [10] discovered the following property of Schur-convexity of the generalized integral quasiarithmetic means.

Theorem D Let f be a real Lebesgue integrable function defined on the interval $I \subseteq \mathbf{R}$, with range J. Let k be a real continuous strictly monotone function on J. Then, for the generalized integral quasiarithmetic mean of function f defined as

$$M_k(f; a, b) = \begin{cases} k^{-1}\left(\frac{1}{b-a}\int_a^b (k \circ f)(t)\mathrm{d}t\right), & a \neq b \\ f(a), & a = b \end{cases} \tag{4}$$

the following hold:

(i) $M_k(f; x, y)$ is Schur-convex on I^2 if $k \circ f$ is convex on I and k is increasing on J or if $k \circ f$ is concave on I and k is decreasing on J;

(ii) $M_k(f; x, y)$ is Schur-concave on I^2 if $k \circ f$ is convex on I and k is decreasing on J or if $k \circ f$ is concave on I and k is increasing on J.

In recent years, Schur-convexity of various functions connected to the Hermite-Hadamard inequality has invoked the interest of many researchers and numerous papers have been dedicated to the investigation of it, see [9-13].

In this paper, comparing (2) with (4), we studied the Schur-convexity of the following two functions:

$$H_{p,q}(f, g; a, b) = \begin{cases} \frac{M_p(f;a,b)}{M_q(g;a,b)}, & a \neq b \\ \frac{f(a)}{g(a)}, & a = b \end{cases} \tag{5}$$

and

$$L_{p,q}(f; g; a, b) = \begin{cases} \left[M_p(f; a, b) - f(\frac{a+b}{2})\right] \cdot \left[g(\frac{a+b}{2}) - M_q(g; a, b)\right], & a \neq b \\ 0, & a = b \end{cases} \tag{6}$$

Our main results are as follows:

Theorem 1 Let f and g be a real Lebesgue integrable function defined on the interval $I \subseteq \mathbf{R}$, with range J_1 and J_2, respectively, p and q be a real continuous strictly increasing function on J_1 and J_2, respectively, and let $M_p(f; a, b) \geq 0$, $M_q(g; a, b) > 0$ and $g\left(\frac{a+b}{2}\right) \neq 0$.

(i) If $p \circ f$ is convex on I, $q \circ g$ is concave on I, then $H_{p,q}(f, g; a, b)$ is Schur-convex on I^2. And then for $a < b$, we have

$$\frac{M_p(f; a, b)}{M_q(g; a, b)} \geq \frac{M_p(f; ta + (1-t)b, tb + (1-t)a)}{M_q(g; ta + (1-t)b, tb + (1-t)a)} \geq \frac{f\left(\frac{a+b}{2}\right)}{g\left(\frac{a+b}{2}\right)} \tag{7}$$

where $\frac{1}{2} \leq t \leq 1$ or $0 \leq t \leq \frac{1}{2}$.

(ii) If $p \circ f$ is concave on I, $q \circ g$ is convex on I, then $H_{p,q}(f, g; a, b)$ is

Schur-concave on I^2. And then the inequality chains (7) reverse hold.

Theorem 2 Let f and g be a real Lebesgue integrable non negative function defined on the interval $I \subseteq \mathbf{R}$, with range J_1 and J_2, respectively, and let $M_p(f; a, b) \geqslant 0$, $M_q(g; a, b) > 0$ and $g\left(\frac{a+b}{2}\right) \neq 0$. If p, q is a real continuous strictly increasing function on J_1 and J_2, respectively, and $p \circ f$ is convex on I, $q \circ g$ is concave on I, then $L_{p,q}(f, g; a, b)$ is Schur-convex on I^2. And then the following inequality chains hold.

$$\frac{M_p(f; a, b)}{M_q(g; a, b)} \geqslant \frac{M_p(f; a, b)}{2M_q(g; a, b)} + \frac{f\left(\frac{a+b}{2}\right)}{2g\left(\frac{a+b}{2}\right)} \geqslant \frac{f\left(\frac{a+b}{2}\right)}{2M_q(g; a, b)} + \frac{M_p(f; a, b)}{2g\left(\frac{a+b}{2}\right)} \geqslant \frac{f\left(\frac{a+b}{2}\right)}{g\left(\frac{a+b}{2}\right)} \tag{8}$$

76.2　Definitions and Lemmas

We need the following definitions and lemmas.

Definition 1 [14–15] Let $\boldsymbol{x} = (x_1, \dots, x_n)$ and $\boldsymbol{y} = (y_1, \dots, y_n) \in \mathbf{R}^n$.

(i) $\boldsymbol{x}$ is said to be majorized by $\boldsymbol{y}$ (in symbols $\boldsymbol{x} \prec \boldsymbol{y}$) if $\sum\limits_{i=1}^{k} x_{[i]} \leqslant \sum\limits_{i=1}^{k} y_{[i]}$

for $k = 1, 2, \dots, n-1$ and $\sum\limits_{i=1}^{n} x_i = \sum\limits_{i=1}^{n} y_i$, where $x_{[1]} \geqslant \cdots \geqslant x_{[n]}$ and $y_{[1]} \geqslant \cdots \geqslant y_{[n]}$ are of $\boldsymbol{x}$ and $\boldsymbol{y}$ in a descending order.

(ii) Let $\Omega \subseteq \mathbf{R}^n$. The function $\varphi \colon \Omega \to \mathbf{R}$ be said to be a Schur-convex function on Ω if $\boldsymbol{x} \prec \boldsymbol{y}$ on Ω implies $\varphi(\boldsymbol{x}) \leq \varphi(\boldsymbol{y})$. φ is said to be a Schur-concave function on Ω if and only if $-\varphi$ is Schur-convex.

Lemma 1 [14–15] Let $\Omega \subseteq \mathbf{R}^n$ be a symmetric set and with a nonempty interior Ω^0, $\varphi : \Omega \to \mathbf{R}$ be a continuous on Ω and differentiable in Ω^0. Then φ is the $Schur - convex(Schur - concave) function$, if and only if φ is symmetric on Ω and

$$(x_1 - x_2)\left(\frac{\partial \varphi}{\partial x_1} - \frac{\partial \varphi}{\partial x_2}\right) \geqslant 0(\leqslant 0) \tag{9}$$

holds for any $\boldsymbol{x} = (x_1, \cdots, x_n) \in \Omega^0$.

Lemma 2 [16] Let $a \leqslant b$, $u(t) = ta + (1-t)b$, $v(t) = tb + (1-t)a$. If $\frac{1}{2} \leqslant t \leqslant 1$ or $0 \leqslant t \leqslant \frac{1}{2}$, then

$$\left(\frac{a+b}{2}, \frac{a+b}{2}\right) \prec (u(t), v(t)) \prec (a, b) \tag{10}$$

Lemma 3

(i) If $\varphi(x)$ is a convex function defined on the convex set $A \subseteq \mathbf{R}$ and if $h : \mathbf{R} \to \mathbf{R}$ is an increasing convex function, then the function $\psi : \mathbf{R} \to \mathbf{R}$ defined by $\psi(x) = h(\varphi(x))$ is convex on A.

(ii) If $\varphi(x)$ is a concave function defined on the convex set $A \subseteq \mathbf{R}$ and if $h : \mathbf{R} \to \mathbf{R}$ is an increasing concave function, then the function $\psi : \mathbf{R} \to \mathbf{R}$ defined by $\psi(x) = h(\varphi(x))$ is concave on A.

Proof We only give the proof of Lemma 3 (i) in detail. Similar argument leads to the proof of Lemma 3 (ii). If $x, y \in A$, then for all $\alpha \in [0, 1]$

$$
\begin{aligned}
\psi(\alpha x + (1 - \alpha)y) &= h(\varphi(\alpha x + (1 - \alpha)y)) \\
&\leqslant h(\alpha \varphi(x) + (1 - \alpha)\varphi(y)) \\
&\leqslant \alpha h(\varphi(x)) + (1 - \alpha)h(\varphi(y)) \\
&= \alpha \psi(x) + (1 - \alpha)\psi(y)
\end{aligned}
$$

Here the first inequality uses the monotonicity of h together with the convexity of φ; the second inequality uses the convexity of h.

76.3 Proofs of main results

Proof of Theorem 1

(i) It is clear that $H_{p,q}(f, g; a, b)$ is symmetric with a, b. Without loss of generality, we may assume $b \geqslant a$. Directly calculating yields

$$
\frac{\partial H_{p,q}}{\partial a} = \frac{1}{M_q^2(g; a, b)} \left(\frac{\partial M_p}{\partial a} M_q - \frac{\partial M_q}{\partial a} M_p \right)
$$

$$
\frac{\partial H_{p,q}}{\partial b} = \frac{1}{M_q^2(g; a, b)} \left(\frac{\partial M_p}{\partial b} M_q - \frac{\partial M_q}{\partial b} M_p \right)
$$

and then

$$
\begin{aligned}
\Delta :&= (b - a) \left(\frac{\partial H_{p,q}}{\partial b} - \frac{\partial H_{p,q}}{\partial a} \right) \\
&= \frac{M_q}{M_q^2(g; a, b)}(b - a) \left(\frac{\partial M_p}{\partial b} - \frac{\partial M_p}{\partial a} \right) - \frac{M_p}{M_q^2(g; a, b)}(b - a) \left(\frac{\partial M_q}{\partial b} - \frac{\partial M_q}{\partial a} \right)
\end{aligned}
$$

From Theorem D and Lemma 1, it follows that

$$(b-a)\left(\frac{\partial M_p}{\partial b}-\frac{\partial M_p}{\partial a}\right)\geqslant 0,\ (b-a)\left(\frac{\partial M_q}{\partial b}-\frac{\partial M_q}{\partial a}\right)\leqslant 0$$

so $\Delta \geqslant 0$, from Lemma 1, it follows that $H_{p,q}(f,g;a,b)$ is Schur-convex on I^2. And then from Lemma 2, we have

$$H_{p,q}(f,g;a,b)\geqslant H_{p,q}(f,g;ta+(1-t)b,tb+(1-t)a)\geqslant H_{p,q}\left(f,g;\frac{a+b}{2},\frac{a+b}{2}\right)$$

that is the inequalities (7) hold.

By the same arguments, we can carry out the proof of the proposition (ii).

The proof of Theorem 1 is completed.

Proof of Theorem 2

It is clear that $L_{p,q}(f,g;a,b)$ is symmetric with a,b. Without loss of generality, we may assume $b\geqslant a$. Directly calculating yields

$$\frac{\partial L_{p,q}}{\partial a}=\left[\frac{\partial M_p}{\partial a}-\frac{1}{2}f'\left(\frac{a+b}{2}\right)\right]\cdot\left[g\left(\frac{a+b}{2}\right)-M_q(g;a,b)\right]+$$
$$\left[\frac{1}{2}g'\left(\frac{a+b}{2}\right)-\frac{\partial M_q}{\partial a}\right]\cdot\left[M_p(f;a,b)-f\left(\frac{a+b}{2}\right)\right]$$

$$\frac{\partial L_{p,q}}{\partial b}=\left[\frac{\partial M_p}{\partial b}-\frac{1}{2}f'\left(\frac{a+b}{2}\right)\right]\cdot\left[g\left(\frac{a+b}{2}\right)-M_q(g;a,b)\right]+$$
$$\left[\frac{1}{2}g'\left(\frac{a+b}{2}\right)-\frac{\partial M_q}{\partial b}\right]\cdot\left[M_p(f;a,b)-f\left(\frac{a+b}{2}\right)\right]$$

and then

$$\Delta:=(b-a)\left(\frac{\partial L_{p,q}}{\partial b}-\frac{\partial L_{p,q}}{\partial a}\right)$$
$$=\left[g\left(\frac{a+b}{2}\right)-M_q(g;a,b)\right](b-a)\left(\frac{\partial M_p}{\partial b}-\frac{\partial M_p}{\partial a}\right)-$$
$$\left[M_p(f;a,b)-f\left(\frac{a+b}{2}\right)\right](b-a)\left(\frac{\partial M_q}{\partial b}-\frac{\partial M_q}{\partial a}\right)$$

From Theorem D and Lemma 1, it follows that

$$(b-a)\left(\frac{\partial M_p}{\partial b}-\frac{\partial M_p}{\partial a}\right)\geqslant 0,\ (b-a)\left(\frac{\partial M_q}{\partial b}-\frac{\partial M_q}{\partial a}\right)\leqslant 0$$

Since $\left(\frac{a+b}{2}, \frac{a+b}{2}\right) \prec (a, b)$, from (i) and (ii) in Theorem D, we have $g\left(\frac{a+b}{2}\right) \geqslant M_q(g; a, b)$ and $M_p(f; a, b) \geqslant f\left(\frac{a+b}{2}\right)$, respectively, so $\Delta \geqslant 0$, from Lemma 1, it follows that $L_{p,q}(f, g; a, b)$ is Schur-convex on I^2.

And then, we have

$$L_{p,q}(f, g; a, b) \geqslant L_{p,q}\left(f, g; \frac{a+b}{2}, \frac{a+b}{2}\right) = 0$$

namely

$$\left[M_p(f; a, b) - f(\frac{a+b}{2})\right] \cdot \left[g(\frac{a+b}{2}) - M_q(g; a, b)\right] \geqslant 0$$

it is equivalent to

$$g(\frac{a+b}{2})M_p(f; a, b) + f(\frac{a+b}{2})M_q(g; a, b)$$
$$\geqslant f(\frac{a+b}{2})g(\frac{a+b}{2}) + M_p(f; a, b)M_q(g; a, b) \tag{11}$$

Dividing each term of the inequalities (11) by $2M_q(g; a, b)g(\frac{a+b}{2})$, we get second inequality in (8).

From the inequalities (7), it is easy to see that

$$g\left(\frac{a+b}{2}\right) M_p(f; a, b) - f\left(\frac{a+b}{2}\right) M_q(g; a, b) \geqslant 0 \tag{12}$$

Dividing each term of the inequalities (12) by $M_q(g; a, b)$, we obtain

$$2g\left(\frac{a+b}{2}\right) \frac{M_p(f; a, b)}{M_q(g; a, b)} - g\left(\frac{a+b}{2}\right) \frac{M_p(f; a, b)}{M_q(g; a, b)} - f\left(\frac{a+b}{2}\right) \geqslant 0 \tag{13}$$

further, dividing each term of the inequalities (13) by $2g\left(\frac{a+b}{2}\right)$, we get first inequality in (8).

From Theorem D, it follows that

$$M_p(f; a, b) \geqslant M_p\left(f; \frac{a+b}{2}, \frac{a+b}{2}\right)$$

and

$$M_q(g; a, b) \leqslant M_q\left(g; \frac{a+b}{2}, \frac{a+b}{2}\right)$$

namely

$$M_p(f; a, b) - f\left(\frac{a+b}{2}\right) \geqslant 0$$

and

$$g\left(\frac{a+b}{2}\right) - M_q(g; a, b) \geqslant 0$$

and then, we have

$$g\left(\frac{a+b}{2}\right)\left[f\left(\frac{a+b}{2}\right)\left(g\left(\frac{a+b}{2}\right) - M_q(g; a, b)\right)\right.$$
$$\left. + M_q(g; a, b)\left(M_p(f; a, b) - f\left(\frac{a+b}{2}\right)\right)\right] \geqslant 0$$

this is

$$\left(g\left(\frac{a+b}{2}\right)\right)^2 f\left(\frac{a+b}{2}\right) + g\left(\frac{a+b}{2}\right) M_p(f; a, b) M_q(g; a, b)$$
$$\geqslant 2g\left(\frac{a+b}{2}\right) f\left(\frac{a+b}{2}\right) M_q(g; a, b) \tag{14}$$

Dividing each term of the inequalities (14) by $2\left(g\left(\frac{a+b}{2}\right)\right)^2 M_q(g; a, b)$, we get third inequality in (8).

The proof of Theorem 2 is completed.

76.4　Applications

Theorem 3 Let f and g be non negative integrable function on $I = [a, b] \subseteq \mathbf{R}_+$, satisfying $\frac{1}{b-a}\int_a^b (g(t))^s dt > 0$ and $g\left(\frac{a+b}{2}\right) > 0$, for $r \geqslant 1$ and $0 < s \leqslant 1$. If f is convex and g is concave on I, then

$$\frac{\left(\frac{1}{b-a}\int_a^b (f(t))^r dt\right)^{\frac{1}{r}}}{\left(\frac{1}{b-a}\int_a^b (g(t))^s dt\right)^{\frac{1}{s}}} \geqslant \frac{\left(\frac{1}{b-a}\int_{tb+(1-t)a}^{ta+(1-t)b} (f(t))^r dt\right)^{\frac{1}{r}}}{\left(\frac{1}{b-a}\int_{tb+(1-t)a}^{ta+(1-t)b} (g(t))^s dt\right)^{\frac{1}{s}}} \geqslant \frac{f\left(\frac{a+b}{2}\right)}{g\left(\frac{a+b}{2}\right)} \tag{15}$$

where $\frac{1}{2} \leqslant t < 1$ or $0 \leqslant t \leqslant \frac{1}{2}$.

If f is concave and g is convex, then the inequality chains (15) reverse hold.

Proof For $r \geqslant 1$ and $0 < s \leqslant 1$, taking $p(x) = x^r$ and $q(x) = x^s$, then p and q is strictly increasing convex and concave on $\mathbf{R}_+$, respectively, and then from Lemma 3, it follows that $f \circ p$ is convex on $[a, b]$ and $g \circ q$

is concave on $[a, b]$, and then by Theorem 1, it is deduced that inequalities (15) hold.

The proof of Theorem 3 is completed.

By a similar proof of Theorem 1, from Theorem 2, we can obtain the following Theorem.

Theorem 4 Let f and g be non negative integrable function on $I = [a, b] \subseteq \mathbf{R}_+$, satisfying $\frac{1}{b-a} \int_a^b (g(t))^s \mathrm{d}t > 0$ and $g\left(\frac{a+b}{2}\right) > 0$, for $r \geqslant 1$ and $0 < s \leqslant 1$. If f is convex and g is concave on I, then

$$\frac{\left(\frac{1}{b-a} \int_a^b (f(t))^r \mathrm{d}t\right)^{\frac{1}{r}}}{\left(\frac{1}{b-a} \int_a^b (g(t))^s \mathrm{d}t\right)^{\frac{1}{s}}} \geqslant \frac{\left(\frac{1}{b-a} \int_a^b (f(t))^r \mathrm{d}t\right)^{\frac{1}{r}}}{2\left(\frac{1}{b-a} \int_a^b (g(t))^s \mathrm{d}t\right)^{\frac{1}{s}}} + \frac{f\left(\frac{a+b}{2}\right)}{2g\left(\frac{a+b}{2}\right)}$$

$$\geqslant \frac{f\left(\frac{a+b}{2}\right)}{2\left(\frac{1}{b-a} \int_a^b (g(t))^s \mathrm{d}t\right)^{\frac{1}{s}}} + \frac{\left(\frac{1}{b-a} \int_a^b (f(t))^r \mathrm{d}t\right)^{\frac{1}{r}}}{2g\left(\frac{a+b}{2}\right)} \geqslant \frac{f\left(\frac{a+b}{2}\right)}{g\left(\frac{a+b}{2}\right)} \qquad (16)$$

Remark 1 It is obvious that inequalities (15) and (16) are strengthening and extension of the inequality (2).

Theorem 5 Let f and g be positive integrable function on $I = [a, b] \subseteq \mathbf{R}_+$, satisfying $g\left(\frac{a+b}{2}\right) > 0$. If $f(x)$ be log-convex function, and $g''(x) \leqslant 0, x \in I$, then

$$\frac{\exp\{\frac{1}{b-a} \int_a^b \log f(t) \mathrm{d}t\}}{\exp\{\frac{1}{b-a} \int_a^b \log g(t) \mathrm{d}t\}} \geqslant \frac{\exp\{\frac{1}{b-a} \int_{tb+(1-t)a}^{ta+(1-t)b} \log f(t) \mathrm{d}t\}}{\exp\{\frac{1}{b-a} \int_{tb+(1-t)a}^{ta+(1-t)b} \log g(t) \mathrm{d}t\}} \geqslant \frac{f\left(\frac{a+b}{2}\right)}{g\left(\frac{a+b}{2}\right)} \qquad (17)$$

where $\frac{1}{2} \leqslant t < 1$ or $0 \leqslant t \leqslant \frac{1}{2}$.

Proof Taking $p(x) = q(x) = \log x$, since $g''(x) \leqslant 0$, and then $(\log g(x))'' = \frac{g(x)g''(x) - (g'(x))^2}{(g(x))^2} \leqslant 0$, this is $\log g(x)$ is concave. $f(x)$ is a log-convex function, namely, $\log f(x)$ is convex. So from Theorem 1, it is deduced that inequalities (17) hold.

Similar to the proof of Theorem 5, by the theorem 2, we can prove the following theorem.

Theorem 6 Let f and g be positive integrable function on $I = [a, b] \subseteq \mathbf{R}_+$, satisfying $g\left(\frac{a+b}{2}\right) > 0$. If $f(x)$ is a log-convex function, and $g''(x) \leqslant$

$0, x \in I$, then

$$\frac{\exp\{\frac{1}{b-a}\int_a^b \log f(t)\mathrm{d}t\}}{\exp\{\frac{1}{b-a}\int_a^b \log g(t)\mathrm{d}t\}} \geqslant \frac{\exp\{\frac{1}{b-a}\int_a^b \log f(t)\mathrm{d}t\}}{2\exp\{\frac{1}{b-a}\int_a^b \log g(t)\mathrm{d}t\}} + \frac{f\left(\frac{a+b}{2}\right)}{2g\left(\frac{a+b}{2}\right)}$$

$$\geqslant \frac{f\left(\frac{a+b}{2}\right)}{2\exp\{\frac{1}{b-a}\int_a^b \log g(t)\mathrm{d}t\}} + \frac{\exp\{\frac{1}{b-a}\int_a^b \log f(t)\mathrm{d}t\}}{2g\left(\frac{a+b}{2}\right)} \geqslant \frac{f\left(\frac{a+b}{2}\right)}{g\left(\frac{a+b}{2}\right)} \qquad (18)$$

In particular, taking $g(x) = \mathrm{e}, x \in [a, b]$, from Theorem 5, we have the following corollary.

Corollary 1 Let f be positive integrable function on $I = [a, b] \subseteq \mathbf{R}_+$. If $f(x)$ is a log-convex function, then

$$\exp\{\frac{1}{b-a}\int_a^b \log f(t)\mathrm{d}t\} \geqslant \exp\{\frac{1}{b-a}\int_{tb+(1-t)a}^{ta+(1-t)b} \log f(t)\mathrm{d}t\} \geqslant f\left(\frac{a+b}{2}\right)$$

$$(19)$$

where $\frac{1}{2} \leqslant t < 1$ or $0 \leqslant t \leqslant \frac{1}{2}$.

Remark 2 In [17], Dragomir and Mond proved that the following inequalities of Hermite-Hadamard type hold for log-convex functions:

$$f\left(\frac{a+b}{2}\right) \leqslant \exp\{\frac{1}{b-a}\int_a^b \log f(t)\mathrm{d}t\}$$

$$\leqslant \frac{1}{b-a}\int_a^b \sqrt{f(t)f(a+b-t)}\mathrm{d}t$$

$$\leqslant \frac{1}{b-a}\int_a^b \log f(t)\mathrm{d}t$$

$$\leqslant \frac{f(a) - f(b)}{\log f(a) - \log f(b)}$$

$$\leqslant \frac{f(a) + f(b)}{2} \qquad (20)$$

The inequality chain (19) is a refinement of the first inequality in [20].

References

[1]　J HADMARD. Étude sur les propriétés des fonctions entières et en particulier d'une fonction considérée par Riemann[J]. J. Math. Pures Appl., 1893, 58 : 171-215.

[2]　S S DRAGOMIR, Y J CHO, S S KIM. Inequalities of Hadamard's type for Lipschitzian mappings and their applications[J]. J. Math. Anal. Appl., 2000, 245 : 489-501.

[3]　G-S YANG, K-L TSENG. Inequalities of Hadamard's type for Lips-

chitzian mappings[J]. J. Math. Anal. Appl., 2001, 260 : 230-238.

[4] M MATIĆ, J PEČARIĆ. Note on inequalities of Hadamard's type for Lipschitzian mappings[J]. Tamkang J. Math., 2001, 32(2) : 127-130.

[5] S J YANG. A direct proof and extensions of an inequality[J]. J. Math. Res. Exposit., 2004, 24(4) : 649-652.

[6] S S DRAGOMIR, R P AGARWAL. Two new mappings associated with Hadamard's inequalities for convex functions[J]. Appl. Math. Lett., 1998, 11(3) : 33-38.

[7] L-C WANG. Some refinements of Hermite-Hadamard inequalities for convex functions[J]. Univ. Beograd. Publ. Elektrotehn. Fak. Ser. Mat., 2004, 15 : 40-45.

[8] L HE. Two new mappings associated with inequalities of Hadamard-type for convex functions[J]. J. Inequal. Pure and Appl. Math., 2009, 10(3), Art. 81, 5 pp.

[9] HUAN-NAN SHI. Two Schur-convex functions relate to Hadamard-type integral inequalities[J]. Publicationes Mathematicae Debrecen, 2011,78(2): 393-403.

[10] VERAČULJAK, IVA FRANJIĆ, ROQIA GHULAM, JOSIP PEČARIĆ. Schur-convexity of averages of convex functions[J]. Journal of Inequalities and Applications, Volume 2011, Article ID 581918, 25 pages doi:10.1155/2011/581918.

[11] H-N SHI. Schur-convex functions relate to Hadamard-type inequalities[J]. J. Math. Inequal., 2007, 1(1): 127-136.

[12] H-N SHI, D-M LI, CH GU. Schur-convexity of a mean of convex function[J]. Appl. Math. Lett., 2009, 22 : 932-937.

[13] ELEZOVIĆ, N, J E PEČARIĆ. Note on Schur-convex functions[J]. Rocky Mountain J.Math., 1998, 29: 853-856.

[14] A M MARSHALL, I OLKIN. Inequalities:theory of majorization and its application[M]. New York : Academies Press, 1979.

[15] B-Y WANG. Foundations of Majorization Inequalities (Kong zhi bu deng shi ji chu)[M]. Beijing: Beijing Normal Univ. Press, 1990.

[16] H-N SHI, Y-M JIANG, W-D JIANG. Schur-convexity and Schur-geometrically concavity of Gini mean[J]. Comput. Math. Appl., 2009, 57 : 266-274.

[17] S S DRAGOMIR, B MOND. Integral inequalities of Hadamard type for log-convex functions[J]. Demonstratio Mathematica, 1998, 31 (2): 354-364.

[18] MEVLÜT TUNÇ. Some integral inequalities for logarithmically convex functions[J]. Journal of the Egyptian Mathematical Society, 2014, 22(2): 177-181.

第77篇　Schur Convexity of Bonferroni Means

(WANG DONGSHENG, SHI HUANNAN. Advanced Studies in Contemporary Mathematics, 2017(27): 599-608)

Abstract: Schur-convexity, Schur-geometric convexity and Schur-harmonic convexity of the Bonferroni means for n variables are investigated, and some mean value inequalities of n variables are established.

Keywords: Schur-convexity; Schur geometric convexity; Schur harmonic convexity; Bonferroni means; majorization; inequalities

77.1　Introduction

Throughout the article, $\mathbf{R}$ denotes the set of real numbers, $\boldsymbol{x} = (x_1, x_2, \cdots, x_n)$ denotes n-tuple (n-dimensional real vectors), the set of vectors can be written as

$$\mathbf{R}^n = \{\boldsymbol{x} = (x_1, x_2, \cdots, x_n) : x_i \in \mathbf{R}, i = 1, \cdots, n\}$$

$$\mathbf{R}^n_+ = \{\boldsymbol{x} = (x_1, x_2, \cdots, x_n) : x_i > 0, i = 1, \cdots, n\}$$

In particular, the notations $\mathbf{R}$ and $\mathbf{R}_+$ denote $\mathbf{R}^1$ and $\mathbf{R}^1_+$, respectively.

Let $\boldsymbol{x} = (x_1, x_2, \cdots, x_n) \in \mathbf{R}^n_+$ and $p, q \geqslant 0, p + q \neq 0$. The Bonferroni mean was originally introduced by Bonferroni in [1], which was defined as follows

$$B^{p,q}(\boldsymbol{x}) = \left(\frac{1}{n(n-1)} \sum_{i,j=1,i\neq j}^{n} x_i^p x_j^q\right)^{\frac{1}{p+q}} \tag{1}$$

Obviously, the Bonferroni mean has the following properties:

(i) $B^{p,q}(0, 0, \cdots, 0) = 0$;

(ii) $B^{p,q}(x, x, \cdots, x) = x$, if $x_i = x$, for all i;

(iii) $B^{p,q}(\boldsymbol{x}) \geqslant B^{p,q}(\boldsymbol{y})$. i.e., the Bonferroni mean is monotonic, if $x_i \geqslant y_i$, for all i;

(iv) $\min\{x_i\} \leqslant B^{p,q}(\boldsymbol{x}) \leqslant \max\{x_i\}$.

Furthermore, if $q = 0$, then by (1), it follows that

$$B^{p,0}(\boldsymbol{x}) = \left(\frac{1}{n} \sum_{i=1}^{n} x_i^p \left(\frac{1}{n-1} \sum_{i,j=1,i\neq j}^{n} x_j^0 \right) \right)^{\frac{1}{p+0}} = \left(\sum_{i=1}^{n} x_i^p \right)^{\frac{1}{p}} \tag{2}$$

which is power means of n variables.

If $n = 2$, then by (1), it follows that

$$B^{p,q}(x, y) = \left(\frac{x^p y^q + x^q y^p}{2} \right)^{\frac{1}{p+q}} \tag{3}$$

which is the generalized Muirhead mean of two variables $M(p, q; x, y)$ (See [2]).

In 2010, Yu-ming Chu etal.[3] studied the Schur-convexity, Schur geometric and harmonic convexities of the generalized Muirhead mean $M(p, q; x, y)$, obtained the following results.

Theorem A[6] For fixed $(p, q) \in \mathbf{R}^2$.

(i) $M(p, q; x, y)$ is Schur-convex with $(x, y) \in \mathbf{R}_+^2$ if and only if $(p, q) \in \{(p, q) \mid (p - q)^2 \geqslant p + q > 0$ and $pq \leqslant 0\}$;

(ii) $M(p, q; x, y)$ is Schur-concave with $(x, y) \in \mathbf{R}_+^2$ if and only if $(p, q) \in \{(p, q) \mid (p - q)^2 \leqslant p + q, (p, q) \neq (0, 0)\} \cup \{(p, q) \mid p + q < 0 \}$.

Theorem B[6] For fixed $(p, q) \in \mathbf{R}^2$.

(i) $M(p, q; x, y)$ is Schur-geometric convex with $(x, y) \in \mathbf{R}_+^2$ if and only if $(p, q) \in \{(p, q) \mid p + q > 0 \}$;

(ii) $M(p, q; x, y)$ is Schur-geometric concave with $(x, y) \in \mathbf{R}_+^2$ if and only if $(p, q) \in \{(p, q) \mid p + q < 0 \}$.

Theorem C[7] For fixed $(p, q) \in \mathbf{R}^2$.

(i) $M(p, q; x, y)$ is Schur-harmonic convex with $(x, y) \in \mathbf{R}_+^2$ if and only if $(p, q) \in \{(p, q) \mid p + q > 0 \} \cup \{(p, q) \mid p \leqslant 0, q \leqslant 0, (p - q)^2 + p + q \leqslant 0, p^2 + q^2 \neq 0\}$;

(ii) $M(p, q; x, y)$ is Schur-harmonic concave with $(x, y) \in \mathbf{R}_+^2$ if and only if

$$(p, q) \in \{(p, q) \mid p \geqslant 0, p + q < 0, (p - q)^2 + p + q \geqslant 0 \} \cup \{(p, q) \mid q \geqslant 0, p + q < 0, (p - q)^2 + p + q \geqslant 0\}.$$

In recent years, the research on Schur convexity of all kinds of means in n variables is more and more active and fruitful (see [10 - 19]). In this paper, we for the case of $n \geqslant 3$, discuss the Schur-convexity, Schur geometric and harmonic convexities of the Bonferroni mean $B^{p,q}(\boldsymbol{x})$. Our main results are as follows:

Theorem 1 For $n \geqslant 3$ and fixed $(p, q) \in \mathbf{R}^2$.

(i) If $0 \leqslant q \leqslant p \leqslant 1$ and $p - q \leqslant \sqrt{p + q}, p + q \neq 0$, then $B^{p,q}(\boldsymbol{x})$ is Schur-concave with $\boldsymbol{x} \in \mathbf{R}_+^n$;

(ii) If $q \leqslant p \leqslant 0$ and $p + q \neq 0$, then $B^{p,q}(\boldsymbol{x})$ is Schur-concave with $\boldsymbol{x} \in \mathbf{R}_+^n$;

(iii) If $p \geqslant 1, q \leqslant 0$ and $p + q > 0$, then $B^{p,q}(\boldsymbol{x})$ is Schur-convex with $\boldsymbol{x} \in \mathbf{R}_+^n$;

(iv) if $p \geqslant 1, q \leqslant 0$ and $p + q < 0$, then $B^{p,q}(\boldsymbol{x})$ is Schur-concave with $\boldsymbol{x} \in \mathbf{R}_+^n$.

Theorem 2 For $n \geqslant 3$ and fixed $(p, q) \in \mathbf{R}^2$.

(i) If $p + q > 0$, then $B^{p,q}(\boldsymbol{x})$ is Schur-geometric convex with $\boldsymbol{x} \in \mathbf{R}_+^n$;

(ii) If $p + q < 0$, then $B^{p,q}(\boldsymbol{x})$ is Schur-geometric concave with $\boldsymbol{x} \in \mathbf{R}_+^n$.

Theorem 3 For $n \geqslant 3$ and fixed $(p, q) \in \mathbf{R}^2$.

(i) If $p \geqslant q \geqslant 0$ and $p + q \neq 0$, then $B^{p,q}(\boldsymbol{x})$ is Schur-harmonic convex with $\boldsymbol{x} \in \mathbf{R}_+^n$;

(ii) If $0 \geqslant p \geqslant q \geqslant -1$ and $p + q \neq 0, (p - q)^2 + p + q \leqslant 0$, then $B^{p,q}(\boldsymbol{x})$ is Schur-harmonic convex with $\boldsymbol{x} \in \mathbf{R}_+^n$;

(iii) If $p \geqslant 0, q \leqslant -1$ and $p + q > 0$, then $B^{p,q}(\boldsymbol{x})$ is Schur-harmonic convex with $\boldsymbol{x} \in \mathbf{R}_+^n$;

(iv) If $p \geqslant 0, q \leqslant -1$ and $p + q < 0$, then $B^{p,q}(\boldsymbol{x})$ is Schur-harmonic concave with $\boldsymbol{x} \in \mathbf{R}_+^n$.

77.2　Definitions and Lemmas

For convenience, we introduce some definitions as follows.

Definition 1　Let $x = (x_1, x_2, \cdots, x_n)$ and $y = (y_1, y_2, \cdots, y_n) \in \mathbf{R}^n$.

(i) $x \geq y$ means $x_i \geqslant y_i$ for all $i = 1, 2, \cdots, n$.

(ii) Let $\Omega \subset \mathbf{R}^n$, $\varphi: \Omega \to \mathbf{R}$ is said to be increasing if $x \geq y$ implies $\varphi(x) \geqslant \varphi(y)$. φ is said to be decreasing if and only if $-\varphi$ is increasing.

Definition 2 Let $x = (x_1, x_2, \cdots, x_n)$ and $y = (y_1, y_2, \cdots, y_n) \in \mathbf{R}^n$.

(i) x is said to be majorized by y (in symbols $x \prec y$) if $\sum\limits_{i=1}^{k} x_{[i]} \leqslant \sum\limits_{i=1}^{k} y_{[i]}$

for $k = 1, 2, \cdots, n - 1$ and $\sum\limits_{i=1}^{n} x_i = \sum\limits_{i=1}^{n} y_i$, where $x_{[1]} \geqslant \cdots \geqslant x_{[n]}$ and $y_{[1]} \geqslant \cdots \geqslant y_{[n]}$ are rearrangements of x and y in a descending order.

(ii) Let $\Omega \subset \mathbf{R}^n$, the function $\varphi: \Omega \to \mathbf{R}$ is said to be Schur-convex on Ω if $x \prec y$ on Ω implies $\varphi(x) \leq \varphi(y)$. φ is said to be a Schur-concave function on Ω if and only if $-\varphi$ is Schur-convex function on Ω.

Definition 3 Let $x = (x_1, x_2, \cdots, x_n)$ and $y = (y_1, y_2, \cdots, y_n) \in \mathbf{R}^n$.

(i) $\Omega \subset \mathbf{R}^n$ is said to be a convex set if $x, y \in \Omega, 0 \leqslant \alpha \leqslant 1$ implies $\alpha x + (1 - \alpha)y = (\alpha x_1 + (1 - \alpha)y_1, \cdots, \alpha x_n + (1 - \alpha)y_n) \in \Omega$.

(ii) Let $\Omega \subset \mathbf{R}^n$ be convex set. A function $\varphi: \Omega \to \mathbf{R}$ is said to be convex on Ω if

$$\varphi(\alpha x + (1 - \alpha)y) \leqslant \alpha \varphi(x) + (1 - \alpha)\varphi(y)$$

for all $x, y \in \Omega$, and all $\alpha \in [0, 1]$. The function φ is said to be concave on Ω if and only if $-\varphi$ is convex function on Ω.

Definition 4

(i) A set $\Omega \subset \mathbf{R}^n$ is called symmetric, if $x \in \Omega$ implies $xP \in \Omega$ for every $n \times n$ permutation matrix P.

(ii) A function $\varphi: \Omega \to \mathbf{R}$ is called symmetric if for every permutation matrix P, $\varphi(xP) = \varphi(x)$ for all $x \in \Omega$.

Lemma 1[4] Let $\Omega \subset \mathbf{R}^n$ be symmetric and have a nonempty interior convex set. Ω^0 is the interior of Ω. $\varphi: \Omega \to \mathbf{R}$ is continuous on Ω and differentiable in Ω^0. Then φ is the $Schur - convex\ (Schur - concave)\ function$

if and only if φ is symmetric on Ω and

$$(x_1 - x_2)\left(\frac{\partial \varphi}{\partial x_1} - \frac{\partial \varphi}{\partial x_2}\right) \geqslant 0 (\leqslant 0) \tag{4}$$

holds for any $x \in \Omega^0$.

The first systematical study of the functions preserving the ordering of majorization was made by Issai Schur in 1923. In Schur's honor, such functions are said to be "Schur-convex". It can be used extensively in analytic inequalities, combinatorial optimization, quantum physics, information theory, and other related fields. See[11].

Definition 4 [6-7]

Let $x = (x_1, x_2, \cdots, x_n) \in \mathbf{R}_+^n$ and $y = (y_1, y_2, \cdots, y_n) \in \mathbf{R}_+^n$.

(i) $\Omega \subset \mathbf{R}_+^n$ is called a geometrically convex set if $(x_1^\alpha y_1^\beta, x_2^\alpha y_2^\beta, \cdots, x_n^\alpha y_n^\beta) \in \Omega$ for all $x, y \in \Omega$ and $\alpha, \beta \in [0, 1]$ such that $\alpha + \beta = 1$.

(ii) Let $\Omega \subset \mathbf{R}_+^n$. The function $\varphi\colon \Omega \to \mathbf{R}_+$ is said to be Schur geometrically convex function on Ω if

$$(\log x_1, \log x_2, \cdots, \log x_n) \prec (\log y_1, \log y_2, \cdots, \log y_n)$$

on Ω implies $\varphi(x) \leqslant \varphi(y)$. The function φ is said to be a Schur geometrically concave function on Ω if and only if $-\varphi$ is Schur geometrically convex function.

Lemma 2 [6] Let $\Omega \subset \mathbf{R}_+^n$ be a symmetric and geometrically convex set with a nonempty interior Ω^0. Let $\varphi : \Omega \to \mathbf{R}_+$ be continuous on Ω and differentiable in Ω^0. If φ is symmetric on Ω and

$$(\log x_1 - \log x_2)\left(x_1 \frac{\partial \varphi}{\partial x_1} - x_2 \frac{\partial \varphi}{\partial x_2}\right) \geqslant 0 \quad (\leqslant 0) \tag{5}$$

holds for any $x = (x_1, x_2, \cdots, x_n) \in \Omega^0$, then φ is a Schur geometrically convex (Schur geometrically concave) function.

The Schur geometric convexity was proposed by Zhang[6] in 2004, we also note that some authors use the term "Schur multiplicative convexity".

In 2009, Chu[8-9] introduced the notion of Schur harmonically convex function, and some interesting inequalities were obtained.

Definition 6 [8] Let $\Omega \subset \mathbf{R}_+^n$.

(i) A set Ω is said to be harmonically convex if $\dfrac{xy}{\lambda x + (1-\lambda)y} \in \Omega$

for every $x, y \in \Omega$ and $\lambda \in [0,1]$, where $xy = \sum\limits_{i=1}^{n} x_i y_i$ and $\dfrac{1}{x} = \left(\dfrac{1}{x_1}, \dfrac{1}{x_2}, \cdots, \dfrac{1}{x_n}\right).$

(ii) A function $\varphi : \Omega \to \mathbf{R}_+$ is said to be Schur harmonically convex on Ω if $\dfrac{1}{x} \prec \dfrac{1}{y}$ implies $\varphi(x) \le \varphi(y)$. A function φ is said to be a Schur harmonically concave function on Ω if and only if $-\varphi$ is a Schur harmonically convex function.

Lemma 2 [8] Let $\Omega \subset \mathbf{R}_+^n$ be a symmetric and harmonically convex set with inner points and let $\varphi : \Omega \to \mathbf{R}_+$ be a continuously symmetric function which is differentiable on Ω^0. Then φ is Schur harmonically convex (Schur harmonically concave) on Ω if and only if

$$(x_1 - x_2)\left(x_1^2 \dfrac{\partial \varphi(x)}{\partial x_1} - x_2^2 \dfrac{\partial \varphi(x)}{\partial x_2}\right) \ge 0 \quad (\le 0), \quad x \in \Omega^0 \tag{6}$$

Lemma 4 For $z \ge 1$, let

$$f(z) = -qz^{p-q+1} + pz^{p-q} - pz + q \tag{7}$$

(i) If $p \ge q \ge 0$ and $(p-q)^2 \le p+q$, then $f(z) \le 0$;

(ii) If $q \le p \le 0$, then $f(z) \ge 0$;

(iii) If $p \ge 0, q \le 0$ and $(p-q)^2 \ge p+q$, then $f(z) \ge 0$;

(iv) If $p \ge 0, q \le 0$ and $p+q \le 0$, then $f(z) \ge 0$.

Proof

$$f'(z) = -q(p-q+1)z^{p-q} + p(p-q)z^{p-q-1} - p$$

$$f'(1) = -q(p-q+1) + p(p-q) - p = (p-q)^2 - (p+q)$$

$$f''(z) = -q(p-q+1)(p-q)z^{p-q-1} + p(p-q)(p-q-1)z^{p-q-2}$$
$$= z^{p-q-2}h(z)$$

where

$$h(z) = -q(p-q+1)(p-q)z + p(p-q)(p-q-1) \tag{8}$$

$$h(1) = -q(p-q+1)(p-q) + p(p-q)(p-q-1) = (p-q)[(p-q)^2 - (p+q)]$$

$$h'(z) = -q(p-q+1)(p-q)$$

(i) If $p \geqslant q \geqslant 0$ and $(p-q)^2 \leqslant p+q$, then $h'(z) \leqslant 0$ and $h(1) \leqslant 0$, therefore $h(z) \leqslant 0$ for $z \geqslant 1$, meanwhile $f''(z) \leqslant 0$, but $f'(1) \leqslant 0$, so that $f'(z) \leqslant 0$, from $f(1) = 0$, it follows that $f(z) \leqslant 0$ for $z \geqslant 1$.

(ii) If $q \leqslant p \leqslant 0$, then $h'(z) \geqslant 0$ and $h(1) \geqslant 0$, therefore $h(z) \geqslant 0$ for $z \geqslant 1$, meanwhile $f''(z) \geqslant 0$, but $f'(1) \geqslant 0$, so that $f'(z) \geqslant 0$, from $f(1) = 0$, it follows that $f(z) \geqslant 0$ for $z \geqslant 1$.

Proving propositions (iii) and (iv) is similar to proposition (ii), so it is omitted.

The proof of lemma 4 is complete.

Lemma 5 For $z \geqslant 1$, let

$$g(z) = pz^{p-q+1} - qz^{p-q} + qz - p \tag{9}$$

(i) If $p \geqslant q \geqslant 0$, then $g(z) \geqslant 0$;

(ii) If $0 \geqslant p \geqslant q$ and $(p-q)^2 + p + q \leqslant 0$, then $g(z) \leqslant 0$;

(iii) If $p \geqslant 0 \geqslant q$ and $(p-q)^2 + p + q \geqslant 0$, then $g(z) \geqslant 0$.

Proof

$$g'(z) = p(p-q+1)z^{p-q} - q(p-q)z^{p-q-1} + q$$

$$g'(1) = p(p-q+1) - q(p-q) + q = (p-q)^2 + p + q$$

$$g''(z) = p(p-q+1)(p-q)z^{p-q-1} - q(p-q)(p-q-1)z^{p-q-2}$$
$$= (p-q)z^{p-q-2}m(z)$$

where

$$m(z) = p(p-q+1)z - q(p-q-1) \tag{10}$$

$$m(1) = (p-q)^2 + p + q$$

$$m'(z) = p(p-q+1)$$

(i) If $p \geqslant q \geqslant 0$, then it is easy to see that $m'(z) \geqslant 0$ for $z \geqslant 1$, but $m(1) \geqslant 0$, therefore $m(z) \geqslant 0$ for $z \geqslant 1$, meanwhile $g''(z) \leqslant 0$, but

$g'(1) > 0$, so that $g'(z) \geqslant 0$, from $g(1) = 0$, it follows that $g(z) \geqslant 0$ for $z \geqslant 1$.

(ii) If $0 \geqslant p \geqslant q$ and $(p - q)^2 + p + q \leqslant 0$, then $m'(z) \leqslant 0$ and $m(1) \leqslant 0$, therefore $m(z) \leqslant 0$ for $z \geqslant 1$, meanwhile $g''(z) \leqslant 0$, but $g'(1) \leqslant 0$, so that $g'(z) \leqslant 0$, from $g(1) = 0$, it follows that $g(z) \leqslant 0$ for $z \geqslant 1$.

Proving propositions (iii) is similar to proposition (i), so it is omitted. The proof of lemma 5 is complete.

Lemma 6 [4] Let $\boldsymbol{x} = (x_1, x_2, \cdots, x_n) \in \mathbf{R}_+^n$ and $A_n(\boldsymbol{x}) = \frac{1}{n} \sum_{i=1}^{n} x_i$. Then

$$\boldsymbol{u} = \left(\underbrace{A_n(\boldsymbol{x}), A_n(\boldsymbol{x}), \cdots, A_n(\boldsymbol{x})}_{n} \right) \prec (x_1, x_2, \cdots, x_n) = \boldsymbol{x} \qquad (11)$$

Lemma 7 [4] If $x_i > 0, i = 1, 2, \ldots, n$, then for all nonnegative constants c satisfying $0 < c < \frac{1}{n} \sum_{i=1}^{n} x_i$

$$\left(\frac{x_1}{\sum\limits_{j=1}^{n} x_j}, \cdots, \frac{x_n}{\sum\limits_{j=1}^{n} x_j} \right) \prec \left(\frac{x_1 - c}{\sum\limits_{j=1}^{n} (x_j - c)}, \cdots, \frac{x_n - c}{\sum\limits_{j=1}^{n} (x_j - c)} \right) \qquad (12)$$

77.3　Proof of Theorems

Proof of Theorem 1 Let

$$b(\boldsymbol{x}) = \frac{1}{n(n-1)} \sum_{i,j=1,i \neq j}^{n} x_i^p x_j^q \qquad (13)$$

Then

$$\frac{\partial B^{p,q}(\boldsymbol{x})}{\partial x_1} = \frac{1}{p+q} (b(\boldsymbol{x}))^{\frac{1}{p+q}-1} \frac{1}{n(n-1)} \cdot$$
$$\left[p x_1^{p-1} (x_2^q + x_3^q + \cdots + x_n^q) + q x_1^{q-1} (x_2^p + x_3^p + \cdots + x_n^p) \right]$$

$$\frac{\partial B^{p,q}(\boldsymbol{x})}{\partial x_2} = \frac{1}{p+q} (b(\boldsymbol{x}))^{\frac{1}{p+q}-1} \frac{1}{n(n-1)} \cdot$$
$$\left[p x_2^{p-1} (x_1^q + x_3^q + \cdots + x_n^q) + p x_2^{q-1} (x_1^p + x_3^p + \cdots + x_n^p) \right]$$

It is easy to see that $B^{p,q}(\boldsymbol{x})$ is symmetric on $\mathbf{R}_+^n$, without loss of generality, we may assume that $x_1 \geqslant x_2 > 0$. Let $z = \frac{x_1}{x_2}$. Then $z \geqslant 1$.

$$
\begin{aligned}
\Delta_1 :=& (x_1 - x_2)\left(\frac{\partial B^{p,q}(\boldsymbol{x})}{\partial x_1} - \frac{\partial B^{p,q}(\boldsymbol{x})}{\partial x_2}\right) \\
=& (x_1 - x_2)\frac{1}{p+q}(b(\boldsymbol{x}))^{\frac{1}{p+q}-1}\frac{1}{n(n-1)}[p(x_3^q + \cdots + x_n^q)(x_1^{p-1} - x_2^{p-1})+ \\
& q(x_3^p + \cdots + x_n^p)(x_1^{q-1} - x_2^{q-1})+ \\
& x_1^{p-1}x_2^{q-1}(px_2 - qx_1) + x_1^{q-1}x_2^{p-1}(qx_2 - px_1)] \\
=& (x_1 - x_2)\frac{1}{p+q}(b(\boldsymbol{x}))^{\frac{1}{p+q}-1}\frac{1}{n(n-1)}[p(x_3^q + \cdots + x_n^q)(x_1^{p-1} - x_2^{p-1})+ \\
& q(x_3^p + \cdots + x_n^p)(x_1^{q-1} - x_2^{q-1}) + x_2^{p+q-1}(z^{p-1}(p-qz) + z^{q-1}(q-pz))] \\
=& (x_1 - x_2)\frac{1}{p+q}(b(\boldsymbol{x}))^{\frac{1}{p+q}-1}\frac{1}{n(n-1)}[p(x_3^q + \cdots + x_n^q)(x_1^{p-1} - x_2^{p-1})+ \\
& q(x_3^p + \cdots + x_n^p)(x_1^{q-1} - x_2^{q-1}) + x_2^{p+q-1}z^{q-1}f(z)]
\end{aligned} \tag{14}
$$

For $n \geqslant 3$ and fixed $(p,q) \in \mathbf{R}^2$.

(i) If $0 \leqslant q \leqslant p \leqslant 1$ and $p - q \leqslant \sqrt{p+q}, p + q \neq 0$, then by (i) in Lemma 4, it follows $f(z) \leqslant 0$. Furthermore, from $x_1 \geqslant x_2 > 0$, we have $x_1^{p-1} - x_2^{p-1} \leqslant 0$ and $x_1^{q-1} - x_2^{q-1} \leqslant 0$. Hence from (12), we conclude that $\Delta_1 \leqslant 0$, by Lemma 1, it follows that $B^{p,q}(\boldsymbol{x})$ is Schur-concave with $\boldsymbol{x} \in \mathbf{R}_+^n$.

(ii) If $q \leqslant p \leqslant 0$ and $p + q \neq 0$, then by (ii) in Lemma 4, it follows $f(z) \geqslant 0$. Furthermore, from $x_1 \geqslant x_2 > 0$, we have $p(x_1^{p-1} - x_2^{p-1}) \geqslant 0$ and $q(x_1^{q-1} - x_2^{q-1}) \geqslant 0$. Notice that $\frac{1}{p+q} < 0$, from (14), we conclude that $\Delta_1 \leqslant 0$, by Lemma 1, it follows that $B^{p,q}(\boldsymbol{x})$ is Schur-concave with $\boldsymbol{x} \in \mathbf{R}_+^n$. then $B^{p,q}(\boldsymbol{x})$ is Schur-concave with $\boldsymbol{x} \in \mathbf{R}_+^n$

(iii) If $p \geqslant 1, q \leqslant 0$ and $p + q > 0$, then $(p-q)^2 \geqslant p - q \geqslant p \geqslant p + q$, and then by (iii) in Lemma 4, it follows $f(z) \geqslant 0$. Furthermore, from $x_1 \geqslant x_2 > 0$, we have $p(x_1^{p-1} - x_2^{p-1}) \geqslant 0$ and $q(x_1^{q-1} - x_2^{q-1}) \geqslant 0$. Notice that $\frac{1}{p+q} > 0$, from (14), we conclude that $\Delta_1 \geqslant 0$, by Lemma 1, it follows that $B^{p,q}(\boldsymbol{x})$ is Schur-convex with $\boldsymbol{x} \in \mathbf{R}_+^n$.

(iv) If $p \geqslant 1, q \leqslant 0$ and $p + q < 0$, then by (iv) in Lemma 4, it follows $f(z) \geqslant 0$. Furthermore, from $x_1 \geqslant x_2 > 0$, we have $p(x_1^{p-1} - x_2^{p-1}) \geqslant 0$ and $q(x_1^{q-1} - x_2^{q-1}) \geqslant 0$. Notice that $\frac{1}{p+q} < 0$, from (14), we conclude that $\Delta_1 \leqslant 0$, by Lemma 1, it follows that $B^{p,q}(\boldsymbol{x})$ is Schur-concave with $\boldsymbol{x} \in R_+^n$.

The proof of Theorem 1 is completed.

Proof of Theorem 2

$$x_1 \frac{\partial B^{p,q}(\boldsymbol{x})}{\partial x_1} = \frac{1}{p+q} \left(b(\boldsymbol{x})\right)^{\frac{1}{p+q}-1} \frac{1}{n(n-1)} \cdot$$

$$[px_1^p \left(x_2^q + x_3^q + \cdots + x_n^q\right) + qx_1^q \left(x_2^p + x_3^p + \cdots + x_n^p\right)]$$

$$x_2 \frac{\partial B^{p,q}(\boldsymbol{x})}{\partial x_2} = \frac{1}{p+q} \left(b(\boldsymbol{x})\right)^{\frac{1}{p+q}-1} \frac{1}{n(n-1)} \cdot$$

$$[px_2^p \left(x_1^q + x_3^q + \cdots + x_n^q\right) + qx_2^q \left(x_1^p + x_3^p + \cdots + x_n^p\right)]$$

Without loss of generality, we may assume that $x_1 \geqslant x_2 > 0$. Let $z = \frac{x_1}{x_2}$. Then $z \geqslant 1$.

$$\Delta_2 := (x_1 - x_2) \left(x_1 \frac{\partial B^{p,q}(\boldsymbol{x})}{\partial x_1} - x_2 \frac{\partial B^{p,q}(\boldsymbol{x})}{\partial x_2} \right)$$

$$= (x_1 - x_2) \frac{1}{p+q} \left(b(\boldsymbol{x})\right)^{\frac{1}{p+q}-1} \frac{1}{n(n-1)} [p(x_3^q + \cdots + x_n^q)(x_1^p - x_2^p) +$$

$$q(x_3^p + \cdots + x_n^p)(x_1^q - x_2^q) + px_1^p x_2^q - px_1^q x_2^p + qx_1^q x_2^p - qx_1^p x_2^q)]$$

$$= (x_1 - x_2) \frac{1}{p+q} \left(b(\boldsymbol{x})\right)^{\frac{1}{p+q}-1} \frac{1}{n(n-1)} [p(x_3^q + \cdots + x_n^q)(x_1^p - x_2^p)x +$$

$$q(x_3^p + \cdots + x_n^p)(x_1^q - x_2^q) + (p-q)x_2^q x_2^p(z^p - z^q)] \tag{15}$$

Note that there are always $p(x_1^p - x_2^p) \geqslant 0$ and $q(x_1^q - x_2^q) \geqslant 0$. For $z > 1$, the function z^t is increasing with t, so $(p-q)(z^p - z^q) \geqslant 0$. Thus if $p+q > 0$, then from (15), we conclude that $\Delta_2 \geqslant 0$, by Lemma 2, it follows that $B^{p,q}(\boldsymbol{x})$ is Schur-geometric convex with $\boldsymbol{x} \in \mathbf{R}_+^n$, if $p+q < 0$, then from (15), we conclude that $\Delta_2 \leqslant 0$, by Lemma 2, it follows that $B^{p,q}(\boldsymbol{x})$ is Schur-geometric concave with $\boldsymbol{x} \in \mathbf{R}_+^n$.

The proof of Theorem 2 is completed.

Proof of Theorem 3

$$x_1^2 \frac{\partial B^{p,q}(\boldsymbol{x})}{\partial x_1} = \frac{1}{p+q} \left(b(\boldsymbol{x})\right)^{\frac{1}{p+q}-1} \frac{1}{n(n-1)} \cdot$$

$$\left[px_1^{p+1} \left(x_2^q + x_3^q + \cdots + x_n^q\right) + qx_1^{p+1} \left(x_2^p + x_3^p + \cdots + x_n^p\right)\right]$$

$$x_2^2 \frac{\partial B^{p,q}(\boldsymbol{x})}{\partial x_2} = \frac{1}{p+q} \left(b(\boldsymbol{x})\right)^{\frac{1}{p+q}-1} \frac{1}{n(n-1)} \cdot$$

$$\left[px_2^{p+1} \left(x_1^q + x_3^q + \cdots + x_n^q\right) + qx_2^{p+1} \left(x_1^p + x_3^p + \cdots + x_n^p\right)\right]$$

Without loss of generality, we may assume that $x_1 \geqslant x_2 > 0$. Let $z = \frac{x_1}{x_2}$. Then $z \geqslant 1$.

$$\Delta_3 := (x_1 - x_2) \left(x_1^2 \frac{\partial B^{p,q}(\boldsymbol{x})}{\partial x_1} - x_2^2 \frac{\partial B^{p,q}(\boldsymbol{x})}{\partial x_2} \right)$$

$$= (x_1 - x_2) \frac{1}{p+q} (b(\boldsymbol{x}))^{\frac{1}{p+q}-1} \frac{1}{n(n-1)} [p(x_3^q + \cdots + x_n^q)(x_1^{p+1} - x_2^{p+1}) +$$

$$q(x_3^p + \cdots + x_n^p)(x_1^{q+1} - x_2^{q+1}) +$$

$$p x_1^{p+1} x_2^q - p x_1^q x_2^{p+1} + q x_1^{q+1} x_2^p - q x_1^p x_2^{q+1})]$$

$$= (x_1 - x_2) \frac{1}{p+q} (b(\boldsymbol{x}))^{\frac{1}{p+q}-1} \frac{1}{n(n-1)} [p(x_3^q + \cdots + x_n^q)(x_1^{p+1} - x_2^{p+1}) +$$

$$q(x_3^p + \cdots + x_n^p)(x_1^{q+1} - x_2^{q+1}) + x_2^{p+1} x_2^q z^q (z^{p-q}(pz - q) - (p - qz))]$$

$$= (x_1 - x_2) \frac{1}{p+q} (b(\boldsymbol{x}))^{\frac{1}{p+q}-1} \frac{1}{n(n-1)} [p(x_3^q + \cdots + x_n^q)(x_1^{p+1} - x_2^{p+1}) +$$

$$q(x_3^p + \cdots + x_n^p)(x_1^{q+1} - x_2^{q+1}) + x_2^{p+1} x_2^q z^q g(z)] \tag{16}$$

For $n \geqslant 3$ and fixed $(p, q) \in \mathbf{R}^2$.

(i) If $p \geqslant q \geqslant 0$ and $p + q \neq 0$, then by (i) in Lemma 5, it follows $g(z) \geqslant 0$. Furthermore, from $x_1 \geqslant x_2 > 0$, we have $p(x_1^{p+1} - x_2^{p+1}) \geqslant 0$ and $q(x_1^{q+1} - x_2^{q+1}) \geqslant 0$. Hence from (16), we conclude that $\Delta_3 \geqslant 0$, by Lemma 3, it follows that $B^{p,q}(\boldsymbol{x})$ is Schur-harmonic convex with $\boldsymbol{x} \in \mathbf{R}_+^n$;

(ii) If $0 \geqslant p \geqslant q \geqslant -1$ and $p + q \neq 0$, $(p - q)^2 + p + q \leqslant 0$ then by (ii) in Lemma 5, it follows $g(z) \leqslant 0$. Furthermore, from $x_1 \geqslant x_2 > 0$, we have $p(x_1^{p+1} - x_2^{p+1}) \leqslant 0$ and $q(x_1^{q+1} - x_2^{q+1}) \leqslant 0$. Notice that $\frac{1}{p+q} < 0$, from (16), we conclude that $\Delta_3 \geqslant 0$, by Lemma 3, it follows that $B^{p,q}(\boldsymbol{x})$ is Schur-harmonic convex with $\boldsymbol{x} \in \mathbf{R}_+^n$;

(iii) If $p \geqslant 0, q \leqslant -1$ and $p + q > 0$, then $(p - q)^2 + p + q \geqslant 0$, and then by (iii) in Lemma 5, it follows $g(z) \geqslant 0$. Furthermore, from $x_1 \geqslant x_2 > 0$, we have $p(x_1^{p+1} - x_2^{p+1}) \geqslant 0$ and $q(x_1^{q+1} - x_2^{q+1}) \geqslant 0$. Notice that $\frac{1}{p+q} > 0$, from (16), we conclude that $\Delta_3 \geqslant 0$, by Lemma 3, it follows that $B^{p,q}(\mathbf{x})$ is Schur-harmonic convex with $\boldsymbol{x} \in \mathbf{R}_+^n$.

(iv) If $p \geqslant 0, q \leqslant -1$ and $p + q < 0$, then $(p - q)^2 + p + q = p^2 + q(q - p + 1) + p(1 - q) \geqslant 0$, by (iv) in Lemma 5, it follows $g(z) \geqslant 0$. Furthermore, from $x_1 \geqslant x_2 > 0$, we have $p(x_1^{p+1} - x_2^{p+1}) \geqslant 0$ and $q(x_1^{q+1} - x_2^{q+1}) \geqslant 0$. Notice that $\frac{1}{p+q} < 0$, from (16), we conclude that $\Delta_3 \leqslant 0$, by Lemma 3, it follows that $B^{p,q}(\boldsymbol{x})$ is Schur-harmonic concave with $\boldsymbol{x} \in \mathbf{R}_+^n$.

The proof of Theorem 3 is completed.

77.4　Applications

Theorem 4 Let $n \geqslant 3$ and $(p,q) \in \mathbf{R}^2$. If $(p,q) \in \{(p,q)|0 \leqslant q \leqslant p \leqslant 1$ and $p+q \neq 0, p-q \leqslant \sqrt{p+q}\} \cup \{(p,q)|p+q < 0, p \geqslant 1, q \leqslant 0\}$, then for $\boldsymbol{x} \in \mathbf{R}_+^n$, we have

$$B^{p,q}(\boldsymbol{x}) \leqslant A_n(\boldsymbol{x}) \tag{17}$$

if $(p,q) \in \{(p,q)|p \geqslant 1, q \leqslant 0$ and $p+q > 0\}$, then the inequality (17) is reversed.

Proof If $(p,q) \in \{(p,q)|0 \leqslant q \leqslant p \leqslant 1$ and $p+q \neq 0, p-q \leqslant \sqrt{p+q}\} \cup \{(p,q)|p+q < 0, p \geqslant 1, q \leqslant 0\}$, then by Theorem 1, from Lemma 6, we have

$$B^{p,q}(\boldsymbol{u}) \geqslant B^{p,q}(\boldsymbol{x})$$

rearranging gives (17), if $(p,q) \in \{(p,q)|p \geqslant 1, q \leqslant 0$ and $p+q > 0\}$, then the inequality (17) is reversed.

The proof is complete.

Theorem 5 Let $n \geqslant 3$ and $(p,q) \in \mathbf{R}^2$, $\boldsymbol{x} \in \mathbf{R}_+^n$, and the constant c satisfying $0 < c < \min\{x_i\}, i = 1,2,\ldots,n$. If $(p,q) \in \{(p,q)|0 \leqslant q \leqslant p \leqslant 1$ and $p+q \neq 0, p-q \leqslant \sqrt{p+q}\} \cup \{(p,q)|p+q < 0, p \geqslant 1, q \leqslant 0\}$, then we have

$$B^{p,q}(x_1 - c, x_2 - c, \ldots, x_n - c) \leqslant \left(1 - \frac{c}{A_n(\boldsymbol{x})}\right) B^{p,q}(x_1, x_2, \ldots, x_n) \tag{18}$$

if $(p,q) \in \{(p,q)|p \geqslant 1, q \leqslant 0$ and $p+q > 0\}$, then the inequality (18) is reversed.

Proof Note that $0 < c < \min\{x_i\} < \frac{1}{n}\sum_{i=1}^{n} x_i$, if $(p,q) \in \{(p,q)|0 \leqslant q \leqslant p \leqslant 1$ and $p+q \neq 0, p-q \leqslant \sqrt{p+q}\} \cup \{(p,q)|p+q < 0, p \geqslant 1, q \leqslant 0\}$, then by Theorem 1, from Lemma 7, we have

$$B^{p,q}\left(\frac{x_1}{\sum\limits_{j=1}^{n} x_j}, \ldots, \frac{x_n}{\sum\limits_{j=1}^{n} x_j}\right) \geqslant B^{p,q}\left(\frac{x_1 - c}{\sum\limits_{j=1}^{n}(x_j - c)}, \ldots, \frac{x_n - c}{\sum\limits_{j=1}^{n}(x_j - c)}\right)$$

rearranging gives (18), if $(p,q) \in \{(p,q)|p \geqslant 1, q \leqslant 0$ and $p+q > 0\}$, then the inequality (18) is reversed.

The proof is complete.

Theorem 6 Let $n \geqslant 3$ and $(p,q) \in \mathbf{R}^2$. If $p+q > 0$, then for $\boldsymbol{x} \in \mathbf{R}_+^n$,

we have

$$B^{p,q}(\boldsymbol{x}) \geqslant G_n(\boldsymbol{x}) \tag{19}$$

where $G_n(\boldsymbol{x}) = \sqrt[n]{\prod\limits_{i=1}^{n} x_i}$. If $p + q < 0$, then the inequality (19) is reversed.

Proof From Lemma 6, we have

$$(\log G_n(\boldsymbol{x}), \log G_n(\boldsymbol{x}), \ldots, \log G_n(\boldsymbol{x})) \prec (\log x_1, \log x_2, \ldots, \log x_n)$$

If $p + q < 0$, then by Theorem 2, we have

$$G_n(\boldsymbol{x}) = B^{p,q}\left(G_n(\boldsymbol{x}), G_n(\boldsymbol{x}), \ldots, G_n(\boldsymbol{x})\right) \leqslant B^{p,q}\left(x_1, x_2, \ldots, x_n\right) = B^{p,q}(\boldsymbol{x})$$

If $p + q < 0$, then the inequality (19) is reversed.

The proof is complete.

Theorem 7 Let $n \geqslant 3$ and $(p, q) \in \mathbf{R}^2$. If $p \geqslant q \geqslant 0$ and $p + q \neq 0$, then for $\boldsymbol{x} \in \mathbf{R}_+^n$, we have

$$B^{p,q}(\boldsymbol{x}) \geqslant H_n(\boldsymbol{x}) \tag{20}$$

where $H_n(\boldsymbol{x}) = \dfrac{n}{\sum\limits_{i=1}^{n} x_i^{-1}}$.

From Lemma 6, we have

$$\left(\frac{1}{H_n(\boldsymbol{x})}, \frac{1}{H_n(\boldsymbol{x})}, \ldots, \frac{1}{H_n(\boldsymbol{x})}\right) \prec \left(\frac{1}{x_1}, \frac{1}{x_2}, \ldots, \frac{1}{x_n}\right).$$

If $p \geqslant q \geqslant 0$ and $p + q \neq 0$, then by Theorem 3, we have

$$H_n(\boldsymbol{x}) = B^{p,q}\left(H_n(\boldsymbol{x}), H_n(\boldsymbol{x}), \ldots, H_n(\boldsymbol{x})\right) \leqslant B^{p,q}\left(x_1, x_2, \ldots, x_n\right) = B^{p,q}(\boldsymbol{x})$$

The proof is complete.

References

[1] C BONFERRONI. Sulle medie multiple di potenze[J]. Bolletino Matematica Italiana, 1950, 5: 267–270.

[2] W-M GONG, W-F XIA, Y-M CHU. The Schur convexity for the generalized Muirhead mean[J]. J. Wath. Inequal., 2014, 8: 855–862.

[3] Y M Chu, W F Xia. Necessary and sufficient conditions for the Schur harmonic convexity of the generalized Muirhead mean[J]. Proceedings of a Razmadze Mathematical Institute, 2010, 152: 19–27.

[4]　A W MARSHALL, I OLKIN, B C ARNOLD. Inequalities: theory of majorization and its application[M]. 2nd ed. New York: Springer Press, 2011.

[5]　B-Y WANG. Foundations of majorization inequalities[M]. Beijing: Beijing Normal Univ. Press, 1990.

[6]　X-M ZHANG. Geometrically convex functions[M]. Hefei: Anhui University Press, 2004.

[7]　C P NICULESCU. Convexity according to the geometric mean[J]. Mathematical Inequalities & Applications, 2000, 3 (2): 155–167.

[8]　Y-M CHU, G-D WANG, X-H ZHANG. The Schur multiplicative and harmonic convexities of the complete symmetric function[J]. Mathematische Nachrichten, 2011, 284 (5-6): 653–663.

[9]　J-X MENG, Y M CHU, X-M TANG. The Schur-harmonic-convexity of dual form of the Hamy symmetric function[J]. Matematički Vesnik, 2010, 62: 37–46.

[10]　K Z GUAN, J-H SHAN. Schur‑convexity for a class of symmetric function and its applications[J]. Math. Inequal. Appl., 2006, 9: 199–210.

[11]　K Z GUAN. Some inequalities for a class of generalized means[J]. J. Inequal. Pure Appl. Math., 2004, 5 (3), Article 69.

[12]　N-G ZHENG, Z-H ZHANG, X-M ZHANG. Schur-convexity of two types of one-parameter mean values in n variables[J]. J. Inequal. Appl., Volume 2007, Article ID 78175, 10 pages doi: 10. 1155/2007/78175.

[13]　W F XIA, Y M CHU. The Schur multiplicative convexity of the weighted generalized logarithmic mean in n variables[J]. International Mathematical Forum, 2009, 4: 1229–1236.

[14]　W-F XIA, Y-M CHU. The Schur convexity of the weighted generalized logarithmic mean values according to harmonic mean[J]. International Journal of Modern Mathematics, 2009, 4: 225–233.

[15]　N-G ZHENG, Z-H ZHANG, X-M ZHANG. The Schur- harmonic-convexity of two types of one-parameter mean values in n variables[J]. Journal of Inequalities and Applications Volume 2007, Article ID 78175, 10 pages. doi:10.1155/2007/78175.

[16]　QIAN XU. Research on schur-p power-convexity of the quotient of arithmetic mean and geometric mean[J]. Journal of Fudan University(Natural Science), 2015, 54: 299–295.

[17] N-G ZHENG, X-M ZHANG, Y-M CHU. Convexity and Geometrical Convexity of the Identic and Logarithmic Means in N Variables[J]. Acta Mathematica Scientia, 2008, 28A: 1173 - 1180.

[18] D S WANG, C-R FU, H-N SHI. Schur-convexity for a mean of n variables with three parameters[J]. Publications de l'Institut Mathématique (Beograd), in Press.

[19] H-N SHI. Majorization Theory and Analytic Inequality[M]. Harbin: Publishing House of Harbin Institute of Technology, China 2012.

第78篇　Schur-m Power Convexity of Geometric Bonferroni Means

(SHI HUANNAN, WU SHANHE. Italian Journal of Pure and Applied

Mathematics, N. 2017, 38: 769 ‑ 776)

Abstract: Schur-convexity, Schur-geometric convexity and Schur-harmonic convexity of the geometric Bonferroni means and the generalized geometric Bonferroni mean for n variables are investigated, and some mean value inequalities of n variables are established.

Keywords: Schur-convexity; Schur m-power convexity; geometric Bonferroni means; majorization

78.1　Introduction

Throughout the article, $\mathbf{R}$ denotes the set of real numbers, $\boldsymbol{x} = (x_1, \cdots, x_n)$ denotes n-tuple (n-dimensional real vectors), the set of vectors can be written as

$$\mathbf{R}^n = \{\boldsymbol{x} = (x_1, \cdots, x_n) : x_i \in \mathbf{R}, i = 1, \cdots, n\}$$

$$\mathbf{R}_+^n = \{\boldsymbol{x} = (x_1, \cdots, x_n) : x_i \geqslant 0, i = 1, \cdots, n\}$$

$$\mathbf{R}_{++}^n = \{\boldsymbol{x} = (x_1, \cdots, x_n) : x_i > 0, i = 1, \cdots, n\}$$

In particular, the notations $\mathbf{R}$, $\mathbf{R}_+$ and $\mathbf{R}_{++}$ denote $\mathbf{R}^1$, $\mathbf{R}_+^1$ and $\mathbf{R}_{++}^1$, respectively.

The Bonferroni mean was initially proposed by Bonferroni [1] and was also investigated intensively by Yager[2]. It has important application in multi criteria decision making(see[2-7]).

637

Definition 1 Let $x = (x_1, x_2, \cdots, x_n) \in \mathbf{R}_+^n$ and $p, q \geqslant 0, p + q \neq 0$. If

$$B^{p,q}(x) = \left(\frac{1}{n(n-1)} \sum_{i,j=1,i\neq j}^{n} x_i^p x_j^q \right)^{\frac{1}{p+q}} \tag{1}$$

$B^{p,q}(x)$ is called the Bonferroni mean.

Based on the usual geometric mean $G(x) = \prod_{i=1}^{n} (x_i)^{\frac{1}{n}}$ and $B^{p,q}(x)$, M. Xia et al [4] introduce the geometric Bonferroni mean such as:

Definition 2 Let $x = (x_1, x_2, \cdots, x_n) \in \mathbf{R}_+^n$ and $(p, q) \in \mathbf{R}_{++}^2$. If

$$GB^{p,q}(x) = \frac{1}{p+q} \prod_{i,j=1,i\neq j}^{n} (px_i + qx_j)^{\frac{1}{n(n-1)}} \tag{2}$$

$GB^{p,q}(x)$ is called the geometric Bonferroni mean.

Obviously, the geometric Bonferroni mean has the following properties:

(i) $GB^{p,q}(0, 0, \cdots, 0) = 0$;

(ii) $GB^{p,q}(x, x, \cdots, x) = x$, if $x_i = x$, for all i;

(iii) $GB^{p,q}(x) \geqslant GB^{p,q}(x)$. i.e., the geometric Bonferroni mean is monotonic, if $x_i \geqslant y_i$, for all i;

(iv) $\min\{x_i\} \leqslant GB^{p,q}(x) \leqslant \max\{x_i\}$.

Furthermore, if $q = 0$, then by Eq.(3), it reduces to the geometric mean

$$GB^{p,0}(x) = \frac{1}{p} \prod_{i,j=1,i\neq j}^{n} (px_i)^{\frac{1}{n(n-1)}} = \prod_{i=1}^{n} (x_i)^{\frac{1}{n}} = G(x)$$

In recent years, the study on the properties of the mean by using theory of majorization is unusually active. Yang [8-10] generalized the notion of Schur convexity to Schur f-convexity. Moreover, he discussed Schur m-power convexity of Stolarsky means[8], Gini means [9] and Daróczy means [10]. Subsequently, many scholars have aroused the interest of Schur m-power convexity (see [11-14]).

In this paper, we discuss the Schur m-power convexity of the geometric Bonferroni mean $B^{p,q}(x)$, Our main results are as follows:

Theorem 1 For fixed $(p, q) \in \mathbf{R}_{++}^2$ and $n \geqslant 3$.

(i) If $m = 0$, then $B^{p,q}(x)$ is Schur m-power concave;

(ii) If $m < 0$, then $B^{p,q}(\boldsymbol{x})$ is Schur m-power convex;

(iii) If $m = 1$, then $B^{p,q}(\boldsymbol{x})$ is Schur m-power concave;

(iv) If $m \geqslant 2$, then $B^{p,q}(\boldsymbol{x})$ is Schur m-power concave.

78.2　Preliminaries

For convenience, we introduce some definitions as follows.

Definition 3 [15−16] Let $\boldsymbol{x} = (x_1, \cdots, x_n)$ and $\boldsymbol{y} = (y_1, \cdots, y_n) \in \mathbf{R}^n$.

(i) $\boldsymbol{x} \geq \boldsymbol{y}$ means $x_i \geqslant y_i$ for all $i = 1, 2, \cdots, n$.

(ii) Let $\Omega \subset \mathbf{R}^n$, $\varphi \colon \Omega \to \mathbf{R}$ is said to be increasing if $\boldsymbol{x} \geq \boldsymbol{y}$ implies $\varphi(\boldsymbol{x}) \geqslant \varphi(\boldsymbol{y})$. φ is said to be decreasing if and only if $-\varphi$ is increasing.

Definition 4 [15−16] Let $\boldsymbol{x} = (x_1, \cdots, x_n)$ and $\boldsymbol{y} = (y_1, \cdots, y_n) \in \mathbf{R}^n$.

(i) $\boldsymbol{x}$ is said to be majorized by $\boldsymbol{y}$ (in symbols $\boldsymbol{x} \prec \boldsymbol{y}$) if $\sum_{i=1}^{k} x_{[i]} \leqslant \sum_{i=1}^{k} y_{[i]}$ for $k = 1, 2, \cdots, n-1$ and $\sum_{i=1}^{n} x_i = \sum_{i=1}^{n} y_i$, where $x_{[1]} \geqslant \cdots \geqslant x_{[n]}$ and $y_{[1]} \geqslant \cdots \geqslant y_{[n]}$ are rearrangements of $\boldsymbol{x}$ and $\boldsymbol{y}$ in a descending order.

(ii) Let $\Omega \subset \mathbf{R}^n$, the function $\varphi \colon \Omega \to \mathbf{R}$ is said to be Schur-convex on Ω if $\boldsymbol{x} \prec \boldsymbol{y}$ on Ω implies $\varphi(\boldsymbol{x}) \leq \varphi(\boldsymbol{y})$. φ is said to be a Schur-concave function on Ω if and only if $-\varphi$ is Schur-convex function on Ω.

Definition 5 [15−16] Let $\boldsymbol{x} = (x_1, \cdots, x_n)$ and $\boldsymbol{y} = (y_1, \cdots, y_n) \in \mathbf{R}^n$.

(i) $\Omega \subset \mathbf{R}^n$ is said to be a convex set if $\boldsymbol{x}, \boldsymbol{y} \in \Omega, 0 \leqslant \alpha \leqslant 1$ implies $\alpha\boldsymbol{x} + (1-\alpha)\boldsymbol{y} = (\alpha x_1 + (1-\alpha)y_1, \cdots, \alpha x_n + (1-\alpha)y_n) \in \Omega$.

(ii) Let $\Omega \subset \mathbf{R}^n$ be convex set. A function $\varphi \colon \Omega \to \mathbf{R}$ is said to be convex on Ω if

$$\varphi(\alpha\boldsymbol{x} + (1-\alpha)\boldsymbol{y}) \leqslant \alpha\varphi(\boldsymbol{x}) + (1-\alpha)\varphi(\boldsymbol{y})$$

for all $\boldsymbol{x}, \boldsymbol{y} \in \Omega$, and all $\alpha \in [0, 1]$. The function φ is said to be concave on Ω if and only if $-\varphi$ is convex function on Ω.

Definition 6 [15−16]

(i) A set $\Omega \subset \mathbf{R}^n$ is called symmetric, if $\boldsymbol{x} \in \Omega$ implies $\boldsymbol{x}\boldsymbol{P} \in \Omega$ for every $n \times n$ permutation matrix $\boldsymbol{P}$.

(ii) A function $\varphi : \Omega \to \mathbf{R}$ is called symmetric if for every permutation matrix $\boldsymbol{P}$, $\varphi(\boldsymbol{x}\boldsymbol{P}) = \varphi(\boldsymbol{x})$ for all $\boldsymbol{x} \in \Omega$.

Lemma 1 [15-16] Let $\Omega \subset \mathbf{R}^n$ be symmetric and have a nonempty interior convex set. Ω^0 is the interior of Ω. $\varphi : \Omega \to \mathbf{R}$ is continuous on Ω and differentiable in Ω^0. Then φ is the $Schur - convex$ ($Schur - concave$) $function$ if and only if φ is symmetric on Ω and

$$(x_1 - x_2) \left(\frac{\partial \varphi}{\partial x_1} - \frac{\partial \varphi}{\partial x_2} \right) \geqslant 0 (\leqslant 0) \tag{3}$$

holds for any $\boldsymbol{x} \in \Omega^0$.

The first systematical study of the functions preserving the ordering of majorization was made by Issai Schur in 1923. In Schur's honor, such functions are said to be "Schur-convex". It can be used extensively in analytic inequalities, combinatorial optimization, quantum physics, information theory, and other related fields. See [11].

Definition 7 [8] Let $f : \mathbf{R}_+ \to \mathbf{R}$ be defined by

$$f(x) = \begin{cases} \dfrac{x^m - 1}{m}, & m \neq 0 \\ \ln x, & m = 0 \end{cases} \tag{4}$$

Then a function $\phi : \Omega \subset \mathbf{R}_+^n \to \mathbf{R}$ is said to be Schur m-power convex on Ω if

$$(f(x_1), f(x_2), \ldots, f(x_n)) \prec (f(y_1), f(y_2), \ldots, f(y_n))$$

for all $(x_1, x_2, \ldots, x_n) \in \Omega$ and $(y_1, y_2, \ldots, y_n) \in \Omega$ implies $\phi(x) \leq \phi(y)$.

If $-\phi$ is Schur m-power convex, then we say that ϕ is Schur m-power concave.

Lemma 2 [8] Let $\Omega \subset \mathbf{R}_+^n$ be a symmetric set with nonempty interior Ω^0 and $\varphi : \Omega \to \mathbf{R}_+$ be continuous on Ω and differentiable in Ω^0. Then φ is Schur m-power convex on Ω if and only if φ is symmetric on Ω and

$$\frac{x_1^m - x_2^m}{m} \left[x_1^{1-m} \frac{\partial \varphi(\boldsymbol{x})}{\partial x_1} - x_2^{1-m} \frac{\partial \varphi(\boldsymbol{x})}{\partial x_2} \right] \geqslant 0, \quad \text{if } m \neq 0 \tag{5}$$

and

$$(\log x_1 - \log x_2) \left[x_1 \frac{\partial \varphi(\boldsymbol{x})}{\partial x_1} - x_2 \frac{\partial \varphi(\boldsymbol{x})}{\partial x_2} \right] \geqslant 0 \quad \text{if } m = 0 \tag{6}$$

for all $\boldsymbol{x} \in \Omega^0$.

78.3　Proof of Theorem

Proof of Theorem 1

$$\log GB^{p,q}(\boldsymbol{x}) = \log \frac{1}{p+q} + \frac{1}{n(n-1)}Q$$

where

$$Q = \sum_{j=3}^{n} [\log(px_1 + qx_j) + \log(px_2 + qx_j)] +$$

$$\sum_{i=3}^{n} [\log(px_i + qx_1) + \log(px_i + qx_2)] +$$

$$\log(px_1 + qx_2) + \log(px_2 + qx_1)$$

$$\frac{\partial GB^{p,q}(\boldsymbol{x})}{\partial x_1} = \frac{GB^{p,q}(\boldsymbol{x})}{n(n-1)} \left(\sum_{j=3}^{n} \frac{p}{px_1+qx_j} + \sum_{i=3}^{n} \frac{q}{px_i+qx_1} + \frac{p}{px_1+qx_2} + \frac{q}{px_2+qx_1} \right)$$

$$\frac{\partial GB^{p,q}(\boldsymbol{x})}{\partial x_2} = \frac{GB^{p,q}(\boldsymbol{x})}{n(n-1)} \left(\sum_{j=3}^{n} \frac{p}{px_2+qx_j} + \sum_{i=3}^{n} \frac{q}{px_i+qx_2} + \frac{q}{px_1+qx_2} + \frac{p}{px_2+qx_1} \right)$$

It is easy to see that $GB^{p,q}(\boldsymbol{x})$ is symmetric on $\mathbf{R}_+^n$. Without loss of generality, we may assume that $x_1 \geqslant x_2$.

$$\Delta := \frac{x_1^m - x_2^m}{m} \left(x_1^{1-m} \frac{GB^{p,q}(\boldsymbol{x})}{\partial x_1} - x_2^{1-m} \frac{\partial GB^{p,q}(\boldsymbol{x})}{\partial a_2} \right)$$

$$= \frac{(x_1^m - x_2^m)GB^{p,q}(\boldsymbol{x})}{mn(n-1)} [p \sum_{j=3}^{n} (\frac{x_1^{1-m}}{px_1 + qx_j} - \frac{x_2^{1-m}}{px_2 + qx_j}) +$$

$$q \sum_{i=3}^{n} (\frac{x_1^{1-m}}{px_i + qx_1} - \frac{x_2^{1-m}}{px_i + qx_2}) +$$

$$\frac{px_1^{1-m} - qx_2^{1-m}}{px_1 + qx_2} + \frac{qx_1^{1-m} - px_2^{1-m}}{px_2 + qx_1}]$$

$$= \frac{(x_1^m - x_2^m)GB^{p,q}(\boldsymbol{x})}{mn(n-1)} [p \sum_{j=3}^{n} \frac{px_1x_2(x_1^{-m} - x_2^{-m}) + qx_j(x_1^{1-m} - x_2^{1-m})}{(px_1 + qx_j)(px_2 + qx_j)} +$$

$$q \sum_{i=3}^{n} \frac{qx_1x_2(x_1^{-m} - x_2^{-m}) + px_i(x_1^{1-m} - x_2^{1-m})}{(px_i + qx_1)(px_i + qx_2)} +$$

$$\frac{x_1x_2(p^2 + q^2)(x_1^{-m} - x_2^{-m}) + 2pq(x_1^{2-m} - x_2^{2-m})}{(px_1 + qx_2)(px_2 + qx_1)}]$$

If $m < 0$, then $x_1^m - x_2^m \leqslant 0$, $x_1^{-m} - x_2^{-m} \geqslant 0$, $x_1^{1-m} - x_2^{1-m} \geqslant 0$ and $x_1^{2-m} - x_2^{2-m} \geqslant 0$. Thus, $\Delta \geqslant 0$, by Lemma 2, it follows that $GB^{p,q}(\boldsymbol{x})$ is Schur m-power convex with $\boldsymbol{x} \in \mathbf{R}_{++}^n$.

If $m \geqslant 2$, then $x_1^m - x_2^m \geqslant 0$, $x_1^{-m} - x_2^{-m} \leqslant 0$, $x_1^{1-m} - x_2^{1-m} \leqslant 0$ and $x_1^{2-m} - x_2^{2-m} \leqslant 0$. Thus, $\Delta \leqslant 0$, by Lemma 2, it follows that $GB^{p,q}(\boldsymbol{x})$ is Schur m-power concave with $\boldsymbol{x} \in \mathbf{R}_{++}^n$.

If $m = 1$, then

$$
\Delta = -\frac{(x_1 - x_2)^2 GB^{p,q}(\boldsymbol{x})}{n(n-1)} \Big[\sum_{j=3}^{n} \frac{p^2}{(px_1 + qx_j)(px_2 + qx_j)} +
$$

$$
\sum_{i=3}^{n} \frac{q^2}{(px_i + qx_1)(px_i + qx_2)} + \frac{(p-q)^2}{(px_1 + qx_2)(px_2 + qx_1)}\Big] \leqslant 0
$$

by Lemma 2, it follows that $GB^{p,q}(\boldsymbol{x})$ is Schur m-power concave with $\boldsymbol{x} \in \mathbf{R}_{++}^n$.

If $m = 0$, then

$$
\Delta = (\log x_1 - \log x_2)\left(x_1 \frac{GB^{p,q}(\boldsymbol{x})}{\partial x_1} - x_2 \frac{\partial GB^{p,q}(\boldsymbol{x})}{\partial x_2}\right)
$$

$$
= \frac{(\log x_1 - \log x_2)GB^{p,q}(\boldsymbol{x})}{n(n-1)}\Big[p \sum_{j=3}^{n}\Big(\frac{x_1}{px_1 + qx_j} - \frac{x_2}{px_2 + qx_j}\Big) +
$$

$$
q \sum_{i=3}^{n}\Big(\frac{x_1}{px_i + qx_1} - \frac{x_2}{px_i + qx_2}\Big) + \frac{px_1 - qx_2}{px_1 + qx_2} + \frac{qx_1 - px_2}{px_2 + qx_1}\Big]
$$

$$
= \frac{(x_1 - x_2)(\log x_1 - \log x_2)GB^{p,q}(\boldsymbol{x})}{n(n-1)}\Big[p \sum_{j=3}^{n} \frac{qx_j}{(px_1 + qx_j)(px_2 + qx_j)} +
$$

$$
q \sum_{i=3}^{n} \frac{px_i}{(px_i + qx_1)(px_i + qx_2)} + \frac{2pq(x_1 + x_2)}{(px_1 + qx_2)(px_2 + qx_1)}\Big] \geqslant 0
$$

By Lemma 2, it follows that $GB^{p,q}(\boldsymbol{x})$ is Schur-m-power convex with $\boldsymbol{x} \in \mathbf{R}_{++}^n$.

The proof of Theorem 1 is completed.

Problem When $0 < m < 1$ or $1 < m < 2$, how the Schur-m-power convexity of the geometric Bonferroni mean $GB^{p,q}(\boldsymbol{x})$?

References

[1] C BONFERRONI. Sulle medie multiple di potenze[J]. Bolletino Matematica Italiana, 1950, 5: 267–270.

[2] R R YAGER. On generalized Bonferroni mean operators for multi-

criteria aggregation[J]. International Journal of Approximate Reasoning, 2009, 50: 1279–1286.

[3]　JIN HAN PARK, EUN JIN PARK. Generalized fuzzy Bonferroni harmonic mean operators and their applications in group decision making[J]. Journal of Applied Mathematics Volume 2013, Article ID 604029, 14 pages.

[4]　MEIMEI XIA, ZESHUI XU, BIN ZHU. Generalized intuitionistic fuzzy Bonferroni means[J] International Journal of Intelligent Systems, January 2012 DOI: 10.1002/int.20515 • Source: DBLP

[5]　JIN HAN PARK, JI YU KIM. Intuitionistic fuzzy optimized weighted geometric Bonferroni means and their applications in Group Decision Making[J]. Fundamenta Informaticae, 2016, 144: 363–381. DOI 10.3233/FI-2016-1341.

[6]　G BELIAKOV, S JAMES, J MORDELOV′ A, T RÜCKSCHLOSSOV′ A, R R YAGER. Generalized Bonferroni mean operators in multicriteria aggregation[J]. Fuzzy Sets and Systems, 2010,161: 2227–2242.

[7]　ZESHUI XU, RONALD R YAGER. Intuitionistic Fuzzy Bonferroni Means[J]. IEEE Transactions on Systems, Man, and Cybernetics, Part B (Cybernetics) 2011, 41 (2).

[8]　Z-H YANG. Schur power convexity of Stolarsky means[J]. Publ. Math. Debrecen, 2012, 80 (1-2): 43–66.

[9]　Z-H YANG. Schur power convexity of Gini means[J]. Bull. Korean Math. Soc., 2013, 50 (2): 485–498.

[10]　Z-H YANG. Schur power comvexity of the daróczy means[J]. Math. Inequal. Appl. 2013, 16 (3): 751–762.

[11]　Y-P DENG, S-H WU, Y-M CHU, D HE. The Schur convexity of the generalized Muirhead-Heronian means[J]. Abstract and Applied Analysis, Volume 2014 (2014), Article ID 706518, 11 pages.

[12]　W WANG, S-G YANG. Schur m-power convexity of a class of multiplicatively convex functions and applications[J]. Abstract and Applied Analysis Volume 2014 (2014), Article ID 258108, 12 pages.

[13]　Q XU. Reserch on Schur p power-convexity of the quotient of arithmetic mean and geometric mean[J]. Journal of Fudan University (Natural Science), 2015, 54 (3): 299–295.

[14]　H-P YIN, H-N SHI, F QI. On Schur m-power convexity for ratios of some means[J]. J. Math. Inequal., 2015, 9 (1): 145–153.

[15] A W MARSHALL, I OLKIN, B C ARNOLD. Inequalities: theory of majorization and its application[M]. 2nd ed. New York: Springer Press, 2011.

[16] B-Y WANG. Foundations of majorization inequalities[M]. Beijing: Beijing Normal Univ. Press, 1990.

第79篇　Schur Convexity of Generalized Geometric Bonferroni Means and the Relevant Inequalities

(Huan-Nan Shi and Shan-He Wu. Journal of Inequalities and Applications 20182018:8 https://doi.org/10.1186/s13660-017-1605-7)

Abstract: In this paper, we discuss the Schur convexity, Schur geometric convexity and Schur harmonic convexity of the generalized geometric Bonferroni mean. Some inequalities related to the generalized geometric Bonferroni mean are established to illustrate the applications of the obtained results.

Keywords: geometric Bonferroni mean; Schur's condition; majorization relationship; inequality

79.1　Introduction

The Schur convexity of functions relating to special means is a very significant research subject and has attracted the interest of many mathematicians. There are numerous articles written on this topic in recent years; see [1-2] and the references therein. As supplements to the Schur convexity of functions, the Schur geometrically convex functions and Schur harmonically convex functions were investigated by Zhang and Yang [3], Chu, Zhang and Wang [4], Chu and Xia [5], Chu, Wang and Zhang [6], Shi and Zhang [7-8], Meng, Chu and Tang [9], Zheng, Zhang and Zhang [10]. These properties of functions have been found to be useful in discovering and proving the inequalities for specialmeans (see [11 - 14]).

Recently, it has come to our attention that a type of means which is symmetrical on n variables $x_1, x_2, \ldots, x_n$ and involves two parameters, it was initially proposed by Bonferroni [15], as follows

$$B^{p,q}(\boldsymbol{x}) = \left(\frac{1}{n(n-1)} \sum_{i,j=1, i \neq j}^{n} x_i^p x_j^q \right)^{\frac{1}{p+q}} \tag{1}$$

where $\boldsymbol{x} = (x_1, x_2, \ldots, x_n), x_i \geqslant 0, i = 1, 2, \ldots, n, p, q \geqslant 0$ and $p + q \neq 0$.

$B^{p,q}(\boldsymbol{x})$ is called the Bonferroni mean. It has important application in multi criteria decision-making (see [16 ~ 21]).

Beliakov, James and Mordelová et al. [22] generalized the Bonferroni mean by introducing three parameters p, q, r, i.e.

$$GB^{p,q,r}(\boldsymbol{x}) = \left(\frac{1}{n(n-1)(n-2)} \sum_{i,j,k=1, i \neq j \neq k}^{n} x_i^p x_j^q x_k^r \right)^{\frac{1}{p+q+r}} \tag{2}$$

where $\boldsymbol{x} = (x_1, x_2, \ldots, x_n), x_i \geqslant 0, i = 1, 2, \ldots, n, p, q, r \geqslant 0$ and $p+q+r \neq 0$.

Motivated by the Bonferroni mean $B^{p,q}(\boldsymbol{x})$ and the geometric mean $G(\boldsymbol{x}) = \prod_{i=1}^{n} (x_i)^{\frac{1}{n}}$, Xia, Xu and Zhu [23] introduced a new mean which is called the geometric Bonferroni mean, as follows

$$GB^{p,q}(\boldsymbol{x}) = \frac{1}{p+q} \prod_{i,j=1, i \neq j}^{n} (px_i + qx_j)^{\frac{1}{n(n-1)}} \tag{3}$$

where $\boldsymbol{x} = (x_1, x_2, \ldots, x_n), x_i > 0, i = 1, 2, \ldots, n, p, q \geqslant 0$ and $p + q \neq 0$.

An extension of the geometric Bonferroni mean was given by Park and Kim in [19], which is called the generalized geometric Bonferroni mean, i.e.

$$GB^{p,q,r}(\boldsymbol{x}) = \frac{1}{p+q+r} \prod_{i,j,k=1, i \neq j \neq k}^{n} (px_i + qx_j + rx_k)^{\frac{1}{n(n-1)(n-2)}} \tag{4}$$

where $\boldsymbol{x} = (x_1, x_2, \ldots, x_n), x_i > 0, i = 1, 2, \ldots, n, p, qr \geqslant 0$ and $p+q+r \neq 0$.

Remark 1 For $r = 0$, it is easy to observe that

$$GB^{p,q,0}(\boldsymbol{x}) = \frac{1}{p+q+0} \prod_{i,j=1, i \neq j}^{n} \left[\prod_{k=1, i \neq j \neq k}^{n} (px_i + qx_j + 0 \times x_k) \right]^{\frac{1}{n(n-1)(n-2)}}$$

$$= \frac{1}{p+q} \sum_{i,j=1, i \neq j}^{n} \left[(px_i + qx_j)^{(n-2)} \right]^{\frac{1}{n(n1)(n2)}}$$

$$= \frac{1}{p+q} \sum_{i,j=1,i\neq j}^{n} (px_i + qx_j)^{\frac{1}{n(n1)}}$$

$$= GB^{p,q}(\boldsymbol{x})$$

Remark 2　If $q = 0, r = 0$, then the generalized geometric Bonferroni mean reduces to the geometric mean, i.e.

$$GB^{p,0,0}(\boldsymbol{x}) = GB^{p,0}(\boldsymbol{x}) = \frac{1}{p} \prod_{i,j=1,i\neq j}^{n} (px_i)^{\frac{1}{n(n-1)}} = \prod_{i=1}^{n}(x_i)^{\frac{1}{n}} = G(\boldsymbol{x})$$

Remark 3　If $\boldsymbol{x} = (x, x, \ldots, x)$, then

$$GB^{p,q,r}(\boldsymbol{x}) = GB^{p,q,r}(x, x, \cdots, x) = x$$

For convenience, throughout the paper $\mathbf{R}$ denotes the set of real numbers, $\boldsymbol{x} = (x_1, x_2, \ldots, x_n)$ denotes n-tuple (n-dimensional real vectors), the set of vectors can be written as

$$\mathbf{R}^n = \{\boldsymbol{x} = (x_1, \cdots, x_n) : x_i \in \mathbf{R}, i = 1, \cdots, n\}$$

$$\mathbf{R}_+^n = \{\boldsymbol{x} = (x_1, \cdots, x_n) : x_i \geqslant 0, i = 1, \cdots, n\}$$

$$\mathbf{R}_{++}^n = \{\boldsymbol{x} = (x_1, \cdots, x_n) : x_i > 0, i = 1, \cdots, n\}$$

In a recent paper [24], Shi and Wu investigated the Schur m-power convexity of the geometric Bonferroni mean $GB^{p,q}(\boldsymbol{x})$. The definition of Schur m-power convex function is as follows

$$f(x) = \begin{cases} \dfrac{x^m - 1}{m}, & m \neq 0 \\ \ln x, & m = 0 \end{cases} \tag{5}$$

Then a function $\phi : \Omega \subset \mathbf{R}_+^n \to \mathbf{R}$ is said to be Schur m-power convex on Ω if

$$(f(x_1), f(x_2), \ldots, f(x_n)) \prec (f(y_1), f(y_2), \ldots, f(y_n))$$

for all $(x_1, x_2, \ldots, x_n) \in \Omega$ and $(y_1, y_2, \ldots, y_n) \in \Omega$ implies $\phi(x) \leq \phi(y)$.

If $-\phi$ is Schur m-power convex, then we say that ϕ is Schur m-power concave.

Shi and Wu [24] obtained the following result.

Proposition 1 For fixed positive real numbers p, q, (i) if $m < 0$ or $m = 0$, then $GB^{p,q}(\boldsymbol{x})$ is Schur m-power convex on $\mathbf{R}_{++}^n$; (ii) if $m = 1$ or $m \geqslant 2$, then $GB^{p,q}(\boldsymbol{x})$ is Schur m-power concave on $\mathbf{R}_{++}^n$.

In this paper we discuss the Schur convexity, Schur geometric convexity and Schur harmonic convexity of the generalized geometric Bonferroni mean $GB^{p,q,r}(\boldsymbol{x})$. Our main results are as follows.

Theorem 1 For fixed non-negative real numbers p, q, r with $p+q+r \neq 0$, if $\boldsymbol{x} = (x_1, x_2, \dots, x_n), n \geqslant 3$, then $GB^{p,q,r}(\boldsymbol{x})$ is Schur concave, Schur geometric convex and Schur harmonic convex on $\mathbf{R}_{++}^n$.

Proposition 2 For fixed non-negative real numbers p, q with $p+q \neq 0$, if $\boldsymbol{x} = (x_1, x_2, \dots, x_n), n \geqslant 3$, then $GB^{p,q}(\boldsymbol{x})$ is Schur concave, Schur geometric convex and Schur harmonic convex on $\mathbf{R}_{++}^n$.

79.2 Preliminaries

We introduce some definitions, lemmas and propositions, which will be used in the proofs of the main results in subsequent sections.

Definition 1 Let $\boldsymbol{x} = (x_1, \cdots, x_n)$ and $\boldsymbol{y} = (y_1, \cdots, y_n) \in \mathbf{R}^n$.

(i) $\boldsymbol{x}$ is said to be majorized by $\boldsymbol{y}$ (in symbols $\boldsymbol{x} \prec \boldsymbol{y}$) if $\sum_{i=1}^{k} x_{[i]} \leqslant \sum_{i=1}^{k} y_{[i]}$ for $k = 1, 2, \cdots, n-1$ and $\sum_{i=1}^{n} x_i = \sum_{i=1}^{n} y_i$, where $x_{[1]} \geqslant \cdots \geqslant x_{[n]}$ and $y_{[1]} \geqslant \cdots \geqslant y_{[n]}$ are rearrangements of $\boldsymbol{x}$ and $\boldsymbol{y}$ in a descending order.

(ii) Let $\Omega \subset \mathbf{R}^n$, the function $\varphi \colon \Omega \to \mathbf{R}$ is said to be Schur-convex on Ω if $\boldsymbol{x} \prec \boldsymbol{y}$ on Ω implies $\varphi(\boldsymbol{x}) \leq \varphi(\boldsymbol{y})$. φ is said to be a Schur-concave function on Ω if and only if $-\varphi$ is Schur-convex function on Ω.

Definition 2 Let $\boldsymbol{x} = (x_1, \cdots, x_n)$ and $\boldsymbol{y} = (y_1, \cdots, y_n) \in \mathbf{R}^n$.

(i) $\Omega \subset \mathbf{R}^n$ is said to be a convex set if $\boldsymbol{x}, \boldsymbol{y} \in \Omega, 0 \leqslant \alpha \leqslant 1$ implies $\alpha \boldsymbol{x} + (1-\alpha)\boldsymbol{y} = (\alpha x_1 + (1-\alpha)y_1, \cdots, \alpha x_n + (1-\alpha)y_n) \in \Omega$.

(ii) Let $\Omega \subset \mathbf{R}^n$ be convex set. A function $\varphi \colon \Omega \to \mathbf{R}$ is said to be convex on Ω if

$$\varphi(\alpha \boldsymbol{x} + (1-\alpha)\boldsymbol{y}) \leqslant \alpha\varphi(\boldsymbol{x}) + (1-\alpha)\varphi(\boldsymbol{y})$$

for all $\boldsymbol{x}, \boldsymbol{y} \in \Omega$, and all $\alpha \in [0, 1]$. The function φ is said to be concave on Ω if and only if $-\varphi$ is convex function on Ω.

Definition 3

(i) A set $\Omega \subset \mathbf{R}^n$ is called symmetric, if $\boldsymbol{x} \in \Omega$ implies $\boldsymbol{x}P \in \Omega$ for every $n \times n$ permutation matrix $\boldsymbol{P}$.

(ii) A function $\varphi : \Omega \to \mathbf{R}$ is called symmetric if for every permutation matrix $\boldsymbol{P}$, $\varphi(\boldsymbol{x}P) = \varphi(\boldsymbol{x})$ for all $\boldsymbol{x} \in \Omega$.

The following proposition is called Schur's condition. It provides an approach for testing whether a vector valued function is Schur convex or not.

Proposition 3[4]　Let $\Omega \subset \mathbf{R}^n$ be symmetric and have a nonempty interior convex set. Ω^0 is the interior of Ω. $\varphi : \Omega \to \mathbf{R}$ is continuous on Ω and differentiable in Ω^0. Then φ is the $Schur - convex$ $(Schur - concave)$ $function$ if and only if φ is symmetric on Ω and

$$(x_1 - x_2)\left(\frac{\partial \varphi}{\partial x_1} - \frac{\partial \varphi}{\partial x_2}\right) \geqslant 0 (\leqslant 0) \tag{6}$$

holds for any $\boldsymbol{x} \in \Omega^0$.

Definition 4[6-7]　Let $\boldsymbol{x} = (x_1, \cdots, x_n) \in \mathbf{R}_+^n$ and $\boldsymbol{y} = (y_1, \cdots, y_n) \in \mathbf{R}_+^n$.

(i) $\Omega \subset \mathbf{R}_+^n$ is called a geometrically convex set if $(x_1^\alpha y_1^\beta, \cdots, x_n^\alpha y_n^\beta) \in \Omega$ for all $\boldsymbol{x}, \boldsymbol{y} \in \Omega$ and $\alpha, \beta \in [0, 1]$ such that $\alpha + \beta = 1$.

(ii) Let $\Omega \subset \mathbf{R}_+^n$. The function $\varphi : \Omega \to \mathbf{R}_+$ is said to be Schur geometrically convex function on Ω if $(\log x_1, \cdots, \log x_n) \prec (\log y_1, \cdots, \log y_n)$ on Ω implies $\varphi(\boldsymbol{x}) \leqslant \varphi(\boldsymbol{y})$. The function φ is said to be a Schur geometrically concave function on Ω if and only if $-\varphi$ is Schur geometrically convex function.

Proposition 4 [6]　Let $\Omega \subset \mathbf{R}_+^n$ be a symmetric and geometrically convex set with a nonempty interior Ω^0. Let $\varphi : \Omega \to \mathbf{R}_+$ be continuous on Ω and differentiable in Ω^0. If φ is symmetric on Ω and

$$(\log x_1 - \log x_2)\left(x_1 \frac{\partial \varphi}{\partial x_1} - x_2 \frac{\partial \varphi}{\partial x_2}\right) \geqslant 0 \quad (\leqslant 0) \tag{7}$$

holds for any $\boldsymbol{x} = (x_1, \cdots, x_n) \in \Omega^0$, then φ is a Schur geometrically convex (Schur geometrically concave) function.

Definition 5[8] Let $\Omega \subset \mathbf{R}_+^n$.

(i) A set Ω is said to be harmonically convex if $\dfrac{xy}{\lambda x + (1-\lambda)y} \in \Omega$ for every $x, y \in \Omega$ and $\lambda \in [0,1]$, where $xy = \sum\limits_{i=1}^{n} x_i y_i$ and $\dfrac{1}{x} = (\dfrac{1}{x_1}, \cdots, \dfrac{1}{x_n})$.

(ii) A function $\varphi : \Omega \to \mathbf{R}_+$ is said to be Schur harmonically convex on Ω if $\dfrac{1}{x} \prec \dfrac{1}{y}$ implies $\varphi(x) \le \varphi(y)$. A function φ is said to be a Schur harmonically concave function on Ω if and only if $-\varphi$ is a Schur harmonically convex function.

Proposition 5 [8] Let $\Omega \subset \mathbf{R}_+^n$ be a symmetric and harmonically convex set with inner points and let $\varphi : \Omega \to \mathbf{R}_+$ be a continuously symmetric function which is differentiable on Ω^0. Then φ is Schur harmonically convex (Schur harmonically concave) on Ω if and only if

$$(x_1 - x_2)\left(x_1^2 \frac{\partial \varphi(x)}{\partial x_1} - x_2^2 \frac{\partial \varphi(x)}{\partial x_2}\right) \geqslant 0 \quad (\leqslant 0), \quad x \in \Omega^0 \tag{8}$$

Remark 1 Propositions 3 and 4 provide analogous Schur's conditions for determining Schur geometrically convex functions and Schur harmonically convex functions, respectively.

Lemma 1 [4] Let $x = (x_1, x_2, \cdots, x_n) \in \mathbf{R}_+^n$ and $A_n(x) = \frac{1}{n}\sum_{i=1}^{n} x_i$. Then

$$u = \Big(\underbrace{A_n(x), A_n(x), \cdots, A_n(x)}_{n}\Big) \prec (x_1, x_2, \cdots, x_n) = x \tag{9}$$

Lemma 2 [4] If $x_i > 0, i = 1, 2, \ldots, n$, then for all nonnegative constants c satisfying $0 < c < \frac{1}{n}\sum\limits_{i=1}^{n} x_i$

$$\left(\frac{x_1}{\sum\limits_{j=1}^{n} x_j}, \cdots, \frac{x_n}{\sum\limits_{j=1}^{n} x_j}\right) \prec \left(\frac{x_1 - c}{\sum\limits_{j=1}^{n}(x_j - c)}, \cdots, \frac{x_n - c}{\sum\limits_{j=1}^{n}(x_j - c)}\right) \tag{10}$$

79.3　Proof of Theorems

Proof of Theorem 1

Write $\varphi(\boldsymbol{x}) := GB^{p,q,r}(\boldsymbol{x})$. Then

$$\log \varphi(\boldsymbol{x}) = \log \frac{1}{p+q+r} + \frac{1}{n(n-1)(n-2)} D$$

where

$$
\begin{aligned}
D = & \sum_{j,k=3,j\neq k}^{n} [\log(px_1 + qx_j + rx_k) + \log(px_2 + qx_j + rx_k)] + \\
& \sum_{i,k=3,i\neq k}^{n} [\log(px_i + qx_1 + rx_k) + \log(px_i + qx_2 + rx_k)] + \\
& \sum_{i,j=3,i\neq j}^{n} [\log(px_i + qx_j + rx_1) + \log(px_i + qx_j + rx_2)] + \\
& \sum_{k=3}^{n} [\log(px_1 + qx_2 + rx_k) + \log(px_2 + qx_1 + rx_k)] + \\
& \sum_{j=3}^{n} [\log(px_1 + qx_j + rx_2) + \log(px_2 + qx_j + rx_1)] + \\
& \sum_{i=3}^{n} [\log(px_i + qx_1 + rx_2) + \log(px_i + qx_2 + rx_1)]
\end{aligned}
$$

$$
\begin{aligned}
\frac{\partial \varphi(\boldsymbol{x})}{\partial x_1} = & \frac{\varphi(\boldsymbol{x})}{n(n-1)(n-2)} \cdot \\
& \Big[\sum_{j,k=3,j\neq k}^{n} \frac{p}{px_1 + qx_j + rx_k} + \sum_{i,k=3,i\neq k}^{n} \frac{q}{px_i + qx_1 + rx_k} + \\
& \sum_{i,j=3,i\neq j}^{n} \frac{r}{px_i + qx_j + rx_1} + \sum_{k=3}^{n} \Big(\frac{p}{px_1 + qx_2 + rx_k} + \frac{q}{px_2 + qx_1 + rx_k} \Big) + \\
& \sum_{j=3}^{n} \Big(\frac{p}{px_1 + qx_j + rx_2} + \frac{r}{px_2 + qx_j + rx_1} \Big) + \\
& \sum_{i=3}^{n} \Big(\frac{q}{px_i + qx_1 + rx_2} + \frac{r}{px_i + qx_2 + rx_1} \Big) \Big]
\end{aligned}
$$

$$
\begin{aligned}
\frac{\partial \varphi(\boldsymbol{x})}{\partial x_2} = & \frac{\varphi(\boldsymbol{x})}{n(n-1)(n-2)} \cdot \\
& \Big[\sum_{j,k=3,j\neq k}^{n} \frac{p}{px_2 + qx_j + rx_k} + \sum_{i,k=3,i\neq k}^{n} \frac{q}{px_i + qx_2 + rx_k} +
\end{aligned}
$$

$$\sum_{i,j=3,i\neq j}^{n}\frac{r}{px_i+qx_j+rx_2}+\sum_{k=3}^{n}(\frac{q}{px_1+qx_2+rx_k}+\frac{p}{px_2+qx_1+rx_k})+$$

$$\sum_{j=3}^{n}(\frac{r}{px_1+qx_j+rx_2}+\frac{p}{px_2+qx_j+rx_1})+$$

$$\sum_{i=3}^{n}(\frac{r}{px_i+qx_1+rx_2}+\frac{q}{px_i+qx_2+rx_1})]$$

It is easy to see that $GB^{p,q,r}(\boldsymbol{x})$ is symmetric on $\mathbf{R}^n_{++}$. For $n\geqslant 3$, we have

$$\Delta_1:=(x_1-x_2)\left(\frac{\partial\varphi(\boldsymbol{x})}{\partial x_1}-\frac{\partial\varphi(\boldsymbol{x})}{\partial x_2}\right)$$

$$=\frac{(x_1-x_2)\varphi(\boldsymbol{x})}{n(n-1)(n-2)}\cdot[p\sum_{j,k=3,j\neq k}^{n}(\frac{1}{px_1+qx_j+rx_k}-\frac{1}{px_2+qx_j+rx_k})+$$

$$q\sum_{i,k=3,i\neq k}^{n}(\frac{1}{px_i+qx_1+rx_k}-\frac{1}{px_i+qx_2+rx_k})+$$

$$r\sum_{i,j=3,i\neq j}^{n}(\frac{1}{px_i+qx_j+rx_1}-\frac{1}{px_i+qx_j+rx_2})+$$

$$\sum_{k=3}^{n}(\frac{p-q}{px_1+qx_2+rx_k}+\frac{q-p}{px_2+qx_1+rx_k})+$$

$$\sum_{j=3}^{n}(\frac{p-r}{px_1+qx_j+rx_2}+\frac{r-p}{px_2+qx_j+rx_1})+$$

$$\sum_{i=3}^{n}(\frac{q-r}{px_i+qx_1+rx_2}+\frac{r-q}{px_i+qx_2+rx_1})]$$

$$=-\frac{(x_1-x_2)^2\varphi(\boldsymbol{x})}{n(n-1)(n-2)}\cdot[\sum_{j,k=3,j\neq k}^{n}\frac{p^2}{(px_1+qx_j+rx_k)(px_2+qx_j+rx_k)}+$$

$$\sum_{i,k=3,i\neq k}^{n}\frac{q^2}{(px_i+qx_1+rx_k)(px_i+qx_2+rx_k)}+$$

$$\sum_{i,j=3,i\neq j}^{n}\frac{r^2}{(px_i+qx_j+rx_1)(px_i+qx_j+rx_2)}+$$

$$\sum_{k=3}^{n}\frac{(p-q)^2}{(px_1+qx_2+rx_k)(px_2+qx_1+rx_k)}+$$

$$\sum_{j=3}^{n}\frac{(p-r)^2}{(px_1+qx_j+rx_2)(px_2+qx_j+rx_1)}+$$

$$\sum_{i=3}^{n} \frac{(q-r)^2}{(px_i + qx_1 + rx_2)(px_i + qx_2 + rx_1)}] \leqslant 0$$

By Lemma 1, it follows that $GB^{p,q,r}(\boldsymbol{x})$ is Schur-concave with $\boldsymbol{x} \in \mathbf{R}_{++}^n$.

For $n \geqslant 3$, we have

$$\Delta_2 := (x_1 - x_2)\left(x_1 \frac{\partial \varphi(\boldsymbol{x})}{\partial x_1} - x_2 \frac{\partial \varphi(\boldsymbol{x})}{\partial x_2}\right)$$

$$= \frac{(x_1 - x_2)\varphi(\boldsymbol{x})}{n(n-1)(n-2)} \cdot [p \sum_{j,k=3,j\neq k}^{n} (\frac{x_1}{px_1 + qx_j + rx_k} - \frac{x_2}{px_2 + qx_j + rx_k}) +$$

$$q \sum_{i,k=3,i\neq k}^{n} (\frac{x_1}{px_i + qx_1 + rx_k} - \frac{x_2}{px_i + qx_2 + rx_k}) +$$

$$r \sum_{i,j=3,i\neq j}^{n} (\frac{x_1}{px_i + qx_j + rx_1} - \frac{x_2}{px_i + qx_j + rx_2}) +$$

$$\sum_{k=3}^{n} (\frac{px_1 - qx_2}{px_1 + qx_2 + rx_k} + \frac{qx_1 - px_2}{px_2 + qx_1 + rx_k}) +$$

$$\sum_{j=3}^{n} (\frac{px_1 - rx_2}{px_1 + qx_j + rx_2} + \frac{rx_1 - px_2}{px_2 + qx_j + rx_1}) +$$

$$\sum_{i=3}^{n} (\frac{qx_1 - rx_2}{px_i + qx_1 + rx_2} + \frac{rx_1 - qx_2}{px_i + qx_2 + rx_1})]$$

$$= \frac{(x_1 - x_2)^2 \varphi(\boldsymbol{x})}{n(n-1)(n-2)} \cdot [\sum_{j,k=3,j\neq k}^{n} \frac{qx_j + rx_k}{(px_1 + qx_j + rx_k)(px_2 + qx_j + rx_k)} +$$

$$\sum_{i,k=3,i\neq k}^{n} \frac{px_i + rx_k}{(px_i + qx_1 + rx_k)(px_i + qx_2 + rx_k)} +$$

$$\sum_{i,j=3,i\neq j}^{n} \frac{px_i + qx_j}{(px_i + qx_j + rx_1)(px_i + qx_j + rx_2)} +$$

$$\sum_{k=3}^{n} \frac{2pq(x_1 + x_2) + rx_k(p+q)}{(px_1 + qx_2 + rx_k)(px_2 + qx_1 + rx_k)} +$$

$$\sum_{j=3}^{n} \frac{2rp(x_1 + x_2) + qx_j(p+r)}{(px_1 + qx_j + rx_2)(px_2 + qx_j + rx_1)} +$$

$$\sum_{i=3}^{n} \frac{2qr(x_1 + x_2) + px_i(q+r)}{(px_i + qx_1 + rx_2)(px_i + qx_2 + rx_1)}] \geqslant 0$$

By Lemma 2, it follows that $GB^{p,q,r}(\boldsymbol{x})$ is Schur-geometric convex with

$\boldsymbol{x} \in \mathbf{R}_{++}^n.$

For $n \geqslant 3$, we have

$$\Delta_3 := (x_1 - x_2)\left(x_1^2 \frac{\partial\varphi(\boldsymbol{x})}{\partial x_1} - x_2^2 \frac{\partial\varphi(\boldsymbol{x})}{\partial x_2}\right)$$

$$= \frac{(x_1 - x_2)\varphi(\boldsymbol{x})}{n(n-1)(n-2)} \cdot [p \sum_{j,k=3,j\neq k}^n (\frac{x_1^2}{px_1 + qx_j + rx_k} - \frac{x_2^2}{px_2 + qx_j + rx_k}) +$$

$$q \sum_{i,k=3,i\neq k}^n (\frac{x_1^2}{px_i + qx_1 + rx_k} - \frac{x_2^2}{px_i + qx_2 + rx_k}) +$$

$$r \sum_{i,j=3,i\neq j}^n (\frac{x_1^2}{px_i + qx_j + rx_1} - \frac{x_2^2}{px_i + qx_j + rx_2}) +$$

$$\sum_{k=3}^n (\frac{px_1^2 - qx_2^2}{px_1 + qx_2 + rx_k} + \frac{qx_1^2 - px_2^2}{px_2 + qx_1 + rx_k}) +$$

$$\sum_{j=3}^n (\frac{px_1^2 - rx_2^2}{px_1 + qx_j + rx_2} + \frac{rx_1^2 - px_2^2}{px_2 + qx_j + rx_1}) +$$

$$\sum_{i=3}^n (\frac{qx_1^2 - rx_2^2}{px_i + qx_1 + rx_2} + \frac{rx_1^2 - qx_2^2}{px_i + qx_2 + rx_1})]$$

$$= \frac{(x_1 - x_2)^2 \varphi(\boldsymbol{x})}{n(n-1)(n-2)} \cdot [\sum_{j,k=3,j\neq k}^n \frac{(x_1 + x_2)(qx_j + rx_k) + px_1 x_2}{(px_1 + qx_j + rx_k)(px_2 + qx_j + rx_k)} +$$

$$\sum_{i,k=3,i\neq k}^n \frac{(x_1 + x_2)(px_i + rx_k) + qx_1 x_2}{(px_i + qx_1 + rx_k)(px_i + qx_2 + rx_k)} +$$

$$\sum_{i,j=3,i\neq j}^n \frac{(x_1 + x_2)(px_i + qx_j) + rx_1 x_2}{(px_i + qx_j + rx_1)(px_i + qx_j + rx_2)} +$$

$$\sum_{k=3}^n \frac{2pq(x_1^2 + x_2^2) + rx_k(x_1 + x_2)(p+q) + x_1 x_2(p+q)^2}{(px_1 + qx_2 + rx_k)(px_2 + qx_1 + rx_k)} +$$

$$\sum_{j=3}^n \frac{2pr(x_1^2 + x_2^2) + qx_j(x_1 + x_2)(p+r) + x_1 x_2(p+r)^2}{(px_1 + qx_j + rx_2)(px_2 + qx_j + rx_1)} +$$

$$\sum_{i=3}^n \frac{2qr(x_1^2 + x_2^2) + px_i(x_1 + x_2)(q+r) + x_1 x_2(q+r)^2}{(px_i + qx_1 + rx_2)(px_i + qx_2 + rx_1)}] \geqslant 0$$

By Lemma 3, it follows that $GB^{p,q,r}(\boldsymbol{x})$ is Schur-harmonic convex with $\boldsymbol{x} \in \mathbf{R}_{++}^n.$

The proof of Theoremn 1 is completed.

79.4 Applications

Theorem 2 Let $x \in \mathbf{R}_{++}^n$ and $(p,q,r) \in \mathbf{R}_+^3$, p,q,r cannot take the value 0 at the same time and $\geqslant 3$. Then

$$GB^{p,q,r}(x) \leqslant A_n(x) \tag{11}$$

Proof By Theorem 1, from Lemma 4, we have

$$GB^{p,q,r}(u) \geqslant GB^{p,q,r}(x)$$

rearranging gives (10).

The proof is complete.

Theorem 3 Let $x \in \mathbf{R}_{++}^n$, $n \geqslant 3$ $(p,q,r) \in \mathbf{R}_{++}^3$, p,q,r cannot take the value 0 at the same time and the constant c satisfying $0 < c < \min\{x_i\}$, $i = 1,2,\ldots,n$. Then

$$GB^{p,q,r}(x-c) := GB^{p,q,r}(x_1-c, x_2-c, \ldots, x_n-c) \leqslant \left(1 - \frac{c}{nA_n(x)}\right) GB^{p,q,r}(x) \tag{12}$$

Proof By Theorem 1, from Lemma 5, we have

$$GB^{p,q}\left(\frac{x_1}{\sum\limits_{j=1}^n x_j}, \ldots, \frac{x_n}{\sum\limits_{j=1}^n x_j}\right) \geqslant GB^{p,q}\left(\frac{x_1-c}{\sum\limits_{j=1}^n (x_j-c)}, \ldots, \frac{x_n-c}{\sum\limits_{j=1}^n (x_j-c)}\right)$$

rearranging gives (11).

The proof is complete.

References

[1] C BONFERRONI. Sulle medie multiple di potenze[J]. Bolletino Matematica Italiana, 1950, 5: 267–270.

[2] W-M GONG, W-F XIA, Y-M CHU. The Schur convexity for the generalized Muirhead mean[J]. J. Wath. Inequal., 2014, 8: 855–862.

[3] Y-M CHU, W-F XIA. Necessary and sufficient conditions for the Schur harmonic convexity of the generalized Muirhead mean[J]. Proceedings of a Razmadze Mathematical Institute, 2010, 152 :19–27.

[4] A W MARSHALL, I OLKIN, B C ARNOLD. Inequalities: theory of majorization and its application[M]. 2nd ed. New York: Springer

Press, 2011.

[5] B-Y WANG. Foundations of majorization inequalities[M].Beijing: Beijing Normal Univ. Press, 1990.

[6] X-M ZHANG. Geometrically convex functions [M]. Hefei: Anhui University Press, 2004.

[7] C P NICULESCU. Convexity according to the geometric mean[J]. Mathematical Inequalities & Applications, 2000, 3: 155–167.

[8] Y-M CHU, G-D WANG, X-H ZHANG. The Schur multiplicative and harmonic convexities of the complete symmetric function[J]. Mathematische Nachrichten, 2011, 284 (5-6): 653–663.

[9] J-X MENG, Y M CHU, X-M TANG. The Schur-harmonic-convexity of dual form of the Hamy symmetric function[J]. Matematički Vesnik, 2010, 62: 37–46.

[10] K Z GUAN, J-H SHEN. Schur‐convexity for a class of symmetric function and its applications[J]. Math. Inequal. Appl., 2006, 9: 199–210.

[11] K Z GUAN. Some inequalities for a class of generalized means[J]. J. Inequal. Pure Appl. Math., 2004,5(3), Article 69.

[12] N-G ZHENG, Z-H ZHANG, X-M ZHANG. Schur-convexity of two types of one-parameter mean values in n variables[J]. J. Inequal. Appl., Volume 2007, Article ID 78175, 10 pages doi: 10. 1155/2007/78175.

[13] W F XIA, Y M CHU. The Schur multiplicative convexity of the weighted generalized logarithmic mean in n variables[J]. International Mathematical Forum, 2009, 4: 1229–1236.

[14] W-F XIA, Y-M CHU. The Schur convexity of the weighted generalized logarithmic mean values according to harmonic mean[J]. International Journal of Modern Mathematics, 2009, 4: 225–233.

[15] N-. ZHENG, Z-H ZHANG, X-M ZHANG. The Schur- harmonic-convexity of two types of one-parameter mean values in n variables[J]. Journal of Inequalities and Applications Volume 2007, Article ID 78175, 10 pages, doi:10.1155/2007/78175.

[16] QIAN XU. Research on schur-p power-convexity of the quotient of arithmetic mean and geometric mean[J]. Journal of Fudan University(Natural Science), 2015, 54: 299–295.

[17] N-G ZHENG, X-M ZHANG, Y-M CHU. Convexity and Geometrical Convexity of the Identic and Logarithmic Means in N Variables[J]. Acta Mathematica Scientia, 2008, 28A: 1173–1180.

[18] D SH WANG, C-R FU, H-N SHI. Schur-convexity for a mean of n variables with three parameters[J].Publications de l'Institut Mathématique (Beograd), in Press.

[19] H-N SHI. Majorization Theory and Analytic Inequality[M]. Harbin: Publishing House of Harbin Institute of Technology, 2012.

[20] MEIMEI XIA, ZESHUI XU, BIN ZHU. Geometric Bonferroni means with their application in multi-criteria decision making[J]. Knowledge-Based Systems, 2013, 40: 88–100.

[21] JIN HAN PARK, EUN JIN PARK. Generalized fuzzy Bonferroni harmonic mean operators and their applications in group decision making[J]. Journal of Applied Mathematics Volume 2013, Article ID 604029, 14 pages.

[22] MEIMEI XIA, ZESHUI XU, BIN ZHU. Generalized intuitionistic fuzzy Bonferroni means[J]. International Journal of Intelligent Systems, January 2012 DOI: 10.1002/int.20515 • Source: DBLP

[23] JIN HAN PARK, JI YU KIM. Intuitionistic fuzzy optimized weighted geometric Bonferroni means and their applications in Group Decision Making[J]. Fundamenta Informaticae, 2016, 144: 363–381. DOI 10.3233/FI-2016-1341.

[24] R R YAGER. On generalized Bonferroni mean operators for multi-criteria aggregation[J]. International Journal of Approximate Reasoning, 2009, 50: 1279–1286.

[25] G BELIAKOV, S JAMES, J MORDELOV' A, T RÜckschlossov'A, R R YAGER. Generalized Bonferroni mean operators in multicriteria aggregation[J]. Fuzzy Sets and Systems, 2010, 161: 2227–2242.

[26] ZESHUI XU, RONALD R YAGER, Intuitionistic Fuzzy Bonferroni Means[J].IEEE Transactions on Systems, Man, and Cybernetics, Part B (Cybernetics), 2011, 41 (2), April.

第80篇　Schur-Convexity for a Mean of Two Variables with Three Parameters

(SHI HUAN-NAN, ZHANG JING. Journal of Computational Analysis

and Applications, 2017, 22 (5): 907-922)

Abstract: Schur-convexity, Schur-geometric convexity and Schur-harmonic convexity for a mean of two variables with three parameters are investigated, and some mean value inequalities of two variables are established.

Keywords: mean of two variables; Schur-convexity; Schur-geometric convexity; Schur-harmonic convexity; majorization; inequalities

80.1　Introduction

Throughout the paper we assume that the set of n-dimensional row vector on the real number field by $\mathbf{R}^n$

$$\mathbf{R}_+^n = \{\boldsymbol{x} = (x_1,\ldots,x_n) \in \mathbf{R}^n : x_i > 0, i = 1,\ldots,n\}$$

In particular, $\mathbf{R}^1$ and $\mathbf{R}_+^1$ denoted by $\mathbf{R}$ and $\mathbf{R}_+$ respectively.

In 2009, Kuang[1] defined a mean of two variables with three parameters as follows

$$K(\omega_1,\omega_2,p;a,b) = \left[\frac{\omega_1 A(a^p,b^p) + \omega_2 G(a^p,b^p)}{\omega_1 + \omega_2}\right]^{\frac{1}{p}} \tag{1}$$

where $A(a,b) = \frac{a+b}{2}$ and $G(a,b) = \sqrt{ab}$ respectively is the arithmetic mean and geometric mean of two positive numbers a and b, parameters $p \neq 0$,

$\omega_1, \omega_2 \geqslant 0$ with $\omega_1 + \omega_2 \neq 0$.

For simplicity, sometimes we will be $K(\omega_1, \omega_2, p; a, b)$ for $K(\omega_1, \omega_2, p)$ or $K(a, b)$.

In particular

$$K\left(1, \frac{\omega}{2}, 1\right) = \frac{a + \omega\sqrt{ab} + b}{\omega + 2}$$

is the generalized Heron mean, which was introduced by Janous [2] in 2001.

$$K\left(1, \frac{\omega}{2}, p\right) = \frac{a^p + \omega(ab)^{p/2} + b^p}{\omega + 2}$$

is the generalized Heron mean with parameter.

In recent years, the study on the properties of the mean with two variables by using theory of majorization is unusually active (see references [10-35]).

In this paper, Schur-convexity, Schur-geometric convexity and Schur-harmonic convexity of $K(\omega_1, \omega_2, p)$ are discussed. As consequences, some interesting inequalities are obtained.

Our main results are as follows:

Theorem 1

(i) When $\omega_1 \omega_2 \neq 0$, if $p \geqslant 2$ and $p(\omega_1 - \frac{\omega_2}{2}) - \omega_1 \geqslant 0$, then $K(\omega_1, \omega_2, p)$ is Schur-convex with $(a, b) \in \mathbf{R}_+^2$; if $1 \leqslant p < 2$ and $p(\omega_1 - \frac{\omega_2}{2}) - \omega_1 \leqslant 0$, then $K(\omega_1, \omega_2, p)$ is Schur-concave with $(a, b) \in \mathbf{R}_+^2$; if $p < 1$, then $K(\omega_1, \omega\omega_2, p)$ is Schur-concave with $(a, b) \in \mathbf{R}_+^2$.

(ii) When $\omega_1 = 0, \omega_2 \neq 0$, $K(\omega_1, \omega_2, p)$ is Schur-concave with $(a, b) \in \mathbf{R}_+^2$.

(iii) When $\omega_1 \neq 0, \omega_2 = 0$, if $p \geqslant 2$, then $K(\omega_1, \omega_2, p)$ is the Schur-convex with $(a, b) \in \mathbf{R}_+^2$; if $p < 2$, then $K(\omega_1, \omega_2, p)$ is the Schur-concave with $(a, b) \in \mathbf{R}_+^2$.

Theorem 2 If $p \geqslant 0$, then $K(\omega_1, \omega_2, p)$ is Schur-geometrically convex with $(a, b) \in \mathbf{R}_+^2$. If $p < 0$, then $K(\omega_1, \omega_2, p)$ is Schur-geometrically concave with $(a, b) \in \mathbf{R}_+^2$.

Theorem 3 If $p \geqslant -1$, then $K(\omega_1, \omega_2, p)$ is Schur-harmonically convex with $(a, b) \in \mathbf{R}_+^2$. If $-2 < p < -1$ and $\omega_1(p + 1) + \omega_2(\frac{p}{2} + 1) \geqslant 0$, then $K(\omega_1, \omega_2, p)$ is Schur-harmonically convex with $(a, b) \in \mathbf{R}_+^2$. If $p \leqslant -2$ and $\omega_1(\frac{p}{2} + 1) + \omega_2 = 0$, then $K(\omega_1, \omega_2, p)$ is Schur-harmonically concave with $(a, b) \in \mathbf{R}_+^2$.

80.2 Definitions and Lemmas

We need the following definitions and lemmas.

Definition 1 [3-4] Let $\boldsymbol{x} = (x_1, \ldots, x_n)$ and $\boldsymbol{y} = (y_1, \ldots, y_n) \in \mathbf{R}^n$.

(i) $\boldsymbol{x}$ is said to be majorized by $\boldsymbol{y}$ (in symbols $\boldsymbol{x} \prec \boldsymbol{y}$) if $\sum\limits_{i=1}^{k} x_{[i]} \leqslant \sum\limits_{i=1}^{k} y_{[i]}$ for $k = 1, 2, \ldots, n-1$ and $\sum\limits_{i=1}^{n} x_i = \sum\limits_{i=1}^{n} y_i$, where $x_{[1]} \geqslant \cdots \geqslant x_{[n]}$ and $y_{[1]} \geqslant \cdots \geqslant y_{[n]}$ are rearrangements of $\boldsymbol{x}$ and $\boldsymbol{y}$ in a descending order.

(ii) $\Omega \subset \mathbf{R}^n$ is called a convex set if $(\alpha x_1 + \beta y_1, \ldots, \alpha x_n + \beta y_n) \in \Omega$ for any $\boldsymbol{x}$ and $\boldsymbol{y} \in \Omega$, where α and $\beta \in [0, 1]$ with $\alpha + \beta = 1$.

(iii) Let $\Omega \subset \mathbf{R}^n$, $\varphi \colon \Omega \to \mathbf{R}$ is said to be a Schur-convex function on Ω if $\boldsymbol{x} \prec \boldsymbol{y}$ on Ω implies $\varphi(\boldsymbol{x}) \leqslant \varphi(\boldsymbol{y})$. φ is said to be a Schur-concave function on Ω if and only if $-\varphi$ is Schur-convex function.

Definition 2 [5-6] Let $\boldsymbol{x} = (x_1, \ldots, x_n)$ and $\boldsymbol{y} = (y_1, \ldots, y_n) \in \mathbf{R}_+^n$.

(i) $\Omega \subset \mathbf{R}_+^n$ is called a geometrically convex set if $(x_1^\alpha y_1^\beta, \ldots, x_n^\alpha y_n^\beta) \in \Omega$ for any $\boldsymbol{x}$ and $\boldsymbol{y} \in \Omega$, where α and $\beta \in [0, 1]$ with $\alpha + \beta = 1$.

(ii) let $\Omega \subset \mathbf{R}_+^n$, $\varphi \colon \Omega \to \mathbf{R}_+$ is said to be a Schur-geometrically convex function on Ω if $(\ln x_1, \ldots, \ln x_n) \prec (\ln y_1, \ldots, \ln y_n)$ on Ω implies $\varphi(\boldsymbol{x}) \leqslant \varphi(\boldsymbol{y})$. φ is said to be a Schur-geometrically concave function on Ω if and only if $-\varphi$ is Schur-geometrically convex function.

Definition 3 [7-8] Let $\Omega \subset \mathbf{R}_+^n$.

(i) A set Ω is said to be a harmonically convex set if $\frac{\boldsymbol{x}\boldsymbol{y}}{\lambda \boldsymbol{x} + (1-\lambda)\boldsymbol{y}} \in \Omega$ for every $\boldsymbol{x}, \boldsymbol{y} \in \Omega$ and $\lambda \in [0, 1]$, where $\boldsymbol{x}\boldsymbol{y} = \sum\limits_{i=1}^{n} x_i y_i$ and $\frac{1}{\boldsymbol{x}} = \left(\frac{1}{x_1}, \cdots, \frac{1}{x_n}\right)$.

(ii) A function $\varphi \colon \Omega \to \mathbf{R}_+$ is said to be a Schur harmonically convex function on Ω if $\frac{1}{\boldsymbol{x}} \prec \frac{1}{\boldsymbol{y}}$ implies $\varphi(\boldsymbol{x}) \leq \varphi(\boldsymbol{y})$. A function φ is said to be a Schur harmonically concave function on Ω if and only if $-\varphi$ is a Schur harmonically convex function.

Lemma 1 [3-4] Let $\Omega \subset \mathbf{R}^n$ is convex set, and has a nonempty interior set Ω^0. Let $\varphi \colon \Omega \to \mathbf{R}$ is continuous on Ω and differentiable in Ω^0. Then

φ is the $Schur-convex(Schur-concave)function$, if and only if it is symmetric on Ω and if

$$(x_1 - x_2)\left(\frac{\partial\varphi}{\partial x_1} - \frac{\partial\varphi}{\partial x_2}\right) \geqslant 0(\leqslant 0)$$

holds for any $\boldsymbol{x} = (x_1, x_2, \cdots, x_n) \in \Omega^0$.

Lemma 2 [5-6] Let $\Omega \subset \mathbf{R}_+^n$ be a symmetric geometrically convex set with a nonempty interior Ω^0. Let $\varphi : \Omega \to \mathbf{R}_+$ be continuous on Ω and differentiable on Ω^0. Then φ is a Schur geometrically convex (Schur geometrically concave) function if and only if φ is symmetric on Ω and

$$(x_1 - x_2)\left(x_1\frac{\partial\varphi}{\partial x_1} - x_2\frac{\partial\varphi}{\partial x_2}\right) \geqslant 0 \quad (\leqslant 0) \tag{2}$$

holds for any $\boldsymbol{x} = (x_1, \cdots, x_n) \in \Omega^0$.

Lemma 3 [7-8] Let $\Omega \subset \mathbf{R}_+^n$ be a symmetric harmonically convex set with a nonempty interior Ω^0. Let $\varphi : \Omega \to \mathbf{R}_+$ be continuous on Ω and differentiable on Ω^0. Then φ is a Schur harmonically convex (Schur harmonically concave) function if and only if φ is symmetric on Ω and

$$(x_1 - x_2)\left(x_1^2\frac{\partial\varphi}{\partial x_1} - x_2^2\frac{\partial\varphi}{\partial x_2}\right) \geqslant 0 \quad (\leqslant 0) \tag{3}$$

holds for any $\boldsymbol{x} = (x_1, \cdots, x_n) \in \Omega^0$.

Lemma 4 [9] Let $a \leqslant b, u(t) = tb + (1 - t)a, v(t) = ta + (1 - t)b$. If $\frac{1}{2} \leqslant t_2 \leqslant t_1 \leqslant 1$ or $0 \leqslant t_1 \leqslant t_2 \leqslant \frac{1}{2}$, then

$$(u(t_2), v(t_2)) \prec (u(t_1), v(t_1)) \prec (a, b) \tag{4}$$

Lemma 5 Let

$$f(x) = \omega_1(p + 1)x^{\frac{p}{2}+1} + \omega_2(\frac{p}{2} + 1)x - \omega_2\frac{p}{2}, \quad x \in [1, \infty)$$

where $\omega_1, \omega_2 \geqslant 0, \omega_1^2 + \omega_2^2 \neq 0$.

If $-2 < p < -1$ and $\omega_1(p+1) + \omega_2(\frac{p}{2}+1) \geqslant 0$, then $f(x) \geqslant 0$, if $p \leqslant -2$ and $\omega_1(p+1) + \omega_2 = 0$, then $f(x) \leqslant 0$.

Proof If $-2 < p < -1$ and $\omega_1(p+1) + \omega_2(\frac{p}{2}+1) \geqslant 0$, then

$$g(x) := \omega_1(p+1)x^{\frac{p}{2}+1} + \omega_2(\frac{p}{2}+1)x$$

$$\geqslant \omega_1(p+1)x^{\frac{p}{2}+1} + \omega_2(\frac{p}{2}+1)x^{\frac{p}{2}+1}$$

$$= x^{\frac{p}{2}+1}[\omega_1(p+1) + \omega_2(\frac{p}{2}+1)]$$

$$\geqslant 0$$

and then $f(x) = g(x) - \omega_2\frac{p}{2} \geqslant 0$.

If $p \leqslant -2$ and $\omega_1(p+1) + \omega_2 = 0$, then

$$f'(x) = \omega_1(p+1)(\frac{p}{2}+1)x^{\frac{p}{2}} + \omega_2(\frac{p}{2}+1)$$

and

$$f''(x) = \omega_1\frac{p}{2}(p+1)(\frac{p}{2}+1)x^{\frac{p}{2}-1} \leqslant 0$$

so $f'(x)$ is decreasing, but

$$f'(1) = \omega_1(p+1)(\frac{p}{2}+1) + \omega_2(\frac{p}{2}+1) = (\frac{p}{2}+1)[\omega_1(p+1) + \omega_2] = 0$$

then $f'(x) \leqslant 0$, so $f(x)$ is decreasing, furthermore

$$f(1) = \omega_1(p+1) + \omega_2(\frac{p}{2}+1) - \omega_2\frac{p}{2} = \omega_1(p+1) + \omega_2 = 0$$

thus $f(x) \leqslant 0$.

80.3　Proofs of Main results

From the definition of $K(\omega_1, \omega_2, p)$, we have

$$K(\omega_1, \omega_2, p) = \left(\frac{\omega_1\frac{a^p+b^p}{2} + \omega_2 a^{\frac{p}{2}}b^{\frac{p}{2}}}{\omega_1 + \omega_2} \right)^{\frac{1}{p}}$$

It is clear that $K(\omega_1, \omega_2, p)$ is symmetric with $(a, b) \in \mathbf{R}_+^2$.

Write

$$m(a, b) := \left[\frac{\omega_1(a^p + b^p) + 2\omega_2 a^{\frac{p}{2}}b^{\frac{p}{2}}}{2(\omega_1 + \omega_2)} \right]^{\frac{1}{p}-1}$$

Proof of Theorem 1 (i) When $\omega_1\omega_2 \neq 0$

$$\frac{\partial K}{\partial a} = m(a,b)\left(\frac{\omega_1 a^{p-1} + \omega_2 a^{\frac{p}{2}-1}b^{\frac{p}{2}}}{\omega_1 + \omega_2}\right)$$

$$\frac{\partial K}{\partial b} = m(a,b)\left(\frac{\omega_1 b^{p-1} + \omega_2 a^{\frac{p}{2}}b^{\frac{p}{2}-1}}{\omega_1 + \omega_2}\right)$$

and then

$$\Delta_1 := (a-b)\left(\frac{\partial K}{\partial a} - \frac{\partial K}{\partial b}\right)$$

$$= \frac{a-b}{2(\omega_1 + \omega_2)}m(a,b)\left[\omega_1(a^{p-1} - b^{p-1}) - \omega_2(a-b)a^{\frac{p}{2}-1}b^{\frac{p}{2}-1}\right]$$

Without loss of generality, we may assume that $a \geqslant b$, then $z := \frac{a}{b} \geqslant 1$,
and then

$$\Delta_1 = \frac{a-b}{2(\omega_1 + \omega_2)}m(a,b)b^{p-1}f(z)$$

where

$$f(z) = \omega_1(z^{p-1} - 1) - \omega_2(z-1)z^{\frac{p}{2}-1}, \quad z \geqslant 1$$

$$f'(z) = \omega_1(p-1)z^{p-2} - \omega_2 z^{\frac{p}{2}-1} - \omega_2\left(\frac{p}{2} - 1\right)(z-1)z^{\frac{p}{2}-2}$$

$$= z^{\frac{p}{2}-2}\left[\omega_1(p-1)z^{\frac{p}{2}} - \omega_2\frac{p}{2}z + \omega_2\left(\frac{p}{2} - 1\right)\right]$$

If $p \geqslant 2$ and $p(\omega_1 - \frac{\omega_2}{2}) - \omega_1 \geqslant 0$, then

$$\omega_1(p-1)z^{\frac{p}{2}} - \omega_2\frac{p}{2}z = z\left[\omega_1(p-1)z^{\frac{p}{2}-1} - \omega_2\frac{p}{2}\right] \geqslant z\left[\omega_1(p-1) - \omega_2\frac{p}{2}\right]$$

Notice that

$$\omega_1(p-1) - \omega_2\frac{p}{2} \geqslant 0 \Leftrightarrow p\left(\omega_1 - \frac{\omega_2}{2}\right) - \omega_1 \geqslant 0,$$

we have $f'(z) \geqslant 0$, for $z \in [1, \infty)$, but $f(1) = 0$, so $f(z) \geqslant 0$, further $\Delta_1 \geqslant 0$.
By Lemma 1, it follows that $K(\omega_1, \omega_2, p)$ is Schur-convex with $(a,b) \in \mathbf{R}_+^2$.

If $1 \leqslant p < 2$ and $p(\omega_1 - \frac{\omega_2}{2}) - \omega_1 \leqslant 0$, then

$$\omega_1(p-1)z^{\frac{p}{2}} - \omega_2\frac{p}{2}z = z\left[\omega_1(p-1)z^{\frac{p}{2}-1} - \omega_2\frac{p}{2}\right] \leqslant z\left[\omega_1(p-1) - \omega_2\frac{p}{2}\right]$$

Notice that

$$\omega_1(p-1) - \omega_2\frac{p}{2} \geqslant 0 \Leftrightarrow p(\omega_1 - \frac{\omega_2}{2}) - \omega_1 \leqslant 0$$

we have $f'(z) \leqslant 0$, for $z \in [1, \infty)$, but $f(1) = 0$, so $f(z) \leqslant 0$, further $\Delta_1 \leqslant 0$. By Lemma 1, it follows that $K(w_1, w_2, p)$ is Schur-concave with $(a, b) \in \mathbf{R}_+^2$.

If $p < 1$, then

$$f(z) = \omega_1(z^{p-1} - 1) - \omega_2(z-1)z^{\frac{p}{2}-1} \leqslant \omega_1(1-1) - \omega_2(z-1)z^{\frac{p}{2}-1} \leqslant 0$$

and then $\Delta_1 \leqslant 0$. By Lemma 1, it follows that $K(w_1, w_2, p)$ is Schur-concave with $(a, b) \in \mathbf{R}_+^2$.

(ii) When $\omega_1 = 0, \omega_2 \neq 0$, $K(w_1, w_2, p) = \sqrt{ab}$, then

$$\Delta_1 := (a-b)\left(\frac{\partial K}{\partial a} - \frac{\partial K}{\partial b}\right) = -\frac{1}{2}\frac{(a-b)^2}{\sqrt{ab}} \leqslant 0$$

By Lemma 1, it follows that $K(w_1, w_2, p)$ is Schur-concave with $(a, b) \in \mathbb{R}_+^2$.

(iii) When $\omega_1 \neq 0, \omega_2 = 0$, $K(w_1, w_2, p) = \left(\frac{a^p + b^p}{2}\right)^{\frac{1}{p}}$, then

$$\Delta_1 := (a-b)\left(\frac{\partial K}{\partial a} - \frac{\partial K}{\partial b}\right) = \frac{1}{2}(a-b)(a^p + b^p)^{\frac{1}{p}-1}(a^{p-1} + b^{p-1})$$

If $p \geqslant 2$, then $a - b$ and $a^{p-1} - b^{p-1}$ has the same sign, so $\Delta_1 \geqslant 0$. By Lemma 1, it follows that $K(w_1, w_2, p)$ is Schur-convex with $(a, b) \in \mathbf{R}_+^2$. If $p < 2$, then $a - b$ and $a^{p-1} - b^{p-1}$ has the opposite sign, so $\Delta_1 \leqslant 0$. By Lemma 1, it follows that $K(w_1, w_2, p)$ is Schur-concave with $(a, b) \in \mathbf{R}_+^2$.

The proof of Theorem 1 is complete.

Proof of Theorem 2 It is easy to see that

$$a\frac{\partial K}{\partial a} = m(a,b)\left(\frac{\omega_1 a^p + \omega_2 a^{\frac{p}{2}}b^{\frac{p}{2}}}{\omega_1 + \omega_2}\right)$$

$$b\frac{\partial K}{\partial b} = m(a,b)\left(\frac{\omega_1 b^p + \omega_2 a^{\frac{p}{2}}b^{\frac{p}{2}}}{\omega_1 + \omega_2}\right)$$

and then

$$\Delta_2 := (a-b)\left(a\frac{\partial K}{\partial a} - b\frac{\partial K}{\partial b}\right) = \frac{(a-b)m(a,b)\omega_1(a^p - b^p)}{2(\omega_1 + \omega_2)}$$

If $p \geqslant 0$, then $a - b$ and $a^p - b^p$ has the same sign, so $\Delta_2 \geqslant 0$. By

Lemma 2, it follows that $K(\omega_1, \omega_2, p)$ is Schur-geometrically convex with $(a, b) \in \mathbf{R}_+^2$. If $p < 0$, then $a - b$ and $a^p - b^p$ has the opposite sign, so $\Delta_2 \leqslant 0$. By Lemma 2, it follows that $K(\omega_1, \omega_2, p)$ is Schur-geometrically concave with $(a, b) \in \mathbf{R}_+^2$.

The proof of Theorem 2 is complete.

Proof of Theorem 3 It is easy to see that

$$a^2 \frac{\partial K}{\partial a} = \frac{m(a, b)}{2(\omega_1 + \omega_2)} \left(\omega_1 a^{p+1} + \omega_2 a^{\frac{p}{2}+1} b^{\frac{p}{2}}\right)$$

$$b^2 \frac{\partial K}{\partial b} = \frac{m(a, b)}{2(\omega_1 + \omega_2)} \left(\omega_1 b^{p+1} + \omega_2 a^{\frac{p}{2}} b^{\frac{p}{2}+1}\right)$$

and then

$$\Delta_3 := (a - b) \left(a^2 \frac{\partial K}{\partial a} - b^2 \frac{\partial K}{\partial b}\right) = \frac{m(a, b)}{2(\omega_1 + \omega_2)} f(x, y)$$

where

$$f(a, b) := \omega_1(a - b)(a^{p+1} - b^{p+1}) + \omega_2 a^{\frac{p}{2}} b^{\frac{p}{2}} (a - b)^2$$

If $p \geqslant -1$, then $a - b$ and $a^{p+1} - b^{p+1}$ has the same sign, so $f(a, b) \geqslant 0$, and then $\Delta_3 \geqslant 0$. By Lemma 3, it follows that $K(\omega_1, \omega_2, p)$ is Schur-harmonically convex with $(a, b) \in \mathbf{R}_+^2$.

Without loss of generality, we may assume that $a \geqslant b$, then $z := \frac{a}{b} \geqslant 1$, and then

$$f(a, b) = b^{p+2}(z - 1)g(z)$$

where

$$g(z) = \omega_1(z^{p+1} - 1) + \omega_2 z^{\frac{p}{2}}(z - 1)$$

$$g'(z) = z^{\frac{p}{2}-1} s(z)$$

where

$$s(z) = \omega_1(p + 1)(z^{\frac{p}{2}+1} + \omega_2(\frac{p}{2} + 1)z - \omega_2 \frac{p}{2}$$

If $-2 < p < -1$ and $\omega_1(p+1) + \omega_2(\frac{p}{2}+1) \geqslant 0$, from Lemma 5, it follows $s(z) \geqslant 0$, and then $g'(z) \geqslant 0$, but $g(1) = 0$, so $g(z) \geqslant 0$ and $f(a, b) \geqslant 0$. Thus $\Delta_3 \geqslant 0$, by Lemma 3, it follows that $K(\omega_1, \omega_2, p)$ is Schur-harmonically convex with $(a, b) \in \mathbf{R}_+^2$.

If f $p \leqslant -2$ and $\omega_1(\frac{p}{2} + 1) + \omega_2 = 0$, from Lemma 5, it follows $s(z) \leqslant 0$, and then $g'(z) \leqslant 0$, but $g(1) = 0$, so $g(z) \leqslant 0$ and $f(a, b) \leqslant 0$. Thus $\Delta_3 \leqslant 0$, by Lemma 3, it follows that $K(\omega_1, \omega_2, p)$ is Schur-harmonically concave with $(a, b) \in \mathbf{R}_+^2$.

The proof of Theorem 3 is complete.

80.4 Applications

Theorem 4 Let $(a, b) \in \mathbf{R}_+^2$, $u(t) = tb + (1 - t)a$, $v(t) = ta + (1 - t)b$. Assume also that $\frac{1}{2} \leqslant t_2 \leqslant t_1 \leqslant 1$ or $0 \leqslant t_1 \leqslant t_2 \leqslant \frac{1}{2}$.

If $\omega_1 \omega_2 \neq 0$, $p \geqslant 2$ and $p(\omega_1 - \frac{\omega_2}{2}) - \omega_1 \geqslant 0$, then we have we have

$$K\left(\omega_1, \omega_2, p; \frac{a + b}{2}, \frac{a + b}{2}\right) \leqslant K(\omega_1, \omega_2, p; u(t_2), v(t_2))$$

$$\leqslant K(\omega_1, \omega_2, p; u(t_1), v(t_1)) \leqslant K(\omega_1, \omega_2, p; a, b) \leqslant G(\omega_1, \omega_2, p; a + b, 0) \qquad (5)$$

If $\omega_1 \omega_2 \neq 0$, $1 \leqslant p < 2$ and $p(\omega_1 - \frac{\omega_2}{2}) - \omega_1 \leqslant 0$, then inequalities in (5) are all reversed.

Proof From Lemma 4, we have

$$\left(\frac{a + b}{2}, \frac{a + b}{2}\right) \prec (u(t_2), v(t_2)) \prec (u(t_1), v(t_1)) \prec (a, b)$$

and it is clear that $(a, b) \prec (a + b - \varepsilon, \varepsilon)$, where ε is enough small positive number.

If $\omega_1 \omega_2 \neq 0$, $p \geqslant 2$ and $p(\omega_1 - \frac{\omega_2}{2}) - \omega_1 \geqslant 0$, by Theorem 1, and let $\varepsilon \to 0$, it follows that (5) are holds. If $\omega_1 \omega_2 \neq 0$, $1 \leqslant p < 2$ and $p(\omega_1 - \frac{\omega_2}{2}) - \omega_1 \leqslant 0$, then inequalities in (5) are all reversed.

The proof is complete.

Theorem 4 enable us to obtain a large number of refined inequalities by assigning appropriate values to the parameters $\omega_1, \omega_2, p, t_1$ and t_2.

For example, putting $\omega_1 = \omega_2 = 1$ in (5), we can get

Corollary 1 Let $p \geqslant 2$. Then for $(a, b) \in \mathbf{R}_+^2$, we have

$$A(a^p, b^p) + G(a^p, b^p) \geqslant 2(A(a, b))^p \qquad (6)$$

Putting $p = \frac{1}{2}, \omega_1 = 2, \omega_2 = 1$ and $t_1 = \frac{3}{4}, t_2 = \frac{1}{2}$ in (5), we can get

Corollary 2 Let $(a, b) \in \mathbf{R}_+^2$. Then

$$\frac{a+b}{2} \geqslant \frac{1}{36}\left[\sqrt{a+3b} + \sqrt[4]{(a+3b)(3a+b)} + \sqrt{3a+b}\right]^2 \geqslant \frac{1}{9}\left(\sqrt{a} + \sqrt[4]{ab} + \sqrt{b}\right)^2 \tag{7}$$

Theorem 5 Let $(a, b) \in \mathbf{R}_+^2$. If $p \geqslant 0(< 0)$, we have

$$G(a, b) \leqslant (\geqslant)K(\omega_1, \omega_2, p; a, b) \tag{8}$$

Proof Since $(\log \sqrt{ab}, \log \sqrt{ab}) \prec (\log a, \log b)$, if $p \geqslant 0(< 0)$, by Theorem 2, it follows

$$G(a, b) = K\left(\omega_1, \omega_2, p; \sqrt{ab}, \sqrt{ab}\right) \leqslant (\geqslant)K(\omega_1, \omega_2, p; a, b)$$

The proof is complete.

For example, putting $\omega_1 = \omega_2 = 1$ in (8), we can get

Corollary 3 Let $(a, b) \in \mathbf{R}_+^2$. If $p \geqslant 0(< 0)$, then

$$A(a^p, b^p) + G(a^p, b^p) \leqslant (\geqslant)2(G(a, b))^p \tag{9}$$

Theorem 6 Let $(a, b) \in \mathbf{R}_+^2$. If $p \geqslant -1$ or if $-2 < p < -1$ and $\omega_1(p+1) + \omega_2(\frac{p}{2}+1) \geqslant 0$, then

$$H(a, b) \leqslant K\left(\omega_1, \omega_2, p; \frac{ab}{tb+(1-t)a}, \frac{ab}{ta+(1-t)b}\right) \leqslant K(\omega_1, \omega_2, p; a, b) \tag{10}$$

where $H(a, b) = \frac{2}{a^{-1}+b^{-1}}$ is the harmonic mean.

If $p \leqslant -2$ and $\omega_1(\frac{p}{2}+1) + \omega_2 = 0$, then inequalities in (10) are all reversed.

Proof By Lemma 4, we have

$$\left(\frac{a^{-1}+b^{-1}}{2}, \frac{a^{-1}+b^{-1}}{2}\right) \prec \left(ta^{-1}+(1-t)b^{-1}, tb^{-1}+(1-t)a^{-1}\right) \prec (a^{-1}, b^{-1})$$

If $p \geqslant -1$ or if $-2 < p < -1$ and $\omega_1(p+1) + \omega_2(\frac{p}{2}+1) \geqslant 0$, then by Theorem 3, it follows

$$H(a, b) = K\left(\omega_1, \omega_2, p; \frac{2}{a^{-1}+b^{-1}}, \frac{2}{a^{-1}+b^{-1}}\right)$$

$$\leqslant K\left(\omega_1, \omega_2, p; \frac{ab}{tb+(1-t)a}, \frac{ab}{ta+(1-t)b}\right)$$

$$\leqslant K(\omega_1, \omega_2, p; a, b)$$

If $p \leqslant -2$ and $\omega_1(\frac{p}{2} + 1) + \omega_2 = 0$, then inequalities in (10) are all reversed. The proof is complete.

Putting $\omega_1 = \omega_2 = 1$ in (10), we can get

Corollary 4 Let $(a, b) \in \mathbf{R}_+^2$. If $p \geqslant -1$ or $-\frac{4}{3} < p < -1$, then

$$A(a^p, b^p) + G(a^p, b^p) \geqslant 2(H(a, b))^p \tag{11}$$

If $p = -4$, then the inequality in (11) is reversed.

References

[1] J-C KUANG. Applied Inequalities (Chang yong bu deng shi[M]. 4th ed. Jinan: Shandong Press of Science and Technology, 2010.

[2] W JANOUS. A note on generalized Heronian means[J]. Mathematical Inequalities & Applications, 2001,4 (3) : 369–375.

[3] B-Y WANG. Foundations of majorization inequalities[M].Beijing: Beijing Normal Univ. Press, 1990.

[4] A W MARSHALL, I OLKIN. Inequalities:theory of majorization and its application[M]. New York : Academies Press, 1979.

[5] X-M ZHANG. Geometrically convex functions [M]. Hefei: Anhui University Press, 2004.

[6] C P NICULESCU. Convexity According to the Geometric Mean, Mathematical Inequalities & Applications[J]. 2000, 3(2):155–167.

[7] Y-M CHU, G-D WANG, X-H ZHANG. The Schur multiplicative and harmonic convexities of the complete symmetric function[J]. Mathematische Nachrichten, 2011, 284 (5-6): 653–663.

[8] J-X MENG, Y M CHU, X-M TANG. The Schur-harmonic-convexity of dual form of the Hamy symmetric function[J]. Matematički Vesnik, 2010, 62 (1): 37–46.

[9] H-N SHI, Y-M JIANG, W-D JIANG. Schur-convexity and Schur-geometrically concavity of Gini mean[J]. Computers and Mathematics with Applications, 2009, 57 : 266–274.

[10] A WITKOWSKI. On Schur convexity and Schur-geometrical convexity of four-parameter family of means[J]. Math. Inequal. Appl., 2011,14 (4): 897–903.

[11] J SÁNDOR. The Schur-convexity of Stolarsky and Gini means[J]. Banach J. Math. Anal. , 2007, 1(2): 212–215.

[12] Y-M CHU, X-M ZHANG. Necessary and sufficient conditions such that extended mean values are Schur-convex or Schur-concave[J]. Journal of Mathematics of Kyoto University, 2008, 48(1): 229–238.

[13] Y-M CHU, X-M ZHANG. The Schur geometrical convexity of the extended mean values[J]. Journal of Convex Analysis, 2008, 15(4): 869–890.

[14] W-F XIA, Y-M CHU. The Schur convexity of Gini mean values in the sense of harmonic mean[J]. Acta Mathematica Scientia 2011, 31B(3): 1103–1112.

[15] H-N SHI, B MIHALY, S-HA WU, D-M LI. Schur convexity of generalized Heronian means involving two parameters[J]. J. Inequal. Appl., vol.2008, Article ID 879273, 9 pages.

[16] W-F XIA, Y-M CHU. The Schur multiplicative convexity of the generalized Muirhead mean, International Journal of Functional Analysis[J]. Operator Theory and Applications, 2009, 1(1): 1–8.

[17] Y-M CHU, W-F XIA. Necessary and sufficient conditions for the Schur harmonic Convexity of the Generalized Muirhead Mean[J]. Proceedings of A. Razmadze Mathematical Institute, 2010, 152 : 19–27.

[18] Z-H YANG. Necessary and sufficient conditions for Schur geometrical convexity of the four-parameter homogeneous means[J]. Abstr. Appl. Anal., Volume 2010, Article ID 830163, 16 pages, doi:10.1155/2010/ 830163.

[19] W-F XIA, Y-M CHU. The Schur convexity of the weighted generalized logarithmic mean values according to harmonic mean[J]. International Journal of Modern Mathematics, 2009, 4(3): 225–233.

[20] A WITRKOWSKI. On Schur-convexity and Schur-geometric convexity of four-parameter family of means[J]. Math. Inequal. Appl., 2011, 14 (4): 897 - -903.

[21] Z-H YANG. Schur harmonic convexity of Gini means[J]. International Mathematical Forum, 2011, 6, (16): 747– 762.

[22] Z-H YANG. Schur power convexity of Stolarsky means[J]. Publ. Math. Debrecen, 2012, 80 (1-2): 43–66.

[23] F QI, J SÁNDOR, S S DRAGOMIR, A SOFO. Notes on the Schur-convexity of the extended mean values[J]. Taiwanese J. Math., 2005, 9(3): 411–420.

[24] L -L FU, B-Y XI, H M SRIVASTAVA. Schur-convexity of the generalized Heronian means involving two positive numbers[J]. Taiwanese Journal of Mathematics, 2011, 15(6): 2721–2731.

[25] T-Y ZHANG, A-P JI, Schur-convexity of generalized Heronian mean[J]. Communications in Computer and Information Science, 1, Volume 244, Information Computing and Applications, Part 1, Pages 25–33.

[26] W-F XIA, Y-M CHU, G-D WANG. Necessary and sufficient conditions for the Schur harmonic convexity or concavity of the extended mean values[J]. Revista De La Uniòn Matemática Argentina, 2010, 51(2) : 121–132.

[27] Y WU, F QI. Schur-harmonic convexity for differences of some means, Analysis, 2012, 32: 1001–1008.

[28] V LOKESHA, K M NAGARAJA, B NAVEEN KUMAR, Y-D WU. Schur convexity of Gnan mean for two variables[J]. NNTDM, 2011, 17 (4): 37–41.

[29] Y WU, F QI, H-N SHI. Schur-harmonic convexity for differences of some special means in two variables[J]. J.Math. Inequal., J.Math. Inequal., 2014, 8 (2): 321–330.

[30] W-M GONG, X SHEN, Y-M CHU. The Schur convexity for the generalized Muirhead mean[J]. J. Math. Inequal., 2014, 8 (4): 855–862.

[31] K M NAGARAJA, SUDHIR KUMAR SAHU. Schur harmonic convexity of Stolarsky extended mean values[J]. Scientia Magna, 2013, 9 (2): 18–29.

[32] K M NAGARAJA, B NAVAVEEN KUMAR, K M NAGARAJA, S PADMANABHAN. Schur geometric convexity for ratio of difference of means[J]. Journal of Scientific Research & Reports, 2014, 3(9): 1211–1219. Article no. JSRR.2014.9.008.

[33] Y-P DEANG, S-H WU, Y-M CHU, D HE. The Schur convexity of the generalized Muirhead-Heronian means[J]. Abstract and Applied Analysis, Volume 2014 (2014), Article ID 706518, 11 pages.

[34] W-M GONG, H SUN, Y-M CHU. The Schur convexity for the generalized muirhead mean, J.Math. Inequal., 2014, 8 (4): 855–862.

[35] H-P YIN, H-N SHI, F QI. On Schur m‐power convexity for ratios of
some means[J]. J.Math. Inequal., 2015, 9 (1): 145–153.

第81篇　Schur-Convexity for a Mean of n Variables with Three Parameters

(WANG DONGSHENG, FU CUN-RU, SHI HUANNAN. Publications

de l'Institut Mathématique (Beograd))

Abstract: Schur-convexity, Schur-geometric convexity and Schur-harmonic convexity for a mean of n variables with three parameters are investigated, and some mean value inequalities of n variables are established.

Keywords: mean of n variables; Schur-convexity; Schur-geometric convexity; Schur-harmonic convexity; majorization, inequalities

81.1　Introduction

Throughout the paper we assume that the set of n-dimensional row vector on the real number field by $\mathbf{R}^n$

$$\mathbf{R}_+^n = \{\boldsymbol{x} = (x_1, \ldots, x_n) \in \mathbf{R}^n : x_i > 0, i = 1, \ldots, n\}$$

In particular, $\mathbf{R}^1$ and $\mathbf{R}_+^1$ denoted by $\mathbf{R}$ and $\mathbf{R}_+$ respectively.

In 2009, Kuang [1] defined a mean of two variables with three parameters as follows

$$K(\omega_1, \omega_2, p; a, b) = \left[\frac{\omega_1 A(a^p, b^p) + \omega_2 G(a^p, b^p)}{\omega_1 + \omega_2} \right]^{\frac{1}{p}} \tag{1}$$

where $A(a, b) = \frac{a+b}{2}$ and $G(a, b) = \sqrt{ab}$ respectively is the arithmetic mean and geometric mean of two positive numbers a and b, parameters $p \neq 0$, $\omega_1, \omega_2 \geqslant 0$ with $\omega_1 + \omega_2 \neq 0$.

In particular

$$K\left(1, 0, p; a, b\right) = M_p(a, b) = \frac{a^p + b^p}{2}$$

is the power mean.

$$K\left(1, \frac{\omega}{2}, 1; a, b\right) = \frac{a + \omega\sqrt{ab} + b}{\omega + 2}$$

is the generalized Heron mean, which was introduced by Janous [2] in 2001.

$$K\left(1, \frac{\omega}{2}, p; a, b\right) = \frac{a^p + \omega(ab)^{p/2} + b^p}{\omega + 2}$$

is the generalized Heron mean with parameter.

Obviously, for $\boldsymbol{a} = (a_1, \ldots, a_n) \in \mathbf{R}_+^n$, $K(\omega_1, \omega_2, p; a, b)$ can be generalized as follows

$$K_n(\omega_1, \omega_2, p; \boldsymbol{a}) = \left[\frac{\omega_1 A_n(a_1^p, \cdots, a_n^p) + \omega_2 G_n(a_1^p, \cdots, a_n^p)}{\omega_1 + \omega_2}\right]^{\frac{1}{p}} \qquad (2)$$

where $A(\boldsymbol{a}) = \frac{a_1 + \cdots + a_n}{n}$ and $G(\boldsymbol{a}) = \sqrt[n]{a_1 \cdots a_n}$ respectively is the arithmetic mean and geometric mean of $\boldsymbol{a}$, parameters $p \neq 0$, $\omega_1, \omega_2 \geqslant 0$ with $\omega_1 + \omega_2 \neq 0$.

For simplicity, sometimes we will be $K_n(\omega_1, \omega_2, p; \boldsymbol{a})$ for $K_n(\omega_1, \omega_2, p)$ or $K_n(a_1, \cdots, a_n)$.

In recent years, the study on the properties of the mean by using theory of majorization is unusually active (see references [10-35]).

In this paper, Schur-convexity, Schur-geometric convexity and Schur-harmonic convexity of $K_n(\omega_1, \omega_2, p; \boldsymbol{a})$ are discussed. As consequences, some interesting inequalities are obtained.

Our main results are as follows:

Theorem 1 Let $\omega_1, \omega_2 \geqslant 0$ with $\omega_1 + \omega_2 \neq 0$.

(i) If $p \leqslant 1$, then $K_n(\omega_1, \omega_2, p; \boldsymbol{a})$ is Schur-concave with $\boldsymbol{a} \in \mathbf{R}_+^n$;

(ii) If $p \geqslant 1$, $\omega_2 = 0$, then $K_n(\omega_1, \omega_2, p; \boldsymbol{a})$ is Schur-convex with $\boldsymbol{a} \in \mathbf{R}_+^n$;

(iii) If $p \geqslant 1$, $\omega_1 = 0$, then $K_n(\omega_1, \omega_2, p; \boldsymbol{a})$ is Schur-concave with $\boldsymbol{a} \in \mathbf{R}_+^n$;

(iv) Let $B = \{\boldsymbol{a} = (a_1, \ldots, a_n) | a_i \geqslant 1, i = 1, \ldots, n$ and $a_1 + \cdots + a_n \leqslant s\}$. If $p \geqslant 2$ and $\omega_1 \geqslant \frac{\omega_2 s^p}{n^p}$, then $K_n(\omega_1, \omega_2, p; \boldsymbol{a})$ is Schur-convex with $\boldsymbol{a} \in B$;

(v) Let $C = \{\boldsymbol{a} = (a_1, \ldots, a_n) | a_i \leqslant 1, i = 1, \ldots, n$ and $\frac{1}{a_1} + \cdots + \frac{1}{a_n} \geqslant r\}$. If $2 \geqslant p \geqslant 1$ and $\omega_1 \leqslant \frac{\omega_2 n^p}{r^p}$, then $K_n(\omega_1, \omega_2, p; \boldsymbol{a})$ is Schur-concave with $\boldsymbol{a} \in C$.

Theorem 2 Let $\omega_1, \omega_2 \geqslant 0$ with $\omega_1 + \omega_2 \neq 0$. If $p > 0$, then $K_n(\omega_1, \omega_2, p; \boldsymbol{a})$ is Schur-geometrically convex with $\boldsymbol{a} \in \mathbf{R}_+^n$. If $p < 0$, then $K_n(\omega_1, \omega_2, p; \boldsymbol{a})$ is Schur-geometrically concave with $\boldsymbol{a} \in \mathbf{R}_+^n$.

Theorem 3 Let $\omega_1, \omega_2 \geqslant 0$ with $\omega_1 + \omega_2 \neq 0$. If $p \geqslant -1$, then $K_n(\omega_1, \omega_2, p; \boldsymbol{a})$ is Schur-harmonically convex with $\boldsymbol{a} \in \mathbf{R}_+^n$.

81.2   Definitions and Lemmas

We need the following definitions and lemmas.

Definition 1 [3-4] Let $\boldsymbol{x} = (x_1, \ldots, x_n)$ and $\boldsymbol{y} = (y_1, \ldots, y_n) \in \mathbf{R}^n$.

(i) $\boldsymbol{x}$ is said to be majorized by $\boldsymbol{y}$ (in symbols $\boldsymbol{x} \prec \boldsymbol{y}$) if $\sum\limits_{i=1}^{k} x_{[i]} \leqslant \sum\limits_{i=1}^{k} y_{[i]}$ for $k = 1, 2, \ldots, n-1$ and $\sum\limits_{i=1}^{n} x_i = \sum\limits_{i=1}^{n} y_i$, where $x_{[1]} \geqslant \cdots \geqslant x_{[n]}$ and $y_{[1]} \geqslant \cdots \geqslant y_{[n]}$ are rearrangements of $\boldsymbol{x}$ and $\boldsymbol{y}$ in a descending order.

(ii) $\Omega \subset \mathbf{R}^n$ is called a convex set if $(\alpha x_1 + \beta y_1, \ldots, \alpha x_n + \beta y_n) \in \Omega$ for any $\boldsymbol{x}$ and $\boldsymbol{y} \in \Omega$, where α and $\beta \in [0, 1]$ with $\alpha + \beta = 1$.

(iii) Let $\Omega \subset \mathbf{R}^n$, $\varphi \colon \Omega \to \mathbf{R}$ is said to be a Schur-convex function on Ω if $\boldsymbol{x} \prec \boldsymbol{y}$ on Ω implies $\varphi(\boldsymbol{x}) \leq \varphi(\boldsymbol{y})$. φ is said to be a Schur-concave function on Ω if and only if $-\varphi$ is Schur-convex function.

Definition 2 [5-6] Let $\boldsymbol{x} = (x_1, \ldots, x_n)$ and $\boldsymbol{y} = (y_1, \ldots, y_n) \in \mathbf{R}_+^n$.

(i) $\Omega \subset \mathbf{R}_+^n$ is called a geometrically convex set if $(x_1^\alpha y_1^\beta, \ldots, x_n^\alpha y_n^\beta) \in \Omega$ for any $\boldsymbol{x}$ and $\boldsymbol{y} \in \Omega$, where α and $\beta \in [0, 1]$ with $\alpha + \beta = 1$.

(ii) Let $\Omega \subset \mathbf{R}_+^n$, $\varphi \colon \Omega \to \mathbf{R}_+$ is said to be a Schur-geometrically convex function on Ω if $(\ln x_1, \ldots, \ln x_n) \prec (\ln y_1, \ldots, \ln y_n)$ on Ω implies $\varphi(\boldsymbol{x}) \leqslant \varphi(\boldsymbol{y})$. φ is said to be a Schur-geometrically concave function on Ω if and only if $-\varphi$ is Schur-geometrically convex function.

Definition 3 [7-8] Let $\Omega \subset \mathbf{R}_+^n$.

(i) A set Ω is said to be a harmonically convex set if $\frac{\boldsymbol{xy}}{\lambda \boldsymbol{x} + (1-\lambda)\boldsymbol{y}} \in \Omega$ for every $\boldsymbol{x}, \boldsymbol{y} \in \Omega$ and $\lambda \in [0, 1]$, where $\boldsymbol{xy} = \sum\limits_{i=1}^{n} x_i y_i$ and $\frac{1}{\boldsymbol{x}} = \left(\frac{1}{x_1}, \cdots, \frac{1}{x_n}\right)$.

(ii) A function $\varphi \colon \Omega \to \mathbf{R}_+$ is said to be a Schur harmonically convex function on Ω if $\frac{1}{\boldsymbol{x}} \prec \frac{1}{\boldsymbol{y}}$ implies $\varphi(\boldsymbol{x}) \leq \varphi(\boldsymbol{y})$. A function φ is said to be a Schur harmonically concave function on Ω if and only if $-\varphi$ is a Schur harmonically convex function.

Lemma 1 [3-4] Let $\Omega \subset \mathbf{R}^n$ is convex set, and has a nonempty interior set Ω^0. Let $\varphi \colon \Omega \to \mathbf{R}$ is continuous on Ω and differentiable in Ω^0. Then φ is the $Schur - convex(Schur - concave) function$, if and only if it is symmetric on Ω and if

$$(x_1 - x_2)\left(\frac{\partial \varphi}{\partial x_1} - \frac{\partial \varphi}{\partial x_2}\right) \geqslant 0 (\leqslant 0)$$

holds for any $\boldsymbol{x} = (x_1, x_2, \cdots, x_n) \in \Omega^0$.

Lemma 2 [5-6] Let $\Omega \subset \mathbf{R}_+^n$ be a symmetric geometrically convex set with a nonempty interior Ω^0. Let $\varphi : \Omega \to \mathbf{R}_+$ be continuous on Ω and differentiable on Ω^0. Then φ is a Schur geometrically convex (Schur geometrically concave) function if and only if φ is symmetric on Ω and

$$(x_1 - x_2)\left(x_1\frac{\partial \varphi}{\partial x_1} - x_2\frac{\partial \varphi}{\partial x_2}\right) \geqslant 0 \quad (\leqslant 0) \tag{3}$$

holds for any $\boldsymbol{x} = (x_1, \cdots, x_n) \in \Omega^0$.

Lemma 3 [7-8] Let $\Omega \subset \mathbf{R}_+^n$ be a symmetric harmonically convex set with a nonempty interior Ω^0. Let $\varphi : \Omega \to \mathbf{R}_+$ be continuous on Ω and differentiable on Ω^0. Then φ is a Schur harmonically convex (Schur harmonically concave) function if and only if φ is symmetric on Ω and

$$(x_1 - x_2)\left(x_1^2\frac{\partial \varphi}{\partial x_1} - x_2^2\frac{\partial \varphi}{\partial x_2}\right) \geqslant 0 \quad (\leqslant 0) \tag{4}$$

holds for any $\boldsymbol{x} = (x_1, \cdots, x_n) \in \Omega^0$.

Lemma 4 [4] Let $\boldsymbol{a} = (a_1, \cdots, a_n) \in \mathbf{R}_+^n$ and $A_n(\boldsymbol{a}) = \frac{1}{n}\sum_{i=1}^{n} a_i$. Then

$$\boldsymbol{u} = \left(\underbrace{A_n(\boldsymbol{a}), \cdots, A_n(\boldsymbol{a})}_{n}\right) \prec (a_1, \cdots, a_n) = \boldsymbol{a} \tag{5}$$

Lemma 5 [4] Let $\boldsymbol{a} = (a_1, \cdots, a_n) \in \mathbf{R}_+^n$. Then for any $c \geqslant 0$, we have

$$\boldsymbol{v} := \left(\frac{a_1 + c}{\sum\limits_{i=1}^{n} a_i + nc}, \cdots, \frac{a_n + c}{\sum\limits_{i=1}^{n} a_i + nc}\right) \prec \left(\frac{a_1}{\sum\limits_{i=1}^{n} a_i}, \cdots, \frac{a_n}{\sum\limits_{i=1}^{n} a_i}\right) =: \boldsymbol{w} \tag{6}$$

Lemma 6 For $x, y \in [1, +\infty)$, if $p \geqslant 2$, then

$$\frac{x^{p-1} - y^{p-1}}{x - y} \geqslant \frac{1}{xy} \tag{7}$$

and for $x, y \in (0, 1]$, if $1 \leqslant p \leqslant 2$, then inequality (7) is reversed.

Proof Without loss of generality, we may assume that $x \geqslant y$. It is easy to see that inequality (7) is equivalent to

$$x^{p-1} + x^{-1} \geqslant y^{p-1} + y^{-1} \tag{8}$$

Let $f(t) = t^{p-1} + t^{-1}$. Then $f'(t) = t^{-2}((p-1)t^p - 1)$. For $x, y \in [1, +\infty)$, if $p \geqslant 2$, then $f'(t) \geqslant 0$, so that the inequality (8) holds. For $x, y \in (0, 1]$, if $1 \leqslant p \leqslant 2$, then $f'(t) \leqslant 0$, which implies (8) is reversed.

81.3 Proofs of Theorems

From the definition of $K_n(\omega_1, \omega_2, p; \boldsymbol{a})$, we have

$$K_n(\omega_1, \omega_2, p; \boldsymbol{a}) = (m(\omega_1, \omega_2, p))^{\frac{1}{p}}$$

where

$$m(\omega_1, \omega_2, p) = \frac{\omega_1 \sum\limits_{i=1}^{n} a_i^p + n\omega_2 \prod\limits_{i=1}^{n} a_i^{\frac{p}{n}}}{n(\omega_1 + \omega_2)}$$

Proofs of Theorems 1 It is clear that $K_n(\omega_1, \omega_2, p; \boldsymbol{a})$ is symmetric with $\boldsymbol{a} \in \mathbf{R}_+^n$.

$$\frac{\partial K_n}{\partial a_1} = \frac{1}{p}(m(\omega_1, \omega_2, p))^{\frac{1}{p}-1} \cdot \frac{\omega_1 p a_1^{p-1} + p\omega_2 a_1^{\frac{p}{n}-1} a_2^{\frac{p}{n}} \prod\limits_{i=3}^{n} a_i^{\frac{p}{n}}}{n(\omega_1 + \omega_2)}$$

$$\frac{\partial K_n}{\partial a_2} = \frac{1}{p}(m(\omega_1, \omega_2, p))^{\frac{1}{p}-1} \cdot \frac{\omega_1 p a_2^{p-1} + p\omega_2 a_1^{\frac{p}{n}} a_2^{\frac{p}{n}-1} \prod\limits_{i=3}^{n} a_i^{\frac{p}{n}}}{n(\omega_1 + \omega_2)}$$

and then

$$\Delta := (a_1 - a_2)\left(\frac{\partial K_n}{\partial a_1} - \frac{\partial K_n}{\partial a_2}\right) = \frac{(m(\omega_1, \omega_2, p))^{\frac{1}{p}-1}}{n(\omega_1 + \omega_2)}(a_1 - a_2)^2 f(\boldsymbol{a})$$

where

$$f(\boldsymbol{a}) = \omega_1 \frac{a_1^{p-1} - a_2^{p-1}}{a_1 - a_2} - \omega_2 \frac{1}{a_1 a_2} \prod\limits_{i=1}^{n} a_i^{\frac{p}{n}} \tag{9}$$

(i) For $p \leqslant 1$, obvious $f(\boldsymbol{a}) \leqslant 0$, and then $\Delta \leqslant 0$, by Lemma 1, it follows that $K_n(\omega_1, \omega_2, p; \boldsymbol{a})$ is Schur-concave with $\boldsymbol{a} \in \mathbf{R}_+^n$;

(ii) From $p \geqslant 1$, $\omega_2 = 0$, it is easy to see $f(\boldsymbol{a}) \geqslant 0$, and then $\Delta \geqslant 0$, by

Lemma 1, it follows that $K_n(\omega_1, \omega_2, p; \boldsymbol{a})$ is Schur-convex with $\boldsymbol{a} \in \mathbf{R}_+^n$;

(iii) If $p \geqslant 1$, $\omega_1 = 0$, then $f(\boldsymbol{a}) \leqslant 0$, and then $\Delta \leqslant 0$, by Lemma 1, it follows that $K_n(\omega_1, \omega_2, p; \boldsymbol{a})$ is Schur-concave with $\boldsymbol{a} \in \mathbf{R}_+^n$;

(iv) Obviously, $K_n(\omega_1, \omega_2, p; \boldsymbol{a})$ is symmetry function on the symmetry convex set B. By the arithmetic-geometric mean value inequality and $a_1 + \cdots + a_n \leqslant s$, we have

$$f(\boldsymbol{a}) \geqslant \omega_1 \frac{a_1^{p-1} - a_2^{p-1}}{a_1 - a_2} - \frac{s^p \omega_2}{n^p a_1 a_2}$$

By Lemma 6, $\frac{a_1^{p-1} - a_2^{p-1}}{a_1 - a_2} \geqslant \frac{1}{a_1 a_2}$, and combining the condition $\omega_1 \geqslant \frac{\omega_2 s^p}{n^p}$, we get $f(\boldsymbol{a}) \geqslant 0$, and then $\Delta \geqslant 0$, by Lemma 1, it follows that $K_n(\omega_1, \omega_2, p; \boldsymbol{a})$ is Schur-convex with $\boldsymbol{a} \in B$;

(v) Obviously, $K_n(\omega_1, \omega_2, p; \boldsymbol{a})$ is symmetry function on the symmetry convex set C. By the geometric-harmonic mean inequality and $\frac{1}{a_1} + \cdots + \frac{1}{a_n} \geqslant r$, we have

$$f(\boldsymbol{a}) \geqslant \omega_1 \frac{a_1^{p-1} - a_2^{p-1}}{a_1 - a_2} - \omega_2 \frac{n^p}{r^p a_1 a_2}$$

By the Lemma 6, $\frac{a_1^{p-1} - a_2^{p-1}}{a_1 - a_2} \leqslant \frac{1}{a_1 a_2}$, and combining the condition $\omega_1 \leqslant \frac{\omega_2 n^p}{r^p}$, we get $f(\boldsymbol{a}) \leqslant 0$, and then $\Delta \leqslant 0$, by Lemma 1, it follows that $K_n(\omega_1, \omega_2, p; \boldsymbol{a})$ is Schur-concave with $\boldsymbol{a} \in C$.

The proof of Theorem 1 is complete.

Proofs of Theorems 2 It is easy to see that

$$a_1 \frac{\partial K_n}{\partial a_1} = \frac{1}{p} (m(\omega_1, \omega_2, p))^{\frac{1}{p} - 1} \cdot \frac{\omega_1 p a_1^p + p \omega_2 \prod\limits_{i=1}^{n} a_i^{\frac{p}{n}}}{n(\omega_1 + \omega_2)}$$

$$a_2 \frac{\partial K_n}{\partial a_2} = \frac{1}{p} (m(\omega_1, \omega_2, p))^{\frac{1}{p} - 1} \cdot \frac{\omega_1 p a_2^p + p \omega_2 \prod\limits_{i=1}^{n} a_i^{\frac{p}{n}}}{n(\omega_1 + \omega_2)}$$

and then

$$\Delta_1 := (a_1 - a_2)\left(a_1 \frac{\partial K_n}{\partial a_1} - a_2 \frac{\partial K_n}{\partial a_2}\right) = \frac{(m(\omega_1, \omega_2, p))^{\frac{1}{p} - 1}}{n(\omega_1 + \omega_2)} \omega_1 (a_1^p - a_2^p)(a_1 - a_2)$$

If $p > 0$, then $a_1 - a_2$ and $a_1^p - a_2^p$ has the same sign, so $\Delta_1 \geqslant 0$. By Lemma 2, it follows that $K_n(\omega_1, \omega_2, p; \boldsymbol{a})$ is Schur-geometrically convex with

$\boldsymbol{a} \in \mathbf{R}_+^n$. If $p < 0$, then $a_1 - a_2$ and $a_1^p - a_2^p$ has the opposite sign, so $\Delta_1 \leqslant 0$. By Lemma 2, it follows that $K_n(\omega_1, \omega_2, p; \boldsymbol{a})$ is Schur-geometrically concave with $\boldsymbol{a} \in \mathbf{R}_+^n$.

The proof of Theorem 2 is complete.

Proofs of Theorems 3 It is easy to see that

$$a_1^2 \frac{\partial K_n}{\partial a_1} = \frac{1}{p}(m(\omega_1, \omega_2, p))^{\frac{1}{p}-1} \cdot \frac{\omega_1 p a_1^{p+1} + p\omega_2 a_1^{\frac{p}{n}+1} a_2^{\frac{p}{n}} \prod_{i=3}^n a_i^{\frac{p}{n}}}{n(\omega_1 + \omega_2)}$$

$$a_2^2 \frac{\partial K_n}{\partial a_2} = \frac{1}{p}(m(\omega_1, \omega_2, p))^{\frac{1}{p}-1} \cdot \frac{\omega_1 p a_2^{p+1} + p\omega_2 a_1^{\frac{p}{n}} a_2^{\frac{p}{n}+1} \prod_{i=3}^n a_i^{\frac{p}{n}}}{n(\omega_1 + \omega_2)}$$

and then

$$\Delta_2 := (a_1 - a_2)\left(a_1^2 \frac{\partial K_n}{\partial a_1} - a_2^2 \frac{\partial K_n}{\partial a_2}\right) = \frac{(m(\omega_1, \omega_2, p))^{\frac{1}{p}-1}}{n(\omega_1 + \omega_2)}(a_1 - a_2)^2 f(\boldsymbol{a})$$

where

$$f(\boldsymbol{a}) = \omega_1 \frac{a_1^{p+1} - a_2^{p+1}}{a_1 - a_2} + \omega_2 \prod_{i=1}^n a_i^{\frac{p}{n}} \tag{10}$$

If $p \geqslant -1$, then $a_1 - a_2$ and $a_1^{p+1} - a_2^{p+1}$ has the same sign, so $f(\boldsymbol{a}) \geqslant 0$, and then $\Delta_2 \geqslant 0$. By Lemma 3, it follows that $K_n(\omega_1, \omega_2, p; \boldsymbol{a})$ is Schur-harmonically convex with $\boldsymbol{a} \in \mathbf{R}_+^n$.

The proof of Theorem 3 is complete.

81.4 Applications

Theorem 4 For $\boldsymbol{a} = (a_1, \ldots, a_n) \in \mathbf{R}_+^n$, if $p \leqslant 1$ and $\omega_1, \omega_2 \geqslant 0$ with $\omega_1 + \omega_2 \neq 0$, then we have

$$A_n(\boldsymbol{a}) \geqslant \left(\frac{\omega_1 A_n(\boldsymbol{a}^p) + \omega_2 G_n(\boldsymbol{a}^p)}{\omega_1 + \omega_2}\right)^{\frac{1}{p}} \tag{11}$$

where $A_n(\boldsymbol{a}^p) = \frac{a_1^p + \cdots + a_n^p}{n}$ and $G_n(\boldsymbol{a}^p) = \sqrt[n]{a_1^p \cdots a_n^p}$.

Proof

If $p \leqslant 1$, $\omega_1 \omega_2 \neq 0$, then by Theorem 1, from Lemma 4, we have

$$K_n(\omega_1, \omega_2, p; \boldsymbol{u}) \geqslant K_n(\omega_1, \omega_2, p; \boldsymbol{a})$$

rearranging gives (12).

The proof is complete.

Theorem 5 For $\boldsymbol{a} = (a_1, \ldots, a_n) \in \mathbf{R}_+^n$, if $p > 0 (< 0)$, then we have

$$G_n(\boldsymbol{a}) \leqslant (\geqslant) M_p(\boldsymbol{a}) \tag{12}$$

Proof By Lemma 4, we have

$$\left(\underbrace{\log G_n(\boldsymbol{a}), \cdots, \log G_n(\boldsymbol{a})}_{n} \right) \prec (\log a_1, \cdots, \log a_n)$$

if $p > 0 (< 0)$, by Theorem 2, it follows

$$K_n \left(\underbrace{G_n(\boldsymbol{a}), \cdots, G_n(\boldsymbol{a})}_{n} \right) \leqslant (\geqslant) K_n (a_1, \cdots, a_n)$$

rearranging gives (12).

The proof is complete.

Theorem 6 For $\boldsymbol{a} = (a_1, \ldots, a_n) \in \mathbf{R}_+^n$, if $p \geqslant -1$ and $\omega_1, \omega_2 \geqslant 0$ with $\omega_1 + \omega_2 \neq 0$, then we have

$$H_n(\boldsymbol{a}) \leqslant \left(\frac{\omega_1 A_n(\boldsymbol{a}^p) + \omega_2 G_n(\boldsymbol{a}^p)}{\omega_1 + \omega_2} \right)^{\frac{1}{p}} \tag{13}$$

where $H_n(\boldsymbol{a}) = \dfrac{n}{\sum\limits_{i=1}^{n} \frac{1}{a_i}}$ is the harmonic mean.

By Lemma 4, we have

$$\left(\underbrace{\frac{1}{H_n(\boldsymbol{a})}, \cdots, \frac{1}{H_n(\boldsymbol{a})}}_{n} \right) \prec \left(\frac{1}{a_1}, \cdots, \frac{1}{a_n} \right)$$

If $p \geqslant -1$, by Theorem 3, it follows

$$K_n \left(\underbrace{H_n(\boldsymbol{a}), \cdots, H_n(\boldsymbol{a})}_{n} \right) \leqslant K_n (a_1, \cdots, a_n)$$

rearranging gives (13).

The proof is complete.

Theorem 7 Let $a = (a_1, \ldots, a_n) \in \mathbf{R}_+^n$ and $\omega_1, \omega_2 \geqslant 0$ with $\omega_1 + \omega_2 \neq 0$. If $2 \geqslant p \geqslant 1$ and $\omega_1 \leqslant \frac{\omega_2}{n^p}$, then for any $c \geqslant 0$, we have

$$\left(\frac{\omega_1 A_n((a+c)^p) + \omega_2 G_n((a+c)^p)}{\omega_1 A_n(a^p)) + \omega_2 G_n(a^p)} \right)^{\frac{1}{p}} \geqslant \frac{A_n(a+c)}{A_n(a)} \tag{14}$$

where $A_n((a+c)^p) = \frac{1}{n} \sum\limits_{i=1}^{n} (a_i + c)^p$ and $G_n((a+c)^p) = \sqrt[n]{\prod\limits_{i=1}^{n} (a_i + c)^p}$.

Proof For v and w in (6), it is clear that $v_i \leqslant 1$ and $w_i \leqslant 1, i = 1, \ldots, n$.

$$\sum_{i=1}^{n} \frac{1}{v_i} = \sum_{i=1}^{n} (x_i + c) \sum_{i=1}^{n} \frac{1}{x_i + c} \geqslant n^2$$

and

$$\sum_{i=1}^{n} \frac{1}{w_i} = \sum_{i=1}^{n} x_i \sum_{i=1}^{n} \frac{1}{x_i} \geqslant n^2$$

If $2 \geqslant p \geqslant 1$ and $\omega_1 \leqslant \frac{\omega_2 n^p}{n^2 p} = \frac{\omega_2}{n^p}$, then by (v) in Theorem 1 and Lemma 5, it follows

$$K_n(\omega_1, \omega_2, p; v) \geqslant K_n(\omega_1, \omega_2, p; w),$$

rearranging gives (14).

The proof is complete.

References

[1] J-C KUANG. Applied Inequalities (Chang yong bu deng shi)[M]. 4th ed. Ji'nan: Shandong Press of Science and Technology, 2010.

[2] W JANOUS. A note on generalized Heronian means[J]. Mathematical Inequalities & Applications, 2001,4 (3) : 369–375.

[3] B-Y WANG. Foundations of majorization inequalities[M].Beijing: Beijing Normal Univ. Press, 1990.

[4] A W MARSHALL, I OLKIN. Inequalities:theory of majorization and its application[M]. New York : Academies Press, 1979.

[5] X-M ZHANG. Geometrically convex functions [M]. Hefei: Anhui University Press, 2004.

[6] C P NICULESCU. Convexity According to the Geometric Mean[J]. Mathematical Inequalities & Applications. 2000, 3(2):155–167.

[7] Y-M CHU, G-D WANG, X-H ZHANG. The Schur multiplicative and harmonic convexities of the complete symmetric function[J]. Mathematische Nachrichten, 2011, 284 (5-6): 653–663.

[8]　J-X MENG, Y M CHU, X-M TANG. The Schur-harmonic-convexity of dual form of the Hamy symmetric function[J]. Matematički Vesnik, 2010, 62 (1): 37-46.

[9]　H-N SHI, Y-M JIANG, W-D JIANG. Schur-convexity and Schur-geometrically concavity of Gini mean[J]. Computers and Mathematics with Applications, 2009, 57 : 266–274.

[10]　A WITKOWSKI. On Schur convexity and Schur-geometrical convexity of four-parameter family of means[J]. Math. Inequal. Appl., 2011,14 (4): 897–903.

[11]　J SÁNDOR. The Schur-convexity of Stolarsky and Gini means[J]. Banach J. Math. Anal. , 2007, 1, No. 2: 212–215.

[12]　Y-M CHU, X-M ZHANG. Necessary and sufficient conditions such that extended mean values are Schur-convex or Schur-concave[J]. Journal of Mathematics of Kyoto University, 2008, 48(1): 229–238.

[13]　Y-M CHU, X-M ZHANG. The Schur geometrical convexity of the extended mean values[J]. Journal of Convex Analysis, 2008, 15, No. 4: 869–890.

[14]　W-F XIA, Y-M CHU. The Schur convexity of Gini mean values in the sense of harmonic mean[J]. Acta Mathematica Scientia 2011, 31B (3) :1103–1112.

[15]　H-N SHI, B MIHALY, S-HA WU, D-M LI. Schur convexity of generalized Heronian means involving two parameters[J]. J. Inequal. Appl., vol.2008, Article ID 879273, 9 pages.

[16]　W-F XIA, Y-M CHU. The Schur multiplicative convexity of the generalized Muirhead mean[J]. International Journal of Functional Analysis, Operator Theory and Applications, 2009, 1(1): 1–8.

[17]　Y-M CHU, W-F XIA. Necessary and sufficient conditions for the Schur harmonic Convexity of the Generalized Muirhead Mean[J]. Proceedings of A. Razmadze Mathematical Institute, 2010, 152 : 19–27.

[18]　Z-H Yang. Necessary and sufficient conditions for Schur geometrical convexity of the four-parameter homogeneous means[J]. Abstr. Appl. Anal., Volume 2010, Article ID 830163, 16 pages, doi:10.1155/2010/ 830163.

[19]　W-F XIA, Y-M CHU. The Schur convexity of the weighted generalized logarithmic mean values according to harmonic mean[J]. International Journal of Modern Mathematics, 2009, 4(3): 225–233.

[20] A WITRKOWSKI. On Schur-convexity and Schur-geometric convexity of four-parameter family of means[J]. Math. Inequal. Appl., 2011, 14 (4): 897 - -903.

[21] Z-H YANG. Schur harmonic convexity of Gini means[J]. International Mathematical Forum, 2011, 6 (16): 747– 762.

[22] Z-H YANG. Schur power convexity of Stolarsky means[J]. Publ. Math. Debrecen, 2012, 80 (1-2): 43–66.

[23] F QI, J SÁNDOR, S S DRAGOMIR, A SOFO. Notes on the Schur-convexity of the extended mean values[J]. Taiwanese J. Math., 2005, 9(3): 411–420.

[24] L -L FU, B-Y XI, H M SRIVASTAVA. Schur-convexity of the generalized Heronian means involving two positive numbers[J]. Taiwanese Journal of Mathematics, 2011, 15(6): 2721–2731.

[25] T-Y ZHANG, A-P JI. Schur-convexity of generalized Heronian mean[J]. Communications in Computer and Information Science, 1, Volume 244, Information Computing and Applications, Part 1, Pages 25–33.

[26] W-F XIA, Y-M CHU, G-D WANG. Necessary and sufficient conditions for the Schur harmonic convexity or concavity of the extended mean values[J]. Revista De La Uniòn Matemática Argentina, 2010, 51(2) : 121–132.

[27] Y WU, F QI. Schur-harmonic convexity for differences of some means [J]. Analysis, 2012, 32: 1001–1008.

[28] V LOKESHA, K M NAGARAJA, B NAVEEN KUMAR, Y-D WU. Schur convexity of Gnan mean for two variables[J]. NNTDM, 2011, 17 (4): 37–41.

[29] Y WU, F QI, H-N SHI. Schur-harmonic convexity for differences of some special means in two variables[J]. J.Math. Inequal., 2014, 8 (2): 321–330.

[30] W-M GONG, X SHEN, Y-M CHU. The Schur convexity for the generalized Muirhead mean[J]. J. Math. Inequal., 2014, 8 (4): 855–862.

[31] K M NAGARAJA, SUDHIR KUMAR SAHU. Schur harmonic convexity of Stolarsky extended mean values[J]. Scientia Magna, 2013, 9 (2): 18–29.

[32] K M NAGARAJA, B NAVAVEEN KUMAR, K M NAGARAJA, S PADMANABHAN. Schur geometric convexity for ratio of difference of

means[J]. Journal of Scientific Research & Reports, 2014, 3(9): 1211–1219.

[33] Y-P DEANG, S-H WU, Y-M CHU, D HE. The Schur convexity of the generalized Muirhead-Heronian means[J]. Abstract and Applied Analysis, 2014 (2014), Article ID 706518, 11 pages.

[34] W-M GONG, H SUN, Y-M CHU. The Schur convexity for the generalized muirhead mean[J]. J.Math. Inequal., 2014, 8 (4): 855–862.

[35] H-P YIN, H-N SHI, F QI. On Schur m - power convexity for ratios of some means[J]. J.Math. Inequal., 2015, 9 (1): 145–153.

第82篇 国内Schur-凸函数研究综述

(石焕南. 广东第二师范学院学报, 2017, 37 (5): 6-12)

摘 要: 本文回顾2003 年以来国内Schur-凸函数研究的进展, 着重介绍国内学者应用Schur-凸函数理论于解析不等式研究所取得的成绩, 并对今后的研究提出几点建议.

关键词: Schur-凸函数; 解析不等式; Schur-几何凸函数; Schur-调和凸函数; Schur-幂凸函数

82.1 引言

1979年, Marshall 和Olkin 合作出版了《Inequalities: Theory of Majorization and Its Application》[1] 一书, 以此为标志, 受控理论(亦称控制不等式理论)成为数学的一门独立的新兴学科. 1979 年9 月至1981 年9 月, 北京师范大学的王伯英教授在美国加州大学(UCSB)做访问学者, 期间学习了这一理论. 回国后于1984 年在国内率先开设了有关受控理论的硕士研究生课程《矩阵与控制不等式》. 1990年, 王伯英教授编著的《控制不等式基础》[2] 一书正式出版, 该书除精选了Marshall 和Olkin 一书中的经典基础理论以外, 还包含了不少王伯英教授精彩的独创内容. 在应用部分, 该书着重讨论了受控理论在矩阵上的应用.

如王伯英教授所言 "控制不等式几乎渗入到各个数学领域, 而且处处扮演着精彩角色, 原因是它常能深刻地描述许多数学量之间的内在关系, 从而便于推得所需的结论; 它还能把许多已有的从不同方法得来的不等式用一种统一的方法简便地推导出来, 它更是推广已有的不等式, 发现新的不等式的一种强有力手段, 控制不等式的理论和应用有着美好的发展前景".

《控制不等式基础》一书的出版极大地推动我国受控理论研究的发展. 截至目前, 我国学者在国内外已发表了三百多篇有关受控理论与解析不等式方面的的研究论文, 绝大多数是2003 年后发表的,其中近百篇刊于SCI期刊. 已形成了一支在国际上具有一定影响的研究队伍, 其中包括王挽澜、续铁权、石焕南、文家金、张小明、褚玉明、关开中、吴善和、杨镇杭、姜卫东、杨定华、席博彦、马统一、李大矛、夏卫锋、张静、何灯、许谦、王文、龙波涌、王东生、傅春茹等. 2011 年, B. C. Arnold, A. M. Marshall 和 I. Olkin 合作的《Inequalities: Theory of Majorization and Its Application》(第二版)[3] 引用了不少国内学者的论文.

2012年, 石焕南的《受控理论与解析不等式》[4] 一书由哈尔滨工业大学出版社出版,该书介绍了受控理论的新推广及受控理论在解析不等式(包括对称函数不等式, 序列不等式, 积分不等式, 平均值不等式等)方面的应用, 并附有400多篇参考文献. 该书出版后的五年间,书中所涉及的几乎所有问题都有了后续的研究成果.

2017年, 受刘培杰数学工作室的举荐,得国家出版基金的赞助,石焕南的《Schur凸函数与不等式》[5] 一书由哈尔滨工业大学出版社出版, 与《受控理论与解析不等式》比较, 该书的参考文献新增了近160 余篇, 还新增了"Schur凸函数与几何不等式" 等章节.

Schur-凸函数是受控理论的核心概念, 本文回顾自2003 年始, 15年间国内Schur-凸函数研究的进展,着重介绍国内学者应用Schur凸函数理论于解析不等式研究所取得的成绩, 并对今后的研究提出几点建议.

82.2　Schur-凸函数及其推广

在本文中, $\mathbf{R}^n$, $\mathbf{R}_+^n$ 和 $\mathbf{R}_{++}^n$ 分别表示 n 维实数集, n 维非负实数集和 n 维正实数集,并记 $\mathbf{R}^1 = \mathbf{R}, \mathbf{R}_+^1 = \mathbf{R}_+$ 和 $\mathbf{R}_{++}^1 = \mathbf{R}_{++}$.

定义 1[1-2]　对于 $\boldsymbol{x} = (x_1, ..., x_n) \in \mathbf{R}^n$, 将 $\boldsymbol{x}$ 的分量排成递减的次序后, 记作 $x_{[1]} \geqslant x_2 \geqslant \cdots \geqslant x_{[n]}$. 设 $\boldsymbol{x}, \boldsymbol{y} \in \mathbf{R}^n$ 满足

(1) $\sum_{i=1}^k x_{[i]} \leqslant \sum_{i=1}^k y_{[i]}, k = 1, 2, ..., n-1$; (2) $\sum_{i=1}^n x_i = \sum_{i=1}^n y_i$. 则称 $\boldsymbol{x}$ 被 $\boldsymbol{y}$ 所控制, 记作 $\boldsymbol{x} \prec \boldsymbol{y}$.

$\boldsymbol{x} \leqslant \boldsymbol{y}$ 表示 $x_i \leqslant y_i, i = 1, ..., n$

定义 2[1-2]　设 $\Omega \subset \mathbf{R}^n, \varphi : \Omega \to \mathbf{R}$, 若在 Ω 上 $\boldsymbol{x} \prec \boldsymbol{y} \Rightarrow \varphi(\boldsymbol{x}) \leqslant \varphi(\boldsymbol{y})$, 则称 φ 为 Ω 上的Schur-凸函数(简称S- 凸函数); 若 $-\varphi$ 是 Ω 上的S-凸函数, 则称 φ 为 Ω 上的S- 凹函数.

若在 Ω 上 $\boldsymbol{x} \leqslant \boldsymbol{y} \Rightarrow \varphi(\boldsymbol{x}) \leqslant \varphi(\boldsymbol{y})$, 则称 φ 为 Ω 上的增函数; 若 $-\varphi$ 是 Ω 上的增函数, 则称 φ 为 Ω 上的减函数.

定理 1[1-2]　(Schur-凸函数判定定理) 设 $\Omega \subset \mathbf{R}^n$ 是有内点的对称凸集, $\varphi : \Omega \to \mathbf{R}$ 在 Ω 上连续, 在 Ω 的内部 Ω^o 可微, 则在 Ω 上S-凸(S-凹)的充

要条件是 φ 在 Ω 上对称且 $\forall \boldsymbol{x} \in \Omega$, 有

$$(x_1 - x_2) \left(\frac{\partial \varphi}{\partial x_1} - \frac{\partial \varphi}{\partial x_2} \right) \geqslant 0 (\leqslant 0).$$

关于Schur-凸函数的推广, 国内学者做了不少开拓性的工作, 并得到国外学者的认可与应用. 2003 年, 张小明首先提出并建立了Schur-几何凸函数的定义及判定定理.

定义 3[6-7]　设 $\Omega \subset \mathbf{R}_{++}^n$, $f : \Omega \to \mathbf{R}_+$, 对于任意 $\boldsymbol{x} = (x_1, \ldots, x_n), \boldsymbol{y} = (y_1, \ldots, y_n) \in \Omega$, 若

$$\ln \boldsymbol{x} := (\ln x_1, \ldots, \ln x_n) \prec (\ln y_1, \ldots, \ln y_n) =: \ln \boldsymbol{y}$$

有 $f(\boldsymbol{x}) \leqslant f(\boldsymbol{y})$, 则称为 f 为 Ω 上的S-几何凸函数. 若 $\ln \boldsymbol{x} \prec \ln \boldsymbol{y}$ 有 $f(\boldsymbol{x}) \geqslant f(\boldsymbol{y})$, 则称 f 为 Ω 上的S-几何凹函数.

定理 2[6-7]　(Schur-几何凸函数判定定理) 设 $\Omega \subset \mathbf{R}_{++}^n$ 是有内点的对称凸集, $\varphi : \Omega \to \mathbf{R}$ 在 Ω 上连续, 在 Ω 的内部 Ω^0 可微, 则在 Ω 上S-几何凸(S-凹)的充要条件是 φ 在 Ω 上对称且 $\forall \boldsymbol{x} \in \Omega$, 有

$$(\ln x_1 - \ln x_2) \left(x_1 \frac{\partial \varphi}{\partial x_1} - x_2 \frac{\partial \varphi}{\partial x_2} \right) \geqslant 0 (\leqslant 0)$$

张小明将其对几何凸函数的研究成果写入其专著[6]和[7].

2008 年, 褚玉明等人首先提出并建立了Schur-调和凸函数的定义及判定定理.

定义 4[8]　设 $\Omega \subset \mathbf{R}_+^n$, $\varphi : \Omega \to \mathbf{R}_+$. $\forall \boldsymbol{x} = (x_1, \cdots, x_n), \boldsymbol{y} = (y_1, \cdots, y_n) \in \Omega$, 若 $(1/x_1, \cdots, 1/x_n) \prec (1/y_1, \cdots, 1/y_n) \Rightarrow \varphi(\boldsymbol{x}) \leqslant \varphi(\boldsymbol{y})$, 则称 φ 为 Ω 上的Schur-调和凸函数; 若 $-\varphi$ 是 Ω 上 S-调和凸函数, 则称 φ 为 Ω 上的 S-调和凹函数.

定理 3[8] (调和凸函数判定定理) 设 $\Omega \subset \mathbf{R}^n$ 是有内点的对称调和凸集, $\varphi : \Omega \to \mathbf{R}_+$ 于 Ω 上连续, 在 Ω 的内部 Ω^0 一阶可微. 若 φ 在 Ω 上对称, 且对于任意 $\boldsymbol{x} = (x_1, \cdots, x_n) \in \Omega^0$, 有

$$(x_1 - x_2) \left(x_1^2 \frac{\partial \varphi}{\partial x_1} - x_2^2 \frac{\partial \varphi}{\partial x_2} \right) \geqslant 0 (\leqslant 0)$$

则 φ 是 Ω 上 S-调和凸(凹)函数.

作为Schur-凸函数, Schur-几何凸函数, Schur-调和凸函数等概念的统一推广, 2010 年杨镇杭定义了Schur-f 凸函数及Schur-幂凸函数, 并建立了相应的判定定理.

定义 5[9]　设 $f : \mathbf{R}_+ \to \mathbf{R}$ 定义如下

$$f(x) = \begin{cases} \dfrac{x^m - 1}{m}, & m \neq 0 \\ \ln x, & m = 0 \end{cases}$$

对于任意 $\boldsymbol{x}, \boldsymbol{y} \in \Omega$, 若 $f(\boldsymbol{x}) \prec f(\boldsymbol{y})$, 有 $\phi(\boldsymbol{x}) \leqslant \phi(\boldsymbol{y})$, 则称 $\phi : \Omega \subset \mathbf{R}_+^n \to \mathbf{R}$ 是 Ω 上的 Schur m- 幂凸函数. 若 $-\phi$ 是 Schur m-幂凸函数, 则称 ϕ 是 Ω 上的 Schur m-幂凹函数.

在定义 5 中, 若分别置 $f(x)$ 为 $x, \ln x$ 和 $\frac{1}{x}$, 则定义 5 就化为 Schur- 凸, Schur-几何凸和 Schur-调和凸函数的定义.

定理 14[9]　(Schur 幂凸函数的判定定理) 设 $\Omega \subset \mathbf{R}^n$ 是有内点的对称凸集, $\varphi : \Omega \to \mathbf{R}$ 在 Ω 上连续, 在 Ω 的内部 Ω^0 可微, 则在 Ω 上 Schur-凸(Schur-凹)的充要条件是 φ 在 Ω 上对称且 $\forall \boldsymbol{x} \in \Omega$, 有

$$\frac{x_1^m - x_2^m}{m} \left(x_1^{1-m} \frac{\partial \varphi}{\partial x_1} - x_2^{1-m} \frac{\partial \varphi}{\partial x_2} \right) \geqslant 0 (\leqslant 0)$$

2009 年, 杨定华[10] 用公理化的方法, 提出了抽象平均、抽象凸函数和抽象受控等概念, 它们分别是平均、凸函数和受控等概念的相应推广. 通过逻辑演绎, 建立了抽象受控不等式的基本定理.

2014 年, 在上述工作的基础上, 杨定华[11] 考虑范畴论的思想和方法: 通过考察对象之间的态射反映对象本身的性质在抽象平均的基础平台上, 基于映射的观点, 首先提出抽象平均、抽象凸函数、抽象控制和抽象受控不等式等同构映射的概念, 建立了抽象凸函数同构映射的基本定理.

82.3　Schur-凸函数性质的移植

将 Schur-凸函数的定义推广到 Schur-几何凸, Schur-调和凸函和 Schur-幂凸函数后, 自然可进一步考虑 Schur-凸函数的某些性质可否向 Schur-几何凸, Schur-调和凸函和 Schur- 幂凸函数移植? 石焕南和张静在这方面做了下述工作.

定义 6[12]　设区间 $I \subset \mathbf{R}_+$, 函数 $\varphi : I \to \mathbf{R}_+$ 连续.

(1) 函数 φ 说是 I 上的 GA 凸(凹) 函数, 若对于 $x, y \in I$, 有

$$\varphi\left(\sqrt{xy}\right) \leqslant (\geqslant) \frac{\varphi(x) + \varphi(y)}{2}$$

(2) 函数 φ 说是 I 上的 HA 凸(凹) 函数, 若对于 $x, y \in I$, 有

$$\varphi\left(\frac{2xy}{x+y}\right) \leqslant (\geqslant) \frac{\varphi(x) + \varphi(y)}{2}$$

设 $\pi = (\pi(1), \cdots, \pi(n))$ 是 $(1, \cdots, n)$ 的任意置换, 关于 Schur-凸函数有如下判定定理:

定理 5[1] 设 $A \subset \mathbf{R}^k$ 是一个对称凸集, φ 是 A 上的 S-凸函数, 具有性质: 对每一个固定的 $x_2, \cdots, x_k, \varphi(z, x_2, \cdots, x_k)$ 关于 z 在 $\{z : (z, x_2, \cdots, x_k) \in A\}$ 上凸, 则对于任何 $n > k$, $\psi(x_1, \cdots, x_n) = \sum\limits_{\pi} \varphi\left(x_{\pi(1)}, \cdots, x_{\pi(k)}\right)$ 在 B 上 S-凸, 其中

$$B = \left\{(x_1, \cdots, x_n) : \left(x_{\pi(1)}, \cdots, x_{\pi(k)}\right) \in A, \text{对于所有的排列 } \pi\right\}$$

继而

$$\overline{\psi}(\boldsymbol{x}) = \sum\limits_{1 \leqslant i_1 < \cdots < i_k \leqslant n} \varphi\left(x_{i_1}, \cdots, x_{i_k}\right)$$

在 B 上 S-凸.

2013年, 石焕南和张静[13] 将定理5 移植到 Schur-几何凸和 Schur-调和凸的情形.

定理 6[13] 设 $A \subset \mathbf{R}^k$ 是一个对称凸集, φ 是 A 上的 S-几何凸函数, 具有性质: 对每一个固定的 $x_2, \cdots, x_k, \varphi(z, x_2, \cdots, x_k)$ 关于 z 在 $\{z : (z, x_2, \cdots, x_k) \in A\}$ 上 GA 凸, 则对于任何 $n > k$,

$$\psi(x_1, \cdots, x_n) = \sum\limits_{\pi} \varphi\left(x_{\pi(1)}, \cdots, x_{\pi(k)}\right)$$

在 B 上 S-几何凸, 其中

$$B = \left\{(x_1, \cdots, x_n) : \left(x_{\pi(1)}, \cdots, x_{\pi(k)}\right) \in A, \text{ 对于所有的排列 } \pi\right\}$$

继而

$$\overline{\psi}(\boldsymbol{x}) = \sum\limits_{1 \leqslant i_1 < \cdots < i_k \leqslant n} \varphi\left(x_{i_1}, \cdots, x_{i_k}\right)$$

在 B 上 S-几何凸.

定理 7[13] 设 $A \subset \mathbf{R}^k$ 是一个对称凸集, φ 是 A 上的 S-调和凸函数, 具有性质: 对每一个固定的 $x_2, \cdots, x_k, \varphi(z, x_2, \cdots, x_k)$ 关于 z 在 $\{z : (z, x_2, \cdots, x_k) \in$

A} 上HA凸, 则对于任何 $n > k$

$$\psi(x_1, \cdots, x_n) = \sum_\pi \varphi\left(x_{\pi(1)}, \cdots, x_{\pi(k)}\right)$$

在 B 上S-调和凸, 其中

$$B = \left\{(x_1, \cdots, x_n) : \left(x_{\pi(1)}, \cdots, x_{\pi(k)}\right) \in A, \text{ 对于所有的排列 } \pi\right\}$$

继而

$$\overline{\psi}(\boldsymbol{x}) = \sum_{1 \leqslant i_1 < \cdots < i_k \leqslant n} \varphi\left(x_{i_1}, \cdots, x_{i_k}\right)$$

在 B 上S-调和凸.

设区间 $I \subset \mathbf{R}$, $\varphi : I \to \mathbf{R}_+$ 是一个对数凸函数. 定义对称函数

$$F_k(\boldsymbol{x}) = \sum_{1 \leqslant i_1 < \ldots < i_k \leqslant n} \prod_{j=1}^{k} f(x_{i_j}), \quad k = 1, \ldots, n$$

王淑红等人[14] 利用Schur 凸函数判定定理, 即定理1证得:

定理 8[14] 设 $I \subset \mathbf{R}$ 是具有非空内部的对称凸集, 函数 $f : I \to \mathbf{R}_{++}$ 在 I 上连续, 在 I 的内部可微且对数凸, 则 $F_k(\boldsymbol{x})$ 在 I^n 上S-凸. 若 f 还是递增函数, 则 $F_k(\boldsymbol{x})$ 在 I^n 上S - 几何凸且S- 调和凸.

张静和石焕南[15] 利用定理5, 定理6 和定理7 简洁地证明了定理8.

2014年, 对于 $F_k(\boldsymbol{x})$ 的对偶函数

$$F_k^*(\boldsymbol{x}) = \prod_{1 \leqslant i_1 < \ldots < i_k \leqslant n} \sum_{j=1}^{k} f(x_{i_j}), \quad k = 1, \ldots, n$$

石焕南和张静[16] 利用定理5, 定理6 和定理7 证得

定理 9[16] 设 $I \subset \mathbf{R}$ 是具有非空内部的对称凸集, 函数 $f : I \to \mathbf{R}_{++}$ 在 I 上连续, 在 I 的内部可微且对数凸, 则 $F_k^*(\boldsymbol{x})$ 在 I^n 上S-凸. 若 f 还是递增函数, 则 $F_k^*(\boldsymbol{x})$ 在 I^n 上S-几何凸且S- 调和凸.

关于复合函数的Schur-凸性, 有如下结论:

定理 10[1] 设区间 $[a, b] \subset \mathbf{R}$, $\varphi : \mathbf{R}^n \to \mathbf{R}$, $f : [a, b] \to \mathbf{R}$, $\psi(x_1, \cdots, x_n) = \varphi(f(x_1), \cdots, f(x_n)) : [a, b]^n \to \mathbf{R}$.

(1) 若 φ 增且S-凸, f 凸, 则 ψ S-凸;

(2) 若 φ 增且S-凹, f 凹, 则 ψ S-凹;

(3) 若 φ 减且S-凸, f 凹, 则 ψ S-凸;

(4) 若 φ 增且S-凸, f 增且凸, 则 ψ 增且S-凸;

(5) 若 φ 减且S-凸, f 减且凹, 则 ψ 增且S-凸;

(6) 若 φ 增且S-凸, f 减且凸, 则 ψ 减且S-凸;

(7) 若 φ 减且S-凸, f 增且凸, 则 ψ 减且S-凸;

(8) 若 φ 减且S-凹, f 减且凹, 则 ψ 增且S-凹.

2015 年, 石焕南和张静[17] 将定理10 移植到Schur-几何凸的情形.

定理 11[17] 设区间$[a,b] \subset \mathbf{R}, \varphi : \mathbf{R}^n \to \mathbf{R}, f : [a,b] \to \mathbf{R}, \psi(x_1, \cdots, x_n) = \varphi(f(x_1), \cdots, f(x_n)) : [a,b]^n \to \mathbf{R}.$

(1) 若 φ 增且S-几何凸, f 几何凸, 则 ψ S-几何凸;

(2) 若 φ 增且S-几何凹, f 几何凹, 则 ψ S-几何凹;

(3) 若 φ 减且S-几何凸, f 几何凹, 则 ψ S-几何凸;

(4) 若 φ 增且S-几何凸, f 增且几何凸, 则 ψ 增且S-几何凸;

(5) 若 φ 减且S-几何凸, f 减且几何凹, 则 ψ 增且S-几何凸;

(6) 若 φ 增且S-几何凸, f 减且几何凸, 则 ψ 减且S-几何凸;

(7) 若 φ 减且S-几何凸, f 增且几何凹, 则 ψ 减且S-几何凸;

(8) 若 φ 减且S-几何凹, f 减且几何凹, 则 ψ 增且S-几何凹.

2017 年, 石焕南和张静[18] 又将定理10 移植到Schur-调和凸的情形.

定理 12[18] 设区间$[a,b] \subset \mathbf{R}, \varphi : \mathbf{R}^n \to \mathbf{R}, f : [a,b] \to \mathbf{R}, \psi(x_1, \cdots, x_n) = \varphi(f(x_1), \cdots, f(x_n)) : [a,b]^n \to \mathbf{R}.$

(1) 若 φ 增且S-调和凸, f 调和凸, 则 ψ S-调和凸;

(2) 若 φ 增且S-调和凹, f 调和凹, 则 ψ S-调和凹;

(3)若 φ 减且S-调和凸, f 调和凹, 则 ψ S-调和凸;

(4) 若 φ 增且S-调和凸, f 增且调和凸, 则 ψ 增且S-调和凸;

(5) 若 φ 减且S-调和凸, f 减且调和凹, 则 ψ 增且S-调和凸;

(6) 若 φ 增且S-调和凸, f 减且调和凸, 则 ψ 减且S-调和凸;

(7) 若 φ 减且S-调和凸, f 增且调和凹, 则 ψ 减且S-调和凸;

(8) 若 φ 减且S-调和凹, f 减且调和凹, 则 ψ 增且S-调和凹.

82.4　Schur-凸函数与解析不等式

近些年国内学者应用Schur-凸函数理论研究解析不等式的成果颇丰, 本文不可能一一赘述, 仅在本节介绍某些笔者感觉比较新颖或比较漂亮的结果.

82.4.1　Schur-凸函数与多元对称函数不等式

一个对称凸集上的S-凸函数必是对称函数,这意味着控制不等式理论最适宜处理对称函数不等式问题,因此研究多元对称函数的Schur-凸性一直是受控理论研究的热点. 近几年, 国内学者得到一些不错的结果,这里介绍数例.

对于下述完全对称函数的复合函数

$$F_n(\boldsymbol{x}, r) = \sum_{i_1+i_2+\cdots+i_n=r} \left(\frac{x_1}{1-x_1}\right)^{i_1} \left(\frac{x_2}{1-x_2}\right)^{i_2} \cdots \left(\frac{x_n}{1-x_n}\right)^{i_n}$$

2014 年, 孙明保等人[19] 考察了它的S-凸性, S-几何凸性和S-调和凸性, 分别利用这三种凸性的判定定理, 即定理1, 定理2 和定理3 证得如下三个定理.

定理 13[19] 对于 $\boldsymbol{x} = (x_1,\ldots,x_n) \in [0,1)^n \cup (1,+\infty)^n$ 和 $r \in \mathbf{N}$.

(1) $F_n(\boldsymbol{x}, r)$ 在$[0,1)^n$ 上递增且Schur-凸;

(2) 若r 是偶数(奇数), 则$F_n(\boldsymbol{x}, r)$ 在$(1,+\infty)^n$ 上递减且Schur- 凸(递增且Schur- 凹)

定理 14[19] 对于 $\boldsymbol{x} = (x_1,\ldots,x_n) \in [0,1)^n \cup (1,+\infty)^n$ 和 $r \in \mathbf{N}$.

(1) $F_n(\boldsymbol{x}, r)$ 在$[0,1)^n$ 上Schur-几何凸;

(2) 若 r 是偶数(奇数), 则$F_n(\boldsymbol{x}, r)$ 在$(1,+\infty)^n$ 上Schur-几何凸(Schur-几何凹).

定理 15[19] 对于 $\boldsymbol{x} = (x_1,\ldots,x_n) \in [0,1)^n \cup (1,+\infty)^n$ 和 $r \in \mathbf{N}$.

(1) $F_n(\boldsymbol{x}, r)$ 在$[0,1)^n$ 上Schur-调和凸;

(2) 若r 是偶数(奇数), 则$F_n(\boldsymbol{x}, r)$ 在$(1,+\infty)^n$ 上调和凸(调和凹).

注 石焕南等人[20] 分别利用S-凸, S-几何凸和S-调和凸的有关性质给出了上述三个定理的简单证明.

关开中[21] 将Hamy对称函数推广为

$$\sum_n^k (f(\boldsymbol{x})) = \sum_{1\leqslant i_1<\ldots<i_k\leqslant n} f\left(\prod_{j=1}^k x_{i_j}^{\frac{1}{k}}\right), \ k = 1,\ldots,n$$

2011 年, 关开中和关汝柯[22] 研究了当f 为MN 凸函数时$\sum_n^k (f(\boldsymbol{x}))$ 的S-凸性, 得到如下结果.

定理 16[22] 设$I \subset \mathbf{R}_{++}, f : I \to \mathbf{R}_{++}$ 连续, 则:

(1) 若f 在I 上递减且AA 凸, 则$\sum_n^k (f(\boldsymbol{x}))$ 在I^n 上S - 凸;

(2) 若f 在I 上递增且AA 凹,则$\sum_n^k (f(\boldsymbol{x}))$ 在I^n 上S - 凹.

定理 17[22] 设 $I \subset \mathbf{R}_{++}, f : I \to \mathbf{R}_{++}$ 连续, 则:

(1) 若 f 在 I 上递减且 GA 凸, 则 $\sum\limits_{n}^{k} (f(\boldsymbol{x}))$ 在 I^n 上 S - 几何凸;

(2) 若 f 在 I 上递增且 GA 凹, 则 $\sum\limits_{n}^{k} (f(\boldsymbol{x}))$ 在 I^n 上 S - 几何凹.

定理 18[22] 设 $I \subset \mathbf{R}_{++}, f : I \to \mathbf{R}_{++}$ 连续, 则:

(1) 若 f 在 I 上递减且 HA 凸, 则 $\sum\limits_{n}^{k} (f(\boldsymbol{x}))$ 在 I^n 上 S - 调和凸;

(2) 若 f 在 I 上递增且 HA 凹, 则 $\sum\limits_{n}^{k} (f(\boldsymbol{x}))$ 在 I^n 上 S - 调和凹.

2014 年, 王文和杨世国[23] 研究了 $\sum\limits_{n}^{k} (f(\boldsymbol{x}))$ 的 S- 幂凸性, 得到如下结果.

定理 19 设 $I \subset \mathbf{R}_+, f : I \to \mathbf{R}_+$ 在 I 内有二阶连续的偏导数. 若 f 在 I 上是单调递增的几何凸函数, 则

(1) 对于 $m \leqslant 0, \sum\limits_{n}^{k} (f(\boldsymbol{x}))$ 在 I^n 上 m 阶 S-幂凸;

(2) 当 $r = 2, n = 2$ 时, 对于 $m > 0, \sum\limits_{n}^{k} (f(\boldsymbol{x}))$ 在 I^n 上 m 阶 S - 幂凹.

注 定理19 是经张鑑等人[24]修正后的结果, 且给出结论(1) 一个简单证明.

王文和杨世国[25] 定义了如下对称函数

$$F_{n,k}(\boldsymbol{x}, r) = \prod_{1 \leqslant i_1 < \ldots < i_k \leqslant n} f\left(\sum_{j=1}^{k} x_{i_j}^r\right)^{\frac{1}{r}}, \quad k = 1, \ldots, n$$

并利用 S-幂凸函数的判定定理证得下述结果.

定理 20[25] 设 $\Omega \subset \mathbf{R}_{++}$ 是一个具有非空内点的对称凸集, $f : \Omega \to \mathbf{R}_{++}$ 在 Ω 上连续, 在 Ω^0 可微. 若 f 是递增的几何凸函数, 则对于 $m \leqslant 0, r > 0, k = 1, 2 \ldots, n, F_{n,k}(\boldsymbol{x}, r)$ 是 Ω 上的 S - 幂凸函数.

注 石焕南[5] 利用 S-几何凸函数和 S-幂凸函数的性质给出了一个简单的证明.

82.4.2 Schur-凸函数与平均值不等式

设 $(r, s) \in \mathbf{R}^2, (x, y) \in \mathbf{R}_{++}^2$. Stolarsky平均定义如下

$$
E(r,s;x,y)=\begin{cases}\left(\dfrac{s}{r}\cdot\dfrac{y^r-x^r}{y^s-x^s}\right)^{1/(r-s)}, & rs(r-s)(x-y)\neq0\\[2ex]\left(\dfrac{1}{r}\cdot\dfrac{y^r-x^r}{\ln y-\ln x}\right)^{1/r}, & r(x-y)\neq0,s=0\\[2ex]\dfrac{1}{e^{1/r}}\left(\dfrac{x^{x^r}}{y^{y^r}}\right)^{1/(x^r-y^r)}, & r(x-y)\neq0,r=s\\[2ex]\sqrt{xy}, & r=s=0,x\neq y\\[1ex]x, & x=y\end{cases}
$$

2012 年, 杨镇杭[26] 考察了$E(r,s;x,y)$的S-幂凸性, 得到:

定理 21[26] 对于固定的$(r,s)\in\mathbf{R}^2$:

(1) 若$m>0$, 则$E(r,s;x,y)$ 关于(x,y) 在$\mathbf{R}^2_{++}$ 上m 阶S - 幂凸(S- 幂凹)当且仅当$r+s\geqslant(\leqslant)3m$ 且$\min\{r,s\}\geqslant m(\leqslant m)$;

(2) 若$m<0$, 则$E(r,s;x,y)$ 关于(x,y) 在$\mathbf{R}^2_{++}$上m 阶S-幂凸(S-幂凹)当且仅当$r+s\geqslant(\leqslant)3m$ 且$\max\{r,s\}\geqslant m(\leqslant m)$;

(3) 若$m=0$, 则$E(r,s;x,y)$ 关于(x,y) 在$\mathbf{R}^2_{++}$ 上S- 幂凸(S-幂凹)当且仅当$r+s\geqslant0(\leqslant0)$.

设$(r,s)\in\mathbf{R}^2,(x,y)\in\mathbf{R}^2_{++}$. Gini平均定义如下

$$
G(r,s;x,y)=\begin{cases}\left(\dfrac{x^s+y^s}{x^r+y^r}\right)^{1/(s-r)}, & r\neq s\\[2ex]\exp\left(\dfrac{x^s\ln x+y^s\ln y}{x^r+y^r}\right), & r=s\end{cases}
$$

2013 年, 杨镇杭[9] 考察了$G(r,s;x,y)$的S-幂凸性,得到:

定理 22[9] 对于固定的$(r,s)\in\mathbf{R}^2$:

(1) 若$m>0$, 则$G(r,s;x,y)$ 关于(x,y) 在$\mathbf{R}^2_{++}$ 上m 阶S - 幂凸(S- 幂凹)当且仅当$r+s\geqslant(\leqslant)m$ 且$\min\{r,s\}\geqslant0(\leqslant0)$;

(2) 若$m<0$, 则$G(r,s;x,y)$ 关于(x,y) 在$\mathbf{R}^2_{++}$上m 阶S-幂凸(S-幂凹)当且仅当$r+s\geqslant(\leqslant)m$ 且$\max\{r,s\}\geqslant0(\leqslant0)$;

(3) 若$m=0$, 则$G(r,s;x,y)$ 关于(x,y) 在$\mathbf{R}^2_{++}$ 上S- 幂凸(S-幂凹)当且仅当$r+s\geqslant0(\leqslant0)$.

广义Heron平均定义如下

$$
H_{p,w}(x,y)=\begin{cases}\left[\dfrac{x^p+w(xy)^{p/2}+y^p}{w+2}\right]^{\frac{1}{p}}, & p\neq0\\[2ex]\sqrt{xy}, & p=0\end{cases}
$$

where $w \geqslant 0$.

杨镇杭[27] 将广义Heron平均$H_{w,p}(x,y)$ 参数w 的取值范围$w \geqslant 0$ 扩展为$w > -2$, 并称此时的广义Heron平均为Daróczy 平均, 得到如下结果.

定理 23[27] 对于固定的$p \in \mathbf{R}, m > 0$ 和$w > -2$, $H_{w,p}(x,y)$ 关于(x,y) 在$\mathbf{R}_{++}^2$ 上m 阶S- 幂凸(S- 幂凹)当且仅当$(w,p) \in \Omega_1((w,p) \in \Omega_2)$, 其中

$$\Omega_1 = \left\{ -2 < w \leqslant 0, p \geqslant \frac{(w+2)m}{2} \right\}$$

$$\cup \left\{ w > 0, p \geqslant \max\left(\frac{(w+2)m}{2}, 2m \right) \right\}$$

$$\Omega_2 = \{ -2 < w < 0, p < 0 \} \cup \left\{ w \geqslant 0, p \leqslant \min\left(\frac{(w+2)m}{2}, 2m \right) \right\}$$

定理 24[27] 对于固定的$p \in \mathbf{R}, m < 0$ 和$w > -2$, $H_{w,p}(x,y)$ 关于(x,y) 在$\mathbf{R}_{++}^2$ 上m 阶S- 幂凸(S- 幂凹)当且仅当$(w,p) \in E_1((w,p) \in E_2)$, 其中

$$E_1 = \{ -2 < w < 0, p > 0 \} \cup \left\{ w \geqslant 0, p \geqslant \max(\frac{w+2}{2}m, 2m) \right\}$$

$$E_2 = \left\{ -2 < w \leqslant 0, p \leqslant \frac{w+2}{2}m \right\}$$

$$\cup \left\{ w > 0, p \leqslant \min(\frac{w+2}{2}m, 2m) \right\}$$

定理 25[27] 对于固定的$p \in \mathbf{R}, m = 0$ 和$w > -2$, $H_{w,p}(x,y)$ 关于(x,y) 在$\mathbf{R}_{++}^2$ 上m 阶S- 幂凸(S- 幂凹)当且仅当$p \geqslant (\leqslant)0$.

2014 年, 邓勇平等人[28] 将广义Heron 平均$H_{w,p}(x,y)$ 和Gini平均$G(r,s;x,y)$ 统一推广为如下含有三个参数的广义Gini-Heron平均并考察了它的S-几何凸性.

$$H_{p,w}(x,y) = \begin{cases} \left[\dfrac{x^p + w(xy)^{p/2} + y^p}{x^q + w(xy)^{q/2} + y^q} \right]^{\frac{1}{p-q}}, & p \neq q \\ \exp\left\{ \dfrac{x^p \ln x + w(xy)^{p/2} \ln xy + y^p \ln y}{x^p + w(xy)^{p/2} + y^p} \right\}, & p = q \end{cases}$$

其中$(p,q) \in \mathbf{R}^2, (x,y) \in \mathbf{R}_{++}^2$.

定理 26[28] 对于固定的$(p,q,w) \in \mathbf{R}^3$:

(1) 若$p + q \geqslant 0$ 且$w \geqslant 0$,则$H_{p,w}(x,y)$ 关于(x,y) 在$\mathbf{R}_{++}^2$ 上S-几何凸;

(2) 若$p + q \leqslant 0$ 且$w \geqslant 0$,则$H_{p,w}(x,y)$ 关于(x,y) 在$\mathbf{R}_{++}^2$ 上S-几何凹.

吴裕东等人[29] 还考察了Gnan平均和它的对偶式的S-凸性和S-几何凸性.

问题 Gnan平均及其对偶式的调和凸性和S-幂凸性如何?

2009年, 顾春, 石焕南[30] 证得如下结果:

定理 27[30] 设$\boldsymbol{x} \in \mathbf{R}_{++}^2$, 则$n$ 元Lehme平均

$$L_p(\boldsymbol{x}) = L_p(x_1, x_2, \ldots, x_n) = \frac{\sum\limits_{i=1}^{n} x_i^p}{\sum\limits_{i=1}^{n} x_i^{p-1}}$$

当$1 \leqslant p \leqslant 2$ 时, 在$\mathbf{R}_{++}^2$ 上S-凸; 而当$0 \leqslant p \leqslant 1$ 时, 在$\mathbf{R}_{++}^2$ 上S-凹.

并猜想: 当$p \geqslant 2$ 时, 在$\mathbf{R}_{++}^2$ 上S-凸; 而当$p \leqslant 0$ 时, 在$\mathbf{R}_{++}^2$ 上S-凹.

2016 年, 傅春茹等人[31] 指出此猜想不成立, 得到如下结果.

定理 28[31] (1) 当$p \geqslant 2$ 时, 对任意的$a > 0$, $L_p(\boldsymbol{x})$ 在$\left[\frac{(p-2)a}{p}, a\right]^n$ 上Schur-凸;

(2) 当$p < 0$ 时, 对任意的$a > 0$, $L_p(\boldsymbol{x})$ 在$\left[a, \frac{(p-2)a}{p}\right]^n$ 上Schur-凹.

定理 29[31] (1) 当$p < \frac{1}{2}$ 且$p \neq 0$ 时, 对任意的$a > 0$, $L_p(\boldsymbol{x})$ 在$\left[a, (\frac{(p-1)}{p})^2 a, \right]^n$ 上S-几何凹;

(2) 当$p > \frac{1}{2}$ 时, 对任意的$a > 0$, $L_p(\boldsymbol{x})$ 在$\left[\frac{(p-1)^2}{p})a, a\right]^n$ 上S-几何凸;

(3) 当$p = 0$ 时, $L_p(\boldsymbol{x})$ 在$\mathbf{R}_{++}^n$ 上S-几何凹.

定理 30[31] (1) 当$0 \leqslant p \leqslant 1$ 时, $L_p(\boldsymbol{x})$ 在$\mathbf{R}_{++}^n$ 上S-调和凸, 当$-1 \leqslant p \leqslant 0$ 时, $L_p(\boldsymbol{x})$ 在$\mathbf{R}_{++}^n$ 上S-调和凹;

(2) 当$p > 1$ 时, 对任意的$a > 0$, $L_p(\boldsymbol{x})$ 在$\left[\frac{p-1}{p+1}a, a\right]^n$ 上S-调和凸;

(3) 当$p < -1$ 时, 对任意的$a > 0$, $L_p(\boldsymbol{x})$ 在$\left[a, \frac{p-1}{p+1}a\right]^n$ 上S-调和凹.

设$\boldsymbol{x} = (x_1, x_2, \cdots, x_n) \in \mathbf{R}_+^n, p, q \geqslant 0, p + q \neq 0$. Bonferroni平均定义如下

$$B^{p,q}(\boldsymbol{x}) = \left(\frac{1}{n(n-1)} \sum_{i,j=1, i \neq j}^{n} x_i^p x_j^q\right)^{\frac{1}{p+q}}$$

Bonferroni平均是二元Muirhead平均的推广. 最近王东生,石焕南[32] 研究了Bonferroni 平均的S-凸性, S-几何凸性和S-调和凸性, 得到如下结果.

定理 31[32] 对于$n \geqslant 3$和固定的$(p, q) \in \mathbf{R}^2$.

(1) 若$0 \leqslant q \leqslant p \leqslant 1$ 且$p - q \leqslant \sqrt{p+q}, p + q \neq 0$, 则$B^{p,q}(\boldsymbol{x})$ 在R_+^n 上S-凹;

(2) 若$q \leqslant p \leqslant 0$ 且$p + q \neq 0$, 则$B^{p,q}(\boldsymbol{x})$ 在$\mathbf{R}_+^n$ 上S- 凹;

(3) 若$p \geqslant 1, q \leqslant 0$ 且$p + q > 0$, 则$B^{p,q}(\boldsymbol{x})$ 在$\mathbf{R}_+^n$ 上S- 凸;

(4) 若$p \geqslant 1, q \leqslant 0$ 且$p + q < 0$, 则$B^{p,q}(\boldsymbol{x})$ 在$\mathbf{R}_+^n$ 上S- 凹.

定理 32[32] 对于$n \geqslant 3$ 和固定的$(p, q) \in \mathbf{R}^2$.

(1) 若 $p + q > 0$, 则 $B^{p,q}(\boldsymbol{x})$ 在 $\mathbf{R}_+^n$ 上S-几何凸;

(2) 若 $p + q < 0$, 则 $B^{p,q}(\boldsymbol{x})$ 在 $\mathbf{R}_+^n$ 上S-几何凹.

定理 33[32] 对于 $n \geqslant 3$ 和固定的 $(p, q) \in \mathbf{R}^2$.

(1) 若 $p \geqslant q \geqslant 0$ 且 $p + q \neq 0$, 则 $B^{p,q}(\boldsymbol{x})$ 在 $\mathbf{R}_+^n$ 上S- 调和凸;

(2) 若 $0 \geqslant p \geqslant q \geqslant -1$ 且 $p + q \neq 0, (p-q)^2 + p + q \leqslant 0$, 则 $B^{p,q}(\boldsymbol{x})$ 在 $\mathbf{R}_+^n$ 上S-调和凸;

(3) 若 $p \geqslant 0, q \leqslant -1$ 且 $p + q > 0$, 则 $B^{p,q}(\boldsymbol{x})$ 在 $\mathbf{R}_+^n$ 上S- 调和凸;

(4) 若 $p \geqslant 0, q \leqslant -1$ 且 $p + q < 0$, 则 $B^{p,q}(\boldsymbol{x})$ 在 $\mathbf{R}_+^n$ 上S- 调和凹.

82.4.3　Schur-凸函数与特殊函数不等式

对于 $x \in \mathbf{C}$ 且 $\mathrm{Re}(x) > 0$, 定义

$$\Gamma_k(x) = \int_0^\infty \mathrm{e}^{-\frac{t^k}{k}} t^{x-1} \mathrm{d}t$$

和

$$\zeta_k(x) = \frac{1}{\Gamma_k(x)} \int_0^\infty \frac{t^{x-k}}{\mathrm{e}^t - 1} \mathrm{d}t, \ x > k$$

$\Gamma_k(x)$ 和 $\zeta_k(x)$ 分别是伽马函数 $\Gamma(x)$ 和黎曼Zeta函数 $\zeta(x)$ 的推广. 张静和石焕南[33] 利用受控理论证得如下结果:

定理 34[33]

$$\frac{\prod_{i=1}^n \Gamma_k(1 + \alpha_i)}{\Gamma_k(\beta + \sum_{i=1}^n \alpha_i)} \leqslant \frac{\prod_{i=1}^n \Gamma_k(1 + \alpha_i x)}{\Gamma_k(\beta + (\sum_{i=1}^n \alpha_i)x)} \leqslant \frac{1}{\Gamma_k(\beta)}$$

其中 $x \in [0, 1], \beta \geqslant 1, \alpha_i > 0, n \in \mathbf{N}$.

定理 35[33]

$$\frac{\prod_{i=1}^n \zeta_k(k + 1 + \alpha_i)\Gamma_k(k + 1 + \alpha_i)}{\zeta_k(k + 1 + \sum_{i=1}^n \alpha_i)\Gamma_k(\beta + k + \sum_{i=1}^n \alpha_i)}$$
$$\leqslant \frac{\prod_{i=1}^n \zeta_k(k + 1 + \alpha_i)\Gamma_k(k + 1 + \alpha_i)x}{\zeta_k(k + 1 + (\sum_{i=1}^n \alpha_i)x)\Gamma_k(\beta + k + (\sum_{i=1}^n \alpha_i)x)}$$
$$\leqslant \frac{\left(\frac{\pi^2}{6}\right)^n}{\zeta_k(\beta + k)\Gamma_k(\beta + k)}$$

其中 $x \in [0, 1], \beta \geqslant 1, \alpha_i > 0, n \in \mathbf{N}$.

卡塔兰数(Catalan numbers)C_n 是组合数学中一类重要的自然数数列,

其计算公式可用伽马函数表为

$$C_n = \frac{4^n \Gamma \left(n + \frac{1}{2}\right)}{\sqrt{\pi} \Gamma(n+2)}$$

祁锋等人[34] 给出一个推广形式

$$C(a,b;x) = \frac{\Gamma(b)}{\Gamma(a)} \left(\frac{b}{a}\right)^x \frac{\Gamma(x+a)}{\Gamma(x+b)}$$

其中$a, b > 0, x \geqslant 0$, 称 $(a,b;x)$ 为卡塔兰-祁函数(或广义卡塔兰函数).

祁锋等人研究了此函数的S-凸性, 得到如下结果:

定理 36[35]　对于$a, b > 0, x \geqslant 0$, 令$F_x(a,b) = |\ln C(a,b,x)|$, 则对于$x \geqslant 0$, $F_x(a,b)$ 关于(a,b) 在$\mathbf{R}^2_{++}$ 上S-凸.

82.5　几点建议

最后, 关于Schur-凸函数的研究, 笔者提如下几点建议:

(1) 在受控理论的研究中, 有两项工作是重要而基础的, 一是发现和建立向量间的控制关系, 二是发现和证明各种Schur-凸函数. 利用Schur-凸函数判定定理是判定Schur-凸函数的主要方法, 但注意有时利用Schur-凸函数的性质判定会简单些.

(2) 进一步完善我们所创立的Schur-几何凸, Schur-调和凸和Schur-幂凸理论, 例如移植Schur-凸函数性质, 还有许多工作可作. 凸函数的各种推广层出不穷, 除了已有的几种推广, Schur-凸函数是不是还可以做一些推广?

(3) 利用受控理论证明不等式主要适用于对称函数不等式, 这是它的局限性, 但对称函数是非常广泛且十分重要的函数, 所以受控理论大有用武之地, 有大量课题有待我们挖掘.

(4) Arnold, Marshall 和Olkin 合作出版的《Inequalities: Theory of Majorization and Its Application》(第二版)中的应用部分涉及组合分析, 几何不等式, 矩阵理论, 数值分析, 概率统计等领域. 目前国内受控理论研究者主要利用受控理论研究解析不等式和矩阵不等式, 努力开拓在其他领域的应用, 可能别有洞天.

参考文献

[1]　MARSHALL A W, OLKIN I. Inequalities: theory of majorization and its application [M]. New York : Academies Press, 1979.

[2]　王伯英. 控制不等式基础[M]. 北京: 北京师范大学出版社, 1990.

[3]   MARSHALL A M, OLKIN I, ARNOLD B C. Inequalities: Theory of majorization and its application[M]. 2nd ed. New York Dordrecht Heidelberg London: Springer 2011.

[4]   石焕南.受控理论与解析不等式[M]. 哈尔滨:哈尔滨工业大学出版社, 2012.

[5]   石焕南.Schur-凸函数与不等式[M]. 哈尔滨:哈尔滨工业大学出版社, 2017.

[6]   张小明. Schur几何凸函数[M]. 合肥：安徽大学出版社, 2004.

[7]   张小明, 褚玉明. 解析不等式新论[M]. 哈尔滨：哈尔滨工业大学出版社, 2009.

[8]   YU-MING CHU, YU-PEI LÜ. The Schur harmonic convexity of the Hamy symmetric function and its applications[J]. Journal of Inequalities and Applications, Volume 2009, Article ID 838529, 10 pages, doi: 10.1155/2009/838529.

[9]   YANG ZHEN-HANG. Schur power convexity of Gini means [J]. Bull. Korean Math. Soc., 2013, 50 (2): 485‑498.

[10]   杨定华. 抽象控制不等式的理论基础[J]. 中国科学A辑：数学, 2009, 39 (7):873-891.

[11]   杨定华. 抽象受控不等式的同构映射[J]. 数学进展, 2014, 43 (5):741-760.

[12]   ANDERSON G D, VAMANAMURTHY M K, VUORINEN M. Generalized convexity and inequalities[J]. J. Math. Anal. Appl., 2007, 335(2): 1294-1308.

[13]   SHI HUAN-NAN, ZHANG JING. Some new judgment theorems of Schur geometric and Schur harmonic convexities for a class of symmetric functions[J]. J. Inequal. Appl., 2013, 2013:527. doi:10.1186/ 1029-242X-2013-527.

[14]   王淑红, 张天宇, 华志强. 一类对称函数的Schur-几何凸性及Schur- 调和凸性[J]. 内蒙古民族大学学报(自然科学版), 2011, 26 (4): 387-390.

[15]   张静, 石焕南. 关于一类对称函数的Schur凸性[J]. 数学的实践与认识, 2013, 43 (19): 292-296.

[16]   SHI HUAN-NAN, ZHANG JING. Schur-convexity, Schur-geometric and harmonic convexities of dual form of a class symmetric functions [J]. J.Math. Inequal., 2014, 8 (2): 349‑358.

[17] SHI HUAN-NAN, ZHANG JING. Compositions involving Schur geometrically convex functions[J]. J. Inequal. Appl., 2015 (2015): 320.

[18] SHI HUAN-NAN, ZHANG JING. Compositions involving Schur harmonically convex functions[J]. Journal of Computational Analysis and Applications, 2017, 22 (5): 907-922.

[19] Sun M-B, CHEN N-B, LI S-H. Some properties of a class of symmetric functions and its applications [J]. Math Nachr, 2014, 287(13): 1530‑1544. doi:10.1002/ mana.201300073

[20] SHI HUAN-NAN, ZHANG JING, MA QINUA. Schur-convexity, Schur-geometric and Schur-harmonic convexity for a composite function of complete symmetric function[J]. SpringerPlus, 2016 (5): 296. DOI 10.1186/s40064-016-1940-z.

[21] GUAN KAI-ZHONG. A class of symmetric functions for multiplicatively convex function [J]. Math. Inequal. Appl., 2007, 10 (4): 745-753.

[22] GUAN KAI-ZHONG, GUAN RUKE. Some properties of a generalized Hamy symmetric function and its applications[J]. J. Math. Anal. Appl., 2011, 376: 494-505.

[23] 王文, 杨世国. 一类对称函数的m-指数凸性[J]. 系统科学与数学, 2014, 34 (3): 367-375.

[24] 张鑑, 顾春, 石焕南. 一类对称函数的Schur -指数凸性的简单证明[J]. 系统科学与数学, 2016, 136 10:1779-1782.

[25] WEN WANG, SHIGUO YANG. Schur m-power convexity of a class of multiplicatively convex functions and applications[J]. Abstract and Applied Analysis, Volume 2014, Article ID 258108, 12 pages.

[26] YANG ZHEN-HANG. Schur power convexity of Stolarsky means[J]. Publ. Math. Debrecen, 2012, 80 (1-2): 43-66.

[27] YANG ZHEN-HANG. Schur power convexity of the Daróczy means[J]. Math. Inequal. Appl., 2013, 16 (3): 751‑762.

[28] DENG YONG-PING, CHU YU-MING, WU SHAN-HE, HE DENG. Schur-geometric convexity of the generalized Gini-Heronian means involving three parameters[J]. J.Inequal. Appl., 2014, 2014:413

[29] LOKESHA V, NAGARAJA K M, NAVEEN KUMAR B, WU Y-D. Schur convexity of Gnan mean for two variables [J]. NNTDM, 2011, 17 (4): 37-41.

[30] 顾春, 石焕南. Lehme平均的Schur凸性和Schur几何凸性[J]. 数学的实践与认识, 2009，39 (12)：183-188.

[31] FU CHUN-RU, WANG DONGSHENG, SHI HUAN-NAN. Schur-convexity for Lehmer mean of n variables[J]. J. Nonlinear Sci. Appl., 2016, 9: 5510-5520.

[32] DONGSHENG WANG, HUAN-NAN SHI. Schur convexity of Bonferroni means[J]. Advanced Studies in Contemporary Mathematics, 待发.

[33] ZHANG JING, SHI HUAN-NAN. Two double inequalities for k-gamma and k-Riemann zeta functions[J]. J. Inequal. Appl., 2014, 2014:191.

[34] QI F, SHI X-T, LIU F-F. An exponential representation for a function involving the gamma function and originating from the Catalan numbers[EB/OL]. http://dx.doi.org/ 10.13140/RG.2.1.1086.4486.

[35] QI F, SHI X-T, MANSOUR M, LIU F-F. Schur-convexity of the catalan-Qi function [EB/OL]. http://www.researchgate. net/ publication/ 281448338)

第83篇　Schur convexity of the generalized geometric Bonferroni mean and the relevant inequalities

(SHI HUANNAN, WU SHANHE. Journal of Inequalities and Applications, (2018) 2018:8)

Abstract: In this paper,we discuss the Schur convexity, Schur geometric convexity and Schur harmonic convexity of the generalized geometric Bonferroni mean. Some inequalities related to the generalized geometric Bonferroni mean are established to illustrate the applications of the obtained results.

Keywords: geometric Bonferroni mean; Schur's condition; majorization relationship; inequality

83.1　Introduction

The Schur convexity of functions relating to special means is a very significant research subject and has attracted the interest of many mathematicians. There are numerous articles written on this topic in recent years; see [1, 2] and the references therein. As supplements to the Schur convexity of functions, the Schur geometrically convex functions and Schur harmonically convex functions were investigated by Zhang and Yang[3], Chu, Zhang and Wang[4], Chu and Xia[5], Chu, Wang and Zhang[6], Shi and Zhang[7,8], Meng, Chu and Tang[9], Zheng, Zhang and Zhang[10]. These properties of functions have been found to be useful in discovering and proving the inequalities for special means(see [11 - 14]).

Recently, it has come to our attention that a type of means which is

701

symmetrical on n variables $x_1, x_2, \ldots, x_n$ and involves two parameters, it was initially proposed by Bonferroni[15], as follows

$$B^{p,q}(\boldsymbol{x}) = \left(\frac{1}{n(n-1)} \sum_{i,j=1, i \neq j}^{n} x_i^p x_j^q \right)^{\frac{1}{p+q}} \tag{1}$$

where $\boldsymbol{x} = (x_1, x_2, \cdots, x_n) \in \mathbf{R}_+^n$ and $p, q \geqslant 0, p + q \neq 0$.

$B^{p,q}(\boldsymbol{x})$ is called the Bonferroni mean. It has important application in multi criteria decision-making (see [16 – 21]).

Beliakov, James and Mordelová et al.[22] generalized the Bonferroni mean by introducing three parameters p, q, r, i.e.

$$GB^{p,q,r}(\boldsymbol{x}) = \left(\frac{1}{n(n-1)(n-2)} \sum_{i,j,k=1, i \neq j \neq k}^{n} x_i^p x_j^q x_k^r \right)^{\frac{1}{p+q+r}} \tag{2}$$

where $\boldsymbol{x} = (x_1, x_2, \cdots, x_n) \in \mathbf{R}_+^n$ and $p, q, r \geqslant 0, p + q + r \neq 0$.

Motivated by the Bonferroni mean $B^{p,q}(\boldsymbol{x})$ and the geometric mean $G(\boldsymbol{x}) = \prod_{i=1}^{n} (x_i)^{\frac{1}{n}}$, Xia, Xu and Zhu[23] introduced a new mean which is called the geometric Bonferroni mean, as follows

$$GB^{p,q}(\boldsymbol{x}) = \frac{1}{p+q} \prod_{i,j=1, i \neq j}^{n} (px_i + qx_j)^{\frac{1}{n(n-1)}} \tag{3}$$

where $\boldsymbol{x} = (x_1, x_2, \ldots, x_n), x_i > 0, i = 1, 2, \ldots, n, p, q \geqslant 0$ and $p + q \neq 0$.

An extension of the geometric Bonferroni mean was given by Park and Kim in [19], which is called the generalized geometric Bonferroni mean, i.e.

$$GGB^{p,q,r}(\boldsymbol{x}) = \frac{1}{p+q+r} \prod_{i,j,k=1, i \neq j \neq k}^{n} (px_i + qx_j + rx_k)^{\frac{1}{n(n-1)(n-2)}} \tag{4}$$

where $\boldsymbol{x} = (x_1, x_2, \cdots, x_n), x_i > 0, i = 1, 2, \ldots, n, p, q, r \geqslant 0$ and $p + q + r \neq 0$.

Remark 1 For $r = 0$, it is easy to observe that

$$GGB^{p,q,0}(\boldsymbol{x}) = \frac{1}{p+q+0} \prod_{i,j=1, i \neq j}^{n} \left[\prod_{k=1, i \neq j \neq k}^{n} (px_i + qx_j + 0 \times x_k) \right]^{\frac{1}{n(n-1)(n-2)}}$$

$$= \frac{1}{p+q} \prod_{i,j=1, i \neq j}^{n} \left[(px_i + qx_j)^{(2-2)} \right]^{\frac{1}{n(n-1)(n-2)}}$$

$$= \frac{1}{p+q} \prod_{i,j=1,i\neq j}^{n} (px_i + qx_j)^{\frac{1}{n(n-1)}}$$

$$= GB^{p,q}(\boldsymbol{x})$$

Remark 2 If $q = 0, r = 0$, then the generalized geometric Bonferroni mean reduces to the geometric mean, i.e.

$$GB^{p,0,0}(\boldsymbol{x}) = GB^{p,0}(\boldsymbol{x}) = \frac{1}{p} \prod_{i,j=1,i\neq j}^{n} (px_i)^{\frac{1}{n(n-1)}} = \prod_{i=1}^{n}(x_i)^{\frac{1}{n}} = G(\boldsymbol{x})$$

Remark 3 If $\boldsymbol{x} = (x, x, \ldots, x)$, then

$$GB^{p,q,r}(\boldsymbol{x}) = GB^{p,q,r}(x, x, \ldots, x) = x$$

For convenience, throughout the paper $\mathbf{R}$ denotes the set of real numbers, $\boldsymbol{x} = (x_1, x_2, \ldots, x_n)$ denotes n-tuple (n-dimensional real vectors), the set of vectors can be written as

$$\mathbf{R}^n = \{\boldsymbol{x} = (x_1, \cdots, x_n) : x_i \in \mathbf{R}, i = 1, \cdots, n\}$$

$$\mathbf{R}^n_+ = \{\boldsymbol{x} = (x_1, \cdots, x_n) : x_i \geqslant 0, i = 1, \cdots, n\}$$

$$\mathbf{R}^n_{++} = \{\boldsymbol{x} = (x_1, \cdots, x_n) : x_i > 0, i = 1, \cdots, n\}$$

In a recent paper [24], Shi and Wu investigated the Schur m-power convexity of the geometric Bonferroni mean $GB^{p,q}(\boldsymbol{x})$. The definition of Schur m-power convex function is as follows

Let $f : \mathbf{R}_+ \to \mathbf{R}$ be defined by

$$F(x) = \begin{cases} f(x) = \dfrac{x^m - 1}{m}, m \neq 0 \\ \ln x, m = 0 \end{cases}$$

Then a function $\phi : \Omega \subset \mathbf{R}^n_+ \to \mathbf{R}$ is said to be Schur m-power convex on Ω if

$$(f(x_1), f(x_2), \ldots, f(x_n)) \prec (f(y_1), f(y_2), \ldots, f(y_n))$$

for all $(x_1, x_2, \ldots, x_n) \in \Omega$ and $(y_1, y_2, \ldots, y_n) \in \Omega$ implies $\phi(x) \leqslant \phi(y)$.

If $-\phi$ is Schur m-power convex, then we say that ϕ is Schur m-power concave.

Shi and Wu[24] obtained the following result.

Proposition 1 For fixed positive real numbers p, q, (i) if $m < 0$ or $m = 0$, then $GB^{p,q}(\boldsymbol{x})$ is Schur m-power convex on $\mathbf{R}_{++}^n$; (ii) if $m = 1$ or $m \geqslant 2$, then $GB^{p,q}(\boldsymbol{x})$ is Schur m-power concave on $\mathbf{R}_{++}^n$.

In this paper we discuss the Schur convexity, Schur geometric convexity and Schur harmonic convexity of the generalized geometric Bonferroni mean $GB^{p,q,r}(\boldsymbol{x})$. Our main results are as follows.

Theorem 1 For fixed non-negative real numbers p, q, r with $p+q+r \neq 0$, if $\boldsymbol{x} = (x_1, x_2, \ldots, x_n), n \geqslant 3$, then $GB^{p,q,r}(\boldsymbol{x})$ is Schur concave, Schur geometric convex and Schur harmonic convex on $\mathbf{R}_{++}^n$.

Corollary 1 For fixed non-negative real numbers p, q with $p + q \neq 0$, if $\boldsymbol{x} = (x_1, x_2, \ldots, x_n)$, $n \geqslant 3$, then $GB^{p,q}(\boldsymbol{x})$ is Schur concave, Schur geometric convex and Schur harmonic convex on $\mathbf{R}_{++}^n$.

83.2　Preliminaries

We introduce some definitions, lemmas and propositions, which will be used in the proofs of the main results in subsequent sections.

Definition 1 (see [1]) Let $\boldsymbol{x} = (x_1, x_2, \cdots, x_n)$ and $\boldsymbol{y} = (y_1, y_2, \cdots, y_n) \in \mathbf{R}^n$.

(i) $\boldsymbol{x}$ is said to be majorized by $\boldsymbol{y}$ (in symbols $\boldsymbol{x} \prec \boldsymbol{y}$) if $\sum_{i=1}^{k} x_{[i]} \leqslant \sum_{i=1}^{k} y_{[i]}$ for $k = 1, 2, \cdots, n-1$ and $\sum_{i=1}^{n} x_i = \sum_{i=1}^{n} y_i$, where $x_{[1]} \geqslant x_{[2]} \geqslant \cdots \geqslant x_{[n]}$ and $y_{[1]} \geqslant y_{[2]} \geqslant \cdots \geqslant y_{[n]}$ are rearrangements of $\boldsymbol{x}$ and $\boldsymbol{y}$ in a descending order.

(ii) Let $\Omega \subset \mathbf{R}^n$, the function $\varphi \colon \Omega \to \mathbf{R}$ is said to be Schur-convex on Ω if $\boldsymbol{x} \prec \boldsymbol{y}$ on Ω implies $\varphi(\boldsymbol{x}) \leq \varphi(\boldsymbol{y})$. φ is said to be a Schur-concave function on Ω if and only if $-\varphi$ is Schur-convex function on Ω.

Definition 2 (see [1]) Let $\boldsymbol{x} = (x_1, x_2, \cdots, x_n)$ and $\boldsymbol{y} = (y_1, y_2, \cdots, y_n) \in \mathbf{R}^n$. $\Omega \subset \mathbf{R}^n$ is said to be a convex set if $\boldsymbol{x}, \boldsymbol{y} \in \Omega, 0 \leqslant \alpha \leqslant 1$ implies

$$\alpha \boldsymbol{x} + (1-\alpha)\boldsymbol{y} = (\alpha x_1 + (1-\alpha)y_1, \alpha x_2 + (1-\alpha)y_2, \cdots, \alpha x_n + (1-\alpha)y_n) \in \Omega$$

Definition 3 (see [1])

(i) A set $\Omega \subset \mathbf{R}^n$ is called symmetric, if $\boldsymbol{x} \in \Omega$ implies $\boldsymbol{x}\boldsymbol{P} \in \Omega$ for every $n \times n$ permutation matrix $\boldsymbol{P}$.

(ii) A function $\varphi : \Omega \to \mathbf{R}$ is called symmetric if for every permutation matrix P and $\varphi(xP) = \varphi(x)$ for all $x \in \Omega$.

The following proposition is called Schur's condition. It provides an approach for testing whether a vector valued function is Schur convex or not.

Proposition 2 (see [1]) Let $\Omega \subset \mathbf{R}^n$ be symmetric and have a nonempty interior convex set. Ω° is the interior of Ω. $\varphi : \Omega \to \mathbf{R}$ is continuous on Ω and differentiable in Ω°. Then φ is the *Schur* $-$ *convex (Schur* $-$ *concave) function* if and only if φ is symmetric on Ω and

$$(x_1 - x_2) \left(\frac{\partial \varphi}{\partial x_1} - \frac{\partial \varphi}{\partial x_2} \right) \geqslant 0 (\leqslant 0) \tag{5}$$

holds for any $x \in \Omega^\circ$.

Definition 4 (see [25]) Let $x = (x_1, x_2, \cdots, x_n)$ and $y = (y_1, y_2, \cdots, y_n) \in \mathbf{R}_+^n$.

(i) $\Omega \subset \mathbf{R}_+^n$ is called a geometrically convex set if $(x_1^\alpha y_1^\beta, x_2^\alpha y_2^\beta, \cdots, x_n^\alpha y_n^\beta) \in \Omega$ for all $x, y \in \Omega$ and $\alpha, \beta \in [0, 1]$ such that $\alpha + \beta = 1$.

(ii) Let $\Omega \subset \mathbf{R}_+^n$. The function $\varphi : \Omega \to \mathbf{R}_+$ is said to be Schur geometrically convex function on Ω if $(\log x_1, \log x_2, \cdots, \log x_n) \prec (\log y_1, \log y_2, \cdots, \log y_n)$ on Ω implies $\varphi(x) \leqslant \varphi(y)$. The function φ is said to be a Schur geometrically concave function on Ω if and only if $-\varphi$ is Schur geometrically convex function.

Proposition 3 (see [25]) Let $\Omega \subset \mathbf{R}_+^n$ be a symmetric and geometrically convex set with a nonempty interior Ω°. Let $\varphi : \Omega \to \mathbf{R}_+$ be continuous on Ω and differentiable in Ω°. If φ is symmetric on Ω and

$$(\log x_1 - \log x_2) \left(x_1 \frac{\partial \varphi}{\partial x_1} - x_2 \frac{\partial \varphi}{\partial x_2} \right) \geqslant 0 \quad (\leqslant 0) \tag{6}$$

holds for any $x = (x_1, x_2, \cdots, x_n) \in \Omega^\circ$, then φ is a Schur geometrically convex (Schur geometrically concave) function.

Definition 5 (see [26]) Let $\Omega \subset \mathbf{R}_+^n$.

(i) A set Ω is said to be harmonically convex if $\dfrac{xy}{\lambda x + (1-\lambda)y} \in \Omega$ for every $\boldsymbol{x}, \boldsymbol{y} \in \Omega$ and $\lambda \in [0,1]$, where $\boldsymbol{xy} = \sum\limits_{i=1}^{n} x_i y_i$ and

$$\frac{1}{\lambda\boldsymbol{x}+(1-\lambda)\boldsymbol{y}} = \left(\frac{1}{\lambda x_1+(1-\lambda)y_1}, \frac{1}{\lambda x_2+(1-\lambda)y_2}, \cdots, \frac{1}{\lambda x_n+(1-\lambda)y_n}\right)$$

(ii) A function $\varphi : \Omega \to \mathbf{R}_+$ is said to be Schur harmonically convex on Ω if $\frac{1}{\boldsymbol{x}} \prec \frac{1}{\boldsymbol{y}}$ implies $\varphi(\boldsymbol{x}) \leqslant \varphi(\boldsymbol{y})$. A function φ is said to be a Schur harmonically concave function on Ω if and only if $-\varphi$ is a Schur harmonically convex function.

Proposition 4 (see [26]) Let $\Omega \subset \mathbf{R}_+^n$ be a symmetric and harmonically convex set with inner points and let $\varphi : \Omega \to \mathbf{R}_+$ be a continuously symmetric function which is differentiable on Ω°. Then φ is Schur harmonically convex (Schur harmonically concave) on Ω if and only if

$$(x_1 - x_2)\left(x_1^2 \frac{\partial\varphi(\boldsymbol{x})}{\partial x_1} - x_2^2 \frac{\partial\varphi(\boldsymbol{x})}{\partial x_2}\right) \geqslant 0 \quad (\leqslant 0), \quad \boldsymbol{x} \in \Omega^\circ \qquad (7)$$

Remark 4 Propositions 3 and 4 provide analogous Schur's conditions for determining Schur geometrically convex functions and Schur harmonically convex functions, respectively.

Lemma 1 (see [1]) Let $\boldsymbol{x} = (x_1, x_2, \cdots, x_n) \in \mathbf{R}_+^n$ and $A_n(\boldsymbol{x}) = \frac{1}{n}\sum\limits_{i=1}^{n} x_i$. Then

$$\left(\underbrace{A_n(\boldsymbol{x}), A_n(\boldsymbol{x}), \cdots, A_n(\boldsymbol{x})}_{n}\right) \prec (x_1, x_2, \cdots, x_n) \qquad (8)$$

Lemma 2 (see [1]) If $x_i > 0, i = 1, 2, \ldots, n$, then for all nonnegative constants c satisfying $0 < c < \frac{1}{n}\sum\limits_{i=1}^{n} x_i$, one has

$$\left(\frac{x_1}{\sum\limits_{j=1}^{n} x_j}, \ldots, \frac{x_n}{\sum\limits_{j=1}^{n} x_j}\right) \prec \left(\frac{x_1 - c}{\sum\limits_{j=1}^{n}(x_j - c)}, \ldots, \frac{x_n - c}{\sum\limits_{j=1}^{n}(x_j - c)}\right) \qquad (9)$$

83.3　Proof of main result

Proof of Theorem 1 Note that the generalized geometric Bonferroni mean is defined by

$$GGB^{p,q,r}(\boldsymbol{x}) = \frac{1}{p+q+r} \prod_{i,j,k=1,i\neq j\neq k}^{n} (px_i + qx_j + rx_k)^{\frac{1}{n(n-1)(n-2)}}$$

taking the natural logarithm gives

$$\log \varphi(\boldsymbol{x}) = \log \frac{1}{p+q+r} + \frac{1}{n(n-1)(n-2)}Q$$

where

$$Q = \sum_{j,k=3,j\neq k}^{n} [\log(px_1 + qx_j + rx_k) + \log(px_2 + qx_j + rx_k)] +$$

$$\sum_{i,k=3,i\neq k}^{n} [\log(px_i + qx_1 + rx_k) + \log(px_i + qx_2 + rx_k)] +$$

$$\sum_{i,j=3,i\neq j}^{n} [\log(px_i + qx_j + rx_1) + \log(px_i + qx_j + rx_2)] +$$

$$\sum_{k=3}^{n} [\log(px_1 + qx_2 + rx_k) + \log(px_2 + qx_1 + rx_k)] +$$

$$\sum_{j=3}^{n} [\log(px_1 + qx_j + rx_2) + \log(px_2 + qx_j + rx_1)] +$$

$$\sum_{i=3}^{n} [\log(px_i + qx_1 + rx_2) + \log(px_i + qx_2 + rx_1)] +$$

$$\sum_{i,j,k=3,i\neq j\neq k}^{n} \log(px_i + qx_j + rx_k)$$

Differentiating $GB^{p,q,r}(\boldsymbol{x})$ with respect to x_1 and x_2, respectively, we have

$$\frac{\partial GB^{p,q,r}(\boldsymbol{x})}{\partial x_1} = \frac{GB^{p,q,r}(\boldsymbol{x})}{n(n-1)(n-2)} \frac{\partial Q}{\partial x_1}$$

$$= \frac{GB^{p,q,r}(\boldsymbol{x})}{n(n-1)(n-2)} \cdot \left[\sum_{j,k=3,j\neq k}^{n} \frac{p}{px_1 + qx_j + rx_k} + \sum_{i,k=3,i\neq k}^{n} \frac{q}{px_i + qx_1 + rx_k} + \right.$$

$$\sum_{i,j=3,i\neq j}^{n} \frac{r}{px_i + qx_j + rx_1} + \sum_{k=3}^{n} \left(\frac{p}{px_1 + qx_2 + rx_k} + \frac{q}{px_2 + qx_1 + rx_k} \right) +$$

$$\sum_{j=3}^{n}(\frac{p}{px_1+qx_j+rx_2}+\frac{r}{px_2+qx_j+rx_1})+$$

$$\sum_{i=3}^{n}(\frac{q}{px_i+qx_1+rx_2}+\frac{r}{px_i+qx_2+rx_1})]$$

$$\frac{\partial GB^{p,q,r}(\boldsymbol{x})}{\partial x_2}=\frac{GB^{p,q,r}(\boldsymbol{x})}{n(n-1)(n-2)}\frac{\partial Q}{\partial x_2}$$

$$=\frac{GB^{p,q,r}(\boldsymbol{x})}{n(n-1)(n-2)}\cdot[\sum_{j,k=3,j\neq k}^{n}\frac{p}{px_2+qx_j+rx_k}+\sum_{i,k=3,i\neq k}^{n}\frac{q}{px_i+qx_2+rx_k}+$$

$$\sum_{i,j=3,i\neq j}^{n}\frac{r}{px_i+qx_j+rx_2}+\sum_{k=3}^{n}(\frac{q}{px_1+qx_2+rx_k}+\frac{p}{px_2+qx_1+rx_k})+$$

$$\sum_{j=3}^{n}(\frac{r}{px_1+qx_j+rx_2}+\frac{p}{px_2+qx_j+rx_1})+$$

$$\sum_{i=3}^{n}(\frac{r}{px_i+qx_1+rx_2}+\frac{q}{px_i+qx_2+rx_1})]$$

It is easy to see that $GGB^{p,q,r}(\boldsymbol{x})$ is symmetric on $\mathbf{R}_{++}^{n}$. For $n\geqslant 3$, we have

$$\Delta_1:=(x_1-x_2)\left(\frac{\partial\varphi(\boldsymbol{x})}{\partial x_1}-\frac{\partial\varphi(\boldsymbol{x})}{\partial x_2}\right)$$

$$=\frac{(x_1-x_2)\varphi(\boldsymbol{x})}{n(n-1)(n-2)}\cdot[p\sum_{j,k=3,j\neq k}^{n}(\frac{1}{px_1+qx_j+rx_k}-\frac{1}{px_2+qx_j+rx_k})+$$

$$q\sum_{i,k=3,i\neq k}^{n}(\frac{1}{px_i+qx_1+rx_k}-\frac{1}{px_i+qx_2+rx_k})+$$

$$r\sum_{i,j=3,i\neq j}^{n}(\frac{1}{px_i+qx_j+rx_1}-\frac{1}{px_i+qx_j+rx_2})+$$

$$\sum_{k=3}^{n}(\frac{p-q}{px_1+qx_2+rx_k}+\frac{q-p}{px_2+qx_1+rx_k})+$$

$$\sum_{j=3}^{n}(\frac{p-r}{px_1+qx_j+rx_2}+\frac{r-p}{px_2+qx_j+rx_1})+$$

$$\sum_{i=3}^{n}(\frac{q-r}{px_i+qx_1+rx_2}+\frac{r-q}{px_i+qx_2+rx_1})]$$

$$=-\frac{(x_1-x_2)^2\varphi(\boldsymbol{x})}{n(n-1)(n-2)}\cdot[\sum_{j,k=3,j\neq k}^{n}\frac{p^2}{(px_1+qx_j+rx_k)(px_2+qx_j+rx_k)}+$$

$$\sum_{i,k=3,i\neq k}^{n} \frac{q^2}{(px_i+qx_1+rx_k)(px_i+qx_2+rx_k)}+$$

$$\sum_{i,j=3,i\neq j}^{n} \frac{r^2}{(px_i+qx_j+rx_1)(px_i+qx_j+rx_2)}+$$

$$\sum_{k=3}^{n} \frac{(p-q)^2}{(px_1+qx_2+rx_k)(px_2+qx_1+rx_k)}+$$

$$\sum_{j=3}^{n} \frac{(p-r)^2}{(px_1+qx_j+rx_2)(px_2+qx_j+rx_1)}+$$

$$\sum_{i=3}^{n} \frac{(q-r)^2}{(px_i+qx_1+rx_2)(px_i+qx_2+rx_1)}\Big].$$

This implies that $\Delta_1 \leqslant 0$ for $\boldsymbol{x} \in \mathbf{R}^n_{++}(n \geqslant 3)$. By Proposition 2, we conclude that $GB^{p,q,r}(\boldsymbol{x})$ is Schur concave on $\mathbf{R}^n_{++}$.

In view of the discrimination criterion of Schur geometrically convexity, we start with the following calculations

$$\Delta_2 :=(x_1-x_2)\left(x_1\frac{\partial GB^{p,q,r}(\boldsymbol{x})}{\partial x_1}-x_2\frac{\partial GB^{p,q,r}(\boldsymbol{x})}{\partial x_2}\right)$$

$$=\frac{(x_1-x_2)GB^{p,q,r}(\boldsymbol{x})}{n(n-1)(n-2)}\cdot[p\sum_{j,k=3,j\neq k}^{n}(\frac{x_1}{px_1+qx_j+rx_k}-\frac{x_2}{px_2+qx_j+rx_k})+$$

$$q\sum_{i,k=3,i\neq k}^{n}(\frac{x_1}{px_i+qx_1+rx_k}-\frac{x_2}{px_i+qx_2+rx_k})+$$

$$r\sum_{i,j=3,i\neq j}^{n}(\frac{x_1}{px_i+qx_j+rx_1}-\frac{x_2}{px_i+qx_j+rx_2})+$$

$$\sum_{k=3}^{n}(\frac{px_1-qx_2}{px_1+qx_2+rx_k}+\frac{qx_1-px_2}{px_2+qx_1+rx_k})+$$

$$\sum_{j=3}^{n}(\frac{px_1-rx_2}{px_1+qx_j+rx_2}+\frac{rx_1-px_2}{px_2+qx_j+rx_1})+$$

$$\sum_{i=3}^{n}(\frac{qx_1-rx_2}{px_i+qx_1+rx_2}+\frac{rx_1-qx_2}{px_i+qx_2+rx_1})]$$

$$=\frac{(x_1-x_2)^2\varphi(\boldsymbol{x})}{n(n-1)(n-2)}\cdot[\sum_{j,k=3,j\neq k}^{n}\frac{qx_j+rx_k}{(px_1+qx_j+rx_k)(px_2+qx_j+rx_k)}+$$

$$\sum_{i,k=3,i\neq k}^{n}\frac{px_i+rx_k}{(px_i+qx_1+rx_k)(px_i+qx_2+rx_k)}+$$

$$\sum_{i,j=3,i\neq j}^{n} \frac{px_i + qx_j}{(px_i + qx_j + rx_1)(px_i + qx_j + rx_2)} +$$

$$\sum_{k=3}^{n} \frac{2pq(x_1 + x_2) + rx_k(p + q)}{(px_1 + qx_2 + rx_k)(px_2 + qx_1 + rx_k)} +$$

$$\sum_{j=3}^{n} \frac{2rp(x_1 + x_2) + qx_j(p + r)}{(px_1 + qx_j + rx_2)(px_2 + qx_j + rx_1)} +$$

$$\sum_{i=3}^{n} \frac{2qr(x_1 + x_2) + px_i(q + r)}{(px_i + qx_1 + rx_2)(px_i + qx_2 + rx_1)} \Big].$$

Thus, we have $\Delta_2 \geqslant 0$ for $\boldsymbol{x} \in \mathbf{R}_{++}^n (n \geqslant 3)$. It follows from Proposition 3 that $GB^{p,q,r}(\boldsymbol{x})$ is Schur geometric convex on $\mathbf{R}_{++}^n$.

Finally, we discuss the Schur harmonic convexity of $GB^{p,q,r}(\boldsymbol{x})$. A direct computation gives

$$\Delta_3 := (x_1 - x_2)\left(x_1^2 \frac{\partial GB^{p,q,r}(\boldsymbol{x})}{\partial x_1} - x_2^2 \frac{\partial GB^{p,q,r}(\boldsymbol{x})}{\partial x_2}\right)$$

$$= \frac{(x_1 - x_2)GB^{p,q,r}(\boldsymbol{x})}{n(n-1)(n-2)} \cdot \Big[p \sum_{j,k=3,j\neq k}^{n} \Big(\frac{x_1^2}{px_1 + qx_j + rx_k} - \frac{x_2^2}{px_2 + qx_j + rx_k}\Big) +$$

$$q \sum_{i,k=3,i\neq k}^{n} \Big(\frac{x_1^2}{px_i + qx_1 + rx_k} - \frac{x_2^2}{px_i + qx_2 + rx_k}\Big) +$$

$$r \sum_{i,j=3,i\neq j}^{n} \Big(\frac{x_1^2}{px_i + qx_j + rx_1} - \frac{x_2^2}{px_i + qx_j + rx_2}\Big) +$$

$$\sum_{k=3}^{n} \Big(\frac{px_1^2 - qx_2^2}{px_1 + qx_2 + rx_k} + \frac{qx_1^2 - px_2^2}{px_2 + qx_1 + rx_k}\Big) +$$

$$\sum_{j=3}^{n} \Big(\frac{px_1^2 - rx_2^2}{px_1 + qx_j + rx_2} + \frac{rx_1^2 - px_2^2}{px_2 + qx_j + rx_1}\Big) +$$

$$\sum_{i=3}^{n} \Big(\frac{qx_1^2 - rx_2^2}{px_i + qx_1 + rx_2} + \frac{rx_1^2 - qx_2^2}{px_i + qx_2 + rx_1}\Big)\Big]$$

$$= \frac{(x_1 - x_2)^2 \varphi(\boldsymbol{x})}{n(n-1)(n-2)} \cdot \Big[\sum_{j,k=3,j\neq k}^{n} \frac{(x_1 + x_2)(qx_j + rx_k) + px_1 x_2}{(px_1 + qx_j + rx_k)(px_2 + qx_j + rx_k)} +$$

$$\sum_{i,k=3,i\neq k}^{n} \frac{(x_1 + x_2)(px_i + rx_k) + qx_1 x_2}{(px_i + qx_1 + rx_k)(px_i + qx_2 + rx_k)} +$$

$$\sum_{i,j=3,i\neq j}^{n} \frac{(x_1 + x_2)(px_i + qx_j) + rx_1 x_2}{(px_i + qx_j + rx_1)(px_i + qx_j + rx_2)} +$$

$$\sum_{k=3}^{n} \frac{2pq(x_1^2 + x_2^2) + rx_k(x_1 + x_2)(p + q) + x_1x_2(p + q)^2}{(px_1 + qx_2 + rx_k)(px_2 + qx_1 + rx_k)} +$$

$$\sum_{j=3}^{n} \frac{2pr(x_1^2 + x_2^2) + qx_j(x_1 + x_2)(p + r) + x_1x_2(p + r)^2}{(px_1 + qx_j + rx_2)(px_2 + qx_j + rx_1)} +$$

$$\sum_{i=3}^{n} \frac{2qr(x_1^2 + x_2^2) + px_i(x_1 + x_2)(q + r) + x_1x_2(q + r)^2}{(px_i + qx_1 + rx_2)(px_i + qx_2 + rx_1)}]$$

Hence, we obtain $\Delta_3 \geqslant 0$ for $\boldsymbol{x} \in \mathbf{R}_{++}^n (n \geqslant 3)$. Using Proposition 4 leads to the assertion that $GB^{p,q,r}(\boldsymbol{x})$ is Schur harmonic convex on $\mathbf{R}_{++}^n$.

The proof of Theorem 1 is completed.

Remark 5 As a direct consequence of Theorem 1, taking $r = 0$ in Theorem 1 together with the identity $GB^{p,q,0}(\boldsymbol{x}) = GB^{p,q}(\boldsymbol{x})$, we arrive at the assertion of Corollary 1.

83.4　Applications

As an application of Theorem 1, we establish the following interesting inequalities for generalized geometric Bonferroni mean.

Theorem 2 Let p, q, r be non-negative real numbers with $p + q + r \neq 0$. Then, for arbitrary $\boldsymbol{x} \in \mathbf{R}_{++}^n$, $n \geqslant 3$

$$GB^{p,q,r}(\boldsymbol{x}) \leqslant A_n(\boldsymbol{x}) \tag{10}$$

Proof　It follows from Theorem 1 that $GB^{p,q,r}(\boldsymbol{x})$ is Schur concave on $\mathbf{R}_{++}^n$. Using Lemma 1, one has

$$\left(\underbrace{A_n(\boldsymbol{x}), A_n(\boldsymbol{x}), \cdots, A_n(\boldsymbol{x})}_{n} \right) \prec (x_1, x_2, \cdots, x_n)$$

Thus, we deduce from Definition 1 that By Theorem 1, from Lemma 4, we have

$$GB^{p,q,r} \left(\underbrace{A_n(\boldsymbol{x}), A_n(\boldsymbol{x}), \cdots, A_n(\boldsymbol{x})}_{n} \right) \geqslant GB^{p,q,r}(x_1, x_2, \ldots, x_n)$$

which implies that

$$A_n(\boldsymbol{x}) \geqslant GB^{p,q,r}(\boldsymbol{x})$$

Theorem 2 is proved.

Theorem 3　Let p, q, r be non-negative real numbers with $p+q+r \neq 0$, and let c be a constant satisfying $0 \leqslant c < A_n(\boldsymbol{x})$, $(\boldsymbol{x} - c) = (x_1 - c, x_2 - c, \ldots, x_n - c)$. Then, for arbitrary $\mathbf{R}_{++}^n (n \geqslant 3)$

$$GB^{p,q,r}(\boldsymbol{x} - c) \leqslant \left(1 - \frac{c}{nA_n(\boldsymbol{x})}\right) GB^{p,q,r}(\boldsymbol{x}) \qquad (11)$$

Proof　By the majorization relationship given in Lemma 2,

$$\left(\frac{x_1}{\sum\limits_{j=1}^{n} x_j}, \ldots, \frac{x_n}{\sum\limits_{j=1}^{n} x_j}\right) \prec \left(\frac{x_1 - c}{\sum\limits_{j=1}^{n}(x_j - c)}, \ldots, \frac{x_n - c}{\sum\limits_{j=1}^{n}(x_j - c)}\right)$$

it follows from Theorem 1 that

$$GB^{p,q,r}\left(\frac{x_1}{\sum\limits_{j=1}^{n} x_j}, \ldots, \frac{x_n}{\sum\limits_{j=1}^{n} x_j}\right) \geqslant GB^{p,q,r}\left(\frac{x_1 - c}{\sum\limits_{j=1}^{n}(x_j - c)}, \ldots, \frac{x_n - c}{\sum\limits_{j=1}^{n}(x_j - c)}\right)$$

that is

$$\frac{GB^{p,q,r}(x_1, x_2, \ldots, x_n)}{\sum\limits_{i=1}^{n} x_i} \geqslant \frac{GB^{p,q,r}(x_1 - c, x_2 - c, \ldots, x_n - c)}{\sum\limits_{i=1}^{n}(x_i - c)}$$

which implies that

$$GB^{p,q,r}(\boldsymbol{x} - c) \leqslant \left(1 - \frac{c}{nA_n(\boldsymbol{x})}\right) GB^{p,q,r}(\boldsymbol{x})$$

This completes the proof of Theorem 3.

83.5　Conclusion

This paper is a follow-up study of our recent work [24], we generalize the geometric Bonferroni mean by introducing three non-negative parameters p, q, r, under the condition of $p + q + r \neq 0$, we prove that the generalized geometric Bonferronimean $GB^{p,q,r}(\boldsymbol{x})$ is Schur concave, Schur geometric convex and Schur harmonic convex on $\mathbf{R}_{++}^n$. As an application of the Schur convexity, we establish two inequalities for generalized geometric Bonferroni mean. In fact, there have been a large number inequalities for means which originate from the Schur convexity of functions. For details, we refer the interested reader to [27 – 32] and the references therein.

Acknowledgements

This research was supported partially by the National Natural Sciences Foundation of China (11601214, 11526107) and the Natural Science Foundation of Fujian Province of China (2016J01023). The authors would like to express sincere appreciation to the anonymous reviewers for their helpful comments and suggestions.

References

[1]　Marshall A W, Olkin I, Arnold B C. Inequalities: Theory of Majorization and Its Application[M]. 2nd ed. New York: Springer, 2011.

[2]　Shi H N. Theory of Majorization and Analytic Inequalities[M]. Harbin: Harbin Institute of Technology Press, 2012.

[3]　Zhang X M, Yang Z H. Differential criterion of n-dimensional geometrically convex functions[J]. J. Appl. Anal., 2007, **13**(2): 197-208.

[4]　Chu Y, Zhang X M, Wang G D. The Schur geometrical convexity of the extended mean values[J]. J. Convex Anal., 2008, **15**(4): 707-718.

[5]　Chu Y M, Xia W F. Necessary and sufficient conditions for the Schur harmonic convexity of the generalized Muirhead mean[J]. Proc. A. Razmadze Math. Inst.,2010, **152**: 19-27.

[6]　Chu Y M, Wang G D, Zhang X H. The Schur multiplicative and harmonic convexities of the complete symmetric function[J]. Math. Nachr., 2011, **284**(5-6): 653-663.

[7]　Shi H N, Zhang J. Some new judgement theorems of Schur geometric and Schur harmonic convexities for a class of symmetric functions[J]. J. Inequal. Appl., 2013, **2013**, Article ID 527.

[8]　Shi H N, Zhang J. Schur-convexity, Schur geometric and Schur harmonic convexities of dual form of a class symmetric functions[J]. J. Math. Inequal.,2014, **8**(2): 349-358.

[9]　Meng J X, Chu Y M, Tang X M. The Schur-harmonic-convexity of dual form of the Hamy symmetric function[J]. Mat. Vesn.,2010, **62**(1): 37-46.

[10]　Zheng N G, Zhang Z H, Zhang X M. Schur-convexity of two types of one-parameter mean values in n variables[J]. J. Inequal. Appl.,2007, **2007**, Article ID 78175.

[11] Wu S H, Shi H N. A relation of weak majorization and its applications to certain inequalities for means[J]. Math. Slovaca,2011, **61**(4): 561-570.

[12] Wu S H. Generalization and sharpness of power means inequality and their applications[J]. J. Math. Anal. Appl.,2005, **312**: 637-652.

[13] Shi H N, Wu S H, Qi F. An alternative note on the Schur-convexity of the extended mean values[J]. Math. Inequal. Appl.,2006, **9**: 219-224.

[14] Fu L L, Xi B Y, Srivastava H M. Schur-convexity of the generalized Heronian means involving two positive numbers[J]. Taiwan. J. Math.,2011, **15**(6): 2721-2731.

[15] Bonferroni C. Sulle medie multiple di potenze[J]. Boll. UMI,1950, **5**(3-4): 267-270.

[16] Yager R. On generalized Bonferroni mean operators for multi-criteria aggregation[J]. Int. J. Approx. Reason, 2009, **50**: 1279-1286.

[17] Park J H, Park E J. Generalized fuzzy Bonferroni harmonic mean operators and their applications in group decision making[J]. J. Appl. Math.,2013, **2013**, Article ID 604029

[18] Xia M, Xu Z, Zhu B. Generalized intuitionistic fuzzy Bonferroni means[J]. Int. J. Intell. Syst., 2012, **27**(1): 23-47.

[19] Park J H, Kim J Y. Intuitionistic fuzzy optimized weighted geometric Bonferroni means and their applications in group decision making[J]. Fundam. Inform., 2016, **144**(3-4): 363-381.

[20] Beliakov G, James S, MordelováJ, Rückschlossová T, Yager R R. Generalized Bonferroni mean operators in multicriteria aggregation[J]. Fuzzy Sets Syst., 2010, **161**(17): 2227-2242.

[21] Xu Z, Yager R R. Intuitionistic fuzzy Bonferroni means[J]. IEEE Trans. Syst. Man Cybern. B, 2011, **41**(2): 568-578.

[22] Beliakov G, James S, Mordelová J, Rückschlossová T, Yager R R. Generalized Bonferroni mean operators in multi-criteria aggregation[J]. Fuzzy Sets Syst., 2010,**161**(17), 2227-2242.

[23] Xia M, Xu Z, Zhu B. Generalized intuitionistic fuzzy Bonferroni means[J]. Int. J. Intell. Syst., 2012, **27**(1): 23-47.

[24] Shi H N, Wu S H. Schur *m*-power convexity of geometric Bonferroni mean[J]. Ital. J. Pure Appl. Math., 2017, **38**: 769-776.

[25] Shi H N, Zhang J, Ma Q H. Schur-convexity, Schur-geometric and Schur-harmonic convexity for a composite function of complete symmetric function[J]. SpringerPlus, 2016, **2016**(5), Article ID 296.

[26] Chu Y, Lv Y. The Schur harmonic convexity of the Hamy symmetric function and its applications[J]. J. Inequal. Appl., 2009, **2009**, Article ID 838529.

[27] Qi F. A note on Schur-convexity of extended mean values[J]. Rocky Mt. J. Math., 2005, **35**(5): 1787-1793.

[28] Qi F, Sándor J, Dragomir S S, Sofo A. Notes on the Schur-convexity of the extended mean values[J]. Taiwan. J. Math., 2005, **9**(3): 411-420.

[29] Sun J, Sun Z L, Xi B Y, Qi F. Schur-geometric and Schur-harmonic convexity of an integral mean for convex functions[J]. Turk. J. Anal. Number Theory, 2015, **3**(3): 87-89.

[30] Wu Y, Qi F. Schur-harmonic convexity for differences of some means[J]. Analysis,2012, **32**(4): 263-270.

[31] Yang Z H. Schur power convexity of Stolarsky means[J]. Publ. Math. (Debr.), 2012, **80**(1-2): 43-66.

[32] Yang Z H. Schur power convexity of the Daróczy means[J]. Math. Inequal. Appl., 2013, **16**(3): 751-762.

第84篇　A concise proof of a double inequality involving the exponential and logarithmic functions

(SHI HUANNAN, WU SHANHE. Italian Journal of Pure and Applied

Mathematics, 2019, 41: 284 - 289.)

Abstract: In this note we provide a concise proof for a double inequality involving the exponential and logarithmic functions, our method is based on the usage of the majorization inequalities and the Schur-convexity of a function.

Keywords: inequality, exponential function; logarithmic function; Schur-convexity; majorization inequality; Stolarsky mean

84.1　Introduction

In [4], Guo and Qi presented a double inequality involving the exponential and logarithmic functions, as follows

$$e^{\frac{x+y}{2}} < \frac{e^x - e^y}{x - y} < \frac{e^x + e^y}{2} \tag{1}$$

where x and y are arbitrary real numbers with $x \neq y$.

In [4], the authors also mentioned that the idea of establishing inequality (1) was motivated by the following two inequalities [1, p. 352]:

$$\frac{x+y}{2} < \frac{(x-1)e^x - (y-1)e^y}{e^x - e^y} \tag{2}$$

$$\frac{x+y}{2} < \frac{(x-1)e^x - (y-1)e^y}{e^x - e^y} \tag{3}$$

where x and y are arbitrary real numbers with $x \neq y$.

Guo and Qi[4] proved the inequality (1) by the method of mathematical analysis. In this paper, we give a new proof of inequality (1) using the majorization inequalities (introduced by Hardy et al.[2]) and Schur-convexity (introduced by Schur[3]). As an application of inequality (1), we establish a comparison result of the Stolarsky means for different parameters.

84.2　Definitions and Lemmas

In this section, we need to introduce some definitions and lemmas relating to the theory of majorization inequalities.

Definition 1[5,6]　Let $\boldsymbol{x} = (x_1, x_2, \cdots, x_n)$ and $\boldsymbol{y} = (y_1, y_2, \cdots, y_n) \in \mathbf{R}^n$.

(1) $\boldsymbol{x}$ is said to be majorized by $\boldsymbol{y}$ (in symbols $\boldsymbol{x} \preceq \boldsymbol{y}$) if $\displaystyle\sum_{i=1}^{k} x_{[i]} \leqslant \sum_{i=1}^{k} y_{[i]}$ for $k = 1, 2, \cdots, n-1$ and $\displaystyle\sum_{i=1}^{n} x_i = \sum_{i=1}^{n} y_i$, where $x_{[1]} \geqslant x_{[2]} \geqslant \cdots \geqslant x_{[n]}$ and $y_{[1]} \geqslant y_{[2]} \geqslant \cdots \geqslant y_{[n]}$ are rearrangements of $\boldsymbol{x}$ and $\boldsymbol{y}$ in a descending order. Furthermore, $\boldsymbol{x}$ is said to be strictly majorized by $\boldsymbol{y}$ (in symbols $\boldsymbol{x} \prec \boldsymbol{y}$), if $\boldsymbol{x}$ is not permutation of $\boldsymbol{y}$.

(2) Let $\Omega \subset \mathbf{R}^n$, the function $\varphi \colon \Omega \to \mathbf{R}$ is said be a strictly Schur-convex on Ω if $\boldsymbol{x} \prec \boldsymbol{y}$ on Ω implies $\varphi(\boldsymbol{x}) < \varphi(\boldsymbol{y})$. φ is said to be a strictly Schur-concave function on Ω if and only if $-\varphi$ is Schur-convex function on Ω.

Definition 2[5,6]　Let set $\Omega \subseteq \mathbf{R}^n$. Ω is said to be a convex set if $\boldsymbol{x}, \boldsymbol{y} \in \Omega, 0 \leqslant \alpha \leqslant 1$ implies $\alpha\boldsymbol{x} + (1-\alpha)\boldsymbol{y} = (\alpha x_1 + (1-\alpha)y_1, \ldots, \alpha x_n + (1-\alpha)y_n) \in \Omega$.

Lemma 1[5,6]　Let $\Omega \subset \mathbf{R}^n$ be symmetric and have a nonempty interior convex set. Ω° is the interior of Ω. $\varphi \colon \Omega \to \mathbf{R}$ is continuous on Ω and differentiable in Ω°. Then φ is the strictly $Schur - convex \; (Schur - concave) \; function$ if and only if φ is symmetric on Ω and

$$(x_1 - x_2)\left(\frac{\partial\varphi}{\partial x_1} - \frac{\partial\varphi}{\partial x_2}\right) > 0 (< 0) \tag{4}$$

holds for any $\boldsymbol{x} \in \Omega^\circ$ and $x_1 \neq x_2$.

84.3 The proof of inequality (1)

Let us first deal with the left-hand inequality of (1), which reads as follows:

Proposition 1 For arbitrary real numbers x, y with $x \neq y$, we have

$$\ln \frac{e^x - e^y}{x - y} < \frac{(x-1)e^x - (y-1)e^y}{e^x - e^y} \tag{5}$$

Proof By the L'Hospital rule, it is easy to find that

$$\lim_{x \to y} \left[\ln \frac{e^x - e^y}{x - y} - \frac{(x-1)e^x - (y-1)e^y}{e^x - e^y} \right] = 0$$

Thus, we define a function $f(x, y)$ by

$$f(x, y) = \begin{cases} \ln(e^x - e^y) - \ln(x - y) - \frac{(x-1)e^x - (y-1)e^y}{e^x - e^y}, & x \neq y \\ 0, & x = y. \end{cases}$$

Note that the inequality (5) is symmetrical with respect to variables x and y: To prove inequality (5), it is sufficient to prove that $f(x, y) < 0$ for $x > y$.

Let us now discuss the Schur-convexity of $f(x, y)$ on $\Omega = f(x, y) : x > y, x; y \in \mathbf{R}$.

Differentiating $f(x, y)$ with respect to x gives

$$\begin{aligned} \frac{\partial f}{\partial x} &= \frac{e^x}{e^x - e^y} - \frac{1}{x - y} - \frac{xe^x(e^x - e^y) - e^x[(x-1)e^x - (y-1)e^y]}{(e^x - e^y)^2} \\ &= \frac{e^{2x} - e^{x+y} - e^{2x} - (y - x - 1)e^{x+y}}{(e^x - e^y)^2} - \frac{1}{x - y} \\ &= \frac{(x - y)e^{x+y}}{(e^x - e^y)^2} - \frac{1}{x - y} \end{aligned}$$

Similarly to the above, we have

$$\frac{\partial f}{\partial y} = \frac{(y - x)e^{x+y}}{(e^x - e^y)^2} - \frac{1}{y - x}$$

Hence

$$\Delta_1 := (x - y)\left(\frac{\partial f}{\partial x} - \frac{\partial f}{\partial y} \right) = 2\left[\frac{(x - y)^2 e^{x+y}}{(e^x - e^y)^2} - 1 \right]$$

It is easy to observe that

$$\frac{(x-y)^2 e^{x+y}}{(e^x - e^y)^2} - 1 < 0 \Leftrightarrow e^{x+y} < \left(\frac{e^x - e^y}{x - y}\right)^2$$

which is equivalent to a known result, the left-hand side inequality of (2). Hence, we obtain $\Delta_1 < 0$. By Lemma 1, we conclude that $f(x,y)$ is strictly Schur-concave on Ω. Further, from an evident majorization relationship

$$\left(\frac{x+y}{2}, \frac{x+y}{2}\right) \prec (x, y)$$

along with the definition of Schur-concave function, we deduce that

$$0 = f\left(\frac{x+y}{2}, \frac{x+y}{2}\right) > f(x, y)$$

which implies the desired inequality (5). The proof of Proposition 1 is complete.

Let us now verify the validity of the right-hand inequality of (1), which is stated by Proposition 2 below.

Proposition 2 For arbitrary real numbers x, y with $x \neq y$, we have

$$\frac{(x-1)e^x - (y-1)e^y}{e^x - e^y} < \ln \frac{e^x + e^y}{2} \tag{6}$$

Proof It is not difficult to verify that

$$\lim_{x \to y} \left[\ln \frac{e^x + e^y}{2} - \frac{(x-1)e^x - (y-1)e^y}{e^x - e^y}\right] = 0$$

Thus, we define a function $g(x, y)$ by

$$g(x, y) = \begin{cases} \ln(e^x + e^y) - \ln 2 - \frac{(x-1)e^x - (y-1)e^y}{e^x - e^y}, & x \neq y \\ 0, & x = y \end{cases}$$

Because the inequality (6) is symmetrical with respect to variables x and y in order to prove inequality (6), it is enough to prove that $g(x, y) > 0$ for $x > y$.

In the following, we discuss the Schur-convexity of $g(x, y)$ on $\Omega = f(x, y) : x > y, x, y \in \mathbf{R}$.

Direct computation gives

$$\frac{\partial g}{\partial x} = \frac{e^x}{e^x + e^y} - \frac{xe^x(e^x - e^y) - e^x[(x-1)e^x - (y-1)e^y]}{(e^x - e^y)^2}$$

$$= \frac{e^x}{e^x + e^y} - \frac{e^{2x} - (x-y+1)e^{x+y}}{(e^x - e^y)^2}$$

and

$$\frac{\partial g}{\partial y} = \frac{e^y}{e^x + e^y} - \frac{e^{2y} - (y-x+1)e^{x+y}}{(e^x - e^y)^2}$$

Therefore

$$\Delta_2 := (x-y)\left(\frac{\partial g}{\partial x} - \frac{\partial g}{\partial y}\right)$$

$$= (x-y)\left[\frac{e^x - e^y}{e^x + e^y} + \frac{e^{2x} - e^{2y} + 2(y-x)e^{x+y}}{(e^x - e^y)^2}\right]$$

$$= \frac{4(x-y)^2}{(e^x + e^y)(e^x - e^y)^2}\left(\frac{e^x + e^y}{2} - \frac{e^x - e^y}{x-y}\right)$$

$$= \frac{4(x-y)^2}{(e^x + e^y)(e^x - e^y)^2}\left(\frac{e^x + e^y}{2} - \frac{1}{x-y}\int_y^x e^t dt\right)$$

Recall the well-known Hermite-Hadamard inequality for a convex function on the interval $[x, y]$

$$\frac{1}{x-y}\int_x^y \psi(t)dt \leqslant \frac{\psi(x) + \psi(y)}{2} \tag{7}$$

If we take $\psi(t) = e^t$, then we have a strict inequality of (7), that is

$$\frac{1}{x-y}\int_x^y e^t dt \leqslant \frac{e^x + e^y}{2}$$

Hence, we obtain $\Delta_2 > 0$, this implies that $g(x, y)$ is strictly Schur-convex on Ω. Then, from

$$\left(\frac{x+y}{2}, \frac{x+y}{2}\right) \prec (x, y)$$

it follows that

$$0 = g\left(\frac{x+y}{2}, \frac{x+y}{2}\right) < g(x, y)$$

which implies the required inequality (6). The Proposition 2 is proved.

84.4 An application to the Stolarsky mean

In order to demonstrate the application of inequality (1), we establish a comparison result of theStolarsky means for different parameters.

Let $(x, y) \in \mathbf{R}_+^2$. The extended mean (or Stolarsky mean) of (x, y) is defined in[7]as

$$
E(a, b; x, y) = \begin{cases}
\left(\dfrac{b}{a} \cdot \dfrac{y^a - x^a}{y^b - x^b} \right)^{1/(a-b)}, & ab(a-b)(x-y) \neq 0 \\[2mm]
\left(\dfrac{1}{a} \cdot \dfrac{y^a - x^a}{\ln y - \ln x} \right)^{1/a}, & a(x-y) \neq 0, b = 0 \\[2mm]
\dfrac{1}{e^{1/a}} \left(\dfrac{x^{x^a}}{y^{y^a}} \right)^{1/(x^a - y^a)}, & a(x-y) \neq 0, a = b \\[2mm]
\sqrt{xy}, & a = b = 0, x \neq y \\[2mm]
x, & x = y
\end{cases}
$$

We have the following inequalities for the Stolarsky mean $E(a, b; x, y)$.

Proposition 3 Let u, v be arbitrary positive numbers with $u \neq v$. Then

$$
E(1, 0; u, v) < E(1, 1; u, v) < E(2, 1; u, v) \tag{8}
$$

Proof Taking $e^x = u$ and $e^y = v$ in the inequality (1), we obtain

$$
\ln \left(\frac{u - v}{\ln u - \ln v} \right) < \frac{(\ln u - 1)u - (\ln v - 1)v}{u - v} < \ln \frac{u + v}{2}
$$

$$
\Leftrightarrow \frac{u - v}{\ln u - \ln v} < \frac{1}{e} \left(\frac{u^u}{v^v} \right)^{1/(u-v)} < \frac{u + v}{2} \tag{9}
$$

Obviously, the inequality (9) can be equivalently transformed to the desired inequality (8) according to the definition of $E(a, b; x, y)$ described above. This proves Proposition 3.

Acknowledgements

This research was supported by the Natural Science Foundation of Fujian province of China under Grant No.2016J01023. All authors contributed equally and significantly in writing this paper.

References

[1] J C Kuang. Applied Inequalities (Chang yong bu deng shi)[M]. 4th ed.Ji'nan: Shandong Press of Science and Technology, 2010 (in Chinese).

[2]   Hardy G H , Littlewood J E , Pólya G. Inequalities[M].London: Cambridge Univ. Press, 1952.

[3]   Schur I. Über eine Klasse von Mittelbildungen mit Anwendungen auf die Determinantentheorie[J]. Sitzungsberichte der Berliner Mathematischen Gesselschaft, 1923(22): 9-20.

[4]   Guo B N, F Qi. On inequalities for the exponential and logarithmic functions and means[J]. Malays. J. Math. Sci., 2016(10): 23-34.

[5]   Wang B Y. Foundations of majorization inequalities[M]. Beijing: Beijing Normal Univ. Press, 1990.

[6]   Marshall A M, Olkin I. Inequalities:theory of majorization and its application[M]. New York: Academies Press, 1979.

[7]   Stolarsky K B. Generalizations of the logarithmic mean[J].Math. Mag., 1975(48):87-92.

第85篇　Schur-Power Convexity of a Completely Symmetric Function Dual

(Huan-Nan Shi and Wei-Shih Du. Symmetry, 2019, 11(7), 897)

Abstract: In this paper, by applying the decision theorem of the Schur-power convex function, the Schur-power convexity of a class of complete symmetric functions are studied. As applications, some new inequalities are established.

Keywords: Schur-power convexity; Schur-convexity; Schur-geometric convexity; Schur-harmonic convexity; completely symmetric function; dual form

85.1 Introduction and Preliminaries

Convexity is a natural notion and plays an important and fundamental role in mathematics, physics, chemistry, biology, economics, engineering, and other sciences. To solve practical problems, several interesting concepts of generalized convexity or generalized concavity have been introduced and studied. Recent important investigations and developments in convex analysis have focused on the study of Schur-convexity, and Schur-geometric and Schur-harmonic convexity of various symmetric functions; see, e.g.[1 - 20] and references therein. It is worth mentioning that discovering and judging Schur-convexity of various symmetric functions is an important topic in the study of the majorization theory. A lot of achievements in this field have been investigated by several authors; for more details, see the first author's monographs [21,22].

Throughout this paper, we denote by N and R, the set of positive integers and real numbers, respectively. Let X be a nonempty set. Denote

$\mathbf{R}_+ := (0, +\infty)$ and $\mathbf{R}_- := (-\infty, 0)$. For a positive integer n, the set X^n for the Cartesian product is the collection of all n-tuples of elements of X. Therefore, we can write $\mathbf{R}^n, \mathbf{R}_+^n$ and $\mathbf{R}_-^n$ as follows

$$X^n = \underbrace{X \times X \times \cdots \times X}_{n}$$

where $X \in \{\mathbf{R}, \mathbf{R}_+, \mathbf{R}_-\}$

Let $\boldsymbol{x} = (x_1, \ldots, x_n)$ and $\boldsymbol{y} = (y_1, \ldots, y_n)$ in $\mathbf{R}^n$. A set $D \subset \mathbf{R}^n$ is said to be convex if $\boldsymbol{x}, \boldsymbol{y} \in D$ and $0 \leqslant \alpha \leqslant 1$ imply

$$\alpha \boldsymbol{x} + (1 - \alpha)\boldsymbol{y} = (\alpha x_1 + (1 - \alpha)y_1, \ldots, \alpha x_n + (1 - \alpha)y_n) \in D$$

Let $D \subset \mathbf{R}^n$ be a convex set. A function $f : D \to \mathbf{R}$ is said to be convex on D, if

$$f(\alpha \boldsymbol{x} + (1 - \alpha)\boldsymbol{y}) \leqslant \alpha f(\boldsymbol{x}) + (1 - \alpha)f(\boldsymbol{y})$$

for all $\boldsymbol{x}, \boldsymbol{y} \in D$, and all $\alpha \in [0, 1]$. The function f is said to be concave on D, if and only if $-f$ is convex on D.

For the reader's convenience and explicit later use, we now recall some basic definitions and notation that will be needed in this paper.

Definition 1 (see [23,24]).

(i) A set $\Omega \subset \mathbf{R}^n$ is called symmetric, if $\boldsymbol{x} \in \Omega$ implies $\boldsymbol{x}\boldsymbol{P} \in \Omega$ for every $n \times n$ permutation matrix $\boldsymbol{P}$.

(ii) A function $\varphi : \Omega \to \mathbf{R}$ is called symmetric if for every permutation matrix $\boldsymbol{P}, \varphi(\boldsymbol{x}\boldsymbol{P}) = \varphi(\boldsymbol{x})$ for all $\boldsymbol{x} \in \Omega$.

Definition 2 (see [23,24]). Let $\boldsymbol{x} = (x_1, \cdots, x_n)$ and $\boldsymbol{y} = (y_1, \cdots, y_n) \in \mathbf{R}^n$.

(i) $\boldsymbol{x} \geq \boldsymbol{y}$ means $x_i \geqslant y_i$ for all $i = 1, 2, \cdots, n$.

(ii) Let $\Omega \subset \mathbf{R}^n$, $\varphi \colon \Omega \to \mathbf{R}$ is said to be increasing if $\boldsymbol{x} \geq \boldsymbol{y}$ implies $\varphi(\boldsymbol{x}) \geqslant \varphi(\boldsymbol{y})$. φ is said to be decreasing if and only if $-\varphi$ is increasing.

Definition 3 (see [23,24]). Let $\boldsymbol{x} = (x_1, \cdots, x_n)$ and $\boldsymbol{y} = (y_1, \cdots, y_n) \in \mathbf{R}^n$.

(i) $\boldsymbol{x}$ is said to be majorized by $\boldsymbol{y}$ (in symbols $\boldsymbol{x} \prec \boldsymbol{y}$) if $\sum\limits_{i=1}^{k} x_{[i]} \leqslant \sum\limits_{i=1}^{k} y_{[i]}$ for $k = 1, 2, \cdots, n - 1$ and $\sum\limits_{i=1}^{n} x_i = \sum\limits_{i=1}^{n} y_i$, where $x_{[1]} \geqslant \cdots \geqslant x_{[n]}$ and

$y_{[1]} \geqslant \cdots \geqslant y_{[n]}$ are rearrangements of x and y in a descending order.

(ii) Let $\Omega \subset \mathbf{R}^n$, the function $\varphi: \Omega \to \mathbf{R}$ is said to be Schur-convex on Ω if $x \prec y$ on Ω implies $\varphi(x) \leq \varphi(y)$. φ is said to be a Schur-concave function on Ω if and only if $-\varphi$ is Schur-convex function on Ω.

The following useful characterizations of Schur-convex and Schur-concave functions were established in [23,24].

Lemma 1 (see [23,24])　Let $\Omega \subset \mathbf{R}^n$ be symmetric and have a nonempty interior convex set. Ω° is the interior of Ω. $\varphi : \Omega \to \mathbf{R}$ is continuous on Ω and differentiable in Ω°. Then φ is the $Schur - convex\ (Schur - concave)\ function$, if and only if φ is symmetric on Ω and

$$(x_1 - x_2)\left(\frac{\partial \varphi}{\partial x_1} - \frac{\partial \varphi}{\partial x_2}\right) \geqslant 0 (\leqslant 0) \tag{1}$$

holds for any $x \in \Omega^\circ$.

In 1923, Professor Issai Schur made the first systematic study of the functions preserving the ordering of majorization. In Schur's honor, such functions are said to be "Schur-convex". It is known that Schur-convexity can be applied extensively in analytic inequalities, combinatorial optimization, quantum physics, information theory, and other related fields (see, e.g., [23]).

Definition 4 (see [25,26])　Let $x = (x_1, x_2, \ldots, x_n) \in \mathbf{R}_+^n$ and $y = (y_1, y_2, \ldots, y_n) \in \mathbf{R}_+^n$.

(i) A set $\Omega \subset \mathbf{R}_+^n$ is called a geometrically convex set if $(x_1^\alpha y_1^\beta, x_2^\alpha y_2^\beta, \ldots, x_n^\alpha y_n^\beta) \in \Omega$ for all $x, y \in \Omega$ and $\alpha, \beta \in [0, 1]$ such that $\alpha + \beta = 1$.

(ii) Let $\Omega \subset \mathbf{R}_+^n$. The function $\varphi: \Omega \to \mathbf{R}_+$ is said to be Schur-geometrically convex function on Ω if $(\log x_1, \log x_2, \ldots, \log x_n) \prec (\log y_1, \log y_2, \ldots, \log y_n)$ on Ω implies $\varphi(x) \leqslant \varphi(y)$. The function φ is said to be a Schur-geometrically concave function on Ω, if and only if $-\varphi$ is Schur-geometrically convex function on Ω.

Lemma 2 (see [25,26])　(Schur-geometrically convex function decision theorem) Let $\Omega \subset \mathbf{R}_+^n$ be a symmetric and geometrically convex set with a nonempty interior Ω°. Let $\varphi : \Omega \to \mathbf{R}_+$ be continuous on Ω and differentiable in Ω°. If φ is symmetric on Ω and

$$(\log x_1 - \log x_2)\left(x_1\frac{\partial \varphi}{\partial x_1} - x_2\frac{\partial \varphi}{\partial x_2}\right) \geqslant 0 \quad (or \leqslant 0, respectively) \tag{2}$$

holds for any $\boldsymbol{x} = (x_1, x_2, \ldots, x_n) \in \Omega^\circ$, then φ is a Schur-geometrically convex (or Schur-geometrically concave, respectively) function.

The Schur-geometric convexity was first proposed and studied by Zhang [25] in 2004 and was widely investigated and improved by many authors, see [27 - 29] and references therein. We also note that some authors use the term "Schur multiplicative convexity".

In 2009, Chu [1 - 3] introduced the notion of Schur-harmonically convex function and established some interesting inequalities for Schur-harmonically convex functions.

Definition 5 (see [1]) Let $\Omega \subset \mathbf{R}_+^n$ or $\Omega \subset \mathbf{R}_-^n$.

(i) A set Ω is said to be harmonically convex if $\dfrac{\boldsymbol{xy}}{\lambda \boldsymbol{x} + (1 - \lambda)\boldsymbol{y}} \in \Omega$
for every $\boldsymbol{x}, \boldsymbol{y} \in \Omega$ and $\lambda \in [0, 1]$, where $\boldsymbol{xy} = \sum_{i=1}^n x_i y_i$ and $\dfrac{1}{\boldsymbol{x}} = (\dfrac{1}{x_1}, \dfrac{1}{x_2}, \ldots, \dfrac{1}{x_n})$.

(ii) A function $\varphi : \Omega \to \mathbf{R}_+$ is said to be Schur-harmonically convex on Ω if $\dfrac{1}{\boldsymbol{x}} \prec \dfrac{1}{\boldsymbol{y}}$ implies $\varphi(\boldsymbol{x}) \leq \varphi(\boldsymbol{y})$. A function φ is said to be a Schur-harmonically concave function on Ω if and only if $-\varphi$ is a Schur-harmonically convex function.

Lemma 3 (see [1])(Schur-harmonically convex function decision theorem) Let $\Omega \subset \mathbf{R}_+^n$ or $\Omega \subset \mathbf{R}_-^n$ be a symmetric and harmonically convex set with inner points and let $\varphi : \Omega \to \mathbf{R}$ be a continuously symmetric function which is differentiable on Ω°. Then φ is Schur-harmonically convex (or Schur-harmonically concave, respectively) on Ω if and only if

$$(x_1 - x_2)\left(x_1^2 \frac{\partial \varphi(\boldsymbol{x})}{\partial x_1} - x_2^2 \frac{\partial \varphi(\boldsymbol{x})}{\partial x_2} \right) \geq 0 \quad (or \leq 0, respectively), \quad \boldsymbol{x} \in \Omega^\circ \ (3)$$

In 2010, Yang[30]defined and introduced the concepts of the Schur-f-convex function and Schur-power convex function which are the generalization and unification of the concepts of Schur-convexity, Schur-geometric convexity, and Schur-harmonic convexity. He established useful characterizations of Schur m-power convex functions and presented their important properties; see [30].

Definition 6 (see [30])　Let $f : \mathbf{R}_+ \to \mathbf{R}$ be defined by

$$f(x) = \begin{cases} \dfrac{x^m - 1}{m}, m \neq 0 \\ \ln x, m = 0 \end{cases} \tag{4}$$

Then a function $\varphi : \Omega \subset \mathbf{R}_+^n \to \mathbf{R}$ is said to be Schur m-power convex on Ω if

$$(f(x_1), f(x_2), \ldots, f(x_n)) \prec (f(y_1), f(y_2), \ldots, f(y_n))$$

for all $\boldsymbol{x} = (x_1, x_2, \ldots, x_n) \in \Omega$ and $\boldsymbol{y} = (y_1, y_2, \ldots, y_n) \in \Omega$ implies $\varphi(\boldsymbol{x}) \leqslant \varphi(\boldsymbol{y})$.

If $-\varphi$ is Schur m-power convex, then we say that φ is Schur m-power concave.

Lemma 4 (see [30])　Let $\Omega \subset \mathbf{R}_+^n$ be a symmetric set with nonempty interior Ω° and $\varphi : \Omega \to \mathbf{R}_+$ be continuous on Ω and differentiable in Ω°. Then φ is Schur m-power convex on Ω, if and only if φ is symmetric on Ω and

$$\frac{x_1^m - x_2^m}{m} \left[x_1^{1-m} \frac{\partial \varphi(\boldsymbol{x})}{\partial x_1} - x_2^{1-m} \frac{\partial \varphi(\boldsymbol{x})}{\partial x_2} \right] \geqslant 0, \quad \text{if } m \neq 0 \tag{5}$$

and

$$(\log x_1 - \log x_2) \left[x_1 \frac{\partial \varphi(\boldsymbol{x})}{\partial x_1} - x_2 \frac{\partial \varphi(\boldsymbol{x})}{\partial x_2} \right] \geqslant 0, \quad \text{if } m = 0 \tag{6}$$

for all $\boldsymbol{x} \in \Omega^\circ$.

For $\boldsymbol{x} = (x_1, x_2, \ldots, x_n) \in \mathbf{R}^n$, the complete symmetric function $c_n(\boldsymbol{x}, r)$ is defined as

$$c_n(\boldsymbol{x}, r) = \sum_{i_1 + i_2 + \cdots + i_n = r} x_1^{i_1} x_2^{i_2} \cdots x_n^{i_n} \tag{7}$$

where $c_0(\boldsymbol{x}, r) = 1$, $r \in \{1, 2, \ldots, n\}$, $i_1, i_2, \ldots, i_n$ are nonnegative integers.

The collection of complete symmetric functions is an important class of symmetric functions which has been investigated by many mathematicians and there are many interesting results in the literature.

In 2006, Guan[5] discussed the Schur-convexity of $c_n(\boldsymbol{x}, r)$ and proved the following result.

Proposition 1　$c_n(\boldsymbol{x}, r)$ is increasing and Schur-convex on $\mathbf{R}_+^n$.

Subsequently, Chu et al.[2] established the following proposition.

Proposition 2　$c_n(\boldsymbol{x}, r)$ is Schur-geometrically convex and Schur-harmonically convex on $\mathbf{R}_+^n$.

In 2016, Shi et al.[19] further studied the Schur-convexity of $c_n(\boldsymbol{x}, r)$ on

$\mathbf{R}^n_-$.and presented the following important result.

Proposition 3(see [19])　If r is even integer (or odd integer, respectively), then $c_n(\boldsymbol{x}, r)$ is decreasing and Schur-convex (or increasing and Schur-concave, respectively) on $\mathbf{R}^n_-$.

Recall that the dual form of the complete symmetric function $c_n(\boldsymbol{x}, r)$ is defined by

$$c_n^*(\boldsymbol{x}, r) = \prod_{i_1+i_2+\cdots+i_n=r} \sum_{j=1}^{n} i_j x_j \tag{8}$$

In 2013, Zhang and Shi [18] established the following two interesting propositions.

Proposition 4(see [18])　For $r = 1, 2, \ldots, n$, $c_n^*(\boldsymbol{x}, r)$ is increasing and Schur-concave on $\mathbf{R}^n_+$.

Proposition 5(see [18])　For $r = 1, 2, \ldots, n$, $c_n^*(\boldsymbol{x}, r)$ is Schur-geometrically convex and Schur-harmonically convex on $\mathbf{R}^n_+$.

Notice that

$$c_n^*(-\boldsymbol{x}, r) = (-1)^r c_n^*(\boldsymbol{x}, r)$$

It is not difficult to prove the following result.

Proposition 5(see [18])　If r is even integer (or odd integer, respectively), then $c_n^*(\boldsymbol{x}, r)$ is decreasing and Schur-concave (or increasing and Schur-convex, respectively) on $\mathbf{R}^n_-$.

In 2014, Sun et al.[6]studied the Schur-convexity, Schur-geometric and harmonic convexities of the following composite function of $c_n(\boldsymbol{x}, r)$

$$c_n\left(\frac{\boldsymbol{x}}{1-\boldsymbol{x}}, r\right) = \sum_{i_1+i_2+\cdots+i_n=r} \prod_{j=1}^{n} \left(\frac{x_j}{1-x_j}\right)^{i_j} \tag{9}$$

By using Lemmas 1 - 3, they proved the following Theorems 1 - 3, respectively.

Theorem 1　For $\boldsymbol{x} = (x_1, x_2, \ldots, x_n) \in (0, 1)^n \cup (1, +\infty)^n$ and $r \in \mathbf{N}$.

(i) $c_n\left(\frac{x}{1-x}, r\right)$ is increasing and Schur-convex on $(0, 1)^n$;

(ii) If r is even integer (or odd integer, respectively), then $c_n\left(\frac{x}{1-x}, r\right)$ is Schur-convex (or Schur-concave, respectively) on $(1, +\infty)^n$, and is decreasing (or increasing, respectively).

Theorem 2　For $\boldsymbol{x} = (x_1, x_2, \ldots, x_n) \in (0, 1)^n \cup (1, +\infty)^n$ and $r \in \mathbf{N}$.

(i) $c_n\left(\frac{x}{1-x}, r\right)$ is Schur-geometrically convex on $(0, 1)^n$;

(ii) If r is even integer (or odd integer, respecti- vely), then $c_n\left(\frac{x}{1-x},r\right)$ is Schur-geometrically convex (or Schur-geometrically concave, respectively) on $(1,+\infty)^n$.

Theorem 3 For $x=(x_1,x_2,\ldots,x_n)\in(0,1)^n\cup(1,+\infty)^n$ and $r\in\mathbf{N}$.

(i) $c_n\left(\frac{x}{1-x},r\right)$ is Schur-harmonically convex on $(0,1)^n$;

(ii) If r is even integer (or odd integer, respectively), then $c_n\left(\frac{x}{1-x},r\right)$ is Schur-harmonically convex (or Schur-harmonically concave, respectively) on $(1,+\infty)^n$.

In 2016, Shi et al.[19] applied the properties of Schur-convex, Schur-geometrically convex, and Schur-harmonically convex functions respectively to give simple proofs of Theorems 1 - 3. Recall that the dual form of the function $c_n\left(\frac{x}{1-x},r\right)$ is defined by

$$c_n^*\left(\frac{x}{1-x},r\right)=\prod_{i_1+i_2+\cdots+i_n=r}\sum_{j=1}^{n}i_j\left(\frac{x_j}{1-x_j}\right)\tag{10}$$

A function associated with this function is

$$c_n^*\left(\frac{x}{x-1},r\right)=\prod_{i_1+i_2+\cdots+i_n=r}\sum_{j=1}^{n}i_j\left(\frac{x_j}{x_j-1}\right)\tag{11}$$

In this work, we will establish some important results for the Schur-power convexity of symmetric functions $c_n^*\left(\frac{x}{x-1},r\right)$ and $c_n^*\left(\frac{x}{1-x},r\right)$. As their applications, some new inequalities are obtained in Section 3.

85.2 Main Results

The following lemmas are very crucial for our main results.

Lemma 5 Let $m\geqslant-1$. For $x_1,x_2\in(1,+\infty)$ and $x_1>x_2$, we have

$$x_1(x_1-1)x_2^{1-m}\geqslant x_2(x_2-1)x_1^{1-m}\tag{12}$$

$$(x_1-1)^2x_2^{1-m}\geqslant(x_2-1)^2x_1^{1-m}\tag{13}$$

$$\frac{(x_1-1)^2x_2^{2-m}}{x_2-1}\geqslant\frac{(x_2-1)^2x_1^{2-m}}{x_1-1}\tag{14}$$

Proof Since

$$[((t-1)t^m)]' = t^{m-1}[t(m+1)-m] \geqslant t^{m-1}[(m+1)-m] \geqslant 0, \text{for } t > 1$$

we have

$$(x_1-1)x_1^m \geqslant (x_2-1)x_2^m, \text{for } x_1 > x_2$$

This inequality is equivalent to inequality (12). Since

$$\left[\frac{(t-1)^2}{t^{(1-m)}}\right]' = \frac{(t-1)[t(m+1)+1-m]}{t^m t^{[2(1-m)]}}$$

$$\geqslant \frac{(t-1)[(m+1)+1-m]}{t^m t^{[2(1-m)]}} \geqslant 0, \text{for } t > 1$$

we obtain

$$\frac{(x_1-1)^2}{x_1^{(1-m)}} \geqslant \frac{(x_2-1)^2}{x_2^{(1-m)}}, \text{for } x_1 > x_2$$

This inequality is equivalent to inequality (13). Since

$$\left[\frac{(t-1)^3}{t^{(2-m)}}\right]' = \frac{(t-1)^2 t^{1-m}[t(m+1)+2-m]}{t^{[2(2-m)]}}$$

$$\geqslant \frac{(t-1)^2 t^{1-m}[(m+1)+2-m]}{t^{[2(2-m)]}} \geqslant 0, \text{for } t > 1$$

we get

$$\frac{(x_1-1)^3}{mx_1^{(2-m)}} \geqslant \frac{(x_2-1)^3}{mx_2^{(2-m)}}, \text{for } x_1 > x_2$$

This inequality is equivalent to inequality (14).

Lemma 6 Let $m \leqslant 0$. For $x_1, x_2 \in (0,1)$ and $x_1 > x_2$, we have

$$x_1^{1-m}x_2(1-x_2) \leqslant x_2^{1-m}x_1(1-x_1) \tag{15}$$

$$x_1^{1-m}(1-x_2)^2 \leqslant x_2^{1-m}(1-x_1)^2 \tag{16}$$

$$\frac{(1-x_2)^2 x_1^{2-m}}{1-x_1} \leqslant \frac{(1-x_1)^2 x_2^{2-m}}{1-x_2} \tag{17}$$

Proof Since

$$((1-t)t^m)' = t^{m-1}[m(1-t)-t] = t^{m-1}[m-t(m+1)] \leqslant 0, \text{for } \in (0,1)$$

we get

$$(1-x_1)x_1^m \leqslant (1-x_2)x_2^m, \text{for } x_1 > x_2$$

This inequality is equivalent to inequality (15). Since

$$\left[\frac{(1-t)^2}{t^{(1-m)}}\right]' = \frac{(t-1)[m(1-t)-(t+1)]}{t^m t^{[2(1-m)]}} \leqslant 0, \text{ for } t \in (0,1)$$

we obtain

$$\frac{(1-x_1)^2}{x_1^{(1-m)}} \geqslant \frac{(1-x_2)^2}{x_2^{(1-m)}}, \text{ for } x_1 > x_2$$

This inequality is equivalent to inequality (16). Since

$$\left[\frac{(1-t)^3}{t^{(2-m)}}\right]' = \frac{(1-t)^2 t^{1-m}[m(1-t)-2(1+t)]}{t^{[2(2-m)]}} \leqslant 0, \text{ for } t \in (0,1)$$

we have

$$\frac{(1-x_1)^3}{mx_1^{(2-m)}} \geqslant \frac{(1-x_2)^3}{mx_2^{(2-m)}}, \text{ for } x_1 > x_2$$

This inequality is equivalent to inequality (17).

Now, we establish the following new result for the Schur-power convexity of $c_n^*((\boldsymbol{x}/(\boldsymbol{x}-1)),r)$.

Theorem 4　Let $r \in \mathbf{N}$. If $m \geqslant -1$, then $c_n^*\left(\frac{\boldsymbol{x}}{\boldsymbol{x}-1},r\right)$ is decreasing and Schur m-power convex on $(1,+\infty)^n$.

Proof　Let $q(t) = \frac{t}{t-1}$. Then

$$q'(t) = -\frac{1}{(t-1)^2}, \quad q''(t) = \frac{2}{(t-1)^3} \tag{18}$$

From Proposition 4, we know that $c_n^*(\boldsymbol{x},r)$ is increasing on $\mathbf{R}_+^n$, but $q(t)$ is decreasing on $\mathbf{R}$, therefore, the function $c_n^*\left(\frac{\boldsymbol{x}}{\boldsymbol{x}-1},r\right)$ is decreasing on $(1,+\infty)^n$.

For $r = 1$ and $r = 2$, it is easy to prove that $c_n^*\left(\frac{\boldsymbol{x}}{\boldsymbol{x}-1},r\right)$ is Schur m-power convex on $(1,+\infty)^n$. Now consider the case of $r \geqslant 3$. By the symmetry of $c_n^*\left(\frac{\boldsymbol{x}}{\boldsymbol{x}-1},r\right)$, without loss of generality, we may assume $x_1 > x_2$. So

$$c_n^*\left(\frac{\boldsymbol{x}}{\boldsymbol{x}-1},r\right) = \prod_{\substack{i_1+i_2+\cdots+i_n=r \\ i_1 \neq 0, i_2=0}} \sum_{j=1}^{n} \frac{i_j x_j}{x_j-1} \times \prod_{\substack{i_1+i_2+\cdots+i_n=r \\ i_1=0, i_2 \neq 0}} \sum_{j=1}^{n} \frac{i_j x_j}{x_j-1} \times$$

$$\prod_{\substack{i_1+i_2+\cdots+i_n=r \\ i_1 \neq 0, i_2 \neq 0}} \sum_{j=1}^{n} \frac{i_j x_j}{x_j-1} \times \prod_{\substack{i_1+i_2+\cdots+i_n=r \\ i_1=0, i_2=0}} \sum_{j=1}^{n} \frac{i_j x_j}{x_j-1}$$

Then we have

$$\frac{\partial c_n^*\left(\frac{\boldsymbol{x}}{\boldsymbol{x}-1},r\right)}{\partial x_1} = c_n^*\left(\frac{\boldsymbol{x}}{\boldsymbol{x}-1},r\right) \times$$

$$\left(\sum_{\substack{i_1+i_2+\cdots+i_n=r \\ i_1\neq 0, i_2=0}} \frac{-i_1}{(x_1-1)^2 \sum_{j=1}^{n}\frac{i_j x_j}{x_j-1}} + \sum_{\substack{i_1+i_2+\cdots+i_n=r \\ i_1\neq 0, i_2\neq 0}} \frac{-i_1}{(x_1-1)^2 \sum_{j=1}^{n}\frac{i_j x_j}{x_j-1}} \right)$$

$$= c_n^*\left(\frac{\boldsymbol{x}}{\boldsymbol{x}-1},r\right) \left(\sum_{\substack{k+k_3+\cdots+k_n=r \\ k\neq 0}} \frac{-k}{(x_1-1)^2\left(\frac{kx_1}{x_1-1}+\sum_{j=3}^{n}\frac{k_j x_j}{x_j-1}\right)} + \right.$$

$$\left. \sum_{\substack{k+m+i_3+\cdots+i_n=r \\ k\neq 0, m\neq 0}} \frac{-k}{(x_1-1)^2\left(\frac{kx_1}{x_1-1}+\frac{mx_2}{x_2-1}+\sum_{j=3}^{n}\frac{k_j x_j}{x_j-1}\right)} \right) \qquad (19)$$

By the same arguments, we get

$$\frac{\partial c_n^*\left(\frac{\boldsymbol{x}}{\boldsymbol{x}-1},r\right)}{\partial x_2} = c_n^*\left(\frac{\boldsymbol{x}}{\boldsymbol{x}-1},r\right) \left(\sum_{\substack{k+k_3+\cdots+k_n=r \\ k\neq 0}} \frac{-k}{(x_2-1)^2\left(\frac{kx_2}{x_2-1}+\sum_{j=3}^{n}\frac{k_j x_j}{x_j-1}\right)} + \right.$$

$$\left. \sum_{\substack{k+m+i_3+\cdots+i_n=r \\ k\neq 0, m\neq 0}} \frac{-k}{(x_2-1)^2\left(\frac{kx_2}{x_2-1}+\frac{mx_1}{x_1-1}+\sum_{j=3}^{n}\frac{k_j x_j}{x_j-1}\right)} \right) \qquad (20)$$

then, it follows from [19] and [20] that

$$x_1^{1-m}\frac{\partial c_n^*\left(\frac{\boldsymbol{x}}{\boldsymbol{x}-1},r\right)}{\partial x_1} - x_2^{1-m}\frac{\partial c_n^*\left(\frac{\boldsymbol{x}}{\boldsymbol{x}-1},r\right)}{\partial x_2} = c_n^*\left(\frac{\boldsymbol{x}}{\boldsymbol{x}-1},r\right)(C_1+C_2)$$

where

$$C_1 = \sum_{\substack{k+k_3+\cdots+k_n=r \\ k\neq 0}} \left(\frac{-kx_1^{1-m}}{(x_1-1)^2\left(\frac{kx_1}{x_1-1}+\sum_{j=3}^{n}\frac{k_j x_j}{x_j-1}\right)} - \frac{-kx_2^{1-m}}{(x_2-1)^2\left(\frac{kx_2}{x_2-1}+\sum_{j=3}^{n}\frac{k_j x_j}{x_j-1}\right)} \right)$$

$$= k \sum_{\substack{k+k_3+\cdots+k_n=r \\ k\neq 0}} \frac{\lambda_1}{(x_1-1)^2\left(\frac{kx_1}{x_1}+\sum_{j=3}^{n}\frac{k_j x_j}{x_j-1}\right)(x_2-1)^2\left(\frac{kx_2}{x_2-1}+\sum_{j=3}^{n}\frac{k_j x_j}{x_j-1}\right)}$$

with

$$\lambda_1 = k[x_1(x_1-1)x_2^{1-m} - x_2(x_2-1)x_1^{1-m}]+$$

$$[(x_1 - 1)^2 x_2^{1-m} - (x_2 - 1)^2 x_1^{1-m}] \sum_{j=3}^{n} \frac{k_j x_j}{x_j - 1}$$

and

$$C_2 = \sum_{\substack{k+m+i_3+\cdots+i_n=r \\ k \neq 0, m \neq 0}} \left(\frac{-kx_1^{1-m}}{(x_1-1)^2 \left(\frac{kx_1}{x_1-1} + \frac{mx_2}{x_2-1} + \sum_{j=3}^{n} \frac{k_j x_j}{x_j-1} \right)} - \right.$$

$$\left. \frac{-kx_2^{1-m}}{(x_2-1)^2 \left(\frac{kx_2}{x_2-1} + \frac{mx_1}{x_1-1} + \sum_{j=3}^{n} \frac{k_j x_j}{x_j-1} \right)} \right)$$

$$= k \sum_{\substack{k+m+i_3+\cdots+i_n=r \\ k \neq 0, m \neq 0}} \frac{\lambda_2}{(x_1-1)^2 \left(\frac{kx_1}{x_1-1} + \frac{mx_2}{x_2-1} + \sum_{j=3}^{n} \frac{k_j x_j}{x_j-1} \right)(x_2-1)^2 \left(\frac{kx_2}{x_2-1} + \frac{mx_1}{x_1-1} + \sum_{j=3}^{n} \frac{k_j x_j}{x_j-1} \right)}$$

with

$$\lambda_2 = k[x_1(x_1 - 1)x_2^{1-m} - x_2(x_2 - 1)x_1^{1-m}] +$$

$$m \left[\frac{(x_1 - 1)^2 x_2^{2-m}}{x_2 - 1} - \frac{(x_2 - 1)^2 x_1^{2-m}}{x_1 - 1} \right] +$$

$$[(x_1 - 1)^2 x_2^{1-m} - (x_2 - 1)^2 x_1^{1-m}] \sum_{j=3}^{n} \frac{k_j x_j}{x_j - 1}$$

By Lemma 5, it is easy to see that $C_1 \geqslant 0$ and $C_2 \geqslant 0$ for $x \in (1, +\infty)^n$, so

$$x_1^{1-m} \frac{\partial c_n^* \left(\frac{x}{x-1}, r \right)}{\partial x_1} - x_2^{1-m} \frac{\partial c_n^* \left(\frac{x}{x-1}, r \right)}{\partial x_2} \geqslant 0$$

By Lemma 4, we prove that $c_n^* \left(\frac{x}{x-1}, r \right)$ is Schur m-Power convex on $(1, +\infty)^n$ for $m \geqslant -1$. The proof is completed.

Next, we present some new results for the Schur-power convexity of $c_n^*((x/(1 - x)), r)$.

Theorem 5　Let $r \in \mathbf{N}$.

(i) $c_n^* \left(\frac{x}{1-x}, r \right)$ is increasing on $\mathbf{R}_+^n$ and Schur-convex on $[\frac{1}{2}, 1)^n$;

(ii) If $m \leqslant 0$, then $c_n^* \left(\frac{x}{1-x}, r \right)$ is Schur-m-power convex on $(0, 1)^n$;

(iii) For $m \geqslant -1$, if r is even integer (or odd integer, respectively), then $c_n^* \left(\frac{x}{1-x}, r \right)$ is Schur-m-power convex(or Schur-m-power concave, respectively) on $(1, +\infty)^n$.

Proof　(i) Let $p(t) = \frac{t}{1-t}$. Then

$$p'(t) = \frac{1}{(1-t)^2}, \quad p''(t) = \frac{2}{(1-t)^3} \tag{21}$$

From Proposition 4, we know that $c_n^*(\boldsymbol{x}, r)$ is increasing on $\mathbf{R}_+^n$, but $p(t)$ is increasing on $\mathbf{R}$, therefore, the function $c_n^*\left(\frac{\boldsymbol{x}}{1-\boldsymbol{x}}, r\right)$ is increasing on $\mathbf{R}_+^n$.

For the case of $r = 1$ and $r = 2$, it is easy to prove that $c_n^*\left(\frac{\boldsymbol{x}}{1-\boldsymbol{x}}, r\right)$ is Schur-convex on $[\frac{1}{2}, 1)^n$.

Now consider the case of $r \geqslant 3$. By the symmetry of $c_n^*\left(\frac{\boldsymbol{x}}{1-\boldsymbol{x}}, r\right)$, without loss of generality, we may assume $x_1 > x_2$. So

$$c_n^*\left(\frac{\boldsymbol{x}}{1-\boldsymbol{x}}, r\right) = \prod_{\substack{i_1+i_2+\cdots+i_n=r \\ i_1 \neq 0, i_2=0}} \sum_{j=1}^{n} \frac{i_j x_j}{1-x_j} \times \prod_{\substack{i_1+i_2+\cdots+i_n=r \\ i_1=0, i_2 \neq 0}} \sum_{j=1}^{n} \frac{i_j x_j}{1-x_j} \times$$

$$\prod_{\substack{i_1+i_2+\cdots+i_n=r \\ i_1 \neq 0, i_2 \neq 0}} \sum_{j=1}^{n} \frac{i_j x_j}{1-x_j} \times \prod_{\substack{i_1+i_2+\cdots+i_n=r \\ i_1=0, i_2=0}} \sum_{j=1}^{n} \frac{i_j x_j}{1-x_j}$$

Then we obtain

$$\frac{\partial c_n^*\left(\frac{\boldsymbol{x}}{1-\boldsymbol{x}}, r\right)}{\partial x_1} = c_n^*\left(\frac{\boldsymbol{x}}{1-\boldsymbol{x}}, r\right) \times$$

$$\left(\sum_{\substack{i_1+i_2+\cdots+i_n=r \\ i_1 \neq 0, i_2=0}} \frac{i_1}{(1-x_1)^2 \sum_{j=1}^{n} \frac{i_j x_j}{1-x_j}} + \sum_{\substack{i_1+i_2+\cdots+i_n=r \\ i_1 \neq 0, i_2 \neq 0}} \frac{i_1}{(1-x_1)^2 \sum_{j=1}^{n} \frac{i_j x_j}{1-x_j}} \right) \tag{22}$$

$$= c_n^*\left(\frac{\boldsymbol{x}}{1-\boldsymbol{x}}, r\right) \left(\sum_{\substack{k+k_3+\cdots+k_n=r \\ k \neq 0}} \frac{k}{(1-x_1)^2\left(\frac{kx_1}{1-x_1} + \sum_{j=3}^{n} \frac{k_j x_j}{1-x_j}\right)} + \right.$$

$$\left. \sum_{\substack{k+m+i_3+\cdots+i_n=r \\ k \neq 0, m \neq 0}} \frac{k}{(1-x_1)^2\left(\frac{kx_1}{1-x_1} + \frac{mx_2}{1-x_2} + \sum_{j=3}^{n} \frac{k_j x_j}{1-x_j}\right)} \right) \tag{23}$$

By the same arguments

$$\frac{\partial c_n^*\left(\frac{\boldsymbol{x}}{1-\boldsymbol{x}}, r\right)}{\partial x_2} = c_n^*\left(\frac{\boldsymbol{x}}{1-\boldsymbol{x}}, r\right) \left(\sum_{\substack{k+k_3+\cdots+k_n=r \\ k \neq 0}} \frac{k}{(1-x_2)^2\left(\frac{kx_2}{1-x_2} + \sum_{j=3}^{n} \frac{k_j x_j}{1-x_j}\right)} + \right.$$

$$\left. \sum_{\substack{k+m+i_3+\cdots+i_n=r \\ k \neq 0, m \neq 0}} \frac{k}{(1-x_2)^2\left(\frac{kx_2}{1-x_2} + \frac{mx_1}{1-x_1} + \sum_{j=3}^{n} \frac{k_j x_j}{1-x_j}\right)} \right) \tag{24}$$

$$\frac{\partial c_n^*\left(\frac{x}{1-x},r\right)}{\partial x_1} - \frac{\partial c_n^*\left(\frac{x}{1-x},r\right)}{\partial x_2} = c_n^*\left(\frac{x}{1-x},r\right)(D_1 + D_2)$$

where

$$D_1 = \sum_{\substack{k+k_3+\cdots+k_n=r \\ k\neq 0}} \left(\frac{k}{(1-x_1)^2\left(\frac{kx_1}{1-x_1}+\sum_{j=3}^{n}\frac{k_jx_j}{1-x_j}\right)} - \frac{k}{(1-x_2)^2\left(\frac{kx_2}{1-x_2}+\sum_{j=3}^{n}\frac{k_jx_j}{1-x_j}\right)} \right)$$

$$= k \sum_{\substack{k+k_3+\cdots+k_n=r \\ k\neq 0}} \frac{k(x_1+x_2-1)(x_1-x_2)+(x_1-x_2)(2-x_1-x_2)\sum_{j=3}^{n}\frac{k_jx_j}{1-x_j}}{(1-x_1)^2\left(\frac{kx_1}{1-x_1}+\sum_{j=3}^{n}\frac{k_jx_j}{1-x_j}\right)(1-x_2)^2\left(\frac{kx_2}{1-x_2}+\sum_{j=3}^{n}\frac{k_jx_j}{1-x_j}\right)}$$

and

$$D_2 = \sum_{\substack{k+m+i_3+\cdots+i_n=r \\ k\neq 0, m\neq 0}} \left(\frac{k}{(1-x_1)^2\left(\frac{kx_1}{1-x_1}+\frac{mx_2}{1-x_2}+\sum_{j=3}^{n}\frac{k_jx_j}{1-x_j}\right)} - \right.$$

$$\left. \frac{k}{(1-x_2)^2\left(\frac{kx_2}{1-x_2}+\frac{mx_1}{1-x_1}+\sum_{j=3}^{n}\frac{k_jx_j}{1-x_j}\right)} \right)$$

$$= k \sum_{\substack{k+m+i_3+\cdots+i_n=r \\ k\neq 0, m\neq 0}} \frac{\delta_1}{(1-x_1)^2\left(\frac{kx_1}{1-x_1}+\frac{mx_2}{1-x_2}+\sum_{j=3}^{n}\frac{k_jx_j}{1-x_j}\right)(1-x_2)^2\left(\frac{kx_2}{1-x_2}+\frac{mx_1}{1-x_1}+\sum_{j=3}^{n}\frac{k_jx_j}{1-x_j}\right)}$$

with

$$\delta_1 = k(x_1+x_2-1)(x_1-x_2) + \left(\frac{(1-x_2)^2mx_1}{1-x_1} - \frac{(1-x_1)^2mx_2}{1-x_2}\right) +$$

$$(x_1-x_2)(2-x_1-x_2)\sum_{j=3}^{n}\frac{k_jx_j}{1-x_j}$$

Let $q(t) = \frac{(1-t)^3}{mt}$. Then $q'(t) = -\frac{m(1+2t)(1-t)^2}{m^2t^2} \leqslant 0$ which implies that $q(t)$ is descending on $\mathbf{R}_+$. So that $\frac{(1-x_1)^3}{mx_1} \leqslant \frac{(1-x_2)^3}{mx_2}$, namely $\frac{(1-x_2)^2mx_1}{1-x_1} - \frac{(1-x_1)^2mx_2}{1-x_2} \geqslant 0$. It is easy to see that $D_1 \geqslant 0$ and $D_2 \geqslant 0$ for $\boldsymbol{x} \in [\frac{1}{2},1)^n$, so

$$\frac{\partial c_n^*\left(\frac{x}{1-x},r\right)}{\partial x_1} - \frac{\partial c_n^*\left(\frac{x}{1-x},r\right)}{\partial x_2} \geqslant 0$$

By Lemma 1, we obtain $c_n^*\left(\frac{x}{1-x},r\right)$ is Schur-convex on $[\frac{1}{2},1)^n$.

(ii) For $r=1$ and $r=2$, it is easy to prove that $c_n^*\left(\frac{x}{1-x},r\right)$ is Schur m-power convex on $(0,1)^n$.

Now consider the case of $r \geqslant 3$. By the symmetry of $c_n^* \left(\frac{x}{1-x}, r \right)$, without loss of generality, we may assume $x_1 > x_2$. From [22] and [24], we have

$$x_1^{1-m} \frac{\partial c_n^* \left(\frac{x}{1-x}, r \right)}{\partial x_1} - x_2^{1-m} \frac{\partial c_n^* \left(\frac{x}{1-x}, r \right)}{\partial x_2} = c_n^* \left(\frac{x}{1-x}, r \right) (F_1 + F_2)$$

where

$$F_1 = \sum_{\substack{k+k_3+\cdots+k_n=r \\ k \neq 0}} \left(\frac{kx_1^{1-m}}{(1-x_1)^2 \left(\frac{kx_1}{1-x_1} + \sum_{j=3}^{n} \frac{k_j x_j}{1-x_j} \right)} - \frac{kx_2^{1-m}}{(1-x_2)^2 \left(\frac{kx_2}{1-x_2} + \sum_{j=3}^{n} \frac{k_j x_j}{1-x_j} \right)} \right)$$

$$= k \sum_{\substack{k+k_3+\cdots+k_n=r \\ k \neq 0}} \frac{\delta_1}{(1-x_1)^2 \left(\frac{kx_1}{1-x_1} + \sum_{j=3}^{n} \frac{k_j x_j}{1-x_j} \right)(1-x_2)^2 \left(\frac{kx_2}{1-x_2} + \sum_{j=3}^{n} \frac{k_j x_j}{1-x_j} \right)}$$

with

$$\delta_1 = k[x_1^{1-m} x_2(1-x_2) - x_2^{1-m} x_1(1-x_1)] +$$
$$[x_1^{1-m}(1-x_2)^2 - x_2^{1-m}(1-x_1)^2] \sum_{j=3}^{n} \frac{k_j x_j}{1-x_j}$$

and

$$F_2 = \sum_{\substack{k+m+i_3+\cdots+i_n=r \\ k \neq 0, m \neq 0}} \left(\frac{kx_1^{1-m}}{(1-x_1)^2 \left(\frac{kx_1}{1-x_1} + \frac{mx_2}{1-x_2} + \sum_{j=3}^{n} \frac{k_j x_j}{1-x_j} \right)} - \right.$$

$$\left. \frac{kx_2^{1-m}}{(1-x_2)^2 \left(\frac{kx_2}{1-x_2} + \frac{mx_1}{1-x_1} + \sum_{j=3}^{n} \frac{k_j x_j}{1-x_j} \right)} \right)$$

$$= k \sum_{\substack{k+m+i_3+\cdots+i_n=r \\ k \neq 0, m \neq 0}} \frac{\delta_2}{(1-x_1)^2 \left(\frac{kx_1}{1-x_1} + \frac{mx_2}{1-x_2} + \sum_{j=3}^{n} \frac{k_j x_j}{1-x_j} \right)(1-x_2)^2 \left(\frac{kx_2}{1-x_2} + \frac{mx_1}{1-x_1} + \sum_{j=3}^{n} \frac{k_j x_j}{1-x_j} \right)}$$

with

$$\delta_2 = k[x_1^{1-m} x_2(1-x_2) - x_2^{1-m} x_1(1-x_1)] +$$
$$m \left[\frac{(1-x_2)^2 x_1^{2-m}}{1-x_1} - \frac{(1-x_1)^2 x_2^{2-m}}{1-x_2} \right] +$$
$$([x_1^{1-m}(1-x_2)^2 - x_2^{1-m}(1-x_1)^2] \sum_{j=3}^{n} \frac{k_j x_j}{1-x_j}$$

By Lemma 6, it is easy to see that $F_1 \geqslant 0$ and $F_2 \geqslant 0$ for $\boldsymbol{x} \in (0,1)^n$, and then

$$x_1^{1-m} \frac{\partial c_n^* \left(\frac{\boldsymbol{x}}{1-\boldsymbol{x}}, r \right)}{\partial x_1} - x_2^{1-m} \frac{\partial c_n^* \left(\frac{\boldsymbol{x}}{1-\boldsymbol{x}}, r \right)}{\partial x_2} \geqslant 0$$

By Lemma 4, we show that $c_n^* \left(\frac{\boldsymbol{x}}{1-\boldsymbol{x}}, r \right)$ is Schur-m power convex on $(0,1)^n$.

(iii) Notice that

$$c_n^* \left(\frac{\boldsymbol{x}}{\boldsymbol{x} - 1}, r \right) = (-1)^r c_n^* \left(\frac{\boldsymbol{x}}{1-\boldsymbol{x}}, r \right) \tag{25}$$

and combining with the Schur-power convexity of $c_n^* \left(\frac{\boldsymbol{x}}{\boldsymbol{x}-1}, r \right)$ on $(1, +\infty)^n$ (see Theorem 1), we can prove (iii). The proof is completed.

According to the relationship between the Schur-power convex function and the Schur-convex function, the Schur-geometrically convex function, and the Schur-harmonically function, we can establish the following two corollaries immediately.

Corollary 1　Let $r \in \mathbf{N}$. Then $c_n^* \left(\frac{\boldsymbol{x}}{\boldsymbol{x}-1}, r \right)$ is Schur-convex, Schur-geometrically convex and Schur-harmonically convex on $(1, +\infty)^n$.

Corollary 2　Let $r \in \mathbf{N}$.

(i) The function $c_n^* \left(\frac{\boldsymbol{x}}{1-\boldsymbol{x}}, r \right)$ is Schur-geometrically convex and Schur-harmonically convex on $(0,1)^n$.

(ii) If r is even integer (or odd integer, respectively), then $c_n^* \left(\frac{\boldsymbol{x}}{1-\boldsymbol{x}}, r \right)$ is Schur-convex, Schur-geometrically convex and Schur-harmonically convex (or Schur-concave, Schur-geometrically concave and Schur-harmonically concave, respectively) on $(1, +\infty)^n$.

Finally, an open problem arises naturally at the end of this section.

Problem 1　For $\boldsymbol{x} \in \left(0, \frac{1}{2} \right)^n$, what is the Schur-convexity of $c_n^* \left(\frac{\boldsymbol{x}}{1-\boldsymbol{x}}, r \right)$? Is it Schur-convex or Schur-concave, or is it uncertain?

85.3　Some applications

It is not difficult to prove the following theorem by applying Theorem 2 and the majorizing relation

$$(A_n(\boldsymbol{x}), A_n(\boldsymbol{x}), \ldots, A_n(\boldsymbol{x})) \prec (x_1, x_2, \ldots, x_n)$$

Theorem 6 If $x = (x_1, x_2, \ldots, x_n) \in [\frac{1}{2}, 1)^n$ and $r \in \mathbf{N}$, or r is even integer and $x \in (1, +\infty)^n$, then

$$c_n^* \left(\frac{x}{1-x}, r \right) \geqslant \left(\frac{rA_n(x)}{1 - A_n(x)} \right)^{\binom{n+r-1}{r}} \tag{26}$$

where $A_n(x) = \frac{1}{n} \sum\limits_{i=1}^{n} x_i$ and $\binom{n+r-1}{r} = \frac{(n+r-1)!}{r!((n+r-1)-r)!}$.

If r is odd and $x \in (1, +\infty)^n$, then the inequality (26) is reversed.

By Corollary 2 and the majorizing relation

$$(\log G_n(x), \log G_n(x), \ldots, \log G_n(x)) \prec (\log x_1, \log x_2, \ldots, \log x_n)$$

we can establish the following result.

Theorem 7 If $x = (x_1, x_2, \ldots, x_n) \in (0, 1)^n$ and $r \in \mathbf{N}$ or r is even integer $x \in (1, +\infty)^n$, then

$$c_n^* \left(\frac{x}{1-x}, r \right) \geqslant \left(\frac{rG_n(x)}{1 - G_n(x)} \right)^{\binom{n+r-1}{r}} \tag{27}$$

where $G_n(x) = \sqrt[n]{\prod\limits_{i=1}^{n} x_i}$ and $\binom{n+r-1}{r} = \frac{(n+r-1)!}{r!((n+r-1)-r)!}$.

If r is odd integer and $x \in (1, +\infty)^n$, then the inequality (27) is reversed.

By using Corollary 2 and the majorizing relation

$$\left(\frac{1}{H_n(x)}, \frac{1}{H_n(x)}, \ldots, \frac{1}{H_n(x)} \right) \prec \left(\frac{1}{x_1}, \frac{1}{x_2}, \ldots, \frac{1}{x_n} \right)$$

we obtain the following theorem.

Theorem 8 If $x = (x_1, x_2, \ldots, x_n) \in (0, 1)^n$ and $r \in \mathbf{N}$, or r is even integer and $x \in (1, +\infty)^n$, then

$$c_n^* \left(\frac{x}{1-x}, r \right) \geqslant \left(\frac{rH_n(x)}{1 - H_n(x)} \right)^{\binom{n+r-1}{r}} \tag{28}$$

where $H_n(x) = \frac{n}{\sum\limits_{i=1}^{n} x_i^{-1}}$ and $\binom{n+r-1}{r} = \frac{(n+r-1)!}{r!((n+r-1)-r)!}$.

If r is odd and $x \in (1, +\infty)^n$, then the inequality (28) is reversed.

85.4　Conclusions

In this paper, we establish the following two important main results of this paper for the Schur-power convexity of symmetric functions $c_n^* \left(\frac{x}{x-1}, r \right)$ and $c_n^* \left(\frac{x}{1-x}, r \right)$:

- (see Theorem 4) Let $r \in \mathbf{N}$. If $m \geqslant -1$, then $c_n^* \left(\frac{x}{x-1}, r \right)$ is decreasing and Schur m-power convex on $(1, +\infty)^n$.

- (see Theorem 5) Let $r \in \mathbf{N}$.

(i) $c_n^* \left(\frac{x}{1-x}, r \right)$ is increasing on $\mathbf{R}_+^n$ and Schur-convex on $[\frac{1}{2}, 1)^n$;

(ii) If $m \leqslant 0$, then $c_n^* \left(\frac{x}{1-x}, r \right)$ is Schur-m-power convex on $(0, 1)^n$;

(iii) For $m \geqslant -1$, if r is even integer (or odd integer, respectively), then $c_n^* \left(\frac{x}{1-x}, r \right)$ is Schur-m-power convex (or Schur-m-power concave, respectively) on $(1, +\infty)^n$.

As applications of our new results, some new inequalities are presented in Section 3.

References

[1] Chu Y M, Sun T C. The Schur harmonic convexity for a class of symmetric functions[J]. Acta Math. Sci., **2010**, 30: 1501 - 1506.

[2] Chu Y M, Wang G D, Zhang X H. The Schur multiplicative and harmonic convexities of the complete symmetric function[J]. Math. Nachr., **2011**, 284: 653 - 663.

[3] Chu Y M, Lv Y P. The Schur harmonic convexity of the Hamy symmetric function and its applications[J]. J. Inequal. Appl., **2009**, 2009, 838529.

[4] Xia W F, Chu Y M. Schur-convexity for a class of symmetric functions and its applications[J]. J. Inequal. Appl., **2009**, 2009, 493759

[5] Guan K Z. Schur-convexity of the complete symmetric function[J]. Math. Inequal. Appl., 2006, **9**: 567 - 576.

[6] Sun M B, Chen N B, Li S H. Some properties of a class of symmetric functions and its applications[J]. Math. Nachr., 2014, **287**: 1530 - 1544.

[7] Xia W F, Chu Y M. Schur convexity and Schur multiplicative convexity for a class of symmeric functions with applications[J]. Ukr. Math. J., 2009, **61**: 1541 – 1555.

[8] Roventa I. Schur convexity of a class of symmetric functions[J]. Ann. Univ. Craiova Math. Comput. Sci. Ser., 2010, **37**: 12 – 18.

[9] Xia W F, Chu Y M. On Schur convexity of some symmetric functions[J]. J. Inequal. Appl., **2010**, 2010: 54325.

[10] Meng J X, Chu Y M, Tang X M. The Schur-harmonic-convexity of dual form of the Hamy symmetric function[J]. Mat. Vesn., 2010, **62**: 37 – 46.

[11] Chu Y M, Xia W F, Zhao T H. Some properties for a class of symmetric functions and applications[J]. J. Math. Inequal., 2011, **5**: 1 – 11.

[12] Guan K Z, Guan R K. Some properties of a generalized Hamy symmetric function and its applications[J]. J. Math. Anal. Appl. 2011, **376**: 494 – 505.

[13] Qian W M. Schur convexity for the ratios of the Hamy and generalized Hamy symmetric functions[J]. J. Inequal. Appl., **2011**, 2011: 131.

[14] Chu Y M, Xia W F, Zhang X H. The Schur concavity, Schur multiplicative and harmonic convexities of the second dual form of the Hamy symmetric function with applications[J]. J. Multivar. Anal., 2012, **105**: 412 – 421.

[15] Roventa I. A note on Schur-concave functions[J]. J. Inequal. Appl., **2012**, 2012: 159.

[16] Xia W F, Zhang X H, Wang G D, Chu Y M. Some properties for a class of symmetric functions with applications[J]. Indian J. Pure Appl. Math., 2012, **43**: 227 – 249.

[17] Shi H N, Zhang J. Schur-convexity of dual form of some symmetric functions[J]. J. Inequal. Appl., **2013**, 2013: 295.

[18] Zhang K S, Shi H N. Schur convexity of dual form of the complete symmetric function[J]. Math. Inequal. Appl., 2013, **16**: 963 – 970.

[19] Shi H N, Zhang J, Ma Q H. Schur-convexity, Schur-geometric and Schur-harmonic convexity for a composite function of complete symmetric function[J]. SpringerPlus, 2016, **5**: 296.

[20] Shi H N, Du W S. Schur-convexity for a class of completely symmetric function dual[J]. Adv. Theory Nonlinear Anal. Appl., 2019, **3**: 74 – 89.

[21] Shi H N. Majorization Theory and Analytical Inequalities[M]. Harbin: Harbin Institute of Technology Press, 2012.

[22] Shi H N. Schur-Convex Functions and Inequalities[M]. Harbin: Harbin Institute of Technology Press, 2012.

[23] Marshall A W, Olkin I, Arnold B C. Inequalities: Theory of Majorization and Its Application[M]. 2nd ed. New York: Springer, 2011.

[24] Wang B Y. Foundations of Majorization Inequalities[M]. Beijing: Beijing Normal University Press, 1990.

[25] Zhang X M. Geometrically Convex Functions[M]. Hefei: Anhui University Press, 2004.

[26] Zhang X H, Chu Y M. New discussion to analytic Inequalities[M]. Harbin: Harbin Institute of Technology Press, 2009.

[27] Chu Y M, Zhang X M, Wang G D. The Schur geometrical convexity of the extended mean values[J]. J. Convex Anal., 2008, 15: 707 – 718.

[28] Guan K Z. A class of symmetric functions for multiplicatively convex function[J]. Math. Inequal. Appl., 2007, 10: 745 – 753.

[29] Sun T C, Lv Y P, Chu Y M. Schur multiplicative and harmonic convexities of generalized Heronian mean in n variables and their applications[J]. Int. J. Pure Appl. Math., 2009, 55: 25 – 33.

[30] Yang Z H. Schur power convexity of Stolarsky means[J]. Publ. Math. Debrecen, 2012, 80: 43 – 66.

第86篇 两个分式不等式的控制证明与推广

(石焕南, 广东第二师范学院学报, 2019，39 (5): 11-15.)

摘 要: 利用完全对称函数及其对偶式的Schur几何凸性, 证明并推广了两个分式不等式.

关键词: 分式不等式; Schur-凸函数; Schur-几何凸函数; 完全对称函数; 对偶式

86.1 引 言

在本文中, $\mathbf{R}^n$, $\mathbf{R}_+^n$ 和$\mathbf{R}_-^n$ 分别表示n 维实数集, n 维正实数集和n 维负实数集, 并记$\mathbf{R}^1 = \mathbf{R}, \mathbf{R}_+^1 = \mathbf{R}_+$ 和$\mathbf{R}_-^1 = \mathbf{R}_-$.

2013年, 李阳刚提出了如下两个不等式猜想.

猜想 1[1] 若$a, b, c > 1, 0 < \lambda \leqslant 1$, 则

$$\frac{1}{a+\lambda} + \frac{1}{b+\lambda} + \frac{1}{c+\lambda} \geqslant \frac{3}{abc+\lambda} \tag{1}$$

猜想 2[1] 若$a_1, a_2, \ldots, a_n > 1, n \in \mathbf{N}_+, n \geqslant 2, 0 < \lambda \leqslant 1$, 则

$$\frac{1}{a_1^n+\lambda} + \frac{1}{a_2^n+\lambda} + \cdots + \frac{1}{a_n^n+\lambda} \geqslant \frac{n}{a_1 a_2 \cdots a_n + \lambda} \tag{2}$$

最近石冶郝等[2], 将不等式(1) 和(2)推广为如下两个定理并利用函数的凹凸性, 借助琴生不等式加以证明.

定理 A[2] 若$a_1, a_2, \ldots, a_n > 1, n \in \mathbf{N}_+, -1 \leqslant \lambda \leqslant 1$, 则当$\mu \geqslant 1$ 时

$$\frac{1}{(a_1^n+\lambda)^\mu} + \frac{1}{(a_2^n+\lambda)^\mu} + \cdots + \frac{1}{(a_n^n+\lambda)^\mu} \geqslant \frac{n}{(a_1 a_2 \cdots a_n + \lambda)^\mu} \tag{3}$$

等号成立当且仅当$a_1 = a_2 = \cdots = a_n$.

定理 B[2] 若 $0 < a_1, a_2, \ldots, a_n \leqslant 1, n \in \mathbf{N}_+, \lambda \geqslant 1$, 则当 $0 < \mu \leqslant 1$ 时

$$\frac{1}{(a_1^n + \lambda)^\mu} + \frac{1}{(a_2^n + \lambda)^\mu} + \cdots + \frac{1}{(a_n^n + \lambda)^\mu} \leqslant \frac{n}{(a_1 a_2 \cdots a_n + \lambda)^\mu} \tag{4}$$

等号成立当且仅当 $a_1 = a_2 = \cdots = a_n$.

本文利用受控理论将定理A 和定理B 推广到完全对称函数及其对偶式的形式.

对于 $\boldsymbol{x} = (x_1, \ldots, x_n) \in \mathbf{R}^n$, 定义完全对称函数 $c_n(\boldsymbol{x}, r)$ 如下

$$c_n(\boldsymbol{x}, r) = \sum_{i_1 + i_2 + \cdots + i_n = r} \prod_{j=1}^n x_j^{i_j} \tag{5}$$

其中 $c_0(\boldsymbol{x}, r) = 1$, $r \in \{1, 2, \ldots, n\}$, $i_1, i_2, \ldots, i_n$ 是非负整数.

完全对称函数是一类重要的对称函数, 诸多学者关注此函数, 并获得一些有趣的结果. 例如, 关开中[3] 讨论了 $c_n(\boldsymbol{x}, r)$ 的Schur-凸性并证得如下结论:

命题 1[3] $c_n(\boldsymbol{x}, r)$ 在 $\mathbf{R}_+^n$ 上递增且Schur-凸.

随后, 褚玉明等人[4] 证得:

命题 2[4] $c_n(\boldsymbol{x}, r)$ 在 $\mathbf{R}_+^n$ 上Schur-几何凸和Schur-调和凸.

2016 年, 石焕南等人[5] 进一步考虑了 $c_n(\boldsymbol{x}, r)$ 在 $\mathbf{R}_-^n$ 上的Schur-凸性, 证得:

命题 3[5] 若 r 是偶数(奇数), 则 $c_n(\boldsymbol{x}, r)$ 在 $\mathbf{R}_-^n$ 上递减且Schur-凸(递增且Schur-凹).

完全对称函数对偶式的定义为

$$c_n^*(\boldsymbol{x}, r) = \prod_{i_1 + i_2 + \cdots + i_n = r} \sum_{j=1}^n i_j x_j \tag{6}$$

其中 $c_0^*(\boldsymbol{x}, r) = 1$, $r \in \{1, 2, \ldots, n\}$, $i_1, i_2, \ldots, i_n$ 是非负整数.

张孔生和石焕南[6] 证得如下两个命题:

命题 4 对于 $r = 1, 2, \ldots, n$, $c_n^*(\boldsymbol{x}, r)$ 在 $\mathbf{R}_+^n$ 上递增且Schur- 凹.

命题 5 对于 $r = 1, 2, \ldots, n$, $c_n^*(\boldsymbol{x}, r)$ 在 $\mathbf{R}_+^n$ 上Schur-几何凸和Schur-调和凸.

注意

$$c_n^*(-\boldsymbol{x}, r) = (-1)^r c_n^*(\boldsymbol{x}, r)$$

不难证明

命题 6 若 r 是一个偶整数(或奇整数), 则 $c_n^*(\boldsymbol{x}, r)$ 在 $\mathbf{R}_-^n$ 上递减且Schur-

凹(或递增且Schur-凸).

记

$$c_n\left(\frac{1}{(\boldsymbol{x}+\lambda)^\mu},r\right)=\sum_{i_1+i_2+\cdots+i_n=r}\prod_{j=1}^n\left(\frac{1}{(x_j+\lambda)^\mu}\right)^{i_j} \qquad (7)$$

$$c_n^*\left(\frac{1}{(\boldsymbol{x}+\lambda)^\mu},r\right)=\prod_{i_1+i_2+\cdots+i_n=r}\sum_{j=1}^n i_j\left(\frac{1}{(x_j+\lambda)^\mu}\right) \qquad (8)$$

我们的主要结果是:

定理1 设$\mu\geqslant 1$,且$-1\leqslant\lambda\leqslant 1$,则对称函数$c_n\left(\frac{1}{(\boldsymbol{x}+\lambda)^\mu},r\right)$在$(1,+\infty)^n$上 Schur-几何凸. 并且对于$\boldsymbol{x}\in(1,+\infty)^n$,如下不等式成立

$$\sum_{i_1+i_2+\cdots+i_n=r}\prod_{j=1}^n\left(\frac{1}{(x_j+\lambda)^\mu}\right)^{i_j}\geqslant\binom{n+r-1}{r}\left(\frac{1}{(G_n(\boldsymbol{x})+\lambda)^\mu}\right)^r \qquad (9)$$

其中$G_n(\boldsymbol{x})=\sqrt[n]{\prod_{i=1}^n x_i}$.

定理2 设$0<\mu\leqslant 1$,且$\lambda\geqslant 1$,则对称函数$c_n^*\left(\frac{1}{(\boldsymbol{x}+\lambda)^\mu},r\right)$在$(0,1)^n$上Schur-几何凹. 并且对于$\boldsymbol{x}\in(0,1)^n$,如下不等式成立

$$\prod_{i_1+i_2+\cdots+i_n=r}\sum_{j=1}^n i_j\left(\frac{1}{(x_j+\lambda)^\mu}\right)\leqslant\left(\frac{r}{(G_n(\boldsymbol{x})+\lambda)^\mu}\right)^{\binom{n+r-1}{r}} \qquad (10)$$

其中$G_n(\boldsymbol{x})=\sqrt[n]{\prod_{i=1}^n x_i}$.

注1 取$r=1$,并置$x_i=a_i^n$,由定理1 即可得定理A,所以定理1 推广了定理A.

对于$0<a_1,a_2,\ldots,a_n\leqslant 1,n\in\mathbf{N}_+,\lambda\geqslant 1$,当$0<\mu\leqslant 1$ 时,取$r=1$,并置$x_i=a_i^n$,由定理2 可得

$$\frac{1}{(a_1^n+\lambda)^\mu}+\frac{1}{(a_2^n+\lambda)^\mu}+\cdots+\frac{1}{(a_n^n+\lambda)^\mu}\leqslant\left(\frac{1}{(a_1a_2\cdots a_n+\lambda)^\mu}\right)^n \qquad (11)$$

注意

$$0\leqslant\frac{1}{(a_1a_2\cdots a_n+\lambda)^\mu}\leqslant 1$$

我们有

$$\left(\frac{1}{(a_1a_2\cdots a_n+\lambda)^\mu}\right)^n\leqslant\frac{1}{(a_1a_2\cdots a_n+\lambda)^\mu}\leqslant\frac{n}{(a_1a_2\cdots a_n+\lambda)^\mu}$$

故不等式(11)加强了不等式(4). 这样定理2 不仅推广了而且还加强了定理B.

取 $n=3, r=2$, 并置 $x_i = a_i^3$, 由定理1 可得如下推论.

推论 1　若 $a_1, a_2, a_3 > 1, -1 \leqslant \lambda \leqslant 1$, 则当 $\mu \geqslant 1$ 时, 有

$$\frac{1}{(a_1^3+\lambda)^\mu(a_2^3+\lambda)^\mu} + \frac{1}{(a_2^3+\lambda)^\mu(a_3^3+\lambda)^\mu} + \frac{1}{(a_3^3+\lambda)^\mu(a_1^3+\lambda)^\mu} +$$

$$\frac{1}{(a_1^3+\lambda)^{2\mu}} + \frac{1}{(a_2^3+\lambda)^{2\mu}} + \frac{1}{(a_3^3+\lambda)^{2\mu}} \geqslant 6\left(\frac{1}{(a_1a_2a_3+\lambda)^\mu}\right)^2 \tag{12}$$

取 $n=3, r=2$, 并置 $x_i = a_i^3$, 由定理2 可得如下推论.

推论 2　若 $0 < a_1, a_2, a_3 \leqslant 1, \lambda > 1$, 则当 $\mu \geqslant 1$ 时, 有

$$\left(\frac{1}{(a_1^3+\lambda)^\mu} + \frac{1}{(a_2^3+\lambda)^\mu}\right)\left(\frac{1}{(a_2^3+\lambda)^\mu} + \frac{1}{(a_3^3+\lambda)^\mu}\right)\left(\frac{1}{(a_3^3+\lambda)^\mu} + \frac{1}{(a_1^3+\lambda)^\mu}\right) \times$$

$$\left(\frac{2}{(a_1^3+\lambda)^\mu}\right)\left(\frac{2}{(a_2^3+\lambda)^\mu}\right)\left(\frac{2}{(a_3^3+\lambda)^\mu}\right) \leqslant \left(\frac{2}{(a_1a_2a_3+\lambda)^\mu}\right)^6 \tag{13}$$

86.2　定义和引理

我们需要如下定义和引理.

定义 1[7-8]　对于 $\boldsymbol{x} = (x_1, \ldots, x_n) \in \mathbf{R}^n$, 将 $\boldsymbol{x}$ 的分量排成递减的次序后, 记作 $x_{[1]} \geqslant x_{[2]} \geqslant \cdots \geqslant x_{[n]}$. 设 $\boldsymbol{x}, \boldsymbol{y} \in \mathbf{R}^n$ 满足

(i) $\sum_{i=1}^k x_{[i]} \leqslant \sum_{i=1}^k y_{[i]}, k = 1, 2, \ldots, n-1$; (ii) $\sum_{i=1}^n x_i = \sum_{i=1}^n y_i$. 则称 $\boldsymbol{x}$ 被 $\boldsymbol{y}$ 所控制, 记作 $\boldsymbol{x} \prec \boldsymbol{y}$.

$\boldsymbol{x} \leqslant \boldsymbol{y}$ 表示 $x_i \leqslant y_i, i = 1, \ldots, n$.

定义 2[7-8]　设 $\Omega \subset \mathbf{R}^n, \varphi : \Omega \to \mathbf{R}$, 若在 Ω 上 $\boldsymbol{x} \prec \boldsymbol{y} \Rightarrow \varphi(\boldsymbol{x}) \leqslant \varphi(\boldsymbol{y})$, 则称 φ 为 Ω 上的Schur-凸函数; 若 $-\varphi$ 是 Ω 上的Schur-凸函数, 则称 φ 为 Ω 上的Schur-凹函数.

若在 Ω 上 $\boldsymbol{x} \leqslant \boldsymbol{y} \Rightarrow \varphi(\boldsymbol{x}) \leqslant \varphi(\boldsymbol{y})$, 则称 φ 为 Ω 上的增函数; 若 $-\varphi$ 是 Ω 上的增函数, 则称 φ 为 Ω 上的减函数.

引理 1[7-8]　(Schur-凸函数判定定理) 设 $\Omega \subset \mathbf{R}^n$ 是有内点的对称凸集, $\varphi : \Omega \to \mathbf{R}$ 在 Ω 上连续, 在 Ω 的内部 Ω° 可微, 则在 Ω 上Schur-凸(Schur-凹)的充要条件是 φ 在 Ω 上对称且 $\forall \boldsymbol{x} \in \Omega$, 有

$$(x_1 - x_2)\left(\frac{\partial \varphi}{\partial x_1} - \frac{\partial \varphi}{\partial x_2}\right) \geqslant 0 (\leqslant 0)$$

关于Schur-凸函数的推广, 国内学者做了不少开拓性的工作, 并得到国外学者的认可与应用. 2003 年, 张小明首先提出并建立了Schur-几何凸函数

的定义及判定定理.

定义 3[9-10] 设 $\Omega \subset \mathbf{R}_{++}^n$, $f : \Omega \to \mathbf{R}_+$, 对于任意 $\boldsymbol{x} = (x_1, \ldots, x_n)$, $\boldsymbol{y} = (y_1, \ldots, y_n) \subset \Omega$, 若

$$\ln \boldsymbol{x} := (\ln x_1, \ldots, \ln x_n) \prec (\ln y_1, \ldots, \ln y_n) =: \ln \boldsymbol{y}$$

有 $f(\boldsymbol{x}) \leqslant f(\boldsymbol{y})$, 则称 f 为 Ω 上的 Schur-几何凸函数(也称作乘性凸函数). 若 $\ln \boldsymbol{x} \prec \ln \boldsymbol{y}$ 有 $f(\boldsymbol{x}) \geqslant f(\boldsymbol{y})$, 则称 f 为 Ω 上的 Schur-几何凹函数(也称作乘性凹函数).

引理 2[9-10] (Schur-几何凸函数判定定理) 设 $\Omega \subset \mathbf{R}_{++}^n$ 是有内点的对称凸集, $\varphi : \Omega \to \mathbf{R}$ 在 Ω 上连续, 在 Ω 的内部 Ω° 可微, 则在 Ω 上 Schur-几何凸(Schur-几何凹)的充要条件是 φ 在 Ω 上对称且 $\forall \boldsymbol{x} \in \Omega$, 有

$$(\ln x_1 - \ln x_2) \left(x_1 \frac{\partial \varphi}{\partial x_1} - x_2 \frac{\partial \varphi}{\partial x_2} \right) \geqslant 0 (\leqslant 0)$$

由定义3,下述命题显然成立.

命题 7[9-10] 设 $\Omega \subset \mathbf{R}_+^n$, 且令

$$\log \Omega = \{ (\log x_1, \ldots, \log x_n) : (x_1, \ldots, x_n) \in \Omega \}$$

则 $\varphi : \Omega \to \mathbf{R}_+$ 在 Ω 上 Schur-几何凸(Schur-几何凹)当且仅当 $\varphi(\mathrm{e}^{x_1}, \ldots, \mathrm{e}^{x_n})$ 在 $\log \Omega$ 上 Schur-凸(Schur-凹).

关于复合函数的 Schur-凸性, 有如下结论:

引理 3[7-8] 设区间 $[a, b] \subset \mathbf{R}$, $\varphi : \mathbf{R}^n \to \mathbf{R}$, $f : [a, b] \to \mathbf{R}$, $\psi(x_1, \cdots, x_n) = \varphi(f(x_1), \cdots, f(x_n)) : [a, b]^n \to \mathbf{R}$.

(i) 若 φ 递增且 Schur-凸, f 凸, 则 ψ Schur-凸;

(ii) 若 φ 递减且 Schur-凸, f 递增且凹, 则 ψ 递减且 Schur-凸;

(iii) 若 φ 递增且 Schur-凹, f 递增且凹, 则 ψ 递增且 Schur-凹;

(iv) 若 φ 递减且 Schur-凸, f 递减且凹, 则 ψ 递增且 Schur-凸;

(v) 若 φ 递增且 Schur-凹, f 递减且凹, 则 ψ 递减且 Schur-凹.

引理 4[7-8] 设 $\boldsymbol{x} = (x_1, x_2, \ldots, x_n) \in \mathbf{R}^n$, 则

$$(A(\boldsymbol{x}), A(\boldsymbol{x}), \ldots, A(\boldsymbol{x})) \prec \boldsymbol{x} = (x_1, x_2, \ldots, x_n) \tag{14}$$

其中 $A(\boldsymbol{x}) = \frac{1}{n} \sum_{i}^{n} x_i$.

86.3 主要结果的证明

定理 1 的证明 令 $f(t) = \dfrac{1}{(e^x + \lambda)^\mu}$. 由

$$f'(t) = \frac{-\mu e^x}{(e^x + \lambda)^{\mu+1}}, \quad f''(t) = \frac{\mu e^x(\mu e^x - \lambda)}{(e^x + \lambda)^{\mu+2}} \tag{15}$$

当 $\mu \geqslant 1$ 时, 且 $-1 \leqslant \lambda \leqslant 1, x > 0$, 则 $f''(x) > 0$, 所以 f 是 $\mathbf{R}_+$ 上的凸函数. 据命题1 和引理3 (i) 知 $c_n\left(\frac{1}{(e^x+\lambda)^\mu}, r\right)$ 在 $\mathbf{R}_+$ 上Schur- 凸. 进而由命题7 知 $c_n\left(\frac{1}{(x+\lambda)^\mu}, r\right)$ 在 $(1, +\infty)^n$ 上Schur-几何凸.

据控制关系(14), 下述控制关系成立

$$(\log G_n(\boldsymbol{x}), \log G_n(\boldsymbol{x}), \ldots, \log G_n(\boldsymbol{x})) \prec (\log x_1, \log x_2, \ldots, \log x_n) \tag{16}$$

由此控制关系和 $c_n\left(\frac{1}{(x+\lambda)^\mu}, r\right)$ 在 $(1, +\infty)^n$ 上Schur-几何凸性, 即知式(9)成立.

定理 2 的证明 当 $0 < \mu \leqslant 1$, 且 $\lambda \geqslant 1, x < 0$ 时, 由式(15)可见 $f'(x) < 0, f''(x) < 0$, 所以 f 是 $\mathbf{R}_-$ 上的递减的凹函数.

据命题 4 和引理3 (v) 知 $c_n^*\left(\frac{1}{(e^x+\lambda)^\mu}, r\right)$ 在 $\mathbf{R}_-$ 上Schur-凹, 进而由命题 7 知 $c_n^*\left(\frac{1}{(x+\lambda)^\mu}, r\right)$ 在 $(0,1)^n$ 上Schur-几何凹.

由式(16)和 $c_n\left(\frac{1}{(x+\lambda)^\mu}, r\right)$ 在 $(0,1)^n$ 上Schur-几何凹性, 即知式(10)成立. 定理2 证毕.

对各种对称函数的Schur-凸性的探讨是近年来不等式研究的一个热点, 利用函数的Schur凸性或Schur-几何凸性来证明不等式是一种十分有效的方法, 有关这方面更多的信息请参见专著[11] 和[12].

参考文献

[1] 李阳刚. 关于一类分式型不等式的证明[J]. 黔南民族师范学院学报, 2013, 33 (4)：127–128.

[2] 石冶郝, 林玲, 孙颖. 两个猜想不等式的证明与推广[J]. 首都师范大学学报(自然科学版), 2019, 40 (3): 14-17.

[3] Guan K-Z. Schur-convexity of the complete symmetric function[J]. Math Inequal Appl, 2006, 9(4):567‑576

[4] Chu Y-M, Wang G-D, Zhang X-H. The Schur multiplicative and harmonic convexities of the complete symmetric function[J]. Math Nachr, 2011, 284(5‑6):653‑663.

[5]   Huan-Nan Shi, Jing Zhang, Qing-Hua Ma. Schur-convexity, Schur-geometric and Schur-harmonic convexity for a composite function of complete symmetric function[J]. SpringerPlus, 2016(5): 296.

[6]   K S Zhang, H N Shi. Schur convexity of dual form of the complete symmetric function[J]. Mathematical Inequalities & Applications, 2013, 16(4): 963-970.

[7]   Marshall A W, Olkin I, Arnold B C. Inequalities: theory of majorization and its application[M]. 2nd ed. New York: Springer, 2011.

[8]   王伯英. 控制不等式基础[M]. 北京: 北京师范大学出版社, 1990.

[9]   张小明. Schur几何凸函数[M]. 合肥:安徽大学出版社, 2004.

[10] 张小明, 褚玉明. 解析不等式新论[M]. 哈尔滨：哈尔滨工业大学出版社, 2009.

[11] 石焕南. 受控理论与解析不等式[M]. 哈尔滨:哈尔滨工业大学出版社, 2012.

[12] 石焕南. Schur凸函数与不等式[M]. 哈尔滨:哈尔滨工业大学出版社, 2017.

第87篇　Schur-convexity for compositions of complete symmetric function dual

(Huan-nan Shi, Pei Wang and Jian Zhang. Journal of Inequalities and Applications, (2020)2020:65)

Abstract: The Schur-convexity for certain compound functions involving the dual of the complete symmetric function are studied. As an application, the Schur-convexity of some special symmetric functions is discussed and some inequalities are established.

Keywords: Schur-convexity; Schur-geometric convexity; Schur-harmonic convexity; completely symmetric function; dual form

87.1　Introduction

Throughout the article, $\mathbf{R}$ denotes the set of real numbers, $\boldsymbol{x} = (x_1, x_2, \ldots, x_n)$ denotes n-tuple (n-dimensional real vectors), the set of vectors can be written as

$$\mathbf{R}^n = \{\boldsymbol{x} = (x_1, x_2, \cdots, x_n) : x_i \in \mathbf{R}, i = 1, 2, \ldots, n\}$$

$$\mathbf{R}^n_+ = \{\boldsymbol{x} = (x_1, x_2, \ldots, x_n) : x_i > 0, i = 1, 2, \ldots, n\}$$

$$\mathbf{R}^n_- = \{\boldsymbol{x} = (x_1, x_2, \ldots, x_n) : x_i < 0, i = 1, 2, \ldots, n\}$$

In particular, the notations $\mathbf{R}$ and $\mathbf{R}_+$ denote $\mathbf{R}^1$ and $\mathbf{R}^1_+$, respectively.

In recent years, the Schur-convexity, Schur-geometric and Schur-harmonic convexities of various symmetric functions are hot topic of inequality research ([8]-[37]).

The following complete symmetric function is an important class of symmetric functions.

For $x = (x_1, x_2, \ldots, x_n) \in \mathbf{R}^n$, the $c_n(x, r)$ is defined as

$$c_n(x, r) = \sum_{i_1+i_2+\cdots+i_n=r} x_1^{i_1} x_2^{i_2} \cdots x_n^{i_n} \tag{1}$$

where $c_0(x, r) = 1$, $r \in \{1, 2, \ldots, n\}$, $i_1, i_2, \ldots, i_n$ are non-negative integers.

It has been investigated by many mathematicians and there are many interesting results in the literature.

Guan[11] discussed the Schur-convexity of $c_n(x, r)$ and proved that:

Proposition 1 $c_n(x, r)$ is increasing and Schur-convex on $\mathbf{R}_+^n$.

Subsequently, Chu et al.[8] proved that:

Proposition 2 $c_n(x, r)$ is Schur-geometrically convex and Schur-harmonically convex on $\mathbf{R}_+^n$.

In 2016, Shi et al.[25] further considered the Schur-convexity of $c_n(x, r)$ on $\mathbf{R}_-^n$, which proved the following proposition.

Proposition 3 If r is even integer(or odd integer, respectively), then $c_n(x, r)$ is decreasing and Schur-convex (or increasing and Schur-concave, respectively) on $\mathbf{R}_-^n$.

The dual form of the complete symmetric function $c_n(x, r)$ is defined as

$$c_n^*(x, r) = \prod_{i_1+i_2+\cdots+i_n=r} \sum_{j=1}^{n} i_j x_j \tag{2}$$

where $c_0^*(x, r) = 1$, $r \in \{1, 2, \ldots, n\}$, $i_1, i_2, \ldots, i_n$ are non-negative integers.

Zhang and Shi[24] proved the following two propositions:

Proposition 4 For $r = 1, 2, \ldots, n$, $c_n^*(x, r)$ is increasing and Schur-concave on $\mathbf{R}_+^n$.

Proposition 5 For $r = 1, 2, \ldots, n$, $c_n^*(x, r)$ is Schur-geometrically convex and Schur-harmonically convex on $\mathbf{R}_+^n$.

Notice that

$$c_n^*(-x, r) = (-1)^r c_n^*(x, r)$$

it is not difficult to prove that.

Proposition 6 If r is even integer (or odd integer, respectively), then $c_n^*(x, r)$ is decreasing and Schur-concave (or increasing and Schur-convex, respectively) on $\mathbf{R}_-^n$.

This paper we will study the Schur-convexity, Schur-geometric and

Schur-harmonic convexities of the following composite function of $c_n^* (\boldsymbol{x}, r)$

$$c_n^* (f(\boldsymbol{x}), r) = c_n^* (f(x_1), f(x_2), \dots f(x_n), r) = \prod_{i_1+i_2+\cdots+i_n=r} \sum_{j=1}^{n} i_j (f(x_j))$$

(3)

where f is a positive function which satisfies certain conditions.

Our main results are as follows:

Theorem 1 Let $I \subset \mathbf{R}$ be a symmetric convex set with nonempty interior and let $f : I \to \mathbf{R}_+$ be continuous on I and differentiable in the interior of I.

(a) If f is a log-convex function on I, then for any $r = 1, 2, \dots, n$, $c_n^* (f(\boldsymbol{x}), r)$ is a Schur-convex function on I^n;

(b) If f is a concave function on I, then for any $r = 1, 2, \dots, n$, $c_n^* (f(\boldsymbol{x}), r)$ is a Schur-concave function on I^n.

Theorem 2 Let $I \subset \mathbf{R}_+$ be a symmetric convex set with nonempty interior and let $f : I \to \mathbf{R}_+$ be continuous on I and differentiable in the interior of I.

(a) If f is a increasing and log-convex function on I, then for any $r = 1, 2, \dots, n$, $c_n^* (f(\boldsymbol{x}), r)$ is a Schur-geometrically convex function on I^n.

(b) If f is a descending and concave function on I, then for any $r = 1, 2, \dots, n$, $c_n^* (f(\boldsymbol{x}), r)$ is a Schur-geometrically concave function on I^n.

Theorem 3 Let $I \subset \mathbf{R}_+$ be a symmetric convex set with nonempty interior and let $f : I \to \mathbf{R}_+$ be continuous on I and differentiable in the interior of I.

(a) If f is a increasing and log-convex function on I, then for any $r = 1, 2, \dots, n$, $c_n^* (f(\boldsymbol{x}), r)$ is a Schur-harmonically convex function on I^n.

(b) If f is a descending and concave function on I, then for any $r = 1, 2, \dots, n$, $c_n^* (f(\boldsymbol{x}), r)$ is a Schur-harmonically concave function on I^n.

87.2　Definitions and Lemmas

For convenience, we introduce some definitions as follows.

Definition 1 (see [1, 2]) Let $x = (x_1, x_2, \ldots, x_n)$ and $y = (y_1, y_2, \ldots, y_n) \in \mathbf{R}^n$.

(a) $x \geq y$ means $x_i \geqslant y_i$ for all $i = 1, 2, \ldots, n$.

(b) Let $\Omega \subset \mathbf{R}^n$, $\varphi \colon \Omega \to \mathbf{R}$ is said to be increasing if $x \geq y$ implies $\varphi(x) \geqslant \varphi(y)$. φ is said to be decreasing if and only if $-\varphi$ is increasing.

Definition 2 (see [1, 2]) Let $x = (x_1, x_2, \ldots, x_n)$ and $y = (y_1, y_2, \ldots, y_n) \in \mathbf{R}^n$.

(a) x is said to be majorized by y (in symbols $x \prec y$) if $\sum\limits_{i=1}^{k} x_{[i]} \leqslant \sum\limits_{i=1}^{k} y_{[i]}$ for $k = 1, 2, \ldots, n-1$ and $\sum\limits_{i=1}^{n} x_i = \sum\limits_{i=1}^{n} y_i$, where $x_{[1]} \geqslant x_{[2]} \geqslant \cdots \geqslant x_{[n]}$ and $y_{[1]} \geqslant y_{[2]} \geqslant \cdots \geqslant y_{[n]}$ are rearrangements of x and y in a descending order.

(b) Let $\Omega \subset \mathbf{R}^n$, $\varphi \colon \Omega \to \mathbf{R}$ is said to be a Schur-convex function on Ω if $x \prec y$ on Ω implies $\varphi(x) \leq \varphi(y)$. φ is said to be a Schur-concave function on Ω if and only if $-\varphi$ is Schur-convex function on Ω.

Definition 3 (see [1, 2]) Let $x = (x_1, x_2, \ldots, x_n)$ and $y = (y_1, y_2, \ldots, y_n) \in \mathbf{R}^n$.

(a) $\Omega \subset \mathbf{R}^n$ is said to be a convex set if $x, y \in \Omega, 0 \leqslant \alpha \leqslant 1$, implies $\alpha x + (1 - \alpha)y = (\alpha x_1 + (1 - \alpha)y_1, \alpha x_2 + (1 - \alpha)y_2, \ldots, \alpha x_n + (1 - \alpha)y_n) \in \Omega$.

(b) Let $\Omega \subset \mathbf{R}^n$ be convex set. A function $\varphi \colon \Omega \to \mathbf{R}$ is said to be a convex function on Ω if

$$\varphi(\alpha x + (1 - \alpha)y) \leqslant \alpha\varphi(x) + (1 - \alpha)\varphi(y)$$

for all $x, y \in \Omega$, and all $\alpha \in [0, 1]$. φ is said to be a concave function on Ω if and only if $-\varphi$ is convex function on Ω.

Definition 4 (see [1, 2])

(a) A set $\Omega \subset \mathbf{R}^n$ is called a symmetric set, if $x \in \Omega$ implies $xP \in \Omega$ for every $n \times n$ permutation matrix P.

(b) A function $\varphi \colon \Omega \to \mathbf{R}$ is called symmetric if for every permutation matrix P, $\varphi(xP) = \varphi(x)$ for all $x \in \Omega$.

Lemma 1 (Schur-convex function decision theorem) [1,2] Let $\Omega \subset \mathbf{R}^n$ be symmetric and have a nonempty interior convex set. Ω^0 is the interior of Ω. $\varphi : \Omega \to \mathbf{R}$ is continuous on Ω and differentiable in Ω^0. Then φ is the *Schur — convex (or Schur — concave, respectively) function*, if and only if φ is symmetric on Ω and

$$(x_1 - x_2) \left(\frac{\partial \varphi}{\partial x_1} - \frac{\partial \varphi}{\partial x_2} \right) \geqslant 0 (\text{or} \leqslant 0, \text{ respectively}) \tag{4}$$

holds for any $x \in \Omega^0$.

The first systematical study of the functions preserving the ordering of majorization was made by Issai Schur in 1923. In Schur's honor, such functions are said to be "Schur-convex". It can be used extensively in analytic inequalities, combinatorial optimization, quantum physics, information theory, and other related fields. See [1].

Definition 5 (see [3]) Let $x = (x_1, x_2, \ldots, x_n) \in \mathbf{R}_+^n$ and $y = (y_1, y_2, \ldots, y_n) \in \mathbf{R}_+^n$.

(a) $\Omega \subset \mathbf{R}_+^n$ is called a geometrically convex set if $(x_1^\alpha y_1^\beta, x_2^\alpha y_2^\beta, \ldots, x_n^\alpha y_n^\beta) \in \Omega$ for all $x, y \in \Omega$ and $\alpha, \beta \in [0, 1]$ such that $\alpha + \beta = 1$.

(b) Let $\Omega \subset \mathbf{R}_+^n$. The function $\varphi \colon \Omega \to \mathbf{R}_+$ is said to be Schur-geometrically convex function on Ω if $(\log x_1, \log x_2, \ldots, \log x_n) \prec (\log y_1, \log y_2, \ldots, \log y_n)$ on Ω implies $\varphi(x) \leqslant \varphi(y)$. The function φ is said to be a Schur-geometrically concave function on Ω, if and only if $-\varphi$ is Schur-geometrically convex function on Ω.

The Schur-geometric convexity was proposed by Zhang[3] in 2004, and was investigated by Chu et al.[4], Guan[5], Sun et al.[6], and so on. We also note that some authors use the term "Schur multiplicative convexity".

In 2009, Chu (see [7], [8], [9]) introduced the notion of Schur-harmonically convex function, and some interesting inequalities were obtained.

Definition 6[7]　Let $\Omega \subset \mathbf{R}_+^n$ or $\Omega \subset \mathbf{R}_-^n$.

(a) A set Ω is said to be harmonically convex if $\dfrac{xy}{\lambda x + (1 - \lambda) y} \in \Omega$ for every $x, y \in \Omega$ and $\lambda \in [0, 1]$, where $xy = \sum\limits_{i=1}^{n} x_i y_i$ and $\dfrac{1}{x} = \left(\dfrac{1}{x_1}, \dfrac{1}{x_2}, \ldots, \dfrac{1}{x_n} \right)$.

(b) A function $\varphi : \Omega \to \mathbf{R}_+$ is said to be Schur-harmonically convex on Ω if $\frac{1}{x} \prec \frac{1}{y}$ implies $\varphi(x) \leq \varphi(y)$. A function φ is said to be a Schur-harmonically concave function on Ω, if and only if $-\varphi$ is a Schur-harmonically convex function.

Remark 1 We extend the definition and determination theorem of Schur-harmonically convex function established by Chu as follows:

(a) $\Omega \subset \mathbf{R}_+^n$ is extended to $\Omega \subset \mathbf{R}_+^n$ or $\Omega \subset \mathbf{R}_-^n$;

(b) The function $\varphi : \Omega \to \mathbf{R}$ must not be a positive function.

Lemma 2 [1,2] Let the set $\mathbf{A}, \mathbf{B} \subset \mathbf{R}$, $\varphi : \mathbf{B}^n \to \mathbf{R}$, $f : \mathbf{A} \to \mathbf{B}$ and $\psi(x_1, x_2, \ldots, x_n) = \varphi(f(x_1), f(x_2), \ldots, f(x_n)) : \mathbf{A}^n \to \mathbf{R}$.

(a) If f is convex and φ is increasing and Schur-convex, then ψ is Schur-convex;

(b) If f is concave and φ is increasing and Schur-concave, then ψ is Schur-concave.

Lemma 3 Let the set $\Omega \subset \mathbf{R}_+^n$. The function $\varphi : \Omega \to \mathbf{R}_+$ is differentiable.

(a) If φ is increasing and Schur-convex, then φ is Schur geometrically convex;

(b) If φ is decreasing and Schur-concave, then φ is Schur geometrically concave.

Lemma 4 Let the set $\Omega \subset \mathbf{R}_+^n$. The function $\varphi : \Omega \to \mathbf{R}_+$ is differentiable.

(a) If φ is increasing and Schur-convex, then φ is Schur-harmonically convex;

(b) If φ is decreasing and Schur-concave, then φ is Schur-harmonically concave.

Lemma 5 [1,2] Let $x = (x_1, x_2, \ldots, x_n) \in \mathbf{R}^n$. Then

$$(A(x), A(x), \ldots, A(x)) \prec x = (x_1, x_2, \ldots, x_n) \tag{5}$$

where $A(x) = \frac{1}{n} \sum_i^n x_i$.

Lemma 6 [29] Let

$$q(t) = \frac{u^t - 1}{t}$$

If $u > 1$, then $q(t)$ is a log-convex function on $\mathbf{R}_+$.

87.3　Proof of main results

Proof of Theorem 1

For the case of $r = 1$ and $r = 2$, it is easy to prove that $c_n^* \left(f(\boldsymbol{x}), r \right)$ is Schur-convex on I^n.

Now consider the case of $r \geqslant 3$. By the symmetry of $c_n^* \left(f(\boldsymbol{x}), r \right)$, without loss of generality, we can set $x_1 > x_2$

$$
c_n^* \left((\boldsymbol{x}), r \right) = \prod_{\substack{i_1+i_2+\cdots+i_n=r \\ i_1 \neq 0, i_2=0}} \sum_{j=1}^{n} i_j f(x_j) \times \prod_{\substack{i_1+i_2+\cdots+i_n=r \\ i_1=0, i_2 \neq 0}} \sum_{j=1}^{n} i_j f(x_j) \times
$$

$$
\prod_{\substack{i_1+i_2+\cdots+i_n=r \\ i_1 \neq 0, i_2 \neq 0}} \sum_{j=1}^{n} i_j f(x_j) \times \prod_{\substack{i_1+i_2+\cdots+i_n=r \\ i_1=0, i_2=0}} \sum_{j=1}^{n} i_j f(x_j)
$$

Then

$$
\frac{\partial c_n^* \left(f(\boldsymbol{x}), r \right)}{\partial x_1} = c_n^* \left(f(\boldsymbol{x}), r \right) \times
$$

$$
\left(\sum_{\substack{i_1+i_2+\cdots+i_n=r \\ i_1 \neq 0, i_2=0}} \frac{i_1 f'(x_1)}{\sum_{j=1}^{n} i_j f(x_j)} + \sum_{\substack{i_1+i_2+\cdots+i_n=r \\ i_1 \neq 0, i_2 \neq 0}} \frac{i_1 f'(x_1)}{\sum_{j=1}^{n} i_j f(x_j)} \right) +
$$

$$
= c_n^* \left(f(\boldsymbol{x}), r \right) \left(\sum_{\substack{k+k_3+\cdots+k_n=r \\ k \neq 0}} \frac{k f'(x_1)}{k f(x_1) + \sum_{j=3}^{n} i_j f(x_j)} \right.
$$

$$
\left. \sum_{\substack{k+m+i_3+\cdots+i_n=r \\ k \neq 0, m \neq 0}} \frac{k f'(x_1)}{k f(x_1) + m f(x_2) + \sum_{j=3}^{n} i_j f(x_j)} \right) \tag{6}
$$

By the same arguments

$$
\frac{\partial c_n^* \left(f(\boldsymbol{x}), r \right)}{\partial x_2} = c_n^* \left(f(\boldsymbol{x}), r \right)
$$

$$
= c_n^* \left(f(\boldsymbol{x}), r \right) \left(\sum_{\substack{k+k_3+\cdots+k_n=r \\ k \neq 0}} \frac{k f'(x_2)}{k f(x_2) + \sum_{j=3}^{n} i_j f(x_j)} + \right.
$$

$$
\left. \sum_{\substack{k+m+i_3+\cdots+i_n=r \\ k \neq 0, m \neq 0}} \frac{k f'(x_2)}{k f(x_2) + m f(x_1) + \sum_{j=3}^{n} i_j f(x_j)} \right) \tag{7}
$$

$$
\frac{\partial c_n^* \left(f(\boldsymbol{x}), r \right)}{\partial x_1} - \frac{\partial c_n^* \left(f(\boldsymbol{x}), r \right)}{\partial x_2} = c_n^* \left(f(\boldsymbol{x}), r \right) (A_1 + A_2)
$$

where

$$A_1 = \sum_{\substack{k+k_3+\cdots+k_n=r \\ k\neq 0}} \left(\frac{kf'(x_1)}{kf(x_1) + \sum\limits_{j=3}^{n} i_j f(x_j)} - \frac{kf'(x_2)}{kf(x_2) + \sum\limits_{j=3}^{n} i_j f(x_j)} \right)$$

$$= k \sum_{\substack{k+k_3+\cdots+k_n=r \\ k\neq 0}} \frac{k(f(x_2)f'(x_1)-f(x_1)f'(x_2))+(f'(x_1)-f'(x_2))\sum\limits_{j=3}^{n} i_j f(x_j)}{(kf(x_1)+\sum\limits_{j=3}^{n} i_j f(x_j))(kf(x_2)+\sum\limits_{j=3}^{n} i_j f(x_j))}$$

$$(8)$$

and

$$A_2 = \sum_{\substack{k+m+i_3+\cdots+i_n=r \\ k\neq 0, m\neq 0}} \left(\frac{kf'(x_1)}{kf(x_1) + mf(x_2) + \sum\limits_{j=3}^{n} i_j f(x_j)} - \right.$$

$$\left. \frac{kf'(x_2)}{kf(x_2) + mf(x_1) + \sum\limits_{j=3}^{n} i_j f(x_j)} \right)$$

$$= k \sum_{\substack{k+m+i_3+\cdots+i_n=r \\ k\neq 0, m\neq 0}} \frac{\delta}{(kf(x_1)+mf(x_2)+\sum\limits_{j=3}^{n} i_j f(x_j))(kf(x_2)+mf(x_1)+\sum\limits_{j=3}^{n} i_j f(x_j))}$$

where

$$\delta = k(f(x_2)f'(x_1) - f(x_1)f'(x_2)) + m(f(x_1)f'(x_1)-$$

$$f(x_2)f'(x_2)) + (f'(x_1) - f'(x_2)) \sum_{j=3}^{n} i_j f(x_j)$$

(a) Since the log-convex function must be convex function, so $f'(x_1) - f'(x_2) \geqslant 0$ and $f(x_2)f'(x_1) - f(x_1)f'(x_2) \geqslant 0$, and since $(f(x)f'(x))' = (f'(x))^2 + f(x)f''(x) \geqslant 0$, so $f(x_1)f'(x_1) - f(x_2)f'(x_2) \geqslant 0$, and then $A_1 \geqslant 0$ and $A_2 \geqslant 0$. For $\boldsymbol{x} \in I^n$, we have

$$\frac{\partial c_n^*\,(f(\boldsymbol{x}),r)}{\partial x_1} - \frac{\partial c_n^*\,(f(\boldsymbol{x}),r)}{\partial x_2} \geqslant 0$$

by Lemma 1, it follows that $c_n^*\,(f(\boldsymbol{x}),r)$ is Schur-convex on I^n.

(b) By Proposition 4, we know that $c_n^*(\boldsymbol{x},r)$ is increasing and Schur-concave on $\mathbf{R}_+^n$. Since f is concave, from (b) in Lemma 4, it follows that $c_n^*\,(f(\boldsymbol{x}),r)$ is Schur-concave on I^n.

The proof of Theorem 1 is completed.

Proof of Theorem 2

Theorem 2 can be proved by Theorem 1 combined with Lemma 3.

The proof of Theorem 2 is completed.

Proof of Theorem 3

Theorem 3 can be proved by Theorem 1 combined with Lemma 4.

The proof of Theorem 3 is completed.

87.4　Applications

Let

$$c_n^* \left(\frac{1}{x}, r \right) = \prod_{i_1+i_2+\cdots+i_n=r} \sum_{j=1}^{n} i_j \left(\frac{1}{x_j} \right) \tag{9}$$

Theorem 4 The symmetric function $c_n^* \left(\frac{1}{x}, r \right)$ is Schur-convex on $\mathbf{R}_+^n$. If r is an even integer (or odd integer, respectively), then $c_n^* \left(\frac{1}{x}, r \right)$ is Schur-convex (or Schur-concave, respectively) on $\mathbf{R}_-^n$.

Proof Let $f(x) = \frac{1}{x}$. Then $(\ln f(x))'' = \frac{1}{x^2}$, so $f(x)$ is log-convex on $\mathbf{R}_+$, by (a) in Theorem 1, it follows that $c_n^* \left(\frac{1}{x}, r \right)$ is Schur-convex on $\mathbf{R}_+^n$.

For $x \in \mathbf{R}_-^n$, $-x \in \mathbf{R}_+^n$, so $c_n^* \left(\frac{1}{-x}, r \right)$ is Schur-convex on $\mathbf{R}_-^n$. But

$$c_n^* \left(\frac{1}{-x}, r \right) = (-1)^r c_n^* \left(\frac{1}{x}, r \right)$$

This means that if r is an even integer, then

$$c_n^* \left(\frac{1}{x}, r \right) = c_n^* \left(\frac{1}{-x}, r \right)$$

is Schur-convex on $\mathbf{R}_-^n$.

If r is an odd integer, then

$$c_n^* \left(\frac{1}{x}, r \right) = -c_n^* \left(\frac{1}{-x}, r \right)$$

is Schur-concave on $\mathbf{R}_-^n$.

The proof of Theorem 4 is completed.

By Theorem 4 and majorizing relation (7), it is not difficult to prove the following corollary.

Corollary 1 If $x \in \mathbf{R}_+^n$ or r is an even integer and $x \in \mathbf{R}_-^n$, then we have

$$\prod_{i_1+i_2+\cdots+i_n=r} \sum_{j=1}^{n} i_j \left(\frac{1}{x_j}\right) \geqslant \left(\frac{r}{A_n(\boldsymbol{x})}\right)^{\binom{n+r-1}{r}} \tag{10}$$

where $A_n(\boldsymbol{x}) = \frac{1}{n}\sum_{i=1}^{n} x_i$ and $\binom{n+r-1}{r} = \frac{(n+r-1)!}{r!((n+r-1)-r)!}$. If r is odd and $\boldsymbol{x} \in \mathbf{R}_-^n$, then the inequality (10) is reversed.

Let

$$c_n^* \left(\frac{\boldsymbol{x}}{1-\boldsymbol{x}}, r\right) = \prod_{i_1+i_2+\cdots+i_n=r} \sum_{j=1}^{n} i_j \left(\frac{x_j}{1-x_j}\right) \tag{11}$$

Theorem 5 The symmetric function $c_n^* \left(\frac{\boldsymbol{x}}{1-\boldsymbol{x}}, r\right)$ is Schur-convex, Schur-geometrically convex and Schur-harmonically convex on $[\frac{1}{2}, 1]^n$.

Proof Let $g(x) = \frac{x}{1-x}$. Then $(\ln g(x))'' = \frac{2x-1}{x^2(1-x)^2}$, so $f(x)$ is log-convex on $[\frac{1}{2}, 1]$, by Theorem 1 (a), it follows that $c_n^* \left(\frac{\boldsymbol{x}}{1-\boldsymbol{x}}, r\right)$ is Schur-convex on $[\frac{1}{2}, 1]^n$. Noting that $g(x)$ is increasing on $[\frac{1}{2}, 1]$, by (a) in Theorem 2 and (a) in Theorem 3, it follows that $c_n^* \left(\frac{\boldsymbol{x}}{1-\boldsymbol{x}}, r\right)$ is Schur-geometrically convex and Schur-harmonically convex on $[\frac{1}{2}, 1]^n$.

The proof of Theorem 5 is completed.

From the majorizing relation (7), the following majorizing relation is established.

$$(\log G_n(\boldsymbol{x}), \log G_n(\boldsymbol{x}), \ldots, \log G_n(\boldsymbol{x})) \prec (\log x_1, \log x_2, \ldots, \log x_n)$$

By this majorizing relation and Theorem 5, it is not difficult to prove the following corollary.

Corollary 2 If $\boldsymbol{x} \in [\frac{1}{2}, 1]^n$, then we have

$$\prod_{i_1+i_2+\cdots+i_n=r} \sum_{j=1}^{n} i_j \left(\frac{x_j}{1-x_j}\right) \geqslant \left(\frac{rG_n(\boldsymbol{x})}{1-G_n(\boldsymbol{x})}\right)^{\binom{n+r-1}{r}} \tag{12}$$

where $G_n(\boldsymbol{x}) = \sqrt[n]{\prod_{i=1}^{n} x_i}$.

Let

$$c_n^* \left(\frac{1+\boldsymbol{x}}{1-\boldsymbol{x}}, r\right) = \prod_{i_1+i_2+\cdots+i_n=r} \sum_{j=1}^{n} i_j \left(\frac{1+x_j}{1-x_j}\right) \tag{13}$$

Theorem 6

(a) The symmetric function $c_n^* \left(\frac{1+\boldsymbol{x}}{1-\boldsymbol{x}}, r\right)$ is Schur-convex, Schur-geometrically convex and Schur-harmonically convex on $(0, 1)^n$;

(b) If r is an even integer (or odd integer, respectively), then $c_n^* \left(\frac{1+x}{1-x}, r \right)$
is Schur-convex (or Schur-concave, respectively) on $(1, +\infty)^n$.

Proof (a) Let $h(x) = \frac{1+x}{1-x}$. Then $(\ln h(x))'' = \frac{4x}{(1+x)^2(1-x)^2}$, so $f(x)$ is
log-convex on $(0, 1)$, by Theorem 1 (a), it follows that $c_n^* \left(\frac{1+x}{1-x}, r \right)$ is Schur-
convex on $(0, 1)^n$. Noting that $h(x)$ is increasing on $(0, 1)^n$, by (a) in Theo-
rem 2 and (a) in Theorem 3, it follows that $c_n^* \left(\frac{1+x}{1-x}, r \right)$ is Schur-geometrically
convex and Schur-harmonically convex on $(0, 1)^n$.

(b) For $x \in (1, +\infty)$, we consider

$$c_n^* \left(\frac{1+x}{x-1}, r \right) = \prod_{i_1+i_2+\cdots+i_n=r} \sum_{j=1}^n i_j \left(\frac{1+x_j}{x_j-1} \right) \tag{14}$$

Let $h_1(x) = \frac{1+x}{x-1}$. Then $(\ln h_1(x))'' = \frac{4x}{(1+x)^2(x-1)^2}$, so $f(x)$ is log-convex on
$(1, +\infty)$, by (a) in Theorem 1, it follows that $c_n^* \left(\frac{1+x}{x-1}, r \right)$ is Schur-convex on
$(1, +\infty)^n$.

Noting that

$$c_n^* \left(\frac{1+x}{1-x}, r \right) = (-1)^r c_n^* \left(\frac{1+x}{x-1}, r \right)$$

Combining the Schur-convexity of $c_n^* \left(\frac{1+x}{x-1}, r \right)$, we can get (b) in Theorem
6.

The proof of Theorem 6 is completed.

Let

$$c_n^* \left(\frac{1}{x} - x, r \right) = \prod_{i_1+i_2+\cdots+i_n=r} \sum_{j=1}^n i_j \left(\frac{1}{x_j} - x_j \right) \tag{15}$$

Theorem 7

(a) If r is an even integer (or odd integer, respectively), then $c_n^* \left(\frac{1}{x} - x, r \right)$
is Schur-concave (or Schur-convex, respectively) on $\mathbf{R}_+^n$.

(b) The symmetric function $c_n^* \left(\frac{1}{x} - x, r \right)$ is Schur-concave on $\mathbf{R}_-^n$.

(c) If r is an even integer, then $c_n^* \left(\frac{1}{x} - x, r \right)$ is Schur-geometrically concave and Schur-harmonically concave on $(-\infty, 1]^n$.

Proof First consider

$$c_n^* \left(x - \frac{1}{x}, r \right) = \prod_{i_1+i_2+\cdots+i_n=r} \sum_{j=1}^n i_j \left(x_j - \frac{1}{x_j} \right)$$

(a) Let $p(x) = x - \frac{1}{x}$. Then $p''(x) = -\frac{2}{x^3}$, so $f(x)$ is concave on $\mathbf{R}_+$, by Theorem 1 (b), it follows that $c_n^* \left(x - \frac{1}{x}, r \right)$ is Schur-concave on $\mathbf{R}_+^n$.

Noting that

$$c_n^* \left(\frac{1}{x} - x, r \right) = (-1)^n c_n^* \left(x - \frac{1}{x}, r \right)$$

combining the Schur-concavity of $c_n^* \left(\frac{1}{x} - x, r \right)$, we can get (a) in Theorem 7.

(b) Noting that

$$c_n^* \left(\frac{1}{-x} - (-x), r \right) = (-1)^r c_n^* \left(\frac{1}{x} - x, r \right)$$

combining (a) in Theorem 7, it is not difficult to verify that (b) in Theorem 7 holds.

(c) Not difficult to verify that $p(x) = x - \frac{1}{x}$ is nonnegative and decreasing on $(-\infty, 1]$, by Lemma 5 and Lemma 6, from (a) and (b) in Theorem 7, it following that (c) in Theorem 7 holds.

The proof of Theorem 7 is completed.

For $u > 1$, Let

$$c_n^* \left(\frac{u^x - 1}{x}, r \right) = \prod_{i_1 + i_2 + \cdots + i_n = r} \sum_{j=1}^n i_j \left(\frac{u^{x_j} - 1}{x_j} \right) \tag{16}$$

Theorem 8 The symmetric function $c_n^* \left(\frac{u^x - 1}{x}, r \right)$ is Schur-convex, Schur-geometrically convex and Schur-harmonically convex on $\mathbf{R}_+^n$ for $u > 1$.

Proof Let $q(t) = \frac{u^t - 1}{t}$. Then from Lemma 8 and (a) in Theorem 1, it follows that $c_n^* \left(\frac{u^x - 1}{x}, r \right)$ is Schur-convex on $\mathbf{R}_+^n$ for $u > 1$.

Since

$$q'(t) = \frac{s(t)}{t^2}$$

where $s(t) = u^t(t \log u - 1) + 1$, $s'(t) = u^t \log u \log u^t > 0$, for $u > 1$ and $t > 0$, so $s(t) \geqslant s(0) = 0$, and then $q'(t) \geqslant 0$, this is $q(t)$ is increasing on $\mathbf{R}_+^n$, by (a) in Theorem 2 and (a) in Theorem 3, it follows that $c_n^* \left(\frac{u^x - 1}{x}, r \right)$ is Schur-geometrically convex and Schur-harmonically convex on $\mathbf{R}_+^n$.

The proof of Theorem 8 is completed.

From the majorizing relation (7), the following majorizing relation is established

$$\left(\frac{1}{H_n(\boldsymbol{x})}, \frac{1}{H_n(\boldsymbol{x})}, \cdots, \frac{1}{H_n(\boldsymbol{x})}\right) \prec \left(\frac{1}{x_1}, \frac{1}{x_2}, \cdots, \frac{1}{x_n}\right)$$

By this majorizing relation and Theorem 8, it is not difficult to prove the following corollary.

Corollary 3 If $\boldsymbol{x} = (x_1, x_2, \ldots, x_n) \in \mathbf{R}_+^n$ and $u > 1$, then

$$\prod_{i_1+i_2+\cdots+i_n=r} \sum_{j=1}^{n} i_j \left(\frac{u^{x_j}-1}{x_j}\right) \geqslant \left(\frac{r(u^{H_n(\boldsymbol{x})}-1)}{H_n(\boldsymbol{x})}\right)^{\binom{n+r-1}{r}} \tag{17}$$

where $H_n(\boldsymbol{x}) = \dfrac{n}{\sum\limits_{i=1}^{n} x_i^{-1}}$.

Discovering and judging Schur convexity of various symmetric functions is an important subject in the study of the majorization theory. In recent years, many domestic scholars have made a lot of achievements in this field (see [31]-[37]).

References

[1] Marshall A W , Olkin I , Arnold B C . Inequalities: Theory of Majorization and Its Application[M]. 2nd ed. New York: Springer, 2011.

[2] Wang B Y . Foundations of Majorization Inequalities[M]. Beijing: Beijing Normal University Press, 1990.

[3] Zhang X M . Geometrically Convex Functions[M]. Hefei: Anhui University Press, 2004.

[4] Chu Y M , Zhang X M , Wang G D . The Schur geometrical convexity of the extended mean values[J]. Journal of Convex Analysis, 2008, 15(4): 707-718.

[5] Guan K Z . A class of symmetric functions for multiplicatively convex function[J]. Mathematical Inequalities & Applications, 2007, 10(4): 745-753.

[6] Sun T C , Lv Y P , Chu Y M . Schur multiplicative and harmonic convexities of generalized Heronian mean in n variables and their applications[J]. International Journal of Pure and Applied Mathematics, 2009, 55(1): 25-33.

[7] Chu Y M , Sun T C . The Schur harmonic convexity for a class of symmetric functions[J]. Acta Mathematica Scientia, 2010, 30B(5): 1501-1506.

[8]   Chu Y M , Wang G D , Zhang X H . The Schur multiplicative and harmonic convexities of the complete symmetric function[J]. Mathematische Nachrichten, 2011, 284(5-6): 653-663.

[9]   Chu Y M , Lv Y P. The Schur harmonic convexity of the Hamy symmetric function and its applications[J]. Journal of Inequalities and Applications, 2009, Article ID 838529, 10 pages.

[10] Xia W F , Chu Y M. Schur-convexity for a class of symmetric functions and its applications[J]. Journal of Inequalities and Applications, 2009, Article ID 493759, 15 pages.

[11] Guan K Z. Schur-convexity of the complete symmetric function[J]. Mathematical Inequalities & Applications, 2006, 9(4): 567-576.

[12] Sun M B, Chen N B, S H Li. Some properties of a class of symmetric functions and its applications[J]. Mathematische Nachrichten, 2014, doi: 10.1002/mana.201300073.

[13] Xia W F, Chu Y M. Schur convexity and Schur multiplicative convexity for a class of symmetric functions with applications[J]. Ukrainian Mathematical Journal, 2009, 61(10): 1541-1555.

[14] Ionel Roventa. Schur convexity of a class of symmetric functions[J]. Annals of the University of Craiova, Mathematics and Computer Science Series, 2010, 37(1): 12-18.

[15] Xia W F, Chu Y M. On Schur convexity of some symmetric functions[J]. Journal of Inequalities and Applications, 2010, Article ID 543250, 12 pages.

[16] Meng J X, Chu Y M, Tang X M. The Schur-harmonic-convexity of dual form of the Hamy symmetric function[J]. Matematiqki Vesnik, 2010, 62(1): 37-46.

[17] Chu Y M, Xia W F, Zhao T H. Some properties for a class of symmetric functions and applications[J]. Journal of Mathematical Inequalities, 2011, 5(1): 1-11.

[18] Guan K Z, Guan R K. Some properties of a generalized Hamy symmetric function and its applications[J]. Journal of Mathematical Analysis and Applications, 2011, 376: 494-505.

[19] Qian W M . Schur convexity for the ratios of the Hamy and generalized Hamy symmetric functions[J]. Journal of Inequalities and Applications, 2011, 2011:131, doi:10.1186/1029-242X-2011-131.

[20] Chu Y M , Xia W F , Zhang X H . The Schur concavity, Schur multiplicative and harmonic convexities of the second dual form of the
Hamy symmetric function with applications[J]. Journal of Multivariate Analysis, 2012, 105(1): 412-421.

[21] Ionel Rovenţa. A note on Schur-concave functions[J]. Journal of Inequalities and Applications, 2012, 2012:159, doi:10.1186/1029-242X-
2012-159.

[22] Xia W F, Zhang X H, Wang G D, Chu Y M. Some properties for a
class of symmetric functions with applications[J]. Indian J. Pure Appl.
Math., 2012, 43(3): 227-249.

[23] Shi H N, Zhang J. Schur-convexity of dual form of some symmetric
functions[J]. Journal of Inequalities and Applications, 2013, 2013: 295,
doi:10.1186/1029-242X-2013-295.

[24] Zhang K S, Shi H N. Schur convexity of dual form of the complete symmetric function[J]. Mathematical Inequalities & Applications, 2013,
16(4): 963-970.

[25] Shi H N, Zhang J, Ma Q H. Schur-convexity, Schur-geometric and
Schur-harmonic convexity for a composite function of complete symmetric function[J]. SpringerPlus, 2016(5): 296.

[26] Zhang X H, Chu Y M. New discussion to analytic Inequalities[M].
Harbin: Harbin Institute of Technology Press, 2009.

[27] Shi H N. Majorization Theory and Analytical Inequalities[M]. Harbin:
Harbin Institute of Technology Press, 2012.

[28] Shi H N. Schur-Convex Functions and Inequalities[M]. Harbin: Harbin
Institute of Technology Press, 2012.

[29] Shi H N , Zhang J . Schur-convexity, Schur-geometric and harmonic
convexities of dual form of a class symmetric functions[J]. J.Math.
Inequal., 2014, 8(2): 349-358.

[30] Shi Huan-nan. Schur-Convex Functions and Inequalities: Volume 1:
Concepts, Properties, and Applications in Symmetric Function In-
equalitics[M]. Harbin: Harbin Institute of Technology PressLtd, 2019.

[31] Shi Huan-nan. Schur-Convex Functions and Inequalities: Volume 2:
Applications in Inequalities[M]. Harbin: Harbin Institute of Technology Press, 2019.

[32] Wu S H, Chu Y M. Schur m-power convexity of generalized geometric Bonferroni mean involving three parameters[J]. J. Inequal. Appl., 2019, 2019, Article ID 57, 11 pages.

[33] Xia W F, Chu Y M. The Schur convexity of Gini mean values in the sense of harmonic mean[J]. Acta Math. Sci., 2011, 31B(3): 1103-1112.

[34] Xia W F, Chu Y M, Wang G D. Necessary and sufficient conditions for the Schur harmonic convexity or concavity of the extended mean values[J]. Rev. Un. Mat. Argentina, 2011, 52(1): 121-132.

[35] Chu Y M, Wang G D, Zhang X H. Schur convexity and Hadamard's inequality[J]. Math. Inequal. Appl., 2010, 13(4): 725-731.

[36] Chu Y M, Xia W F, Zhao T H. Schur convexity for a class of symmetric functions[J]. Sci. China Math., 2010, 53(2): 465-474.

[37] Guessab, Allal Schmeisser, Gerhard. Sharp integral inequalities of the Hermite-Hadamard type[J]. J. Approx. Theory, 2002, 115 (2): 260-288.

论文与著作目录

作者在国内外发表论文160篇, 其中半数刊于国内核心期刊或国际期刊, 37篇被 *Science Citation Index* 或 *Engineering Index* 收录.

论文目录:

1. 石焕南. 也谈巧算"百分比"[J]. 数理统计与管理, 1990, (3): 56.

2. 石焕南. 一个简单的证明[J]. 数学通报, 1991 (4): 40.

3. 石焕南. 一个组合恒等式的概率证明[J]. 厦门数学通讯, 1991 (1): 2-3.

4. 概率方法在级数求和中的应用[J]. 数学通报, 1992 (3): 34-35,18.

5. 石焕南,石敏琪. 概率方法在不等式证明中的应用[J]. 数学通报, 1992, (7): 34-37.

6. 石焕南. 一道全俄竞赛题的概率证法[J]. 中学数学, 1992 (11): 35.

7. 石焕南. 关于一类不等式的再推广和引申[J]. 数学通报, 1993 (1): 30-31.

8. 石焕南. 也谈一个不等式命题[J]. 数学通报, 1993 (7): 44-46.

9. 石焕南. 一道IMO试题的新推广[J]. 湖南数学通讯, 1994 (5): 35-36.

10. 刘国瑞, 石焕南. 正选型婚配群体中隐性纯合体频率的估算公式[J]. 北京联合大学学报(自然科学版), 1994, 8 (1): 50-53.

11. 石焕南. 关于对称函数的一类不等式[J]. 数学通报, 1996 (3): 38-40.

12. 石焕南, 石敏琪. 对称平均值基本定理应用数例[J]. 数学通报, 1996, (10): 44-45.

13. 石焕南. 代数不等式概率证法举例[J]. 工科数学, 1996,12 (3): 146-149.

14. 石敏琪, 石焕南. 关于广义对数平均的一个不等式[J]. 数学通报, 1997, (5): 37-38.

15. 石焕南. 一道不等式习题的应用[J]. 数学教学研究, 1997年优秀论文专辑: 56-57.

16. 石焕南. 一个对称函数下界的加强[J]. 数学通报, 1998 (5): 46.

17. 石焕南. 幂方根不等式的推广[J]. 河北理科教学研究, 1998 (2): 20-21.

18. 石焕南. 一类对称函数不等式的控制证明[J].成都大学学报(自然科学版),1998, 17 (4): 22-24

19. 石焕南. 一类对称函数不等式的加细与推广[J]. 数学的实践与认识, 1999, 29 (4): 81-85.

20. 石焕南. 一类对称函数不等式的控制证明[J]. 工科数学, 1999,15 (3): 140-142.

21. 石焕南. 初等对称函数差的Schur凸性[J]. 湖南教育学院学报, 1999, 17 (5): 135-138.

22. 石焕南. 积分不等式概率证法举例[J]. 高等数学研究, 1999 年专刊: 31-33.

23. 石焕南. 一类对称函数不等式的加强、推广及应用[J]. 北京联合大学学报(自然科学版), 1999, 13 (2): 51-55.

24. 石焕南. 与四面体内点有关的一类不等式[J]. 数学通报, 1999 (6): 45-46.

25. 石焕南. 整值随机变量期望的一个表示式的应用与推广[J]. 辽宁师范大学学报(自然科学版), 2000, 23 (1): 102-105.

26. 石焕南. 也谈改变观点的创造性技巧[J]. 数学通报, 2000 (2): 42-43.

27. 石焕南. 一个平均值不等式的别证与推广[J]. 河南师范大学学报(自然科学版), 2000, 20 (专刊): 8-9.

28. 石焕南. Bonferroni不等式的推广及其应用[J]. 北京联合大学学报(自然科学版), 2000, 13 (2): 51-55.

29. 石焕南. 优超理论的一个简单命题及其几何应用[C]// 杨学枝.不等式研. 拉萨: 西藏人民出版社, 2000: 323-327.

30. 石焕南. 两种特殊四面体中的三角不等式[C]// 杨学枝. 不等式研究. 拉萨: 西藏人民出版社, 2000: 343-348.

31. 文家金, 石焕南. Maclaurin不等式的最优化加强[J]. 成都大学学报(自然科学版), 2000, 19 (3): 1-8.

32. 石焕南, 贾玉友. 一个数学问题的推广[J]. 数学通报, 2000 (11):40-41.

33. 贾玉友, 石焕南. 一类函数的最值问题[J]. 中学数学, 2000, (11):32.

34. 石焕南. Klamkin不等式的多边形推广[J]. 安徽教育学院学报, 2000,18(6): 12-14.

35. 石焕南, 文家金, 周步骏. 关于幂平均值的一个不等式[J]. 数学的实践与认识, 2001, 31(2): 327-330.

36. 石焕南. 一个分析不等式的推广[J]. 成都大学学报(自然科学版), 2001, 20(1): 5-7.

37. 石焕南, 李大矛. 一类三角不等式的控制证明[J]. 滨州师专学报, 2001, 17(2): 31-33.

38. 石焕南. 初等对称函数对偶式的Schur-凹性及其应用[J]. 东北师范大学学报(自然科学版), 2001, 33 (增刊): 24-27.

39. 石焕南, 李大矛. 凸数列的一个等价条件及其应用[J]. 曲阜师范大学学报(自然科学版), 2001,27(4): 4-6.

40. 石焕南. 一类无理不等式[J]. 数学通报2001 (12):39-40.

41. 石焕南. Wierstrass不等式的新推广[J]. 数学的实践与认识, 2002, 32 (1):132-135.

42. 石焕南. Turner‐Conway不等式的概率证明[J]. 商丘师范学院学报, 2002, (2): 51-52.

43. 石焕南, 范淑香. 两个组合恒等式的概率证明[J]. 山东师范大学学报(自然科学版), 2002, 17(1): 12-14.

44. 石焕南, 范淑香. 组合数的一项性质的概率证明[J]. 数学通报, 2002(6):43.

45. 石焕南, 石敏琪. 一道新编应用题的概率解法[J]. 数学通讯, 2002(15): 30.

46. 石焕南. Popoviciu不等式的新推广[J]. 四川师范大学学报(自然科学版), 2002, 25(5): 510-511.

47. 石焕南, 石敏琪. 一道分式不等式的概率证明及推广[J]. 中学数学, 2002,(11): 47.

48. 罗钊, 文家金, 石焕南. 含k-Brocard点的一类几何不等式[J]. 四川大学学报(自然科学版), 2002, 39(6)：971-976.

49. 吴善和, 石焕南. Shc66的再加强[J]. 中学数学, 2003 (3):42.

50. 石焕南, 杨蕾. 高中数学中的不等式的概率证明[J]. 数学通讯, 2003 (7): 11-12.

51. 石焕南, 续铁权, 顾春. 整幂函数不等式的控制证明[J]. 商丘师范学院学报，2003, 19 (2):46-48.

52. 吴善和, 石焕南. 平均值不等式的推广及应用[J]. 贵州教育学院学报(自然科学版), 2003, 14 (2):14-16.

53. 吴善和, 石焕南. 再探关于四面体的Nesbitt不等式[J]. 福建中学数学, 2003, (5):20.

54. 吴善和, 石焕南. 一类无理不等式的控制证明[J]. 首都师范大学学报(自然科学版), 2003, 24 (3):13-16.

55. 吴善和, 石焕南. 凸序列不等式的控制证明[J]. 数学的实践与认识, 2003, 33(12)：132-137.

56. 吴善和, 石焕南. 一个分式不等式的多参数推广及应用[J]. 中学数学研究, 2003(11)：16-17.

57. 石焕南, 徐坚, 段红梅. 一类组合题的概率解法[J]. 现代中学数学2003(1)：27-29.

58. 吴善和, 石焕南. 一个无理不等式的简证及类似[J]. 福建中学数学, 2004(2)：20.

59. 石焕南, 李大矛. Extensions and refinements of Adamovic's inequality[J]. 数学季刊, 2004, 19 (1): 35-40.

60. 石焕南. 极限 $\lim\limits_{n\to\infty}(1+1/n)^n$ 存在的控制证明[J]. 云南师范大学学报(自然科学版), 2004, 24 (2)：13-15.

61. 石焕南. 凸数列的一个等价条件及其应用, II [J]. 数学杂志, 2004, 24(4): 390-394.

62. WU SHAN-HE, SHI HUAN-NAN. Generalizations of a class of inequalities for products[J]. J. Ineg. Pure Appl. Math., 2004, 5(3) Article 77.

63. 王挽澜, 文家金, 石焕南. 幂平均不等式的最优值[J]. 数学学报，2004, 47(6)：1053-1062.

64. 石焕南, 张鉴, 徐坚. 一类积分不等式的控制证明[J]. 首都师范大学学报(自然科学版), 2004, 25 (4): 11-13.

65. 石焕南, 李康海, 石敏琪. 一类代数不等式的概率证明[J]. 北京联合大学学报(自然科学版), 2005, 18 (1):42-44

66. 吴善和, 石焕南. 周界中点三角形两个面积不等式的加强[J]. 中学数学教学参考, 2005(4)：61.

67. 吴善和, 石焕南. 两个涉及动点的几何不等式的推广[J]. 福建中学数学, 2005 (6):15-16.

68. SHI HUAN-NAN. Exponential generalization of Newman's inequality and Klamkin's inequality[J]. 东北数学, 2005, 21(4)：431-438.

69. 石焕南, 石敏琪.一道典型概率例题的教学实录[J]. 高等数学研究, 2005, 8 (3)：27-28.

70. 李世杰, 石焕南.凸函数初探[J]. 北京联合大学学报(自然科学版), 2005, 18 (3): 19-24, 36.

71. SHI HUAN-NAN, QI FENG, WU SHAN-HE. An alternative note on the Schur-convexity of the extended mean values[J]. Math. Inequal. Appl., 2006，9 (2)：219-224.

72. 江永明, 石焕南. 一个双参数二元不等式的推广[J]. 北京联合大学学报(自然科学版), 2006, 19 (2): 68-72.

73. SHI HUAN-NAN. Refinements of an inequality for the rational fraction[J]. 纯粹数学与应用数学, 2006，22(2)：256-262.

74. 李大矛，石焕南. 一个二元平均值不等式猜想的新证明[J]. 数学的实践与认识, 2006, 36(4): 278-283

75. 李大矛, 顾春, 石焕南. Heron平均幂型推广的Schur凸性[J]. 数学的实践与认识, 2006, 36 (9): 387-390.

76. SHI HUAN-NAN. Majorized proof and refinement of the discrete Steffensen's inequality[J]. Taiwanese Journal of Mathematics, 2007, 11 (4): 1203-1208.

77. SHI HUAN-NAN, Schur-convex functions relate to Hadamard-type inequalities[J]. J.Math. Inequal., 2007, 1 (1): 127-136.

78. 石焕南, 李世杰.关于$\langle l, t\rangle$对数性凸函数的几个不等式[J]. 浙江万里学院学报, 2007,(20) 2: 6-12.

79. 续铁权, 石焕南. 两个凸函数单调平均不等式的改进[J]. 数学的实践与认识, 2007, 37(19): 150-154.

80. SHI HUAN-NAN, WU SHAN-HE. Refinement of an inequality for the generalized logarithmic mean[J]. 数学季刊, 2008,23(4): 594-599.

81. 石焕南. Bernoulli不等式的控制证明及推广[J]. 北京联合大学学报(自然科学版), 2008, 22 (2): 58-61.

82. 顾春, 石焕南. 反向Chrystal不等式[J]. 数学的实践与认识, 2008, 38 (13):163-167.

83. SHI HUAN-NAN. Generalizations of Bernoulli's Inequality with Applications[J]. J. Math. Inequal.,2008, 2 (1): 101-107.

84. 石焕南. 涉及Schwarz积分不等式的Schur-凸函数[J]. 湖南理工学院学报(自然科学版), 2008, 21 (4): 1-3.

85. SHI HUAN-NAN, MIHALY BENCZE, WU SHAN-HE, LI DA-MAO. Schur convexity of generalized Heronian means involving two parameters[J]. J. Inequal. Appl., Volume 2008, Article ID 879273, 9 pages doi:10.1155/2008/879273.

86. SHI HUAN-NAN, LI DA-MAO, GU CHUN. Schur-convexity of a mean of convex function[J]. Applied Mathematics Letters, 2009, 22: 932-937.

87. SHI HUAN-NAN, JIANG YONG-MING, JIANG WEI-DONG. Schur-convexity and Schur-geometrically concavity of Gini mean[J]. Computers and Mathematics wit Applications,2009, 57 : 266-274.

88. LI DA-MAO,SHI HUAN-NAN.Schur convexity and Schur-geometrically concavity of generalized exponent mean[J]. J. Math. Inequal., 2009, 3 (2): 217－225.

89. 石焕南. 关于二元幂平均的一个不等式[J]. 北京联合大学学报(自然科学版), 2009, 23 (2)：62-64.

90. 顾春, 石焕南. Lehme平均的Schur凸性和Schur几何凸性[J]. 数学的实践与认识, 2009, 39(12)：183-188.

91. 江永明, 石焕南. Stolarsky 与Gini 平均的一个比较[J]. 湖南理工学院学报(自然科学版), 2009, 22 (3)：7-12.

92. 石焕南, 张小明. 一对互补对称函数的Schur凸性[J]. 湖南理工学院学报(自然科学版), 2009, 22 (4)：1-5.

93. 江永明, 石焕南. Jensen-Janous-Klamkin型不等式[J]. 成都大学学报(自然科学版), 2009, 28 (3)：208-214.

94. SHI HUAN-NAN. A generalization of Qi's inequality for sums[J]. Kragujevac J. Math., 2010, 33: 101-106.

95. 石焕南.一类控制不等式及其应用[J]. 北京联合大学学报(自然科学版), 2010, 24 (1)：60-64.

96. 石焕南. 一个对称函数不等式猜想的控制证明[J]. 湖南理工学院学报(自然科学版), 2010, 23 (2)：1-3.

97. SHI HUAN-NAN, ZHANG JIAN, LI DA-MAO. Schur-geometric concavity for difference of some means[J]. Applied Math. E-Notes, 2010, 10: 275-284.

98. 李大矛, 石焕南, 杨志明. Stolarsky单参数不等式的推广[J]. 北京联合大学学报(自然科学版), 2010, 24 (3)：56-58.

99. 张小明, 石焕南. 国内学者对凸函数理论的若干研究成果介绍[J]. 中学教研参考·数学(高中版), 2010, 10: 9-32.

100. SHI HUAN-NAN. Two Schur-convex functions relate to Hadamard-type integral inequalities[J]. Publicationes Mathematicae Debrecen, 2011, 78(2): 393-403.

101. 张鉴, 石焕南. 加权算术-几何平均值不等式的控制证明[J]. 北京联合大学学报(自然科学版), 2011, 25 (4):46-47.

102. 石焕南. 关于三个对称函数的Schur-凹凸性[J]. 河西学院学报, 2011, 28 (2)：13-17.

103. 李大矛, 张鉴, 石焕南. Seiffert平均的Schur凸性和Schur几何凸性[J]. 湖南理工学院学报(自然科学版), 2011, 24 (2)：7-10.

104. WU SHAN-HE, SHI HUAN-NAN. A relation of weak majorization and its applications to certain inequalities for means[J]. Mathematica Slovaca, 2011, 61 (4)：561-570.

105. SHI HUAN-NAN, ZHANG JIAN, GU CHUN. Remarks on:Inequalities: theory of majorization and Its application by A. M. Marshall et al[C]// IEEE.Proceedings of 2011 World Congress on Engineering and Technology. Vol.1,454-456.

106. SHI HUAN-NAN, ZHANG JIAN, GU CHUN. New proofs of Schur-concavity for a class of symmetric functions[J]. J. Inequal. Appl., 2012:12 doi:10.1186/1029-242X-2012-12.

107. 褚玉明, 张小明, 石焕南. Gautschi-型不等式及其应用[J]. 数学物理学报, 2012, 32 A(4): 698-708.

108. 李大矛, 徐蕾, 石焕南. Gauss型函数方程与Moskovitz幂型平均值的新特征[J]. 数学的实践与认识, 2012, 42 (4)：179-185.

109. 石焕南, 何灯. 涉及完全对称函数的对偶不等式链[C]// 杨学枝. 不等式研究(第二辑). 哈尔滨: 哈尔滨工业大学出版社, 2012: 87-93.

110. 李大矛, 石焕南. 拟算术平均的Schur凸性和Schur 几何凸性[C]// 杨学枝. 不等式研究(第二辑). 哈尔滨: 哈尔滨工业大学出版社, 2012: 40-46.

111. 石焕南, 顾春, 张鉴. 一个Schur凸性判定定理的应用[J]. 四川师范大学学报(自然科学版), 2012, 35 (3):345-348.

112. SHI HUAN-NAN, LI DA-MAO, ZHANG JIAN. Refinements of inequalities among difference of means[J]. International Journal of Mathematics and Mathematical Sciences Volume 2012, Article ID 315697, 15 pages doi:10.1155/2012/315697.

113. 石焕南. 一个不等式命题的概率证明[J]. 湖南理工学院学报(自然科学版), 2012, 25 (3): 1-2,72

114. SHI HUAN-NAN, GU CHUN. Sharpening of Kai-lai Zhong's Inequality[J]. Journal of Latex Class Files, 2007, 6 (1): 1-4.

115. SHI HUAN-NAN, ZHANG JING. Schur-convexity of dual form of some symmetric functions[J]. J. Inequal. Appl., 2013, 2013:295.

116. ZHANG KONGSHENG, SHI HUAN-NAN. Schur convexity of dual form of the complete symmetric function[J]. Math. Inequal. Appl., 2013, 16 (4): 963‒970.

117. 张静, 石焕南. 关于一类对称函数的Schur凸性[J]. 数学的实践与认识, 2013, 43 (19): 292-296.

118. 石焕南, 张静. 一类条件不等式的控制证明与应用[J]. 纯粹数学与应用数学, 2013, 29 (5):441-449.

119. SHI HUAN-NAN, ZHANG JING. Some new judgment theorems of Schur geometric and Schur harmonic convexities for a class of symmetric functions[J]. J. Inequal. Appl., 2013, 2013:527 doi:10.1186/1029-242X-2013-527.

120. SHI HUAN-NAN, ZHANG JING. A Reverse Analytic Inequality for the Elementary Symmetric Function with Applications[J]. Journal of Applied Mathematics, vol. 2013, Article ID 674567, 5 pages, 2013. doi:10.1155/2013/674567.

121. WU YING, QI FENG, SHI HUAN-NAN. Schur-harmonic convexity for differences of some special means in two variables[J]. J.Math. Inequal., 2014, 8 (2): 321‒330.

122. ZHANG JING, SHI HUAN-NAN. Two double inequalities for k-gamma and k-Riemann zeta functions[J]. J. Inequal. Appl., 2014, 2014:191

123. SHI HUAN-NAN, ZHANG JING. Schur-convexity, Schur-geometric and harmonic convexities of dual form of a class symmetric functions[J]. J.Math. Inequal., 2014, 8 (2): 349-358.

124. 石焕南, 李明. 等差数列的凸性和对数凸性[J]. 湖南理工学院学报(自然科学版), 2014, 27 (3):1-6.

125. YIN HONG-PING, SHI HUAN-NAN, QI FENG. On Schur m‒power convexity for ratios of some means[J]. J. Math. Inequal., 2015, 9 (1): 145‒153.

126. 石焕南. 等比数列的凸性和对数凸性[J]. 广东第二师范学院学报, 2015,35 (3):9-15.

127. SHI HUAN-NAN, ZHANG JING. Compositions involving Schur geometrically convex functions[J]. J. Inequal. Appl., (2015) 2015:320 DOI 10.1186/s13660-015-0842-x

128. 王东生, 石焕南. 一个有限和不等式及其应用[J]. 湖南理工学院学报(自然科学版), 2015, 28 (3):1-3.

129. SHI HUAN-NAN. Hermite-Hadamard type inequalities for functions with a bounded second derivative[J]. Proceedings of the Jangjeon Mathematical Society, 2016, 19 (1): 135-144.

130. ZHANG JING, SHI HUAN-NAN. Multi-parameter generalization of Rado-Popoviciu inequalities[J]. J. Math. Inequal., 2016,10 (2): 577–582.

131. WANG DONGSHENG, FU CUN-RU, SHI HUAN-NAN. Schur-m power convexity for a mean of two variables with three parameters[J]. Journal of Nonlinear Science and Applications, J. Nonlinear Sci. Appl., 2016, 9 : 2298-2304.

132. SHI HUAN-NAN, ZHANG JING, MA QING-HUA. Schur-convexity, Schur-geometric and Schur- harmonic convexity for a composite function of complete symmetric function[J]. SpringerPlus, 2016(5): 296. DOI 10.1186/s40064-016-1940-z.

133. SHI HUAN-NAN. Majorized proof of arithmetic-geometric-harmonic means inequality[J]. Advanced studies in contemporary mathematics, 2016, 26 (4): 681 - 684.

134. 张鉴, 石焕南, 顾春. 凸函数的两个性质的控制证明[J]. 高等数学研究, 2016, 19 (4):32-33.

135. 张鉴, 顾春, 石焕南. 一类对称函数的Schur-指数凸性的简单证明[J]. 系统科学与数学, 2016, 36 (10):1779-1782.

136. 石焕南, 张鉴, 顾春. 凸数列的几个加权和性质的控制证明[J]. 四川师范大学学报(自然科学版), 2016, 39 (3): 373-376.

137. FU CUN-RU, WANG DONGSHENG, SHI HUAN-NAN. Schur-convexity for Lehmer mean of n variables[J]. J. Nonlinear Sci. Appl., 2016, 9: 5510-5520.

138. SHI HUAN-NAN, ZHANG JING. Compositions involving Schur harmonically convex functions[J].Journal of Computational Analysis and Applications, 2017, 22 (5): 907-922.

139. WANG DONGSHENG, FU CHUN-RU, SHI HUAN-NAN. Schur-convexity for a mean of n variables with three parameters[J]. Publications de l'Institut Mathématique (Beograd), 已录用.

140. SUN YI-JIN, WANG DONG SHENG, SHI HUAN-NAN. Two Schur-convex functions related to the generalized integral quasiarithmetic means[J]. Advances in Inequalities and Applications, Adv. Inequal. Appl. 2017, 2017:7. Available online at http://scik.org.

141. WANG DONGSHENG, SHI HUAN-NAN. Schur convexity of Bonferroni means[J]. Advanced Studies in Contemporary Mathematics, 2017, 27 (4): 599-608.

142. SHI HUAN-NAN, WU SHAN-HE. Schur-m Power convexity of geometric Bonferroni means[J]. Italian Journal of Pure and Applied Mathematics, N. 38-2017, 769 - 776.

143. 石焕南. 国内舒尔凸函数研究综述[J]. 广东第二师范学院学报, 2017, 37 (5): 6-12.

144. SHI HUAN-NAN, WU SHAN-HE. Schur convexity of the generalized geometric Bonferroni mean and the relevant inequalities[J/OL]. Journal of Inequalities and Applications, 2018, 2018: 8.http://doi.org/10.1186/s13660-017-1605-7.

145. 王东升, 石焕南. 一个代数不等式的n元推广[J]. 数学通报, 2018,57(3) : 57-59.

146. TAO ZHANG, HUAN-NAN SHI, BO-YAN XI, ALATANCANG CHEN. Majorization involving the cyclic moving average[J]. Journal of Inequalities and Applications, 2018(2018): 152.

147. CHUN-RU FU, DONGSHENG WANG, HUANNAN SHI. Schur-convexity for a mean of two variables with three parameters[J]. Filomat, 2018, 32(19): 6643 - 6651.

148. HUAN-NAN SHI, SHANHE WU. A concise proof of a double inequality involving the exponential and logarithmic functions[J]. Italian Journal of Pure and Applied Mathematics, 2019, 41: 284 - 289.

149. H N SHI, W S DU. Schur-Convexity for a Class of Completely Symmetric Function Dual[J]. Adv. Theory Nonlinear Anal. Appl., 2019, (3): 74 - 88.

150. BO-YAN XI, YING WU, HUAN-NAN SHI, FENG QI. Generalizations of several inequalities related to multivariate geometric means[J]. Mathematics, 2019(7), no. 6, 15 pages.

151. 石焕南. 关于一个完全对称函数舒尔凸性的注记[J]. 广东第二师范学院学报, 2019，39 (3): 14-17.

152. HUAN-NAN SHI, WEI-SHIH DU. Schur-power convexity of a Completely Symmetric Function Dual[J]. Symmetry, 2019, 11(7): 897.

153. SHAN-HE WU, HUAN-NAN SHI, DONG-SHENG WANG. Schur convexity of generalized geometric Bonferroni mean involving three parameters[J]. Italian Journal of Pure And Applied Mathematics - N.42 - 2019 (196 - 207).

154. 石焕南. 两个分式不等式的控制证明与推广[J]. 广东第二师范学院学报, 2019, 39 (5): 11-15.

155. 王东生, 付春茹, 石焕南. Bonferroni平均的Schur凸性及应用[J]. 兰州理工大学学报, 2019, 45 (5): 153-158.

156. 石焕南, 王飞, 王东生. 一个三角不等式的控制证明与推广[J]. 广东第二师范学院学报,2020，40 (3):12-16.

157. HUAN-NAN SHI, PEI WANG, JIAN ZHANG. Schur-convexity for compositions of complete symmetric function dual[J]. Journal of Inequalities and Applications,(2020)2020:65

158. HUAN-NAN SHI, SHANHE WU. Schur convexity of the dual form of complete symmetric function involving exponent parameter[J]. Italian Journal of Pure and Applied Mathematics, Accepted.

159. DONG-SHENG WANG, CHUNRU FU, HUANNAN SHI. Schur-m power convexity of Cauchy means and its application[J]. Rocky Mountain J. Math., Accepted.

160. FU CHUN-RU, WANG DONGSHENG, SHI HUAN-NAN. Schur-convexity for a mean of two variables with three parameters[J]. Filomat, 已录用.

著作目录:

1. 石焕南. 受控理论与解析不等式[M]. 哈尔滨: 哈尔滨工业大学出版社, 2012.

2. 石焕南. Schur凸函数与不等式[M]. 哈尔滨: 哈尔滨工业大学出版社, 2017.

3. HUAN-NAN SHI. Schur-Convex Functions and Inequalities: Volume 1: Concepts, Properties, and Applications in Symmetric Function Inequalities[M]. Harbin: Harbin Institute of Technology Press. Berlin: Walter de Gruyter GmbH.

4. HUAN-NAN SHI. Schur-Convex Functions and Inequalities: Volume 2: Applications in Inequalities[M]. Harbin: Harbin Institute of Technology Press. Berlin: Walter de Gruyter GmbH.

附录1　追念胡克教授[①]

　　2009年8月，我参加了在浙江海宁召开的全国第三届不等式年会，从匡继昌教授处得知著名数学家、江西师范大学教授胡克教授已去世，后从网上确定胡教授于2009年2月7日病故，享年88岁（许多文献资料写胡教授出生于1925年，胡教授来信讲这是错误的，应是1921年）。胡克出生于江西奉新，曾任江西省省政协常委，江西师范大学数学与信息管理学院数学研究所所长。1998年4月离休。胡克1946年入国立中正大学数学系学习，1948年参加解放军，1952-1957年于复旦大学工作。受到著名数学家陈建功先生的指导与赞赏，1957年被错划为右派，接着下放农村劳动，1960年回乡务农。1979年恢复名誉，到江西师范大学工作，1981年提升为教授。1991年享受国务院颁发的特殊津贴。他长期致力于单复函数理论和应用的研究与教学工作，治学严谨，学术造诣很深，是国内外著名的单叶函数专家之一。1979年以来，他在中外杂志上发表论文80多篇，在单叶函数$S(a)$族研究取得了公认的"一些极为重要的结果"，解决了奇单叶函数相邻系数模平方差Duren猜测。胡克教授是国内解析不等式研究的老前辈。他在《中国科学》上发表的"一个不等式及若干应用"论文，被美国数学评论评价为"一个卓越的，非凡的不等式"。著有《解析不等式的若干问题》、《单叶函数的若干问题》等。胡克教授还是中国气功学术研究会名誉理事，江西气功学术研究会副理事长。在人体功能与数学理论的研究上，胡教授提出"人体存在两个信息处理系统"，得到有关专家的重视和肯定，中国科学院学部委员胡海昌教授认为是"很有意思"的工作。我与胡克教授的交往始于1998年。那年我写信与胡教授，询购他的著作《基础不等式创建、改进与应用》，想不到不久老人家就馈赠了我一本。为了寻求胡克对中国不等式小组的指导和帮助，我向他介绍了中国不等式小组的发展状况并复印了《不等式研究通讯》的一些文章（如匡继昌教授的"向不等式研究的广度和深度进军"一文）以及我的一些论文寄于他。老人家很感兴趣。以后我就托付褚小光将每期《不等式研究通讯》寄与他。从2001年始，胡教授给我写过18封信。在此交往中，有两件事令我意外且感动。一是胡教

　　①选自《不等式研究》，石焕南，2009,16(4)：498.

授曾有意要以我为主将《基础不等式创建、改进与应用》一书修订为《解析不等式的若干问题》。因当时我身体欠佳，更因我对胡克教授研究的内容不甚熟悉，怕写不好有损他的大作，便婉言谢绝了他。胡教授不仅不怪罪我，还鼓励我将来写出自己的专著，而他以83岁的高龄很快独立完成了修订工作。二是在书信往来中，我曾谈及我于1992年因脑溢血做了开颅手术。想不到引得胡教授的关注，并给予我深切的关怀。他从我的信中提取信息，指出我身上可能存在的其他疾病，甚至指出我爱人所患疾病。他来信讲："我1983年生病，中西医都无法治疗之下，不能工作，才开始练气功。后来因练功逐渐好转，也就迷上气功。对气功道理进行科学分析，运用中医的经络理论解释。自己也因此出现了一点特殊功能而写了一本'气功学'的书。觉得中国几千年传留下来的养生学即现在所谓的'气功学'是值得科学宣扬的。因此我为学生开了十多年'气功选修课'，用点穴按摩方法也救助了一些医院无法治疗的病人。"他还讲到："你动过手术，希望多保重身体。海带炖肉汤有助心血管病，荞麦做稀饭有助降血压，愿你经常服用。气功锻炼对身体锻炼很有好处。但搞迷信，搞政治，江湖味很重的人教，千万不能去。"胡教授还给我推荐了一些气功动作，并详细解释了其中的科学道理。例如，他解释"双手合十于胸"这一练功动作的科学道理：因为双手手温有差别，一般经常用右手，右手温度高于左手。合十后，由热力学定律"高温向低温流"，因此右手的温度向左手流，引起全身温度流动。物理上又讲，有温度流动，就有电流动，因此人身上的生物微电也流动。人身上有温度流动，有微电流动，那么人体微血管循环，新陈代谢也会适当调整，病情也就会适当调整好转，很多事例证明了这一结论。胡教授差不多每次来信总要问及我的身体状况，对我这样的一个远在千里之外的无名晚辈如此关怀入微，令我深切地感受到他的高尚人格和慈悲心怀。2005年，胡教授不顾年事已高，只身参加了在广州召开的第三届不等式年会，我得以与老人家谋面。因胡克教授不会上网，我每年都寄张邮政贺卡给他拜年，他也随即回赠贺卡与我。去年春节前，我给胡克教授寄了贺卡后，迟迟未见他的回复，我预感他身体不妙，便打电话与他，他的家人接的电话，说胡教授已糊涂了，无法接电话。我默默地祈祷，希望老人家早日康复并能再次见到他。想不到没过不久他就离世了。

2009年8月30日
于北京

附录2　我与胡克教授的两封通信

(石焕南教授2001年11月10日写给胡克教授的信)

胡教授：您好！

10月21日和29日来信均收到，谢谢您对我的关心，为我提供营养食谱。《中国数学家与数学英才大辞典》(第一卷)(海南出版社)以较长的辞条介绍了您的事迹，从中了解了您的坎坷经历及在数学、气功学、医学上的诸多成就，令晚生景仰。

我患脑溢血前后，血压、血脂均不高。医生说可能是脑血管畸形，以及用脑不当等诸多因素所致。我现在每天服用"脑白金"，感觉效果不错，头脑比以前清醒。经历了那场生死劫难，我领悟了两点：一是要爱惜身体。这不是个人的事，是对父母，妻儿的一份责任，一份沉甸甸的责任，要科学地对待这个"本钱"。二是要珍惜生命，从死亡线上走回来，倍感生命之可贵，而生命又是短暂的，因此要活得充实一些，"含金量"高一点，不枉在世上走一遭。

褚小光来电话讲，早些时，他从上海(他在上海一家公司工作，家在苏州)给您寄去近几期的《不等式研究通讯》，后被退了回来，他又第二次寄去了，您至今没收到，不知哪个环节出了差错。不过，以后每期出版都会继续给您寄去。我手头没有多余的研究通讯，寄去的是以前让我校对的部分清样。

祝康健！

学生　石焕南

2001.11.10

(胡克教授2005年2月12日写给石焕南教授的信)

石教授：

很高兴收到了您关于修改增添《解析不等式的若干问题》的回信。我想您一定会做得很好。我写此书前也想了一段时间，决定从十个数学研究注意要点出发归纳来写。因为年纪大了，心力不足，而自己不是研究"不等式"的专家, 只是因为研究复分析的需要, 而研究一些需用的不等式和关连的不等式, 不敢多擅尚不熟悉的东西. 我们通了很久的信, 有了相互了解的基础, 所以请阁下为主将来完成增加篇章修改. 以我亲身经历和想法, 认为首要(1)搞好身体健康;(2) 坚强意志; (3) 活泼思想;(4) 避免过多社会活动, 千万要注意. 我建议阁下要注意下面三点, (A) 暑假要到青山绿水美景去休闲一时, 特别是一些著名的佛道寺院所在山地。我的经验是山川之灵气, 日月的精华确使人神往. (B) 家中平常培植一些盆景,有空多观察其劣缺, 分散我们集中研究时的疲劳. 有时修补盆景劣缺, 创新了盆景, 均会激发我们研究工作的联想, 破难创新出现. (C) 平常修炼身体, 现在国家已决定有几种气功功法可以教练学习的, 如《五禽戏》《八段锦》《六字诀》《易筋经》《瑜伽功》, 你选择一种最简的学. 其实《南无阿弥陀佛》也是一种六字诀. 不是菩萨, 更不是迷信. 它是梵音不是汉字音, 是用声波来调理人体功能的. 多念它有一股气会绕周天运行. 所以佛教净土宗无文化的僧、尼, 只需多念《南无阿弥陀佛》, 不须念其他经语. 有的会有很好的功能.

我看 "不等式" 书, 到现在还没见过有哪一本书有《Hardy-Littlewood-Polya不等式》那样引人入胜. 它是三个人合写的, 真是不朽之作.

徐利治, 王兴华合写的《数学分析的方法和例题选讲》是一本好书. 原为徐单独写, 后来(廿年后)加入王, 变成了更好的教学参考书, 并获国家优秀成果奖.

我不希望《不等式》写的很深, 范围很广, 但希望数学系高年级学生和研究生能看懂, 能启发其创新思维就好, 所以我想总结技巧、创新证法及结论能收入一些更好. 例如

1. $\sum\limits_{n=1}^{\infty} \frac{\sin nx}{n} > 0, 0 < x < \pi$;
2. Hardy-Littlewood极大值.

两定理技巧证明, 化难为易, 删繁就简是值得效法. 不知阁下以为如何?

我实在年老了, 只希望阁下早日能为主将合作写出修订本, 造福年轻数学工作者. 请写一个框架吧? 增加什么章节? 我写的有的定理我将改为更好的. 因已公开在数学国际会议讲过.

好, 再叙. 祝节日快乐, 身体健康!

胡克

2005.2.12

北京
联合大学 **职业技术师范学院**
北京师范大学分校

胡教授：您好！

10月29日来信敬悉，谢谢您对我事情的关心。从《中国现代名人教育荟萃大辞典》（第一卷）（海南出版社）以较长的篇幅介绍了您的事迹，从中了解了您的坎坷经历及在教学、气功学、医学上的诸多成就，令晚生钦佩。

我患脑溢血前后，血压、血脂均不高，医生说可能是脑血管畸型以及用脑过度等诸多因素所致。我后来服用"脑心舒"，疗效效果也好，现血脂也比以前缓解。经历那场生死劫难，我今领悟两点：一是珍惜身体，这不是个人的事，是对父母、妻小的一份责任，一份沉甸甸的责任，要珍惜身体，好好活这个年龄。二是要珍惜生命，从死亡线上走回来，倍感生命之可贵，而生命又是短暂的，因此要活得充实一些，令"含金量"高一些，不枉在世上走一遭。

您的大作已收讫，早些时候他从上海（他在上海一家公司工作，家在南方）给您寄过几期《公式研究通讯》，后来退了回来，他又寄第二次寄去了，您至今没收到，不知哪个环节出了差错。不过，以后每期出版都会继续给您寄去。祝

顺健！

晚生 石焕南
2001.11.10

R1

江西师范大学

石教授：

很高兴收到了您关于修改增添《解析不等式的若干问题》的回信，我想您一定做得很好。我写此书也想了一段时间，决从十个数学研究注意要点出发归纳来做。因为年纪已大了心力不足，而自己不是研究"不等式"，只是因为研究复分析的需要而研究一些需用的不等式和牵连的不等式。我们相识很久了，有了相互了解的基础，所以请阁下为主来完成修增加篇章修改。以我亲身经历和想法，认为首要①搞好身体②坚定意志③活泼思想④避免过多社会活动。我还议阁下要注意下面几点：①暑假要到青山绿水美景去休闲一些，特别是一些著名佛道都要去，我的经验是山川之灵气，日月精华，确使人神往。②平常增植一些盆景，因常有空去观美好，久之搞我们案中研究时疲劳，有时修剪盆景去，创新了盆景也有会激发你研究工作联想，破难创新出现。③平常修炼身体，现在国家已决定有几种气功功法可以教练习如《五禽戏》《八段锦》《六字诀》以筋经》《简功》，你选择一种最简的练。其实南乒所注意：现在念要音节《南无阿弥陀佛》有的不是梵音。　　　　　P.2.

江西师范大学

弥陀佛》也是一种六字诀夹，不是菩萨，更不是迷信，咒也是梵音不是语言，是用声波来调理人体之功能的。多念念有一股气会绕周天运行，所以净土宗有无文化的僧尼，只念《南无阿弥陀佛》，不读会其它经语有的会有很好功能。

我看不"等式"书，到现在还没见有哪一本有巴那样引人入胜。它是三个人合写的，真是不朽之作。　　　　　　　　　　　　Hardy-Littlewood-Pólya 不等式

徐利治、王兴华先生的《数学分析的方法及例题选讲》，建议先为徐单独写后来加入了，变成更好教学参考书，改为获国家优秀成果奖。

我不希望把不"等式"写得很深范围很广，但希望我学完高年学生和研究生能看懂能启发其创新思维就好，所以我想强调技巧创新证法及绪论能收入更好。例如

$1. \quad \sum_{n=1}^{\infty} \frac{\sin nx}{n} > 0, \quad 0 < x < \pi$

$2. \quad \text{Hardy-Littlewood 极大值}$

用技巧证明只使用初等方法，不知阁下以为如何？

我实在年老了。只希望阁下能为之将合作写出够为这福年青数学工作者。请定一个框架吧？增加什么章节？我写的有的我修改为更好的。它公开在数学国际会议讲过。

　　　　　　　　　　　　　　　　　　　致团叙，春节以快乐，身体健康！

　　　　　　　　　　　　　　　　　　　　　　胡克 2005.2.12

索　　引

刘培杰数学工作室
已出版(即将出版)图书目录——高等数学

书　　名	出版时间	定　价	编号
距离几何分析导引	2015—02	68.00	446
大学几何学	2017—01	78.00	688
关于曲面的一般研究	2016—11	48.00	690
近世纯粹几何学初论	2017—01	58.00	711
拓扑学与几何学基础讲义	2017—04	58.00	756
物理学中的几何方法	2017—06	88.00	767
几何学简史	2017—08	28.00	833
微分几何学历史概要	2020—07	58.00	1194
复变函数引论	2013—10	68.00	269
伸缩变换与抛物旋转	2015—01	38.00	449
无穷分析引论(上)	2013—04	88.00	247
无穷分析引论(下)	2013—04	98.00	245
数学分析	2014—04	28.00	338
数学分析中的一个新方法及其应用	2013—01	38.00	231
数学分析例选:通过范例学技巧	2013—01	88.00	243
高等代数例选:通过范例学技巧	2015—06	88.00	475
基础数论例选:通过范例学技巧	2018—09	58.00	978
三角级数论(上册)(陈建功)	2013—01	38.00	232
三角级数论(下册)(陈建功)	2013—01	48.00	233
三角级数论(哈代)	2013—06	48.00	254
三角级数	2015—07	28.00	263
超越数	2011—03	18.00	109
三角和方法	2011—03	18.00	112
随机过程(Ⅰ)	2014—01	78.00	224
随机过程(Ⅱ)	2014—01	68.00	235
算术探索	2011—12	158.00	148
组合数学	2012—04	28.00	178
组合数学浅谈	2012—03	28.00	159
丢番图方程引论	2012—03	48.00	172
拉普拉斯变换及其应用	2015—02	38.00	447
高等代数.上	2016—01	38.00	548
高等代数.下	2016—01	38.00	549
高等代数教程	2016—01	58.00	579
高等代数引论	2020—07	48.00	1174
数学解析教程.上卷.1	2016—01	58.00	546
数学解析教程.上卷.2	2016—01	38.00	553
数学解析教程.下卷.1	2017—04	48.00	781
数学解析教程.下卷.2	2017—06	48.00	782
函数构造论.上	2016—01	38.00	554
函数构造论.中	2017—06	48.00	555
函数构造论.下	2016—09	48.00	680
函数逼近论(上)	2019—02	98.00	1014
概周期函数	2016—01	48.00	572
变叙的项的极限分布律	2016—01	18.00	573
整函数	2012—08	18.00	161
近代拓扑学研究	2013—04	38.00	239
多项式和无理数	2008—01	68.00	22

刘培杰数学工作室
已出版(即将出版)图书目录——高等数学

书　　名	出版时间	定　价	编号
模糊数据统计学	2008—03	48.00	31
模糊分析学与特殊泛函空间	2013—01	68.00	241
常微分方程	2016—01	58.00	586
平稳随机函数导论	2016—03	48.00	587
量子力学原理.上	2016—01	38.00	588
图与矩阵	2014—08	40.00	644
钢丝绳原理:第二版	2017—01	78.00	745
代数拓扑和微分拓扑简史	2017—06	68.00	791
半序空间泛函分析.上	2018—06	48.00	924
半序空间泛函分析.下	2018—06	68.00	925
概率分布的部分识别	2018—07	68.00	929
Cartan 型单模李超代数的上同调及极大子代数	2018—07	38.00	932
纯数学与应用数学若干问题研究	2019—03	98.00	1017
数理金融学与数理经济学若干问题研究	2020—07	98.00	1180
受控理论与解析不等式	2012—05	78.00	165
不等式的分拆降维降幂方法与可读证明(第2版)	2020—07	78.00	1184
石焕南文集:受控理论与不等式研究	2020—09	198.00	1198
实变函数论	2012—06	78.00	181
复变函数论	2015—08	38.00	504
非光滑优化及其变分分析	2014—01	48.00	230
疏散的马尔科夫链	2014—01	58.00	266
马尔科夫过程论基础	2015—01	28.00	433
初等微分拓扑学	2012—07	18.00	182
方程式论	2011—03	38.00	105
Galois 理论	2011—03	18.00	107
古典数学难题与伽罗瓦理论	2012—11	58.00	223
伽罗华与群论	2014—01	28.00	290
代数方程的根式解及伽罗瓦理论	2011—03	28.00	108
代数方程的根式解及伽罗瓦理论(第二版)	2015—01	28.00	423
线性偏微分方程讲义	2011—03	18.00	110
几类微分方程数值方法的研究	2015—05	38.00	485
分数阶微分方程理论与应用	2020—05	95.00	1182
N 体问题的周期解	2011—03	28.00	111
代数方程式论	2011—05	18.00	121
线性代数与几何:英文	2016—06	58.00	578
动力系统的不变量与函数方程	2011—07	48.00	137
基于短语评价的翻译知识获取	2012—02	48.00	168
应用随机过程	2012—04	48.00	187
概率论导引	2012—04	18.00	179
矩阵论(上)	2013—06	58.00	250
矩阵论(下)	2013—06	48.00	251
对称锥互补问题的内点法:理论分析与算法实现	2014—08	68.00	368
抽象代数:方法导引	2013—06	38.00	257
集论	2016—01	48.00	576
多项式理论研究综述	2016—01	38.00	577
函数论	2014—11	78.00	395
反问题的计算方法及应用	2011—11	28.00	147
数阵及其应用	2012—02	28.00	164
绝对值方程—折边与组合图形的解析研究	2012—07	48.00	186
代数函数论(上)	2015—07	38.00	494
代数函数论(下)	2015—07	38.00	495

刘培杰数学工作室

已出版(即将出版)图书目录——高等数学

书　　名	出版时间	定　价	编号
偏微分方程论:法文	2015—10	48.00	533
时标动力学方程的指数型二分性与周期解	2016—04	48.00	606
重刚体绕不动点运动方程的积分法	2016—05	68.00	608
水轮机水力稳定性	2016—05	48.00	620
Lévy噪音驱动的传染病模型的动力学行为	2016—05	48.00	667
铣加工动力学系统稳定性研究的数学方法	2016—11	28.00	710
时滞系统:Lyapunov泛函和矩阵	2017—05	68.00	784
粒子图像测速仪实用指南:第二版	2017—08	78.00	790
数域的上同调	2017—08	98.00	799
图的正交因子分解(英文)	2018—01	38.00	881
图的度因子和分支因子:英文	2019—09	88.00	1108
点云模型的优化配准方法研究	2018—07	58.00	927
锥形波入射粗糙表面反散射问题理论与算法	2018—03	68.00	936
广义逆的理论与计算	2018—07	58.00	973
不定方程及其应用	2018—12	58.00	998
几类椭圆型偏微分方程高效数值算法研究	2018—08	48.00	1025
现代密码算法概论	2019—05	98.00	1061
模形式的 p-进性质	2019—06	78.00	1088
混沌动力学:分形、平铺、代换	2019—09	48.00	1109
微分方程,动力系统与混沌引论:第3版	2020—05	65.00	1144
Galois上同调	2020—04	138.00	1131
毕达哥拉斯定理:英文	2020—03	38.00	1133
吴振奎高等数学解题真经(概率统计卷)	2012—01	38.00	149
吴振奎高等数学解题真经(微积分卷)	2012—01	68.00	150
吴振奎高等数学解题真经(线性代数卷)	2012—01	58.00	151
高等数学解题全攻略(上卷)	2013—06	58.00	252
高等数学解题全攻略(下卷)	2013—06	58.00	253
高等数学复习纲要	2014—01	18.00	384
超越吉米多维奇.数列的极限	2009—11	48.00	58
超越普里瓦洛夫.留数卷	2015—01	28.00	437
超越普里瓦洛夫.无穷乘积与它对解析函数的应用卷	2015—05	28.00	477
超越普里瓦洛夫.积分卷	2015—06	18.00	481
超越普里瓦洛夫.基础知识卷	2015—06	28.00	482
超越普里瓦洛夫.数项级数卷	2015—07	38.00	489
超越普里瓦洛夫.微分、解析函数、导数卷	2018—01	48.00	852
统计学专业英语	2007—03	28.00	16
统计学专业英语(第二版)	2012—07	48.00	176
统计学专业英语(第三版)	2015—04	68.00	465
代换分析:英文	2015—07	38.00	499
历届美国大学生数学竞赛试题集.第一卷(1938—1949)	2015—01	28.00	397
历届美国大学生数学竞赛试题集.第二卷(1950—1959)	2015—01	28.00	398
历届美国大学生数学竞赛试题集.第三卷(1960—1969)	2015—01	28.00	399
历届美国大学生数学竞赛试题集.第四卷(1970—1979)	2015—01	18.00	400
历届美国大学生数学竞赛试题集.第五卷(1980—1989)	2015—01	28.00	401
历届美国大学生数学竞赛试题集.第六卷(1990—1999)	2015—01	28.00	402
历届美国大学生数学竞赛试题集.第七卷(2000—2009)	2015—08	18.00	403
历届美国大学生数学竞赛试题集.第八卷(2010—2012)	2015—01	18.00	404
超越普特南试题:大学数学竞赛中的方法与技巧	2017—04	98.00	758
历届国际大学生数学竞赛试题集(1994—2010)	2012—01	28.00	143

刘培杰数学工作室
已出版(即将出版)图书目录——高等数学

书　名	出版时间	定　价	编号
全国大学生数学夏令营数学竞赛试题及解答	2007－03	28.00	15
全国大学生数学竞赛辅导教程	2012－07	28.00	189
全国大学生数学竞赛复习全书(第2版)	2017－05	58.00	787
历届美国大学生数学竞赛试题集	2009－03	88.00	43
前苏联大学生数学奥林匹克竞赛题解(上编)	2012－04	28.00	169
前苏联大学生数学奥林匹克竞赛题解(下编)	2012－04	38.00	170
大学生数学竞赛讲义	2014－09	28.00	371
大学生数学竞赛教程——高等数学(基础篇、提高篇)	2018－09	128.00	968
普林斯顿大学数学竞赛	2016－06	38.00	669
越过211,刷到985:考研数学二	2019－10	68.00	1115
初等数论难题集(第一卷)	2009－05	68.00	44
初等数论难题集(第二卷)(上、下)	2011－02	128.00	82,83
数论概貌	2011－03	18.00	93
代数数论(第二版)	2013－08	58.00	94
代数多项式	2014－06	38.00	289
初等数论的知识与问题	2011－02	28.00	95
超越数论基础	2011－03	28.00	96
数论初等教程	2011－03	28.00	97
数论基础	2011－03	18.00	98
数论基础与维诺格拉多夫	2014－03	18.00	292
解析数论基础	2012－08	28.00	216
解析数论基础(第二版)	2014－01	48.00	287
解析数论问题集(第二版)(原版引进)	2014－05	88.00	343
解析数论问题集(第二版)(中译本)	2016－04	88.00	607
解析数论基础(潘承洞,潘承彪著)	2016－07	98.00	673
解析数论导引	2016－07	58.00	674
数论入门	2011－03	38.00	99
代数数论入门	2015－03	38.00	448
数论开篇	2012－07	28.00	194
解析数论引论	2011－03	48.00	100
Barban Davenport Halberstam 均值和	2009－01	40.00	33
基础数论	2011－03	28.00	101
初等数论100例	2011－05	18.00	122
初等数论经典例题	2012－07	18.00	204
最新世界各国数学奥林匹克中的初等数论试题(上、下)	2012－01	138.00	144,145
初等数论(Ⅰ)	2012－01	18.00	156
初等数论(Ⅱ)	2012－01	18.00	157
初等数论(Ⅲ)	2012－01	28.00	158
平面几何与数论中未解决的新老问题	2013－01	68.00	229
代数数论简史	2014－11	28.00	408
代数数论	2015－09	88.00	532
代数、数论及分析习题集	2016－11	98.00	695
数论导引提要及习题解答	2016－01	48.00	559
素数定理的初等证明. 第2版	2016－09	48.00	686
数论中的模函数与狄利克雷级数(第二版)	2017－11	78.00	837
数论:数学导引	2018－01	68.00	849
域论	2018－04	68.00	884
代数数论(冯克勤 编著)	2018－04	68.00	885
范氏大代数	2019－02	98.00	1016

刘培杰数学工作室
已出版(即将出版)图书目录——高等数学

书　　名	出版时间	定　价	编号
新编 640 个世界著名数学智力趣题	2014—01	88.00	242
500 个最新世界著名数学智力趣题	2008—06	48.00	3
400 个最新世界著名数学最值问题	2008—09	48.00	36
500 个世界著名数学征解问题	2009—06	48.00	52
400 个中国最佳初等数学征解老问题	2010—01	48.00	60
500 个俄罗斯数学经典老题	2011—01	28.00	81
1000 个国外中学物理好题	2012—04	48.00	174
300 个日本高考数学题	2012—05	38.00	142
700 个早期日本高考数学试题	2017—02	88.00	752
500 个前苏联早期高考数学试题及解答	2012—05	28.00	185
546 个早期俄罗斯大学生数学竞赛题	2014—03	38.00	285
548 个来自美苏的数学好问题	2014—11	28.00	396
20 所苏联著名大学早期入学试题	2015—02	18.00	452
161 道德国工科大学生必做的微分方程习题	2015—05	28.00	469
500 个德国工科大学生必做的高数习题	2015—06	28.00	478
360 个数学竞赛问题	2016—08	58.00	677
德国讲义日本考题.微积分卷	2015—04	48.00	456
德国讲义日本考题.微分方程卷	2015—04	38.00	457
二十世纪中叶中、英、美、日、法、俄高考数学试题精选	2017—06	38.00	783

博弈论精粹	2008—03	58.00	30
博弈论精粹.第二版(精装)	2015—01	88.00	461
数学 我爱你	2008—01	28.00	20
精神的圣徒　别样的人生——60 位中国数学家成长的历程	2008—09	48.00	39
数学史概论	2009—06	78.00	50
数学史概论(精装)	2013—03	158.00	272
数学史选讲	2016—01	48.00	544
斐波那契数列	2010—02	28.00	65
数学拼盘和斐波那契魔方	2010—07	38.00	72
斐波那契数列欣赏	2011—01	28.00	160
数学的创造	2011—02	48.00	85
数学美与创造力	2016—01	48.00	595
数海拾贝	2016—01	48.00	590
数学中的美	2011—02	38.00	84
数论中的美学	2014—12	38.00	351
数学王者　科学巨人——高斯	2015—01	28.00	428
振兴祖国数学的圆梦之旅:中国初等数学研究史话	2015—06	98.00	490
二十世纪中国数学史料研究	2015—10	48.00	536
数字谜、数阵图与棋盘覆盖	2016—01	58.00	298
时间的形状	2016—01	38.00	556
数学发现的艺术:数学探索中的合情推理	2016—07	58.00	671
活跃在数学中的参数	2016—07	48.00	675

书 名	出版时间	定 价	编号
格点和面积	2012—07	18.00	191
射影几何趣谈	2012—04	28.00	175
斯潘纳尔引理——从一道加拿大数学奥林匹克试题谈起	2014—01	28.00	228
李普希兹条件——从几道近年高考数学试题谈起	2012—10	18.00	221
拉格朗日中值定理——从一道北京高考试题的解法谈起	2015—10	18.00	197
闵科夫斯基定理——从一道清华大学自主招生试题谈起	2014—01	28.00	198
哈尔测度——从一道冬令营试题的背景谈起	2012—08	28.00	202
切比雪夫逼近问题——从一道中国台北数学奥林匹克试题谈起	2013—04	38.00	238
伯恩斯坦多项式与贝齐尔曲面——从一道全国高中数学联赛试题谈起	2013—03	38.00	236
卡塔兰猜想——从一道普特南竞赛试题谈起	2013—06	18.00	256
麦卡锡函数和阿克曼函数——从一道前南斯拉夫数学奥林匹克试题谈起	2012—08	18.00	201
贝蒂定理与拉姆贝克莫斯尔定理——从一个拣石子游戏谈起	2012—08	18.00	217
皮亚诺曲线和豪斯道夫分球定理——从无限集谈起	2012—08	18.00	211
平面凸图形与凸多面体	2012—10	28.00	218
斯坦因豪斯问题——从一道二十五省市自治区中学数学竞赛试题谈起	2012—07	18.00	196
纽结理论中的亚历山大多项式与琼斯多项式——从一道北京市高一数学竞赛试题谈起	2012—07	28.00	195
原则与策略——从波利亚"解题表"谈起	2013—04	38.00	244
转化与化归——从三大尺规作图不能问题谈起	2012—08	28.00	214
代数几何中的贝祖定理(第一版)——从一道IMO试题的解法谈起	2013—08	18.00	193
成功连贯理论与约当块理论——从一道比利时数学竞赛试题谈起	2012—04	18.00	180
素数判定与大数分解	2014—08	18.00	199
置换多项式及其应用	2012—10	18.00	220
椭圆函数与模函数——从一道美国加州大学洛杉矶分校(UCLA)博士资格考题谈起	2012—10	28.00	219
差分方程的拉格朗日方法——从一道2011年全国高考理科试题的解法谈起	2012—08	28.00	200
力学在几何中的一些应用	2013—01	38.00	240
高斯散度定理、斯托克斯定理和平面格林定理——从一道国际大学生数学竞赛试题谈起	即将出版		
康托维奇不等式——从一道全国高中联赛试题谈起	2013—03	28.00	337
西格尔引理——从一道第18届IMO试题的解法谈起	即将出版		
罗斯定理——从一道前苏联数学竞赛试题谈起	即将出版		
拉克斯定理和阿廷定理——从一道IMO试题的解法谈起	2014—01	58.00	246
毕卡大定理——从一道美国大学数学竞赛试题谈起	2014—07	18.00	350
贝齐尔曲线——从一道全国高中联赛试题谈起	即将出版		
拉格朗日乘子定理——从一道2005年全国高中联赛试题的高等数学解法谈起	2015—05	28.00	480
雅可比定理——从一道日本数学奥林匹克试题谈起	2013—04	48.00	249
李天岩—约克定理——从一道波兰数学竞赛试题谈起	2014—06	28.00	349
整系数多项式因式分解的一般方法——从克朗耐克算法谈起	即将出版		

刘培杰数学工作室

已出版(即将出版)图书目录——高等数学

书　名	出版时间	定　价	编号
布劳维不动点定理——从一道前苏联数学奥林匹克试题谈起	2014—01	38.00	273
伯恩赛德定理——从一道英国数学奥林匹克试题谈起	即将出版		
布查特－莫斯特定理——从一道上海市初中竞赛试题谈起	即将出版		
数论中的同余数问题——从一道普林南竞赛试题谈起	即将出版		
范·德蒙行列式——从一道美国数学奥林匹克试题谈起	即将出版		
中国剩余定理:总数法构建中国历史年表	2015—01	28.00	430
牛顿程序与方程求根——从一道全国高考试题解法谈起	即将出版		
库默尔定理——从一道IMO预选试题谈起	即将出版		
卢丁定理——从一道冬令营试题的解法谈起	即将出版		
沃斯滕霍姆定理——从一道IMO预选试题谈起	即将出版		
卡尔松不等式——从一道莫斯科数学奥林匹克试题谈起	即将出版		
信息论中的香农熵——从一道近年高考压轴题谈起	即将出版		
约当不等式——从一道希望杯竞赛试题谈起	即将出版		
拉比诺维奇定理	即将出版		
刘维尔定理——从一道《美国数学月刊》征解问题的解法谈起	即将出版		
卡塔兰恒等式与级数求和——从一道IMO试题的解法谈起	即将出版		
勒让德猜想与素数分布——从一道爱尔兰竞赛试题谈起	即将出版		
天平称重与信息论——从一道基辅市数学奥林匹克试题谈起	即将出版		
哈密尔顿－凯莱定理:从一道高中数学联赛试题的解法谈起	2014—09	18.00	376
艾思特曼定理——从一道CMO试题的解法谈起	即将出版		
一个爱尔特希问题——从一道西德数学奥林匹克试题谈起	即将出版		
有限群中的爱丁格尔问题——从一道北京市初中二年级数学竞赛试题谈起	即将出版		
糖水中的不等式——从初等数学到高等数学	2019—07	48.00	1093
帕斯卡三角形	2014—03	18.00	294
蒲丰投针问题——从2009年清华大学的一道自主招生试题谈起	2014—01	38.00	295
斯图姆定理——从一道"华约"自主招生试题的解法谈起	2014—01	18.00	296
许瓦兹引理——从一道加利福尼亚大学伯克利分校数学系博士生试题谈起	2014—08	18.00	297
拉姆塞定理——从王诗宬院士的一个问题谈起	2016—04	48.00	299
坐标法	2013—12	28.00	332
数论三角形	2014—04	38.00	341
毕克定理	2014—07	18.00	352
数林掠影	2014—09	48.00	389
我们周围的概率	2014—10	38.00	390
凸函数最值定理:从一道华约自主招生题的解法谈起	2014—10	28.00	391
易学与数学奥林匹克	2014—10	38.00	392
生物数学趣谈	2015—01	18.00	409
反演	2015—01	28.00	420
因式分解与圆锥曲线	2015—01	18.00	426
轨迹	2015—01	28.00	427
面积原理:从常庚哲命的一道CMO试题的积分解法谈起	2015—01	48.00	431
形形色色的不动点定理:从一道28届IMO试题谈起	2015—01	38.00	439
柯西函数方程:从一道上海交大自主招生的试题谈起	2015—02	28.00	440

书　名	出版时间	定　价	编号
三角恒等式	2015—02	28.00	442
无理性判定:从一道2014年"北约"自主招生试题谈起	2015—01	38.00	443
数学归纳法	2015—03	18.00	451
极端原理与解题	2015—04	28.00	464
法雷级数	2014—08	18.00	367
摆线族	2015—01	38.00	438
函数方程及其解法	2015—05	38.00	470
含参数的方程和不等式	2012—09	28.00	213
希尔伯特第十问题	2016—01	38.00	543
无穷小量的求和	2016—01	28.00	545
切比雪夫多项式:从一道清华大学金秋营试题谈起	2016—01	38.00	583
泽肯多夫定理	2016—03	38.00	599
代数等式证题法	2016—01	28.00	600
三角等式证题法	2016—01	28.00	601
吴大任教授藏书中的一个因式分解公式:从一道美国数学邀请赛试题的解法谈起	2016—06	28.00	656
易卦——类万物的数学模型	2017—08	68.00	838
"不可思议"的数与数系可持续发展	2018—01	38.00	878
最短线	2018—01	38.00	879
从毕达哥拉斯到怀尔斯	2007—10	48.00	9
从迪利克雷到维斯卡尔迪	2008—01	48.00	21
从哥德巴赫到陈景润	2008—05	98.00	35
从庞加莱到佩雷尔曼	2011—08	138.00	136
从费马到怀尔斯——费马大定理的历史	2013—10	198.00	I
从庞加莱到佩雷尔曼——庞加莱猜想的历史	2013—10	298.00	II
从切比雪夫到爱尔特希(上)——素数定理的初等证明	2013—07	48.00	III
从切比雪夫到爱尔特希(下)——素数定理100年	2012—12	98.00	III
从高斯到盖尔方特——二次域的高斯猜想	2013—10	198.00	IV
从库默尔到朗兰兹——朗兰兹猜想的历史	2014—01	98.00	V
从比勃巴赫到德布朗斯——比勃巴赫猜想的历史	2014—02	298.00	VI
从麦比乌斯到陈省身——麦比乌斯变换与麦比乌斯带	2014—02	298.00	VII
从布尔到豪斯道夫——布尔方程与格论漫谈	2013—10	198.00	VIII
从开普勒到阿诺德——三体问题的历史	2014—05	298.00	IX
从华林到华罗庚——华林问题的历史	2013—10	298.00	X
数学物理大百科全书.第1卷	2016—01	418.00	508
数学物理大百科全书.第2卷	2016—01	408.00	509
数学物理大百科全书.第3卷	2016—01	396.00	510
数学物理大百科全书.第4卷	2016—01	408.00	511
数学物理大百科全书.第5卷	2016—01	368.00	512
朱德祥代数与几何讲义.第1卷	2017—01	38.00	697
朱德祥代数与几何讲义.第2卷	2017—01	28.00	698
朱德祥代数与几何讲义.第3卷	2017—01	28.00	699

刘培杰数学工作室
已出版(即将出版)图书目录——高等数学

书 名	出版时间	定 价	编号
闵嗣鹤文集	2011—03	98.00	102
吴从炘数学活动三十年(1951~1980)	2010—07	99.00	32
吴从炘数学活动又三十年(1981~2010)	2015—07	98.00	491
斯米尔诺夫高等数学.第一卷	2018—03	88.00	770
斯米尔诺夫高等数学.第二卷.第一分册	2018—03	68.00	771
斯米尔诺夫高等数学.第二卷.第二分册	2018—03	68.00	772
斯米尔诺夫高等数学.第二卷.第三分册	2018—03	48.00	773
斯米尔诺夫高等数学.第三卷.第一分册	2018—03	58.00	774
斯米尔诺夫高等数学.第三卷.第二分册	2018—03	58.00	775
斯米尔诺夫高等数学.第三卷.第三分册	2018—03	68.00	776
斯米尔诺夫高等数学.第四卷.第一分册	2018—03	48.00	777
斯米尔诺夫高等数学.第四卷.第二分册	2018—03	88.00	778
斯米尔诺夫高等数学.第五卷.第一分册	2018—03	58.00	779
斯米尔诺夫高等数学.第五卷.第二分册	2018—03	68.00	780
zeta 函数,q-zeta 函数,相伴级数与积分	2015—08	88.00	513
微分形式:理论与练习	2015—08	58.00	514
离散与微分包含的逼近和优化	2015—08	58.00	515
艾伦·图灵:他的工作与影响	2016—01	98.00	560
测度理论概率导论,第 2 版	2016—01	88.00	561
带有潜在故障恢复系统的半马尔柯夫模型控制	2016—01	98.00	562
数学分析原理	2016—01	88.00	563
随机偏微分方程的有效动力学	2016—01	88.00	564
图的谱半径	2016—01	58.00	565
量子机器学习中数据挖掘的量子计算方法	2016—01	98.00	566
量子物理的非常规方法	2016—01	118.00	567
运输过程的统一非局部理论:广义波尔兹曼物理动力学,第2版	2016—01	198.00	568
量子力学与经典力学之间的联系在原子、分子及电动力学系统建模中的应用	2016—01	58.00	569
算术域	2018—01	158.00	821
高等数学竞赛:1962—1991 年的米洛克斯·史怀哲竞赛	2018—01	128.00	822
用数学奥林匹克精神解决数论问题	2018—01	108.00	823
代数几何(德语)	2018—04	68.00	824
丢番图逼近论	2018—01	78.00	825
代数几何学基础教程	2018—01	98.00	826
解析数论入门课程	2018—01	78.00	827
数论中的丢番图问题	2018—01	78.00	829
数论(梦幻之旅):第五届中日数论研讨会演讲集	2018—01	68.00	830
数论新应用	2018—01	68.00	831
数论	2018—01	78.00	832
测度与积分	2019—04	68.00	1059
卡塔兰数入门	2019—05	68.00	1060

刘培杰数学工作室
已出版(即将出版)图书目录——高等数学

书　　名	出版时间	定　价	编号
湍流十讲	2018—04	108.00	886
无穷维李代数:第3版	2018—04	98.00	887
等值、不变量和对称性:英文	2018—04	78.00	888
解析数论	2018—09	78.00	889
《数学原理》的演化:伯特兰·罗素撰写第二版时的手稿与笔记	2018—04	108.00	890
哈密尔顿数学论文集(第4卷):几何学、分析学、天文学、概率和有限差分等	2019—05	108.00	891
数学王子——高斯	2018—01	48.00	858
坎坷奇星——阿贝尔	2018—01	48.00	859
闪烁奇星——伽罗瓦	2018—01	58.00	860
无穷统帅——康托尔	2018—01	48.00	861
科学公主——柯瓦列夫斯卡娅	2018—01	48.00	862
抽象代数之母——埃米·诺特	2018—01	48.00	863
电脑先驱——图灵	2018—01	58.00	864
昔日神童——维纳	2018—01	48.00	865
数坛怪侠——爱尔特希	2018—01	68.00	866
当代世界中的数学.数学思想与数学基础	2019—01	38.00	892
当代世界中的数学.数学问题	2019—01	38.00	893
当代世界中的数学.应用数学与数学应用	2019—01	38.00	894
当代世界中的数学.数学王国的新疆域(一)	2019—01	38.00	895
当代世界中的数学.数学王国的新疆域(二)	2019—01	38.00	896
当代世界中的数学.数林撷英(一)	2019—01	38.00	897
当代世界中的数学.数林撷英(二)	2019—01	48.00	898
当代世界中的数学.数学之路	2019—01	38.00	899
偏微分方程全局吸引子的特性:英文	2018—09	108.00	979
整函数与下调和函数:英文	2018—09	118.00	980
幂等分析:英文	2018—09	118.00	981
李群,离散子群与不变量理论:英文	2018—09	108.00	982
动力系统与统计力学:英文	2018—09	118.00	983
表示论与动力系统:英文	2018—09	118.00	984
初级统计学:循序渐进的方法:第10版	2019—05	68.00	1067
工程师与科学家微分方程用书:第4版	2019—07	58.00	1068
大学代数与三角学	2019—06	78.00	1069
培养数学能力的途径	2019—07	38.00	1070
工程师与科学家统计学:第4版	2019—06	58.00	1071
贸易与经济中的应用统计学:第6版	2019—06	58.00	1072
傅立叶级数和边值问题:第8版	2019—05	48.00	1073
通往天文学的途径:第5版	2019—05	58.00	1074

刘培杰数学工作室
已出版（即将出版）图书目录——高等数学

书　名	出版时间	定　价	编号
拉马努金笔记.第1卷	2019—06	165.00	1078
拉马努金笔记.第2卷	2019—06	165.00	1079
拉马努金笔记.第3卷	2019—06	165.00	1080
拉马努金笔记.第4卷	2019—06	165.00	1081
拉马努金笔记.第5卷	2019—06	165.00	1082
拉马努金遗失笔记.第1卷	2019—06	109.00	1083
拉马努金遗失笔记.第2卷	2019—06	109.00	1084
拉马努金遗失笔记.第3卷	2019—06	109.00	1085
拉马努金遗失笔记.第4卷	2019—06	109.00	1086
数论:1976年纽约洛克菲勒大学数论会议记录	2020—06	68.00	1145
数论:卡本代尔1979:1979年在南伊利诺伊卡本代尔大学举行的数论会议记录	2020—06	78.00	1146
数论:诺德韦克豪特1983:1983年在诺德韦克豪特举行的Journees Arithmetiques数论大会会议记录	2020—06	68.00	1147
数论:1985—1988年在纽约城市大学研究生院和大学中心举办的研讨会	2020—06	68.00	1148
数论:1987年在乌尔姆举行的Journees Arithmetiques数论大会会议记录	2020—06	68.00	1149
数论:马德拉斯1987:1987年在马德拉斯安娜大学举行的国际拉马努金百年纪念大会会议记录	2020—06	68.00	1150
解析数论:1988年在东京举行的日法研讨会会议记录	2020—06	68.00	1151
解析数论:2002年在意大利切特拉罗举行的C.I.M.E.暑期班演讲集	2020—06	68.00	1152
量子世界中的蝴蝶:最迷人的量子分形故事	2020—06	118.00	1157
走进量子力学	2020—06	118.00	1158
计算物理学概论	2020—06	48.00	1159
物质,空间和时间的理论:量子理论	即将出版		1160
物质,空间和时间的理论:经典理论	即将出版		1161
量子场理论:解释世界的神秘背景	2020—07	38.00	1162
计算物理学概论	即将出版		1163
行星状星云	即将出版		1164
基本宇宙学:从亚里士多德的宇宙到大爆炸	2020—08	58.00	1165
数学磁流体力学	2020—07	58.00	1166
计算科学:第1卷,计算的科学(日文)	2020—07	88.00	1167
计算科学:第2卷,计算与宇宙(日文)	2020—07	88.00	1168
计算科学:第3卷,计算与物质(日文)	2020—07	88.00	1169
计算科学:第4卷,计算与生命(日文)	2020—07	88.00	1170
计算科学:第5卷,计算与地球环境(日文)	2020—07	88.00	1171
计算科学:第6卷,计算与社会(日文)	2020—07	88.00	1172
计算科学.别卷,超级计算机(日文)	2020—07	88.00	1173

刘培杰数学工作室
已出版(即将出版)图书目录——高等数学

书　名	出版时间	定　价	编号
代数与数论:综合方法	即将出版		1185
复分析:现代函数理论第一课	2020—07	58.00	1186
斐波那契数列和卡特兰数:导论	即将出版		1187
组合推理:计数艺术介绍	2020—07	88.00	1188
二次互反律的傅里叶分析证明	2020—07	48.00	1189
旋瓦兹分布的希尔伯特变换与应用	2020—07	58.00	1190
泛函分析:巴拿赫空间理论入门	即将出版		1191

联系地址:哈尔滨市南岗区复华四道街 10 号　哈尔滨工业大学出版社刘培杰数学工作室
网　　址:http://lpj.hit.edu.cn/
邮　　编:150006
联系电话:0451—86281378　　13904613167
E-mail:lpj1378@163.com